INVESTIGATING BIOLOGY

Laboratory Manual

Ninth Edition

Judith Giles Morgan
Emory University

M. Eloise Brown Carter
Oxford College of Emory University

330 Hudson Street, NY, NY 10013

The authors and publisher believe that the lab experiments described in this publication, when conducted in conformity with the safety precautions described herein and according to the school's laboratory safety procedures, are reasonably safe for the students to whom the manual is directed. Nonetheless, many of the described experiments are accompanied by some degree of risk, including human error, the failure or misuse of laboratory or electrical equipment, mismeasurement, spills of chemicals, and exposure to sharp objects, heat, bodily fluids, blood, or other biologics. The authors and publisher disclaim any liability arising from such risks in connection with any of the experiments contained in this manual. If students have questions or problems with materials, procedures, or instructions on any experiment, they should always ask their instructor for help before proceeding.

Courseware Portfolio Management Specialist: Josh Frost
Courseware Director, Content Development: Ginnie Simione Jutson
Managing Producer, Science: Michael Early
Content Producer, Science: Margaret Young
Production Management and Composition: iEnergizer Aptara®, Ltd.
Interior Design: iEnergizer Aptara®, Ltd.
Illustrators: Lachina
Rights & Permissions Manager: Ben Ferrini
Rights & Permissions Project Manager: Laura Murray, Cenveo
Manufacturing Buyer: Stacey J. Weinberger
Photo Researcher: Maureen Spuhler
VP Product Marketing: Christy Lesko
Product Marketing Manager: Christa Pesek Pelaez

Cover Photo Credit: Radius Images/Getty Images

Library of Congress Cataloging-in-Publication Data
Names: Morgan, Judith Giles, author. | Carter, M. Eloise Brown, author.
Title: Investigating biology laboratory manual / Judith Giles Morgan, Emory University, M. Eloise Brown Carter, Oxford College of Emory University.
Description: Ninth edition. | San Francisco: Pearson, 2017.
Identifiers: LCCN 2016043895| ISBN 9780134519227 (annotated instructor's edition) | ISBN 9780134473468 (student edition)
Subjects: LCSH: Biology--Laboratory manuals.
Classification: LCC QH317 .M74 2016 | DDC 570.78--dc23 LC record available at https://lccn.loc.gov/2016043895d

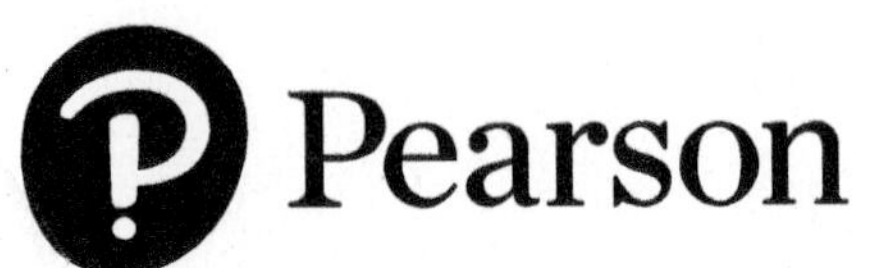

ISBN-13: 978-0-134-47346-8 ISBN-10: 0-134-47346-9 Student Edition
ISBN-13: 978-0-134-51922-7 ISBN-10: 0-134-51922-1 Annotated Instructor's Edition

www.pearsonhighered.com

16 2022

The most remarkable discovery made by scientists is science itself.

JACOB BRONOWSKI

Preface

Our knowledge and understanding of the biological world is based on the scientific enterprise of asking questions and formulating and testing hypotheses. Scientists gather data, then evaluate and interpret their results, communicating their findings through papers and presentations. An important aspect of learning biology is participating in the process of science and developing creative and critical reasoning skills. Our goal in writing this laboratory manual is to present a laboratory program that engages students in the scientific process and encourages scientific thinking. We want students to experience the excitement of discovery and the satisfaction of solving problems and connecting concepts. For us, investigating biology is more than just doing experiments; it is an approach to teaching and learning that promotes inquiry.

The laboratory exercises are designed to encourage students to ask questions, to pose hypotheses, and to make predictions before they initiate laboratory work. Students are asked to synthesize results from their observations and experiments, then draw conclusions based on evidence. Whenever possible, students apply their results to new problems and case studies. *Investigating Biology* provides students with many opportunities to design their own open-inquiry investigations as part of the laboratory. In addition, students can pursue independent investigations using the suggestions and extensions provided at the end of lab topics. Scientific writing and communication are emphasized throughout the laboratory manual and supported by an appendix that includes instructions for writing each section of a scientific paper, obtaining information from primary sources, and preparing oral and poster presentations. Instructors are given suggestions for organizing a laboratory writing program.

Investigating Biology uses a stepwise approach to developing scientific knowledge and skills, as students in early lab topics practice asking questions, developing hypotheses, and designing experiments. Early laboratory experiences build on knowledge (e.g., how cells and enzymes function, genetics, and then biodiversity as an expression of variation and evolution). At the same time, students are developing laboratory and thinking skills, such as pipetting, using instruments, analyzing results, organizing and managing collaborative teams, and developing systematic approaches to experimental design and problem solving. Lab topics are a mixture of directed-inquiry and open-inquiry, with more directed investigations in the initial lab topics and increasing opportunities for open-inquiry in subsequent lab topics. In the open-inquiry investigations, students are presented with a topic and preliminary experiment, then encouraged to develop their own questions, design their investigations, and write their proposals before implementing their own investigations. They must then collect and interpret their data. We use a similar approach for scientific writing. Students write individual sections of a scientific paper for initial lab topics, receiving feedback and the opportunity to revise. Then, for open-inquiry investigations, they are prepared to use their writing expertise to write a complete scientific paper. This incremental approach to knowledge, skills, and disciplined ways of thinking is a hallmark of our approach throughout the lab manual and is adaptable to programs with large laboratory sections or small groups of students. *Investigating Biology* provides a comprehensive introduction to the diverse topics and subdisciplines in the biological sciences, always with an emphasis on scientific investigation.

We are convinced that involving students in the process of science through investigating biological phenomena is the best way to teach. The organization of this laboratory manual with a mix of directed-inquiry and open-inquiry investigations complements this approach to teaching and learning.

New in the Ninth Edition

The ninth edition of *Investigating Biology* offers expanded opportunities for students to participate in science. Adopters of the ninth edition will notice a continued emphasis on recurring themes in biology, including structure and function, unity and diversity, transmission of genetic information, energy transformations, and the overarching theme of evolution. These themes are developed across hierarchical levels from cells to organisms to ecosystems.

Scientists and students in the biological sciences strive to understand complex systems, structures, and processes that range from the molecular to the landscape. Often their inquiries are enhanced by visualization through images, models, and digital media. Building on the addition of full color in the eighth edition of the manual, in this edition we have revised selected pieces of art and added new photos to accompany art in many lab topics, including enhancing dissections, investigating biological diversity and phylogenetic relationships, and illustrating procedures and experimental design. We continue to update and expand media options in the resources sections at the end of each lab topic, including new videos from Biointeractive (HHMI).

We have revised procedures and added tables to clarify methodology. In Lab Topic 10, we have incorporated a new system to visualize gels using SYBR® Safe and blue light illumination because of the high quality results, ease of use, safety, and reduced waste. In Appendix C, we have updated instructions for the two most widely used spectrophotometers and for vernier calipers.

In this edition you will find expanded coverage of vertebrate anatomy with photographs to accompany drawings and, in response to current interests in brain function and suggestions by reviewers, a new investigation of the sheep brain in Lab Topic 24. The fungi lab topic introduced in the eighth edition has been expanded to include additional life cycles and investigations. We continue updating the phylogeny of the protists and invertebrate phyla to emphasize the latest published hypotheses for phylogenetic relationships based on the explosion of molecular research in this area.

In all of the lab topics, we include connections with advances in current research. In the lab topics with opportunities for open-inquiry, we include suggestions to prompt student questions that will lead to independent investigations and to collaborative team research. The opportunities to extend research in the laboratory and through the scientific literature have been enriched with new suggestions for Investigative Extensions and Applying Your Knowledge questions. Additional support for investigations is provided in the *Preparation Guide for Investigating Biology* available to instructors.

Writing and scientific communication continue to be a strong component of the laboratory program. In the ninth edition, we revised Appendix A Scientific Writing and Communication to provide resources on how to read and summarize scientific literature and to reduce plagiarism. We provide tips for preparing and presenting oral papers and posters and extended resources and websites that provide instructions, checklists, and examples. The writing program is supported by materials in the Preparation Guide, including new templates for proposal writing to coordinate with open-inquiry investigations.

For instructors, we have modified the AIE Teaching Plans to provide clear objectives and support for teaching using inquiry-driven approaches. We include recommendations for organization, student development, and evaluation. No other laboratory manual is supported by a comprehensive laboratory Preparation Guide that includes detailed information on planning, ordering, materials preparation, organization, and care and culture of organisms. The catalog numbers have been updated for all lab topics, and new vendors and sources are included.

In all of these changes and modifications, our objective has been to provide laboratory experiences that are challenging to students and allow them to participate in scientific investigations. We are keenly aware of the constraints on laboratory programs, including number of students and sections, preparation of laboratories and instructors, and the expense and mentoring that are required to teach inquiring young scientists. Our objective is to provide the options and resources that will assist you in meeting the goals for your laboratory program.

Laboratory Topics

The laboratory topics build on information and techniques in previous exercises. Various laboratory exercises incorporate a combination of directed and open-ended procedures. There are basically three types of lab topics included in the manual:

1. *Directed-Inquiry Investigations,* in which exercises have been constructed to involve students in the process of science. We have organized these lab topics to include introductory information from which students develop hypotheses and then predict the results of their experiments. They collect their data and summarize the data in tables and figures of their own construction. The students must then accept or reject their hypotheses, based on their results. Examples of these directed-inquiry investigations include Lab Topic 3 Diffusion and Osmosis, Lab Topic 4 Enzymes, and Lab Topic 6 Photosynthesis.
2. *Key Theme Investigations,* in which laboratory exercises have been designed and reorganized with a focus on key themes of biology, for example, the unity and diversity of life. In these thematic exercises, students summarize and synthesize their results and observations connecting to the key themes. They use their observations as evidence in support of these major concepts and apply their understanding to new problems. Examples of these laboratories (and their underlying themes) include Lab Topic 2 Microscopes and Cells (unity and diversity of life); Lab Topics 14 and 15

Plant Diversity I and II (adaptation to the land environment); and Lab Topics 22 to 24 Vertebrate Anatomy I, II, and III (structure and function).

3. *Open-Inquiry Investigations,* in which students generate their own hypotheses and design their own experiments. These exercises begin with an introduction and a simple experiment that demonstrates procedures. Then students are given suggestions and encouraged to develop their own questions and methodologies for further investigation. Examples of these open-inquiry investigations include Lab Topic 5 Cellular Respiration and Fermentation, Lab Topic 11 Population Genetics, Lab Topic 13 Protists, Lab Topic 17 The Kingdom Fungi, Lab Topic 21 Plant Growth, Lab Topic 26 Animal Behavior, and Lab Topic 28 Ecology II.

Lab topics are designated as *Directed-Inquiry, Key Theme, or Open-Inquiry Investigations* in the table of contents.

Scientific Communication: Writing and Presenting

Scientists communicate their results in writing and in presentations to research groups and at meetings. Undergraduates need instruction in writing and an opportunity to practice these skills; however, instructors do not have the time to critique hundreds of student research reports for each exercise. Throughout this lab manual, teams of students work together on improving their skills. They are asked to organize and present their results to their peers during the discussion and summary sessions in the laboratory. Students are also required to write as part of each laboratory. They summarize and discuss their results and then apply information to new problems in the questions at the end of the laboratory.

We have also incorporated a scientific writing program into our lab manual in a stepwise fashion. Students must answer questions and summarize results within the context of the laboratory exercises. For directed-inquiry investigations, students are required to submit one section of a scientific paper. For example, they might submit the Results section for one experiment in Lab Topic 3, and the Discussion section for one experiment in Lab Topic 4. Once students have experience writing each section, they write at least one complete scientific paper for an open-inquiry investigation, for example, Lab Topics 13, 17, 21, and 26. Instructions for writing a scientific paper, developing an oral presentation, and creating a poster are included in Appendix A, which also contains suggestions for developing a systematic writing program. Instructors may choose to organize a session similar to scientific meetings for oral or poster presentations.

See Appendix A and the Teaching Plans in the Annotated Instructor's Edition for additional information on scientific writing and presentations.

Special Features

Reviewing Your Knowledge: Students recall terminology and content by describing and explaining fundamental concepts. Students examine their results and then use evidence as they evaluate their understanding and knowledge.

Applying Your Knowledge: As instructors, we want our students to be challenged to think and to develop critical thinking skills. Throughout this manual, students are asked to work logically through problems, critique results, and modify hypotheses. To emphasize these skills further, we have developed a section in each laboratory topic called Applying Your Knowledge, in which students are asked to apply their knowledge to new problems and to make connections between topics. The questions encourage students to use their knowledge as they solve problems developed from current research as well as issues in science, medicine, and society.

Excel Tables: All data tables used for recording and analyzing results are available in Excel format for modification and use in the laboratory. These tables are designated in the table title and can be downloaded to laboratory computers for student use at www.masteringbiology.com in the Study Area.

Images and Data: Full-color images and figures are integrated throughout the lab manual. Lab Topic 16 Bioinformatics: Molecular Phylogeny of Plants is supported by images in the text as well as the masteringbiology.com website. Folders found under the Study Area contain images of the plants used in the investigation along with the edited and ready-to-use nucleotide sequences for each species. These can be downloaded to laptops or accessed on the website.

Investigative Extensions and Case Studies: Students and instructors have expressed interest in extending laboratory topics to open-inquiry investigations or simply to pursue additional questions that connect the topic to current research and issues. At the end of most lab topics, we have provided questions to prompt student-designed investigations. For a few, we provide case studies that build on the lab topic and require additional reading and research.

Student Media and Web Resources: Providing media and Web resources can enhance teaching and learning in the laboratory. For students, we have included references to videos that connect to their laboratory activities. These are designated in the text with a media icon. Also, at the end of each lab topic, we have included the section **Student Media: BioFlix, Activities, Investigations, Videos, and Data Tables,** which directs students to the website for **activities** and **investigations** that can be used to prepare for the laboratory or review and practice after the laboratory. These media resources are available at www.masteringbiology.com in the Study Area.

Safety Considerations: Safety concerns are noted in the text by the use of icons for general safety and for biohazards. Laboratory safety is also addressed in the Teaching Plan at the end of each lab topic in the Instructor's Edition. Note the **Laboratory Safety Guidelines** printed on the inside front cover.

Notes to Students: To assure student success, cautionary reminders and notes of special interest are also highlighted in the text.

Appendixes: Information needed in several laboratory topics is included in the appendixes: scientific writing and communication, the metric system, instrumentation and techniques, using chi-square analysis, and dissection terminology and techniques.

Instructional Support

The Preparation Guide and the Teaching Plans in the Annotated Instructor's Edition provide valuable suggestions and essential information for the successful implementation of the laboratory topics.

Preparation Guide: A detailed *Preparation Guide for Investigating Biology* accompanies the laboratory manual. It contains materials lists, suggested vendors, instructions for preparing solutions and constructing materials, schedules for planning advance preparation, and suggestions for organizing materials in the lab. The Preparation Guide is essential for successfully preparing and teaching these investigative laboratories. It is now available in hard copy and electronically at masteringbiology.com in Instructor Resources, Instructor Guides for Supplements. Instructors can download preparation and ordering lists as needed and customize these for their program.

Annotated Instructor's Edition: Teaching biology using an investigative approach requires that instructors guide students in posing questions and hypotheses from which they can predict the results of their experiments. We have included additional support for the instructor in the form of instructor's annotations in each lab topic. These annotations are intended to guide the instructor in responding to students, not to provide the right answers to every student question. We encourage instructors to become the guide to discovery rather than the repository of correct answers. Features of the Annotated Instructor's Edition include:

- **Margin notes** with simple suggestions, such as accepting hypotheses as long as they are testable, hints for success, and additional explanations appropriate for the instructor.
- **Suggested answers** to student questions.
- **Typical results**
- **Explanatory figures**
- **Teaching Plans**

Teaching Plans: The teaching plans are the instructor's guide to organizing and teaching each laboratory, and they reflect our objective to systematically develop more effective ways to engage students in the study of biology. The teaching plans have been particularly useful for instructors initiating investigative approaches to lab programs. Instructors should feel free to modify these plans to meet their specific needs. The Teaching Plan for each lab topic includes:

- **Detailed objectives** both for content and for development of skills in problem solving and scientific methodology.
- **Suggested order of the lab.**
- **Estimated time requirements** for each portion of the lab.
- **Suggested options for organizing the activities for a 2-hour lab period.**
- **Hints on how to manage groups of students** and involve them in investigations that might otherwise become passive learning experiences.
- Information about **lab safety precautions.**

Acknowledgments

The development of our ideas, the realization of those ideas in laboratory investigations, and the preparation of this laboratory manual are the result of collaborations with many colleagues over the years. We are indebted to our teaching assistants, whose critical evaluations and insightful suggestions helped shape the exercises. Several colleagues made especially helpful or critical contributions to our efforts, including Evelyn Bailey, Steve Baker, Joy Budensiek, Sarah Fankhauser, Nitya Jacob, LaTonia Taliaferro-Smith, and Theodosia Wade. We appreciate the technological support and advice of Scott Foster, Rob Morgan, Kyle Nelson, and Cleave Pierce. We are grateful to Jacobus de Roode, Emory University, for sharing photographs and new research on the behavior of monarch butterflies, as well as suggestions for the exercise in Lab Topic 1 Scientific Investigation. We are grateful for the support and guidance provided by the editorial and production group at Pearson. Our thanks to Beth Wilbur, Josh Frost, Margaret Young, Lauren Harp, Mike Early, Maureen Spuhler, Janet Wehner, and Monica Moosang. We are especially indebted to all the laboratory educators who have shared their ideas, hints for success, and philosophies of teaching with us, particularly our friends in the Association for Biology Laboratory Education (ABLE). We are grateful to our reviewers for sharing their words of encouragement and criticism, which were essential to the success of our work. We thank our students who, over the years, have provided the ultimate test for these investigations and our ideas. Finally, our deep appreciation to all members of our families for their good humor, patience, and encouragement, with special thanks to Bill Morgan for keeping us supplied with coffee, water, and carbs during our extended and intense working sessions.

Reviewers

Mark Chiappone, *Miami Dade*
Andrew David, *Clarkson University*
Jennifer Ellie, *Wichita State*
Eugenia Hurley, *Fordham University*
Kristin Latham, *Oregon State*
Abby Levitt, *North Central State*
Ana Ribeiro, *College of Mount Saint Vincent*
Bin Shuai, *Wichita State*

Reviewers for Previous Editions

Mitch Albers, *Minneapolis Community College*
Carol Alia, *Sarah Lawrence College*
Connie L. Allen, *Edison Community College*
Phyllis Baudoin, *Xavier University of Louisiana*
Mariette Baxendale, *University of Missouri, St Louis*
Carol Bernson, *University of Illinois*
Charles Biggers, *Memphis State University*
Sandra Bobick, *Community College of Allegheny County*
Amanda Boose, *Gonzaga University*
Jerry Bricker, *Cameron University*
Jeffrey T. Burkhart, *University of LaVerne*
Linda Burroughs, *Rider University*
Thomas Butler, *SUNY Rockland*
Neil Campbell, *University of California, Riverside*
Nickie Cauthen, *Clayton College and State University*
Deborah Clark, *Oregon State University*
Sunita Cooke, *Montgomery College*
Charles Creutz, *University of Toledo*
Diana R. Cundell, *Philadelphia University*
Jean DeSaix, *University of North Carolina, Chapel Hill*
Rebecca Diladdo, *Suffolk University*
John Doherty, *Villanova University*
Susan Dunford, *University of Cincinnati*
Sharon Fugate, *Madisonville Community College*
April Ann Fong, *Portland Community College, Sylvania Campus*
Caitlin Gabor, *Southwest Texas State University*
William Garnett, *Raymond Walters College*

Troy Giambernardi, *Grand Rapids Community College*
Patricia S. Glas, *The Citadel*
Michelle F. Guadette, *Tufts University*
Regina Guiliani, *Saint Peter's College*
Tracy Halward, *Colorado State University*
Teresa Hanlon, *Washington and Lee University*
John Hare, *Linfield College*
Laurie Henderson, *Sam Houston State University*
Daniel Hoffman, *Bucknell University*
Alan Jaworski, *University of Georgia*
Virginia A. Johnson, *Towson State University*
Tamas Kapros, *University of Missouri, Kansas City*
Douglas Kane, *Defiance College*
Ann Kircher, *College of Alameda*
Nikolai Kirov, *New York University*
George Krasilovsky, *Rockland Community College*
Laura G. Leff, *Kent State University*
Brian T. Livingston, *University of Missouri, Kansas City*
Raymond Lynn, *Utah State University*
Presley Martin, *Drexel University*
Stephanie Maruhnich, *Florida State College at Jacksonville*
Barbara J. Maynard, *Colorado State University*
Dawn McGill, *Virginia Western Community College*
Jacqueline McLaughlin, *Pennsylvania State University, Lehigh Campus*
Michael McVay, *Green River Community College*
Charles W. Mims, *The University of Georgia*
Alison Mostrom, *Philadelphia College of Pharmacy and Science*
Margaret Olney, *Saint Martin's University*
Greg Paulson, *Washington State University*
Lynn Petrullo, *College of New Rochelle*
Shelley A. Phelan, *Fairfield University*
Marcia Pierce, *Eastern Kentucky University*
Catherine Purzycki, *University of the Sciences*
Jane Rasco, *University of Alabama*
Carolyn Rasor, *Southern Methodist University*
John Rousseau, *Whatcom Community College*
Ana Ribiero, *College of Mount Saint Vincent*
Eugenia Ribiero-Hurley, *Fordham University*
Michael R. Schaefer, *University of Missouri, Kansas City*
Daniel C. Scheirer, *Northeastern University*
Bin Shuai, *Wichita State University*
Sandra Slivka, *San Diego Miramar College*
Nanci Smith, *University of Texas, San Antonio*
Jenise Snyder, *Ursuline College*
Gerald Summers, *University of Missouri*
Sheryl Swartz Soukup, *Illinois Wesleylan University*
Nina Theis, *Elms University*
Michael Toliver, *Eureka College*
Geraldine W. Twitty, *Howard University*
Shauna Weyrauch, *Central Ohio Technical College*
Roberta Williams, *University of Nevada, Las Vegas*
Dan Wivagg, *Baylor University*
Carol Yeager, *The University of Alabama*

M. Eloise Brown Carter Judith Giles Morgan

To Mary, Bill, Rob, Laura, Tori, Kate, and Will, with love
J.G.M.

To Stefanie and Cyndi, with love
M.E.B.C.

About the Authors

Judith Giles Morgan received her M.A. degree from the University of Virginia and her Ph.D. from the University of Texas, Austin. She is Professor Emeritus of Biology at Emory University, where she developed a general biology laboratory curriculum for majors to incorporate an investigative approach and a TA training program for multisection investigative laboratories. Dr. Morgan directed and taught Emory's Coastal Biology Program for many years. She served as an officer and board member of the Association for Biology Laboratory Education. Dr. Morgan has worked to improve science achievement for promising area high school students, as she directed science summer programs at Emory University, and by leading science laboratories for elementary classrooms.

M. Eloise Brown Carter earned her M.S. and Ph.D. from Emory University and is Professor of Biology at Oxford College of Emory University. With expertise in plant ecology, she has incorporated independent and team research into introductory and advanced biology courses. Dr. Carter served as President of the Association of Southeastern Biologists. She has received several teaching and service awards, including Oxford's Phi Theta Kappa and Fleming Awards, and Emory's Williams Award and Thomas Jefferson Award. Dr. Carter taught in the Oxford Institute for Environmental Education, a program for precollege teachers to improve science education through development of schoolyard investigations.

Contents

Credits

Photo Credits

Lab Topic 1

1.1: Jaap de Roode; 1.10: Barry Mansell/Nature Picture Library.

Lab Topic 2

2.1: Judith G. Morgan; 2.2: Eloise Carter; 2.3: Judith G. Morgan; 2.5a: Biophoto Associates/Science Source; 2.5b: Steve Gschmeissner/Science Source; 2.6: Melba/Age Fotostock; 2.7: Richard L. Howey; 2.8: Yuuii Tsukii/Protist Information Server; 2.9: Dr. Ralf Wagner; 2.10: Manfred Kage/Science Source; 2.11: Perennou Nuridsany/Science Source; 2.12: Bruce Iverson/SPL/Science Source; 2.13: Yuuii Tsukii/Protist Information Server.

Lab Topic 3

3.3: Photo InsoliteRealite/Science Source; 3.7: Eloise Carter.

Lab Topic 4

109: Steve Shott/Dorling Kindersley, Ltd.

Lab Topic 5

5.7: L. Brent Selinger/Pearson Education, Inc.

Lab Topic 6

6.3: Eloise Carter; 6.4: Judith Morgan; 6.7: L. Brent Selinger/Pearson Education, Inc.

Lab Topic 7

7.3: Biophoto Associates/Science Source; 7.9a: Don W. Fawcett/Science Source; 7.9c: B. A. Palevitz, Courtesy of E. H. Newcomb/University of Wisconsin; 7.11: Pearson Education, Inc.; 7.12: Pearson Education, Inc.; 7.13: Pearson Education, Inc.; 7.14: Pearson Education, Inc.; 7.15: Ed Reschke/Photolibrary/Getty Images; 7.16: Pearson Education, Inc.; 7.17a, 7.17b, 7.17c, 7.17e: Ed Reschke/Photolibrary/Getty Images; 7.17d: Michael Abbey/Science Source; 7.18: Judith G. Morgan; 7.21g1: Michael Franklin Micro Imaging; 7.21g2: Dr. George L. Barron/University of Guelph; 7.23: Michael Clayton; 182: Elisabeth S. Pierson/Pearson Education, Inc.

Lab Topic 8

8.3: JS Photo/Alamy Stock Photo.

Lab Topic 9

9.4a, b: Judith G. Morgan; 9.5a, b: Gunter Reuter.

Lab Topic 11

11.5a-d: The genetic basis of adaptive melanism in pocket mice. M. W. Nachman, H.E. Hoekstra, and S. L. D'Agostino. Proc Natl Acad Sci U S A. 2003 Apr 29; 100(9):5268-73, fig. 1B.; 11.6: Eye of Science/Science Source.

Lab Topic 12

12.2a: Christine Case; 12.2b: Centers for Disease Control and Prevention (CDC); 12.5: ASM/Science Source; 12.10: L. Brent Selinger/Pearson Education, Inc.; 12.11: L. Brent Selinger/Pearson Education, Inc.

Lab Topic 13

13.1b: John Mansfield; 13.2: Michael Abbey/Science Source; 13.3a: Frank Fox/Science Source; 13.3b: Eric V. Grave/Science Source; 13.4a: Colin Bates/Coastal Imageworks (coastalimageworks.com); 13.4b: David Hall/Science Source; 13.4c: Eloise Carter; 13.5: Fred M. Rhoades; 13.6b: Virginia Institute of Marine Science; 13.7a: David M. Phillips/Science Source; 13.7b: DP Wilson/Frank Lane Picture Agency Limited; 13.8: Biophoto Associates/Science Source; 13.9: Claude Carre/Science Source; 13.10: Image Quest 3.D/NHPA/Photoshot; 13.11: Melba/AGE Fotostock; 13.12: William H. Crowder/National Geographic Creative/Alamy Stock Photo; 13.13: Stephen Sharnoff; 13.14: Bruno in Columbus; 13.15a: Ed Reschke/Photolibrary/Getty Images; 13.15b: D. P. Wilson/Eric & David Hoskins/Science Source; 13.15c: Wayne Armstrong; 13.16a: Wolfgang Bettinghofer/David Patterson, micro*scope; 13.16b: Blickwinkel/NaturimBild/Alamy Stock Photo; 13.16c: Laurie Campbell/NHPA/Photoshot; 13.17: Center for Aquatic and Invasive Plants, Institute of Food and Agricultural Sciences (IFAS), University of Florida.

Lab Topic 14

14.4: Tony Wharton/Frank Lane Picture Agency Limited; 14.6: Judith G. Morgan; 14.7a: Patrick J. Lynch/Science Source; 14.7b: Jody Banks, Purdue University; 14.8b: Steven P. Lynch, Lynch Images; 14.9: Gordon Dickson; 14.10: Shari L. Morris/Age Fotostock; 14.11a: Eloise Carter; 14.11b: Judith G. Morgan; 14.12: Eloise Carter; 14.14a: The Open University.

Lab Topic 15

15.1a1, 15.1c: Eloise Carter; 15.1a2, 15.1b: Judith G. Morgan; 15.1d: Biophoto Associates/Science Source; 15.1e1: Bob Gibbons/Photoshot; 15.1e2: Al Schneider; 15.4a: F. Rauschenbach/F1online/Age Fotostock; 15.4b: Darlyne A. Murawski/National Geographic Creative/Alamy Stock Photo; 15.4c: Anthony Mercieca/Science Source; 15.4d: Merlin D. Tuttle/Science Source; 15.6: Ed Reschke; 15.7: Eloise Carter.

Lab Topic 16

432 1, 2: Eloise Carter; 433 1, 3: Eloise Carter; 433 2: Judith G. Morgan; 434 1: Eloise Carter; 434 2-4: Judith G. Morgan. 16.4-11: San Diego Supercomputer Center. Used by permission.

Lab Topic 17

17.2: Darlyne A. Murawski/National Geographic/Getty Images; 17.3a: Comar/Fotolia; 17.3b: Maureen Spuhler;17.5: Judith Morgan;17.6 inset: Biophoto Associates/Science Source; 17.7a: Hecker/Blickwinkel/Alamy Stock Photo;17.7b: Jubal Harshaw/Shutterstock;17.7c: Biology Pics/Science Source;17.8 inset: Judith G. Morgan; 17.8a, 17.8c: Judith G. Morgan; 17.8b: Photo by Joe H. Taft; 17.8d: Nancy Rose; 17.9a-c: Judith G. Morgan.

Lab Topic 18

18.3: Wilfried Bay-Nouailhat/Mer-Et-Littoral;18.4: Paddy Ryan Photographic;18.6: Edward L. Snow/Photoshot;18.8b: Rick Gillis, University of Wisconsin-La Crosse Biology Department;18.9: Kjell Sandved/Science Source; 18.10: Robert L. Dunne/Science Source.

Lab Topic 19

19.2a: Kjell Sandved/Photoshot; 19.3a: Judith G. Morgan; 19.4a: Luis Castaneda Inc./The Image Bank/Getty Images;19.6: Samuel Morisoli/Alamy Stock Photo;19.7a: Jeff Rotman/Nature Picture Library; 19.7b: Eugene Sim/Fotolia; 19.8: Heather Angel/Natural Visions/Alamy Stock Photo.

Lab Topic 20

20.2a, 20.2c, 20.2e: Judith G. Morgan; 20.2b: Eric Grave/Science Source; 20.2d: Gene Weller; 20.4b: Dorling Kindersley Ltd./Alamy Stock Photo; 20.6b, 20.6c: Ed Reschke/Photolibrary/Getty Images;

20.7b, 20.7c: Ed Reschke/Photolibrary/Getty Images; 20.9b: Professor Ray F. Evert; 20.9c: Michael Clayton; 20.11b: Ed Reschke/Photolibrary/Getty Images.

Lab Topic 21

21.7: Martin Shields/Alamy Stock Photo; 21.8: Burkhard Schulz-PSLA Dept., University of Maryland.

Lab Topic 22

22.1: Marian Rice; 22.2a, 22.2c, 22.2d: Marian Rice; 22.2b: Nina Zanetti/Pearson Education, Inc.; 22.2e: Biophoto Associates/Science Source; 22.3:Marian Rice; 22.4b: Victor P. Eroschenko; 22.5a: Judith G. Morgan; 22.7: John Hamilton/Pearson Education, Inc.; 22.9c: Anne Geller; 22.10a: LUMEN Histology.

Lab Topic 23

23.1: Charles Venglarik/Pearson Education, Inc.; 23.2b: Charles Venglarik/Pearson Education, Inc.; 23.6: Charles Venglarik/Pearson Education, Inc.; 23.9: Biophoto Associates/Science Source; 23.10c: John Hamilton/Pearson Education, Inc.

Lab Topic 24

24.2b: John Hamilton/Pearson Education, Inc.; 24.3b: Elena Dorfman/Pearson Education, Inc.; 24.7a: Pearson Education, Inc.; 24.7b: Pearson Education, Inc.; 24.7c: Pearson Education, Inc.; 24.8b: Judith G. Morgan.

Lab Topic 25

25.5: George von Dassow; 25.6: George von Dassow; 25.7: George von Dassow; 25.8: George von Dassow; 25.9: George von Dassow; 25.10: George von Dassow; 25.11: George von Dassow; 25.13a, 25.13b, 25.13c, 25.13d, 25.13f: Mary Martin/Science Source; 25.13e: Andrew Ewald; 25.14: Mike O Carroll/Alamy Stock Photo; 25.15: © 2003-Steven J. Baskauf bioimages.vanderbilt.edu; 25.16: © 2003-Steven J. Baskauf bioimages.vanderbilt.edu; 25.17: Judith G. Morgan; 25.18: © 2003-Steven J. Baskauf bioimages.vanderbilt.edu; 25.19: Judith G. Morgan; 25.20: Judith G. Morgan; 25.21: Judith G. Morgan; 25.22: © 2003-Steven J. Baskauf bioimages.vanderbilt.edu; 25.23: Judith G. Morgan; 25.24 (inset): Judith G. Morgan; 25.25a: Pearson Education, Inc.; 25.26: Pearson Education, Inc; 25.27: Pearson Education, Inc; 25.28: Oxford Scientific/Getty Images; 25.29: Pearson Education, Inc.

Lab Topic 26

26.1: Blickwinkel/Hartl/Alamy Stock Photo; 26.2: Danny Smythe/Alamy Stock Photo; 26.3: Narin Sapaisarn/Fotolia.

Appendix C

C-1a: Eloise Carter; C-1b: Eloise Carter; C-2: L. Brent Selinger/Pearson Education, Inc.; C-3: Thermo-Fisher Scientific; C-4a: Eloise Carter; C-4b: Eloise Carter.

Appendix E

E-1, 2, 3: Eric Isselee/Shutterstock.

Text Credits

Lab Topic 1

1.4: Based on the Isle Royale Wolf-Moose Study;1.5: National Center for Health Statistics Biology & Microbiology;1.7: Based on Lefèvre, T., A. Chiang, M. Kelavkar, H. Li, J. Li, C. Lopez, F. deCastillejo, L. Oliver, Y. Potini, M. Hunter, and J. C. de Roode. "Behavioural Resistance Against a Protozoan Parasite in the Monarch Butterfly." Journal of Animal Ecology, 2012, vol. 81, pp. 70–79.; 1.8: Based on Lefèvre, T., A. Chiang, M. Kelavkar, H. Li, J. Li, C. Lopez, F. deCastillejo, L. Oliver, Y. Potini, M. Hunter, and J. C. de Roode. "Behavioural Resistance Against a Protozoan Parasite in the Monarch Butterfly." Journal of Animal Ecology, 2012, vol. 81, pp. 70–79; 1.9: Based on Dietary lipid quality and mitochondrial membrane composition in trout: Responses of membrane enzymes and oxidative capacities by N. Martin, D. P. Bureau, Y. Marty, E. Kraffe & H. Guderley, published by Springer Verlag Berlin Heidelberg, 2012.

Lab Topic 28

PG # 781: Cartoon by www.jimhunt.us.

LAB TOPIC 1

Scientific Investigation

Before going to lab, read the Introduction and Exercises 1.1 and 1.2. Be prepared to answer all questions and contribute your ideas in a class discussion.

Laboratory Objectives

After completing this lab topic, you should be able to:

1. Identify and characterize questions that can be answered through scientific investigation.
2. Define *hypothesis* and explain what characterizes a good scientific hypothesis.
3. Identify and describe the components of a scientific experiment.
4. Develop hypotheses and predictions for a scientific investigation.
5. Design and conduct an experiment to test the hypothesis.
6. Summarize and present results in tables and graphs.
7. Discuss results and critique experiments.
8. Interpret and communicate results.

Introduction

Biology is the study of the phenomena of life, and biological scientists—researchers, teachers, and students—observe living systems and organisms, ask questions, and propose explanations for those observations. Scientific investigation is a way of testing those explanations. Science assumes that biological systems are understandable and can be explained by fundamental rules or laws. Scientific investigations share some common elements and procedures, which are referred to as the *scientific method*. Not all scientists follow these procedures in a strict fashion, but each of the elements is usually present. Science is a creative human endeavor that involves asking questions, making observations, developing explanatory hypotheses, and testing those hypotheses. Scientists closely scrutinize investigations in their field, and each scientist presents his or her work at scientific meetings or in professional publications, providing evidence from observations and experiments that supports the scientist's explanations of biological phenomena.

In this lab topic, you will not only review the process that scientists generally use to ask and answer questions about the living world, but you will develop the skills to conduct and critique scientific investigations. Like scientists, you will work in research teams in this laboratory and others, collaborating as you ask questions and solve problems. Throughout this laboratory manual, you will be investigating biology using the methodology of scientists, asking questions, proposing explanations, designing experiments, predicting results, collecting and analyzing data, and interpreting your results in light of your hypotheses.

EXERCISE 1.1

Questions and Hypotheses

This exercise explores the nature of scientific questions and hypotheses. Before going to lab, read the explanatory paragraphs, answer the questions, and then be prepared to present your ideas in the class discussion.

Lab Study A. Asking Questions

Scientists are characteristically curious and creative individuals whose curiosity is directed toward understanding the natural world. They use their study of previous research or personal observations of natural phenomena as a basis for asking questions about the underlying causes or reasons for these phenomena. *For a question to be pursued by scientists, the phenomenon must be well defined and testable. The elements must be measurable and controllable.*

There are limits to the ability of science to answer questions. Science is only one of many ways of knowing about the world in which we live. Consider, for example, this question: Do excessively high temperatures cause people to behave immorally? Can a scientist investigate this question? Temperature is certainly a well-defined, measurable, and controllable factor, but morality of behavior is not scientifically measurable. We probably could not even reach a consensus on the definition. Thus, there is no experiment that can be performed to test the question, and this question cannot be pursued through the scientific process. Which of the following questions do you think can be answered scientifically?

1. Are domesticated dogs as smart as wolves?
2. Will drinking milk from arthritic goats reduce HIV levels in infected individuals?
3. Do urban honey bees prefer processed sugar over flower nectar?
4. How effective are marigold and rosemary as insect repellants?
5. Should governments prohibit distribution of GM food during severe famine?

How did you decide which questions can be answered scientifically?

Lab Study B. Developing Hypotheses

As questions are asked, scientists attempt to answer them by proposing possible explanations. Those proposed explanations are called **hypotheses**. A hypothesis tentatively explains something observed. It proposes an answer to a question. Consider the preceding question 4 about insect repellents. One hypothesis based on this question might be "Marigold and rosemary extracts are more effective than DEET in repelling insects." The hypothesis has suggested a possible explanation that compares the difference in efficacy between these plant extracts and DEET.

A scientifically useful hypothesis must be testable and falsifiable (able to be proved false). To satisfy the requirement that a hypothesis be falsifiable, it must be possible that the test results do not support the explanation. In our example, the experiment might be to spray one arm with plant extracts and the other with DEET. Then place both arms in a chamber with mosquitoes. If mosquitoes bite both arms with equal frequency or the DEET arm has fewer bites, then the hypothesis has been falsified. *Even though the hypothesis can be falsified, it can never be proved true.* The evidence from an investigation can only *provide support* for the hypothesis. In our example, if there are fewer bites on the arm with plant extracts than on the DEET-treated arm, then the hypothesis has not been proved, but has been supported by the evidence. Other explanations still must be excluded, and new evidence from additional experiments and observations might falsify this hypothesis at a later date. In science, seldom does a single test provide results that clearly support or falsify a hypothesis. In most cases, the evidence serves to modify the hypothesis or the conditions of the experiment.

Science is a way of knowing about the natural world (Moore, 1993) that involves testing hypotheses or explanations. Students often think that controlled experiments are the only way to test a hypothesis. The test of a hypothesis may include experimentation, additional observations, or the synthesis of information from a variety of sources. Many scientific advances have relied on other procedures and information to test hypotheses. For example, James Watson and Francis Crick developed a model that was their hypothesis for the structure of DNA. Their model could only be supported if the accumulated data from a number of other scientists were consistent with the model. Actually, their first model (hypothesis) was falsified by the work of Rosalind Franklin. Their final model was tested and supported not only by the ongoing work of Franklin and Maurice Wilkins but also by research previously published by Erwin Chargaff and others. Watson and Crick won the Nobel Prize for their scientific work. They did not perform a controlled experiment in the laboratory but tested their powerful hypothesis through the use of existing evidence from other research. Methods other than experimentation are acceptable in testing hypotheses. Think about other areas of science that require comparative observations and the accumulation of data from a variety of sources, all of which must be consistent with and support hypotheses or else be inconsistent and falsify hypotheses.

The information in your biology textbook is often thought of as a collection of facts, well understood and correct. It is true that much of the knowledge of biology has been derived through scientific investigations, has been thoroughly tested, and is supported by strong evidence. However, scientific knowledge is always subject to novel experiments and new technology, any aspect of which may result in modification of our ideas and a better understanding of biological phenomena. The self-correcting nature of science can be seen in newly proposed phylogenetic trees illustrating evolutionary relationships among groups of animals. For many years these relationships were based on studies of animal structure and development. Recently, new tools in molecular biology have provided scientists with the ability to sequence genes and even whole genomes. Results from molecular biology reveal new relationships among animal groups that are the basis for reconstructing evolutionary trees.

Application

Before scientific questions can be answered, they must first be converted to hypotheses, which can be tested. For each of the following questions, write an explanatory hypothesis. Recall that the hypothesis is a statement that explains the phenomenon you are interested in investigating.

1. Can human dander cause allergic reactions in cats?

2. Are teens who take vitamin B_{12} supplements more likely to have acne outbreaks?

Scientists often propose and reject a variety of hypotheses before they design a single test. Discuss with your class which of the following statements would be useful as scientific hypotheses and could be investigated using scientific procedures. *Give the reason for each answer by stating whether it could possibly be falsified and what factors are measurable and controllable.*

1. A high-fat diet in pregnant women may increase chances of ADHD in offspring.
2. Living on a farm protects children from developing asthma and hayfever.
3. The Y chromosome in modern humans is the same as the Y chromosome in Neanderthals.
4. Genetically modified foods have caused an increase in gluten intolerance.
5. "Snake oil" from pythons can be used to build heart muscle.

EXERCISE 1.2

Designing Experiments to Test Hypotheses

The most creative aspect of science is designing a test of your hypothesis that will provide unambiguous evidence to falsify or support a particular explanation. Scientists often design, critique, and modify a variety of experiments and other tests before they commit the time and resources to perform a single experiment. In this exercise, you will follow the procedure for experimentally testing hypotheses, but it is important to remember that other methods, including observation and the synthesis of other sources of data, are acceptable in scientific investigations. An experiment involves defining variables, outlining a procedure, and determining controls to be used as the experiment is performed. Once the experiment is defined, the investigator predicts the outcome of the experiment based on the hypothesis.

Read the following case study of a scientific investigation of the disease defense mechanisms in monarch butterflies. You can hear Dr. de Roode describe his research in a recent TED Talk, How butterflies self-medicate: https://www.ted.com/talks/jaap_de_roode_how_butterflies_self_medicate. In Lab Study A you will determine the types of variables involved, and in Lab Study B, you will determine the experimental procedure for this experiment and for others.

Case Study: Investigating Adult Monarch Butterfly Preference for Food Plants That Reduce Parasitic Infection in Their Offspring (Lefèvre et al., 2012)

In nature, free-living organisms are constantly infected with a wide range of parasites, and it is not uncommon to discover that many organisms have evolved defensive mechanisms to reduce the negative effects of parasitic infection. The monarch butterfly (*Danaus plexippus*) is regularly infected by the protozoan parasite *Ophryocystis elektroscirrha*, and this infection negatively affects the growth and survivability of these butterflies.

In the life cycle of a monarch butterfly, an adult female lays eggs on a milkweed plant, and after hatching, the larvae feed on the plant (Figure 1.1). Each larva pupates to an immobile chrysalis, from which a butterfly later emerges. Scientists in the laboratory of Jacobus de Roode of Emory University are studying if monarch butterflies have evolved defense mechanisms to protect offspring or adults against these protozoan parasites. In one experiment performed in de Roode's lab, the researchers asked if adult female butterflies infected with the protozoan parasite *preferentially* lay their eggs on milkweed plants that reduce parasite growth in their offspring, or do they *randomly* deposit eggs, regardless of the type of milkweed plant? *They hypothesized that female butterflies would lay more eggs on plants that reduce parasite growth.*

FIGURE 1.1
Butterfly and milkweed experiment. Monarch butterfly laying eggs on a milkweed plant.

To test this hypothesis, they bred monarchs in the lab to obtain larvae. They then infected some larvae with the protozoan parasite, while leaving other larvae uninfected. All larvae were reared in a similar environment. As adult butterflies emerged, infected females were transferred to a mating cage. Control female monarchs (*uninfected*) were transferred to another cage. Uninfected males were released into each cage to serve as mating partners. Three days after mating (average time required for egg maturation), 10 infected and 10 uninfected females were released one at a time into an egg-laying cage containing two milkweed plants. One plant, *Asclepias curassavica*, is known to contain high concentrations of chemicals that reduce parasite infection in offspring, and the second plant, *Asclepias incarnata*, contains the chemicals but in lower concentrations. After allowing the females to lay eggs on the milkweed plants for 1 hour, they were returned to their holding cage. The investigators then counted the number of eggs laid on each milkweed plant and calculated the proportion of eggs laid on *A. curassavica* (based on the total number of eggs laid on both species). Two days later the experiment was repeated with the same female butterflies (Figure 1.2).

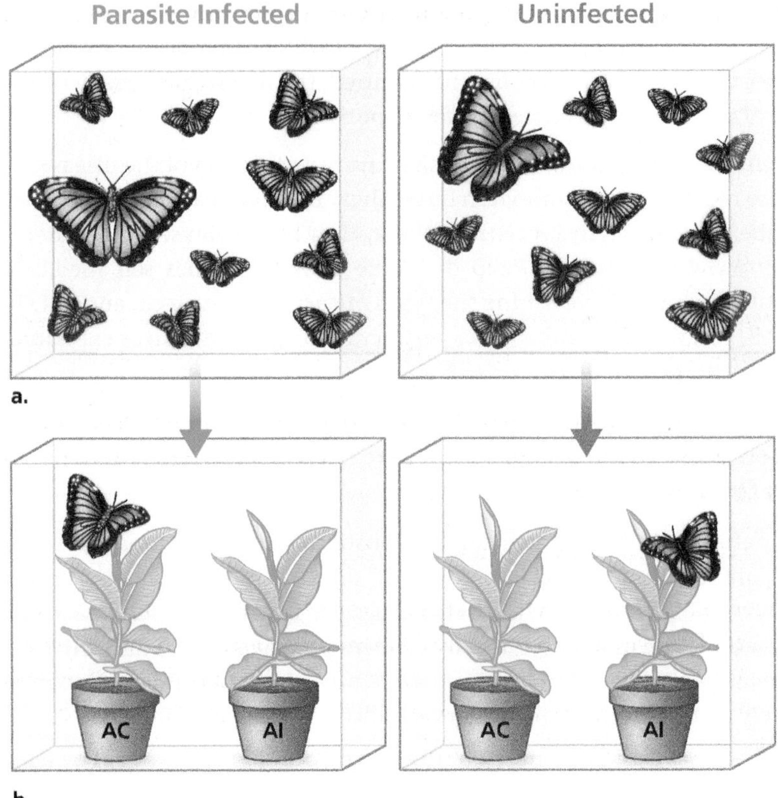

FIGURE 1.2
Experimental design for investigating parasite defense in monarch butterflies. (a) Parasite infected and uninfected female butterflies in mating cages. (b) Butterflies lay eggs in separate cages containing two milkweed plants with different levels of anti-parasite chemicals: *A. curassavica* (AC) with high levels and *A. incarnata* (AI) with low levels. Number of eggs laid is counted. The experiment is repeated two days later.

Hypothesis

State the hypothesis for this study investigating the effect of parasite infection on egg laying in monarch butterflies:

Lab Study A. Determining the Variables

Read the description of each category of variable, and then identify the variable described in the preceding investigation. The variables in an experiment must be clearly defined and measurable. The investigator must identify and define *dependent, independent,* and *controlled* variables for a particular experiment.

The Dependent Variable

Within the experiment, one variable will be measured, counted, or observed in response to the experimental conditions. This variable is the **dependent variable**. For the monarch butterfly experiment as described, one dependent variable was measured. What is this variable?

Although only one dependent variable was measured in this experiment, it is acceptable to measure several dependent variables in an experiment. What additional dependent variables might be measured in this experiment, and why is this acceptable?

Scientists often include more than one dependent variable because different measurements may give additional information about the hypothesis being investigated. In this case, the question is whether females prefer to lay their eggs on plants that reduce parasitic infection in their offspring. Therefore, measuring characteristics that indicate the survivability of the next generation might add more evidence to support the hypothesis.

The Independent Variable

The scientist will choose one variable or experimental condition to manipulate or change. This variable is considered the most important variable by which to test the investigator's hypothesis, and is called the **independent variable**. What was the independent variable in the investigation of female butterfly preference of milkweed plants?

Can you suggest other variables that the investigators might have changed that would have had an effect on the dependent variable?

Although other factors such as light, humidity, the source of the larvae, the natural habitat of the butterflies, differences in test plants—might affect the dependent variable—only one independent variable is usually chosen. Why is it important to have only one independent variable?

The Controlled Variable

Consider the variables that you identified as alternative independent variables. Although they are not part of the hypothesis being tested in this investigation, they could have significant effects on the outcome of this experiment. These variables must therefore be kept constant during the course of the experiment. They are known as the **controlled variables**. The underlying assumption in experimental design is that the selected independent variable is the one affecting the dependent variable. This is only true if all other variables are controlled. What are the controlled variables in this experiment?

Lab Study B. Choosing or Designing the Procedure

The **procedure** is the stepwise method, or sequence of steps, to be performed for the experiment. It should be recorded in a laboratory notebook before initiating the experiment, and any exceptions or modifications should be noted during the experiment. The procedures may be designed from research published in scientific journals, through collaboration with colleagues in the lab or other institutions, or by means of one's own novel and creative ideas. The process of outlining the procedure includes determining levels of treatments, numbers of replications, and control treatment(s).

Level of Treatment

The value set for the independent variable is called the **level of treatment**. This is based on knowledge of the system and the biological significance of the treatment level. For example, if you were investigating the effect of sulfur dioxide (the independent variable) on plants growing near a coal-fired power plant, you would treat plants in your experiment with concentrations of sulfur dioxide that fall below, throughout, and above the range found around the power plant. In some experiments, however, independent variables represent categories that do not have a level of treatment. This is the case in the butterfly experiment, where the independent variable is the presence of parasites in the adult butterflies.

Replication

Scientific investigations are not valid if the conclusions drawn from them are based on only one experiment with one or two individuals. Generally, the same procedure will be repeated several times (replication) and many individuals must be used. Describe replication in this experiment.

Control

The experimental design includes a control in which the independent variable is held at an established level or is omitted. The control or control treatment serves as a benchmark that allows the scientist to decide whether the predicted effect is really due to the independent variable. What was the control in this experiment?

What is the difference between the control and the controlled variables discussed previously?

Lab Study C. Making Predictions

Once the investigator has developed a testable hypothesis and a way to test that hypothesis, then he or she can clearly state the predicted outcome that will either support or falsify the hypothesis. The **prediction** is always based on the particular experiment designed to test a specific hypothesis. Predictions are written in the form of if/then statements: "If the hypothesis is true, then the results of the experiment will be ..."; for example, "**If** extracts of marigold and rosemary are more effective than DEET in repelling insects, **then** there will be fewer bites on the arm sprayed with the plant extract compared to the arm sprayed with DEET after a 5-minute exposure to mosquitoes." Making a prediction provides a critical analysis of the experimental design. If the predictions are not clear, the procedure can be modified before beginning the experiment. For the butterfly experiment, the hypothesis was: "Infected monarch butterflies preferentially lay their eggs on plants that reduce parasite infection in their offspring." *What should the prediction be? State your prediction.*

To evaluate the results of the experiment, the investigator always returns to the prediction. If the results match the prediction, then the hypothesis is supported. If the results do not match the prediction, then the hypothesis is falsified. Either way, the scientist has increased knowledge of the process being studied. Many times the falsification of a hypothesis can provide more information than confirmation, as the ideas and data must be critically evaluated in light of new information. In the butterfly experiment, the scientist may learn that the prediction is supported—there is a greater proportion of eggs on the milkweed plant with higher levels of anti-parasite chemicals *(A. curassavica).* As a next step, the investigator may wish to examine the longevity of infected monarchs reared on *A. curassavica* compared with those reared on *A. incarnata.*

Return to page 4 and review your hypotheses for the numbered questions. Consider how you might design an experiment to test the first hypothesis. For example, you might separate a litter of kittens into two groups and raise them in two environments—one with human dander present and another where the air is filtered, removing human dander. The prediction might be:

> **If** human dander causes allergic reactions in cats *(a restatement of the hypothesis),* **then** the cats raised in the environment with human dander will develop more allergies than those cats reared in a dander-free environment *(predicting the results of the experiment).*

Now consider an experiment you might design to test the second hypothesis. How will you measure "acne outbreak"?

State a prediction for this hypothesis and experiment. Use the if/then format.

The actual test of the prediction is one of the great moments in research. No matter the results, the scientist is not just following a procedure but truly testing a creative explanation derived from an interesting question.

Discussion

1. From this exercise, list the essential components of scientific investigations from asking a question to carrying out an experiment.

2. From this exercise, list the variables that must be identified in designing an experiment.

3. What are the components of an experimental procedure?

EXERCISE 1.3

Designing an Experiment

Materials

steps or platform, 8 to 14 inches high
metronome
clock or stopwatch with seconds
optional: smartphones or tablets with a metronome app (See Website resources for suggestions)

Introduction

The Centers for Disease Control and Prevention (CDC) as well as the U.S. Department of Health and Human Services (HHS) have developed guidelines to improve health and fitness in response to continued concerns about chronic disease as a result of sedentary lifestyles and obesity. One measure of physical fitness is cardiovascular fitness, the body's ability to provide oxygen-rich blood to actively working tissues during exercise. Both the lungs and heart contribute to cardiovascular fitness, which can be affected by a number of factors, including types or extent of exercise, smoking, weight, and other physiological indicators. The "2008 Physical Activity Guidelines for Americans" (HHS, 2008) recommends 150 minutes a week of moderate (or 75 minutes of vigorous) physical activity, as well as strength-building exercises two or more times a week. A number of questions about the factors that affect cardiovascular fitness, particularly in college students, remain unanswered. Are the young adults who follow the HHS guidelines more fit than those who are sedentary? Can you be fit and obese or thin and less fit? How does sleep affect cardiovascular fitness? Can smokers regain the cardiovascular fitness levels of nonsmokers? If you are a smoker, how long before your cardiovascular fitness is affected?

Cardiovascular fitness can be determined by measuring a person's pulse rate and respiration rate before and after a period of exercise. A person who is more fit may have a relatively slower pulse rate and lower respiratory rate after exercise, and his or her pulse rate should return to the normal (resting rate) more quickly than that of a person who is less fit. Several tests of cardiovascular fitness have been developed, but the easiest to use and for comparing the results with other studies, is the Harvard Step Test (Simon, 2005). In this test, the subject steps up and down at a constant rate for 3 to 5 minutes. The pulse rate is measured before and after exercise, as well as the recovery rate at 1, 2, and 3 minutes after exercise. These values can then be compared directly or used to calculate a fitness index.

In this exercise you will brainstorm questions about the factors that affect cardiovascular fitness in college students. As a class you will select one of these questions to pursue in today's laboratory. You will develop a hypothesis, design an experiment using the step test, and state your prediction. In the following exercises you will collect data, analyze and discuss your results, and critique your experiment.

In your research teams, take a few minutes to discuss several *specific* questions that you are interested in asking about factors that affect cardiovascular fitness in college students. Write your questions in the margin of the lab manual. Discuss with your teammates a testable hypothesis for each question. *Choose one question and hypothesis from your group to add to a class list recorded by the instructor. Consider the characteristics of a question that can be pursued scientifically as you select the class question.*

The class selects one question and hypothesis to investigate and then develops the experimental design using the basic step test as a starting point. They then predict the results of the experiments. All teams carry out the same experiment, pooling and analyzing the class data.

Record the ***question*** chosen by the class.

Hypothesis

Record a hypothesis for the question chosen by the class. Consider the characteristics of a testable hypothesis.

The Experiment

A form of the step test has been used to evaluate cardiovascular fitness for decades. There are several different versions of the test that allow the subject to exercise for a specified amount of time, at a given rate, using a standard-height step. The rate and duration of exercise should increase the heart rate without stressing the subject. Heart rate (beats per minute) before and after exercise can be measured. In some procedures the recovery heart rate is also measured at 1, 2, and 3 minutes after exercise. Recovery heart rates can be used to determine a fitness index. You may modify the design for your experiment. The basic elements of the step test are provided below along with possible measurements. *Do not increase the rate or duration beyond those suggested.*

1. *Team Organization:* Select the subjects for the two test groups who will perform the step test. Designate other students to measure pulse rates and record data. *Each team of students should ideally have four students: one student in each test group, with the other students designated as recorders to measure pulse rates and record data for the test subjects.*

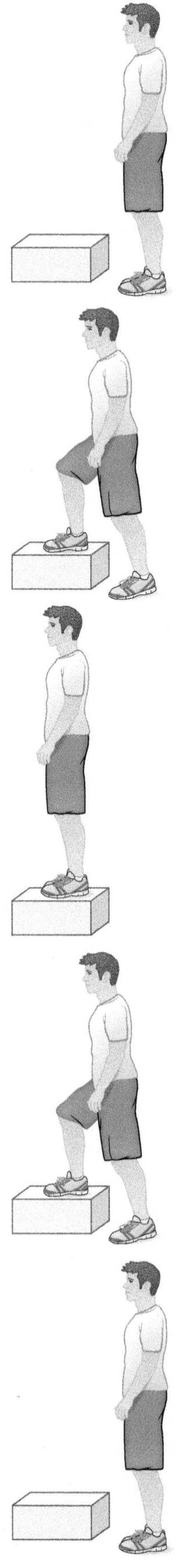

FIGURE 1.3 **The Step Test.** The subject steps up on a platform and then down again, keeping the rate constant.

> ⚠ Students with respiratory or circulatory disorders should not participate as test subjects in the experiment. Do not exceed the rate or duration beyond those suggested.

2. *Step Test Basics:* The subject steps up and down on a platform or step, approximately 20–36 cm (8 to 14 inches) in height, for 3 to 4 minutes at a rate of 24 to 30 steps per minute. Begin by stepping up with one foot and follow with the other. The sequence for one step will be "up, up" and then "down, down" (Figure 1.3). Use a metronome to count steps to ensure that all subjects maintain a constant step rate. For a rate of 30 steps per minute, set the metronome to 120 beats per minute to coincide with the four steps—up and down. Metronome apps are available for smartphones and tablets (see Websites at end of the lab topic for suggestions).
3. *Pulse Rate:* The subject's pulse rate is measured before the test (resting rate) and 1 minute after the test. Have the subject sit quietly in a chair when the pulse is counted. To take a manual pulse, the recorder should use three fingers to find the pulse in the radial artery (the artery on the underside of the wrist above the thumb) or in the carotid artery (under the edge of the jaw near the neck). Count the beats per minute (bpm). (Count the beats for 30 seconds and multiply by 2). You can compare pulse rate before and after exercise, and also calculate the difference in pulse rate.
4. *Recovery Rate:* Additionally, the pulse rate may be measured at 1, 2, and 3 minutes after the test to determine the recovery rate (measure of the return to resting heart rate). Count the pulse from 1 to 1.5 minutes, then 2 to 2.5 minutes, and then 3 to 3.5 minutes. Multiply the minutes by 2 for beats per minute (bpm).
5. *Fitness Index:* You may choose to calculate the fitness index from the recovery pulse rates at 1, 2, and 3 minutes. These can then be compared to the Fitness Index Ratings in Table 1.1 or graphed to compare the two experimental groups (Bird et al., 1998).

   Calculate the fitness index (FI) using the following equation:

$$FI = \frac{\text{Duration of test (sec)}}{(t_1 + t_2 + t_3)} \times 100$$

   t_1 = bpm 1 min. after exercise; t_2 = bpm at 2 min.; t_3 = bpm at 3 min.

List the details of the step test procedure designed by the class. Be sure all team members understand and follow the same procedure.

Number and names of subjects in each test group:

Names of students (recorders) designated to measure and record data:

Step rate:

Duration of test:

Height of step:

When are pulse rates (bpm) measured?

How are pulse rates (bpm) measured?

TABLE 1.1 Fitness Index for Step Test Indicating Relative Fitness

Harvard Step Test Ratings	
Fitness Index	**Rating**
<55	Poor
55–64	Low Average
65–79	High Average
80–89	Good
>90	Excellent

(Bird et al., 1998)

What is (are) the dependent variable(s) in your experiment?

What is the independent variable?

Controlled variables:

Control:

Level of treatment:

Replication:

Prediction

Predict the results of each experiment based on your hypotheses (if/then).

Performing the Experiment

Following the procedures established by your class for the step test, perform the experiment and record the results for your team in Table 1.2.

A = Athletic
NA = Non-Athletic
Stair height = 3.5m

TABLE 1.2 Results of Step Test for Team Members. *Modify the Table for the Measurements Recorded in Your Experiment.*

Download an Excel version of this table from www.masteringbiology.com in the Study Area under Lab Media.

Measurements	Test Group 1: A	Test Group 2: NA
Before step test **Pulse rate (bpm)**	80	64
1 minute after step test **Pulse rate (bpm) = t_1**	96	88
2 minutes after step test **Pulse rate (bpm) = t_2**	64	66
3 minutes after step test **Pulse rate (bpm) = t_3**	60	64
Fitness Index	109.09	110.09
$FI = \dfrac{\text{Duration of test (sec)}}{(t_1 + t_2 + t_3)} \times 100$		

A: $\frac{240}{96+64+60} \times 100 = \frac{240}{220} \times 100 = 109.09$

NA: $\frac{240}{88+66+64} = 110.09$

Results

1. Customize Table 1.2 to record the results for your team. List the test groups (for example, athlete, nonathlete) above the columns at the top of the table. Cross through or add to the rows to indicate the dependent variables for your experiment.
2. Record the results for your team in Table 1.2.
3. Calculate the fitness index (if selected for your experiment) and record in Table 1.2.
4. Customize Table 1.3 for the *total class data* to include the test groups and the dependent variables measured.
5. Record the class results for all subjects in Table 1.3.
6. Calculate the class averages.
7. Compare your results with the fitness index and ratings in Table 1.1.
8. Analyze and present your results following the directions in Exercise 1.4.

EXERCISE 1.4

Presenting and Analyzing Results

Once the data are collected, they must be organized and summarized so that scientists can determine if the hypothesis has been supported or falsified. In this exercise, you will design **tables** and graphs; the latter are also called **figures**. Tables and figures have two primary functions. They are used *(1) to help you analyze and interpret your results and (2) to enhance the clarity with which you present the work to a reader or viewer.*

TABLE 1.3 Summary of Class Data for All Subjects. *Modify the Table and Record the Results from the Experiment by Test Groups 1 and 2.*

Download an Excel version of this table from www.masteringbiology.com in the Study Area under Lab Media.

Results for Test Group 1: Athletic

Subject	1	2	3	4	5	6	Average
Before step test Pulse rate (bpm) 40x2	57.3	80	72	80	64	70.66	
1 minute after step test Pulse rate (bpm) 48	80	120	112	96	88	99.20	
2 minutes after step test Pulse rate (bpm) 32	72	96	92	64	88	82.40	
3 minutes after step test Pulse rate (bpm) 30	64	84	120	60	76	80.80	
Fitness Index	110.49	180	74.1	109.091	95.2	93.90	

Results for Test Group 2: Non Athletic

Subject	1	2	3	4	5	6	Average
Before step test Pulse rate (bpm) 32	73.3	84	100	64	80	80.26	
1 minute after step test Pulse rate (bpm) 44	108	92	172	88	104	112.80	
2 minutes after step test Pulse rate (bpm) 33	88	80	152	66	72	91.60	
3 minutes after step test Pulse rate (bpm) 32	84	80	136	64	88	90.40	
Fitness Index	85.7	95.23	52.2	110.09	90.9	86.82	

Lab Study A. Tables

You have collected data from your experiment, and the information may appear at first glance to have little meaning. Look at your data. How could you organize the data set to make it easier to interpret? You could *average* the data set for each test group, but even averages can be rather uninformative. Could you use a summary table to convey the data (in this case, averages)?

Table 1.4 is an example of a table using data from an experiment investigating the effects of sulfur dioxide on soybean production. In this experiment, 48 soybean plants in flower were divided into two groups, experimental and control. One group (the experimental group) was exposed to 0.6 ppm (parts per million) of sulfur dioxide for 4 hours. The other group was exposed to filtered air for the same amount of time. The plants were then maintained in the greenhouse, and when the seed pods matured, the investigators counted the number of pods per plant and the number of seeds per pod. Note that the number of replicates and the units of measurement are provided in the table title.

Tables often are used to present results that have many data points. They are also useful for displaying several dependent variables and when the quantitative values rather than the trends are the focus. For example, average number of bean pods, average number of seeds per pod, and average weight of pods per plant for treated and untreated plants could all be presented in one table.

The following guidelines will help you construct a table.

- All values of the same kind should read down the column, not across a row. Include only data that are important in presenting the results and for further discussion.
- Information and results that are not essential (for example: test-tube number, simple calculations, or data with no differences) should be omitted.
- The heading of each column should include units of measurement, if appropriate.
- Tables are numbered consecutively throughout a lab report or scientific paper. For example: Table 1.4 would be the fourth table in your report.
- The **title**, which is located at the top of the table, should be clear and concise, with enough information to allow the table to be understandable apart from the text. Capitalize the first and important words in the title. Do not capitalize articles (*a, an, the*), short prepositions, and conjunctions. The title does not need a period at the end.
- Refer to each table in the written text. Summarize the data and refer to the table—for example, "The plants treated with sulfur dioxide produced an average of 1.96 seeds per pod (Table 1.4)." Do not write, "See the results in Table 1.4."
- If you are using a database program, such as Excel, the first step should be to sketch your table on paper before constructing it on the computer.

TABLE 1.4 Effects of 4-Hour Exposure to 0.6 ppm Sulfur Dioxide on Average Seed and Pod Production in Soybeans (24 Replicates)

Treatment	Seeds per Pod	Pods per Plant
Control	3.26	16
SO_2	1.96	13

Application

1. Using the data from your experiment, design a summary table on the next page to present the results for one of your dependent variables (for example, pulse rate before and one minute after the step test).
2. Label this Table 1.5. Compose a title for your table. Include the number of replications used to calculate the averages. Refer to the guidelines in the previous section of this lab topic for composing titles.

Lab Study B. Figures

Graphs, diagrams, drawings, and photographs are all called *figures*. The results of an experiment usually are presented graphically, showing the relationships among the independent and dependent variable(s). A graph or figure provides a visual summary of the results. Often, characteristics of the data are not apparent in a table but may become clear in a graph. By looking at a graph, then, you can visualize the effect that the independent variable has on the dependent variable and detect trends in your data. Making a graph may be one of the first steps in analyzing your results.

The presentation of your data in a graph will assist you in interpreting and communicating your results. In the final steps of a scientific investigation, you must be able to construct a logical argument based on your results that either supports or falsifies your starting hypothesis. Your graph should be accurately and clearly constructed, easily interpreted, and well annotated.

The following guidelines will help you to construct such a graph:

- Use graph paper and a ruler to plot the values accurately. If using a database program, you should first sketch your axes and data points before constructing the figure on the computer.
- *The independent variable is graphed on the* x-axis *(horizontal axis, or abscissa), and the dependent variable on the* y-axis *(vertical axis, or ordinate).*
- The numerical range for each axis should be appropriate for the data being plotted. Generally, begin both axes of the graph at zero (the extreme left corner). Then choose your intervals and range to maximize the use of the graph space. Choose intervals that are logically spaced and therefore will allow easy interpretation of the graph, for example, intervals of 5s or 10s. To avoid generating graphs with wasted space, you may signify unused graph space by two perpendicular tic marks between the zero and your lowest number on one or both axes.
- Label the axes to indicate the variable and the units of measurement. Include a legend if colors or shading is used to indicate different aspects of the experiment.
- Choose the type of graph that best presents your data. Line graphs and bar graphs are most frequently used. The choice of graph type depends on the nature of the variable being graphed.
- Compose a **title** for your figure and write it below your graph. Figures should be numbered consecutively throughout a lab report or scientific paper. Each figure is given a caption or title that describes its contents, giving enough information to allow the figure to be self-contained. Capitalize only the first word in a figure title and place a period at the end.

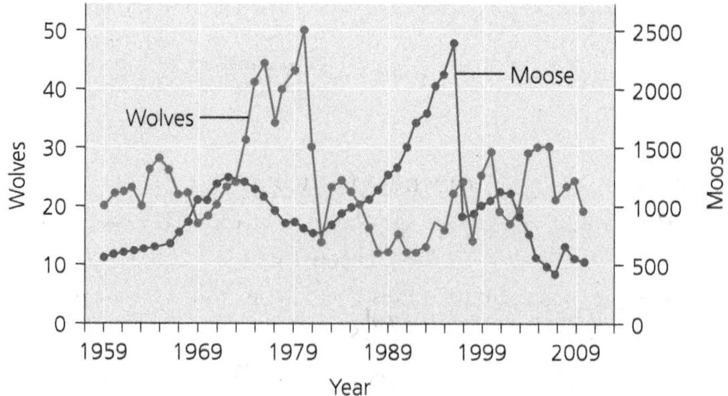

FIGURE 1.4
Five decades of fluctuating wolf and moose populations on Isle Royale, a remote wilderness island in Lake Superior.
Source: From the Isle Royale Wolf-Moose Study

The Line Graph

Line graphs show changes in the quantity of the chosen variable and emphasize the rise and fall of the values over their range. Use a line graph to present continuous quantitative data. For example, changes in a dependent variable such as changes in weight measured over time would be depicted best in a line graph.

- Whether to connect the dots or draw a best fit curve depends on the type of data and how they were collected. To show trends, draw smooth curves or straight lines to fit the values plotted for any one data set. Connect the points dot to dot when emphasizing meaningful changes in values on the *x*-axis.
- If more than one set of data is presented on a graph, use different colors or symbols and provide a key or legend to indicate which set is which.
- A boxed graph, instead of one with only two sides, makes it easier to see the values on the right side of the graph.

Note the features of a line graph in Figure 1.4, which shows the relationship between wolf and moose abundance over time on Isle Royale in Lake Superior.

The Bar Graph

Bar graphs are constructed following the same principles as for line graphs, except that vertical bars, in a series, are drawn down to the horizontal axis. Bar graphs are often used for data that represent separate or discontinuous groups or nonnumerical categories, thus emphasizing the discrete differences between the groups. For example, a bar graph might be used to depict differences in the proportion of eggs laid by monarch butterflies on milkweeds with and without chemicals that reduce parasite infection. Bar graphs are also used when the values on the *x*-axis are numerical but grouped together. These graphs are called histograms.

Note the features of a bar graph in Figure 1.5, which shows the percentage of calories from fast food among adults aged 20 and over.

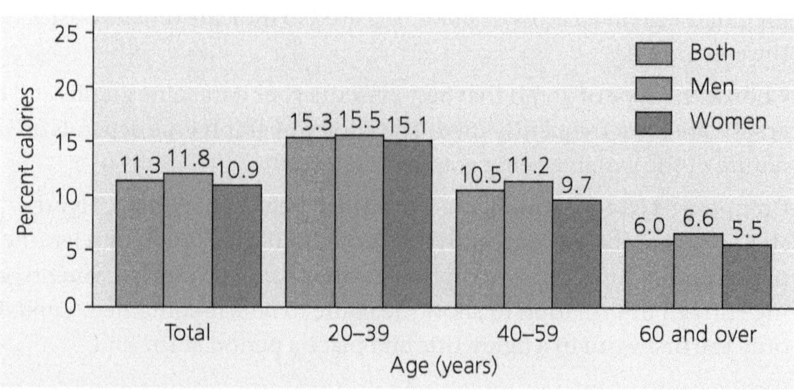

FIGURE 1.5
Percentage of calories from fast food among adults aged 20 and over, by gender and age: United States, 2007–2010.
Source: National Health and Nutrition Survey, 2007–2010

You will be asked to design graphs throughout this laboratory manual. Remember, the primary function of the figure is to present your results in the clearest manner to enhance the interpretation and presentation of your data.

Application

1. Using data from your experiments and the grid provided below, design a *bar graph* that shows the relationship between the dependent and independent variables in your experiment. Discuss with your teammates how to design one figure so that it includes the data for the independent variable and one or more dependent variables.
   a. What was the independent variable for your experiment? On which axis would you graph this?

   b. What was the dependent variable? Write this on the appropriate axis.

2. Add a *legend* to your figure to distinguish the two test groups.
3. Draw, label, and compose a *title* for your figure.
4. Imagine an experiment similar to the one you have performed where it would be appropriate to use a line graph.

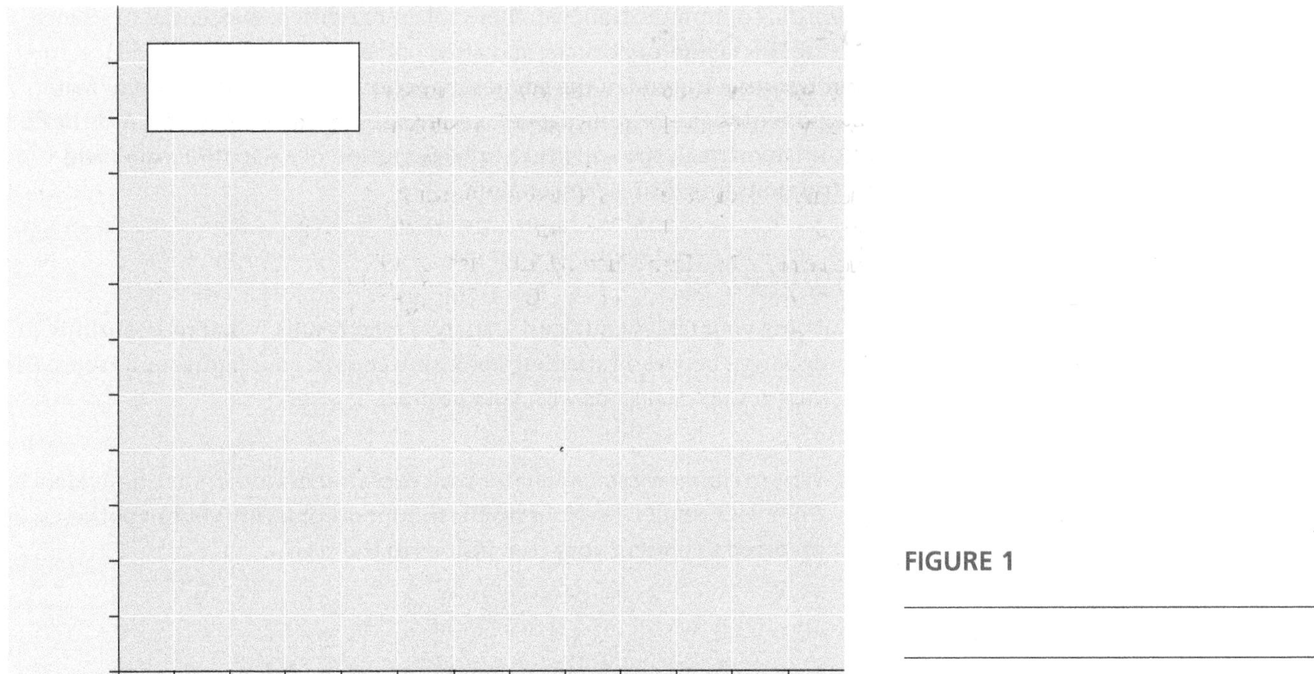

FIGURE 1

EXERCISE 1.5

Interpreting and Communicating Results

The last component of a scientific investigation is to interpret the results and discuss their implications in light of the hypothesis and supporting literature. The investigator studies the results, including tables and figures, and determines if the hypothesis has been supported or falsified. If the hypothesis has been falsified, the investigator must suggest alternate hypotheses for testing. If the hypothesis has been supported, the investigator suggests additional experiments to strengthen the hypothesis, using the same or alternate methods.

Scientists will thoroughly investigate a scientific question: testing hypotheses, collecting data, and analyzing results. In the early stages of a scientific study, scientists review the scientific literature relevant to their topic. They continue to review related published research as they interpret their results and develop conclusions. The final phase of a scientific investigation is the communication of the results to other scientists. Preliminary results may be presented within a laboratory research group and at scientific meetings where the findings can be discussed. Ultimately, the completed project is presented in the form of a scientific paper that is reviewed by scientists within the field and then published in a scientific journal. Research that is peer reviewed and published in a scientific journal is an example of a *primary reference* that can be cited by other scientists. The ideas, procedures, results, analyses, and conclusions of all scientific investigations are critically scrutinized by other scientists. Because of this, science is sometimes described as *self-correcting*, meaning that the work is subject to being repeated and critiqued by other scientists, who may have new evidence or different results that will then modify our understanding of the phenomenon under investigation. *Reproducibility* is a critical requirement of scientific inquiry.

Scientific communication, whether spoken or written, is essential to science. During this laboratory course, you often will be asked to present and interpret your results at the end of the laboratory period. Additionally, you will write components of a scientific paper for many lab topics. In Appendix A at the end of the lab manual, you will find a full description of a scientific paper and instructions for writing each section.

Application

1. Using your tables and figures, analyze your results. What relationships are apparent between variables? Look for trends in your figures and tables. Discuss your conclusions with your group.

2. Write a summary statement for your experiments incorporating evidence from your results. Use your results to support or falsify your hypotheses. Be prepared to present your conclusions to the class.

3. Critique your experiment. List the weaknesses and suggested improvements in the following table.

Weaknesses in Experiment	Improvement

4. Suggest additional and modified hypotheses that might be tested in the future. Briefly describe your next experiment.

5. Refer to Appendix A "Scientific Writing and Communication," at the end of your lab manual . Briefly describe the four major parts of a scientific paper. What is the abstract? What information is found in a References Cited section? What sections of a scientific paper always include references?

REVIEWING YOUR KNOWLEDGE

1. Review the major components of an experiment by matching the following terms to the correct definition: *control, controlled variables, level of treatment, dependent variable, replication, procedure, prediction, hypothesis, independent variable.*
   a. Variables that are kept constant during the experiment (variables not being manipulated)

   b. Tentative explanation for an observation

   c. What the investigator varies in the experiment (for example, time, pH, temperature, concentration)

   d. Process used to measure the dependent variable

e. Appropriate values to use for the independent variable

f. Treatment that eliminates the independent variable or sets it at a standard value

g. What the investigator measures, counts, or records; what is being affected in the experiment

h. Number of times the experiment is repeated

i. Statement of the expected results of an experiment based on the hypothesis

2. Identify the dependent and independent variables in the following investigations. (*Circle* the dependent variable and *underline* the independent variable.)

   a. Scientists inject 50 people with an experimental vaccine immediately after exposure to Ebola and 50 different people after a 21-day delay. The number of people who contract Ebola are counted.

   b. Coral reefs compete for light and space with seaweeds that may cover the coral like a lawn. Scientists measure the percent of coral covered with seaweed and the rate of coral growth in test plots with and without goby fish present.

   c. Researchers use a molecular tool called CRISPR to modify a gene in 35 beagle embryos to test if the modified gene will increase the amount of muscle produced by the dogs.

3. Suggest a control treatment for each of the following experiments.

   a. To investigate if spots on butterfly wings scare off songbirds, scientists expose birds to butterflies with large spots.

b. To investigate if night lights foster depression, a group of hamsters is kept in chronic dim light throughout the night.

4. Propose an experiment and suggest a control treatment for each of the following questions.

a. Do bearded dragons that are genetically male change into females if eggs are incubated at temperatures above 32°C?

Experiment:

Control:

b. Is the cause of colic in newborns related to the microbial community in their intestines?

Experiment:

Control:

c. Female waterbugs lay their eggs on the backs of male waterbugs who care for the eggs. Do female waterbugs prefer to mate with male waterbugs with large numbers of eggs?

Experiment:

Control:

5. A recent study of 59,806 Caucasian women reports that women who take aspirin for five or more years are 30% less likely to develop melanoma than women who do not use aspirin. List other variables that would be important to control in this study (controlled variables).

6. A study of the effects of vitamin D in the body has shown that persons with lower levels of vitamin D are more likely to have chronic diseases such as cardiovascular disease, breast cancer, colds, and flu. Can you conclude that vitamin D prevents these chronic diseases? What other possible explanations for this correlation can you suggest?

7. What is the essential feature of science that makes it different from other ways of understanding the natural world?

APPLYING YOUR KNOWLEDGE

Interpreting Graphed and Tabular Data

1. The use of DDT for malaria control stopped being funded by the World Health Organization (WHO) in the 1980s. The World Bank required a ban on DDT for developing countries seeking loans. At the same time there has been an increase in the resistance of the malarial parasite to the most common antimalarial drugs. Since 2001, WHO has allowed the spraying of DDT in Africa on interior walls to kill mosquitoes. Review the graph in Figure 1.6 and information provided and then answer the following questions.

   What is the independent variable?

   What is the dependent variable?

   Why was a bar graph selected to present these data? Could the authors have used a line graph?

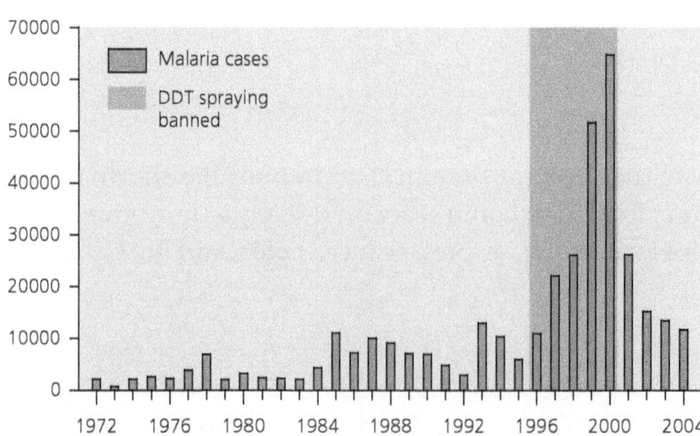

FIGURE 1.6
Malaria cases in South Africa before, during, and after banned DDT spraying.
Source: After Opar, 2006

Write a statement summarizing the results. Specifically address trends from 1972–1992, 1995–2000, and 2001–2004.

2. Review the experimental conditions for the monarch butterfly and milkweed experiment in Exercise 1.2. Recall that adult butterflies lay their eggs on milkweed plants and that the caterpillars that hatch feed on the milkweed leaves. Adult monarch butterflies are often infected by a protozoan parasite that affects their survivability and growth. Some milkweeds (*A. curassavica)* produce high concentrations of anti-parasitic compounds that reduce parasitic infections in monarch butterfly offspring.

   Scientists hypothesized that infected monarch butterflies *preferentially* lay their eggs on milkweeds that reduce parasite infection (*A. curassavica*). In the experiment, the investigators measured the proportion of eggs laid on *A. curassavica* by infected and uninfected adult monarch butterflies.

   If the butterflies showed no preference for the *A. curassavica* plants, what would be the expected proportion?

   The results of this experiment are shown in Figure 1.7. Do the results support or falsify the hypothesis? Explain your answer using the data presented.

   In a second experiment scientists hypothesized that parasite-infected caterpillars would consume a greater proportion of their diet from the anti-parasite (*A. curassavica*) milkweed. In the experiment, caterpillars were given the choice of feeding on milkweed leaves with high concentrations

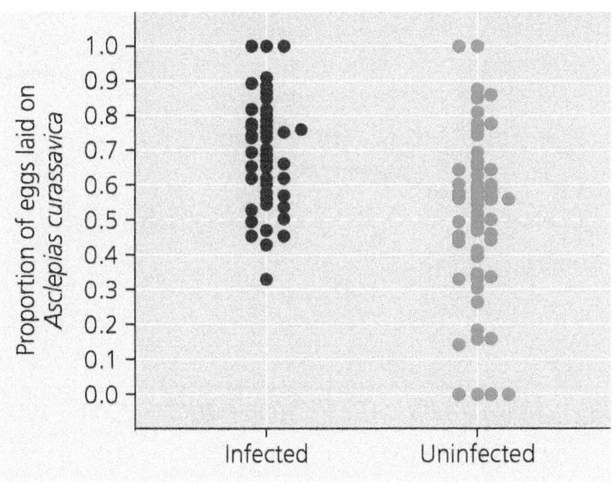

FIGURE 1.7
Proportion of eggs laid on anti-parasitic milkweed (*A. curassavica*) by infected and uninfected monarch butterflies.
Source: Lefèvre et al., 2012

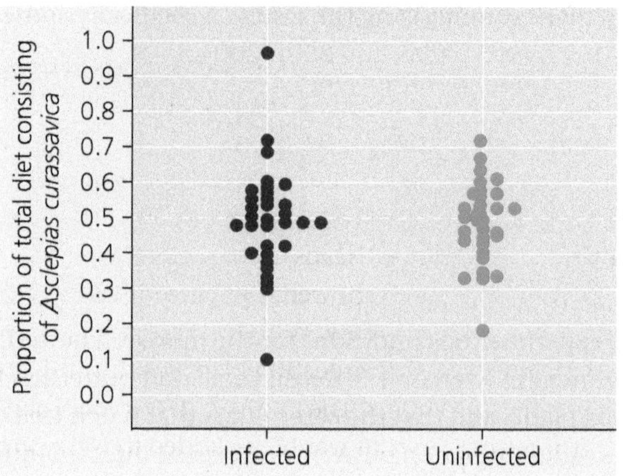

FIGURE 1.8
Proportion of diet from anti-parasitic milkweed (*A. curassavica*) consumed by infected and uninfected caterpillars.
Source: Lefèvre et al., 2012

of anti-parasite compounds (*A. curassavica*) or another milkweed with low concentrations.

The results of the caterpillar feeding experiment are shown in Figure 1.8. Do these results support of falsify the hypothesis? Explain your answer using the data presented.

3. Review the guidelines for graphs on pages 17–19 earlier in the chapter and critique Figure 1.9. This figure illustrates the number and birth rate for twins born in the United States between 1980 and 2009. Suggest changes that you could make to improve this figure.

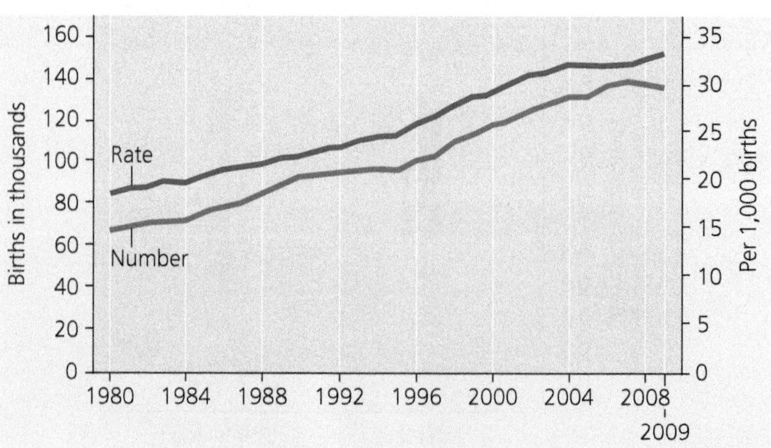

FIGURE 1.9
Number and Rate of Twin Births.
Source: Modified from Martin et al., 2012

Practicing Experimental Design

1. The Zika virus that is carried by mosquitoes has been detected in Africa and Asia for over 50 years, but only in recent years has this virus been implicated as the cause of birth defects in humans. In late 2014, public health officials in Brazil noted a significant increase in cases of Zika virus infection, and by the fall of 2015, physicians noted a similar increase in numbers of babies born with small heads, a condition known as microcephaly. By the next year a large study showed a 30 percent increase in the number of babies born with microcephaly. Was the increase in Zika virus infection the cause of the increase in microcephaly, or was this just a correlation? Some scientists speculated that the increase in microcephaly was due to the pesticide pyriproxyfen that was being added to water supplies in Brazil in an effort to control mosquito populations. Still others proposed other causes, including genetic mutations, exposure to alcohol and other drugs, malnutrition, or reduced blood supply to the fetus brain in pregnancy. You are a researcher recruited by the CDC to investigate the problem of increased microcephaly in babies in Brazil in 2016. Propose a hypothesis to explain this problem, and then design an experiment to test your hypothesis. How can you be sure that your observations are not just correlations?

   Proposed hypothesis:

   Experiment to test the hypothesis:

2. "Grains of paradise" plants grow in the swampy region inhabited by the western lowland gorilla and make up 80–90% of the gorillas' diet, and are even utilized in constructing their nests each night. Captive lowland gorillas in zoos are not fed grains of paradise, but rather have a complex diet of processed vitamin-rich food plus fruits and vegetables available in the marketplace. Recently, scientists identified potent anti-inflammatory and antimicrobial compounds in grains of paradise that may hold the key to a puzzling question: "Why do western lowland gorillas in zoos have an alarmingly high rate of cardiomyopathy (a type of heart disease)?" Hypothesize about the effect of diet (grains of paradise) on rates of heart disease. Describe a simple preliminary experiment to test your hypothesis, and state a prediction.

   Hypothesis:

   Experiment:

Prediction:

FIGURE 1.10
Red-cockaded woodpecker. A federally endangered species that lives in old-growth longleaf pine forests of the southeastern United States.

3. Scientists have been studying ways to increase numbers of new populations of the federally endangered red-cockaded woodpecker (Figure 1.10). Previous research has shown that woodpeckers prefer to colonize forests with *naturally* existing nesting holes, rarely moving into new territories with no holes. Scientists wondered if birds would move into new sites if they included trees with *artificially* constructed nesting cavities. They hypothesized that birds *would* colonize these forests with artificially constructed cavities. Describe a possible experiment to test the hypothesis, including a control. State a prediction.

Experiment:

Control:

Prediction:

STUDENT MEDIA: BioFlix, Activities, Investigations, Videos, and Data Tables

www.masteringbiology.com (select Study Area)

Activities—Ch. 1: Graph It! An Introduction to Graphing

Investigations—Ch. 1: How Does Acid Precipitation Affect Trees?

Data Tables—Table 1.2 and 1.3 can be downloaded in Excel format. Look in the Study Area under Lab Media.

REFERENCES

Barnard, C., F. Gilbert, and P. McGregor. *Asking Questions in Biology*, 4th ed. Harlow, England: Pearson, 2011.

Bird, S. R., A. Smith, and K. James. *Exercise Benefits and Prescriptions*, Cheltenham, UK: Nelson Thornes Ltd., 1998.

Burch, Druin. "Eat Dirt? Allergies, Autoimmune Disease, and the Hygiene Hypothesis," *Natural History*, 2012, vol. 120, pp. 12–15. Good discussion of correlation vs. causation in scientific research.

Knisely, K. *A Student Handbook for Writing in Biology*, 4th ed. Sunderland, MA: Sinauer Associates, 2013. (Instructions for making graphs using Excel, Appendix 2.)

Lefèvre, T., A. Chiang, M. Kelavkar, H. Li, J. Li, C. Lopez, F. deCastillejo, L. Oliver, Y. Potini, M. Hunter, and J. C. de Roode. "Behavioural Resistance Against a Protozoan Parasite in the Monarch Butterfly." *Journal of Animal Ecology*, 2012, vol. 81, pp. 70–79.

Martin, J. A., B. E. Hamilton, and M. J. K. Osterman. "Three Decades of Twin Births in the United States, 1980–2009." *NCHS Data Brief*, 2012, no. 80. http://www.cdc.gov/nchs/data/databriefs/db80.htm. Accessed March, 2013.

Moore, J. *Science as a Way of Knowing*. Cambridge, MA: Harvard University Press, 1993.

Opar, A. "The Return of DDT." *SEED*, 2006, vol. 2, no. 8, p. 20.

Pechenik, J. *A Short Guide to Writing About Biology*, 9th ed. San Francisco, CA: Pearson, 2016.

Rosen, M. "Rapid Spread of Zika Virus Raises Alarm." *Science News*, 2016, vol. 189, no. 4, pp. 16–18.

Simon, H. B. *The No Sweat Exercise Plan: Lose Weight, Get Healthy, and Live Longer*. New York: McGraw-Hill, 2005.

Urry, L., M. Cain, S. Wasserman, P. Minorsky, and J. Reece. *Campbell Biology*, 11th ed. San Francisco, CA: Pearson, 2017.

U.S. Department of Health and Human Services. "2008 Physical Activity Guidelines for Americans." http://health.gov/paguidelines/factsheetprof.aspx

Walters, J. R., C. K. Copeyon, and J. H. Carter. "Test of the Ecological Basis of Cooperative Breeding in Red-cockaded Woodpeckers." *The Auk*, 1992, vol. 1009, no. 1, pp. 90–97.

WEBSITES

Download smartphone and tablet apps for the metronome:

Pro Metronome—Beat with sound and light. Easy to use and works well. Free.

Isle Royale Wolf-Moose Study. For information on the longest-running study of predator/prey interactions: http://www.isleroyalewolf.org

"2008 Physical Activity Guidelines for Americans" by U.S. Department of Health and Human Services: http://health.gov/paguidelines/factsheetprof.aspx

Recommendations for physical activity and diet from the Centers for Disease Control and Prevention: http://www.cdc.gov/physicalactivity

TED Talk, "How Butterflies Self-Medicate," Jaap de Roode: https://www.ted.com/talks/jaap_de_roode_how_butterflies_self_medicate

LAB TOPIC 2

Microscopes and Cells

Laboratory Objectives

After completing this lab topic, you should be able to:

1. Identify the parts of compound and stereoscopic microscopes and be proficient in their correct use in biological studies.
2. Compare the transmission electron microscope and light microscope. Describe procedures used in preparing materials for electron microscopy and compare these with procedures used in light microscopy.
3. Identify cell structures and organelles from electron micrographs and state the functions of each.
4. Describe features of specific cells and determine characteristics shared by all cells studied.
5. Compare the structure of animal and plant cells as seen in both light and electron microscopy.
6. Distinguish between eukaryotic and prokaryotic cells.
7. Discuss the evolutionary significance of increasing complexity from unicellular to multicellular organization and provide examples from the lab.

Introduction

According to cell theory, the *cell* is the fundamental biological unit, the smallest and simplest biological structure possessing all the characteristics of the living condition. All living organisms are composed of one or more cells, and every activity taking place in a living organism is ultimately related to metabolic activities in cells. Thus, understanding the processes of life necessitates an understanding of the structure and function of the cell.

The earliest known cells found in fossilized sediments 3.5 billion years old (called **prokaryotic** cells) lack nuclei and membrane-bound organelles. Cells with a membrane-bound nucleus and organelles (**eukaryotic** cells) do not appear in the fossil record for another 2 billion years. But the eventual evolution of the eukaryotic cell and its internal compartmentalization led to enormous biological diversity in single cells. The evolution of loose aggregates of cells ultimately to colonies of connected cells provided for specialization, so that groups of cells had specific and different functions. This early division of labor included cells whose primary function was locomotion or reproduction. The evolution of multicellularity appears to have originated more than once in eukaryotes and provided an opportunity for extensive adaptive radiation as organisms specialized and diversified, eventually giving rise to fungi, plants, and animals. This general trend in increasing complexity and specialization seen in the history of life will be illustrated in Lab Topic 2.

Given the fundamental role played by cells in the organization of life, one can readily understand why the study of cells is essential to the study of life. Cells, however, are below the limit of resolution of the human eye. We cannot study them without using a microscope. The microscope has probably contributed more than any other instrument to the development of biology as a science and continues today to be the principal tool used in medical and biological research. There are four types of microscopes commonly used by biologists. You will learn how to use two of these microscopes, the compound microscope and the stereoscopic microscope, in today's laboratory. Both of these microscopes use visible light as the source of illumination and are called light microscopes. Two other microscopes, the scanning electron microscope and the transmission electron microscope, use electrons as the source of illumination. Electron microscopes are able to view objects much smaller than those seen in a light microscope. Although these microscopes are not used in this laboratory, you will be given the opportunity to learn more about them in Exercise 2.4.

Microscopes are used by biologists in numerous subdisciplines: genetics, molecular biology, neurobiology, cell biology, evolution, and ecology. The knowledge and skills you develop today will be used and enhanced throughout this course and throughout your career in biology. It is important, therefore, that you take the time to master these exercises thoroughly.

EXERCISE 2.1

The Compound Light Microscope

Materials

compound microscope

Introduction

The microscope is designed to make objects visible that are too difficult or too small to see with the unaided eye. There are many variations of light microscopes, including phase-contrast, darkfield, polarizing, and UV. These differ primarily in the source and manner in which light is passed through the specimen to be viewed.

The microscopes in biology lab are usually compound binocular or monocular light microscopes, some of which may have phase-contrast attachments. **Compound** means that the scopes have a minimum of two magnifying lenses (the ocular and the objective lenses). **Binocular microscopes** have two eyepieces, **monoculars** have only one eyepiece, and **light** refers to the type of illumination used, that is, visible light from a lamp.

Your success in and enjoyment of a large portion of the laboratory work in introductory biology will depend on how proficient you become in the use of the microscope. When used and maintained correctly, these precision instruments are capable of producing images of the highest quality.

Although there are many variations in the features of microscopes, they are all constructed on a similar plan (Figure 2.1). In this exercise you will be introduced to the common variations found in different models of compound microscopes and asked to identify those features found on your microscope.

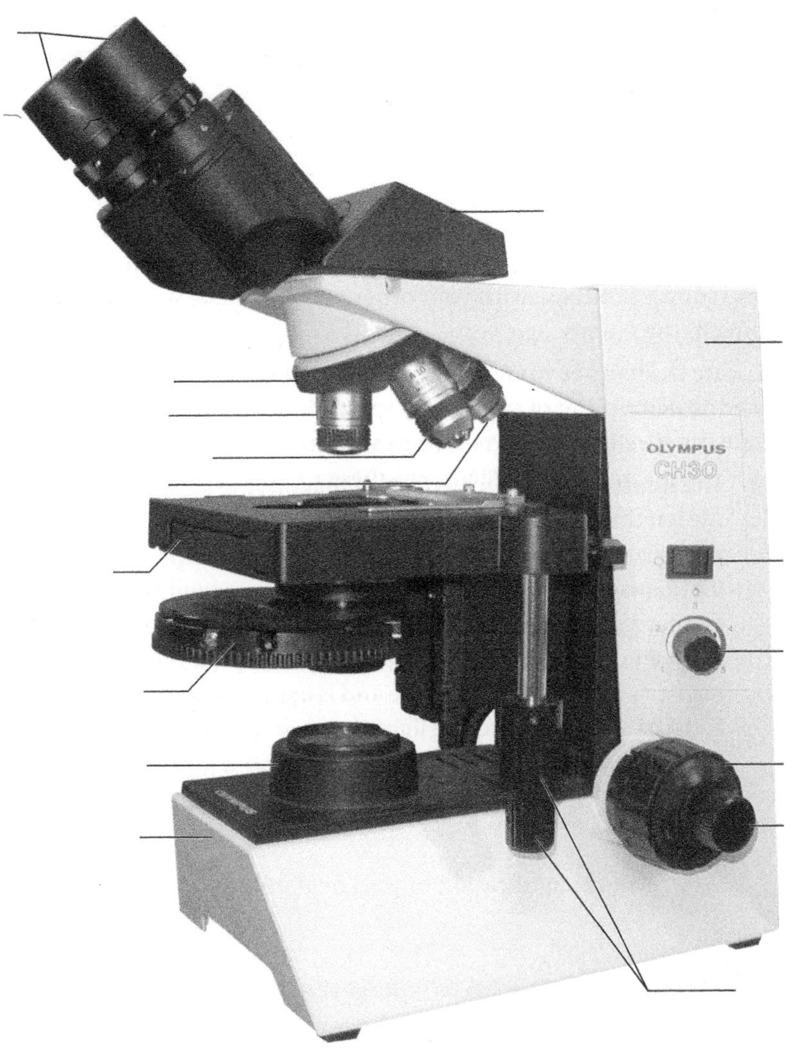

FIGURE 2.1a

The compound binocular light microscope. Locate the parts of your microscope described in Exercise 2.1 and label this photograph. Indicate in the margin of your lab manual any features unique to your microscope.

FIGURE 2.1b

Enlarged photo of compound light microscope as viewed from under the stage. This microscope is equipped with phase-contrast optics. Locate the condenser, condenser adjustment knob, phase-contrast revolving turret, and iris diaphragm on your microscope (if present) and label them on the diagram.

Please treat these microscopes with the greatest care!

Procedure

1. Obtain a compound light microscope, following directions from your instructor. To carry the microscope correctly, hold the arm with one hand, and support the base with your other hand. Remove the cover, but do not plug in the microscope.
2. Locate the parts of your microscope, and label Figure 2.1. Refer to the following description of a typical microscope. In the spaces provided, indicate the specific features related to your microscope.
   a. The **head** supports the two sets of magnifying lenses. The **ocular** is the lens in the eyepiece, which typically has a magnification of 10×. If your microscope is binocular, the distance between the eyepieces (**interpupillary distance**) can be adjusted to suit your eyes. Move the eyepieces apart, and look for the scale used to indicate the distance between the eyepieces. Do not adjust the eyepieces at this time. A pointer has been placed in the eyepiece and is used to point to an object in the **field of view**, the circle of light that one sees in the microscope.

      Is your microscope monocular (one eyepiece) or binocular (two eyepieces)?

      What is the magnification of your ocular(s)?

Although the eyepiece may be removable, it should not be removed from the microscope.

   b. **Objectives** are the three lenses on the **revolving nosepiece**. The shortest lens typically has a magnifying power of 4× and is called the **scanning lens**. The **intermediate lens** is 10×, and the longest, the **high-power lens**, is 40×. You may have a fourth position in the nosepiece on your microscope which may be empty or it may have an oil immersion objective (100×). It is important to clean both the objective and ocular lenses before each use. Dirty lenses will cause a blurring or fogging of the image. Always use lens paper for cleaning! Any other material (including Kimwipes®) may scratch the lenses.

      Typically the magnifying power of each objective is written on the objective as 4×, 10×, or 40×. What is the magnification of each of your objectives? List them in order of increasing magnification.

   c. The **arm** supports the stage and condenser lens. The **condenser lens** is used to focus the light from the **lamp** through the specimen to be viewed. The height of the condenser can be adjusted by an **adjustment knob**. The **iris diaphragm** controls the width of the circle of light and, therefore, the amount of light passing through the specimen.

If your microscope has phase-contrast optics, the condenser may be housed in a **revolving turret**. When the turret is set on 0, the normal optical arrangement is in place. This condition is called **brightfield microscopy**. Another position of the turret sets phase-contrast optics in place. To use phase-contrast, the turret setting must correspond to the magnifying power of the objective being used.

Is your microscope equipped with phase-contrast optics?

The **stage** supports the specimen to be viewed. A mechanical stage can be moved right and left and back and forth by two **stage adjustment knobs**. With a stationary stage, the slide is secured under stage clips and moved slightly by hand while viewing the slide. The distance between the stage and the objective can be adjusted with the **coarse** and **fine focus adjustment knobs**.
Does your microscope have a mechanical or stationary stage?

d. The **base** acts as a stand for the microscope and houses the lamp. In some microscopes, the intensity of the light that passes through the specimen can be adjusted with the **light intensity lever**. Generally, more light is needed when using high magnification than when using low magnification. Describe the light system for your microscope.

EXERCISE 2.2

Basic Microscope Techniques

Materials

clear ruler
coverslips
prepared slides:
  letter and crossed thread
lens paper
blank slides
Kimwipes®
dropper bottle with
  distilled water

Introduction

In this exercise, you will learn to use the microscope to examine a recognizable object, a slide of the letter *e*. Recall that microscopes vary, so you may have to omit steps that refer to features not available on your microscope. The following procedure will allow you to practice adjusting your microscope to become proficient in locating a specimen, focusing clearly, and adjusting the light for the best contrast.

Procedure

1. Clean microscope lenses.

   Each time you use the microscope, you should begin by cleaning the lenses. Using lens paper moistened with a drop of distilled water, wipe the ocular, objective, and condenser lenses. Wipe them again with a piece of dry lens paper.

> ⓘ Use only lens paper on microscope lenses. Do not use Kimwipes®, tissues, or other papers.

2. Adjust the focus on your microscope.
   a. Plug your microscope into the outlet.
   b. Turn on the light. Adjust the light intensity to mid-range if your microscope has that feature.
   c. Rotate the 4× objective into position using the revolving nosepiece ring, not the objective itself.
   d. Take the letter slide and wipe it with a Kimwipes® tissue. Each time you study a prepared slide, you should first wipe it clean. Place the letter slide on the stage, and center it over the stage opening.

> Slides should be placed on and removed from the stage only when the 4× objective is in place. Removing a slide when the higher objectives are in position may scratch the lenses.

   e. Look through the ocular and bring the letter into rough focus by slowly focusing upward using the coarse adjustment.
   f. For binocular microscopes, looking through the oculars, move the oculars until you see only one image of the letter *e*. In this position, the oculars should be aligned with your pupils. In the margin of your lab manual, make a note of the **interpupillary distance** on the scale between the oculars. Each new lab day, before you begin to use the microscope, set this distance.
   g. Raise the condenser to its highest position, and fully close the iris diaphragm.
   h. Looking through the ocular, slowly lower the condenser just until the graininess disappears. Slowly open the iris diaphragm just until the entire field of view is illuminated. This is the correct position for both the condenser and the iris diaphragm.
   i. Rotate the 10× objective into position.
   j. Look through the ocular and slowly focus upward with the coarse adjustment knob until the image is in rough focus. Sharpen the focus using the fine adjustment knob.

> Do not turn the fine adjustment knob more than two revolutions in either direction. If the image does not come into focus, return to 10× and refocus using the coarse adjustment.

k. For binocular microscopes, cover your left eye and use the fine adjustment knob to focus the fixed (right) ocular until the letter *e* is in maximum focus. Now cover the right eye and, using the diopter ring on the left ocular, bring the image into focus. The letter *e* should now be in focus for both of your eyes. Each new lab day, as you begin to study your first slide, repeat this procedure.

l. You can increase or decrease the contrast by adjusting the iris diaphragm opening. Note that the maximum amount of light provides little contrast. Adjust the aperture until the image is sharp.

m. Move the slide slowly to the right. In what direction does the image in the ocular move?

n. Is the image in the ocular inverted relative to the specimen on the stage?

o. Center the specimen in the field of view; then rotate the 40× objective into position while watching from the side. *If it appears that the objective will hit the slide, stop and ask for assistance.*

> Most of the microscopes have **parfocal** lenses, which means that little refocusing is required when moving from one lens to another. If your scope is *not* parfocal, ask your instructor for assistance.

p. After the 40× objective is in place, focus using the fine adjustment knob.

> Never focus with the coarse adjustment knob when you are using the high-power objective.

q. The distance between the specimen and the objective lens is called the **working distance**. Is this distance greater with the 40× or the 10× objective?

3. Compute the total magnification of the specimen being viewed. To do so, multiply the magnification of the ocular lens by that of the objective lens.

a. What is the total magnification of the letter as the microscope is now set?

b. What would be the total magnification if the ocular were 20× and the objective were 100× (oil immersion)? This magnification approaches the upper limit for seeing fine details in a specimen using a light microscope.

4. Measure the diameter of the **field of view**. Once you determine the size of the field of view for any combination of ocular and objective lenses, you can determine the size of any structure within that field.
   a. Rotate the 4× objective into position and remove the letter slide.
   b. Place a clear ruler on the stage, and focus on its edge.
   c. The distance between two lines on the ruler is 1 mm. What is the diameter (mm) of the field of view?

   d. Convert this measurement to micrometers (μm), a more commonly used unit of measurement in microscopy (1 mm = 1,000 μm).
   e. Measure the diameters of the field of view for the 10× and 40× objectives, and enter all three in the spaces below to be used for future reference.

      4× =　　　　10× =　　　　40× =

   f. What is the relationship between the size of the field of view and magnification?

5. Determine spatial relationships. The **depth of field** is the thickness of the specimen that may be seen in focus at one time. Because the depth of focus is very short in the compound microscope, focus up and down to clearly view all planes of a specimen.
   a. Rotate the 4× objective into position and remove the ruler. Take a slide of crossed threads, wipe it with a Kimwipe®, and place the slide on the stage. Center the slide so that the region where the two threads cross is in the center of the stage opening.
   b. Focus on the region where the threads cross. Are both threads in focus at the same time?

   c. Rotate the 10× objective into position and focus on the cross. Are both threads in focus at the same time?

      Does the 4× or the 10× objective have a shorter depth of field?

d. Focus upward (move the stage up) with the coarse adjustment until both threads are just out of focus. Slowly focus down using the fine adjustment. Which thread comes into focus first? Is this thread lying under or over the other thread?

e. Rotate the 40× objective into position and slowly focus up and down, using the fine adjustment only. Does the 10× or the 40× objective have a shorter depth of field?

6. At the end of your microscope session, use the following procedures to store your microscope.
   a. Rotate the 4× objective into position.
   b. Remove the slide from the stage.
   c. Return the phase-contrast condenser to the 0 setting if you have used phase-contrast.
   d. Set the light intensity to its lowest setting and turn off the power.
   e. Unplug the cord and wrap it around the base of the microscope.
   f. Replace the dust cover.
   g. Return the microscope to the cabinet using two hands; one hand should hold the arm, and the other should support the base.

These steps should be followed every time you store your microscope.

EXERCISE 2.3

The Stereoscopic Microscope

Materials

stereoscopic microscope
dissecting needles
living *Elodea*
microscope slides
droppers of water
coverslips

Introduction

The stereoscopic (dissecting) microscope has relatively low magnification, 7× to 30×, and is used for viewing and manipulating relatively large objects. The binocular feature creates the stereoscopic effect. The stereoscopic microscope is similar to the compound microscope except in the following ways: (1) The depth of field is much greater than with the compound microscope, so objects are seen in three dimensions, and (2) the light source can be directed down onto as well as up through an object, which permits the viewing of objects too thick to transmit light. Light directed down on the object is called **reflected** or **incident** light. Light passing through the object is called **transmitted light**.

Procedure

1. Remove your stereoscopic microscope from the cabinet and locate the parts labeled in Figure 2.2. Locate the switches for both incident and transmitted light. In the margin of your lab manual, note any features of your microscope that are not shown in the figure. What is the range of magnification for your microscope?
2. Observe an object of your choice at increasing magnification. Select an object that fits easily on the stage (e.g., ring, coin, fingertip, pen, ruler).
   a. Place the object on the stage and adjust the interpupillary distance (distance between the oculars) by gently pushing or pulling the oculars until you can see the object as a single image.
   b. Change the magnification and note the three-dimensional characteristics of your object.
   c. Adjust the lights, both reflected and transmitted. Which light gives you the best view of your object?

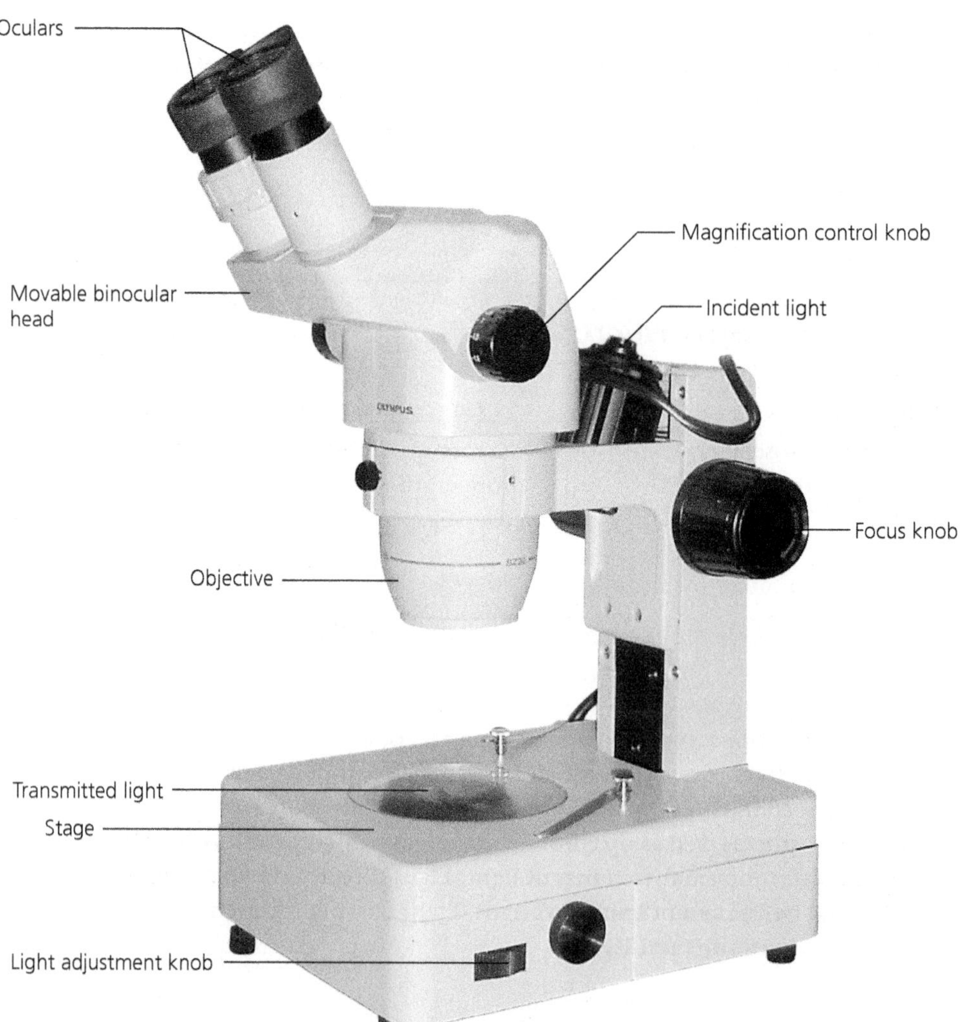

FIGURE 2.2
The stereoscopic (dissecting) microscope. Locate the parts of your microscope by referring to this photograph. Note in the margin any features of your microscope that are not shown in the photograph.

3. Prepare a **wet mount** of *Elodea*. Living material is often prepared for observation using a wet mount. (The material is either in water or covered with water prior to adding a coverslip.) You will use this technique to view living material under the dissecting and compound microscopes (Figure 2.3).
   a. Place a drop of water in the center of a clean microscope slide.
   b. Remove a single leaf of *Elodea,* and place it in the drop of water.
   c. Using a dissecting needle, place a coverslip at a 45° angle above the slide with one edge of the coverslip in contact with the edge of the water droplet, as shown.
   d. Lower the coverslip slowly onto the slide, being careful not to trap air bubbles in the droplet. The function of the coverslip is threefold: (1) to flatten the preparation, (2) to keep the preparation from drying out, and (3) to protect the objective lenses. Over long periods of time, the preparation may dry out, at which point water can be added to one edge of the coverslip.

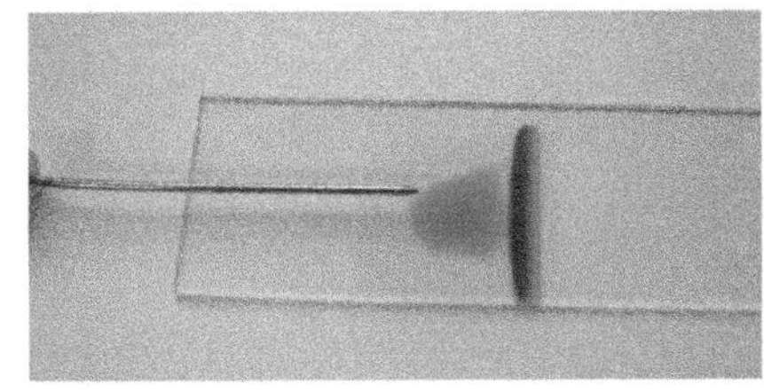

FIGURE 2.3
Preparation of a wet mount.
Place a drop of water and your specimen on the slide. Using a dissecting needle, slowly lower a coverslip onto the slide, being careful not to trap air bubbles in the droplet.

Specimens can be viewed without a coverslip using the stereoscopic microscope, but a coverslip must always be used with the compound microscope.

4. Observe the structure of the *Elodea* leaf at increasing magnification.
   a. Place the leaf slide on the stage and adjust the focus. Change the magnification and note the characteristics of the leaf at increased magnification.
   b. Sketch the leaf in the margin of your lab manual and list, in the space below, the structures that are visible at low and high magnification.

      Low:

      High:

      Is it possible to see cells in the leaf using the stereoscopic microscope?

      Organelles?

   c. Save your slide for later study. In Exercise 2.5, Lab Study C, you will be asked to compare these observations of *Elodea* with those made while using the compound microscope.

EXERCISE 2.4

The Transmission Electron Microscope

Materials

demonstration resources for the electron microscope
electron micrographs

Introduction

The transmission electron microscope (TEM) magnifies objects approximately 1,000× larger than a light microscope can (up to 1,000,000×). This difference depends on the **resolving power** of the electron microscope, which allows the viewer to see two objects of comparable size that are close together and still be able to recognize that they are two objects rather than one. Resolving power, in turn, depends on the wavelength of light (or electrons) passed through the specimen: the shorter the wavelength, the greater the resolution. Electron microscopes focus a beam of electrons through a specimen to illuminate it, rather than a beam of light, and electrons have a much shorter wavelength than does visible light. For that reason, the resolving power of electron microscopes is much greater than that of light microscopes. Both the electron and light microscopes can be equipped with lenses that allow for tremendous magnification, but only the electron microscope has sufficient resolving power to make these lenses useful.

Procedure

1. Compare the features of the light and electron microscopes (Figure 2.4).
   a. Name three structures found in both microscopes.

   b. What is the source of illumination in the electron microscope?

      For the compound microscope?

   c. Describe how the lenses differ for the two microscopes.

2. Using the resources provided by your instructor, review the procedures and materials used to prepare specimens for electron microscopy. Websites that describe electron microscopy are listed at the end of this lab topic.
3. Although the magnifying power of an electron microscope is much greater than that of a light microscope, one important disadvantage of studying cells with the transmission electron microscope is that the process for preparing cells and tissues kills the cells. Define the following terms associated with electron microscopy. As you define these terms, you will understand why living cells cannot be studied with a transmission electron microscope.

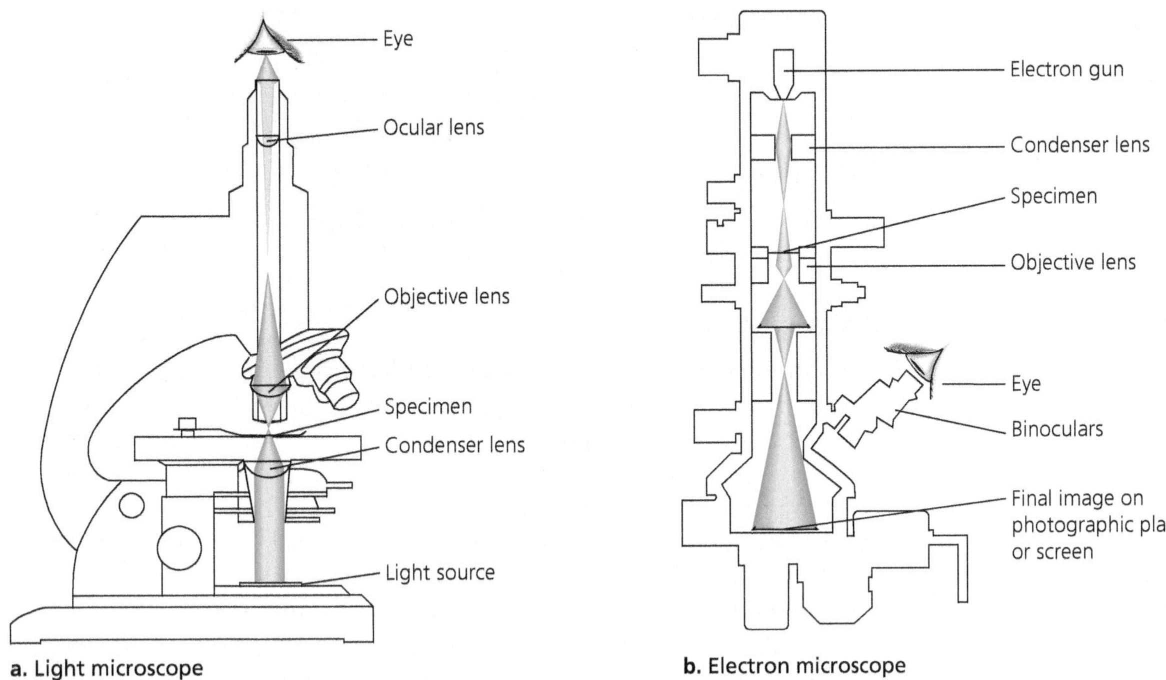

FIGURE 2.4

Comparison of light microscope and electron microscope. The source of illumination is light for the light microscope and electrons for the electron microscope. A condenser lens focuses the light or the electrons on the specimen being viewed. The image is magnified by glass objectives in light microscopy and by electromagnets in electron microscopy.

fixation:

embedding in plastic:

staining with heavy metals:

glass or diamond knife:

ultramicrotome:

fluorescent screen:

4. In addition to the *transmission electron microscope* (TEM), there is an additional type of electron microscope that is frequently used for specific purposes when studying cells and tissues, and even whole organisms. This microscope is called a *scanning electron microscope* (SEM). Using your text or the Web, investigate the use of scanning electron microscopy for biological applications.

   When would it be appropriate to use the SEM rather than the TEM?

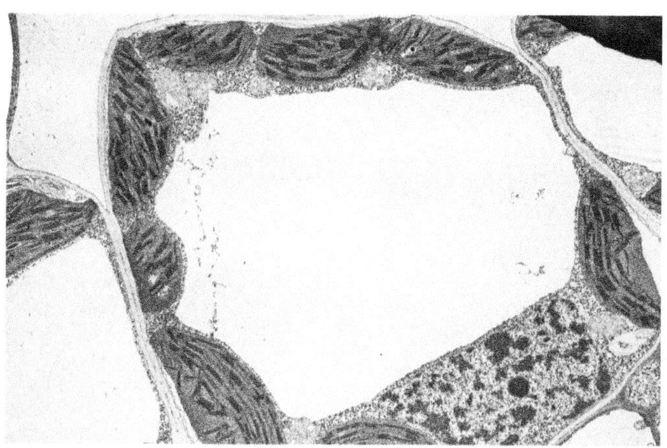

a.

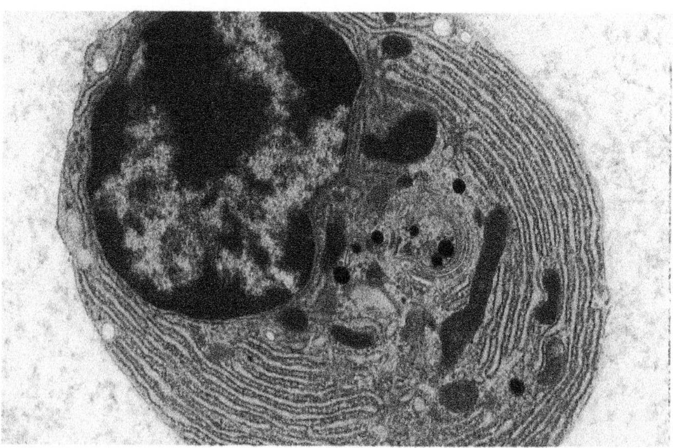

b.

FIGURE 2.5
Cells as seen in a transmission electron microscope. (a) Electron micrograph of a plant cell with chloroplasts in the cytoplasm around the large central vacuole. (b) Electron micrograph of a white blood cell with mitochondria, a Golgi apparatus, and extensive rough endoplasmic reticulum in the cytoplasm around the large nucleus containing dark stained chromatin.

5. When a transmission electron microscope is used, cells are usually studied using electron micrographs, photographs taken of the image seen on the fluorescent screen. Observe the electron micrographs in Figure 2.5a and b, respectively, a plant cell and an animal cell. Other micrographs may be on demonstration in the laboratory, and also check your lecture text for examples. Working with your lab partner, see if you can identify and label the following organelles and structures in Figure 2.5 and in other micrographs on demonstration in the laboratory.

   plasma membrane, cell wall, nucleus, chloroplast, mitochondria, large central vacuole, Golgi apparatus, lysosome, endoplasmic reticulum, ribosome

   Predict which of these organelles will *not* be visible when studying plant and animal cells using the *light* microscope. *Underline* those structures in the above list. Return to this activity after you have completed Exercise 2.5, Lab Study C, to confirm your predictions.

EXERCISE 2.5

The Organization of Cells

In this exercise, you will examine the features common to all *eukaryotic cells* that are indicative of their common ancestry. All eukaryotic cells are surrounded by a cell membrane (also called a plasma membrane) and contain a nucleus, a structure that controls cell metabolism and division. Cytoplasm, a semifluid substance that contains organelles, fills the cell. Organelles perform specific functions for the cell. However, you will observe that all cells are not the same. Some organisms are **unicellular** (single-celled), with all living functions (respiration, digestion, reproduction, and excretion) handled by that one cell. Others form random, temporary **aggregates**, or clusters, of cells. Clusters composed of a consistent and predictable number of cells are called **colonies**. Simple colonies are clusters of cells of similar types with a predictable structure. They are united, but the cells have no physiological connections. More complex colonies have cells of different types. In some colonial algae the cells are called *somatic cells* (cells that are not reproductive) and *reproductive cells* (cells that specialize in reproduction). In these colonies, if *either* type of cell is isolated from the colony, it may be reproductive, dividing and producing new colonies.

Other algae may contain both cell types, somatic and reproductive, but their somatic cells *never* become reproductive, even when isolated. Furthermore, their reproductive cells cannot persist independently, but must be associated with somatic cells to live. These algae are described as **multicellular**. They demonstrate the following two defining features:

- Multicellular organisms consist of two or more types of cells with specialized structure and function.
- If any one of the cell types of the organism is isolated, it is not capable of perpetuating the species in nature.

In more complex algae, fungi, plants, and animals, specialized cell types may be organized into *tissues* that perform particular functions for the organism. Tissues, in turn, may combine to form *organs*, and tissues and organs combine to form a coordinated single *organism*.

In this exercise, you will examine selected unicellular, aggregate, colonial, and multicellular organisms. You will identify many of the organelles that they have in common and some organelles that are unique.

Lab Study A. Unicellular Organisms

Materials

microscope slides
culture of *Amoeba*
living termites
forceps
coverslips
dissecting needles
insect Ringers
pipettes

Introduction

Unicellular eukaryotic organisms may be **autotrophic** (photosynthetic) or **heterotrophic** (deriving food from other organisms or their by-products). These diverse organisms, called protists, will be studied in detail in Lab Topic 13.

Procedure

1. Examine a living *Amoeba* (Figure 2.6) under the compound microscope. Amoebas are aquatic organisms commonly found in ponds. To transfer a specimen to your slide, follow these procedures:
   a. Place the culture dish containing the amoeba under the dissecting microscope, and focus on the bottom of the dish. The amoeba will appear as a whitish, irregularly shaped organism attached to the bottom.
   b. Using a clean pipette (it is important not to interchange pipettes between culture dishes), transfer a drop with several amoebas to your microscope slide. To do this, squeeze the pipette bulb *before* you place

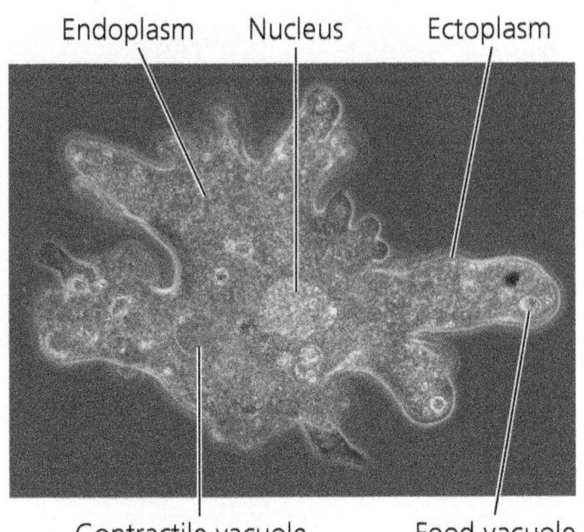

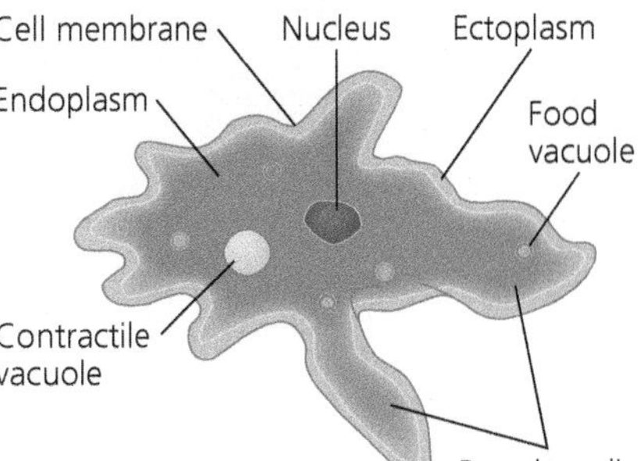

FIGURE 2.6
Amoeba. An amoeba moves using pseudopodia. Observe the living organisms using the compound microscope.

the tip under the surface of the water. Disturbing the culture as little as possible, pipette a drop of water with debris from the *bottom* of the culture dish. You may use your stereoscopic microscope to scan the slide to locate amoebas before continuing.

c. Cover your preparation with a clean coverslip.

d. Under low power on the compound scope, scan the slide to locate an amoeba. Center the specimen in your field of view; then switch to higher powers.

e. Identify the following structures in the amoeba:

The **cell membrane** is the boundary that separates the organism from its surroundings.

Ectoplasm is the thin, transparent layer of cytoplasm directly beneath the cell membrane.

Endoplasm is the granular cytoplasm containing the cell organelles.

The **nucleus** is the grayish, football-shaped body that is somewhat granular in appearance. This organelle, which directs the cellular activities, will often be seen moving within the endoplasm.

Contractile vacuoles are clear, spherical vesicles of varying sizes that gradually enlarge as they fill with excess water. Once you've located a vacuole, watch it fill and then empty its contents into the surrounding environment. These vacuoles serve an excretory function for the amoeba.

Food vacuoles are small, dark, irregularly shaped vesicles within the endoplasm. They contain undigested food particles.

Pseudopodia ("false feet") are fingerlike projections of the cytoplasm. They are used for locomotion as well as for trapping and engulfing food in a process called **phagocytosis**.

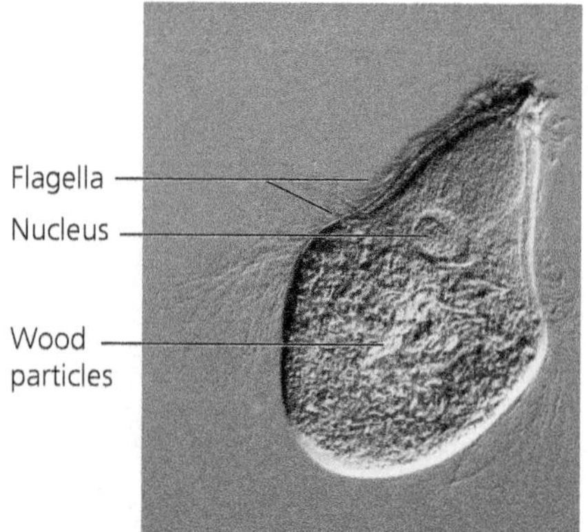

Student Media Videos—Ch. 28: Amoeba; Amoeba Pseudopodia

2. Examine *Trichonympha* (Figure 2.7) using your compound microscope. This unicellular organism and other organisms live in a symbiotic relationship in guts of termites, where they digest wood particles eaten by the insect. Termites lack the enzymes necessary to digest wood and are dependent on *Trichonympha* to make the nutrients in the wood available to them. *Trichonympha* has become so well adapted to the environment of the termite's gut that it cannot survive outside of it. First, you will have to separate the *Trichonympha* from the termite.

To obtain a specimen:

a. Place a couple of drops of **insect Ringers** (a saline solution that has the same concentration as the internal environment of insects) on a clean microscope slide.

b. Using forceps or your fingers, transfer a termite into the drop of Ringers.

c. Place the slide under the dissecting microscope.

d. Place the tips of dissecting needles at either end of the termite and pull in opposite directions.

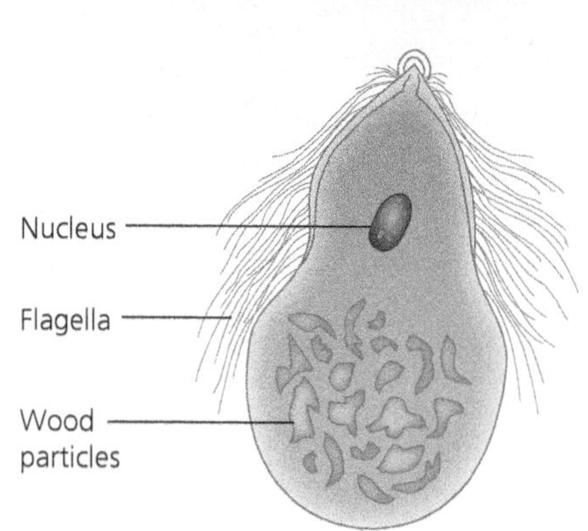

FIGURE 2.7
Trichonympha. A community of microorganisms, including *Trichonympha*, inhabits the intestine of the termite. Following the procedure in Exercise 2.5, Lab Study A, disperse the microorganisms and locate the cellular structures in *Trichonympha*.

e. Locate the long tube that is the termite's intestine. Remove all the larger parts of the insect from the slide.

f. Using a dissecting needle, mash the intestine to release the *Trichonympha* and other protozoa and bacteria.

g. Cover your preparation with a clean coverslip.

h. Transfer your slide to the compound microscope and scan the slide under low power. Center several *Trichonympha* in the field of view and switch to higher powers.

> Several types of protozoans and bacteria will be present in the termite gut.

i. Locate the following structures visible when using the highest power.

Flagella are the long, hairlike structures on the outside of the organism. The function of the flagella is not fully understood. Within the gut of the termite, the organisms live in such high density that movement by flagellar action seems unlikely and perhaps impossible.

The **nucleus** is a somewhat spherical organelle near the middle of the organism.

Wood particles may be located in the posterior region of the organism.

Lab Study B. Aggregate and Colonial Organisms

Materials

microscope slides
dissecting needles
forceps
coverslips
tree bark with *Protococcus*
cultures of *Scenedesmus*
transfer pipettes in cultures

Introduction

Unlike unicellular organisms, which live independently of each other, colonial organisms are cells that live in groups and are to some degree dependent on one another. The organisms studied in this exercise show an increasing degree of interaction among cells.

Procedure

1. Examine *Protococcus* under the compound microscope. *Protococcus* (Figure 2.8) is a terrestrial green algae that grows on the north sides of trees and is often referred to as "moss."

   a. To obtain a specimen, use a dissecting needle to brush off a small amount of the green growth on the piece of tree bark provided into a drop of water on a clean microscope slide. Avoid scraping bark onto the slide. Cover the preparation with a clean coverslip.

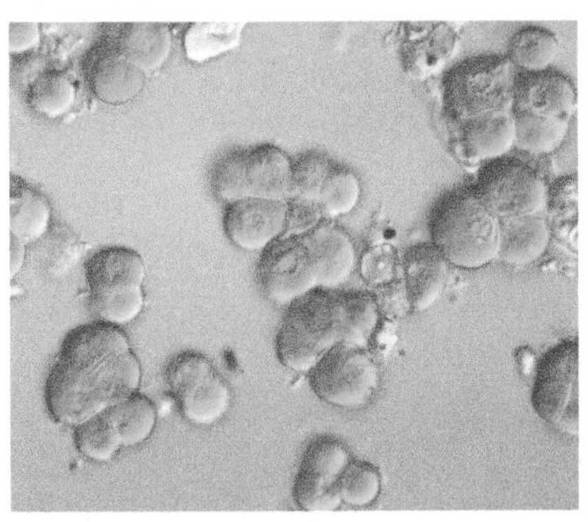

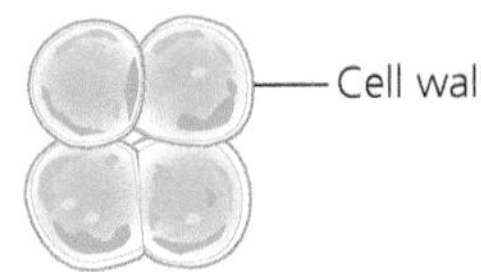

FIGURE 2.8
Protococcus. *Protococcus* is a terrestrial green algae that forms loose aggregates on the bark of trees.

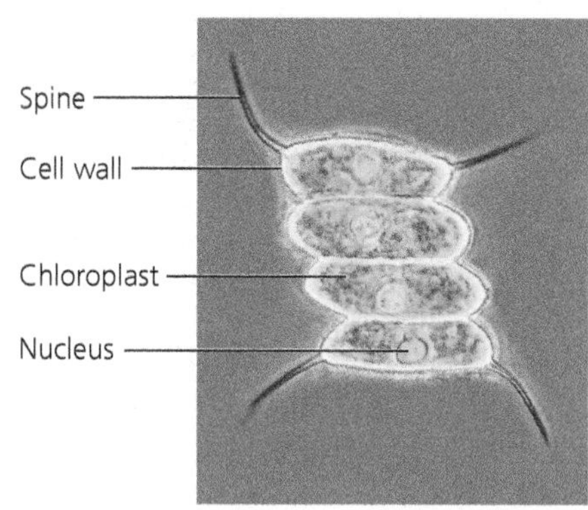

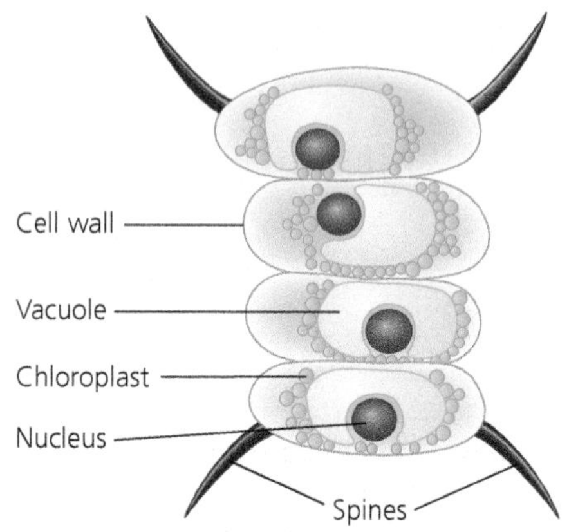

FIGURE 2.9
Scenedesmus. *Scenedesmus* is an aquatic algae that usually occurs in simple colonies of four cells connected by the cell wall.

   b. Observe at highest power that these cells are **aggregates**: The size of the cell groupings is random, and there are no permanent connections between cells. Each cell is surrounded by a cell membrane and an outer **cell wall**.
   c. Observe several small cell groupings and avoid large clumps of cells. Cellular detail may be obscure.
2. Examine living *Scenedesmus* under the compound microscope. *Scenedesmus* (Figure 2.9) is an aquatic green algae that is common in aquaria and polluted water.
   a. To obtain a specimen, place a drop from the culture dish (using a clean pipette) onto a clean microscope slide, and cover it with a clean coverslip.
   b. Observe that the cells of this organism form a **simple colony**: The cells always occur in groups of from four to eight cells, and they are permanently united.
   c. Identify the following structures.

      The **nucleus** is the spherical organelle in the approximate middle of each cell.

      Vacuoles are the transparent spheres that tend to occur at either end of the cells.

      Spines are the transparent projections that occur on the two end cells.

      Cell walls surround each cell.

Lab Study C. Multicellular Organisms

Materials

microscope slides
dropper bottles of water
toothpicks
coverslips
Elodea
methylene blue
finger bowl with disinfectant
broken glass chips
Volvox cultures

Introduction

Review the criteria for characterizing an organism as *multicellular* in the introduction of this exercise. Recall that multicellular organisms have structural and physiological connections with two or more cell types with specialized structure and function. One cell type cannot persist when isolated from other cells in the organism. If these cells are isolated, they are not capable of perpetuating the species. In this lab study, you will examine an example of a green algae, a plant, and an animal to investigate the criteria for multicellularity and observe cells that compose basic tissue types.

Procedure

Volvox

Volvox (Figure 2.10) is an aquatic green algae that is common in aquaria, ponds, and lakes. In older literature this organism was described as colonial and was not considered to be multicellular. Today, however, scientists have concluded that it is more accurate to call *Volvox* multicellular. In this activity you will look for evidence that supports this conclusion.

1. Examine living *Volvox* under the compound microscope. To obtain a specimen, prepare a wet mount as you did for *Scenedesmus* with the following addition: Before placing a drop of the culture on your slide, place several glass chips on the slide. This will keep the coverslip from crushing these spherical organisms.
2. Observe that the cells of this organism lie in a transparent matrix forming a large hollow sphere. The approximately 500 to 50,000 (depending on the species) nonreproductive somatic cells are permanently united by cytoplasmic connections. These cells have chloroplasts for photosynthesis and flagella that beat in a coordinated motion to move the colony like a ball. During asexual reproduction, certain cells in the sphere (reproductive cells) enlarge and migrate inward to become daughter colonies.
3. Identify the following structures: **somatic cells** with **cytoplasmic connections** and **flagella**. Depending on the magnification of your microscope, you may be able to distinguish **cell walls** and **nuclei** in the cells. **Daughter colonies** are smaller spheres within the larger colony. These are released when the parent colony disintegrates.

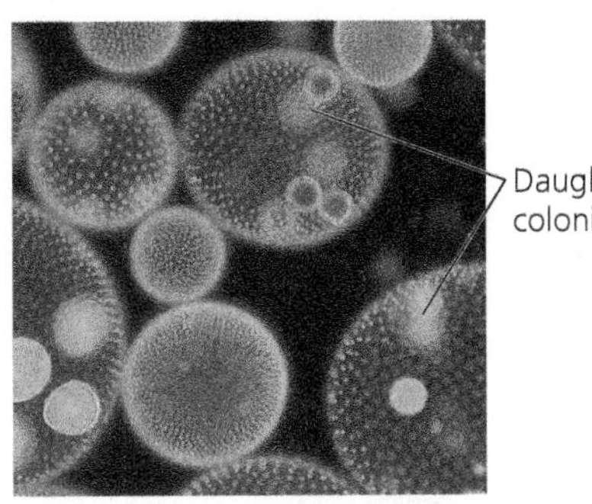

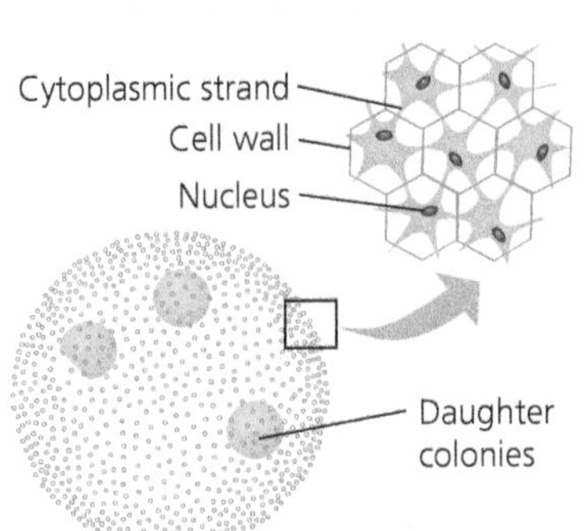

FIGURE 2.10
Volvox. In this organism, the individual cells are interconnected by cytoplasmic strands to form a sphere. Small clusters of cells, called daughter colonies, are specialized for reproduction.

Student Media Video—Ch. 28: *Volvox* Colony

Plant Cells

1. The major characteristics of a typical plant cell are readily seen in the leaf cells of *Elodea,* a common aquatic plant (Figure 2.11). Prepare a wet mount and examine one of the youngest (smallest) leaves from a sprig of *Elodea* under the compound microscope.
2. Identify the following structures.

   The **cell wall** is the rigid outer framework surrounding the cell. This structure gives the cell a definite shape and support. It is not found in animal cells.

   Protoplasm is the organized contents of the cell, exclusive of the cell wall.

   Cytoplasm is the protoplasm of the cell, exclusive of the nucleus.

   The **central vacuole** is a membrane-bound sac within the cytoplasm that is filled with water and dissolved substances. This structure serves to store metabolic wastes and gives the cell support by means of turgor pressure. Animal cells also have vacuoles, but they are not as large and conspicuous as those found in plants.

   Chloroplasts are the green, spherical organelles often seen moving within the cytoplasm. These organelles carry the pigment chlorophyll that is involved in photosynthesis. As the microscope light heats up the cells, cytoplasm and chloroplasts may begin to move around the central vacuole in a process called *cytoplasmic streaming,* or *cyclosis.*

   The **nucleus** is the usually spherical, transparent organelle within the cytoplasm. This structure controls cell metabolism and division.
3. What three structures observed in *Elodea* are unique to plants?

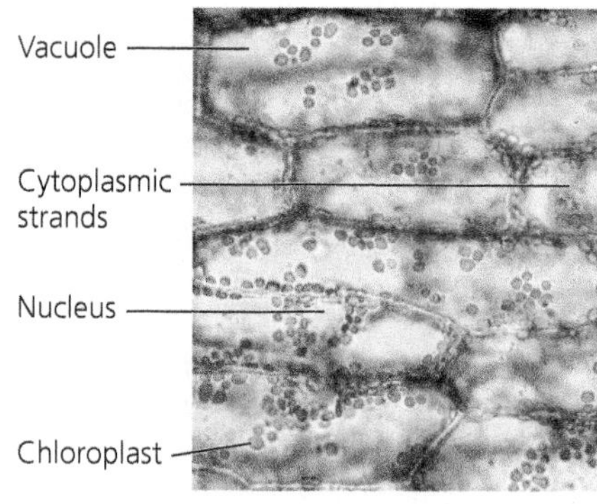

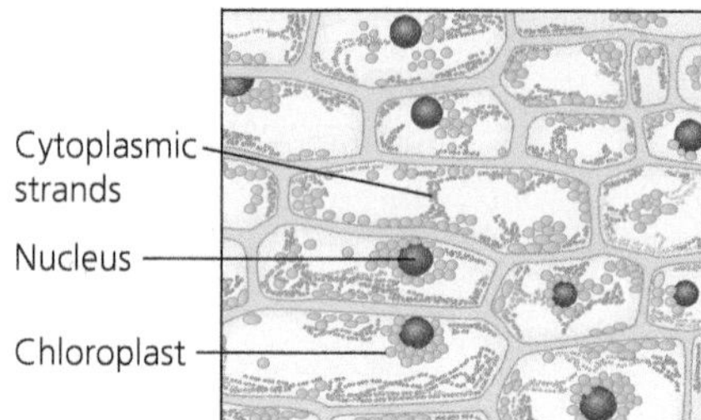

FIGURE 2.11
Elodea. *Elodea* is an aquatic plant commonly grown in freshwater aquaria. The cell structures may be difficult to see because of the three-dimensional cell shape and the presence of a large central vacuole.

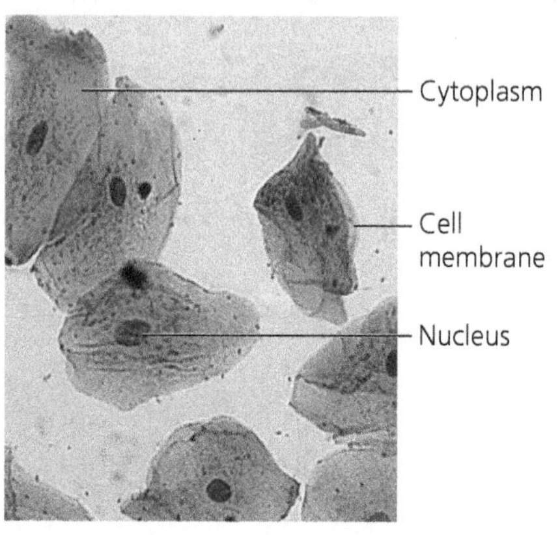

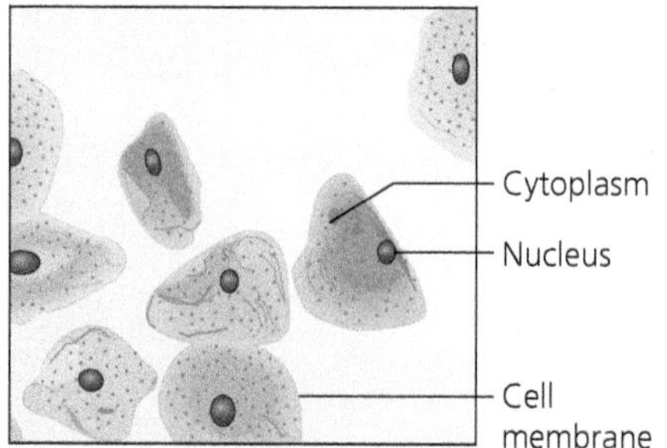

FIGURE 2.12
Human epithelial cells. The epithelial cells that line your cheek are thin, flat cells that you can remove easily from your cheek by scraping it with a toothpick.

4. Compare your observations of *Elodea* using the compound scope with those made in Exercise 2.3 using the stereoscopic scope. List the structures seen with each:

   Stereoscopic:

   Compound:

Student Media Video—Ch. 6: Cytoplasmic Streaming

Animal Cells

1. Animals are multicellular heterotrophic organisms that ingest organic matter. They are composed of cells that can be categorized into four major tissue groups: epithelial, connective, muscle, and nervous tissue. In this lab study, you will examine epithelial cells. Similar to the epidermal cells of plants, **epithelial cells** occur on the outside of animals and serve to protect the animals from water loss, mechanical injury, and foreign invaders. In addition, epithelial cells line interior cavities and ducts in animals. Examine the epithelial cells (Figure 2.12) that form the lining of your inner cheek. To obtain a specimen, follow this procedure:
   a. Add a small drop of water to a clean microscope slide. With a clean toothpick, gently scrape the inside of your cheek several times.
   b. Roll the scraping into the drop of water, add a small drop of methylene blue, and cover with a coverslip. Discard the used toothpick in disinfectant.
   c. Using the compound microscope, view the cells under higher powers.
2. Observe that these cells are extremely flat and so may be folded over on themselves. Attempt to locate several cells that are not badly folded, and study their detail.
3. Identify the following structures.

   The **cell membrane** is the boundary that separates the cell from its surroundings.

   The **nucleus** is the large, circular organelle near the middle of the cell.

   Cytoplasm is the granular contents of the cell, exclusive of the nucleus.

Lab Study D. Unknowns

Materials

microscope slides
coverslips
pond water or culture of unknowns with pipette

Introduction

Use this lab study to see if you have met the objectives of this lab topic. As you carry out this lab study, (1) think carefully about using correct microscopic techniques; (2) distinguish organisms with different cellular organization or configuration (unicellular, colonial, etc.); (3) note how the different organisms are similar yet different; and (4) note cell differences.

All of the cells studied to this point in this lab topic have been examples of **eukaryotic** cells. As you examine drops of pond water as described in the following procedure, you may observe examples of prokaryotic cells in colonies or filaments. Eukaryotic cells have a true nucleus containing chromosomes with genetic material separated from the remainder of the cell by a nuclear envelope. All cellular organelles are also bound by membranes. In prokaryotic cells genetic material is not bound by a nuclear envelope, and no membrane-bound organelles are present. Prokaryotic cells will be studied in more detail in Lab Topic 12 Bacteriology.

Procedure

1. Examine several drops of the culture of pond water that you collected, or examine the unknown culture provided by the instructor.
2. Record in Table 2.1 the characteristics of at least four different organisms.
3. Determine if a well-defined nucleus and organelles are present (eukaryote).

TABLE 2.1 Characteristics of Organisms Found in Pond Water

Unknown	Means of Locomotion	Cell Wall (+/–)	Chloroplasts (+/–)	Cellular Organization	Eukaryote (yes or no)
1					
2					
3					
4					
5					

REVIEWING YOUR KNOWLEDGE

1. Describe at least two types of materials or observations that would necessitate the use of the stereoscopic microscope.

2. State the function of each of these microscope parts and indicate if it is found in the light microscope (l), the electron microscope (e), or both (b):

   a. objective lens

   b. condenser lens

c. ocular

d. electron source (gun)

3. a. What cellular features differentiate plants from animals?

   b. How are the structures that are unique to plants important to their success?

4. Return to Step 5, p. 46. Based on your observations in today's laboratory, *circle* those organelles that are visible in the light microscope. Compare your observations with your initial predictions.
5. Review the criteria used to distinguish between colonial and multicellular organisms. Why is *Volvox* now considered multicellular?

APPLYING YOUR KNOWLEDGE

1. Using examples studied in this laboratory, propose a hypothesis for the evolution of life on Earth from single-celled organisms to multicellular organisms. Reviewing your investigations of organisms in this laboratory, can you suggest the most significant step in the evolution of multicellularity?

2. We often imply that multicellular organisms are more advanced (and therefore more successful) than unicellular or colonial organisms. Explain why this is not true, using examples from this lab or elsewhere.

3. Following is a list of tissues that have specialized functions and demonstrate corresponding specialization of subcellular structure. Match the tissue with the letter of the cell structures and organelles listed to the right that would be abundant in these cells. Use your text for a description of any organelles or structures that are unfamiliar to you.

Tissues	***Cell Structures and Organelles***
• Enzyme (protein)-secreting cells of the pancreas	a. plasma membrane
	b. mitochondria
	c. Golgi apparatus
• Insect flight muscles	d. chloroplast
	e. endoplasmic reticulum
	f. cilia and flagella
• Cells lining the respiratory passages	g. vacuole
	h. ribosome
	i. lysosome
• White blood cells that engulf and destroy invading bacteria	
• Leaf cells of cacti	

4. One organism found in a termite's gut is *Mixotricha paradoxa*. This strange creature looks like a single-celled swimming ciliate under low magnification. However, the electron microscope reveals that it contains spherical bacteria rather than mitochondria and has on its surface, rather than cilia, hundreds of thousands of spirilla and bacilla bacteria. You are the scientist who first observed this organism. How would you describe this organism—single-celled? aggregate? colony? multicellular? Review definitions of these terms in the introduction to Exercise 2.5. Can the structure of this organism give you any insight into the evolution of eukaryotic cells? (*Hint:* See the discussion of the endosymbiosis hypothesis in your text.)

5. *Pleodorina* is an aquatic green algae that is common in ponds, lakes, and roadside ditches (Figure 2.13). This organism is made up of 32 to 128 cells that are embedded in a gel-like matrix. In mature colonies two types of cells can be distinguished, small somatic cells and larger reproductive cells that divide to form new colonies. Somatic cells carry on photosynthesis, but may become reproductive if isolated from the colony.

   Review the criteria used to determine multicellularity, and decide if *Pleodorina* should be classified as multicellular or colonial.

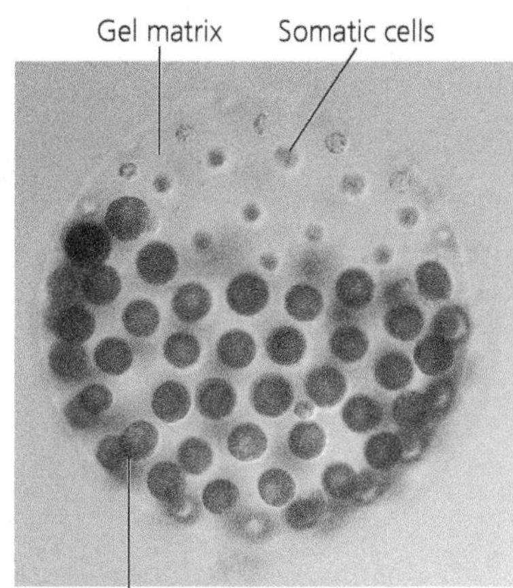

FIGURE 2.13
Pleodorina.

INVESTIGATIVE EXTENSIONS

1. Survey bodies of water surrounding your campus and assess the environmental conditions. Obtain samples of organisms in ponds, lakes, or rivers from several sites with different environmental conditions. Are some sites more polluted than others? Are there temperature differences? If a fountain is available on campus, compare this site with water in a pond that has no fountain (less water movement). You may find a site near the power-generating facility of your campus.

   Measure and note as many independent variables as possible—temperature of the water, relative levels of pollution, surroundings (near an open space, between buildings, etc.).

   List types of organisms observed. Describe the characteristics of organisms. Note differences in density and diversity of organisms in samples taken from different sources.

2. Investigate additional organelles in plants. Plants have specific cellular structures related to their lives as plants. For example, plants use chlorophyll pigments bound in the membranes of **chloroplasts** to absorb wavelengths of light that fuel photosynthesis. The starch synthesized in photosynthesis is later stored in special organelles called **leucoplasts** or **amyloplasts**. Plants are colorful (think of the many functions of color in plants). Some of these colorful pigments are water soluble and stored in vacuoles, but others, including some of the brightly colored carotenoids (reds, yellows, oranges) are membrane bound in organelles called **chromoplasts**. Chloroplasts, chromoplasts, and leucoplasts are three kinds of **plastids**, all of which are similar in structure and typically found in plant cells.

   Consider the structure and function of these three plastids. Where in the plant body would you expect to find them—in roots, stems, leaves, flowers, or fruits? Hypothesize which of these plastids might be found in green pepper skin, tomato skin, potato tubers, or other plant material.

   To test for chromoplasts and chloroplasts, peel the thin epidermis from the outside of the plant or make very thin sections of plant tissue. These can be placed on a slide with water and a coverslip then viewed under high power. To visualize leucoplasts, try adding a drop of iodine to the slide before covering. Iodine stains starch a dark blue or black.

3. In recent years, new technology has improved the quality of images that can be obtained using the light microscope. Although not used in most introductory biology labs, many research laboratories use more sophisticated light microscopes in addition to those that you have used and studied in this lab topic. You have used the conventional optical compound light microscope and the stereoscopic light microscope, and you have learned about the transmission election microscope and the scanning electron microscope. Three other light microscopes that may be used in biological research are the inverted microscope, the fluorescent microscope, and the confocal microscope.

   Using Web resources, investigate the design of these microscopes, the advantages they provide, and their application in a research laboratory.

STUDENT MEDIA: BioFlix, Activities, Investigations, and Videos

www.masteringbiology.com (select Study Area)

BioFlix—Ch. 6: Tour of an Animal Cell; Tour of a Plant Cell

Activities—Ch. 6: Build an Animal Cell and a Plant Cell; Build a Chloroplast and a Mitochondrion; Cilia and Flagella; Review: Animal Cell Structure and Function; Review: Plant Cell Structure and Function

Investigations—Ch. 6: What Is the Size and Scale of Our World?

Videos—Ch. 6: Cytoplasmic Streaming; *Paramecium* Cilia; Ch. 28: *Volvox* Colony; *Volvox* Flagella; Amoeba

REFERENCES

Alberts, B. and A. Johnson. *Molecular Biology of the Cell,* 6th ed. New York: Garland, 2014.

Becker, W. M., L. J. Kleinsmith, and J. Hardin. *The World of the Cell,* 7th ed. Pearson, 2009.

Herron, M. D., J. D. Hackett, F. O. Aylward, and R. E. Michod. "Triassic Origin and Early Radiation of Multicellular Volvocine Algae." *PNAS*, 2009, vol. 106, pp. 3254–3258.

Kirk, D. L. "A Twelve-step Program for Evolving Multicellularity and a Division of Labor." *BioEssays*, 2005, vol. 27, pp. 299–310. (The treatment of *Volvox* as multicellular and the inclusion of *Pleodorina* as a colonial organism was based on discussions with D. L. Kirk.)

Margulis, L. and D. Sagan. "The Beast with Five Genomes," *Natural History,* 2001, vol. 110, pp. 38–41.

Urry, L., M. Cain, S. Wasserman, P. Minorsky, and J. Reece. *Campbell Biology,* 11th ed. San Francisco, CA: Pearson, 2017.

WEBSITES

Cartoon depictions of cellular structures as seen in electron micrographs:
http://www.cellsalive.com

Detailed procedures for fixing, embedding, and staining tissues in both light and electron microscopy:
http://www.bristol.ac.uk/vetpath/cpl/emtechs.htm

Information on collecting and identifying organisms found in a freshwater pond:
www.microscopy-uk.org.uk/pond/index.html

Report of research that suggests that one genetic mutation more than 600 million years ago could have led to multicellularity:
https://around.uoregon.edu/content/mutation-protein-combo-and-life-went-multicellular

Structure and function of the electron microscope; good overview of the electron microscope. Search the phrase "electron microscope." Then choose "Structure and function of the electron microscope." See also "Scanning Electron Microscope," "Images from a Scanning Electron Microscope," and "Amazing Scanning Electron Microscope."
http://www.youtube.com/videos

Ultrastructure of the cell (electron micrographs); excellent electron micrographs of many cells. Click on the listed type of cell to view enlarged portions of different animal cells:
http://www.bu.edu/histology/m/t_electr.htm

LAB TOPIC 3

Diffusion and Osmosis

Laboratory Objectives

After completing this lab topic, you should be able to:

1. Describe the mechanism of diffusion at the molecular level.
2. List several factors that influence the rate of diffusion.
3. Describe a selectively permeable membrane, and explain its role in osmosis.
4. Define *hypotonic, hypertonic,* and *isotonic* in terms of relative concentrations of water and solute (dissolved substance).
5. Discuss the influence of the cell wall on osmotic behavior in cells.
6. Explain how incubating plant tissues in a series of dilutions of sucrose can give an approximate measurement of osmolarity of tissue cells.
7. Explain why diffusion and osmosis are important to cells.
8. Apply principles of osmotic activity to medical, domestic, and environmental activities.
9. Discuss the scientific process, propose questions and hypotheses, and make predictions based on experiments to test hypotheses.
10. Practice scientific persuasion and communication by constructing and interpreting graphs.

Introduction

Maintaining the steady state of a cell is achieved only through regulated movement of materials through cytoplasm, across organelle membranes, and across the plasma membrane. This regulated movement facilitates communication within the cell and between cytoplasm and the external environment. The cytoplasm and extracellular environment of the cell are aqueous solutions. They are composed of water, which is the **solvent**, or dissolving agent, and numerous organic and inorganic molecules, which are the **solutes**, or dissolved substances. Organelle membranes and the plasma membrane are **selectively permeable**, allowing water to freely pass through but regulating the movement of solutes.

The cell actively moves some dissolved substances across membranes, expending adenosine triphosphate (ATP) (biological energy) to accomplish the movement. Other substances move passively, without expenditure of ATP from the cell, but only if the cell membrane is permeable to those substances. Water and selected solutes move passively through the cell and cell membranes by **diffusion**, a physical process in which molecules move from an area where they are in high concentration to one where their concentration is lower. The energy driving diffusion comes only from the intrinsic kinetic energy (energy of motion or thermal energy) in all atoms and molecules. If nothing hinders the movement, a solute will diffuse until it reaches equilibrium.

Osmosis is a type of diffusion; the diffusion of *water* through a selectively permeable membrane from a region where it is highly concentrated to a region where its concentration is lower. The difference in concentration of water occurs if there is an unequal distribution of at least one dissolved substance on either side of a membrane and the membrane is impermeable to that substance. For example, if a membrane that is impermeable to sucrose separates a solution of sucrose from distilled water, water will move from the distilled water, where it is in higher concentration, through the membrane into the sucrose solution, where it is in lower concentration.

Figure 3.1 illustrates this process. This figure shows a U-shaped glass tube with a selectively permeable membrane separating two solutions. In Figure 3.1a, the solution on the right side of the membrane has a greater concentration of solute molecules (that cannot cross the membrane) than the solution on the left. The solution on the right may be described as having a greater **osmolarity** (solute concentration expressed as molarity). The solution on the left side of the membrane has a lower concentration of solute molecules, or a lower osmolarity. The net flow of water will be from left to right—from the solution with a greater concentration of water and lower concentration of solute to the solution with a greater concentration of solute. Diffusion of water will continue until the concentration of solute molecules relative to water molecules on both sides of the membrane is nearly equal (Figure 3.1b).

Cells are separated from their environment by the plasma membrane, a selectively permeable membrane. When referring to cells three terms, **hypertonic**, **hypotonic**, and **isotonic**, are used to describe the **tonicity** (the relative concentration of solutions inside and outside a cell that will determine the direction of water flow). A solution with a greater concentration of solute particles is said to be hypertonic relative to a solution with lower solute particles (hypotonic). When solute concentrations are equal, the solutions are isotonic.

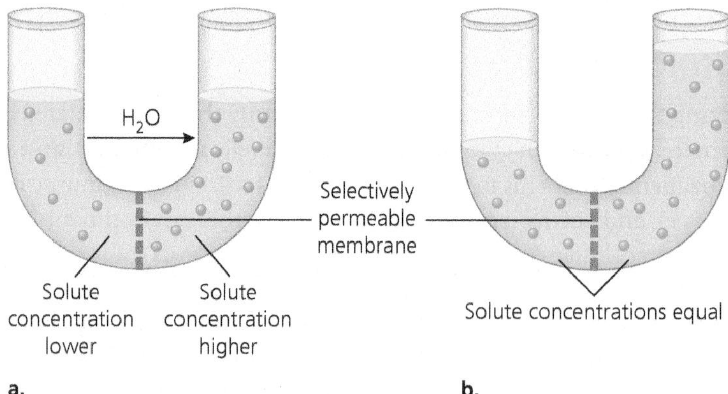

FIGURE 3.1

Osmosis. a. Two solutions separated by a selectively permeable membrane. Water molecules (the solvent) can pass through the membrane pores, but molecules of the solute (dissolved substance) cannot. The solution on the left of the membrane has a *lower* concentration of solute molecules. The solution on the right of the membrane has a *higher* concentration of solute molecules. The net flow of the solvent (water) will be from left to right until the solute concentrations in both solutions are approximately equal.
b. The concentration of solutes on both sides of the membrane is nearly equal. As water molecules diffuse through the membrane, eventually the concentration of the solute will be approximately equal on both sides of the membrane.

EXERCISE 3.1

Diffusion of Molecules

In this exercise you will investigate characteristics of molecules that facilitate diffusion, factors that influence diffusion rates, and diffusion of solutes through a selectively permeable membrane.

Experiment A. Kinetic Energy of Molecules

Materials

dropper bottle of water
carmine powder
dissecting needle
slide and coverslip
compound microscope

Introduction

Molecules of a liquid or gas are constantly in motion because of the intrinsic kinetic energy in all atoms and molecules. In 1827, Robert Brown, a Scottish botanist, noticed that pollen grains suspended in water on a slide appeared to move by a force that he was unable to explain. In 1905, Albert Einstein, searching for evidence that would prove the existence of atoms and molecules, predicted that the motion observed by Brown must exist, although he did not realize that it had been studied for many years. Only after the kinetic energy of molecules was understood did scientists ask if the motion observed by Brown and predicted by Einstein could be the result of molecular kinetic energy being passed to larger particles. We now know that intrinsic molecular kinetic energy is the driving force of diffusion. In this experiment, you will observe large particles suspended in water in motion similar to that observed by Brown, traditionally called **Brownian movement**. You will relate the motion observed to the forces that bring about diffusion.

Procedure

Work in pairs. One person should set up the microscope while the other person makes a slide as follows:

1. Place a drop of water on the slide.
2. Touch the tip of a dissecting needle to the drop of water and then into the dry carmine.
3. Add the carmine on the needle to the drop of water on the slide, mix, cover with a coverslip, and observe under the compound microscope.
4. Observe on low power and then high power. Focus as much as possible on one particle of carmine.
5. Record your findings in the Results section, and draw conclusions based on your results in the Discussion section.

Results

Describe the movement of single carmine particles.

1. Is the movement random or directional?
2. Does the movement ever stop?

3. Do smaller particles move more rapidly than larger particles? Other observations?

Discussion

1. Are you actually observing molecular movement? Explain.

2. How can molecular movement bring about diffusion?

3. List several processes in cell metabolism where diffusion is important.

Experiment B. Diffusion of Molecules Through a Selectively Permeable Membrane

Materials

string or rubber band
wax pencil
30% glucose solution
starch solution
I_2KI solution
Benedict's reagent
hot plate
500-mL beaker one-third filled with water
handheld test tube holder
3 standard test tubes
disposable transfer pipettes
2 400-mL beakers to hold dialysis bag
30-cm strip of moist dialysis tubing

Introduction

Dialysis tubing is a membrane made of regenerated cellulose fibers formed into a flat tube. If two solutions containing dissolved substances of different molecular weights are separated by this membrane, some substances may readily pass through the pores of the membrane, but others may be excluded.

Working in teams of four students, you will investigate the selective permeability of dialysis tubing. You will test the permeability of the tubing to the reducing sugar, glucose (molecular weight 180), starch (a variable-length polymer of glucose), and iodine potassium iodide (I_2KI). You will place a solution of glucose and starch into a dialysis tubing bag and then place this bag into a solution of I_2KI. Sketch and label the design of this experiment in the margin of your lab manual or on separate paper to help you develop your hypotheses.

You will use two tests in your experiment:

1. *I_2KI test for presence of starch*

   When I_2KI is added to the unknown solution, the solution turns purple or black if starch is present. If no starch is present, the solution remains a pale yellow-amber color.

2. *Benedict's test for reducing sugar*

   When Benedict's reagent is added to the unknown solution and the solution is heated, the solution turns green, orange, or orange-red if a reducing sugar is present (the color indicates the sugar concentration). If no reducing sugar is present, the solution remains the color of Benedict's reagent (blue).

Question

Remember that every experiment begins with a question. Review the design of this experiment in the Introduction of this experiment. Formulate a question about the permeability of dialysis tubing. The question may be broad, but it must propose an idea that has measurable and controllable elements.

Hypothesis

Hypothesize about the selective permeability of dialysis tubing to the substances being tested.

Prediction

Predict the results of the I_2KI and Benedict's tests based on your hypothesis (if/then).

Procedure

1. Prepare the dialysis bag with the initial solutions.
   a. Fold over 3 cm at the end of a 25- to 30-cm piece of dialysis tubing that has been soaking in water for a few minutes, pleat the folded end "accordion style," and close the end of the tube with the string or a rubber band, forming a bag. This procedure must secure the end of the bag so that no solution can seep through.
   b. Roll the opposite end of the bag between your fingers until it opens, and add 4 pipettesful of 30% glucose into the bag. Then add 4 pipettesful of starch solution to the glucose in the bag.
   c. Hold the bag closed and mix its contents. Record its color in Table 3.1 in the Results section. Carefully rinse the outside of the bag in tap water.
   d. Add 200 mL of water to a 400- to 500-mL beaker. Add several droppersful of I_2KI solution to the water until it is visibly yellow-amber. Record the color of the H_2O-plus-I_2KI solution in Table 3.1.

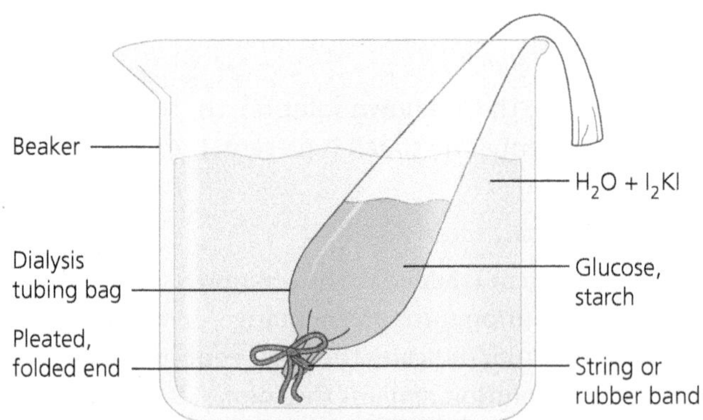

FIGURE 3.2
Setup for Exercise 3.1, Experiment B. The dialysis tubing bag, securely closed at one end, is placed in the beaker of water and I_2KI. The open end of the bag should drape over the edge of the beaker.

e. Place the bag in the beaker so that the untied end of the bag hangs over the edge of the beaker (Figure 3.2). *Do not allow the liquid to spill out of the bag!* If the bag is too full, remove some of the liquid and rinse the outside of the bag again. If needed, place a rubber band around the beaker, holding the bag securely in place. If some of the liquid spills into the beaker, dispose of the beaker water, rinse, and fill again.

2. Leave the bag in the beaker for about 30 minutes. (You should go to another lab activity and then return to check your setup periodically.)
3. After 30 minutes, carefully remove the bag and stand it in a dry beaker.
4. Record in Table 3.1 the final color of the solution in the bag and the final color of the solution in the beaker.
5. Perform the Benedict's test for the presence of sugar in the solutions.
   a. Label three clean test tubes: control, bag, and beaker.
   b. Put 2 pipettesful of water in the control tube.
   c. Put 2 pipettesful of the bag solution in the bag tube.
   d. Put 2 pipettesful of the beaker solution in the beaker tube.
   e. Add 1 dropperful of Benedict's reagent to each tube.
   f. Heat the test tubes in a boiling water bath for about 3 minutes.
   g. Record your results in Table 3.1.
6. Review your results in Table 3.1 and draw your conclusions in the Discussion section.

Results

Complete Table 3.1 as you observe the results of Experiment B.

TABLE 3.1 Results of Experiment Investigating the Permeability of Dialysis Tubing to Glucose, I_2KI, and Starch

Solution Source	Original Contents	Original Color	Final Color	Color After Benedict's Test
Bag 3Pt 3Pt	Glucose starch	Clear	Blue	Dark Yellow
Beaker	water I_2KI	Amber	Amber	Yellow
Control	H20	—	—	Blue

Discussion

1. What is the significance of the final colors and the colors after the Benedict's tests? Did the results support your hypothesis? Explain, giving evidence from the results of your tests.

2. How can you explain your results?

3. From your results, predict the size of I_2KI molecules relative to glucose and starch.

4. What colors would you expect if the experiment started with glucose and I_2KI inside the bag and starch in the beaker? Explain.

I would expect the beaker to turn blue And the bag to stay the same color since the I_2KI goes into the starch

EXERCISE 3.2

Osmotic Activity in Cells

All organisms must maintain an optimum internal osmotic environment. Terrestrial vertebrates must take in and eliminate water using internal regulatory systems to ensure that the environment of tissues and organs remains in osmotic balance. Exchange of waste and nutrients between blood and tissues depends on the maintenance of this condition. Plants and animals living in fresh water must control the osmotic uptake of water into their hypertonic cells.

In this exercise, you will investigate the osmotic behavior of plant and animal cells placed in different molar solutions. What happens to these cells when they are placed in hypotonic or hypertonic solutions? This question will be investigated in the following experiments.

Experiment A. Osmotic Behavior of Animal Cells

Materials

On demonstration:

- test tube rack
- 3 test tubes with screw caps, each containing one of the three solutions of unknown osmolarity (solute concentration)
- ox blood
- newspaper or other printed page

For microscopic observations:

- 4 clean microscope slides and coverslips
- wax pencil
- dropper bottle of ox blood
- dropper bottles with three solutions of unknown osmolarity

Introduction

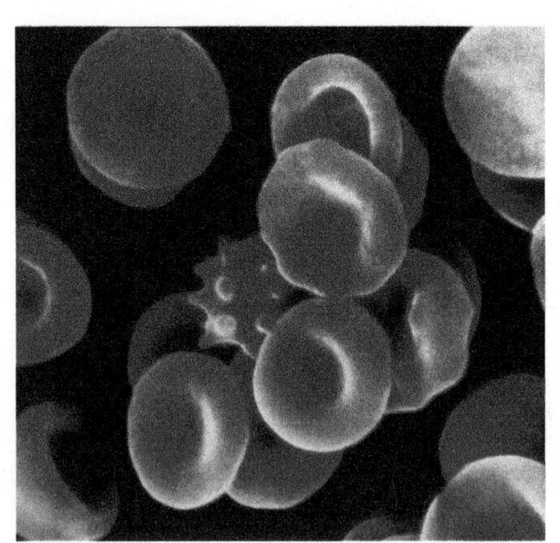

FIGURE 3.3
Human red blood cells. The star-shaped cell in the center has lost water and crenated.

Mature red blood cells (erythrocytes) are little more than packages of hemoglobin bound by a plasma membrane permeable to small molecules, such as oxygen and carbon dioxide, but impermeable to larger molecules, such as proteins, sodium chloride, and sucrose. In mammals these cells even lack nuclei when mature, and as they float in isotonic blood plasma, their shape is flattened and pinched inward into a biconcave disk. Oxygen and carbon dioxide diffuse across the membrane, allowing the cell to carry out its primary function, gas transport, which is enhanced by the increased surface area created by the shape of the cell. Scientists questioning what happens to red blood cells in different molar solutions observed that the cells respond dramatically if they are not in an isotonic environment. When water moves into red blood cells placed in a hypotonic solution, the cells swell and the membranes burst, or undergo **lysis**. When water moves out of red blood cells placed in a hypertonic solution, the cells shrivel and appear bumpy, or **crenate**. In this experiment, you will investigate the behavior of red blood cells when the osmolarity of the environment changes from isotonic to hypertonic or hypotonic. (See Figure 3.3.)

Hypothesis

Hypothesize about the behavior of red blood cells when they are placed in hypertonic or hypotonic environments.

hyper 150 hypo
crenate normal lysed

Prediction

Predict the results of the experiment based on your hypothesis (if/then).

Procedure

1. Observe the three test tubes containing unknown solutions and blood on demonstration. These tubes have been prepared in the following way.

   Test tube 1: 15 mL of unknown solution A

   Test tube 2: 15 mL of unknown solution B

   Test tube 3: 15 mL of unknown solution C

   Your instructor has added 5 drops of ox blood to each test tube.

   Observe the appearance of each test tube. Is it opaque? Is it translucent? Describe your observations in Table 3.2 in the Results section.
2. Be sure each test tube cap is securely tightened, then hold each test tube flat against the printed newspaper article or page of text.
3. Attempt to read the print. Describe in Table 3.2 in the Results section.

   Continue your investigation of the osmotic behavior of animal cells by performing microscopic observations of cells in the three unknown solutions.

> Have your microscope ready, and observe slides immediately after you have prepared them. Do one slide at a time.

4. Label four clean microscope slides A, B, C, and D.
5. Place a drop of blood on slide D, cover with a coverslip, and observe the shape of the red blood cells with no treatment. Record your observations in Table 3.3 in the Results section.
6. Locate the three dropper bottles (A, B, C) containing solutions of unknown osmolarity. Put a drop of solution A on slide A and add a coverslip. Place the slide on the microscope stage and carefully add a small drop of blood to the edge of the coverslip. The blood cells will be drawn under the coverslip by capillary action.
7. As you view through the microscope, carefully watch the cells as they come into contact with solution A; record your observations in Table 3.3.
8. Repeat steps 6 and 7 with solutions B and C.
9. Record your observations in Table 3.3. Draw your conclusions in the Discussion section.

Results

1. Record your observations of the demonstration test tubes in Table 3.2.

TABLE 3.2 Appearance of Unknown Solutions A, B, and C

	Appearance of the Solution	Can You Read the Print?
Test tube 1 (unknown A)	Crenated	No
Test tube 2 (unknown B)	Normal	No
Test tube 3 (unknown C)	lysed	yes

2. Record your microscopic observations of red blood cell behavior in Table 3.3.

TABLE 3.3 Appearance of Red Blood Cells in Test Solutions

Solution	Appearance/Condition of Cells
D (blood only)	Circles
A hyper	crenated
B iso	Normal little dots
C hypo	lysed arms sticking out

Discussion

Explain your results in terms of your hypothesis.

1. Explain the appearance of the three test tubes on demonstration.

2. Based on the demonstration and your microscopic investigation, which of the three solutions is hypotonic to the red blood cells?

   Hypertonic?

   Isotonic?

   Verify your conclusions with the laboratory instructor.

3. What conditions might lead to results other than those expected?

Experiment B. Osmotic Behavior in Cells with a Cell Wall

Materials

On demonstration:

- 2 compound microscopes labeled A and B
- 1 slide of *Elodea* in a hypertonic salt solution
- 1 slide of *Elodea* in distilled water

Introduction

In their natural environment, cells of freshwater plants and algae are bathed in water containing only dilute concentrations of solvents. The net flow of water is from the surrounding medium into the cells. To understand this process, review the structure of *Elodea* cells from Lab Topic 2.

The presence of a cell wall and a large fluid-filled central vacuole in a plant or algal cell will affect the cell's response to solutions of differing molarities. When a plant cell is placed in a hypertonic solution, water moves out of the cell; the protoplast shrinks and may pull away from the cell wall. This process is called **plasmolysis**, and the cell is described as **plasmolyzed** (Figure 3.4). In a hypotonic solution, as water moves into the cell and ultimately into the cell's central vacuole, the cell's **protoplast** (the plant cell exclusive of the cell wall—the cytoplasm enclosed by plasma membrane) expands. The cell wall, however, restricts the expansion and the cell becomes **turgid**, resulting in **turgor pressure** (pressure of the protoplast on the cell wall owing to uptake of water). A high turgor pressure will prevent further movement of water into the cell. This process is a good example of the interaction between pressure and osmolarity in determining the direction of the net movement of water. The hypertonic condition in the cell draws water into the cell until the membrane-enclosed cytoplasm presses against the cell wall. Turgor pressure begins to force water through the membrane and out of the cell, changing the direction of net flow of water (Figure 3.5).

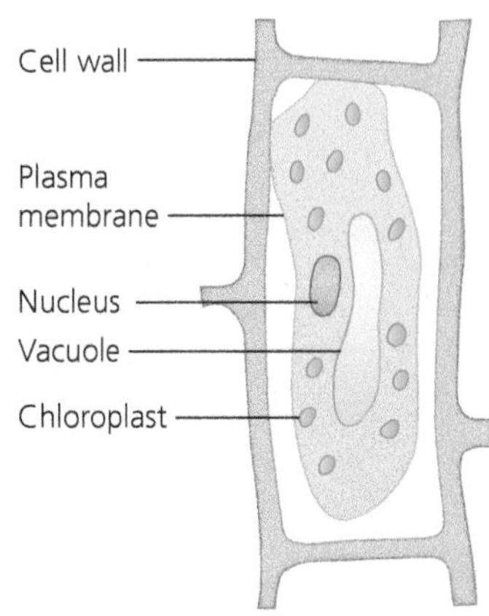

FIGURE 3.4
Plant cell placed in a hypertonic solution. Water leaves the central vacuole and the cytoplasm shrinks, a process called plasmolysis.

Scientists call the combined force created by solute concentration and physical pressure **water potential**. For a detailed explanation of water potential, see a

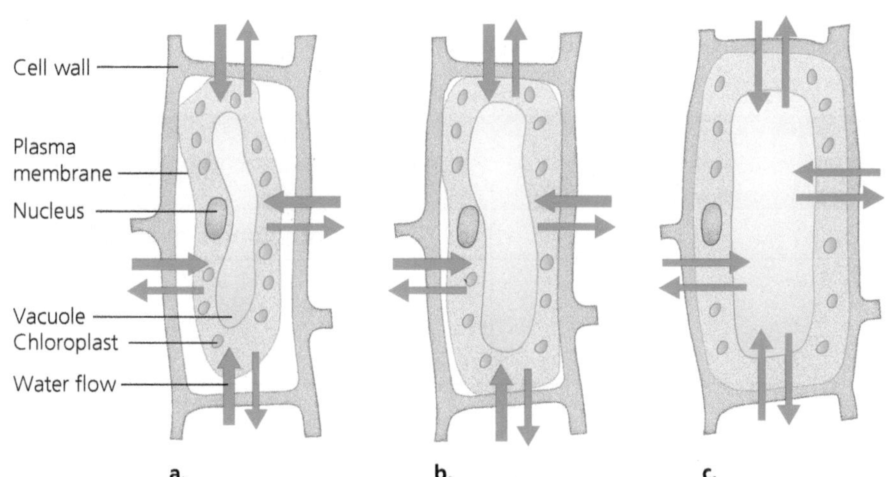

FIGURE 3.5
The effect of turgor pressure on the cell wall and the direction of net flow of water in a plant cell. A plant cell undergoes changes in a hypotonic solution. (a) Low turgor pressure. The net flow of water comes into the cell from the surrounding hypotonic medium. (b) Turgor pressure increases. The protoplast begins to press on the cell wall. (c) Greatest turgor pressure. The tendency to take up water is ultimately restricted by the cell wall, creating a back pressure on the protoplast. Water enters and leaves the cell at the same rate. The cell is turgid.

discussion of plant transport mechanisms in your text (e.g., Chapter 36 in *Campbell Biology,* 11th ed.). In contrast to an animal cell, the ideal state for a plant cell is turgidity. When a plant cell is turgid, it is not isotonic with its surroundings but is hypertonic, having a higher solute concentration than its surroundings. In this state, the plant cell protoplast presses on the cell wall. The pressure of the protoplast on the cell wall is an important force in plant activity. For example, it may cause young cells to "grow" as the elastic cell wall expands.

For this experiment, two slides have been set up on demonstration microscopes. On each slide, *Elodea* has been placed in a different molar solution: One is hypotonic (distilled water) and one is hypertonic (concentrated salt solution).

Question

Propose a question about the movement of water in *Elodea* leaves placed in different molar solutions.

Hypothesis

Hypothesize about the movement of water in cells with a cell wall when they are placed in hypertonic or hypotonic environments.

Prediction

Predict the appearance of *Elodea* cells placed in the two solutions (if/then).

Procedure

1. Observe the two demonstration microscopes with *Elodea* in solutions A and B.
2. Record your observations in Table 3.4 in the Results section, and draw your conclusions in the Discussion section.

Results

In the margin of your manual, sketch the appearance of the *Elodea* leaves and describe the appearance of the *Elodea* cells in Table 3.4.

TABLE 3.4 Appearance of *Elodea* Cells in Unknown Solutions A and B

Solution	Appearance/Condition of Cells
A Pond water	normal/ stacked
B Hyper/salt water	broken down

Discussion

1. Based on your predictions and observations, which solution is hypertonic?

   Hypotonic?

2. Which solution has the greatest osmolarity?

3. Would you expect pond water to be isotonic, hypertonic, or hypotonic to *Elodea* cells? Explain.

4. Verify your conclusions with your laboratory instructor.

Student Media Videos—Ch. 7: Turgid *Elodea;* Plasmolysis

EXERCISE 3.3

Investigating Osmolarity of Plant Cells

Knowing the solute concentration of cells has both medical and agricultural applications. In plants, scientists know that for normal activities to take place, the amount of water relative to solute concentration in cells must be maintained within a reasonable range. If plant cells have a reduced water content, all vital functions slow down.

In the following experiments, you will estimate the osmolarity (solute concentration) of potato tuber cells using two methods, change in weight and change in volume. You will incubate pieces of potato tuber in sucrose solutions of known molarity. The object is to find the molarity at which weight or volume of the potato tuber tissue does not change, indicating that there has been no net loss or gain of water. This molarity is an indirect measure of the solute concentration of the potato tuber. This measure is indirect because water movement in plant cells is also affected by the presence of cell walls (see Figure 3.5c).

Work in teams of four. Each team will measure either weight change or volume change. Time will be available near the end of the laboratory period for each team to present its results to the class for discussion and conclusions.

Experiment A. Estimating Osmolarity by Change in Weight

Materials

1 large potato tuber
7 250-mL beakers (disposable cups may be substituted)
wax marking pencil
forceps
balance that weighs to the nearest 0.01 g
aluminum foil
petri dish
sucrose solutions: 0.1, 0.2, 0.3, 0.4, 0.5, 0.6 molar (*M*)
razor blade
cork borer
deionized (DI) water (0 molar)
paper towels
metric ruler
calculator

Introduction

In this experiment, you will determine the weight of several potato tuber cylinders and incubate them in a series of sucrose solutions. After the cylinders have incubated, you will weigh them and determine if they have gained or lost weight. This information will enable you to estimate the osmolarity of the potato tuber tissue.

Question

What question is being investigated in this experiment?

Hypothesis

Hypothesize about the osmolarity of potato tuber tissue in relation to the sucrose solutions.

Prediction

Predict the results of the experiment based on your hypothesis (if/then).

Procedure

1. Obtain 100 mL of DI water and 100 mL of each of the sucrose solutions. Put each solution in a separate, appropriately labeled 250-mL beaker or paper cup.

Cork borers and razor blades can cut! Use them with extreme care! To use the cork borer, hold the potato in such a way that the borer will not push through the potato into your hand.

2. Use a sharp cork borer to obtain seven cylinders of potato. Push the borer through the length of the potato, twisting it back and forth. When the borer is filled, remove from the potato and push the potato cylinder out of the borer. You must have seven complete, undamaged cylinders at least 5 cm long.
3. Line up the potato cylinders and, using a sharp razor blade, cut all cylinders to a uniform length, about 5 cm, removing the peel from the ends.
4. Place all seven potato samples in a petri dish, and keep them covered to prevent their drying out.

> In subsequent steps, treat each sample individually. Work quickly. To provide consistency, each person should do one task to all cylinders (one person wipe, another weigh, another slice, another record data).

5. Remove a cylinder from the petri dish, and place it between the folds of a paper towel to blot sides and ends.
6. Weigh it to the nearest 0.01 g on the aluminum sheet on the balance. Record the weight in Table 3.5 in the Results section.
7. Immediately cut the cylinder lengthwise into two long halves.
8. Transfer the potato pieces to the water beaker.
9. Note what time the potato pieces are placed in the water beaker. Time: ______________.
10. Repeat steps 5 to 8 with each cylinder, placing potato pieces in the appropriate incubating solution from 0.1 to 0.6 *M*.

> Be sure that the initial weight of the cylinder placed in each test solution is accurately recorded.

11. Incubate 1.5 to 2 hours. (As this takes place, you will be performing other lab activities.)
12. Swirl each beaker every 10 to 15 minutes as the potato pieces incubate.
13. At the end of the incubation period, record the time when the potato pieces are removed. Time: ______________.

    Calculate the approximate incubation time in Table 3.5.
14. Remove the potato pieces from the first sample. Blot the pieces on a paper towel, removing excess solution only.
15. Weigh the potato pieces and record the final weight in Table 3.5.

16. Repeat this procedure until all samples have been weighed in the chronological order in which they were initially placed in the test solutions.
17. Record your data in the Results section, and complete the questions in the Discussion section.

Results

1. Complete Table 3.5. To calculate percentage change in weight, use this formula:

$$\text{Percentage change in weight} = \frac{\text{weight change}}{\text{initial weight}} \times 100$$

If the sample gained in weight, the value should be positive. If it lost in weight, the value should be negative.

2. Plot percentage change in weight as a function of the sucrose molarity in Figure 3.6.
   a. Place a 0 in the *middle of the y-axis*. Choose appropriate scales.
   b. Label the axes of the graph: Determine dependent and independent variables, and place each on the appropriate axis (see Lab Topic 1 for assistance in graphing).
   c. Graph your results. Weight increase (positive values) should be above the zero change line on the "percentage change in weight" axis. Weight decrease should be below the zero change line.
   d. Construct a curve that best fits the data points. Use this curve to estimate the osmolarity of the potato tuber.
   e. Compose an appropriate figure title.

TABLE 3.5 Data for Experiment Estimating Osmolarity by Change in Weight

Download an Excel version from www.masteringbiology.com in the Study Area under Lab Media

Approximate time in solutions: ________

	Sucrose Molarity						
	0.0	**0.1**	**0.2**	**0.3**	**0.4**	**0.5**	**0.6**
Final weight (g)							
Initial weight (g)							
Weight change (g)							
% change in weight							

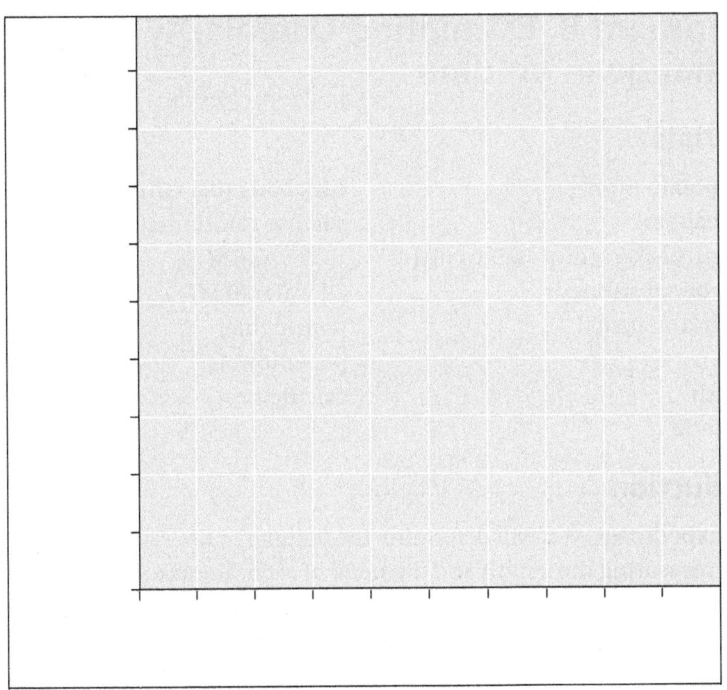

FIGURE 3.6 ______________________

Discussion

1. At what sucrose molarity does the curve cross the zero change line on the graph?

2. Explain how this information can be used to determine the osmolarity of the potato tuber tissue.

3. In more dilute concentrations of sucrose, the weight of the potato pieces ______________ (increases/decreases) after incubation. What forces other than solute concentration will have an impact on the amount of water taken up by the potato pieces (see Figure 3.5c)?

4. Estimate the osmolarity of the potato tuber tissue.

Experiment B. Estimating Osmolarity by Change in Volume

Materials

1 large potato tuber
digital caliper
7 250-mL beakers (disposable cups may be substituted)
wax marking pencil
forceps
petri dish
razor blade
cork borer (0.5-cm diameter)
sucrose solutions: 0.1, 0.2, 0.3, 0.4, 0.5, 0.6 *M*
DI water (0 *M*)
metric ruler
paper towels
calculator

Introduction

In this experiment, you will determine the volume of several potato tuber cylinders by measuring the length and diameter of each. To make these measurements you will use a digital caliper, a tool that is used by scientists, engineers, and precision woodworkers to measure the distance between two opposite sides of an object. The caliper you will use can accurately measure to 0.01 mm. You will then incubate the potato cylinders in a series of sucrose solutions. After the cylinders have incubated, you will again measure their length and diameter and determine if they have increased or decreased in size. This information will enable you to estimate the osmolarity (solute concentration) of the potato tuber tissue.

Question

What question is being investigated in this experiment?

Hypothesis

Hypothesize about the osmolarity of potato tuber tissue.

Prediction

Predict the results of the experiment based on your hypothesis (if/then).

Procedure

1. Practice measuring with the digital caliper (Figure 3.7).
   a. Identify the following parts of the caliper and add these labels to Figure 3.7: *stationary arm, movable arm, digital display, power on/off button, zero button, metric/inch button* above the display in this model (mm in/F), *ruler* with inches scale on bottom, metric on top. Note that the buttons may be in different colors on your caliper. If so, indicate those colors on the figure.
   b. Turn on the caliper. Use the metric/inch button to select metric.

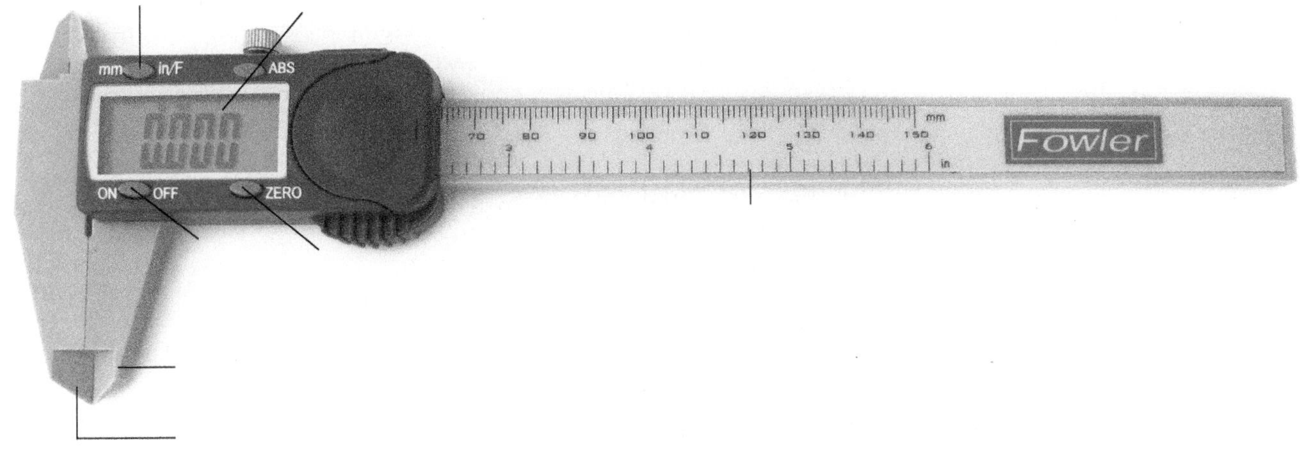

FIGURE 3.7
Digital caliper. Identify the *stationary arm, movable arm, digital display, power on/off button, zero button, metric/inch button,* and *ruler.*

c. Completely close the caliper arms and press the zero button.

d. Choose a small object (a coin will work) and place it between the two arms, adjusting the movable arm until both arms just touch the object.

e. Read the measurement of your object in the screen of the digital display. The measurement is in millimeters (mm) and this instrument accurately measures to 0.01 millimeter. For example, a reading of 50.55 is 50.55mm.

2. Obtain 100 mL of DI water and 100 mL of each of the sucrose solutions. Put each solution in a separate, appropriately labeled 250-mL beaker or paper cup.

Use cork borers and razor blades with extreme care! To use the cork borer, hold the potato in such a way that the borer will not push through the potato into your hand.

3. Use a sharp cork borer to obtain seven cylinders of potato. Push the borer through the length of the potato, twisting it back and forth. When the borer is filled, remove it from the potato and push the potato cylinder out of the borer. You must have seven complete, undamaged cylinders at least 5 cm long.

4. Line up the potato cylinders and, using a sharp razor blade, cut all cylinders to a uniform length, about 5 cm, removing the peel from the ends.

5. Place all seven potato samples in a petri dish, and keep them covered to prevent their drying out.

> In subsequent steps, treat each sample individually. Work quickly. To provide consistency, each person should do one task to all cylinders (one person wipe, another measure, another record data).

6. Remove a cylinder from the petri dish, and place it between the folds of a paper towel to blot sides and ends.
7. Using the caliper, measure the length and diameter of the cylinder to the nearest 0.01 mm, and record these measurements in Table 3.6 in the Results section. To measure, both arms of the caliper should touch but not compress the cylinder.
8. Transfer the cylinder to the 0 *M* (water) beaker.
9. Note the time the cylinder is placed in the 0 *M* beaker. Time: 1.5hrs.
10. Repeat steps 6 to 8 with each cylinder, placing the cylinders in the appropriate incubating solution from 0.1 to 0.6 *M*.

> Be sure that the initial length and diameter of the cylinder placed in each test solution are accurately recorded.

11. Incubate from 1.5 to 2 hours. (During this time period, you will be performing other lab activities.)
12. Swirl each beaker every 10 to 15 minutes as the cylinders incubate.
13. At the end of the incubation period, record the time each cylinder is removed from a solution. Time: _______.

    Calculate the approximate incubation time in Table 3.6.
14. Remove the cylinders in the chronological order in which they were initially placed in the test solutions.
15. Blot each cylinder as it is removed (sides and ends), and use the caliper to measure the length and diameter to the nearest 0.01 mm.
16. Finish recording your data in the Results section, and answer the questions in the Discussion section.

Results

1. Complete Table 3.6. To calculate the volume of a cylinder, use this formula:

$$\text{Volume of a cylinder (mm}^3) = \pi(\text{diameter}/2)^2 \times \text{length}$$

$$(\pi = 3.14)$$

To calculate percentage change in volume, use this formula:

$$\text{Percentage change in volume} = \frac{\text{change in volume}}{\text{initial volume}} \times 100$$

If the sample increases in volume, the value will be positive. If it decreases in volume, the value will be negative.

TABLE 3.6 Data for Experiment Estimating Osmolarity by Change in Volume

Download an Excel version from www.masteringbiology.com in the Study Area under Lab Media

Approximate time in solutions: ________

	Sucrose Molarity						
	0.0	**0.1**	**0.2**	**0.3**	**0.4**	**0.5**	**0.6**
Final diameter (mm)	4.1	3.99	3.98	3.8	3.51	3.5	2.57
Final length (mm)	5.31	5.152	5.00	5.05	4.91	4.92	4.51
Final volume (mm³)	69.97	64.41	62.2	57.27	47.51	47.33	23.39
Initial diameter (mm)	3.52	3.52	3.51	3.49	3.52	3.51	3.52
Initial length (mm)	48.5	48.52	48.52	48.51	49.2	51.9	49.3
Initial volume (mm³)	471.7	466.4	469.1	463.8	480.3	501.9	479.5
Change in volume (mm³)	401.73	401.99	406.9	406.53	432.79	454.57	456.11
% change in volume	85.22	86.18	86.74	87.65	90.10	90.56	95.12

2. Plot percentage change in volume as a function of the sucrose molarity in Figure 3.8.

   a. Place a 0 in the *middle of the y-axis*. Choose appropriate scales.

   b. Label the axes of the graph: Determine dependent and independent variables, and place each on the appropriate axis (see Lab Topic 1).

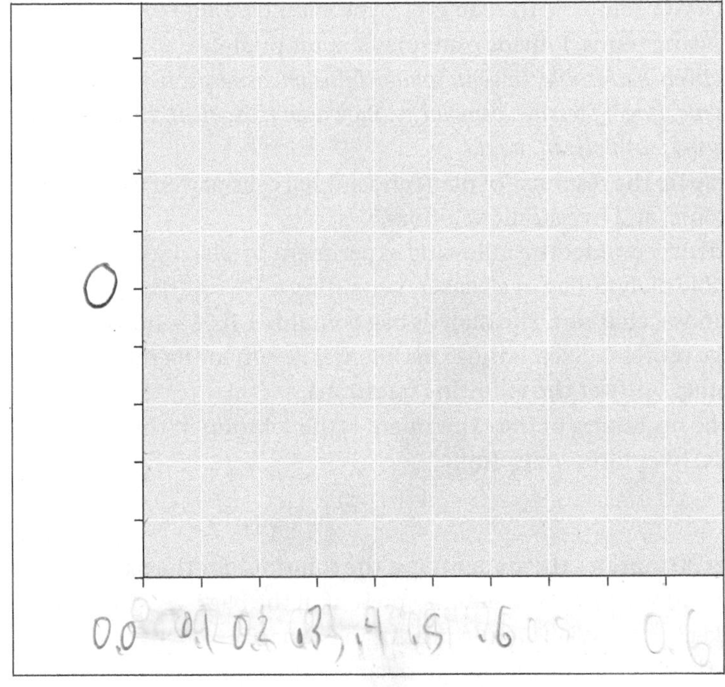

FIGURE 3.8 ______________________

c. Graph your results. Volume increase should be above the zero change line on the "percentage change in volume" axis. Volume decrease should be below the zero change line.

d. Construct a curve that best fits the data points. Use this curve to estimate the osmolarity of the potato tuber.

e. Compose an appropriate figure title.

Discussion

1. At what sucrose molarity does the curve cross the zero change line on the graph?

2. Explain how this information can be used to determine the osmolarity of the potato tuber tissue.

3. In more dilute concentrations of sucrose, the volume of the potato pieces ____________ (increases/decreases) after incubation. What forces other than solute concentration will have an impact on the amount of water taken up by the pieces?

4. Estimate the osmolarity of the potato tuber tissue.

REVIEWING YOUR KNOWLEDGE

1. Once you complete this lab topic, you should be able to define and use the following terms. Provide examples if appropriate.
selectively permeable, solvent, solute, diffusion, osmosis, hypotonic, hypertonic, isotonic, turgor pressure, osmolarity, Brownian movement, lysis, crenate, plasmolysis, plasmolyzed, turgid
2. Compare the response of plant and animal cells placed in hypertonic, isotonic, and hypotonic solutions.
3. Students conduct the following experiment by placing a selectively permeable dialysis bag in a beaker containing a 5% solution of fructose (monosaccharide). The dialysis bag contains a 10% solution of albumin (large protein). Sketch the experimental design in the margin of your lab manual. Answer the following questions:
At the beginning of the experiment is the solution in the beaker hypertonic, hypotonic, or isotonic?

After 30 minutes the students test the solutions for the presence of albumin and fructose. Was albumin present in the bag? In the beaker? Was fructose present in the bag? In the beaker? Did the dialysis bag increase or decrease in volume?

APPLYING YOUR KNOWLEDGE

1. Unlike animals, plants never absorb enough water that their cells burst, but they frequently lose enough water to wilt. Describe plant wilting in terms of turgor pressure. What is the optimum environment for growing plants?

2. The pond water samples you observed in Lab Topic 2, Lab Study D probably contained a variety of multicellular, colonial, and single-celled organisms. Some of these may have had cell walls, but others were lacking cell walls. What adaptations for osmoregulation are found in single-celled organisms, such as the *Amoeba,* and multicellular organisms that lack cell walls but live in a hypotonic environment?

3. There are many applications for the process of osmosis in addition to those examples in living organisms. One interesting application is the use of commercially available so-called "hydration bags" that can be used to produce safe drinking water by campers, hikers, or persons working in an environment where pure water is not available. These bags use "forward osmosis," a process that uses a selectively permeable membrane to separate the potentially polluted water from a hypertonic solution containing ingestible solutes inside the bag. The ingestible solutes are usually glucose or some other sugar. How do you think this works? Use a diagram to illustrate your ideas and use the Web to learn more about the process.

4. Your instructor has given you an assignment to design a system for producing an orange juice concentrate. How could you use information from this lab topic to assist you in this design?

5. Each year in the United States, millions of people become ill and thousands die from eating food contaminated with bacteria. Controlling the growth of microorganisms on food is a particular challenge, and over the course of many centuries various methods have been developed for controlling microorganisms in food. Some of these include preservation by irradiation, drying, freezing, canning, and salting. Preservation by salting is based on the principle of osmosis. One example of this is the preservation of ham by applying large amounts of salt ("salt-cured") or sugar

("sugar-cured") to the meat within 48 hours after slaughter. Using information you learned in this lab, speculate about how this process works to preserve the ham.

INVESTIGATIVE EXTENSIONS

1. Organisms that live in marine environments are described as being *euryhaline* (able to live in waters of a wide range of salinity) or *stenohaline* (unable to withstand wide variation in the salinity of the surrounding water). This characteristic often determines the range of habitat for a marine animal. For example, one would predict that a sessile (attached) marine invertebrate growing on a pier in a tidal river where the salinity changes as the tide floods and ebbs is more euryhaline than an organism living in the open ocean.
   Design an experiment to test the range of tolerance of two or more marine invertebrates (for example, barnacles, sea squirts, small crabs, marine mussels, periwinkles) available from biological supply houses. Determine if they may be described as euryhaline or stenohaline organisms.
2. General science teachers have long known that white vinegar (or a 10% acetic acid solution) will dissolve the shell of a chicken egg, leaving intact the membranes surrounding the albumin and yolk.
   a. Design and perform an experiment to allow you to estimate the *osmolarity* of the chicken egg cell. Hint: It will take 24 to 36 hours to completely remove the shell. If the process goes too slowly, move the eggs to a fresh acid solution after about 12 hours.
   b. Design and perform an experiment to answer the question, "Does the *concentration* of the solute have an effect on the *rate of movement* of the solvent (water) across the membrane of a chicken egg?"
   c. Design and perform an experiment to answer the question, "Does temperature have an effect on the *rate* of osmosis?"

STUDENT MEDIA: BioFlix, Activities, Investigations, Videos, and Data Tables

www.masteringbiology.com (select Study Area)

BioFlix—Ch. 7: Membrane Transport

Activities—Ch. 1: Graph It! An Introduction to Graphing; Ch. 7: Membrane Structure; Selective Permeability of Membranes; Diffusion; Osmosis and Water Balance in Cells

Investigations—Ch. 7: How Do Salt Concentrations Affect Cells?

Videos—Ch. 7: Turgid *Elodea;* Plasmolysis

Data Tables—Table 3.5 and 3.6 can be downloaded in Excel format. Look in the Study Area under Lab Media.

REFERENCES

Lang, F., and S. Waldegger. "Regulating Cell Volume." *American Scientist,* 1997, vol. 85, pp. 456–463.

Urry, L., M. Cain, S. Wasserman, P. Minorsky, and J. Reece. *Campbell Biology*, 11th ed. San Francisco, CA: Pearson, 2017.

WEBSITES

A discussion of osmosis and diffusion in kidney dialysis: http://www.toltec.biz/how_hemodialysis_works.htm

A discussion of reverse osmosis and its applications: http://en.wikipedia.org/wiki/Reverse_osmosis

Includes sections entitled "Real-Life Applications." http://www.scienceclarified.com/everyday/Real-Life-Chemistry-Vol-2/Osmosis.html

NOTES

NOTES

LAB TOPIC 4

Enzymes

Laboratory Objectives

After completing this lab topic, you should be able to:

1. Define *enzyme* and describe the activity of enzymes in cells.
2. Differentiate competitive and noncompetitive inhibition.
3. Discuss the effects of varying environmental conditions such as pH and temperature on the rate of enzyme activity.
4. Discuss the effects of varying enzyme and substrate concentrations on the rate of enzyme activity.
5. Discuss the scientific process, propose questions and hypotheses, and make predictions based on hypotheses and experimental design.
6. Practice scientific thinking and communication by constructing and interpreting graphs of enzyme activity.

Introduction

Living cells perform a multitude of chemical reactions very rapidly because of the participation of enzymes. **Enzymes** are biological **catalysts**, compounds that speed up a chemical reaction without being used up or altered in the reaction. The material with which the catalyst reacts, called the **substrate**, is modified during the reaction to form a new product (see Figure 4.1). But because the enzyme itself emerges from the reaction unchanged and ready to bind with another substrate molecule, a small amount of enzyme can alter a relatively enormous amount of substrate.

The **active site** of an enzyme will bind with the substrate, forming the **enzyme-substrate complex**. It is here that catalysis takes place, and when it is complete, the complex dissociates into enzyme and product or products.

Enzymes are, in part or in whole, proteins and are highly specific in function. Because enzymes lower the **activation energy** needed for reactions to take place, they accelerate the rate of reactions. They do not, however, determine the direction in which a reaction will go or its final equilibrium.

Enzyme activity is influenced by many factors. Varying environmental conditions, such as pH or temperature, may change the three-dimensional shape of an enzyme and alter its rate of activity. Specific chemicals may also bind to an enzyme and modify its shape. Chemicals that must bind for the enzyme to be active are called **activators**. **Cofactors** are nonprotein substances that usually bind to the active site on the enzyme and are essential for the enzyme to work. A cofactor may be as simple as a metal ion, or it may be a more complex organic molecule, called a **coenzyme**. Many vitamins act as coenzymes. Chemicals

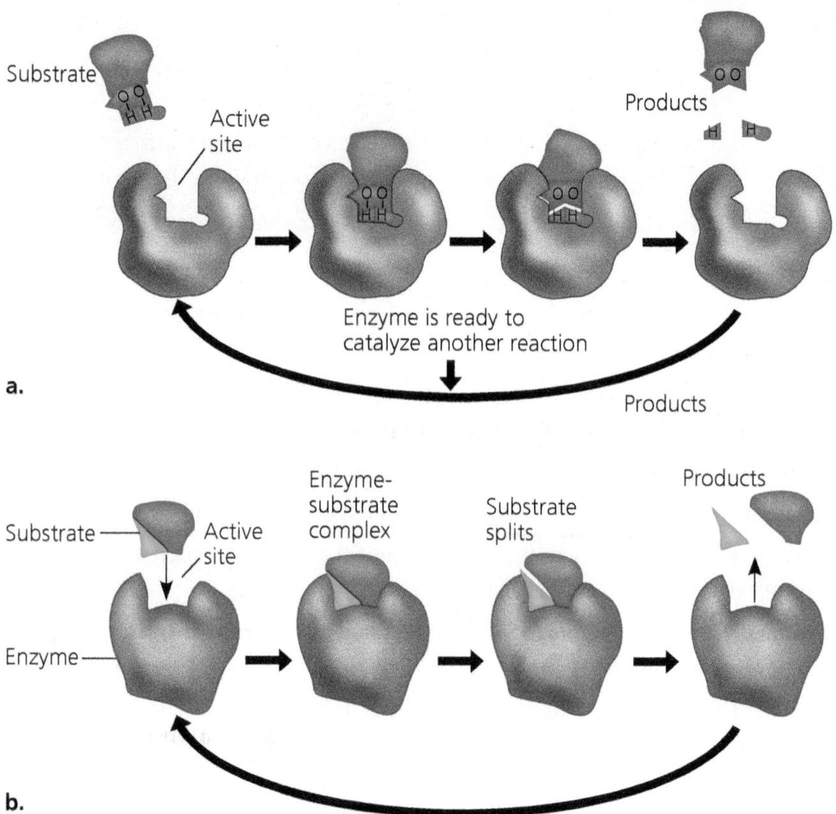

FIGURE 4.1
Enzyme activity. A substrate or substrates bind to the active site of the enzyme, forming the enzyme-substrate complex, which then dissociates into enzyme and product(s). The enzyme may catalyze the addition or removal of a molecule or a portion of a molecule from the substrate to produce the product (a), or the enzyme may catalyze the splitting of a substrate into its component subunits (b).

that shut off the activity of specific enzymes are called **inhibitors**, and their action can be classified as **competitive** or **noncompetitive inhibition**.

Review Figure 4.1 illustrating enzyme activity. There are two ways to measure enzyme activity: (1) Determine the rate of disappearance of the substrate, and (2) determine the rate of appearance of the product.

In this laboratory, you will use both methods to investigate the activity of two enzymes, **catechol oxidase** and **amylase**. (The names of most enzymes are related to their substrates and end in *-ase*.) You will use an inhibitor to influence the activity of catechol oxidase and determine if it is a competitive or noncompetitive inhibitor. Additionally, you will investigate the effect of changing environmental conditions on the rate of amylase activity.

EXERCISE 4.1

Experimental Method and the Action of Catechol Oxidase

Materials

test-tube rack
3 small test tubes
small Parafilm™ squares
calibrated 5-mL pipette
3 calibrated 1-mL pipettes
disposable pasteur pipettes
pipette filler
pipette bulb
distilled or deionized (DI) water
Irish potato extract
catechol
disposable gloves (optional)

Introduction

This exercise will investigate the result of catechol oxidase activity. In the presence of oxygen, catechol oxidase catalyzes the removal of electrons and hydrogens from **catechol**, a phenolic compound found in plant cells. Catechol is converted to benzoquinone. Molecules of benzoquinone subsequently polymerize (combine) to form a dark pigment called catechol melanin to distinguish it from the melanin found in animals. The hydrogens combine with oxygen, forming water (Figure 4.2). The melanin is responsible for the darkening of fruits and vegetables, such as apples and potatoes, after exposure to air.

In this exercise you will use an extract of potato tuber to test for the presence of catechol oxidase and to establish the appearance of the products when the reaction takes place.

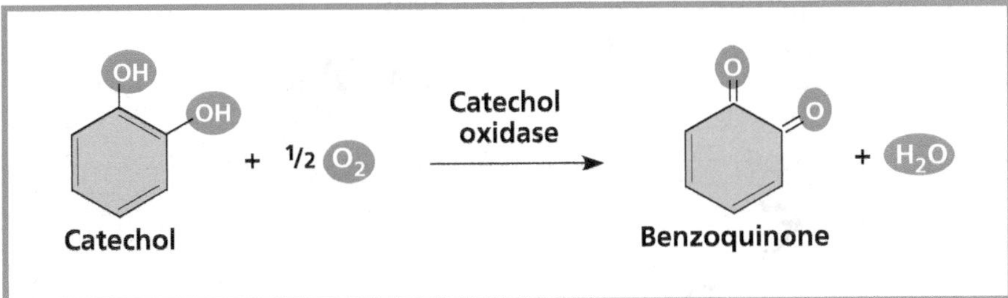

FIGURE 4.2
The oxidation of catechol. In the presence of catechol oxidase, catechol is converted to benzoquinone. Hydrogens removed from catechol combine with oxygen to form water.

Question

Remember that every experiment begins with a question. Review the information given earlier about the activity of catechol oxidase. You will be performing an experiment using potato extract.

Formulate a question about catechol oxidase and potato extract. The question may be broad, but it must propose an idea that has measurable and controllable elements.

Hypothesis

Construct a hypothesis for the presence or absence of catechol oxidase in potato extract. Remember, the hypothesis must be testable. It is possible for you to propose one or more hypotheses, but all must be testable.

Prediction

Predict the result of the experiment based on your hypothesis. To test for the presence or absence of catechol oxidase in potato extract, your prediction would be what you expect to observe as the result of this experiment (if/then).

Catechol is a poison! Avoid contact with all solutions. Do not pipette any solutions by mouth. Wash hands thoroughly after each experiment. If a spill occurs, notify the instructor. If the instructor is unavailable, wear disposable gloves and use dry paper towels to wipe up the spill. Follow dry towels with towels soaked in soap and water. Dispose of all towels in the trash.

Procedure

1. Using Table 4.1, prepare the three experimental tubes. Note that all tubes should contain the same total amount of solution. Do not cross-contaminate pipettes! After each tube is prepared, use your finger to hold a Parafilm™ square securely over the tube mouth and then rotate the tube to mix the contents thoroughly. Use a fresh square for each tube.

TABLE 4.1 Contents of the Three Experimental Tubes

Tube	Distilled Water	Catechol	Distilled Water	Potato Extract
1	5 mL	0.5 mL (10 drops)	0.5 mL (10 drops)	—
2	5 mL	0.5 mL (10 drops)	—	0.5 mL (10 drops)
3	5 mL	—	0.5 mL (10 drops)	0.5 mL (10 drops)

2. Explain the experimental design: What is the purpose of each of the three test tubes? Which is the control tube? Is more than one control tube necessary? Explain. Which is the experimental tube? Why is an additional 0.5 mL of distilled water added to tubes 1 and 3, but not tube 2?

3. Observe the reactions in the tubes, and record your observations in the Results section below. Explain your conclusions in the Discussion section.
4. After recording your results, dispose of the solutions in the test tubes in the waste container indicated by your instructor. Do not pour down the drain.

Results

Design a simple table to record results (Table 4.2).

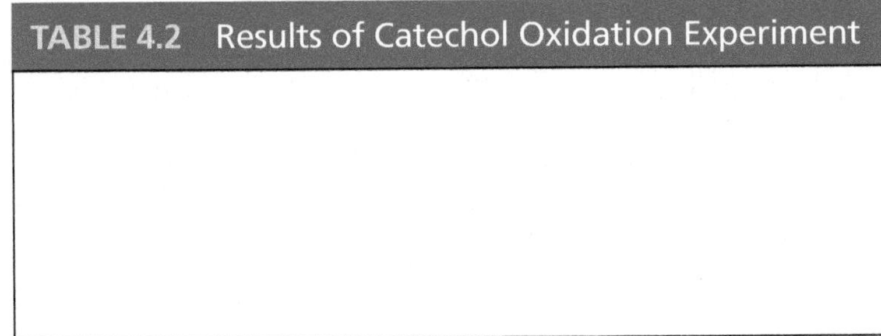

TABLE 4.2 Results of Catechol Oxidation Experiment

Discussion

1. Explain your results in terms of your hypothesis.

2. In this experiment, the enzyme catechol oxidase was extracted from potato. However, this was not a purified preparation; it contained hundreds of enzymes. What evidence supports the assumption that catechol oxidase was the enzyme studied?

EXERCISE 4.2

Inhibiting the Action of Catechol Oxidase

Materials

test-tube rack
3 small test tubes
small Parafilm™ squares
calibrated 5-mL pipette
4 calibrated 1-mL pipettes
disposable pasteur pipettes
pipette bulb
distilled water
potato extract
catechol
phenylthiourea (PTU)
disposable gloves (optional)

Introduction

This exercise will investigate the inhibition of enzyme activity by specific chemicals called **inhibitors**. The specific inhibitor used is **phenylthiourea (PTU)**. To be active, catechol oxidase requires copper as a cofactor. PTU is known to combine with the copper in catechol oxidase and inhibit its enzymatic activity.

An inhibitor molecule affects an enzyme in one of two ways. **Competitive inhibition** takes place when a molecule that is structurally similar to the substrate for a particular reaction competes for a position at the active site on the enzyme. This ties up the enzyme so that it is not available to the substrate. Competitive inhibition can be reversed if the concentration of the substrate is raised to sufficiently high levels while the concentration of the inhibitor is held constant (Figure 4.3a and b).

In **noncompetitive inhibition**, the inhibitor is not structurally similar to the substrate and does not compete for position in the active site. It may physically block the access to the active site or it may bind to a part of the enzyme that is not the active site, causing the conformation of the protein to change

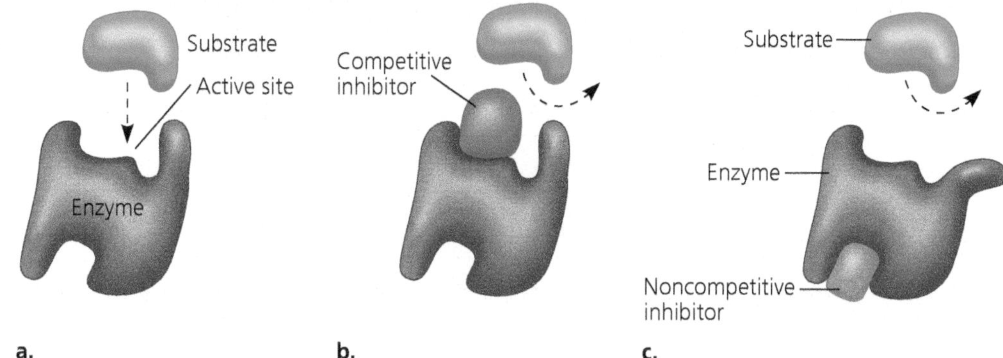

FIGURE 4.3
Action of enzyme inhibitors—competitive and noncompetitive. (a) Without inhibition, the substrate binds to the active site of an enzyme. (b) A **competitive inhibitor** mimics the substrate and competes for the position at the active site on the enzyme. (c) In some examples of noncompetitive inhibition, the inhibitor binds to the enzyme at a location away from the active site, changing the conformation of the enzyme, rendering it inactive.

(Figure 4.3c). A noncompetitive inhibitor may also inhibit the active site by interfering with a cofactor. In noncompetitive inhibition the inhibitor can become unbound, reversing the inhibition. However, unlike in competitive inhibition, *adding additional substrate* will not reverse the inhibition.

In the following experiment, you will determine if PTU is a competitive or noncompetitive inhibitor.

Question

Pose a question about the activity of PTU.

Hypothesis

Hypothesize about the nature of inhibition by PTU.

Prediction

Predict the results of the experiment based on your hypothesis (if/then).

Procedure

PTU and catechol are poisons! Avoid contact with solutions. Do not pipette any solutions by mouth. Wash hands thoroughly after the experiment. If a spill occurs, notify the instructor. If the instructor is unavailable, wear disposable gloves and use dry paper towels to wipe up the spill. Follow dry towels with towels soaked in soap and water. Dispose of all towels in the trash.

TABLE 4.3 Contents of the Three Experimental Tubes

Tube	Distilled Water	Potato Extract	PTU	Distilled Water	Catechol
1	5 mL	0.5 mL	0.5 mL	0.5 mL	0.5 mL
2	5 mL	0.5 mL	0.5 mL	—	1 mL
3	5 mL	0.5 mL	—	1 mL	0.5 mL

1. Using Table 4.3, prepare three experimental tubes. Be sure to add solutions in the sequence given in the table (water first, potato extract next, PTU next, etc.). Cover each tube with a fresh Parafilm™ square and mix.
2. Which test tube is the control? Will the pigment develop in this tube? Why or why not?
3. Why was the concentration of catechol increased in test tube 2?
4. Why should the catechol be added to the test tubes last?
5. Record your observations in the Results section, and explain your results in the Discussion section.
6. After recording your results in Table 4.4, dispose of the solutions in the test tubes in the waste container indicated by your instructor. Do not pour down the drain.

Results

Design a table to record your results (Table 4.4) in the margin.

TABLE 4.4 Results of Inhibition Experiment

Discussion

1. Explain your results in terms of your hypothesis.
2. One member of your team is not convinced that you have adequately tested your hypothesis. How could you expand this experiment to provide additional evidence to strengthen your conclusion?

EXERCISE 4.3

Influence of Enzyme Concentration, pH, and Temperature on the Activity of Amylase

Introduction

In the following exercise, you will investigate the influence of enzyme concentration, pH, and temperature on the activity of the enzyme **amylase**. Amylase is found in the **saliva** of many animals, including humans, that utilize **starch** (amylose) as a source of food. Starch, the principal reserve carbohydrate stores of plants, is a polysaccharide composed of a large number of glucose monomers joined together. Amylase is responsible for the preliminary digestion of starch. In short, amylase breaks up the chains of glucose molecules in starch into maltose, a two-glucose-unit compound. Further digestion of this disaccharide requires other enzymes present in pancreatic and intestinal secretions. To help us follow the digestion of starch into maltose by salivary amylase, we will take advantage of the fact that starch, but not maltose, turns a dark purple color when treated with a solution of I_2KI (this solution is normally yellow-amber in color). Draw equations to help you remember these reactions in the margin of your lab manual.

In the following experiments, the *rate of disappearance* of starch in different amylase concentrations allows a quantitative measurement of reaction rate. Recall that the rate of appearance of the product (in this case, maltose) would give the same information, but the starch test is simpler.

You will be assigned to a team with three or four students. Each team will carry out only one of the experiments. However, each student is responsible for understanding all experiments and results. Be prepared to present your results to the entire class. Your instructor may require you to write a component of a scientific paper. (See Appendix A.)

Experiment A. The Influence of Enzyme Concentration on the Rate of Starch Digestion

Materials

test-tube rack
10 standard test tubes
wax pencil
test plate
flask of distilled or DI water
beaker of distilled or DI rinse water
5-mL graduated cylinder
1 calibrated 1-mL pipette
2 calibrated 5-mL pipettes
disposable pasteur pipettes
pipette bulb
buffer solution (pH = 6.8)
I_2KI solution
1% starch solution
1% amylase solution

Introduction

In this experiment you will vary the concentration of the enzyme amylase to determine what effect the variation will have on the rate of the reaction. You will make serial dilutions of the amylase resulting in a range of enzyme concentrations. For serial dilutions, you will take an aliquot (sample) of the original enzyme and dilute it with an equal amount of water for a 1 : 1 dilution (50% of the original concentration). You will then take an aliquot of the resulting 1 : 1

solution and add an equal amount of water for a 1:3 dilution of the original concentration. You will continue this series of dilutions until you have four different amylase concentrations.

Question

Pose a question about enzyme concentration and reaction rate.

Hypothesis

Hypothesize about the effect of changing the enzyme concentration on the rate of reaction.

Prediction

Predict the results of the experiment based on your hypothesis (if/then).

Procedure

1. Prepare the amylase dilution (test-tube set I):
   a. Number five standard test tubes 1 through 5.
   b. Using the 5-mL graduated pipette, add 5 mL distilled water to each test tube.
   c. Make serial dilutions as follows (use the graduated cylinder):

      Tube 1: Add 5 mL amylase and mix by rolling the tube between your hands. (Dilution: 1:1; 0.5% amylase)

      Tube 2: Add 5 mL amylase solution from tube 1 and mix. (Dilution: 1:3; 0.25% amylase)

      Tube 3: Add 5 mL amylase solution from tube 2 and mix. (Dilution: 1:7; 0.125% amylase)

      Tube 4: Add 5 mL amylase solution from tube 3 and mix. (Dilution: 1:15; 0.063% amylase)

      Tube 5: Add 5 mL amylase solution from tube 4 and mix. (Dilution: 1:31; 0.031% amylase)

      Rinse the graduated cylinder thoroughly.

      Refer to Table 4.5 to confirm the contents of each test tube in set I. Write the final concentrations of the amylase solutions in the last row of this table (see step 1c above). You will use these concentrations as you record and graph your results (see Table 4.6).

TABLE 4.5 Test Tube Set I—Contents After the Amylase Dilution

	Tube 1	Tube 2	Tube 3	Tube 4	Tube 5
Distilled water	5 mL	5 mL	5 mL	5 mL	5 mL
Amylase solution	5 mL	5 ml from tube 1	5 mL from tube 2	5 mL from tube 3	5 mL from tube 4
Final amylase concentration					

2. Prepare the experimental test tubes (test-tube set II):
   a. Number a second set (set II) of five standard test tubes 1 through 5.
   b. Beginning with tube 5 of set I, transfer 2 mL of this enzyme dilution into tube 5 of set II. Use a 5-mL pipette for the transfer. Rinse the pipette in distilled water, and repeat the procedure for tubes 4, 3, 2, and 1, transferring 2 mL of tube 4 (set I) into tube 4 (set II), and so forth. After these transfers have been carried out, test-tube set I will no longer be used and can be set aside.
   c. Add 2 mL of pH 6.8 buffer solution to each of the tubes in the second set. Mix by rolling the tubes between your hands. Set these tubes aside.
   d. Add 1 or 2 drops of I_2KI to each compartment of four rows of a test plate. You will use a separate row for each concentration of amylase.
   e. Using the second set of tubes, proceed with the tests beginning with tube 5.
      (1) Using a clean 1-mL pipette, add 1 mL of the 1% starch solution to tube 5 and mix by rolling the tube between your hands. One team member should immediately record the time. This is time 0.
      (2) Quickly remove 1 drop of the mixture with a disposable pasteur pipette, and add it to a drop of I_2KI in the first compartment on the test plate (time 0).

Remember, when the enzyme and substrate are together, the reaction has begun!

      (3) Sample the reaction mixture at 10-second intervals, each time using a new compartment of the test plate. Continue until a blue color is no longer produced and the I_2KI solution remains yellow-amber (indicating the digestion of all the starch). Record the time required for the digestion of the starch in Table 4.6.
      (4) Repeat steps 1 through 4 for the other four concentrations (tubes 4, 3, 2, and 1 of set II).
3. Finish recording your findings in the Results section, and state your conclusions in the Discussion section.

Results

1. Complete Table 4.6 as you determine rates of digestion (time of starch disappearance) in different enzyme concentrations.

TABLE 4.6 Time of Starch Disappearance in Different Concentrations of the Enzyme Amylase

Download an Excel version from www.masteringbiology.com in the Study Area under Lab Media

Tube	% Amylase	Time of Starch Disappearance (in seconds)
1	0.50	
2	0.25	
3	0.125	
4	0.063	
5	0.031	

2. Construct a graph (Figure 4.4) to illustrate your results. See Lab Topic 1 for assistance in graph construction.
   a. What is the independent variable? Which is the appropriate axis for this variable?

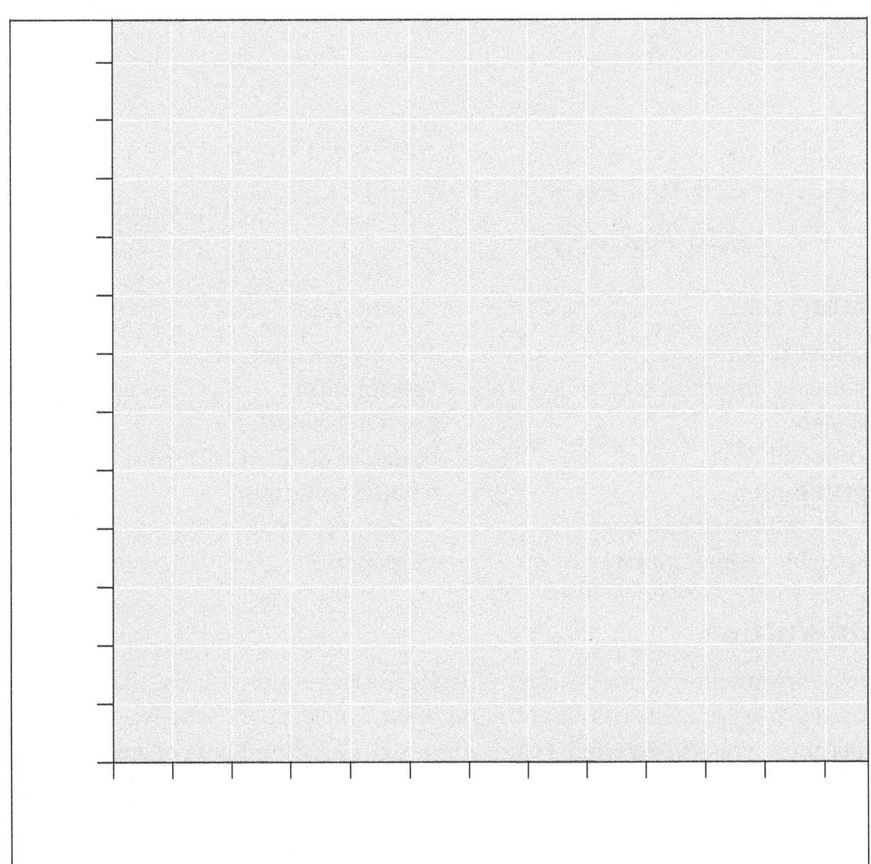

FIGURE 4.4
Enzyme Reaction rate (time of starch disappearance) for different concentrations of amylase.

b. What is the dependent variable? Which is the appropriate axis for this variable?

c. Label the axes of the graph. Using Table 4.6, note the maximum number of seconds in your results, and choose an appropriate scale for the dependent variable. Reaction rate, the dependent variable, was measured as time of starch (product) disappearance. *The data must therefore be graphed in reverse order* because the highest values indicate the slowest reaction rate. You should place "0" at the end of the axis and write "fast" by your 0. Place your highest number near the origin (where the *x*- and *y*-axes cross). Write "slow" near the origin. Choose an appropriate scale for the independent variable (% amylase), and label this axis.

Discussion

1. Explain your results in terms of your hypothesis. Describe the shape of the reaction rate curve as substrate concentration increases. What factors are responsible for the change in reaction rate?

2. Speculate about the shape of a curve measuring reaction rate if you had increased the concentration of enzyme, but held the concentration of substrate constant.

Experiment B. The Effect of pH on Amylase Activity

Materials

test-tube rack
6 standard test tubes
test plate
wax pencil
pipette bulb
3 5-mL calibrated pipettes
disposable pasteur pipettes

1% amylase solution
I_2KI solution
1% starch solution
beaker of distilled or DI rinse water
6 buffer solutions
(pH = 4, 5, 6, 7, 8, 9)
pH paper

Introduction

The environmental factor pH can influence the three-dimensional shape of an enzyme. Every enzyme has an optimum pH at which it is most active. In this experiment you will determine the optimum pH for the activity of amylase. What was the source of the amylase used in this experiment? (Check the Introduction to this exercise.)

Question

Pose a question about pH and reaction rate.

Hypothesis

Hypothesize about the rate of activity of amylase at various pHs. Consider the source of the amylase.

Prediction

Predict the results of the experiment based on your hypothesis (if/then).

Procedure

1. Using a wax pencil, number six standard test tubes 1 through 6. Beginning with tube 1 and pH 4, mark one tube for each pH of buffer (4, 5, 6, 7, 8, 9). After you mark the test tubes, use a 5-mL graduated pipette to add 5 mL of the appropriate buffer to each test tube (5 mL buffer 4 to tube 1, 5 mL buffer 5 to tube 2, etc.). Rinse the pipette with distilled water after dispensing each buffer.

Buffers can burn skin! Avoid contact with all solutions. Do not pipette any solutions by mouth. Wash hands thoroughly after each experiment. If a spill occurs, notify the instructor. If the instructor is unavailable, wear disposable gloves and use dry paper towels to wipe up the spill. Follow dry towels with towels soaked in soap and water. Dispose of all towels in the trash.

2. Using a clean 5-mL graduated pipette, add 1.5 mL amylase solution to each tube and mix by rolling the tubes in your hands.
3. Introduce 1 or 2 drops of I_2KI into the compartments of several rows of the test plate.
4. Using only tube 1, add 2.5 mL of the 1% starch solution with a clean 5-mL pipette. Leave the pipette in the starch solution. Mix by rolling the tube in your hands. One team member should immediately record the time. This is time 0. Start testing immediately (next step).

Remember, when the enzyme and substrate are together, the reaction has begun!

5. Using a disposable pasteur pipette, remove a drop of the reaction mixture from tube 1. Add to a drop of I_2KI on the test plate.
6. Sample the reaction mixture at 10-second intervals, *each time using a new compartment of the test plate*. Continue until a blue color is no longer produced and the I_2KI solution remains yellow-amber (indicating the digestion of all the starch).
7. Record the time required for the digestion of the starch in Table 4.7. If after 7 minutes there is no color change, terminate the experiment with that reaction mixture.
8. Repeat steps 4 through 7 using the other five test tubes. Use separate rows on the test plate for each pH. Rinse the pipette between uses. Record results in Table 4.7.
9. Graph your observations in the Results section, and explain your results in the Discussion section.
10. After recording your results, dispose of the solutions in the test tubes in the waste container indicated by your instructor. Do not pour down the drain.

TABLE 4.7 Time of Starch Disappearance in Different pH Environments for the Enzyme Amylase

Download an Excel version from www.masteringbiology.com in the Study Area under Lab Media

Tube	pH	Time of Starch Disappearance (in minutes)
1	4	
2	5	
3	6	
4	7	
5	8	
6	9	

Results

1. Complete Table 4.7 as rates of digestion (time of starch disappearance) in different pHs are determined.
2. Construct a graph using Figure 4.5 to illustrate your results. See Lab Topic 1 for assistance in graph construction.

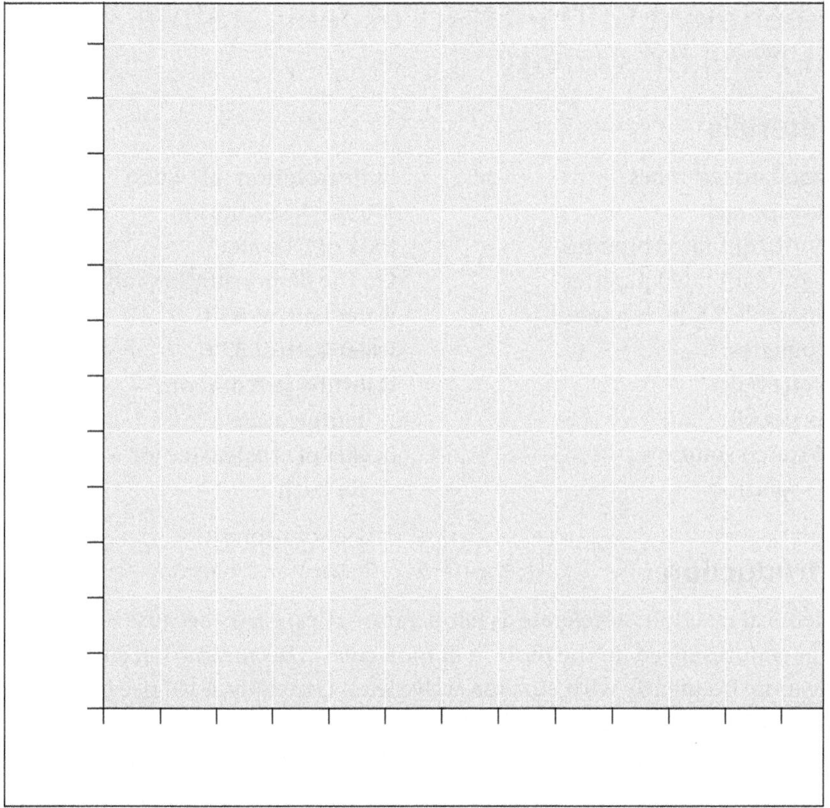

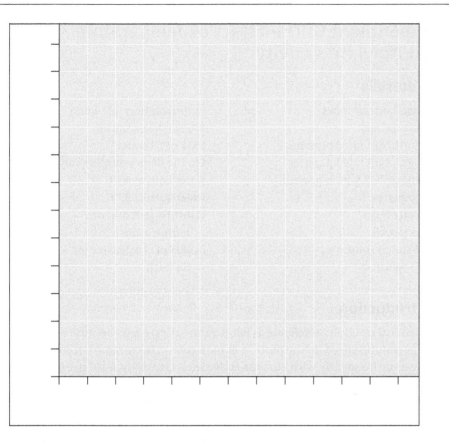

FIGURE 4.5
Reaction rate (time of starch disappearance) of amylase in different pH environments.

a. What is the independent variable? Which is the appropriate axis for this variable?

b. What is the dependent variable? Which is the appropriate axis for this variable?

c. Label the axes of the graph. Using Table 4.7, note the maximum number of minutes in your results, and choose an appropriate scale for the dependent variable. Reaction rate, the dependent variable, was measured as time of starch (product) disappearance. *The data must therefore be graphed in reverse order* because the highest values indicate the slowest reaction rate. You should place "0" at the end of the axis and write "fast" by your 0. Place your highest number near the origin (where the *x*- and *y*-axes cross). Write "slow" near the origin. Choose an appropriate scale for the independent variable (pH) and label the axis.

Discussion

Explain your results in terms of your hypothesis. Describe the shape of the reaction rate curve obtained with change in pH. What factors are responsible for the shape of this curve?

Experiment C. The Effect of Temperature on Amylase Activity

Materials

8 standard test tubes
test-tube rack
2 5-mL calibrated pipettes
2 1-mL calibrated pipettes
disposable 7.5-inch pasteur pipettes
pipette bulb
wax pencil
1% starch solution
I_2KI solution
buffer solution (pH = 6.8)
1% amylase solution
flask of DI water
On the demonstration table:
water bath at 80°C
water bath at 37°C
test-tube rack at room temperature
beaker of crushed ice for ice bath

Introduction

Chemical reactions accelerate as temperature rises, partly because increased temperatures speed up the motion of molecules. This means that substrates collide more frequently with enzyme active sites. Generally, a 10° rise in temperature results in a two- to threefold increase in the rate of a particular reaction. However, at high temperatures, the integrity of proteins can be irreversibly denatured. The activity of enzymes is dependent on the proper tertiary and quaternary structures; the optimum temperature for activity, therefore, may vary, depending on the structure of the enzyme.

What was the source of the amylase used in this experiment? (Check the Introduction to this exercise.)

Question

Pose a question about temperature and reaction rate.

Hypothesis

Hypothesize about the rate of activity of amylase at various temperatures. Consider the source of the amylase.

Prediction

Predict the results of the experiment based on your hypothesis (if/then).

Procedure

1. Number four standard test tubes 1 through 4.
2. Using the 5-mL calibrated pipette, add 2 mL of the 1% starch solution to each tube.
3. Using a clean 5-mL pipette, add 4 mL DI water to each tube.
4. Add 1 mL of 6.8 buffer to each tube.
5. Place the test tubes as follows:

   Tube 1: 80°C water bath
   Tube 2: 37°C water bath
   Tube 3: test-tube rack (room temperature, or about 22°C)
   Tube 4: beaker of crushed ice (4°C)
6. Number and mark a second set of standard test tubes 1A through 4A. Use the 1-mL calibrated pipette to add 1 mL amylase to each tube, and place as follows (do not mix together the solutions in the two sets of tubes until instructed to do so):

   Tube 1A: 80°C water bath
   Tube 2A: 37°C water bath
   Tube 3A: test-tube rack (room temperature, or about 22°C)
   Tube 4A: beaker of crushed ice (4°C)
7. Let all eight tubes sit in the above environments for 10 minutes. You should have one tube of amylase and one of starch at each temperature.
8. Fill several rows of the test plate with 1 or 2 drops of I_2KI per compartment.

> Remember, when the enzyme and substrate are together, the reaction has begun!

9. Leaving the tubes in the above environments as they are being tested, mix tubes 1 and 1A, record the time (this is time 0), and use a disposable pipette to immediately add 1 or 2 drops of the mixture to a drop of I_2KI on the test plate.
10. Continue adding mixture drops to new wells of I_2KI at 30-second intervals until a blue color is no longer produced and the I_2KI solution remains yellow-amber (indicating that all the starch is digested). If within 10 minutes there is no color change, terminate the experiment with that particular reaction mixture. Record your results in Table 4.8.
11. Repeat steps 9 and 10 for the other reaction mixtures (mix tubes 2 and 2A and test, mix 3 and 3A and test, etc.).
12. If time permits, transfer tube 4 (in ice, containing the enzyme and substrate) into the 37°C water bath. After 2 minutes, test the contents of the tube at 10-second intervals. Record these results in the margin of your lab manual and refer to them as you answer Discussion Question 2.
13. Finish recording and graphing your observations in the Results section, and explain your results in the Discussion section.

Results

1. Complete Table 4.8 as rates of digestion (time of starch disappearance) in different temperatures are determined.

TABLE 4.8 Time of Starch Disappearance in Different Temperatures for the Enzyme Amylase

Download an Excel version from www.masteringbiology.com in the Study Area under Lab Media

Tube	Temp. °C	Time of Starch Disappearance (in minutes)
1	80°	
2	37°	
3	22°	
4	4°	

2. Construct a graph using Figure 4.6 to illustrate your results. See Lab Topic 1 for assistance in graph construction.
   a. What is the independent variable? Which is the appropriate axis for this variable?

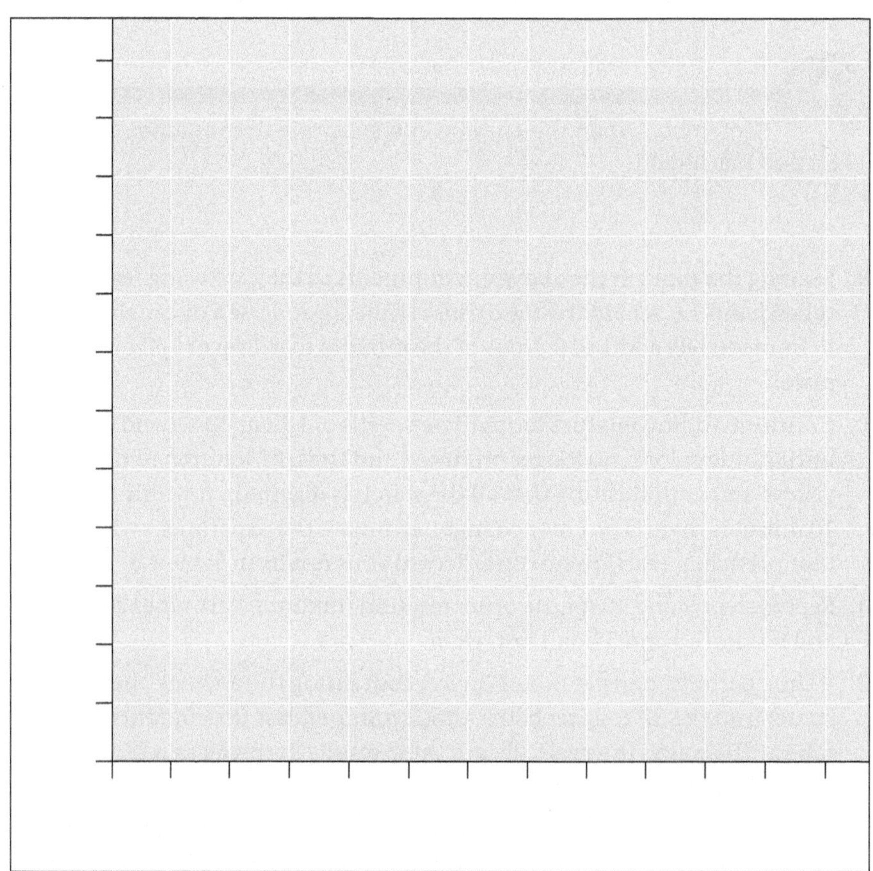

FIGURE 4.6
Reaction rate (time of starch disappearance) of amylase in various temperatures.

b. What is the dependent variable? Which is the appropriate axis for this variable?

c. Label the axes of the graph. Using Table 4.8, note the maximum number of minutes in your results, and choose an appropriate scale for the dependent variable. Reaction rate, the dependent variable, was measured as the time of starch disappearance. *The data must therefore be graphed in reverse order* because the highest values indicate the slowest reaction rate. You should place "0" at the end of the axis and write "fast" by your 0. Place your highest number near the origin (where the *x*- and *y*-axes cross). Write "slow" near the origin. Choose an appropriate scale for the independent variable (temperature) and label the axis.

Discussion

1. Explain your results in terms of your hypothesis. Describe the shape of the reaction rate curve as temperatures increase. What factors are responsible for the shape of this curve?

2. What do you think would happen to the reaction rate in the tube incubated in ice if this tube, with enzyme and substrate already mixed, were placed in the 37°C water bath?

   Explain your answer in terms of the effect of various temperatures on enzyme structure and the rate of enzyme activity.

REVIEWING YOUR KNOWLEDGE

1. Define and use the following terms, providing examples if appropriate: *catalyst, enzyme, substrate, active site, cofactor, coenzyme, competitive inhibition, noncompetitive inhibition.*

   a. Compare and contrast competitive and noncompetitive inhibition.

b. Why does adding additional substrate overcome competitive but not noncompetitive inhibition?

2. Review changes in reaction rate as substrate concentration, pH, and temperature change. Explain the conditions or factors that are responsible for these changes.

APPLYING YOUR KNOWLEDGE

1. Many organisms are able to live in extremely cold or hot temperatures, for example, the thermophilic bacteria that live in the hot springs of Yellowstone National Park. In the margin of your manual, draw a figure representing the predicted activity curve for enzymes in bacteria found in these habitats.
2. pH plays a crucial role in a normally functioning digestive tract. Most digestive enzymes are sensitive to pH and will be most active at a pH specific for the enzyme (see Exercise 4.3, Experiment B). As fluids pass through the digestive tract of humans, the pH abruptly changes from one region to another. Fluid in the mouth and esophagus is about pH 6.8 to 7 and in the stomach is pH 2; the pH of fluid in the small intestine is quickly neutralized from 2 to about 7, and further along in the small intestine, to about 8.

   a. Speculate about possible adaptive advantages for the extremely low pH of 2 in the stomach.

   b. Many people suffer from "heartburn" and take medications such as antacids (which neutralize acids). What could be the possible side effects of taking such medication?

3. Ethylene glycol, the main ingredient in antifreeze, is an odorless, colorless, sweet-tasting water-soluble chemical that, when ingested, is a toxic poison. The treatment for ethylene glycol poisoning involves emergency procedures that include intravenous doses of ethanol for several hours. To understand how this treatment works, you need to know that ethylene glycol is metabolized by the enzyme alcohol dehydrogenase (ADH) into four products, and these products are the actual toxic agents. If this reaction can be inhibited, then the kidneys will eliminate the ethylene glycol intact before the toxins are produced. Chemists know that ADH has 100× greater affinity for ethanol than for ethylene glycol.

Given your knowledge of enzyme activity and inhibitors, speculate about the mechanism by which ethanol prevents ethylene glycol poisoning.

4. There is an enzyme that catalyzes the production of the pigment responsible for dark fur color in Siamese cats and Himalayan rabbits. This enzyme is *thermolabile,* meaning that it does not function at higher temperatures. Rabbits raised at 5°C are all black. If raised at 20°C, they are white with black paws, ears, and noses; they are all white when raised at 35°C. Which of the following best represents an activity curve for this enzyme?

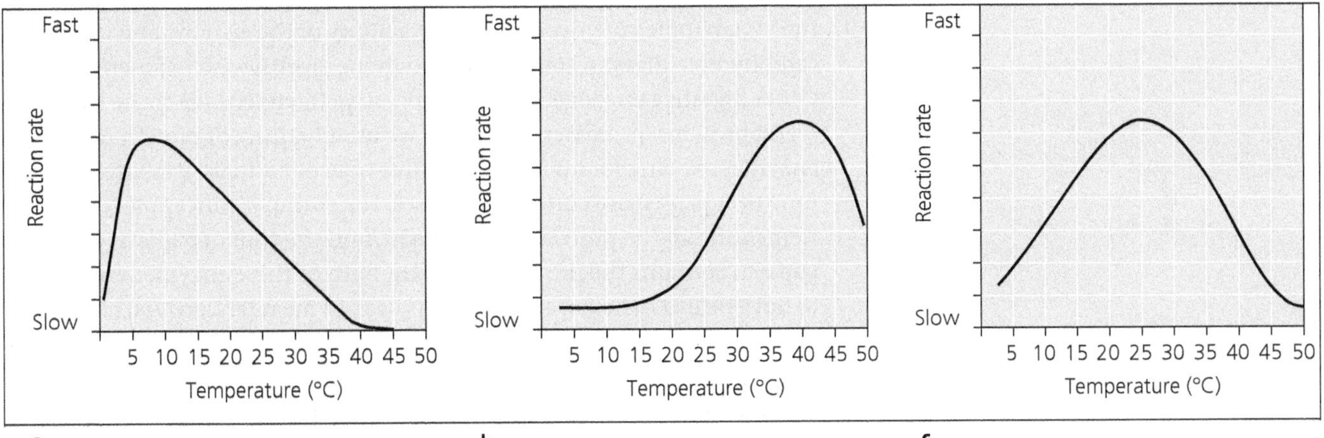

5. You are investigating a specific compound known to inhibit the activity of an enzyme. You hypothesize that the compound is a competitive inhibitor of the enzyme's activity. To test your hypothesis, you design an experiment where you add increasing amounts of the enzyme's substrate to a solution containing the enzyme, and then you test the reaction rate. Which of the following graphs represents the predicted results of your experiment, based on your hypothesis?

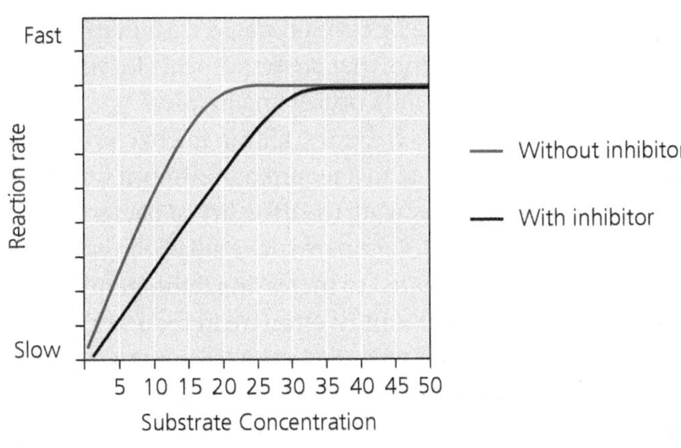

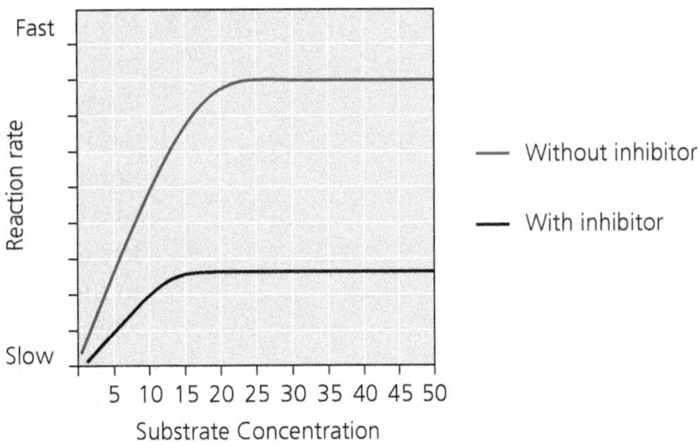

6. A researcher discovers that an antibiotic, sulfanilamide, can be used to treat infections caused by a particular bacterial species. She finds out that the structure of sulfanilamide is almost identical to that of para-amino benzoic acid (PABA), one of the intermediate compounds in the metabolic pathway leading to folic acid, a vitamin required for bacterial growth and survival. Speculate about the mechanism of drug action in controlling bacterial growth. Hint: Enzymes catalyze all reactions in metabolic pathways.

INVESTIGATIVE EXTENSIONS

1. In this lab topic you investigated the activity of the enzyme amylase, a carbohydrate-digesting enzyme found in the digestive tract of many animals. Other digestive enzymes in animal digestive tracts include those that digest proteins, proteases. Although we usually think of protein-digesting enzymes as being found in the digestive tract of animals, it is interesting to note that plants also contain proteases. For example, pineapple fruits and stems contain the protease **bromelain**, and the fruit of papaya (Carica papaya) contains the protease **papain**. Both of these enzymes are reported to have health benefits, and papain is used in meat tenderizers, for example, Adolph's Meat Tenderizer.™
   Using protocols and materials presented in this lab topic, design an experiment to investigate the activity of one of these enzymes. What is the optimum pH for its activity? At what temperature is it denatured? Does it digest a range of common proteins, for example, gelatin or albumin? Use library resources or the Web to research other questions you might have about proteases in plants; for example, what role do they play in plant metabolism?
2. Cottage cheese and many other cheeses are produced by adding the enzyme rennin to milk, producing curds and whey. Using information you learned in this lab, design an experiment to determine the pH at which rennin is most effective. How will you measure the results? Using sources from the Web or library, find tests that can be used to detect protein, sugars, and fats, and use these tests to determine the chemical makeup of curds and whey.
3. It is well known that raw cows' milk can be contaminated with many bacteria, including thermophilic bacteria, that are responsible for many diseases, such as tuberculosis. Modern milk processing involves pasteurization, a process that denatures enzymes, killing the bacteria in liquid food. In this process milk is heated to a required minimum temperature for a specific amount of time. Investigate the diversity of bacteria found in raw milk (fresh from the cow). Compare the results from raw milk to the bacteria present as the milk is heated to increasing temperatures. Ask what is the minimum temperature that is most effective in reducing bacteria diversity? To perform this study, you will need supplies used to culture milk bacteria, including TGY (tryptone, glucose, yeast extract) agar plates, and you will need a way to control the temperature of the milk, such as a water bath.

4. The experiments using catechol oxidase may be quantified by measuring results using a Spectronic 20D+. Modify the instructions in Exercises 4.1 and 4.2 according to the following instructions.
   a. Rather than using three experimental tubes, use four cuvettes, and label them B (the blank), 1, 2, and 3.
   b. Prepare the solutions in cuvettes as directed for test tubes 1, 2, and 3, omitting catechol until ready to take the reading. The blank will contain 5.5 mL of distilled water and 0.5mL of potato extract only. Do not add catechol to the blank. (If using drops, the blank will be 5 mL distilled water, 10 drops of distilled water, and 10 drops of potato extract.)
   c. Design a table to record your data.
   d. Record the absorbance of the solutions in the four cuvettes. (For steps 1–3 following, see the figure for the Spectronic 20D+ and details of its operation in Appendix C. Steps 4–6 are self-explanatory.)
      (1) Set the wavelength at 540 nm.
      (2) Zero the instrument.
      (3) Calibrate the instrument.
      (4) Add the catechol to the appropriate cuvettes.
      (5) Insert cuvette 1 into the sample holder and record absorbance in your data table. Immediately repeat with cuvettes 2 and 3.
      (6) Continue to take readings at 5-minute intervals for 30 minutes or an appropriate time.

STUDENT MEDIA: BioFlix, Activities, Investigations, Videos, and Data Tables

www.masteringbiology.com (select Study Area)

Activities—Ch. 1: Graph It! An Introduction to Graphing; Ch. 8: How Enzymes Work

Data Tables—Tables 4.6, 4.7, and 4.8 can be downloaded in Excel format. Look in the Study Area under Lab Media.

REFERENCES

Cheung, R. "Heat Beaters." *Science News*, 2012, vol. 182, no. 3, pp. 26–27. Reports about the search for enzymes in bacteria that are active at high temperatures to be used for industrial purposes.

Klabunde, T., C. Eicken, J. Sacchettini, and B. Krebs. "Crystal Structure of a Plant Catechol Oxidase Containing a Dicopper Center." *Nature Structural Biology*, 1998, vol. 5, no. 12, pp. 1084–1090. Full text available online. (Discusses activity of catechol oxidase and the specific mode of inhibition by PTU.)

Mathews, C. K., K. E. Van Holde, and K. G. Ahern. *Biochemistry*, 4th ed. San Francisco, CA: Pearson, 2012.

Nelson, D. and M. Cox. *Lehninger Principles of Biochemistry*, 6th ed. New York: Worth W.H. Freeman, Publishers, 2012.

WEBSITES

Describes enzyme preparations approved as food additives:
http://www.fda.gov/Food/IngredientsPackagingLabeling/

Enzymes used in the dairy industry:
http://biotech.about.com/od/casestudies
/a/dairyenzymes.htm

Information and models of alpha-amylase on the Protein Data Bank website. Includes an interesting discussion on the use of glucose isomerase in the production of high-fructose corn syrup.
http://www.rcsb.org/pdb/101/motm.do?momID=74

NOTES

LAB TOPIC 5

Cellular Respiration and Fermentation

This lab topic gives you another opportunity for scientific thinking and to practice the scientific process introduced in Lab Topic 1. Before going to lab, review scientific investigation in Lab Topic 1 and carefully read Lab Topic 5. Be prepared to use this information to design an experiment in fermentation or cellular respiration.

Laboratory Objectives

After completing this lab topic, you should be able to:

1. Describe alcoholic fermentation, naming reactants and products.
2. Describe cellular respiration, naming reactants and products.
3. Explain oxidation-reduction reactions in cellular respiration.
4. Name and describe environmental factors that influence enzymatic activity.
5. Explain spectrophotometry and describe how this process can be used to measure aerobic respiration.
6. Propose hypotheses and make predictions based on them.
7. Develop your own independent questions, hypotheses, and predictions, then execute an experiment testing factors that influence fermentation or cellular respiration.
8. Practice scientific thinking and communication by analyzing, interpreting, and presenting results.

Introduction

You have been investigating cells and their activities in Lab Topics 2–4: cellular structure and evolution (Lab Topic 2), movement across cell membranes (Lab Topic 3), and enzymatic activities (Lab Topic 4). This lab topic and the following one (Photosynthesis, Lab Topic 6) investigate energy transformations in cells. Photosynthesis is the process of transferring the sun's radiant energy to organic molecules, namely, glucose (Figure 5.1). This lab topic investigates **fermentation** and **cellular respiration**, cellular processes that release the energy in glucose to synthesize **adenosine triphosphate** (ATP). The energy in ATP can then be used to perform cellular work. Fermentation is an anaerobic (in the absence of oxygen) process; cellular respiration is aerobic (utilizing oxygen). *All living organisms, including bacteria, protists, plants, and animals, produce ATP in cellular respiration or fermentation and then use ATP in their metabolism.*

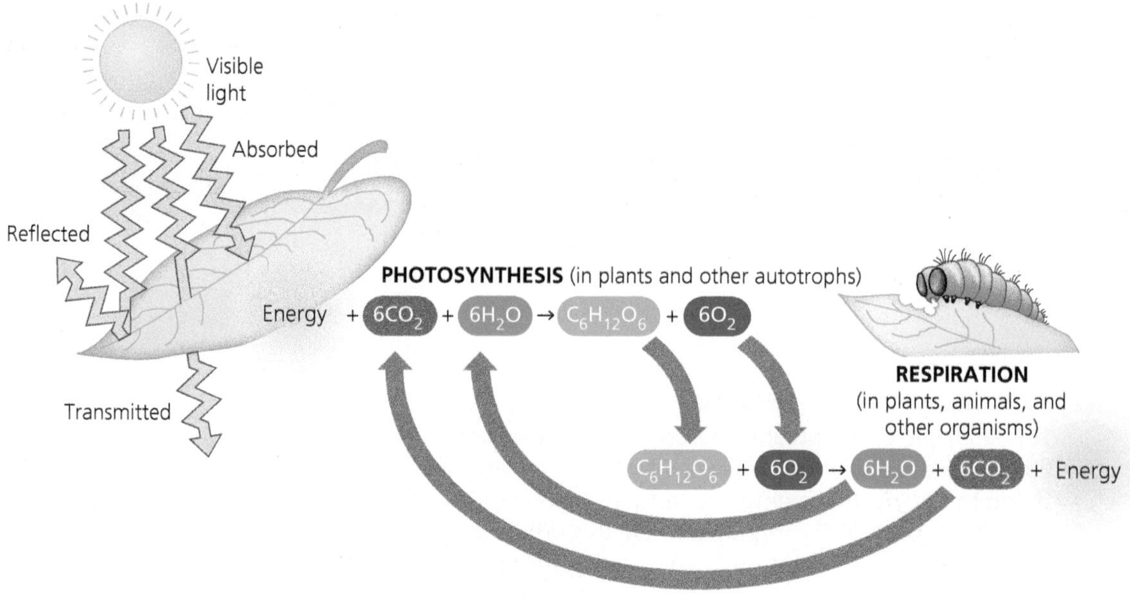

FIGURE 5.1
Energy flow through photosynthesis and cellular respiration. Light energy from the sun is transformed to chemical energy in photosynthesis. This chemical energy is used to synthesize glucose and oxygen from carbon dioxide and water. The energy stored in plant organic molecules—glucose, for example—can be utilized by plants or by consumers. The energy in organic molecules is released to form ATP during cellular respiration in plants, animals, and other organisms.

Fermentation and cellular respiration involve oxidation-reduction reactions (redox reactions). Redox reactions are always defined in terms of electron transfers, oxidation being the *loss* of electrons and reduction the *gain* of electrons. In cellular respiration, two hydrogen atoms are removed from glucose (oxidation) and transferred to a coenzyme called nicotinamide adenine dinucleotide (**NAD+**), reducing this compound to **NADH**. Think of these two hydrogen atoms as 2 electrons and 2 protons. NAD^+ is the oxidizing agent that is reduced to NADH by the addition of 2 electrons and 1 proton. The other proton (H^+) is released into the cell solution. NADH transfers electrons to the electron transport chain. The transfer of electrons from one molecule to another releases energy, and this energy can be used to synthesize ATP.

Cellular respiration is a sequence of three metabolic stages: **glycolysis** in the cytoplasm, and the **citric acid cycle** and the **electron transport chain** in mitochondria (Figure 5.2). Fermentation involves glycolysis but does not involve the citric acid cycle and the electron transport chain, which are inhibited at low oxygen levels. Two common types of fermentation are **alcoholic fermentation** and **lactic acid fermentation**. Animals, certain fungi, and some bacteria convert pyruvate produced in glycolysis to lactate. Plants and some fungi, yeast in particular, convert pyruvate to ethanol and carbon dioxide. Cellular respiration is much more efficient than fermentation in producing ATP. Cellular respiration can produce a maximum of 32 ATP molecules; fermentation produces only 2 ATP molecules.

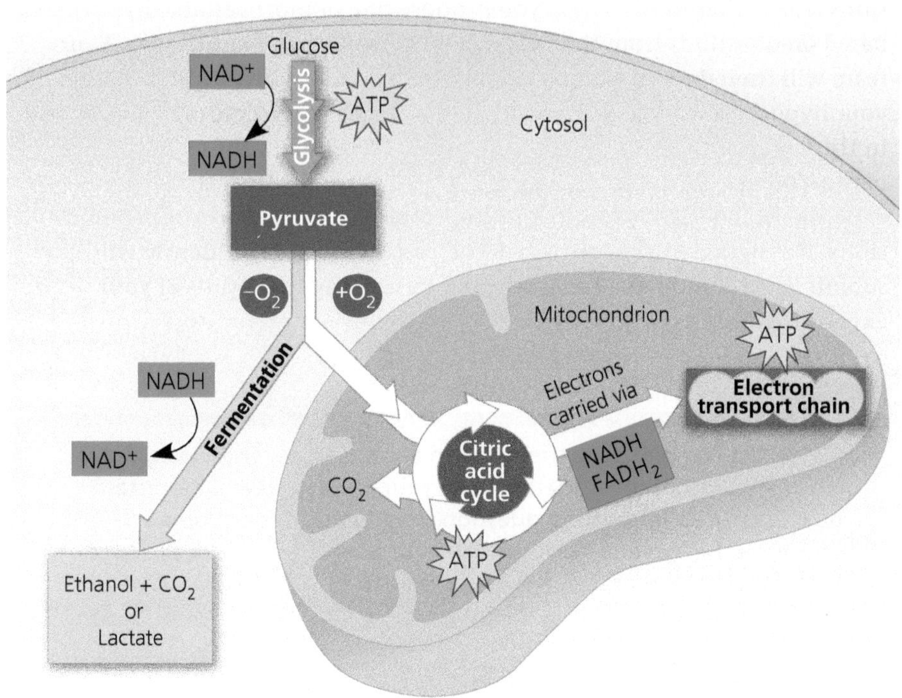

FIGURE 5.2

Stages of cellular respiration and fermentation. Cellular respiration consists of glycolysis, the citric acid cycle, and the electron transport chain. Glycolysis is also a stage in fermentation.

Before you begin today's lab topic, refer to the preceding paragraph and Figure 5.2 as you review major pathways, reactants, and products of fermentation and cellular respiration by answering the following questions.

1. Which processes are anaerobic?

2. Which processes are aerobic?

3. Which processes take place in the cytoplasm of the cell?

4. Which processes take place in mitochondria?

5. What is the initial reactant in cellular respiration?

6. What is (are) the product(s) of the anaerobic processes?

7. What is (are) the product(s) of the aerobic processes?

8. Which gives the greater yield of ATP, alcoholic fermentation or cellular respiration?

In this lab topic you will investigate alcoholic fermentation first and then cellular respiration. Working in teams of two to four students, you will first perform *two* introductory experiments (Experiment A of Exercise 1.1 and Experiment A of Exercise 1.2). Experiment B in each exercise provides questions and background to help you propose one or more testable hypotheses based on questions from the experiments or your prior knowledge. Your team will then design and carry out an independent investigation based on your hypotheses, completing your observations and recording your results in this laboratory period. After discussing the results, your team will prepare an oral presentation in which you will persuade the class that your experimental design is sound and that your results support your conclusions. If assigned by the lab instructor, *each of you* independently will submit Results and Discussion sections describing the results of your experiment (see Appendix A).

First complete Experiment A in each exercise. Then discuss possible questions for investigation with your research team. Be certain you can pose an interesting question from which to develop a testable hypothesis. Design and perform the experiment today. Prepare to report your results in oral and/or written form.

EXERCISE 5.1

Alcoholic Fermentation

For centuries, humans have taken advantage of yeast fermentation to produce alcoholic beverages, bread, and other fermented foods. In 2016, scientists at Stanford University reconstructed the recipe for beer from the residue they discovered on 5000-year-old pottery found in China. Consider the products of fermentation and their roles in making these economically and culturally important foods and beverages. Alcoholic fermentation begins with glycolysis, a series of reactions breaking glucose into two molecules of **pyruvate** with a net yield of 2 ATP and 2 NADH molecules. In anaerobic environments, in two steps the pyruvate (a 3-carbon molecule) is converted to ethyl alcohol (ethanol, a 2-carbon molecule) and CO_2. In this process the 2 NADH molecules are oxidized, replenishing the NAD^+ used in glycolysis (Figure 5.2).

Experiment A. Alcoholic Fermentation in Yeast

Materials

4 respirometers:
  test tubes, 1-mL graduated pipettes, aquarium tubing, flasks, binder clips
pipette pump
3 5-mL graduated pipettes, labeled "DI water," "yeast," and "glucose"
3-inch donut-shaped metal weights
yeast solution
glucose solution
DI water
water bath
wax pencil

Introduction

In this experiment, you will investigate alcoholic fermentation in a yeast (a single-celled fungus), *Saccharomyces cerevisiae,* or "baker's yeast." When oxygen is low, some fungi, including yeast and most plants, switch from cellular respiration to alcoholic fermentation. In bread making, starch in the flour is converted to glucose and fructose, which then serve as the starting compounds for fermentation. The resulting carbon dioxide is trapped in the dough, causing it to rise. Ethanol is also produced in bread making but evaporates during baking. Alcoholic fermentation is also an economically and culturally important process in the production of alcoholic beverages, such as beer and wine. Recently, interest in alternative biofuels has created a new industry in which fermentation is central to the conversion of energy crops (e.g., corn), forestry and municipal waste, and agricultural residues to ethanol.

In this laboratory experiment, the carbon dioxide (CO_2) produced, being a gas, bubbles out of the solution and can be used as an indication of the relative rate of fermentation taking place. Figure 5.3 shows the respirometers you will use to collect CO_2. The rate of fermentation, a series of enzymatic reactions, can be affected by several factors, for example, concentration of yeast, concentration of glucose, or temperature. In this experiment you will investigate *the effects of yeast concentration.* In your independent study, Exercise 5.3, you may choose to investigate other independent variables.

Hypothesis

Hypothesize about the effect of different concentrations of yeast on the rate of fermentation.

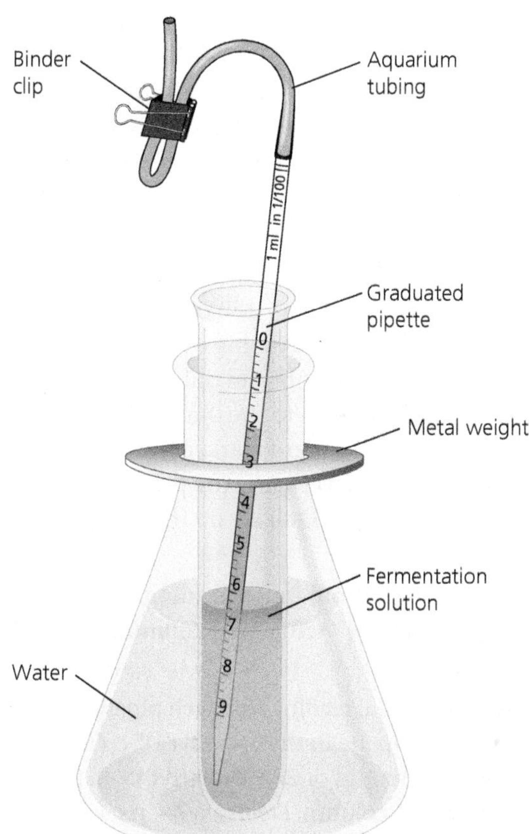

FIGURE 5.3
Respirometer used for yeast fermentation.

Prediction

Predict the results of the experiment based on your hypothesis (if/then).

Procedure

1. Obtain four flasks and add enough tap water to keep them from floating in a water bath (fill to about 5 cm from the top of the flask). Label the flasks 1, 2, 3, and 4. To stabilize the flasks, place a 3-inch donut-shaped metal weight over the neck of the flasks.
2. Obtain four test tubes (fermentation tubes) and label them 1, 2, 3, and 4. Add solutions as in Table 5.1 to the appropriate tubes. Rotate each tube to distribute the yeast evenly in the tube. Place tubes in the corresponding numbered flasks.

TABLE 5.1 Contents of Fermentation Solutions (volumes in mL)

Tube	DI Water	Yeast Suspension	Glucose Solution
1	4	0	3
2	6	1	0
3	3	1	3
4	1	3	3

3. To each tube, add a 1-mL graduated pipette to which a piece of plastic aquarium tubing has been attached.
4. Place the flasks with the test tubes and graduated pipettes in the water bath at 30°C. Allow them to equilibrate for about 5 minutes with the tubing unclamped (Figure 5.3).
5. Attach the pipette pump to the free end of the tubing on the first pipette. Use the pipette pump to draw the fermentation solution up into the pipette. Fill it past the calibrated portion of the tube, but do not draw the solution into the tubing. Fold the tubing over and clamp it shut with the binder clip so the solution does not run out. Open the clip slightly, and allow the solution to drain down to the 0-mL calibration line (or slightly below). If the level is below the zero mark, open the clamp slightly while another student adjusts the level using the pipette pump. Be patient. This may require a couple attempts! Quickly do the same for the other three pipettes.
6. In Table 5.2, quickly record your initial readings for each pipette in the "Initial reading" row in each "Actual (A)" column. This will be the *initial* reading (I).
7. Two minutes after the initial readings for each pipette, record the actual readings (A) in mL for each pipette in the "Actual (A)" column. Subtract I from A to determine the total amount of CO_2 evolved (A – I). Record this value in the "CO_2 Evolved (A – I)" column. *From now on, you will subtract the initial reading from each actual reading to determine the total amount of CO_2 evolved.*

8. Continue taking readings every 2 minutes for each of the solutions for 20 minutes. Avoid moving the pipette once the experiment begins. Remember, take the actual reading from the pipette and subtract the initial reading to get the total amount of CO_2 evolved in each test tube.
9. Record your results in Table 5.2.

Results

1. Complete Table 5.2.
2. Using Figure 5.4, construct a graph to illustrate your results. The graph can also be completed in Excel.
   a. What is (are) the independent variable(s)? Which is the appropriate axis for this variable?

TABLE 5.2 Total CO_2 Evolved by Different Concentrations of Yeast. Actual values are the graduated pipette readings. For CO_2 evolved values, subtract the initial reading from the actual reading. This is the amount of CO_2 accumulated over time.

Download an Excel version from www.masteringbiology.com in the Study Area under Lab Media

	Tube 1		Tube 2		Tube 3		Tube 4	
Time (min)	Actual (A)	CO_2 Evolved (A – I)	Actual (A)	CO_2 Evolved (A – I)	Actual (A)	CO_2 Evolved (A – I)	Actual (A)	CO_2 Evolved (A – I)
Initial reading (I)	.7		0.2		.4		.7	
2	.3	.22	.5	.3	.8	.4	.48	.6
4								
6								
8								
10								
12								
14								
16								
18								
20								

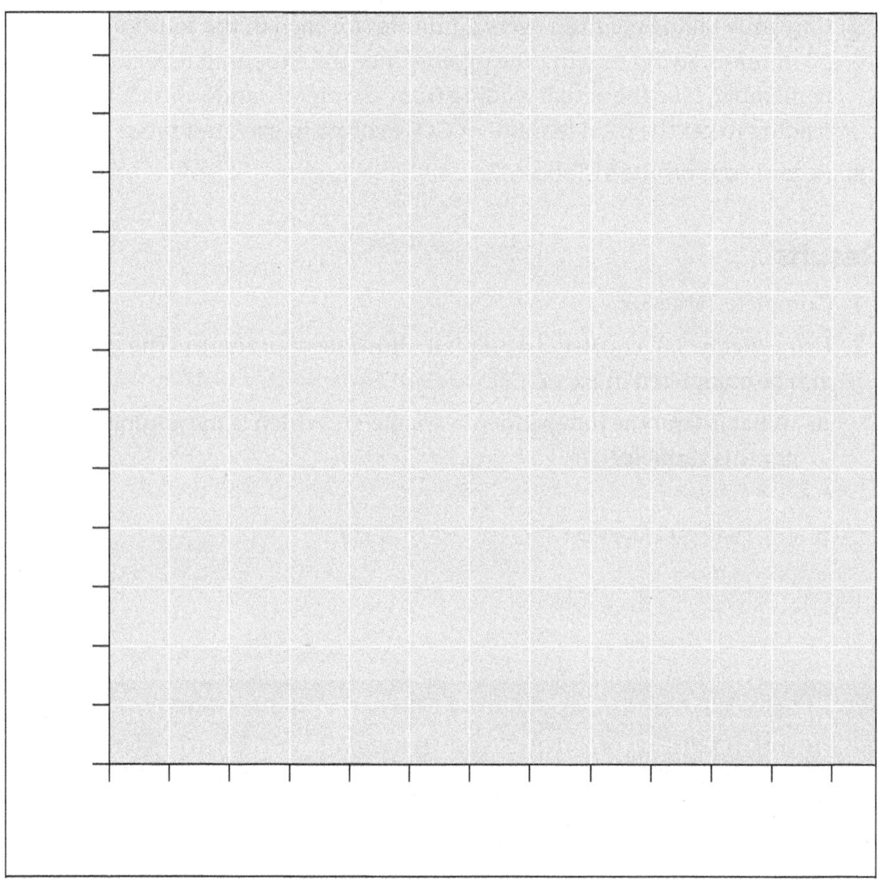

FIGURE 5.4

b. What is the dependent variable? Which is the appropriate axis for this variable?

c. Choose an appropriate scale and label the *x*- and *y*-axes.

d. Should you use a legend? If so, what would this include?

e. Compose a figure title.

Discussion

1. Explain the experimental design. What is the purpose of each test tube? Which is (are) the control tube(s)?

2. Which test tube had the highest rate of fermentation? Explain why.

3. Which test tube had the lowest rate of fermentation? Explain why.

4. Why were different amounts of water added to each fermentation solution?

Experiment B. Student-Designed Investigation of Alcoholic Fermentation

Materials

all materials from Experiment A
beakers
graduated pipettes of various sizes
different substrates: sucrose, saccharin, Nutrasweet™, Splenda™, fructose, starch, glycogen, honey, corn syrup, pyruvate, maltose, stevia, agave nectar
different types of yeast: dry active, quick rise, Pasteur champagne (or other yeast for wine making), sourdough yeast (*Candida milleri*), beer-brewing yeast *(S. cerevisiae and S. pastorianus*—a hybrid or other strains)
various fermentation inhibitors: sodium fluoride, ethyl alcohol, Na benzoate
various salt solutions
various pH buffers
spices: ground cinnamon, cloves, caraway, ginger, cardamom, nutmeg, mace, thyme, dry mustard, hot peppers, ground cayenne pepper, celery seed, turmeric, oregano, sage, fennel, cloves
tannins: oak bark, black walnut bark or husks
disposable gloves
additional glassware
mortar and pestle

Introduction

If your team chooses to study alcoholic fermentation for your independent investigation and report, design a simple experiment to investigate some factor that affects alcoholic fermentation. Use the available materials or ask your instructor about the availability of additional materials.

Procedure

1. Collaborating with your research team, read the following potential questions, and choose a question to investigate using this list or an idea from your prior knowledge. You may want to check your text and other sources for supporting information. You should be able to explain the rationale behind your choice of question. For example, if you choose to investigate *starch* as a substrate, you should be able to explain that the yeast must first digest starch before the glucose can be used in alcoholic fermentation and the impact this might have on the experiment.

   Check with your instructor to confirm how many experimental tubes are available for each team. You may have to limit comparisons to two or three and decide about the number of replicates. In some cases you may select one inhibitor or promoter and test at several concentrations.

a. Would other substrates be as effective as glucose in alcoholic fermentation?

Possible substrates:
sucrose (table sugar—glucose and fructose disaccharide)
honey (mainly glucose and fructose)
corn syrup (fructose and sucrose)
starch (glucose polymer in plants)
saccharin, Equal™, Splenda™
fructose
maltose
pyruvate
stevia
agave nectar

b. Would fermentation rates change with different types of yeasts, for example, quick rise, Champagne yeast used in wine making, and *Candida milleri* used in sourdough?

c. What environmental conditions are optimum for alcoholic fermentation? What temperature ranges? What pH ranges?

Do these environmental conditions have different effects on different strains or species of yeast?

d. What is the maximum amount of ethyl alcohol that can be tolerated by yeast cells?

If you select toxins or fermentation inhibitors for your investigation, ask the instructor about safety procedures. Post safety precautions and follow safety protocol, including wearing gloves, protective eyewear, and proper disposal of materials. Notify the instructor of any spills.

e. Sodium fluoride, commonly used to prevent tooth decay, inhibits an enzyme in glycolysis. At what concentration is it most effective?

f. Would adding $MgSO_4$ enhance glycolysis? $MgSO_4$ provides Mg^{++}, a cofactor necessary to activate some enzymes in glycolysis.

g. Does the amount of capsaicin (bioactive compounds in hot peppers) inhibit or enhance fermentation? Compare peppers with different Scoville ratings, which is the heat scale for peppers.

h. Herbs, spices, and essential oils are used in food science for their flavor, but also for their antimicrobial properties (Juneja et al., 2012). Some spices enhance yeast activity while others inhibit it (Corriher, 1997). Do spices enhance or inhibit fermentation? Are the effects of spice dependent on the concentration, enhancing at low concentrations and inhibiting at high? Try ginger, ground cardamom, caraway, cinnamon, mace, nutmeg, thyme, dry mustard, ground cayenne pepper, hot peppers, turmeric, oregano, celery seed, sage, fennel, cloves, or others.

i. Salt is often used as a food preservative to prevent bacterial and fungal growth (for example, in country ham). But salt is also important to enhance the flavor of bread when added in small amounts. At what concentration does salt begin to inhibit yeast fermentation?

j. Does the food preservative Na benzoate inhibit cellular respiration?

k. How do fermentation rates compare for baker's yeast (*Saccharomyces cervisiae*) and sourdough yeast (*Candida milleri*) in different pH environments?

l. How do fermentation rates compare for yeast used in brewing most beers (*S. cervisiae*) and lager (*S. pastorianus*—a hybrid between *S. cervisiae* and *S. eubayanus*)?

m. Tannins are produced by many woody plants and have antimicrobial properties that can interfere with fermentation in the production of biofuels. Are tannin extracts effective in inhibiting yeast fermentation?

2. Design your experiment, proposing hypotheses, making predictions, and determining procedures as instructed in Exercise 5.3.

EXERCISE 5.2

Cellular Respiration

Most organisms produce ATP using cellular respiration, a process that involves glycolysis, the citric acid cycle, and the electron transport chain. In cellular respiration, many more ATP molecules are produced than were produced in alcoholic fermentation (potentially 30–32 compared to 2). After the series of reactions in the cytoplasm (glycolysis), pyruvate enters the mitochondria, where enzymes for the citric acid cycle and the electron transport chain are located. The citric acid cycle is a series of eight steps, each catalyzed by a specific enzyme. As one compound is converted to another, CO_2 is given off and hydrogen ions and electrons are removed. The electrons and hydrogen ions are passed to NAD^+ and another electron carrier, FAD (flavin adenine dinucleotide). NADH and $FADH_2$ carry the electrons to the electron transport chain, where the electrons pass along the chain to the final electron acceptor, oxygen. In the process, ATP molecules and water are produced (Figure 5.2).

Experiment A. Oxidation-Reduction Reactions in a Mitochondrial Suspension

Materials

mitochondrial suspension
succinate
phosphate buffer
DPIP solution
1-mL graduated pipette
pipette pump
4 cuvettes or small test tubes
Parafilm® squares
Kimwipes®
spectrophotometer
wax pencil

Introduction

In this experiment, you will investigate cellular respiration in isolated mitochondria. Your instructor has prepared a mitochondrial suspension from pulverized lima beans. The suspension has been kept on ice to prevent enzyme degradation, and the citric acid cycle will continue in the isolated mitochondria as in intact cells. Sucrose has been added to the mitochondrial suspension to maintain an osmotic balance.

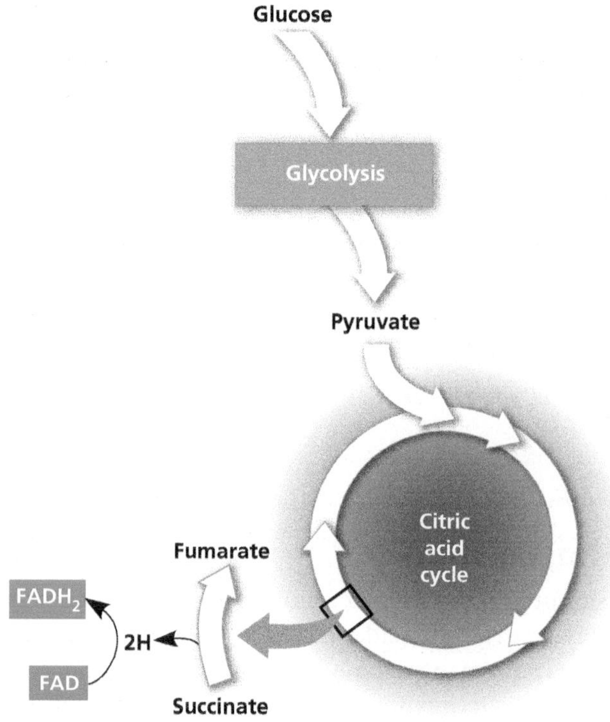

FIGURE 5.5
At one point in the citric acid cycle succinate is converted to fumarate. Hydrogens from succinate pass to FAD, reducing it to $FADH_2$.

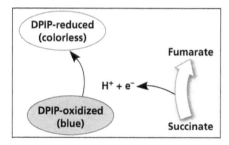

One step in the citric acid cycle is the enzyme-catalyzed conversion of succinate to fumarate in a redox reaction. In intact cells, succinate loses hydrogen ions and electrons to FAD, and, in the process, fumarate is formed (Figure 5.5).

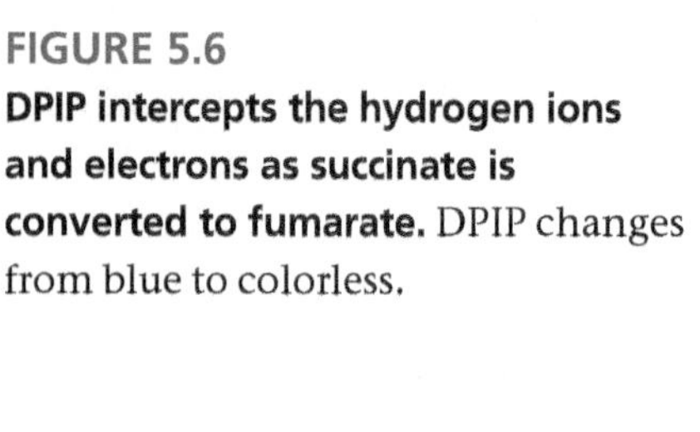

We will utilize this step in the citric acid cycle to investigate the rate of cellular respiration under different conditions. To perform this experiment, we will add a substance called DPIP (di-chlorophenol-indophenol), an electron acceptor that intercepts the hydrogen ions and electrons released from succinate, changing the DPIP from an oxidized to a reduced state. DPIP is *blue* in its oxidized state but changes from blue to *colorless* as it is reduced (Figure 5.6).

DPIP-reduced (colorless)
Fumarate
$H^+ + e^-$
DPIP-oxidized (blue)
Succinate

FIGURE 5.6
DPIP intercepts the hydrogen ions and electrons as succinate is converted to fumarate. DPIP changes from blue to colorless.

We can use this color change to measure the respiration rate. To do this, however, we must have some quantitative means of measuring color change. An instrument called a **spectrophotometer** will allow us to do this. A spectrophotometer measures the amount of light absorbed by a pigment. In the spectrophotometer, a specific wavelength of light (chosen by the operator) passes through the pigment solution being tested—in this case, the blue DPIP. The spectrophotometer then measures the proportion of light *transmitted* by the DPIP and shows a reading on a calibrated scale. As the DPIP changes from blue to clear, it will absorb less light and more light will pass through (be transmitted through) the solution, and the transmittance reading will increase. As aerobic respiration takes place, what should happen to the percent transmittance of light through the DPIP?

Our experiment will involve using succinate as the substrate and investigating the effect that *changing the amount of succinate will have on the cellular respiration rate.*

Hypothesis

Hypothesize about the effect of an increased amount of substrate on the rate of cellular respiration.

Prediction

Predict the results of the experiment based on your hypothesis (if/then).

Procedure

1. Prepare the spectrophotometer.

   The instructions that follow are for a Thermo Scientific Spectronic 20D+ (Figure 5.7). Turn on the machine (power switch C) at least 15 minutes before beginning.(*See* Appendix C *for instructions for the digital Spectronic 200.*)

   a. Using the wavelength control knob (A), select the wavelength: 600 nm. Your instructor has previously determined that this wavelength is absorbed by DPIP.

   b. Press the mode selection button to select "Transmittance."

   c. Zero the instrument by adjusting the control knob (the same as power switch C) so that the meter reads 0% transmittance. There should be no cuvette in the instrument, and the sample holder cover must be closed. Once it is set, do not change this setting.

2. Obtain four cuvettes and label them B, 1, 2, and 3. The B will be the blank.
3. Prepare the blank first by measuring 4.6 mL phosphate buffer, 0.3 mL mitochondrial suspension, and 0.1 mL succinate into the B cuvette. Cover the cuvette tightly with Parafilm and invert it to mix the reactants thoroughly.
4. Calibrate the spectrophotometer as follows: Wipe cuvette B with a Kimwipe and insert it into the sample holder. Be sure you align the etched mark on the cuvette with the line on the sample holder. Close the cover. Adjust the light control (F) until the meter reads 100% transmittance. Remove cuvette B. You are now ready to prepare the experimental cuvettes. The blank corrects for differences in transmittance due to the mitochondrial solution.

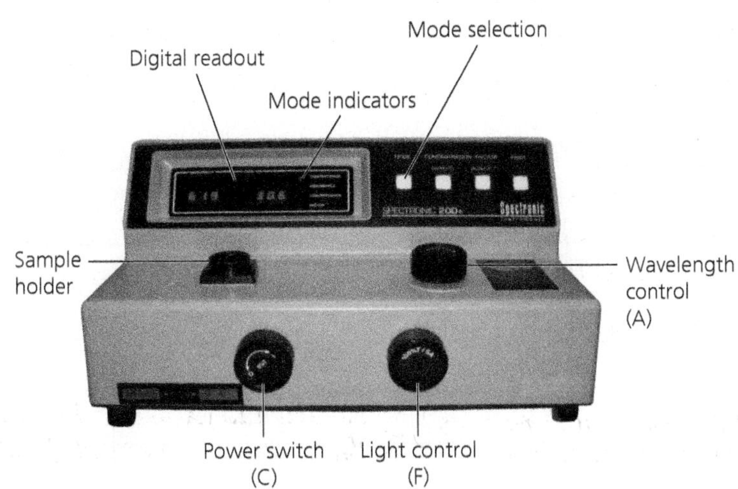

FIGURE 5.7
The Thermo Scientific Spectronic 20D+. A spectrophotometer measures the proportion of light of different wavelengths absorbed and transmitted by a pigment solution. Inside the spectrophotometer, light is separated into its component wavelengths and passed through a sample.

TABLE 5.3 Contents of Experimental Tubes (volumes in mL)

Tube	Phosphate Buffer	DPIP	Mitochondrial Suspension	Succinate (add last)
1	4.4	0.3	0.3	0
2	4.3	0.3	0.3	0.1
3	4.2	0.3	0.3	0.2

5. Measure the phosphate buffer, DPIP, and mitochondrial suspension into cuvettes 1, 2, and 3 as specified in Table 5.3.

Do not add the succinate yet!

6. Perform the next two steps as *quickly* as possible. First, add the succinate to each cuvette.
7. Cover the opening to tube 1 with Parafilm, wipe it with a Kimwipe, immediately insert it into the sample holder, and record the percent transmittance in Table 5.4 in the Results section. Repeat this step for tubes 2 and 3.

> *If the initial reading is higher than 30%, tell your instructor immediately.* You may need to add another drop of DPIP to each tube and repeat step 7. The reading must be low enough (the solution dark enough) to give readings for 20–30 minutes. If the solution is too light (the transmittance is above 30%), the reactions will go to completion too quickly to detect differences in the tubes.

8. Before each set of readings, insert the blank, cuvette B, into the sample holder. Adjust to 100% transmittance if necessary.
9. Continue to take readings at 5-minute intervals for 20–30 minutes. *Each time, before you take a reading, cover the tube opening with Parafilm and invert it to mix the contents.* Record the results in Table 5.4.
10. After recording your results dispose of the solutions in your cuvettes in the waste container indicated by your instructor.

Results

1. Complete Table 5.4. Compose a title for the table.

TABLE 5.4

Download an Excel version from www.masteringbiology.com in the Study Area under Lab Media

	Time (min)						
Tube	0	5	10	15	20	25	30
1	22.0	27.6	30.4	32.3	33.2	34.0	34.4
2	18.4	25.8	28.2	30.8	32.0	33.4	34.4
3	22.0	32.0	37.8	43.0	46.2	50.2	53.8

2. Using Figure 5.8, construct a graph to illustrate your results. The graph can also be completed in Excel.

   a. What is (are) the independent variable(s)? Which is the appropriate axis for this variable?

   b. What is the dependent variable? Which is the appropriate axis for this variable?

   c. Choose an appropriate scale and label the x- and y-axes.

   d. Should you use a legend? If so, what would this include?

   e. Compose a figure title.

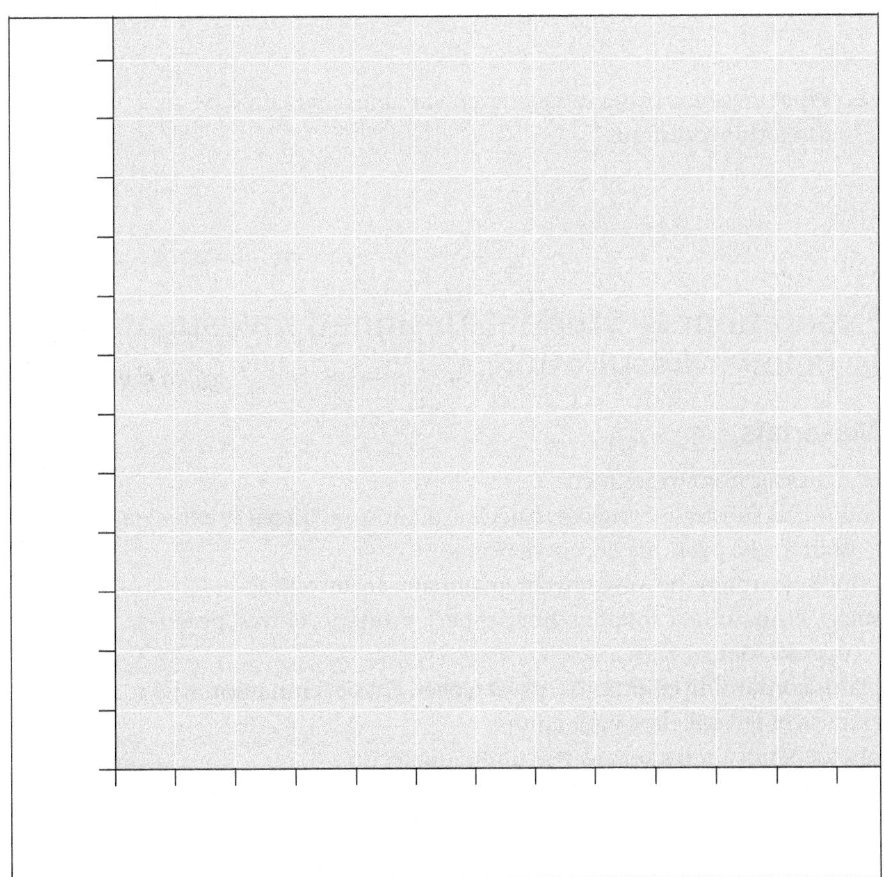

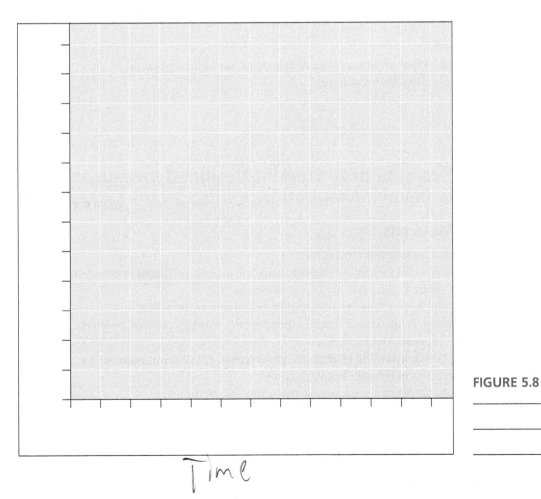

Time

FIGURE 5.8

____

____

____

Discussion

1. Explain the experimental design. What is the role of each of the components of the experimental mixtures?

   lima bean extract

   succinate

   DPIP

   phosphate buffer

2. Which experimental tube is the control?

3. In which experimental tube did transmittance increase more rapidly? Explain.

4. Why should the succinate be added to the reaction tubes last?

5. Was your hypothesis falsified or supported by the results? Use your data to support your answer.

6. What are some other independent variables that could be investigated using this technique?

Experiment B. Student-Designed Investigation in Cellular Respiration

Materials

all materials from Experiment A
additional substrates: glucose, fructose, maltose, artificial sweeteners, starch, glycogen, stevia, agave nectar
inhibitors: rotenone, oligomycin, malonate, antimycin A
spices: cinnamon, capsaicin (hot pepper), turmeric, cloves, peaches, cassava root
plants containing coumarins: sweet clover, Cassia, cinnamon sticks
extracts of red oak bark with lignins
phytochemicals: berberine, curcumin, quercetin
different pH buffers
ice bath
water bath
disposable gloves
hammer
mortar and pestle

Introduction

If your team chooses to study cellular respiration for your independent investigation and report, design a simple experiment to investigate some factor that affects cellular respiration. Use the available materials, or ask your instructor about the availability of additional materials.

If you select toxins or respiratory inhibitors for your investigation, ask the instructor about safety procedures. Post safety precautions and follow safety protocol, including wearing gloves, protective eyewear, and proper disposal of materials. Notify the instructor of any spills.

Procedure

1. Collaborating with your research team, read the following potential questions, and choose a question to investigate using this list or an idea from your prior knowledge. You may want to check your text and other sources for supporting information. You should be able to explain the rationale behind your choice of question. For example, if you choose to investigate *starch* as a substrate, you should be able to explain that the starch must be digested before glucose can enter the pathway and the impact this might have on the experiment.

   Check with your instructor to confirm how many experimental tubes are available for each team. You may have to limit comparisons to two or three and decide about the number of replicates. In some cases, you may select one inhibitor or promoter and test at several concentrations.

   a. Would other substrates be as effective as succinate in cellular respiration? Possible substrates:

      glucose

      sucrose (table sugar—disaccharide of glucose and fructose)

      starch (glucose polymer in plants)

      saccharin, Nutrasweet™, Splenda™, or other artificial sweeteners

      fructose

      agave nectar

      stevia

   b. What environmental conditions are optimum for cellular respiration? What temperature ranges? What pH ranges?

   c. What inhibitors of cellular respiration are most effective? Consider the following list:

      Rotenone, an insecticide, inhibits electron flow in the electron transport chain.

      Oligomycin, an antibiotic, inhibits ATP synthesis.

      Malonate blocks the conversion of succinate to malate. How would you determine if this is competitive or noncompetitive inhibition?

      Antimycin is an antibiotic that inhibits the transfer of electrons to oxygen.

Crushed peach seeds, foxglove (*Digitalis*) seeds, and cassava root contain cyanoglycosides, potential respiratory inhibitors.

Some spices: tumeric (contains curcumin), cloves, hot peppers (contain capsaicin), cinnamon

Plants with a high concentration of coumarin: sweet clover or Cassia cinnamon sticks

Lignins and tannins found in extracts from red oak bark or black walnut bark or husks

d. Phytochemicals are compounds produced by plants that have bioactivity. Do phytochemicals inhibit or enhance respiration in mitochondria? Examples with the plant source in parentheses: berberine (barberry), curcumin (turmeric), and quercetin (colorful fruits).

2. Design your experiment, proposing hypotheses, making predictions, and determining procedures as instructed in Exercise 5.3.

EXERCISE 5.3

Designing and Performing Your Open-Inquiry Investigation

Materials

See each Experiment B Materials list in Exercises 5.1 and 5.2.

Introduction

Now that you have completed both introductory investigations, your research team should decide if you will investigate fermentation or cellular respiration. Return to the investigation of your choice and review the suggestions in Experiment B. Use Lab Topic 1 Scientific Investigation as a reference for designing and performing this independent investigation. You will need to think critically and creatively as you ask questions and formulate your hypothesis. As a team, review and modify the procedures, determine any additional required materials, review the techniques, and assign tasks to all members of your research team. Your experiments will be successful if you plan carefully, think critically, perform lab techniques accurately and systematically, and record and report data accurately. The following outline will assist you in designing and performing your original investigation.

Procedure

1. **Develop a research question to investigate.** Consider one or two potential questions, then as a team select one question. Suggested questions are included in Experiment B as a starting point. (Refer to Lab Topic 1, Exercise 1.1, Lab Study A. Asking Questions.)

Question:

2. **Formulate a testable hypothesis.** (Refer to Lab Topic 1, Exercise 1.1, Lab Study B. Developing Hypotheses.)
   Hypothesis:

3. **Summarize the essential elements of the experiment.** (Use separate paper.)

4. **Predict the results of your experiment based on your hypothesis.** (Refer to Lab Topic 1, Exercise 1.2, Lab Study C. Making Predictions.)
   Prediction: (if/then)

5. **Outline the procedures to be used in the experiment.** (Refer to Lab Topic 1, Exercise 1.2, Lab Study B. Choosing or Designing the Procedure.)
   a. Review and modify the procedures used in Experiment A (see either Fermentation or Cellular Respiration). List each step in your procedure in numerical order.
   b. Create a table with contents of the experimental tubes (model this after either Table 5.1 for fermentation or Table 5.3 for cellular respiration). If computers are available, create or copy your new table into Excel.
   c. Critique your procedure: check for replicates, levels of treatment, controls, time intervals, total volume of experimental tubes, experimental conditions, glassware, and equipment.
   d. If your experiment requires materials other than those provided, ask your laboratory instructor about their availability. If possible, submit requests in advance.
   e. Create a table for data collection, using Table 5.2 (fermentation) or Table 5.4 (cellular respiration) as a model. If computers are available, create or copy your new data table into Excel.

6. **Perform the experiment,** making observations and collecting data for analysis.

If your experiment involves the use of toxins or respiration inhibitors, use them only in liquid form as provided by the instructor. Wear protective gloves and eyewear. Ask your instructor about proper disposal procedures. If a spill occurs, notify your instructor immediately for proper cleanup.

7. **Record results including observations and data** in your data table. Be thorough when collecting data. Make notes about experimental conditions and observations. Do not rely on your memory for information that you will need when reporting your results.

8. **Prepare your discussion.** Discuss your results in light of your hypothesis.
   a. Review your prediction. Did your results correspond to the prediction you made? If not, explain how your results are different from your predictions, and why this might have occurred.
   b. Review your hypothesis. Review your results (tables and graphs). Do your results support or falsify your hypothesis? Explain your answer, using your data for support.
   c. If you had problems with the procedure or questionable results, explain how they might have influenced your conclusion.
   d. If you had an opportunity to repeat and expand this experiment to make your results more convincing, what would you do?
   e. Summarize the conclusion you have drawn from your results.
9. **Be prepared to report your results to the class.** Prepare to persuade your fellow scientists that your experimental design is sound and that your results support your conclusions.
10. If your instructor requires it, **submit Results and Discussion sections** of a scientific paper (see Appendix A). Keep in mind that although you have performed the experiments as a team, you must turn in a lab report of *your original writing*. Your tables and figures may be similar to those of your team members, but your Results and Discussion sections must be the product of your own literature search and creative thinking.

REVIEWING YOUR KNOWLEDGE

1. Having completed this lab topic, you should be able to define, describe, and use the following terms: *aerobic, anaerobic, substrate, reactants, products, spectrophotometer, respirometer, NAD+, NADH, FAD,* $FADH_2$*, ATP.*
2. State the beginning reactants and the end products of glycolysis, alcoholic fermentation, the citric acid cycle, and the electron transport chain. Describe where these processes take place in the cell and the conditions under which they operate (aerobic or anaerobic).

   glycolysis:

   alcoholic fermentation:

   citric acid cycle:

   electron transport chain:

3. Suppose you do another experiment using DPIP to study cellular respiration in isolated mitochondria, and the results using the spectrophotometer show a final percent transmittance reading of 42% in tube 1 and

78% in tube 2. Both tubes had an initial reading of 30%. In which tube did the greater amount of cellular respiration occur? Explain your answer in terms of the changes that take place in DPIP.

4. How do you know that the electrons causing the change in color of DPIP are involved in the succinate–fumarate step?

APPLYING YOUR KNOWLEDGE

1. Two characteristics of natural wines are that they have a maximum alcohol content of 14% and are "sparkling" wines. Apply your understanding of alcoholic fermentation to explain these characteristics.

2. Succinate dehydrogenase is the enzyme that catalyzes the conversion of succinate to fumarate, releasing electrons that are used to reduce FAD to $FADH_2$. If a mutation changed the active site of this enzyme, what do you predict might be the effect on the remaining steps in the citric acid cycle? Would there be any effect on the amount of ATP produced in cellular respiration from 1 molecule of glucose?

3. Increasing concerns about dependence on fossil fuels has generated research and development of renewable fuels, such as ethanol. Ethanol can be produced from grains (for example, corn) or from plant biomass (tree chips and sawdust, composted food waste, or paper products). The production of ethanol is an industrial application of fermentation studied in introductory biology laboratories. Using information from your laboratory experiments, propose the components necessary for converting corn to ethanol. (Consider reactants, products, and environmental conditions.) Corn plants store energy as starch, which cannot enter the fermentation pathway directly. Based on your knowledge of the molecular structure of starch, include components necessary to convert starch to simple sugars.

4. The titan arum (*Amorphophallus titanum*) has an enormous inflorescence up to 3 meters tall. The plant rarely flowers, but when it does visitors flock to botanical gardens, and in the forest pollinators are lured to the flowers in the night. The titan arum is able to generate heat and raise its temperature above ambient temperatures. Botanists have suggested that the ability to produce heat is important in these plants because it enhances the spread of the carrion-like scent to the far reaches of the tropical forest canopy. Clearly, these plants must have a high respiratory rate to produce temperatures as high as 36°C. How could you determine if the temperature is the result of cellular respiration? What features of the plant surface and cell structure might be present if respiration is actively occurring in the flowers?

STUDENT MEDIA: BioFlix, Activities, Investigations, Videos, and Data Tables

www.masteringbiology.com (select Study Area)

BioFlix—Ch. 9: Cellular Respiration

Activities—Ch. 9: Overview of Cellular Respiration; Glycolysis; The Citric Acid Cycle; Electron Transport; Fermentation

Investigations—Ch. 9: How Is the Rate of Cellular Respiration Measured?

Data Tables—Tables 5.2 and 5.4 can be downloaded in Excel format. Look in the Study Area under Lab Media.

REFERENCES

Corriher, S. O. *Cookwise: The Hows and Whys of Successful Cooking.* New York: William Morrow, 1997.

Gänzle, M. G., M. Ehmann, and W. P. Hammes. "Modeling of Growth of *Lactobacillus sanfarciscensis* and *Candida milleri* in Response to Parameters of Sourdough Fermentation." *Appl. Environ. Microbiol.,* 1998, vol. 64, pp. 2616–2623.

Gibson, B. and G. Liti. "*Saccharomyces pastrianus*: Genomic insights inspiring innovation for industry. " *Yeast*, 2015, vol. 32, pp. 17–27.

Heldt, H. W. and B. Piechulla. *Plant Biochemistry*, 4th ed. New York: Academic Press, 2010. Provides information on metabolism, as well as plant extracts and secondary metabolites with antimicrobial activity.

Juneja, V. K., H. P. Dwivedi, and X. Yan. "Novel Natural Food Antimicrobials." *Annual Review of Food Science and Technology*, 2012, vol. 3, pp. 381–403.

Levetin, E. and K. McMahon. *Plants and Society*, 7th ed. New York: McGraw-Hill Publishers, 2015. Includes information on spices, medicinal plants, toxins, and other possible choices for independent investigations.

Nelson, D. L. and M. M. Cox. *Lehninger, Principles of Biochemistry*, 6th ed. New York: W.H. Freeman, 2012.

Seymour, R. S. "Biophysics and Physiology of Temperature Regulation in Thermogenic Flowers." *Bioscience Reports*, 2001, vol. 21, pp. 223–236.

Shelef, L. A. "Antimicrobial Effects of Spices." *Journal of Food Safety*, 1984, vol. 6, pp. 29–44.

Urry, L., M. Cain, S. Wasserman, P. Minorsky, and J. Reece. *Campbell Biology*, 11th ed. San Francisco, CA: Pearson, 2017.

Zhang, Y. and J. Ye. "Mitochondrial inhibitor as a new class of insulin sensitizer." *Acta Pharm Sin B*, 2012, vol. 2(4), pp. 341–349.

The procedure used to assay mitochondrial activity was based on a procedure from "Succinic Acid Dehydrogenase Activity of Plant Mitochondria," in F. Witham, D. Blaydes, and R. Devlin, *Exercises in Plant Physiology*. Boston, MA: Prindle, Weber & Schmidt, 1971.

WEBSITES

Cornell University Poisonous Plants Page:
http://www.ansci.cornell.edu/plants/index.html

Metabolic poisons:
http://www.ruf.rice.edu/~bioslabs/studies/mitochondria/mitopoisons.html

Phaff Yeast Collection at the University of California Davis: http://phaffcollection.ucdavis.edu/
See the Preparation Guide that accompanies this lab manual for information on yeast strains and sources.

Oxidative Phosphorylation: Uncouplers and Inhibitors, University of Leeds, School of Biochemistry and Molecular Biology:
http://www.bmb.leeds.ac.uk/illingworth/oxphos/poisons.htm

Titan arum flowers—timelapse at Kew Gardens:
http://www.youtube.com/watch?v=AD75AA9m03E

U.S. Dept. of Energy, alternative energy, biomass, and biofuels: http://energy.gov/science-innovation/energy-sources/renewable-energy/biomass

LAB TOPIC 6

Photosynthesis

Laboratory Objectives

After completing this lab topic, you should be able to:

1. Describe the roles played by light and pigment in photosynthesis.
2. Name and describe pigments found in photosynthesizing tissues.
3. Explain the separation of pigments by paper chromatography, based on their molecular structure.
4. Demonstrate an understanding of the process of spectrophotometry and the procedure for using the spectrophotometer.

Introduction

Without photosynthesis, there could be no life on Earth as we know it. The Earth is an open system constantly requiring an input of energy to drive the processes of life. All energy entering the biosphere is channeled from the sun into organic molecules via the process of photosynthesis. As the sun's hydrogen is converted to helium, energy in the form of photons is produced. These photons pass to Earth's surface and those with certain wavelengths within the visible light portion of the electromagnetic spectrum are absorbed by pigments in the chloroplasts of plants, initiating the process of photosynthesis.

In photosynthesis, light energy is transformed into chemical energy. That chemical energy is used to synthesize organic compounds (glucose) from CO_2, and in the process water is used and O_2 is released. Glucose, a primary source of energy for all cells, may be converted to sucrose and transported or stored in the polymer starch. The organic compounds produced in photosynthesis can enter the respiratory pathway to provide a source of energy for cellular work, for example, active transport or cell division. These organic molecules also provide the carbon skeletons for the synthesis of proteins, lipids, and other molecules that are necessary for plant growth and development.

Plants (autotrophs) are sometimes called producers because they are the ultimate source of organic compounds and stored energy for all heterotrophs. Animals consume plants and convert the plant molecules into their own organic molecules and energy sources, forming the basis for ecological food webs. Oxygen, also produced by photosynthesis, is necessary for aerobic respiration in the cells of plants, animals, and other organisms (Figure 6.1).

In this laboratory, you will investigate the effects of environmental factors (light) and cellular structures (pigments) on photosynthesis. In several experiments, you will determine photosynthetic activity by testing for the storage product of photosynthesis, starch, by using iodine potassium iodide (I_2KI), which stains starch purple-black. This is the same test for the presence of starch that you used to study starch digestion by amylase in Lab Topic 4 Enzymes.

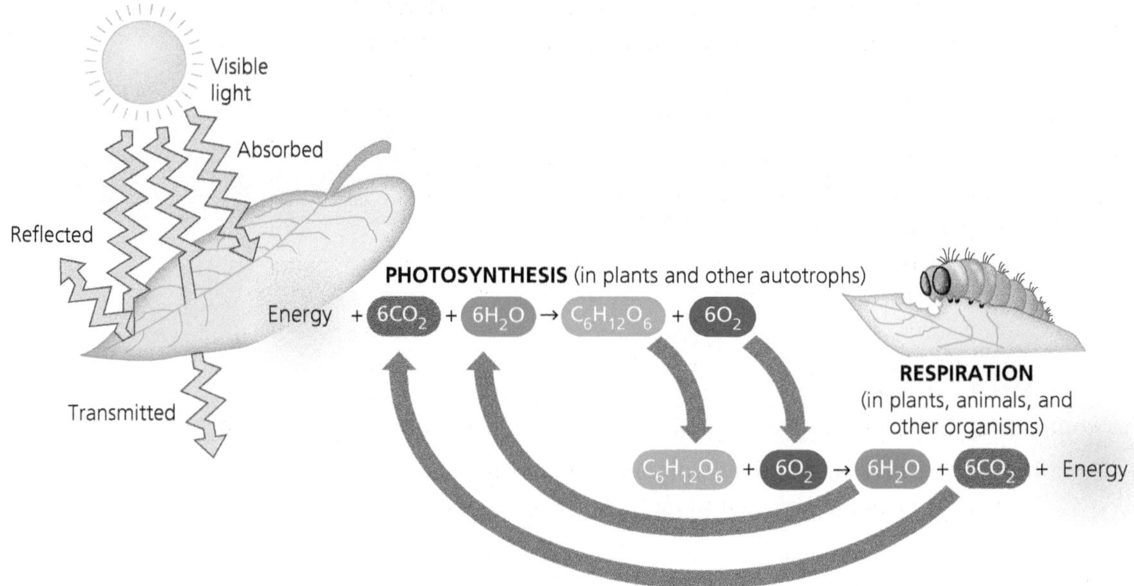

FIGURE 6.1
Energy flow through photosynthesis and respiration. Energy flows from the sun into the biological systems of Earth, and visible light may be reflected, transmitted, or absorbed. Plants absorb light energy and convert it to chemical energy during photosynthesis. In this process, carbon dioxide and water are used to synthesize glucose (and ultimately other organic compounds) and to release oxygen. These organic molecules and the energy stored in them can be utilized by animals and other organisms that consume them. The energy in organic molecules is released to form ATP during cellular respiration in plants, animals, and other organisms.

EXERCISE 6.1

The Wavelengths of Light for Photosynthesis

Materials

black construction paper
green, red, and blue plastic filters
paper clips
forceps
hot plate
petri dish
squirt bottle of water
scissors
1 geranium plant with at least 4 good leaves per 8 students
1 1,000-mL beaker filled with 300 mL of water
1 400-mL beaker filled with 200 mL of 80% ethyl alcohol
dropper bottle with concentrated I_2KI solution

Introduction

Plants produce a variety of pigments, molecules that can absorb specific wavelengths of light. Light that is not absorbed by these pigments is either reflected or transmitted. The colors that we see are a result of the wavelengths of light that are reflected by the pigments. Plants utilize some pigments to capture light energy that is transformed into chemical energy during the light dependent reactions of photosynthesis. These pigments are bound to proteins in the thylakoid membranes of chloroplasts. Other pigments are produced to attract or repel animals or to protect plant tissues from environmental stresses. Some of these pigments are stored in the aqueous solution of the central vacuole.

In this exercise, you will investigate the photosynthetic activity of different wavelengths of light. You will determine if products of photosynthesis are present in leaf tissue that has been exposed to different wavelengths of light for several days. Working with other students in groups of eight, you will cover small portions of different leaves of a geranium plant with pieces of black paper and green, red, and blue plastic filters. Each pair of students will be responsible for one of the four treatments. Later you will determine photosynthetic activity by testing for the presence of starch under the paper or filters. Starch is synthesized from glucose molecules and can be easily tested as an indirect product of photosynthesis. Starch will turn dark in the presence of iodine (I_2KI). A change from the yellow-amber color of the iodine solution to a purple-black color is a positive test for the presence of starch.

Plastic filters used in this exercise are designed to reflect and transmit the appropriate wavelengths of light to correspond to the visible light spectrum. For example, green filters reflect and transmit green light. What wavelengths of light will be absorbed by the green filter? Refer to Figure 6.2, which shows the electromagnetic spectrum. What wavelengths of light will be reflected and transmitted by the black paper and each of the colored filters? Note that the same wavelengths of light are reflected and transmitted by the filters.

Hypothesis

Hypothesize about photosynthetic activity in leaf cells covered with different colored filters as described.

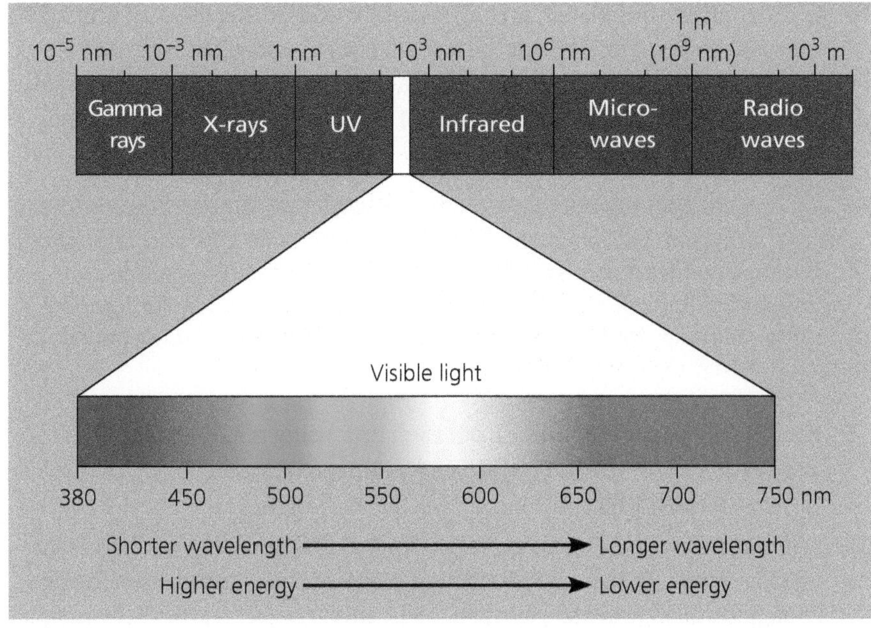

FIGURE 6.2

The electromagnetic spectrum. Visible light is only a small segment of the sun's energy that reaches the Earth. Photosynthetic pigments absorb only specific wavelengths within the visible light range.

Prediction

Predict the results of the experiment based on your hypothesis (if/then).

Procedure

1. Four to five days before the experiment is to be carried out, cut a piece from each color of plastic filter and one from the black construction paper. Each piece should be a rectangle approximately 2.5 cm by 5 cm. Double over the strip and slide the edge of a healthy geranium leaf, still attached to the plant, between the folded edges. Carefully slip a slightly sprung paper clip over the paper or filter, securing the paper or filter to the leaf. The strip should be on both sides of the leaf (Figure 6.3). Follow this procedure with the other colors and the black construction paper, using a different leaf for each strip. Return the plant with treated leaves to bright light. Your instructor may have already carried out this step for you.

FIGURE 6.3
Geranium leaf with filter attached. Fold the filter and slide over the edge of the geranium leaf. Secure with a sprung paper clip.

2. On the day of the lab, carry the plant with leaves covered to your desk. You will have to be able to recognize each leaf after the paper is removed and the leaf is boiled. To facilitate this, with your teammates devise a way to distinguish each leaf, and write the distinction in the space provided. Differences in size or shape may distinguish different leaves, but it may be necessary to introduce distinguishing features, such as by cutting the petioles to different lengths or cutting out small notches in a portion of the leaf not covered by the paper. Record below the distinguishing differences for each treatment.

   black paper:
   green filter:
   red filter:
   blue filter:

3. After you have distinguished each leaf, sketch the leaf in the Results section, showing the position of the paper or filter on the leaf.
4. Set up the boiling alcohol bath. Place a 1,000-mL beaker containing 300 mL of water on the hot plate. Carefully place the 400-mL beaker containing 200 mL of 80% ethyl alcohol into the larger beaker of water. Turn on the hot plate and bring the nested beakers to a boil. Adjust the temperature to maintain slow boiling. Do not place the beaker of alcohol directly on the hot plate.

> Ethyl alcohol is highly flammable! Do not place the beaker of alcohol directly on the hot plate. To bring it to a boil, raise the temperature of the hot plate until the alcohol just boils, and then reduce the temperature to maintain slow boiling. Do not leave boiling alcohol unattended.

5. Remove the paper and filters from each leaf; using forceps, carefully drop all the leaves into the boiling alcohol solution to extract the pigments. Save the plastic filters.
6. When the leaves are almost white, use forceps to remove them from the alcohol. Place them in separate petri dishes, rinse with distilled water, and add enough distilled water to each dish to just cover the leaf. Turn off the hot plate if all teams have completed boiling leaves for this exercise (and Exercise 6.2).

7. Add drops of I_2KI solution to the water until a pale amber color is obtained. I_2KI reacts with starch to produce a purple-black color.
8. Wait about 5 minutes and sketch each leaf in the Results section, indicating which areas of the leaf tested positive for starch.

Results

1. Sketch and label each leaf before boiling, showing the location of the paper or filter.

2. Sketch and label each leaf after staining to show the location of the purple-black color.

Discussion

1. Which treatment allowed the greatest photosynthetic activity? (Explain your results in terms of your hypothesis.)

2. When the red filter is placed on a leaf, what wavelengths of light pass through and reach the leaf cells below? (Check wavelengths in Figure 6.2, which shows the electromagnetic spectrum.)

   Green filter?

   Blue filter?

3. Was starch present under the black construction paper? Explain this in light of the fact that black absorbs all wavelengths of light.

EXERCISE 6.2

Pigments in Photosynthesis

Materials

Coleus plant with multicolored leaves (green, pink, purple, and white or yellow)
forceps
1 1,000-mL beaker filled with 300 mL of water
1 400-mL beaker filled with 200 mL of 80% ethyl alcohol
dropper bottle with concentrated I_2KI solution
hot plate
squirt bottle of water

Introduction

Various pigments are found in plants, as anyone who visits a botanical garden in spring or a deciduous forest in autumn well knows. A pigment is a substance that absorbs light. If a pigment absorbs all wavelengths of visible light, it appears black. The black construction paper used in Exercise 6.1 is colored with such a pigment. Other pigments absorb some wavelengths and reflect others. Yellow pigments, for example, reflect light wavelengths in the yellow portion of the visible light spectrum, green pigments reflect in the green portion, and so on.

Some colors are produced by only one pigment, but an even greater diversity of colors can be produced by the cumulative effects of different pigments in cells. Green colors in plants are produced by the presence of chlorophylls *a* and *b* located in the chloroplasts. Yellow, orange, and bright red colors are produced by carotenoids, also in chloroplasts. Blues, violets, purples, pinks, and dark reds are usually produced by a group of water-soluble pigments, the anthocyanins, that are located in cell vacuoles and do not contribute to photosynthesis. Additional colors may be produced by mixtures of these pigments in cells.

Working with one other student, you will use the I_2KI test for starch as in Exercise 6.1 to determine which pigment(s) in a *Coleus* leaf support photosynthesis. Before beginning the experiment, examine your *Coleus* leaf and hypothesize about the location of photosynthesis based on the leaf colors (Figure 6.4).

Hypothesis

Hypothesize about the location of photosynthesis based on the leaf colors.

Prediction

Predict the results of the experiment based on your hypothesis (if/then).

FIGURE 6.4
***Coleus* leaves with green and pink pigments.**

TABLE 6.1 Predicted and Observed Results for the Presence of Starch in Colored Regions of the *Coleus* Leaf

Color	Pigments	Starch Present (predicted) + or −	Starch Present (actual results) + or −
Green			
Purple			
Pink			
White			
Other			

Procedure

1. Remove a multicolored leaf from a *Coleus* plant that has been in strong light for several hours.
2. In Table 6.1, list the colors of your leaf, predict the pigments present to create that color, and predict the results of the I_2KI starch test in each area of the leaf.
3. Sketch the leaf outline in the margin next to the Results section, mapping the color distribution before the I_2KI test.
4. Extract the pigments as previously described in Exercise 6.1, and test the leaf for photosynthetic activity using I_2KI.

Ethyl alcohol is highly flammable! Do not place the beaker of alcohol directly on the hot plate. To bring it to a boil, raise the temperature of the hot plate until the alcohol just boils, and then reduce the temperature to maintain slow boiling. Do not leave boiling alcohol unattended.

5. Sketch the leaf again in the Results section, outlining the areas showing a positive starch test.

Results

1. Record the results of the I_2KI test in Table 6.1.
2. Compare the sketches of the Coleus leaf before and after the I_2KI test.
3. Which pigments supported photosynthesis? Record your results in Table 6.1.

Before I_2KI Test:

Discussion

Describe and explain your results based on your hypothesis.

After I_2KI Test:

EXERCISE 6.3

Separation and Identification of Plant Pigments by Paper Chromatography

Materials

capillary tube
beakers
extractions of leaf pigments in acetone
forceps
scissors
acetone
chromatography paper stapled into a cylinder marked with a pencil line about 1 cm from one end
1-L jar with lid, containing solvent of petroleum ether and acetone

Introduction

Your instructor has prepared an extract of chloroplast pigments from fresh green grass or fresh spinach. A blender was used to rupture the cells, and the pigments were then extracted with acetone, an organic solvent. Working with one other student, begin this exercise by separating the pigments extracted using paper chromatography. To do this, you will apply the pigment extract to a cylinder of chromatographic paper. You will then place the cylinder in a jar with the organic solvents petroleum ether and acetone. The solvents will move up the paper and carry the pigments along; the pigments will move at different rates, depending on their different solubilities in the solvents used and the degree of attraction to the paper. The leading edge of the solvent is called the **front**. Discrete pigment bands will be formed from the front, back to the point where pigments were added to the paper.

The following information will be helpful to you as you make predictions and interpret results:

1. **Polar molecules** or substances dissolve (or are attracted to) polar molecules.
2. **Nonpolar molecules** are attracted to nonpolar molecules to varying degrees.
3. Chromatography paper (cellulose) is a polar (charged) substance.
4. The solvent, made of petroleum ether and acetone, is relatively nonpolar.
5. The *most nonpolar* substance will dissolve in the nonpolar solvent *first.*
6. The *most polar* substance will be attracted to the polar chromatography paper; therefore, it will move *last.*

Use this information and the molecular structure of major leaf pigments to predict the relative solubilities and separation patterns for the pigments and to identify the pigment bands. Study the molecular structure of the four common plant pigments in Figure 6.5. As you study these diagrams, rank the pigments according to polarity in the space provided. To determine polarity, circle and count the number of polar oxygen groups present in each molecule.

Most polar:

Least polar:

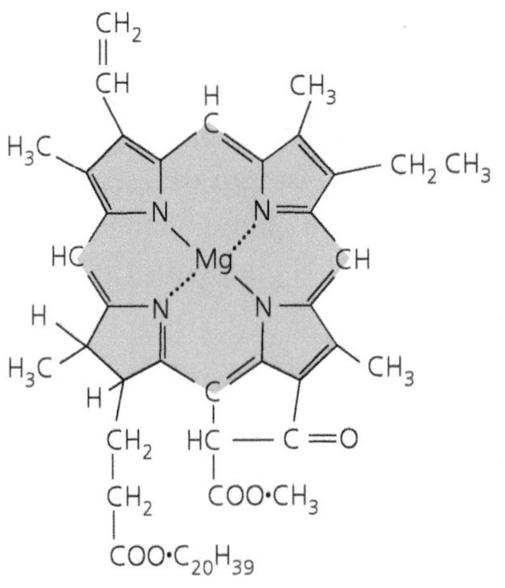

a. Chlorophyll *a*

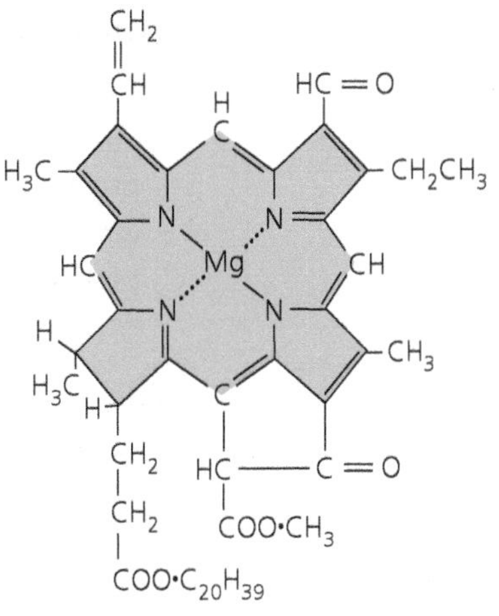

b. Chlorophyll *b*

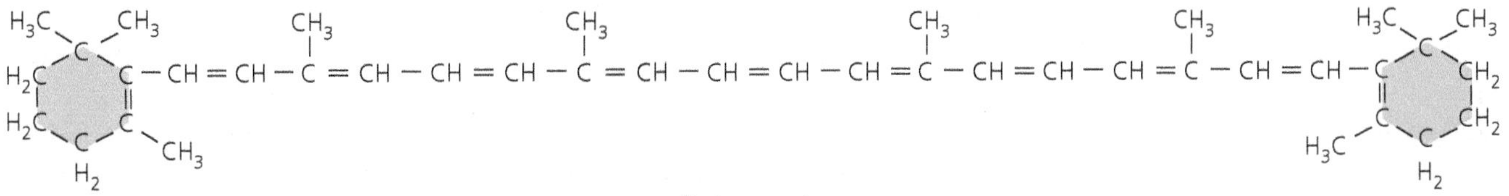

c. Beta carotene

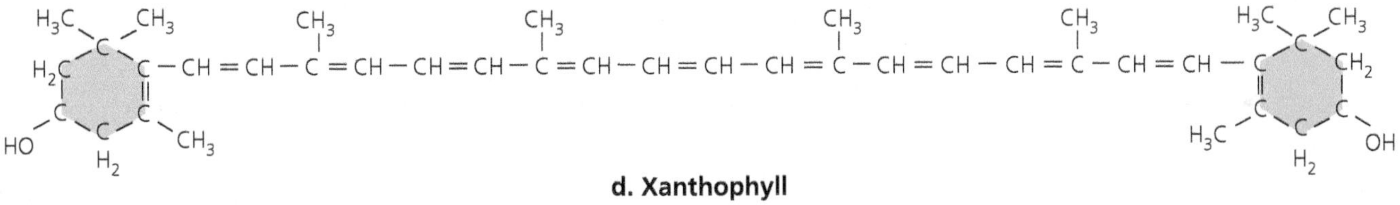

d. Xanthophyll

FIGURE 6.5

Molecular structure of major leaf pigments. The molecular structure of (a) chlorophyll *a*, (b) chlorophyll *b*, (c) beta carotene, and (d) xanthophyll. To determine polarity, circle and count the number of polar oxygen groups present in each molecule.

Hypothesis

State a hypothesis relating polarities and solubilities of pigments.

Beta will be third while chlorophyll b will be the lowest xanthophyll with be the highest ChlorophyllA will be second lowest

Prediction

Predict the results of the experiment based on your hypothesis (if/then).

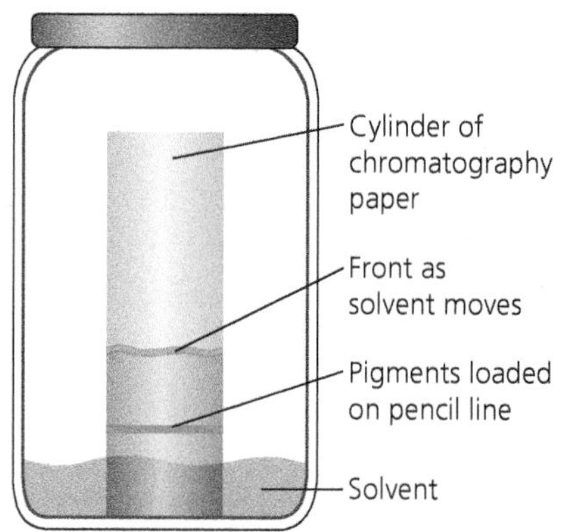

FIGURE 6.6
Paper chromatography of photosynthetic pigments. Add the pigment solution to the paper cylinder along the pencil line. Then carefully place the cylinder into a jar containing a small amount of solvent. Close the lid and watch the pigments separate according to their molecular structures and solubilities.

Procedure

1. Using a capillary tube, streak the leaf pigment extract on a pencil line previously drawn 1 cm from the edge of the paper cylinder. Allow the chlorophyll to dry. Repeat this step three or four times, allowing the extract to air-dry each time. You should have a band of green pigments along the pencil line. The darker your band of pigments, the better the results of your experiment will be.

Perform the next step in a hood or in a well-ventilated room. Do not inhale the fumes of the solvent. *NO SPARKS!* Acetone and petroleum ether are extremely flammable. Avoid contact with all solutions. Wash hands with soap and water. If a spill occurs, notify the instructor. If an instructor is not available, do not attempt to clean up. Leave the room.

2. Obtain the jar containing the petroleum ether and acetone solvent. Using forceps, carefully lower the loaded paper cylinder into the solvent, and quickly cover the jar tightly with the lid (Figure 6.6). *Avoid inhaling the solvent.* The jar should now contain a saturated atmosphere of the solvent. Allow the chromatography to proceed until the solvent front has reached to within 3 cm of the top of the cylinder.
3. Remove the cylinder from the jar, allow it to dry, and remove the staples. Replace the lid on the jar and do not discard the solvent.
4. Save your paper with the separated pigments for the next exercise.

Results

Sketch the chromatography paper. Label the color of the various bands. The front, or leading edge of the paper, should be at the top. The pencil line where pigment was added originally should be at the bottom.

Discussion

Based on your hypothesis and predictions, identify the various pigment bands. *The entire class should come to a consensus about the identifications.* Label your drawing in the Results section above, indicating the correct identification of the pigment bands.

EXERCISE 6.4

Determining the Absorption Spectrum for Leaf Pigments

Materials

spectrophotometer
Kimwipes®
2 cuvettes
20-mL beakers to elute pigments
1 150-mL beaker to hold cuvettes
acetone
cork stoppers for cuvettes

Introduction

In Exercise 6.1, you applied colored plastic filters and black paper to leaves to determine which wavelengths of light would support photosynthesis. Review your conclusions from that exercise and from Exercise 6.2 about pigments used in photosynthesis. Which pigments did you conclude support photosynthesis?

In Exercise 6.4, you will work in teams of four or five students, carrying your investigation a step further by plotting the absorption spectrum of leaf pigments separated by paper chromatography. The **absorption spectrum** is the absorption pattern for a particular pigment, showing relative absorbance at different wavelengths of light. For example, we know that chlorophyll *a* is a green pigment, and we know that it reflects or transmits green wavelengths of light. We do not know, however, the relative proportions of wavelengths of light absorbed by chlorophyll *a*. This information is of interest because it suggests that those wavelengths showing greatest absorbance are important in photosynthesis.

The absorption spectrum can be determined with an instrument called a **spectrophotometer**, or **colorimeter**. A spectrophotometer measures the proportions of light of different wavelengths (colors) absorbed and transmitted by a pigment solution. It does this by passing a beam of light of a particular wavelength (designated by the operator) through the pigment solution being tested. The spectrophotometer then measures the proportion of light transmitted or, conversely, absorbed by that particular pigment and shows the reading on the calibrated scale.

Before measuring the absorption spectrum of the four pigments separated by paper chromatography, consult the diagram of the electromagnetic spectrum (Figure 6.2) and predict the wavelengths of light at which absorption will be greatest for each pigment. Record your predictions in Table 6.2.

TABLE 6.2 Predicted Wavelengths of Greatest Absorption for the Photosynthetic Pigments

Pigment	Wavelengths of Greatest Absorption (predicted)
1. Chlorophyll *a*	
2. Chlorophyll *b*	
3. Carotene	
4. Xanthophyll	

Hypothesis

State a hypothesis that describes the general relationship of each of the pigments to the color of light that it absorbs.

Prediction

Predict the results based on your hypothesis (if/then).

Procedure

1. Cut out the pigments you separated by paper chromatography, and distribute the paper strips as follows:

   Team 1: carotene
   Team 2: xanthophyll
   Team 3: chlorophyll *a*
   Team 4: chlorophyll *b*
   Teams 5 and 6: will determine the absorption spectrum of the total pigment solution

Perform the next three steps in a hood or in a well-ventilated room. Do not inhale the fumes of the solvent. *NO SPARKS!* Acetone is extremely flammable. Avoid contact with all solutions. If a spill occurs, notify the instructor. Wash hands with soap and water.

2. *Teams 1 to 4.* Dilute the pigments as follows: Cut up the chromatography paper with your assigned pigment into a small (20-mL) beaker. Add 10 mL of acetone to the beaker and swirl. This solution containing a single pigment will be your solution B, to be used to determine the absorption spectrum for that pigment. Your reference material will be acetone with no pigments, solution A.

3. *Teams 5 and 6.* Add drops of the original chlorophyll extract solution (acetone pigment mixture) to 10 mL of acetone until it looks pale green. This will be your pigment solution for cuvette B. Your reference material will be acetone with no pigment. This will be in cuvette A.
4. Each team should fill two cuvettes two-thirds full, one (B) with the pigment solution, the other (A) with the reference material (acetone only). Wipe both cuvettes with a Kimwipe to remove fingerprints, and handle cuvettes only with Kimwipes as you proceed.

   What is the purpose of the cuvette with reference material only?

The instructions that follow are for the spectrophotometer, Spectronic 20D+. If using the newer model, Spectronic 200, see the instructions in Appendix C.

5. Measure the absorption spectrum. Record your measurements in Table 6.3. Turn on the machine (power switch C) for at least 15 minutes before beginning (Figure 6.7).
   a. Press the mode selection button to select "Transmittance."
   b. *Select the beginning wavelength* using the wavelength control knob (A). Begin measurements at 400 nanometers (nm).
   c. *Zero the instrument* by adjusting the 0 control knob (same as the power switch C) so that the meter reads 0% transmittance. There should be no cuvette in the instrument, and the sample holder cover must be closed.
   d. *Calibrate the instrument.* Insert cuvette A into the sample holder and close the lid. (Be sure to align the etched mark on the cuvette with the line on the sample holder.) Adjust the light control (F) until the meter reads 100% transmittance, or 0 absorption. You are now ready to make your first reading.
   e. *Change the mode.* Push the mode selection button to change the mode to "Absorbance."

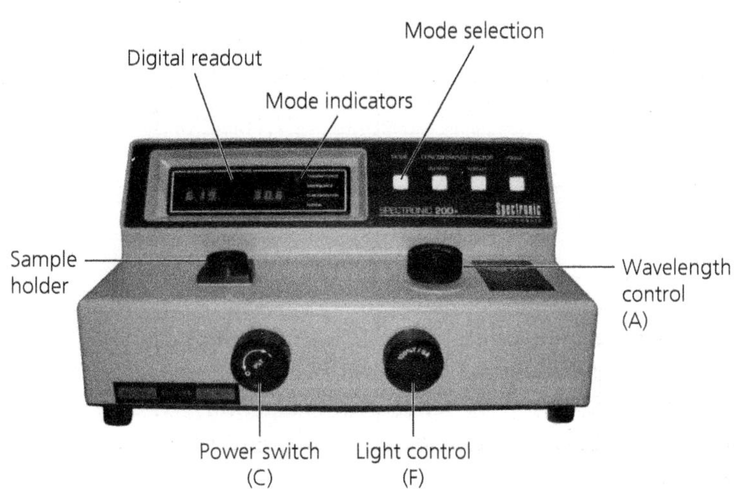

FIGURE 6.7
The Thermo Scientific Spectronic 20D+. A spectrophotometer measures the proportion of light of different wavelengths absorbed and transmitted by a pigment solution. Inside the spectrophotometer, light is separated into its component wavelengths and passed through a sample. The graph of absorption at different wavelengths for a solution is called an *absorption spectrum*.

f. *Begin your readings.* Remove cuvette A and insert cuvette B. (Align the etched mark.) Close the cover. Record the reading on the absorbance scale. Remove cuvette B.

g. *Recalibrate the instrument.* Insert cuvette A into the sample holder, and set the wavelength to 420 nm. Again, calibrate the instrument to 100% transmittance (0 absorption) with cuvette A in place, using the light control (F). *Note: Each time you calibrate to 100% transmittance, use the mode selection button to choose "Transmittance" and then switch back to "Absorbance" for your reading.*

h. *Take the second reading.* Remove cuvette A and insert cuvette B. Record absorbance at 420 nm. Remove cuvette B.

i. *Continue your observations,* increasing the wavelength by 20-nm increments until you reach 720 nm. Be sure to recalibrate each time you change the wavelength.

j. After recording your results in Table 6.3, dispose of the pigment extract in the waste containers provided by your instructor.

6. Pool data from all teams to complete Table 6.3.

TABLE 6.3 Absorbance of Photosynthetic Pigments Extracted from Fresh_______________*

Download an Excel version from www.masteringbiology.com in the Study Area under Lab Media

Wavelength	Chlorophyll *a*	Chlorophyll *b*	Xanthophyll	Carotene	Total Pigment
400					
420					
440					
460					
480					
500					
520					
540					
560					
580					
600					
620					
640					
660					
680					
700					
720					

*Complete title with name of plant used for extract, for example, beans.

FIGURE 6.8

Absorption spectrum for chlorophyll *a*, chlorophyll *b*, carotene, and xanthophyll. Plot your results from Exercise 6.4. Label all axes, and draw smooth curves to fit the data. Label the graph for easy identification of pigments.

Results

1. Using the readings recorded in Table 6.3, plot in Figure 6.8 the absorption spectrum for each pigment.
2. Refer to Lab Topic 1 for information about constructing graphs. Choose appropriate scales for the axes, determine dependent and independent variables, and plot data points. Draw smooth curves to fit the values plotted. Label the graph for easy identification of pigments plotted, or prepare a legend and use colored pencils.

Discussion

1. List in the margin or on another page the pigments extracted and the optimum wavelength(s) of light for absorption for each pigment.
2. Which pigment is most important in the process of photosynthesis? Support your choice with evidence from your results.

3. Chlorophyll *b* and carotenoids are called *accessory pigments.* Using data from your results, speculate about the roles of these pigments in photosynthesis.

REVIEWING YOUR KNOWLEDGE

1. Using your previous knowledge of photosynthesis and the results from today's exercises, explain the role, origin, or fate of each factor involved in the process of photosynthesis.

2. A pigment solution contains compound A with 4 polar groups and compound B with 2 polar groups. You plan to separate these compounds using paper chromatography with a nonpolar solvent. Predict the location of the two bands relative to the solvent front. Explain your answer.

APPLYING YOUR KNOWLEDGE

1. Rice is the number one food crop, feeding over 50% of the world's population. Some scientists estimate that at the current rate of population growth, rice farmers will need to produce 50% more rice per hectare by 2050. Researchers are working to increase the photosynthetic efficiency of rice to meet the concern over food shortage. Using your knowledge of photosynthesis, suggest features of the plant and photosynthetic process that could be modified. Think creatively!

2. In response to shortened day length and cool temperatures in the fall, many trees begin a period of senescence when the breakdown of chlorophyll exceeds chlorophyll production. The leaves of these trees appear to change to yellow and orange. Using your knowledge of photosynthetic pigments, explain the source of these yellow-orange hues.

3. Duckweed is a tiny floating aquatic plant that can reproduce so rapidly that it can completely cover the surface of small fish hatchery ponds. One proposal to limit the growth of this invasive plant is to use colored panels to shield the ponds, inhibit photosynthesis, and kill this pest. What color filters would you order for the panels in this project? What wavelengths of light will be absorbed by these filters? Do you think this proposal will be effective in killing the duckweed?

4. In a classic experiment in photosynthesis performed in 1883 by the German botanist Thomas Engelmann, he surrounded a filament of algae with oxygen-requiring bacteria. He then exposed the algal strand to the visible-light spectrum along its length. In which wavelengths of light along the algal strand would you expect the bacteria to cluster? Explain.

5. An **action spectrum** is a graph that illustrates the efficacy of different wavelengths of light in promoting photosynthesis. Engelmann's elegant experiment in Question 4 is an example of an action spectrum where the efficacy of various wavelengths of light was measured by the production of oxygen and the clustering of bacteria. How does an *action spectrum* differ from the *absorption spectrum* that you created in Exercise 6.4?

6. In 2010, a new form of chlorophyll (chlorophyll *f*) was discovered in cyanobacteria assemblages of stromatolites growing in Shark Bay, Australia. Chlorophyll *f* has been isolated, and scientists are investigating the structure and function of this pigment. Based on your understanding of photosynthesis and your work in the laboratory, how would you determine the wavelengths of light the pigment absorbs? How would you determine if this pigment plays a role in photosynthesis? (Hint: Think about the overall equation for photosynthesis.)

INVESTIGATIVE EXTENSIONS

1. Design and perform comparative investigations of plant pigment systems. The technique in this lab topic can be used to compare plant pigments for cultivated varieties of a species, plants growing in different environments, different species, or even seasonal differences for a single tree or several species.

   Suggestions for comparative studies:

   - Plants grown in shade versus sun
   - Varieties of hostas or other plants that have blue-green, yellow-green, or bright green leaves
   - Variegated plants that have combinations of yellow-green and bright green leaves compared to plants with uniformly green leaves
   - Fresh fall leaves from red maples, hickories, or other species
   - Fresh leaves from hardwoods collected and tested early in the fall while still green, then at 1-week intervals as they change colors

   Note: Remember that anthocyanins are water-soluble pigments located in cell vacuoles and they do not contribute to photosynthesis.

2. Some photosynthetic pigments are produced by all autotrophs; however, other pigments are found in only some plant and algae groups. The photosynthetic pigments found in plants, green algae, diatoms, dinoflagellates, and brown algae are shown in the table below. Which pigments are common to all groups? How do these groups differ for other pigments?

Organism	Chlorophyll *a*	Chlorophyll *b*	Chlorophyll *c*	Carotenoids
Plants	+	+	o	+
Green algae	+	+	o	+
Diatoms	+	o	+	+
Dinoflagellates	+	o	+	+
Brown algae	+	o	+	+

Investigate the molecular structure for chlorophyll *a*, *b*, and *c*, using online sources. Based on your research predict the position of the chlorophyll *c* pigment (relative to *a* and *b*), if separated using paper chromatography.

If you have a source of brown algae or diatoms, collect samples and prepare an acetone extract of the pigments using the techniques from this lab topic. Consult the Preparation Guide if necessary. Separate the pigments using paper chromatography, and then determine the absorption spectrum for the pigments. Compare your results with extracts from plants or green algae, for example, *Ulva*.

Research the molecular and cellular differences, as well as the environmental conditions and evolutionary relationships for these groups of plants and algae. Discuss how these differences are related to differences in pigmentation and photosynthetic activity.

STUDENT MEDIA: BioFlix, Activities, Investigations, Videos, and Data Tables

www.masteringbiology.com (select Study Area)

BioFlix—Ch. 10: Photosynthesis

Activities—Ch. 10: The Sites of Photosynthesis; Overview of Photosynthesis; Light Energy and Pigments; Space-Filling Model of Chlorophyll

Investigations—Ch. 10: How Is the Rate of Photosynthesis Measured? How Does Paper Chromatography Separate Plant Pigments?

Video—Ch. 10: Chloroplast Movement

Data Tables—Table 6.3 can be downloaded in Excel format. Look in the Study Area under Lab Media.

REFERENCES

Ehrenberg, R. "Fifth Form of Chlorophyll Discovered." *Science News,* vol. 178, no. 6, 2010.

Evert, R. F. and S. E. Eichorn. *Raven Biology of Plants,* 8th ed. New York: W. H. Freeman, 2013.

Jabr, F. "A New Form of Chlorophyll." *Scientific American,* August 19, 2010. Accessed at http://www.scientificamerican.com/article.cfm?id=new-form-chlorophyll&page=2

Motten, Alex. "Diversity of Photosynthetic Pigments." *Tested Studies for Laboratory Teaching,* vol. 25, Proceedings of the 25th Workshop/Conference of the Association for Biology Laboratory Education (ABLE), Michael A. O'Donnell, Editor, 2003.

Nelson, D. and M. Cox. *Lehninger Principles of Biochemistry,* 6th ed. New York: W. H. Freeman, 2012.

Taiz, L., E. Zeigler, I. Moller, and A. Murphy. *Plant Physiology and Development,* 6th ed. Sunderland, MA: Sinauer, 2015.

WEBSITES

Arizona State University Center for Bioenergy and Photosynthesis. Includes background information on bioenergy, as well as readings and videos: http://bioenergy.asu.edu/

LAB TOPIC 7

Mitosis and Meiosis

Laboratory Objectives

After completing this lab topic, you should be able to:

1. Describe the activities of chromosomes, centrioles, and microtubules in the cell cycle, including all stages of mitosis and meiosis.
2. Recognize human chromosomes in leukocytes.
3. Identify the stages of mitosis in root tip and whitefish blastula cells.
4. Describe differences in mitosis and cytokinesis in plant and animal cells.
5. Describe differences in mitosis and meiosis.
6. Explain crossing over, and describe how this can bring about particular arrangements of ascospores in the fungus *Sordaria*.

Introduction

The nuclei in cells of eukaryotic organisms contain chromosomes with clusters of **genes**, discrete units of hereditary information consisting of duplicated deoxyribonucleic acid (DNA). Structural proteins in the chromosomes organize the DNA and participate in DNA folding and condensation. When cells divide, chromosomes and genes are duplicated and passed on to daughter cells. Single-celled organisms may divide for reproduction. Multicellular organisms have reproductive cells (eggs or sperm), but they also have somatic (body) cells that divide for growth and development or replacement.

In somatic cells and single-celled organisms, the nucleus divides by **mitosis** into two daughter nuclei, which have the same number of chromosomes and the same genes as the parent cell. For example, the epidermis or outer layer of skin tissue is continuously being replaced through cell reproduction involving mitosis. All of these new skin cells are genetically identical. Yeast and amoeba are both single-celled organisms that can reproduce asexually through mitotic divisions to form additional organisms—genetically identical clones.

Cancerous cells are characterized by uncontrolled mitotic and cell division, and therefore the study of mitosis and its regulation is key to developing new cancer treatments. In 2009, three scientists studying chromosomes and the regulation of mitosis were awarded the Nobel Prize in medicine for their discovery of **telomeres**. Telomeres are DNA sequences on the ends of chromosomes that become shorter during every mitotic cycle of somatic cells. Without telomeres protecting the chromosome ends, important genes located at the ends of chromosomes might be lost in mitosis. The loss of telomeres in each division of somatic cells may be one of many regulatory mechanisms

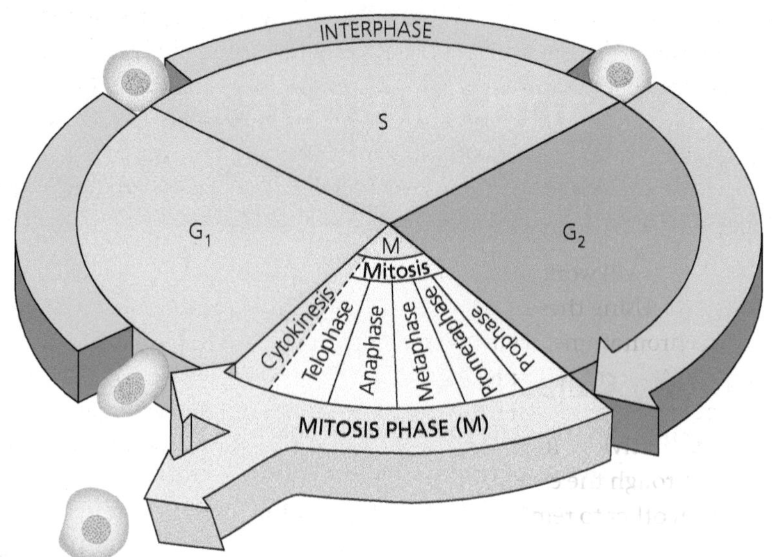

FIGURE 7.1
The cell cycle. In interphase (G_1, S, G_2), DNA replication and most of the cell's growth and biochemical activity take place. In the M phase, the nucleus divides in mitosis, and the cytoplasm divides in cytokinesis.

that limit the number of mitotic cycles that cells can undergo. If telomeres fail to shorten, cells may continue to divide, as in cancer cells.

In multicellular organisms, in preparation for sexual reproduction, a type of nuclear division called **meiosis** takes place. In meiosis, certain cells in ovaries or testes (or sporangia in plants) divide twice, but the chromosomes only replicate once. This process results in the four daughter nuclei with new combinations of chromosomes. Eggs or sperm (or spores in plants) are eventually formed. In contrast to mitosis, the process of meiosis contributes to the genetic variation that is important in sexual reproduction. Generally in both mitosis and meiosis, after nuclear division the cytoplasm divides, a process called **cytokinesis**.

Events from the beginning of one cell division to the beginning of the next are collectively called the **cell cycle**. The cell cycle is divided into two major phases: interphase and the mitotic (M) phase. The M phase represents the division of the nucleus and cytoplasm (Figure 7.1).

EXERCISE 7.1

Modeling the Cell Cycle and Mitosis in an Animal Cell

Materials

60 pop beads of one color
60 pop beads of another color
4 magnetic centromeres
4 centrioles (small plastic cylinders)

Introduction

Scientists use models to represent natural structures and processes that are too small, too large, or too complex to investigate directly. Scientists develop their models from observations and experimental data, usually accumulated from a variety of sources. Building a model can represent the culmination of a body of scientific work, but most models represent a well-developed hypothesis that can then be tested against the natural system and modified.

Linus Pauling's novel and successful technique of building a physical model of hemoglobin was based on available chemical data. This technique was later adopted by Francis Crick and James Watson to elucidate the nature of the hereditary material, DNA. Watson and Crick built a wire model utilizing evidence collected by many scientists. They presented their conclusions about the structure of the DNA helix in the journal *Nature* in April 1953 and were awarded the Nobel Prize for their discovery in 1962.

Today in lab you will work with a partner to build models of cell division: mitosis and meiosis. Using these models will enhance your understanding of the behavior of chromosomes, centrioles, membranes, and microtubules during the cell cycle. After completing your model, you will consider ways in which it is and is not an appropriate model for the cell cycle. You and your partner should discuss activities in each stage of the cell cycle as you build your model. After going through the exercise once together, you will demonstrate the model to each other to reinforce your understanding.

In the model of mitosis that you will build, your cell will be a **diploid** cell ($2n$) with four chromosomes. This means that you will have two homologous pairs of chromosomes. In diploid cells, **homologous chromosomes** are the same length, have the same centromere position, and contain genes for the same characters. One pair will be long chromosomes, and the other pair will be short chromosomes. (**Haploid** cells have only one of each homologous pair of chromosomes, denoted n.)

Lab Study A. Interphase

During interphase, a cell performs its specific functions: Liver cells produce bile; intestinal cells absorb nutrients; pancreatic cells secrete enzymes; skin cells produce keratin. Interphase, a growth phase, begins once a cell completes cell division and consists of three subphases, G_1, S, and G_2. Each new cell has a nucleus that is surrounded by a **nuclear envelope** and that contains chromosomes in an uncoiled, or decondensed, state. In this uncoiled state, the mass of DNA and protein is called **chromatin**. Located outside the nucleus in animal cells is the **centrosome**, a granular region that contains a pair of **centrioles**. The centrosome is the organizing center for microtubules in animal cells (Figure 7.2).

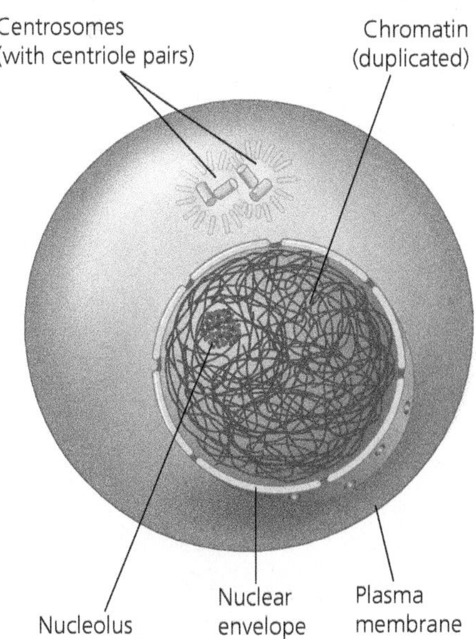

FIGURE 7.2
Interphase.

Procedure

1. Build two pairs of homologous chromosomes, one long pair and one short pair. Construct the first homologous pair of single chromosomes using 10 beads of one color for one member of the long pair and 10 beads of the other color for the other member of the pair. Place the magnetic centromere at any position in the chromosome, but note that it must be in the same position on homologous chromosomes. The centromere appears as a constricted region when chromosomes are condensed. Build the short pair with the same two different colors, but use fewer beads. You should have enough beads left over to duplicate each chromosome later.
2. Model **interphase** of the cell cycle:
   a. Pile all the assembled chromosomes in the center of your work area to represent the decondensed chromosomes as a mass of chromatin in **G_1 (gap 1)**.

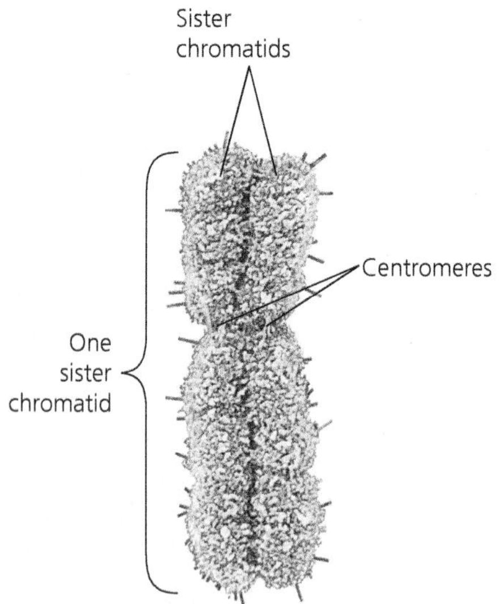

FIGURE 7.3
Duplicated chromosome composed of two sister chromatids held together at the centromeres; condensed as in prometaphase.

b. Position two centrioles as a pair just outside your nucleus. Have the two members of the centriole pair at right angles to each other. (Recall, however, that plant cells with a few exceptions do not have centrioles.)

In the G_1 phase, the cytoplasmic mass increases and will continue to do so throughout interphase. Proteins are synthesized, new organelles are formed, and some organelles such as mitochondria and chloroplasts grow and divide in two. Throughout interphase, one or more dark, round bodies, called **nucleoli** (singular, **nucleolus**), are visible in the nucleus.

c. Duplicate the centrioles: Add a second pair of centrioles to your model; again, have the two centrioles at right angles to each other. **Centriole duplication** begins in late G_1 or early S phase.

d. Duplicate the chromosomes in your model cell to represent DNA replication in the **S (synthesis) phase**: Make a second strand that is identical to the first strand of each chromosome. In duplicating chromosomes, you will use two magnets to form the new centromere. Each sister chromatid will have a centromere represented by the magnet.

Unique activities taking place during the S phase of the cell cycle are the replication of chromosomal DNA and the synthesis of chromosomal proteins. DNA synthesis continues until chromosomes have been duplicated. Each strand of a duplicated chromosome is called a **sister chromatid**. Sister chromatids are identical to each other and are held together most tightly at the **centromere** (Figure 7.3).

e. Do not disturb the chromosomes to represent **G_2 (gap 2)**.

During the G_2 phase, in addition to continuing cell activities, cells prepare for mitosis. Enzymes and other proteins necessary for cell division continue to be synthesized during this phase.

f. Separate your centriole pairs, moving them toward opposite poles of the nucleus to represent that the G_2 phase is coming to an end and mitosis is about to begin.

How many pairs of homologous chromosomes are present in your cell during this stage of the cell cycle?

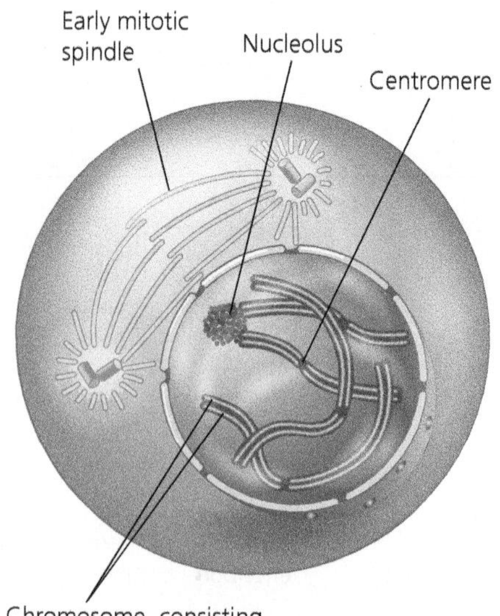

FIGURE 7.4
Prophase mitosis.

Lab Study B. M Phase (Mitosis and Cytokinesis)

In the M phase, the nucleus and cytoplasm divide. Nuclear division is called *mitosis*. Cytoplasmic division is called *cytokinesis*. Mitosis is divided into five stages: prophase, prometaphase, metaphase, anaphase, and telophase.

Procedure

1. To represent **prophase**, leave the chromosomes piled in the center of the work area.

   Prophase begins when chromosomes begin to coil and condense. At this time they become visible in the light microscope. Centrioles continue to move to opposite poles of the nucleus, and as they do so, a fibrous, rounded structure tapering toward each end, called a **spindle**, begins to form between them. Nucleoli begin to disappear (Figure 7.4).

What structures make up the fibers of the spindle? (Check text if necessary.)

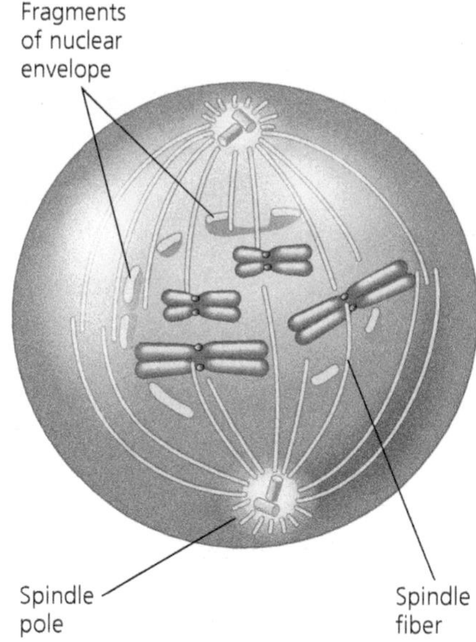

FIGURE 7.5
Prometaphase mitosis.

2. At **prometaphase**, the centrioles are at the poles of the cell. To represent the actions in prometaphase, move the centromeres of your chromosomes to lie on an imaginary plane (the equator) midway between the two poles established by the centrioles.

   During prometaphase, chromosomes continue to condense (Figure 7.3). The nuclear envelope breaks down as the spindle continues to form. Some spindle microtubules become associated with chromosomes at protein structures called **kinetochores**. Each sister chromatid has a kinetochore associated with the centromere. These spindle microtubules now extend from the chromosomes to the centrosomes at the poles. The push and pull of spindle fibers on the chromosomes ultimately leads to their movement to the equator. When the centromeres lie on the equator, prometaphase ends and the next stage begins (Figure 7.5).

   How many duplicated chromosomes are present in your prometaphase nucleus?

3. To represent **metaphase**, a relatively static stage, leave the chromosomes with centromeres lying on the equator.

   In metaphase, duplicated chromosomes lie on the equator (also called the metaphase plate). The two sister chromatids are held together at the centromere region. Metaphase ends as the centromeres separate.

   Label Figure 7.6 with *chromosome, spindle fibers, centrosome, centrioles, kinetochore, equator.*

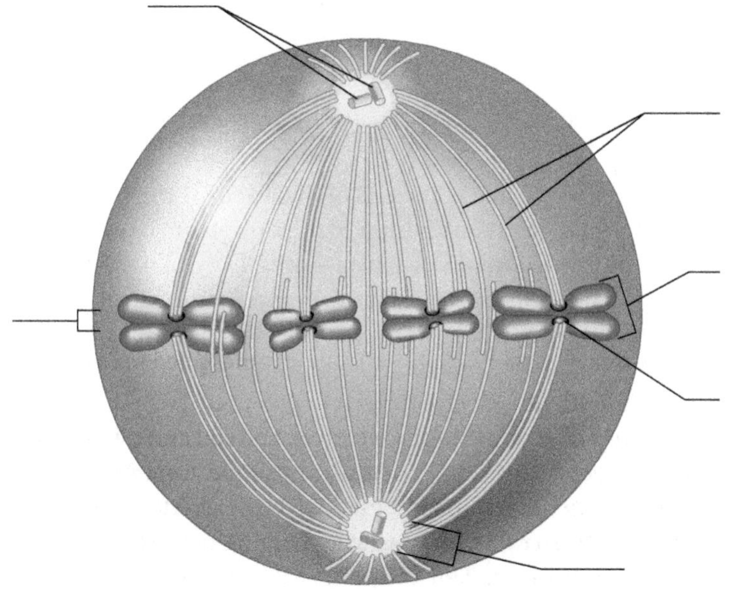

FIGURE 7.6
The mitotic spindle at metaphase.

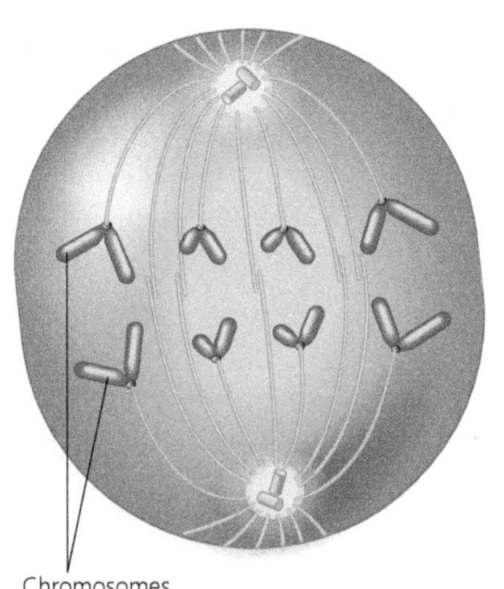

FIGURE 7.7
Anaphase mitosis.

4. Holding on to the centromeres, pull the magnetic centromeres apart and move them toward opposite poles. This action represents **anaphase**.

   After the centromeres separate, sister chromatids move apart and begin to move toward opposite poles. Chromatids are now called **chromosomes**. Anaphase ends as the chromosomes reach the poles (Figure 7.7).

   Describe the movement of the chromosome arms as you move the centromeres to the poles.

   Biologists are currently investigating the role played by spindle fibers in chromosome movement toward the poles. Check your text for a discussion of one hypothesis, and briefly summarize it here.

5. Pile your chromosomes at the poles to represent **telophase**.

   As chromosomes reach the poles, anaphase ends and telophase begins. The spindle begins to break down. Chromosomes begin to uncoil, and nucleoli reappear. A nuclear envelope forms around each new cluster of chromosomes. Telophase ends when the nuclear envelopes are complete (Figure 7.8).

   How many chromosomes are in each new nucleus?

   How many chromosomes were present in the nucleus when the process began?

   How would the condition of telomeres have changed from a previous cell cycle? (See the Introduction to this lab topic.)

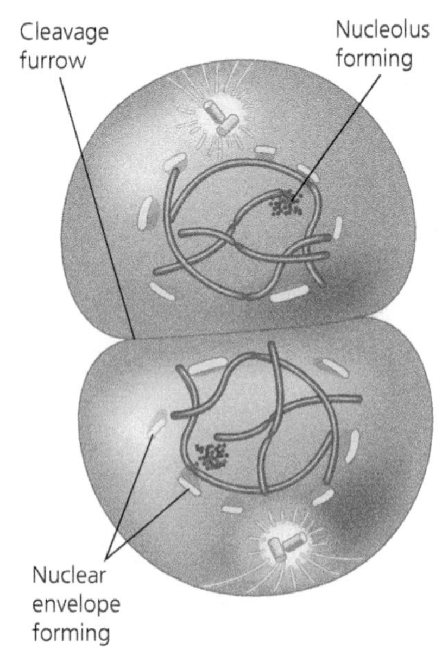

FIGURE 7.8
Telophase mitosis.

6. To represent cytokinesis, leave the two new chromosome masses at the poles.

   The end of telophase marks the end of nuclear division, or mitosis. Sometime during telophase, the division of the cytoplasm, or cytokinesis, results in the formation of two separate cells. In cytokinesis in cells of animals, fungi, and slime molds, a **cleavage furrow** forms at the equator and eventually pinches the parent cell cytoplasm in two (Figure 7.9a). Actin and myosin, the same molecules found in muscle cells, contribute to the formation of the cleavage furrow. In plant cells, membrane-bound vesicles migrate to the center of the equatorial plane and fuse to form the **cell plate**. This eventually extends across the cell, dividing the cytoplasm in two and forming the cell membrane. Cell wall materials in the vesicles are released into the space between the membranes of the cell plate forming the new cell wall (Figure 7.9b and c).

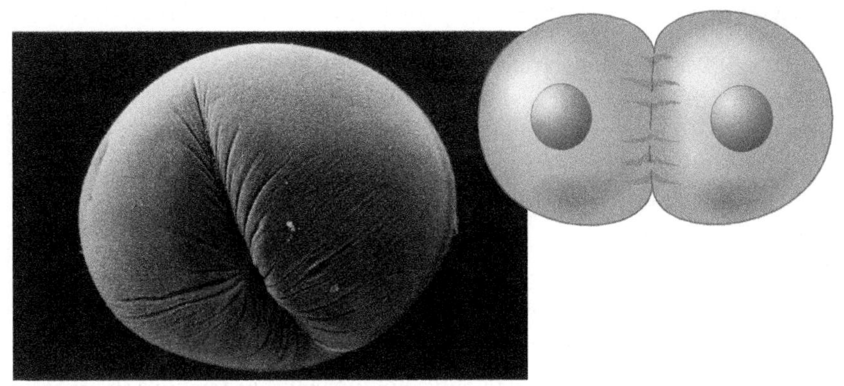

a.

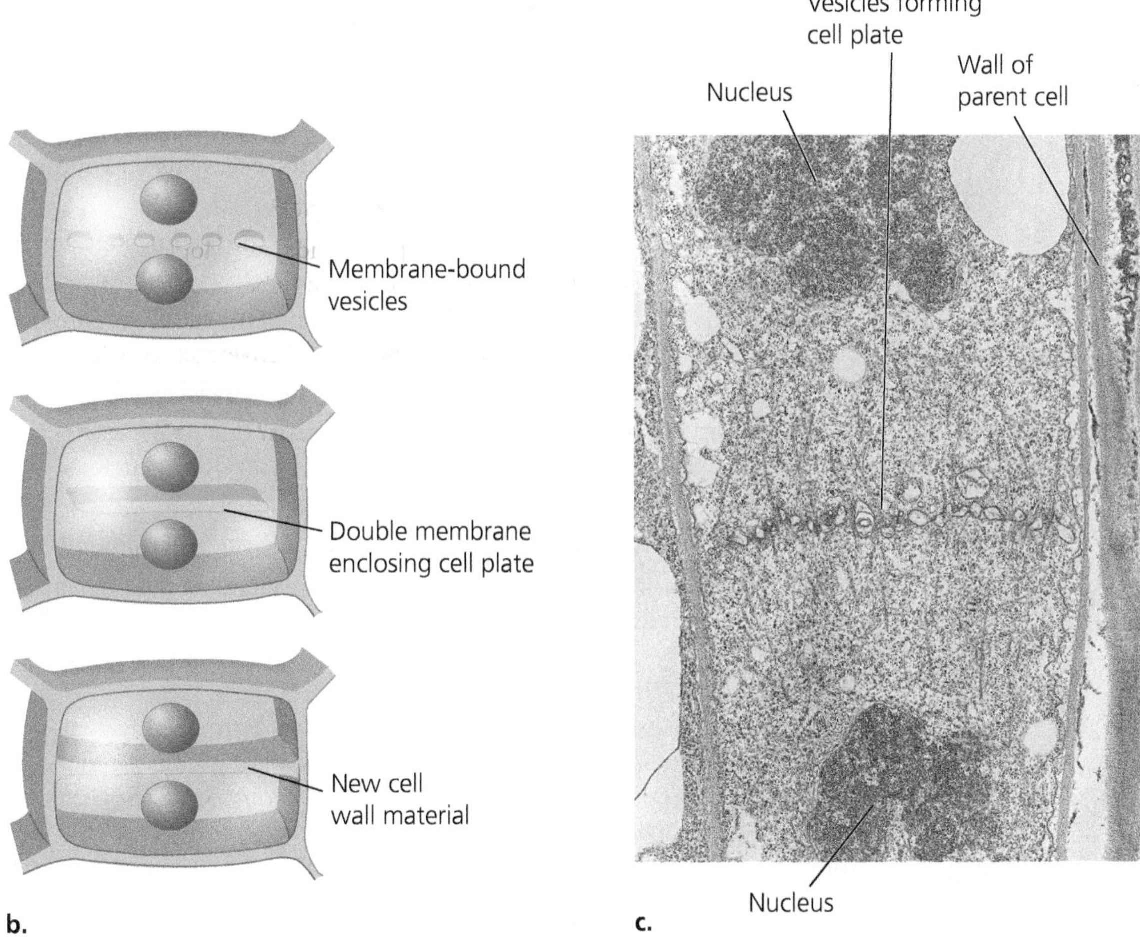

FIGURE 7.9

Cytokinesis in animal and plant cells. (a) In animal cells, a cleavage furrow forms at the equator and pinches the cytoplasm in two. (b) In plants, a cell plate forms in the center of the cell and grows until it divides the cytoplasm in two. (c) Photomicrograph of cytokinesis in a plant cell.

EXERCISE 7.2

Observing Mitosis and Cytokinesis in Plant Cells

Materials

prepared slide of onion root tip
compound microscope

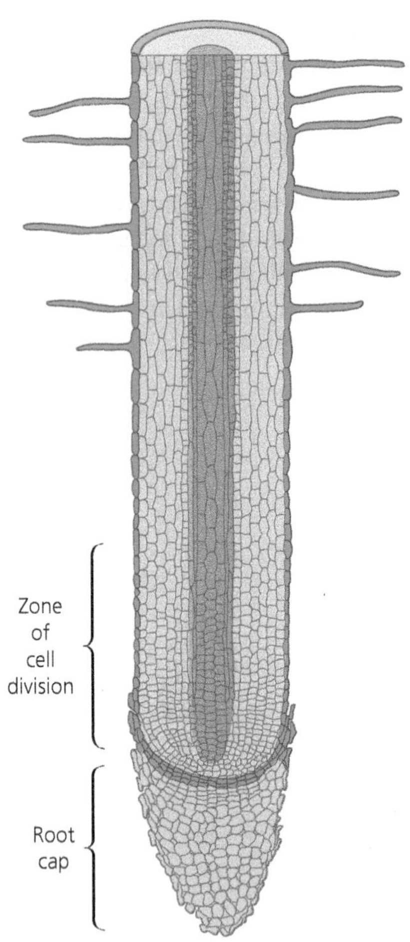

FIGURE 7.10
Longitudinal section through a root tip. Cells are dividing in the zone of cell division just behind the root cap.

Introduction

The behavior of chromosomes during the cell cycle is similar in animal and plant cells. However, differences in cell division do exist. Plant cells have no centrioles, yet they have bundles of microtubules that converge toward the poles at the ends of a spindle. Cell walls in plant cells dictate differences in cytokinesis. In this exercise, you will observe dividing cells in the zone of cell division of a root tip.

Procedure

1. Examine a prepared slide of a longitudinal section through an onion root tip using low power on the compound microscope.
2. Locate the region most likely to have dividing cells, just behind the root cap (Figure 7.10).

   At the tip of the root is a root cap that protects the tender root tip as it grows through the soil. Just behind the root cap is the **zone of cell division**. Notice that rows of cells extend upward from this zone. As cells divide in the zone of cell division, the root tip is pushed farther into the soil. Cells produced by division begin to mature, elongating and differentiating into specialized cells, such as those that conduct water and nutrients throughout the plant.
3. Focus on the zone of cell division. Then switch to the intermediate power, focus, and switch to high power.
4. Survey the zone of cell division and locate stages of the cell cycle: *interphase, prophase, prometaphase, metaphase, anaphase, telophase,* and *cytokinesis.*
5. As you find a dividing cell, speculate about its stage of division, read the following descriptions given of each stage to verify that your guess is correct, and, if necessary, confirm your conclusion with the instructor.
6. Draw the cell in the appropriate boxes provided. Label the *nucleus, nucleolus, chromosome, chromatin, mitotic spindle,* and *cell plate* when appropriate.
7. Compare the drawing of your cell with the corresponding photograph and note any differences. Expect to observe variations in cells.

Interphase (G_1, S, G_2)

Nuclear material is surrounded by a nuclear envelope. Dark-staining bodies, nucleoli, are visible. Chromosomes appear only as dark granules within the nucleus. Collectively, the chromosome mass is called *chromatin.* The chromosomes are not individually distinguishable because they are uncoiled into long, thin strands. Chromosomes are replicated during this stage (Figure 7.11).

FIGURE 7.11
Interphase. Observe the dividing cells under the microscope. Draw and label an interphase cell in the box on the right.

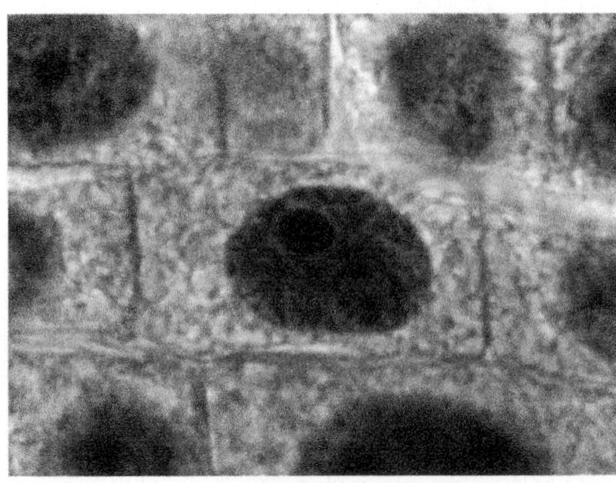

Prophase

Chromosomes begin to coil and become distinguishable thin, threadlike structures, widely dispersed in the nucleus during prophase. Although there are no centrioles in plant cells, a spindle begins to form at regions near the nuclear envelope. Nucleoli begin to disappear. The nuclear envelope is still intact (Figure 7.12).

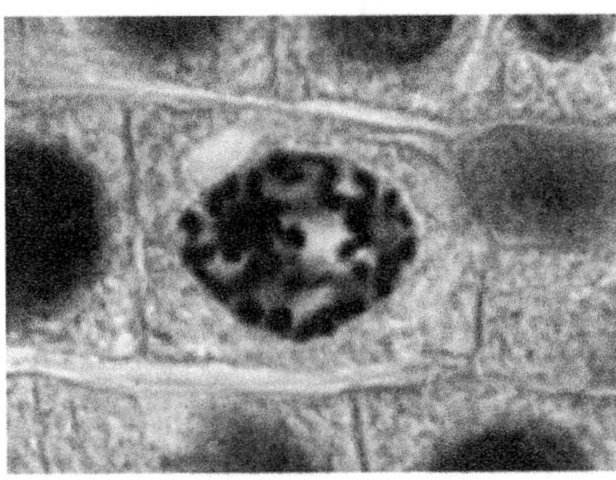

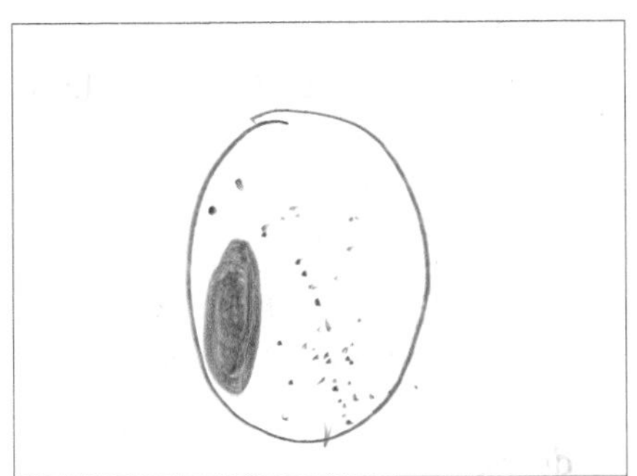

FIGURE 7.12
Prophase.

Prometaphase

By prometaphase, the chromosomes are thick and short. Each chromosome is duplicated, consisting of two chromatids held together at the centromeres. The nuclear envelope and nucleoli break down in prometaphase. Chromosomes are attached to the spindle and move toward the equator (Figure 7.13).

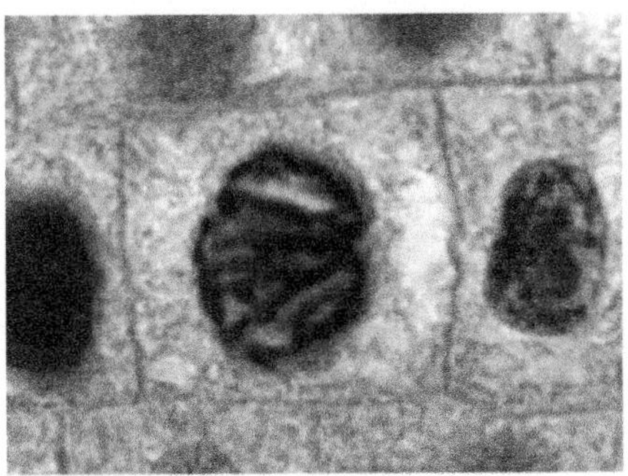

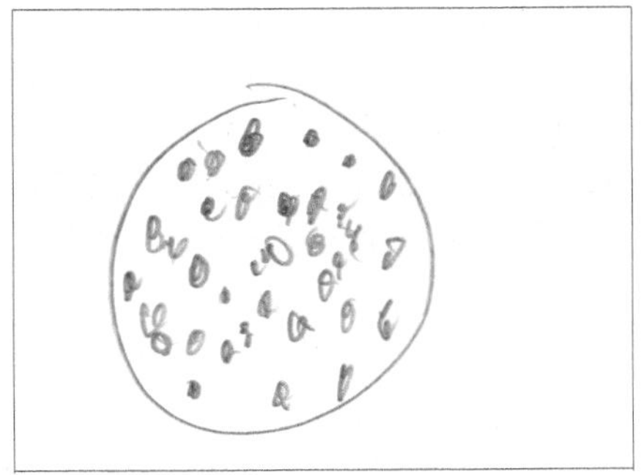

FIGURE 7.13
Prometaphase.

Metaphase

Metaphase begins when the centromeres of the chromosomes lie on the equator of the cell. The arms of the chromatids extend randomly in all directions. A spindle may be apparent. Spindle microtubules are attached to kinetochores at the centromere region and extend to the poles of the cell. As metaphase ends and anaphase begins, the centromeres separate (Figure 7.14).

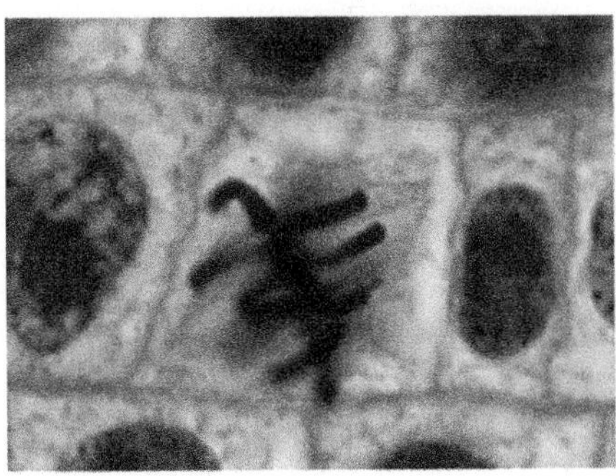

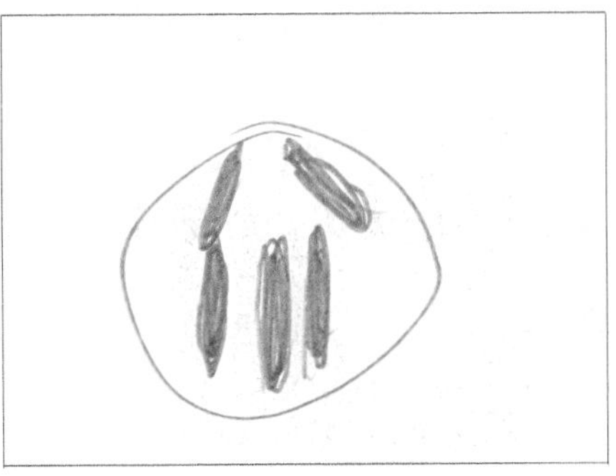

FIGURE 7.14
Metaphase.

Anaphase

The separation of centromeres marks the beginning of anaphase. Each former chromatid is now a new single chromosome. These chromosomes are drawn apart toward opposite poles of the cell. Anaphase ends when the migrating chromosomes reach their respective poles (Figure 7.15).

FIGURE 7.15
Anaphase.

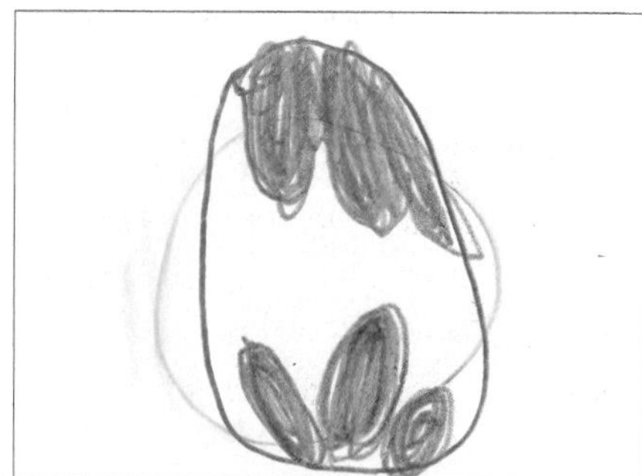

Telophase and Cytokinesis

Chromosomes have now reached the poles. The nuclear envelope re-forms around each compact mass of chromosomes. Nucleoli reappear. Chromosomes begin to uncoil and become indistinct. Cytokinesis is accomplished by the formation of a cell plate that begins in the center of the equatorial plane and grows outward to the cell wall (Figure 7.16).

FIGURE 7.16
Telophase and cytokinesis.

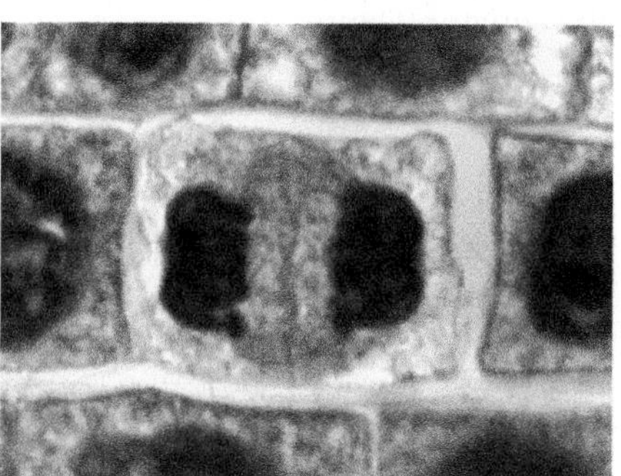

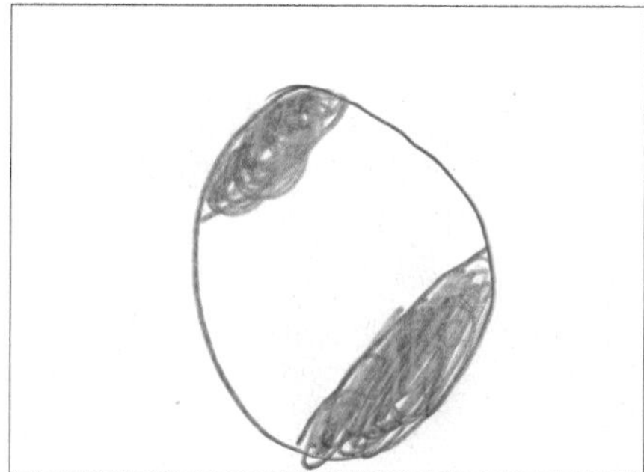

EXERCISE 7.3

Observing Chromosomes, Mitosis, and Cytokinesis in Animal Cells

In this exercise, you will look at the general shape and form of human chromosomes and observe chromosomes and the stages of mitotic division in the whitefish. You will also compare these chromosomes with the plant chromosomes studied in Exercise 7.2. Chromosome structure in animals and plants is basically the same in that both have centromeres and arms. However, plant chromosomes are generally larger than animal chromosomes.

Lab Study A. Mitosis in Whitefish Blastula Cells

Materials

prepared slide of sections of whitefish blastulas
compound microscope

Introduction

The most convenient source of actively dividing cells in animals is the early embryo, where cells are large and divide rapidly with a short interphase. In blastulas (an early embryonic stage), a large percentage of cells will be dividing at any given time. By examining cross sections of whitefish blastulas, you should be able to locate many dividing cells in various stages of mitosis and cytokinesis.

Procedure

1. Examine a prepared slide of whitefish blastula cross sections. Find one of the blastula sections on the lowest power, focus, switch to intermediate power, focus, and switch to high power.
2. As you locate a dividing cell, identify the stage of mitosis (Figure 7.17a–e). You will need to scan several blastulae on your slide to *locate all stages.*
3. Identify the following in several cells:

   nucleus, nuclear envelope, chromosomes, mitotic spindle, cleavage furrow;
   asters—an array of microtubules surrounding each centriole pair at the poles of the spindle
   centrioles—small dots seen at the poles around which the microtubules of the spindle and asters appear to radiate

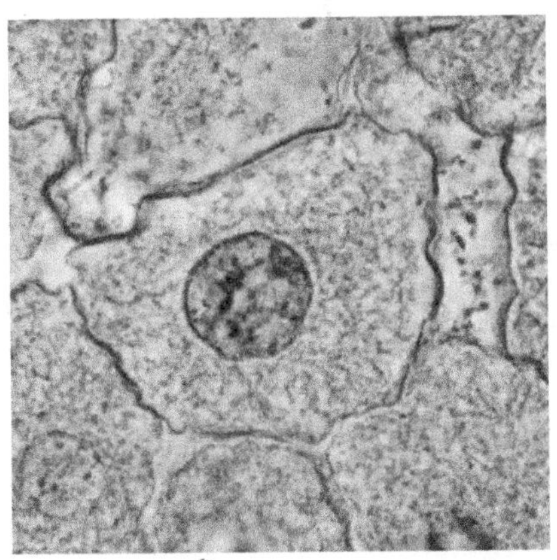

a.

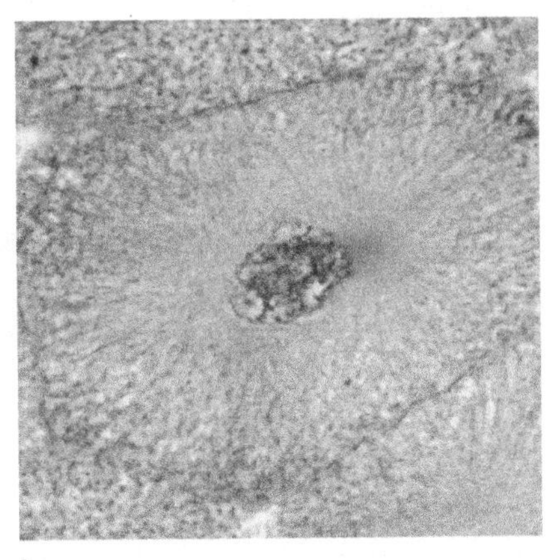

b.

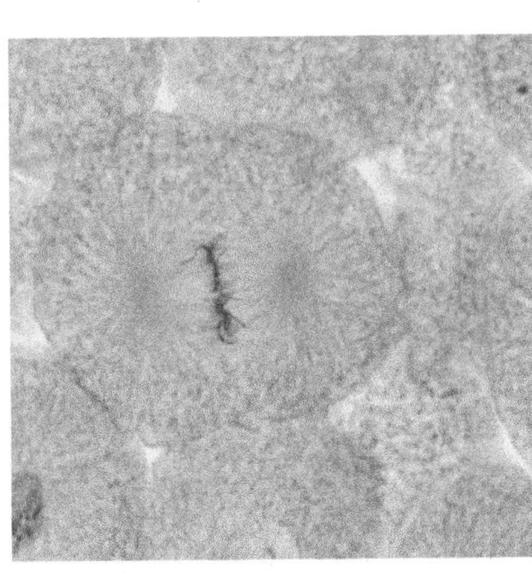

c.

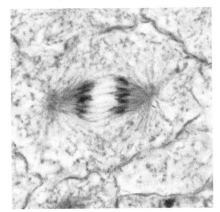

d.

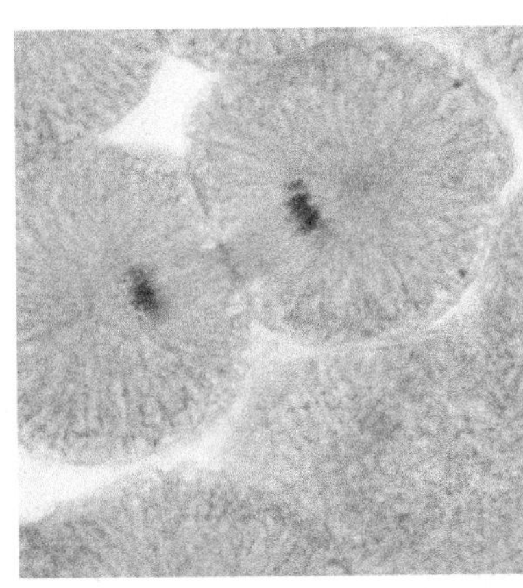

e.

FIGURE 7.17
Mitosis in animal cells, whitefish blastula. (a) Interphase. (b) Prophase. (c) Metaphase. (d) Anaphase. (e) Telophase and cytokinesis.

Results

1. List several major differences you have observed between mitosis in animal cells and mitosis in plant cells:

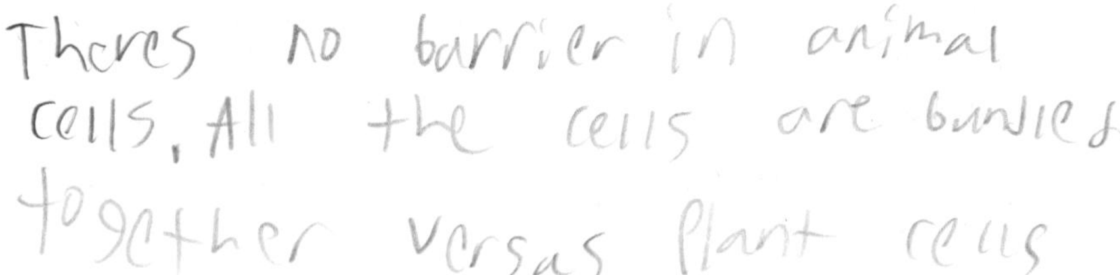

Theres no barrier in animal
cells, All the cells are bundled
together versus plant cells

2. Review your observations of telophase and cytokinesis in plants and animals. Then in the space provided draw a plant cell and an animal cell at cytokinesis. Label the important differences.

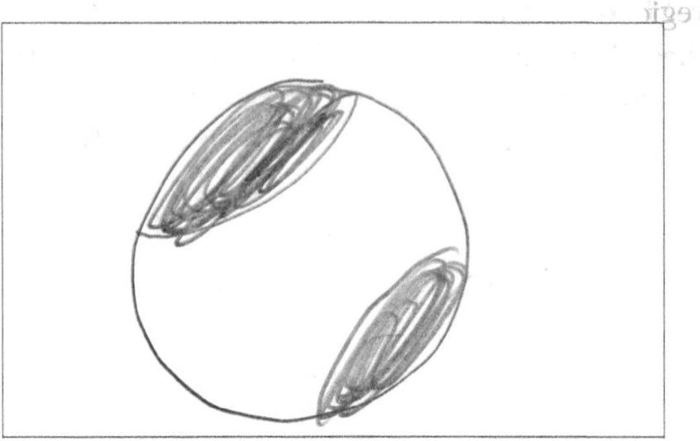

Telophase and cytokinesis in a plant cell

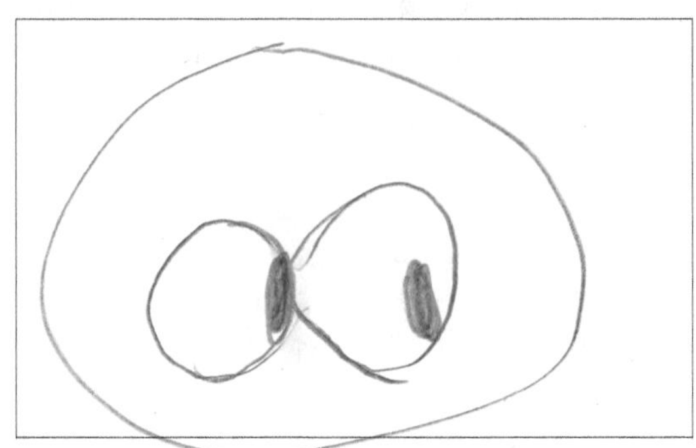

Telophase and cytokinesis in an animal cell

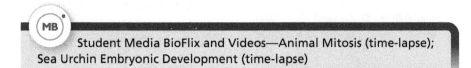

MB Student Media BioFlix and Videos—Animal Mitosis (time-lapse); Sea Urchin Embryonic Development (time-lapse)

Lab Study B. Human Chromosomes in Dividing Leukocytes

Materials

slides of human leukocytes (white blood cells) on demonstration with compound microscopes

Introduction

Cytogeneticists examining dividing cells of humans can frequently detect chromosome abnormalities that lead to intellectual and developmental disability. To examine human chromosomes, leukocytes are isolated from a small sample of the patient's blood and cultured in a medium that inhibits spindle formation during mitosis. As cells begin mitosis, chromosomes condense and become distinct, but in the absence of a spindle they cannot move to the poles in anaphase. You will observe a slide in which many cells have chromosomes condensed as in prometaphase or metaphase, but they are not aligned on a spindle equator (Figure 7.18).

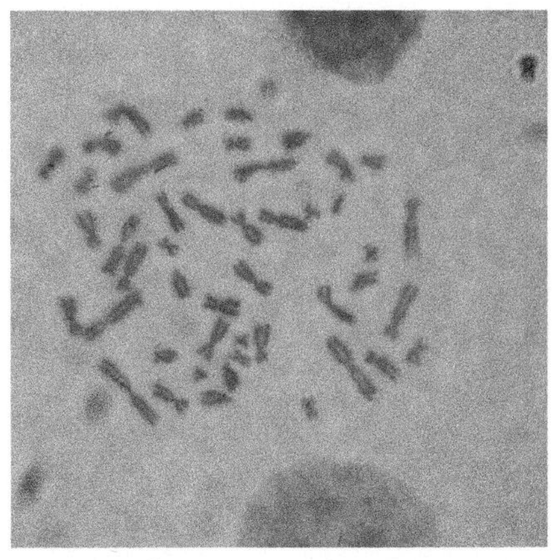

FIGURE 7.18
Human chromosomes. Can you count the number of replicated chromosomes in this photomicrograph?

Procedure

1. Attempt to count the chromosomes in one cell in the field of view. Normally, humans have 46 chromosomes. Persons with trisomy 21 (three copies of chromosome 21), or Down syndrome, have 47 chromosomes. Are the cells on this slide from a person with a normal chromosome number? 47 chromosomes

2. Notice that each chromosome is duplicated, being made up of two sister chromatids held together at the centromere region. At very high magnifications, bands can be seen on the chromosomes. Abnormalities in banding patterns can also be an indication of intellectual and developmental disability.

EXERCISE 7.4

Modeling Meiosis

Materials

60 pop beads of one color
60 pop beads of another color
8 magnetic centromeres
4 centrioles
letters *B*, *D*, *b*, and *d* printed on mailing labels

Introduction

Meiosis takes place in all organisms that reproduce sexually. In animals, meiosis occurs in special cells of the gonads; in plants, in special cells of the sporangia. Meiosis consists of *two* nuclear divisions, **meiosis I** and **II**, with an atypical interphase between the divisions during which cells do not grow and synthesis of DNA does not take place. This means that meiosis I and II result in four cells from each parent cell, each containing half the number of chromosomes, one from each homologous pair. Recall that cells with only one of each homologous pair of chromosomes are haploid (n) cells. The parent cells, with pairs of homologous chromosomes, are diploid ($2n$). The haploid cells become sperm (in males), eggs (in females), or spores (in plants). One advantage of meiosis in sexually reproducing organisms is that it prevents the chromosome number from doubling with every generation when fertilization occurs.

What would be the consequences in successive generations of offspring if the chromosome number were not reduced during meiosis?

Meiosis involves the very precise movement and sorting of chromosomes (in the case of humans, 23 homologous pairs, 46 chromosomes, or 92 chromatids).

This complicated process is not always perfect. Sometimes the homologous pairs do not separate properly or the sister chromatids fail to separate. When this happens the result is that the final nuclei may have either one too many chromosomes or one too few. As you observed in Exercise 7.3 Lab Study B, individuals with Down syndrome have 47 chromosomes as a result of an error in meiosis, in which a gamete from one of the parents had two copies of chromosome 21.

Lab Study A. Interphase

Working with another student, you will build a model of the nucleus of a cell in interphase before meiosis. Nuclear and chromosome activities are similar to those in mitosis. You and your partner should discuss activities in the nucleus and chromosomes in each stage. Go through the exercise once together, and then demonstrate the model to each other to reinforce your understanding. Compare activities in meiosis with those in mitosis as you build your model.

Procedure

1. Build the premeiotic interphase nucleus much as you did the mitotic interphase nucleus. Have two pairs of chromosomes ($2n = 4$) of distinctly different sizes and different centromere positions. Have one member of each pair of homologs be one color, and the other member a different color.
2. To represent G_1 (gap 1), pile your four chromosomes in the center of your work area. The chromosomes are decondensed.

   Cell activities in G_1 are similar to those activities in G_1 of the interphase before mitosis.

   In G_1, are chromosomes single or duplicated?

3. Duplicate the chromosomes to represent DNA duplication in the S (synthesis) phase. Each chromatid will have a centromere. What color should the sister chromatids be for each pair?

4. Duplicate the centriole pair.
5. Leave the chromosomes piled in the center of the work area to represent G_2 (gap 2).

   As in mitosis, in G_2 the cell prepares for meiosis by synthesizing proteins and enzymes necessary for nuclear division.

Lab Study B. Meiosis I

Meiosis consists of two consecutive nuclear divisions, called *meiosis I* and *meiosis II*. As the first division begins, the chromosomes coil and condense, as in mitosis. Meiosis I is radically different from mitosis, however, and the differences immediately become apparent. In your modeling, as you detect the differences, make notes in the margin of your lab manual.

Procedure

1. Meiosis I begins with the chromosomes piled in the center of your work area.

   As chromosomes begin to coil and condense, prophase I begins. Each chromosome is duplicated, made up of two sister chromatids. Two pairs of centrioles are located outside of the nucleus.

2. Separate the two centriole pairs and move them to opposite poles of the nucleus.

   The nuclear envelope breaks down and the spindle begins to form, as in mitosis.

3. Move each homologous chromosome to pair with its partner. You should have four strands together.

   Early in prophase I, each chromosome finds its homolog and pairs in a tight association. The process of homologous pairing is called **synapsis**. Initially, the synapsed homologous pairs are held together by a zipperlike protein called the **synaptonemal complex**. Because the chromosomes are duplicated, this means that each paired duplicated chromosome complex is made of four strands, sometimes referred to as a tetrad.

   How many tetrad complexes do you have in your cell, which is $2n = 4$?

4. Represent the phenomenon of **crossing over** by detaching and exchanging identical segments of any two nonsister chromatids in a tetrad.

   Crossing over takes place between nonsister chromatids in the tetrad. In this process, a segment from one chromatid will break and exchange with the exact same segment on a nonsister chromatid in the tetrad. The points of attachment where crossovers form are called **chiasmata** (singular, **chiasma**).

5. Return the exchanged segments of chromosome to their original chromosomes before performing the crossing-over activity in the next step.

   Genes (traits) are often expressed in different forms. For example, when the gene for seed color is expressed in pea plants, the seed may be green or yellow. Alternative forms of genes are called **alleles**. Green and yellow are alleles of the seed-color gene. It is significant that crossing over produces new allelic combinations among genes along a chromatid. To see how new allelic combinations are produced, proceed to step 6.

6. Using the letters printed on mailing labels, label one bead (gene locus) on each chromatid of one chromosome *B* for brown hair color. Label the beads in the same position on the two chromatids of the other member of the homologous pair *b* for blond hair color.

   The *B* and *b* represent alleles, or alternate forms of the gene for hair color.

   On the chromatids with the *B* allele, label another gene *D* for dark eye color. On the other member of the homologous pair of chromosomes, label the same gene *d* for pale eyes. In other words, one chromosome will have *BD*, the other chromosome, *bd* (Figure 7.19).

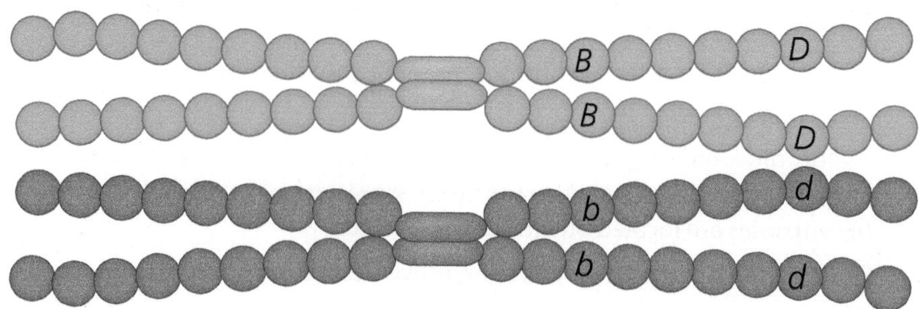

FIGURE 7.19

Arrangement of alleles *B, b, D,* and *d* on chromosome models. One duplicated homologous chromosome has *B* alleles and *D* alleles on each chromatid. The other has *b* and *d* alleles on each chromatid.

7. Have a single crossover take place involving only two of the four chromatids between the loci for hair color and eye color. Remember, the crossover must take place between nonsister chromatids.

   What combinations of alleles are now present on the chromatids?

8. Confirm your results with your laboratory instructor.

> If you are having difficulty envisioning the activities of chromosomes in prophase I and understanding their significance, discuss these events with your lab partner and, if needed, ask questions of your lab instructor before proceeding to the next stage of meiosis I.

9. Move your tetrads to the equator, midway between the two poles.

   Late in prophase I, tetrads move to the equator.

10. To represent metaphase I, leave the tetrads lying at the equator.

    During this stage, tetrads lie on the equatorial plane. *Centromeres do not separate as they do in mitosis.*

11. To represent anaphase I, separate each duplicated chromosome from its homolog, and move one homolog toward each pole.

    How does the structure of chromosomes in anaphase I differ from that in anaphase in mitosis?

12. To represent telophase I, place the chromosomes at the poles. You should have one long and one short chromosome at each pole, representing a homolog from each pair.

    Two nuclei now form, followed by cytokinesis. How many chromosomes are in each nucleus?

    Would you describe the new nuclei as being diploid ($2n$) or haploid (n)?

How has crossing over changed the combination of alleles in the new nuclei?

Are both chromosomes of the same color in the same nucleus? Compare your results with others.

13. To represent meiotic interphase, leave the chromosomes in the two piles formed at the end of meiosis I.

    The interphase between meiosis I and meiosis II is usually short. There is little cell growth and no synthesis of DNA. All the machinery for a second nuclear division is synthesized, however.

14. Duplicate the centriole pairs.

Lab Study C. Meiosis II

The events that take place in meiosis II are similar to the events of mitosis. Meiosis I results in two nuclei with half the number of chromosomes as the parent cell, but the chromosomes are duplicated (made of two chromatids), just as they are at the beginning of mitosis. The events in meiosis II must change duplicated chromosomes into single chromosomes. As meiosis II begins, two new spindles begin to form, establishing the axes for the dispersal of chromosomes to each new nucleus.

Procedure

1. To represent prophase II, separate the centrioles and set up the axes of the two new spindles. Pile the chromosomes in the center of each spindle.

   The events that take place in each of the nuclei in prophase II are similar to those of a mitosis prophase. In each new cell the centrioles move to the poles, nucleoli break down, the nuclear envelope breaks down, and a new spindle forms. The new spindle forms at a right angle to the axis of the spindle in meiosis I.

2. Align the chromosomes at the equator of their respective spindles.

   As the chromosomes reach the equator, prophase II ends and metaphase II begins.

3. Leave the chromosomes on the equator to represent metaphase II.

4. Pull the two magnets of each duplicated chromosome apart.

   As metaphase II ends, the centromeres finally separate and anaphase II begins.

5. Separate sister chromatids (now chromosomes) and move them to opposite poles.

   In anaphase II, single chromosomes move to the poles.

6. Pile the chromosomes at the poles.

As telophase II begins, chromosomes arrive at the poles. Spindles break down. Nucleoli reappear. Nuclear envelopes form around each bunch of chromosomes as the chromosomes uncoil. Cytokinesis follows meiosis II.

a. What is the total number of nuclei and cells now present?

b. How many chromosomes are in each?

c. How many cells were present when the entire process began?

d. How many chromosomes were present per cell when the entire process began?

e. How many of the cells formed by the meiotic division just modeled are genetically identical? (Assume that alternate forms of genes exist on homologs.)

f. Explain your results in terms of independent assortment and crossing over. (Refer to your textbook.)

Results

Summarize the major differences between mitosis and meiosis in Table 7.1.

TABLE 7.1 Comparing Nuclear and Chromosomal Activities in Mitosis and Meiosis

	Mitosis	Meiosis
Synapsis		
Crossing over		
When centromeres separate		
Chromosome structure and movement during anaphase		
Number of divisions		
Number of cells resulting		
Number of chromosomes in daughter cells		
Genetic similarity of daughter cells to parent cells		

EXERCISE 7.5

Meiosis in *Sordaria fimicola:* A Study of Crossing Over

Materials

petri dish containing mycelia resulting from a cross between *Sordaria* with black and tan spores
slides and coverslips
dropper bottles of water
matches
wire bacterial transfer loop
alcohol lamp

Introduction

In the study of meiosis, you demonstrated that genetic recombination may occur as a result of the exchange of genetic material between homologous chromosomes in the process of crossing over. Crossing over occurs during prophase I, when homologous chromosomes synapse. While they are joined in this complex, nonsister chromatids may break at corresponding points and exchange parts. A point at which they appear temporarily joined as a result of this exchange is called a **chiasma** (Figure 7.20).

Sordaria fimicola is a fungus that spends most of its life as a haploid **mycelium**, a mass of cells arranged in filaments. When conditions are favorable, cells of filaments from two different mating types fuse (see Figures 7.21a and b); ultimately, the nuclei fuse (Figure 7.21c), and $2n$ zygotes are produced, each inside a structure called an **ascus** (plural, **asci**) (Figure 7.21d). Asci are protected within fruiting bodies called **ascocarps**. Each $2n$ zygote undergoes meiosis, and the resulting cells (ascospores) remain aligned, the position of an ascospore within the ascus depending on the orientation of separating chromosomes on the equatorial plane of meiosis I. After meiosis, each resulting ascospore divides once by mitosis (Figure 7.21e), resulting in eight ascospores per ascus (Figure 7.21f). This unique sequence of events means that it is easy to detect the occurrence of crossing over involving chromatids carrying alleles that encode for the color of spores and mycelia.

If two mating types of *Sordaria,* one with black spores and the other with tan spores, are grown on the same petri dish (Figure 7.23), mycelia from the two may grow together, and certain cells may fuse. Nuclei from two fused cells then fuse, and the resulting zygote contains one chromosome carrying the allele for black spores and another carrying the allele for tan spores. After meiosis takes place, one mitosis follows, and the result is eight ascospores in one ascus: four black spores and four tan spores. If no crossing over has taken place, the arrangement of spores will appear as in Figure 7.22.

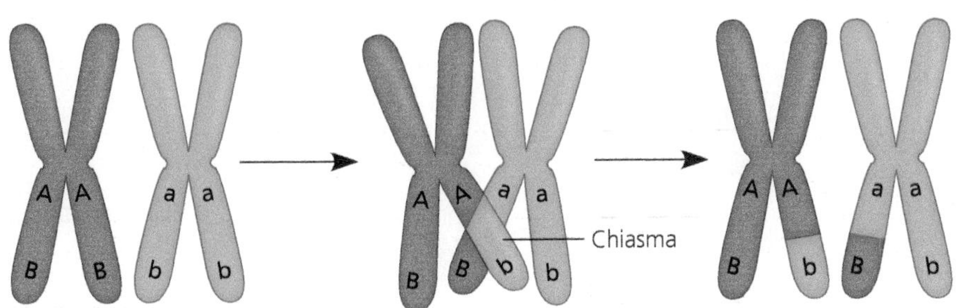

FIGURE 7.20
Crossing over. Chromatid arms break and rejoin with a nonsister member of the tetrad, forming a chiasma between nonsister chromatids. This process results in the exchange of genetic material.

FIGURE 7.21
Abbreviated diagram of the life cycle of *Sordaria fimicola*. (a) Cells from filaments of two different mating types fuse. (b) One cell with two nuclei is formed. (c) The two nuclei fuse, forming a 2*n* zygote. (d) The zygote nucleus begins meiosis, and an ascus begins to form in the ascocarp. (e) Meiosis continues, followed by mitosis. (f) The mature ascus contains eight ascospores. (g) Micrograph of crushed ascocarp with asci containing ascospores.

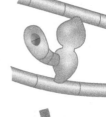

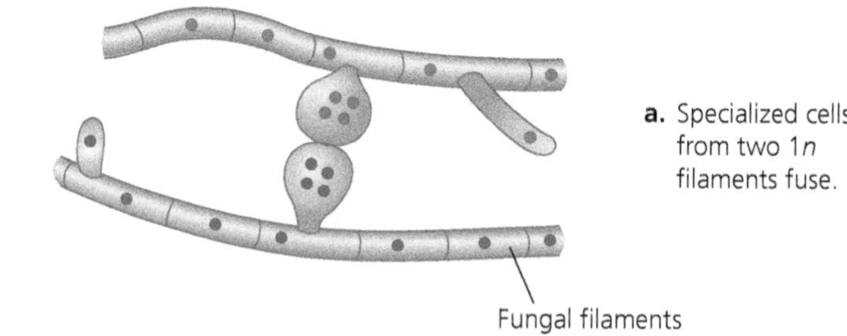

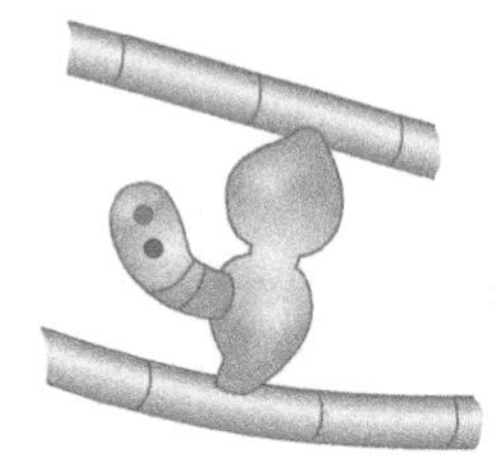

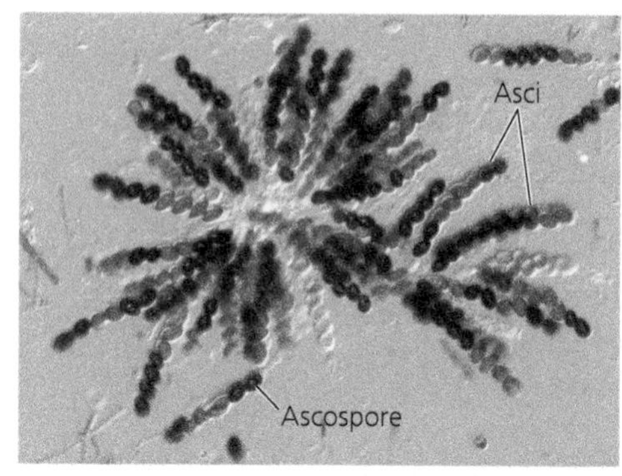

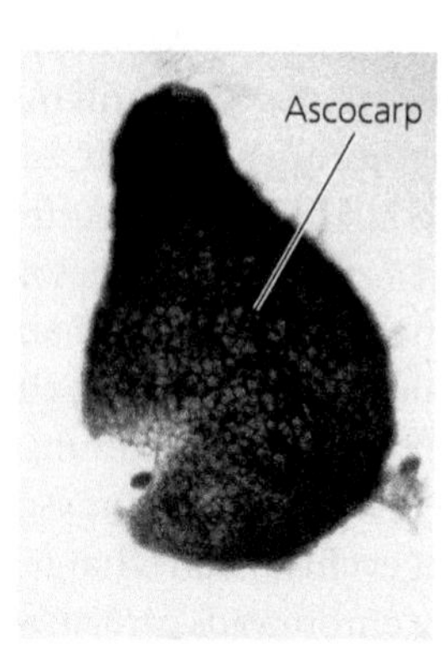

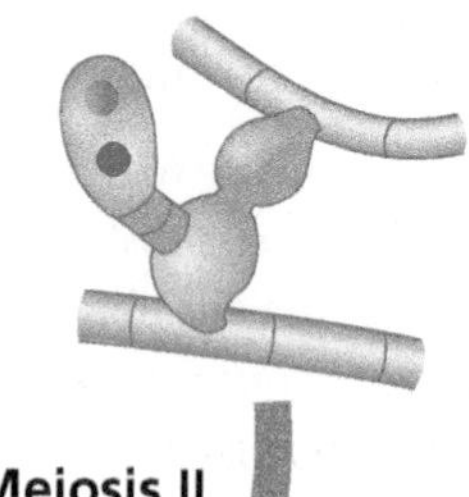

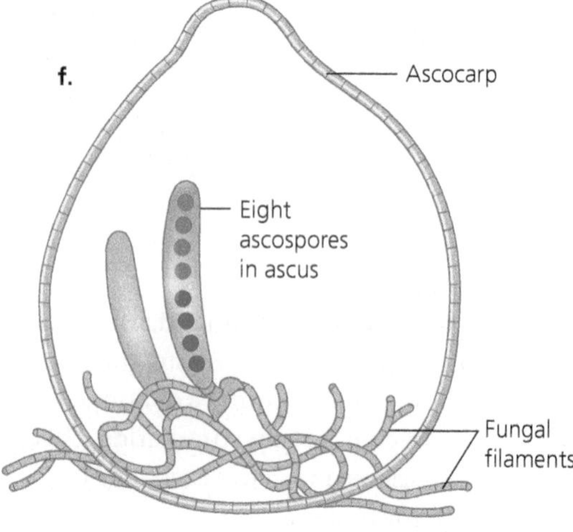

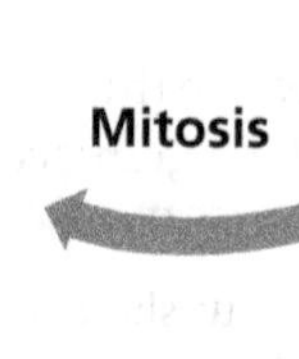

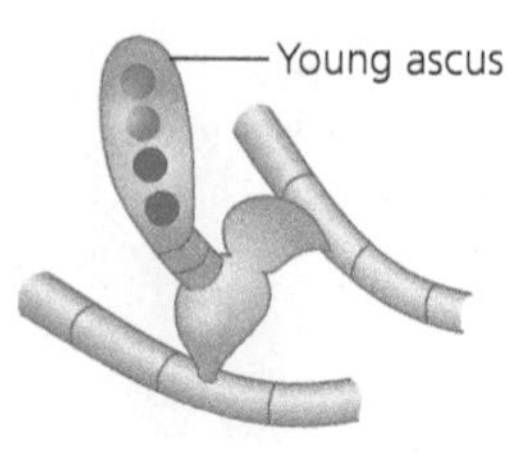

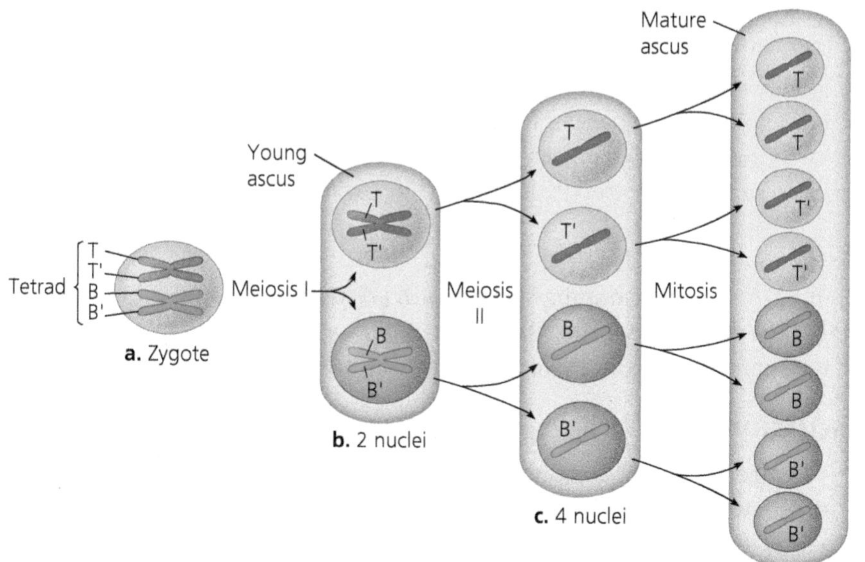

FIGURE 7.22

Arrangement of spores in asci resulting from a cross between fungi with black spores and fungi with tan spores when no crossing over takes place. (a) In the zygote nucleus, the light homologous chromosome has chromatids labeled T and T′. Each chromatid has identical tan alleles for spore color. The dark homologous chromosome (chromatids labeled B and B′) has black alleles. (b) During meiosis I, the two homologous chromosomes separate into two different nuclei retained in one developing ascus. (c) Meiosis II produces four nuclei, two containing a chromosome with the tan allele and two containing a chromosome with the black allele, still within the one ascus. (d) Now each nucleus divides by mitosis, followed by cytokinesis, resulting in eight cells, called ascospores. The ascus now contains eight ascospores. Four of the spores have the tan allele in their nuclei and appear tan. Four ascospores have the black allele and appear black.

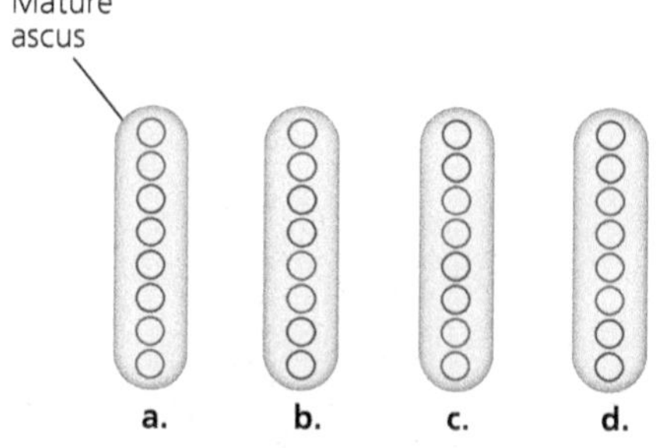

If crossing over does take place, the arrangement of spores will differ. In the spaces provided, using Figure 7.22 as a reference, draw diagrams that illustrate the *predicted* arrangement of spores in the ascus when crossing over takes place between the following chromatids and the alleles for color are exchanged: (a) T′ and B′, (b) T′ and B, (c) T and B′, and (d) T and B.

In lab today, you will observe living cultures of crosses between black and tan *Sordaria.* You will look for asci with spores arranged as in your predictions.

Procedure

1. Place a drop of water on a clean slide, and carry it and a coverslip to the demonstration table.
2. Light the alcohol lamp and flame a transfer loop.
3. Open the lid of the *Sordaria* culture slightly, and use the loop or other instrument to remove several ascocarps from the region near the edge of the dish where the two strains have grown together (Figure 7.23).
4. Place several ascocarps in the drop of water on your slide, and cover them with the coverslip.
5. Return to your work area.

FIGURE 7.23

Two mating types of *Sordaria fimicola* and hybrid zones. Most likely location for ascocarps containing asci with hybrid spores. Following the procedure, collect dark, round ascocarps from along the petri dish perimeter.

6. Using the eraser end of a pencil, tap lightly on the coverslip to break open and flatten out the ascocarps.
7. Systematically scan back and forth across the slide using the intermediate power of the compound microscope. When you locate clusters of asci, focus, switch to high power, count the asci, and determine if crossing over has taken place. Record your numbers in Table 7.2.

Results

In Table 7.2, record the numbers of asci with (a) spores all of one color (indicating that the zygote was formed by fusion of cells of the same strain), (b) black and tan spores with no crossover, and (c) black and tan spores with a crossover.

Discussion

1. What percentage of asci observed resulted from the fusion of cells from different strains?

2. What percentage of those asci resulting from the fusion of different strains demonstrates crossovers?

TABLE 7.2 Microscopic Observations of Crossing Over

Asci Types	Number of Asci in Each Category
Spores all one color	
Crossover absent	
Crossover present	

REVIEWING YOUR KNOWLEDGE

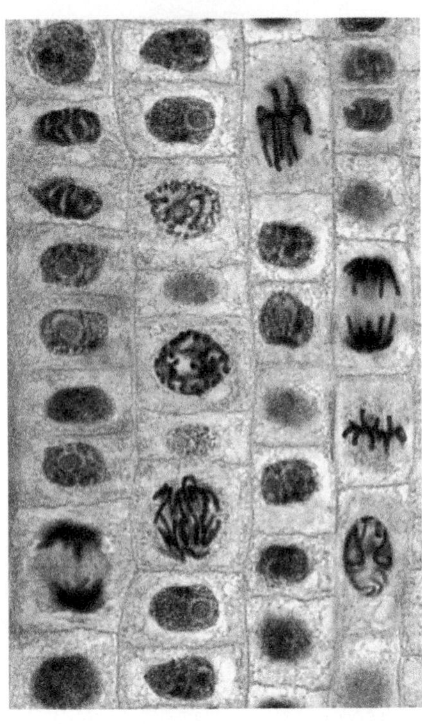

1. Define the following terms and use each in a meaningful sentence. Give examples when appropriate.
   mitosis, meiosis, cytokinesis, chromosome, chromatin, centromere, centriole, centrosome, kinetochore, spindle, aster, homologous chromosome, synaptonemal complex, synapsis, chiasma, sister chromatid, nucleolus, cell plate, cleavage furrow, diploid, haploid, crossing over, mycelium, ascocarp, ascus
2. Describe the activity of chromosomes in each stage of mitosis.

3. In the photomicrograph of dividing root cells in the margin, identify interphase and the following stages of mitosis: prophase, metaphase, anaphase, telophase, and cytokinesis.
4. Describe the activity of chromosomes in each stage of meiosis I and meiosis II.

5. Observe the drawing of several stages of meiosis below.

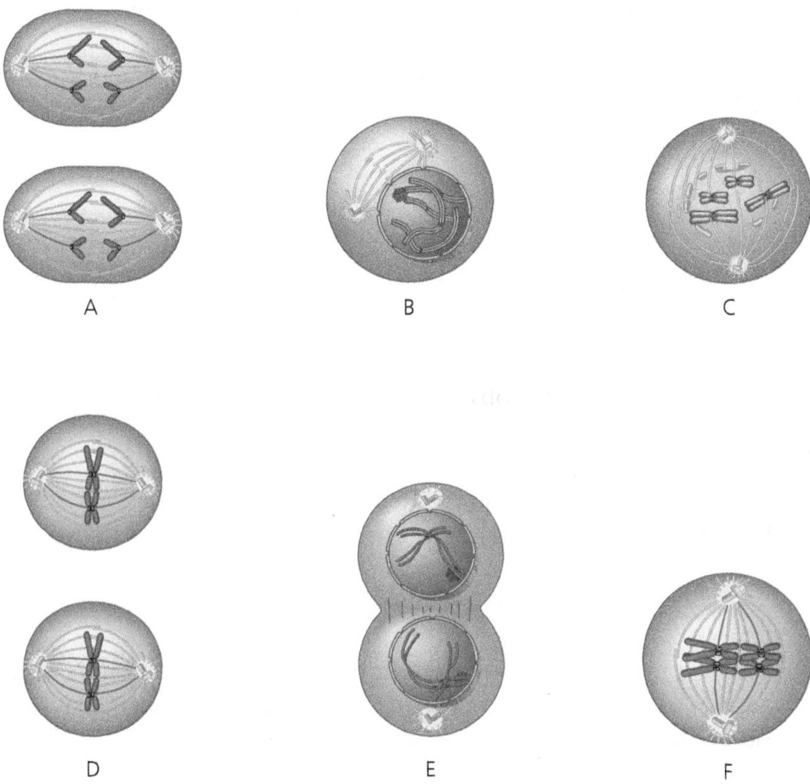

a. Using the designated letters, list the stages of meiosis in sequence.

b. Label each stage (include I or II).

c. At what stage would crossing over occur?

d. What is the diploid number for this organism?

6. Mitosis is important as organisms, both animals and plants, increase in size and grow new tissues and organs. Unlike animals, plants continue to grow throughout their lives. Where would you expect mitosis to be most common in the body of a mature plant?

7. What role would mitosis play in the body of an adult animal?

8. What advantage does the process of crossing over bring to reproduction?

9. Why would the method of cytokinesis in animal cells not work in plant cells?

APPLYING YOUR KNOWLEDGE

1. Explain why models are important to scientific study of biological systems. Provide two examples of models other than those described in the exercises.

2. You have probably heard that the liver of an adult human can "regenerate itself." How is the process of regeneration related to mitosis? Why is it possible to have a living donor for a transplanted liver?

3. Identical twins Jan and Fran were very close sisters. So, when Jan died suddenly, Fran moved in to help take care of Jan's daughter (her niece), Millie. Some time later Fran married her brother-in-law and became Millie's stepmother. When Fran announced that she was pregnant, poor Millie became confused and curious. "So," Millie asked, "who is this baby? Will she be my twin? Will she be my sister, my stepsister, my cousin?" Can you answer her questions? What is the genetic relationship between Millie and the baby? What processes are involved in the formation of gametes, and how do they affect genetic variation?

4. Two natural plant products, vinblastine from the rosy periwinkle and paclitaxel (taxol) from the Pacific yew, have been used successfully in the treatment of a wide range of cancers. These chemicals work by interfering with mitosis but by different methods. Vinblastine inhibits mitosis by preventing the assembly of the spindle. Paclitaxel promotes microtubule synthesis and binds to the microtubules, preventing the depolymerization (disassembly) of the spindle during mitosis. Based on your knowledge of mitosis, at what stage do you expect cell division to be interrupted by each

of these cancer-fighting compounds? Explain your answer based on the activities occurring at the mitotic stage.

5. The sheep "Dolly" was the first mammal cloned by transplanting a nucleus from a cell in an adult sheep into an egg of a donor sheep. Dolly appeared to grow normally for several years, but at age 6, she developed complications from several physical conditions normally seen only in older sheep. There are many hypotheses about causes for Dolly's deterioration. Propose one hypothesis about the role of telomeres in dividing cells as she aged.

INVESTIGATIVE EXTENSIONS

1. What determines when cells divide or do not divide? How do tissues detect when new cells are necessary? What factors prevent or control cell division? What is the sequence of events that affect cell division? These questions are crucial to understanding normal cell regeneration, tissue repair, growth, and development. These are also the central questions for cancer researchers. The cell control system depends on the actions of cyclin and CDK (cyclin-dependent kinase). You can explore these questions in a simulation of the cell control system at the Nobel Prize website http://www.nobelprize.org/educational/medicine/2001/cellcycle.html

   You will enter the cell nucleus and determine the events of cell division, in the correct order, to replace dying cells. In this simulation you must make the right decisions or the cell will be destroyed, and you will have to start over. For additional instructions, click on the "Help" button on the bottom of the page.

   At this site you can also see a video of a cell dividing and read about the 2001 Nobel Prize for research on control of the cell cycle.

2. Plants can be used to investigate similarities and differences among species for chromosome numbers and morphology during mitosis (root tips) and meiosis (anthers). Onions (*Allium* sp.), spiderwort (*Tradescantia* spp.), and society garlic (*Tulbaghia violacea*) are readily available cultivated species that are well suited for these studies.

Procedure for Preparing Root Tips to Observe Chromosomes in Mitosis

a. Obtain actively growing root tips. For onion use toothpicks to suspend an onion bulb on a beaker so that the root area remains in water until the roots begin to grow. *Tulbaghia violacea* has fewer ($2n = 10$) and larger chromosomes than onion. To obtain these root tips, remove the plant from the pot and remove the youngest translucent root tips. Gently wash the cut root tips to remove soil.

b. Use scissors to trim several young, translucent root tips about 3 mm long from the bulb or plant.

c. Soak the root tips in a few drops of 1 *N* HCl in a watch glass for about 10 min.

d. Add a drop of acetocarmine to one or two clean microscope slides.

e. Transfer the tips to the drop of acetocarmine and stain for about 15 min. Alternatively, you may place all of the tips in acetocarmine in a watch glass.

f. Draw off the acetocarmine using a paper towel and add a few drops of water or, if using the watch glass, transfer one tip into a few drops of water on a slide.

g. Lower a coverslip onto the tip, cover the slide and coverslip with a paper towel and press *straight* down using your thumbs or a pencil eraser to spread the tip. Be careful not to slide the coverslip.

h. Examine the tip using first intermediate power and then high power on a compound microscope. If the tip does not appear sufficiently spread out, remove the slide, add more water to the edge of the coverslip, and press again. If the cells are too pale, remove the slide and add a drop of acetocarmine to the edge of the coverslip. After a few minutes draw off the acetocarmine and add a drop of water to the edge of the coverslip.

Procedure for Preparing Anthers to Observe Chromosomes in Meiosis

a. When flowers are available, prepare anther squashes to obtain chromosomes undergoing meiosis. *Tradescantia* sp. and *Tulbaghia violacea* are easily grown in the greenhouse and give excellent results.

b. Add a drop of acetocarmine to a clean slide.

c. Use a dissecting needle to remove young anthers during the time that meiosis is taking place, before the pollen is mature. Place the anthers in the acetocarmine drop on the slide.

d. Carefully add a coverslip, draw off any excess acetocarmine, and tap the coverslip with a pencil eraser to break open the anthers.

e. Using low power on the microscope, examine the slide to locate meiotic figures. If cells in the anthers are still clumped, remove the slide and tap the coverslip again. The cells should be spread to one or two layers thick. Search on intermediate and then high powers to locate meiotic figures.

STUDENT MEDIA: BioFlix, Activities, Investigations, and Videos

www.masteringbiology.com (select Study Area)

BioFlix—Ch. 12: Mitosis

Ch. 13: Meiosis

Activities—Ch. 12: Roles of Cell Division; The Cell Cycle; Mitosis and Cytokinesis Animation; Causes of Cancer

Ch. 13: Asexual and Sexual Life Cycles; Meiosis Animation; Origins of Genetic Variation

Ch. 15: Mistakes in Meiosis

Investigations—Ch. 12: How Much Time Do Cells Spend in Each Phase of Mitosis?

Ch. 13: How Can the Frequency of Crossing Over Be Estimated?

Videos—Ch. 12: Animal Mitosis; Sea Urchin Embryonic Development

REFERENCES

Bold, H. C., C. J. Alexopoulos, and T. Delevoryas. *Morphology of Plants and Fungi.* New York: Harper & Row, 1980.

Hardin, J., G. Bertoni, and L. Kleinsmith. *Becker's World of the Cell*, 9th ed. San Francisco, CA: Pearson, 2016.

Olive, L. S. "Genetics of *Sordaria fimicola.* I. Ascospore Color Mutants." *American Journal of Botany,* 1956, vol. 43, p. 97.

Snyder, Lucy. "Pharmacology of Vinblastine, Vincristine Vindesine and Vinorelbine." *Cyberbotanica*, 2004. [online] Available at http://http://life.nthu.edu.tw/~g864204/botany.html.

van Wyck, B. and M. Wink. *Medicinal Plants of the World.* Portland: Timber Press, 2004.

WEBSITES

Karyotyping investigation and other resources for cell reproduction at the University of Arizona Biology Project:
http://www.biology.arizona.edu

Animation of meiosis and independent assortment: Google youtube.com and search for meiosis animation

"Control of the Cell Cycle," Nobelprize.org simulation of control of the cell cycle:
http://www.nobelprize.org.educational/medicine/2001/cellcycle.html

Current research on mitosis and videos of plant and animal cells dividing:
http://www.bio.unc.edu/faculty/salmon/lab/mitosis

Description of the cell cycle and mitosis, relating these processes to cancer:
http://cancerquest.emory.edu

Normal and abnormal karyotypes may be downloaded for teaching purposes from this site:
http://worms.zoology.wisc.edu/zooweb/Phelps/karyotype.html

Photos of live cancer cells dividing in real time:
http://www.cellsalive.com/cam1.htm

See this website for animations of the cell cycle, mitosis, and meiosis:
http://www.cellsalive.com

NOTES

LAB TOPIC 8

Mendelian Genetics I: Fast Plants

Laboratory Objectives

After completing this lab topic, you should be able to:

1. Describe the mode of inheritance of three traits in *Brassica rapa* Wisconsin Fast Plants™.
2. Use the Mendelian model to test for patterns of inheritance.
3. Describe and use the chi-square test to compare observed and expected results from genetic crosses.
4. Describe at least one type of non-Mendelian inheritance.
5. Define terminology used in the study of genetics.

Introduction

Gregor Mendel's study of inheritance in garden peas provided evidence that traits were inherited as "particles" rather than by blending, which was the generally accepted model of that time. Although the Mendelian model does not explain every type of inheritance, it remains the cornerstone of genetics because it provides a testable model from which a scientist can generate hypotheses with predictable results.

In this lab topic, you will investigate the inheritance of traits in a small plant, *Brassica rapa,* using the Mendelian model to develop your hypotheses. You will actually follow many of Mendel's procedures as you work with a **monohybrid** (single trait) **cross** and a **dihybrid** (two traits) **cross**. Each cross begins with true-breeding parents, each of which is **homozygous** (has two identical alleles for a given trait). If parents with different traits (for example, yellow and green pea pods) are cross-pollinated, under the Mendelian model the offspring will be **heterozygous** (receiving one allele from each parent). However, the offspring will express only one allele, which Mendel described as **dominant**; that is, only one allele needs to be present to be expressed. The masked trait is **recessive**; two alleles must be present for a recessive trait to be expressed. The parents in this cross are referred to as the **P generation**, and the hybrid offspring as the **F_1** (first filial or hybrid) **generation**. Two F_1 hybrids are then crossed to produce the next group of offspring, or **F_2 generation**. The predicted Mendelian results for the F_2 generation from a monohybrid cross would be one-fourth of the offspring with the recessive trait and three-fourths with the dominant trait (Figure 8.1). To explain these results, Mendel formulated the **law of segregation**, which states that allele pairs separate during the formation of gametes (during meiosis), with the paired condition being restored during fertilization to form the zygote.

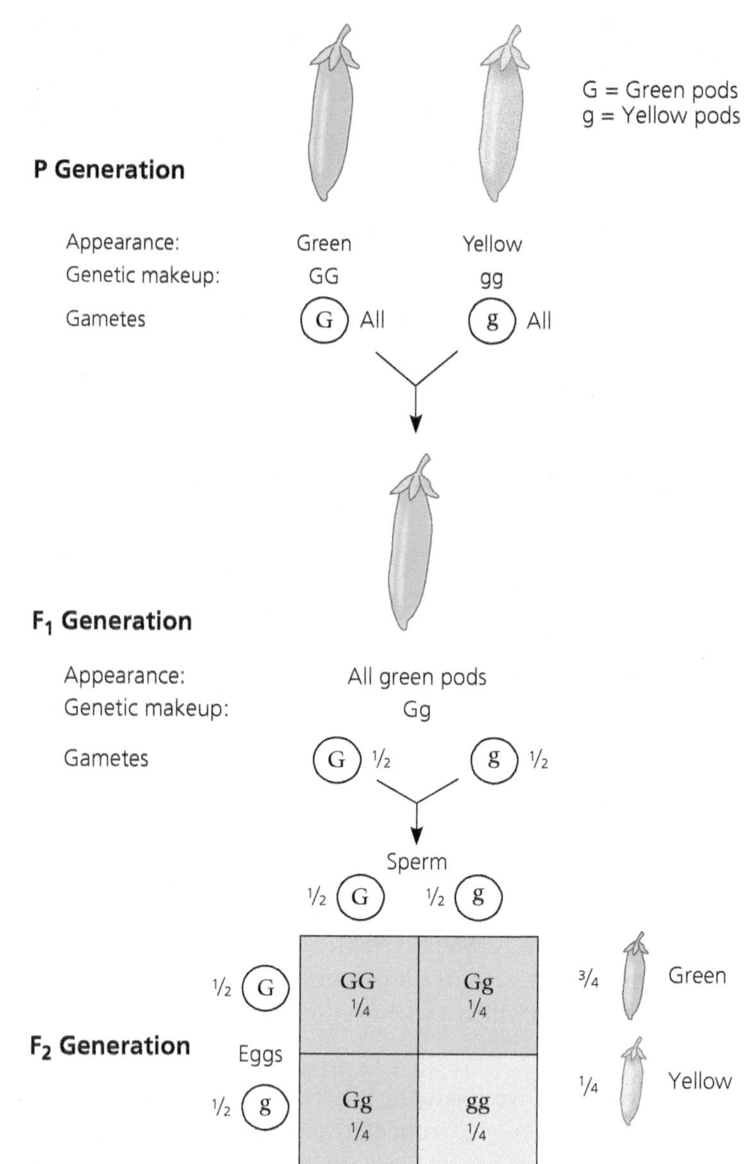

FIGURE 8.1
Mendel's law of segregation. Mendel proposed that allele pairs separate during the formation of gametes and that the paired condition is restored during fertilization. The expected results for the F_2 generation of a monohybrid cross would be 75% dominant (green pea pods) and 25% recessive (yellow pea pods).

The results of Mendel's dihybrid crosses indicated that traits did segregate and that pairs of alleles assorted themselves independently of other pairs of alleles during meiosis. This is Mendel's **law of independent assortment**, which integrates the processes studied in Lab Topic 7 Mitosis and Meiosis, with the inheritance of alleles that are carried on chromosomes (Figure 8.2).

Part of the intrigue of genetics comes from the non-Mendelian results of experiments, which require modified hypotheses and novel procedures. Imagine the excitement in Thomas Hunt Morgan's lab when crosses involving white-eyed fruit flies suggested that the inheritance of eye color might somehow be related to the sex of the fly! (This is an example of what is called **sex linkage**.) Mendel could not have investigated sex linkage in peas because peas, like most plants, do not have sex chromosomes but produce both sexes in one flower.

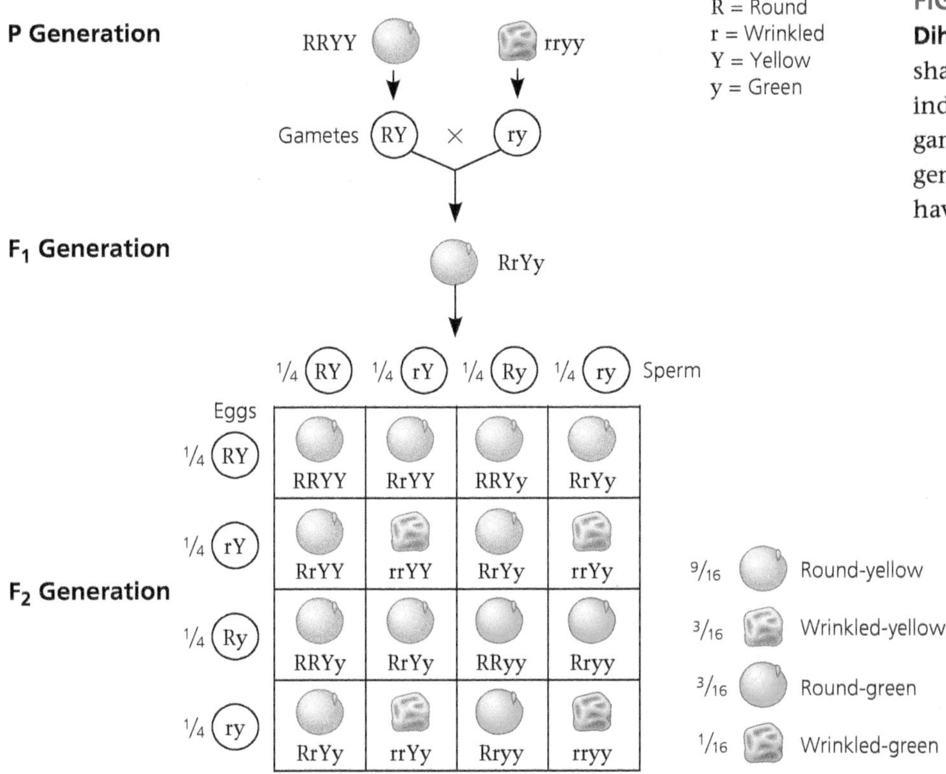

FIGURE 8.2
Dihybrid cross. If two traits (seed shape and seed color) segregate independently, then four kinds of gametes will be produced by the F_1 generation, and the offspring will have a 9:3:3:1 phenotypic ratio.

Cytoplasmic inheritance is another example of non-Mendelian inheritance. Not all genetic information is determined by nuclear genes. Both the mitochondria and chloroplasts carry genetic information, which is passed to the next generation in the cytoplasm rather than in the nucleus. These genes are inherited from the maternal parent because the cytoplasm of the zygote comes entirely from the egg. The sperm contributes only nuclear material. Each person carries the mitochondrial DNA of his or her mother, and, likewise, mitochondrial and chloroplast DNA in plants is maternally inherited.

Genetics has a unique vocabulary that you must master if you are to understand and communicate the concepts. You began developing this vocabulary when you studied mitosis and meiosis. Before you begin today's lab, review definitions of *trait, gene, allele,* and *chromosome* from Lab Topic 7; use your textbook to define any terms you do not recognize in Table 8.1 on the next page.

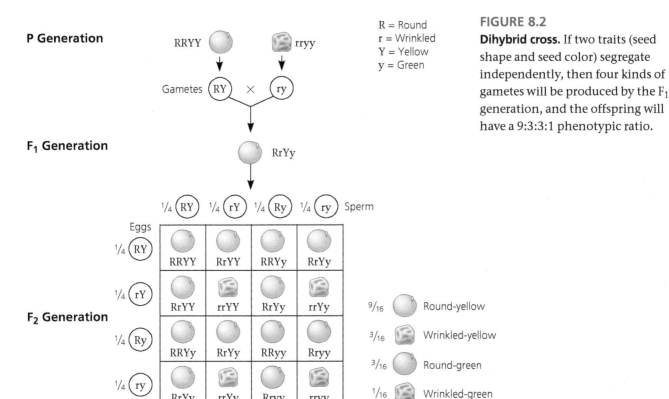

In this laboratory, you will investigate inheritance of three traits in Wisconsin Fast Plants. These plants, developed by Dr. Paul Williams and a team of scientists in the Department of Plant Pathology, University of Wisconsin, Madison, are strains of *Brassica rapa* (RCBr, rapid-cycling brassicas) that complete their entire breeding cycle from seed to seed in 35 days (Figure 8.3 and Figure 8.4). Because of their rapid breeding cycle, plants in the Brassica or mustard family are now utilized as model plants for teaching and research, much as *Drosophila,* the fruit fly, has historically been the animal of choice for research in

TABLE 8.1 Genetic Terminology

Term	Definition
Genotype	
Phenotype	
Wild-type trait	
Mutant trait	
F_1	
F_2	
Dominant trait	
Recessive trait	
Hybrid	
Homozygous	
Heterozygous	

eukaryotic genetics. The Wisconsin scientists have developed several strains of plants with strongly contrasting traits. In this lab topic, you will investigate the inheritance of three of these traits: presence or absence of anthocyanin, yielding green or purple plants; color of leaves—yellow-green or green; and presence or absence of variegated (green and white) leaves (Figure 8.3).

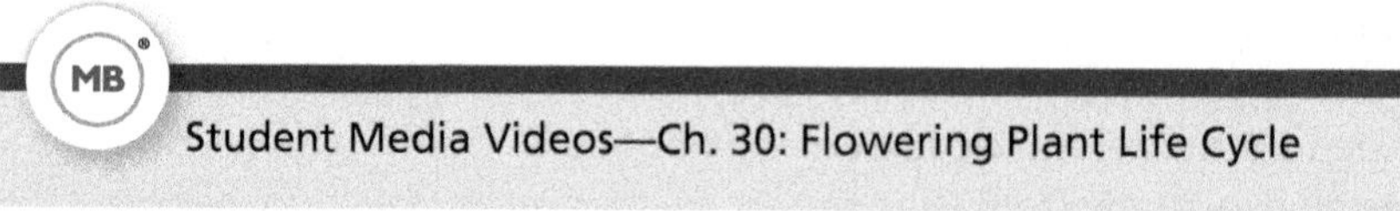

MB
Student Media Videos—Ch. 30: Flowering Plant Life Cycle

The Crucifer Genetics Cooperative for *Brassica* has developed guidelines for designating symbols to represent particular genotypes of *Brassica* that are similar to symbols used in other model systems, such as *Drosophila* (see Lab Topic 9, Mendelian Genetics II: *Drosophila*). Students who wish to perform extended

FIGURE 8.3
***Brassica rapa* seedlings.** Yellow-green seedlings in the left half of the quad and wild type with anthocyanin in the right half.

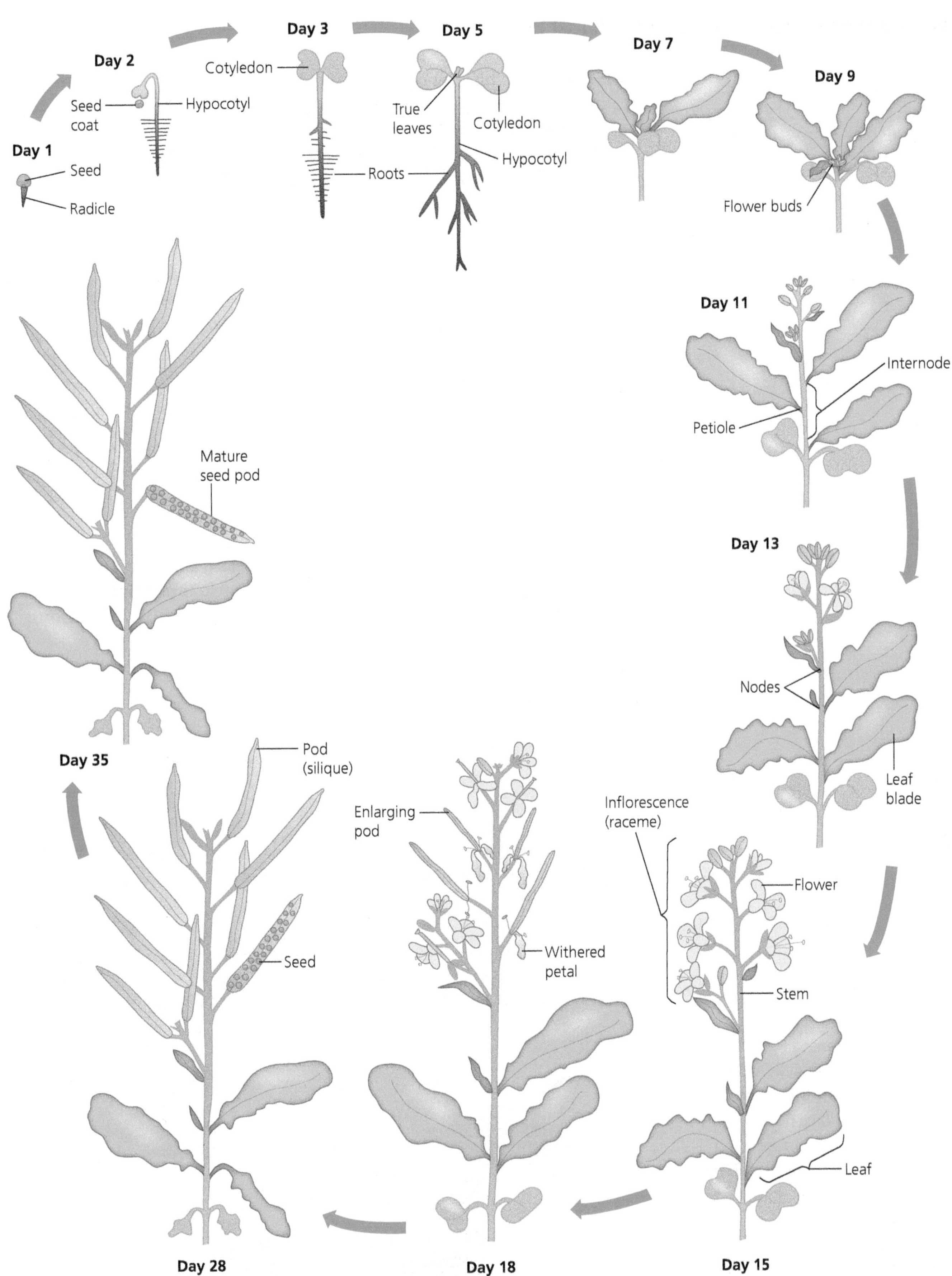

FIGURE 8.4

Life cycle of *Brassica rapa*, Fast Plant. The entire life cycle of this Fast Plant is completed from seed to seed in 35 days. Note the stages of development and the day in the life cycle.

investigations using *Brassica* should refer to the Wisconsin Fast Plants website for guidelines for allelic and genotypic symbols. However, for the purposes of this lab, you should use the common symbols of Mendelian inheritance, in which the dominant allele is designated by the capitalized first letter of the trait name, and the recessive allele by the lowercase letter.

Materials for All Exercises

- seeds of designated phenotypes
- seed-collecting pan
- small envelopes
- wicks
- label tape
- Styrofoam "quads" used to germinate seeds (see Figure 8.5)
- fluorescent light bank (see Figure 8.6)
- water reservoir for petri dish (plastic dish of appropriate size or cut-off base of 2-L bottle)
- water and dropper
- potting mix
- fertilizer pellets
- watering tray
- 20% $CuSO_4$ or anti-algae squares
- petri dish with filter paper

Overview of Exercises

The crosses for each exercise are briefly outlined below, along with the general procedures that will be used in all exercises. These include planting seeds, pollinating flowers, and harvesting and germinating seeds. Specific instructions are provided within each exercise.

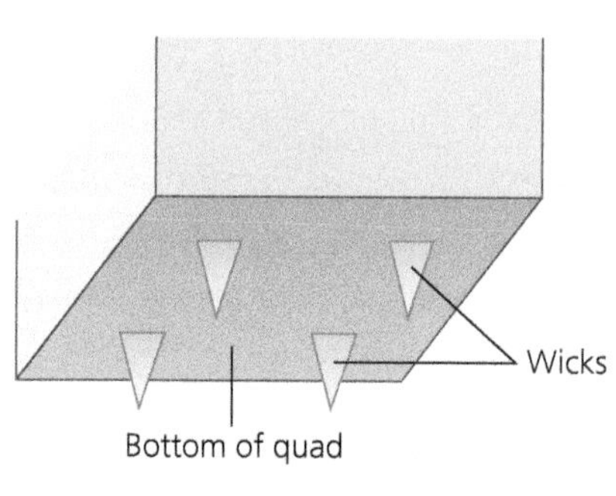

a. Quad and wicks

Potting soil in a cell
Seeds
Fertilizer pellets
Quad
Wick

b. Planting

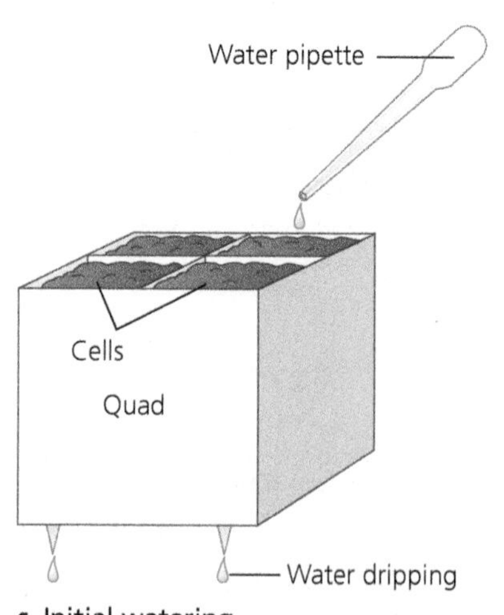

c. Initial watering

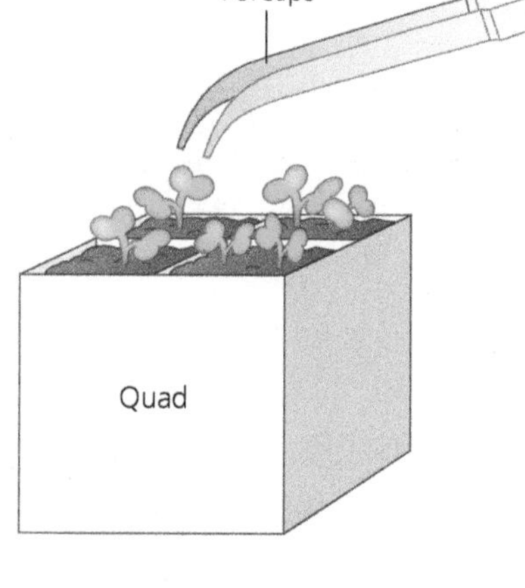

d. Thinning

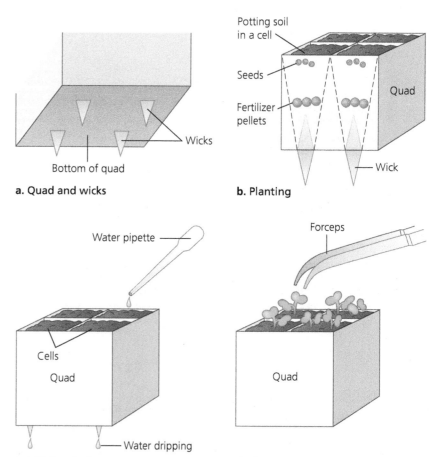

FIGURE 8.5
Planting and thinning Fast Plants.
(a) Place absorbent wicks in the quad, with the tips extending from holes in the bottom. (b) Place potting soil, fertilizer pellets, and more soil in each cell. Make a small depression, and plant two or three seeds in each cell; then cover with soil. (c) Water seeds with a dropper until water drips from the wicks. Place on watering tray. (d) After recording seedling phenotypes, remove all but one large healthy seedling in each cell.

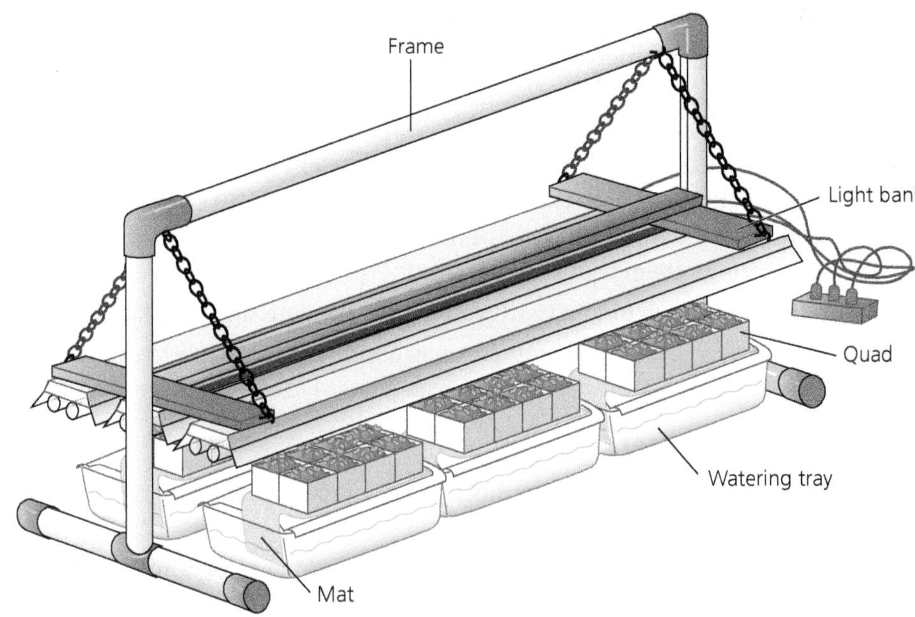

FIGURE 8.6

Growing system for Fast Plants. Fast Plants require high light to complete their life cycle in 35 days. The light bank should be adjusted to remain 2–3 inches above plants. The light should be on 24 hours a day.

Exercises 8.1 and 8.2

You will begin with F_1 hybrid seeds resulting from a cross previously made between homozygous wild-type plants and homozygous mutant plants. You will germinate these F_1 seeds, cross-pollinate the flowers, allow the flowers to mature and produce F_2 seeds, collect the mature F_2 seeds, germinate the F_2 seeds, and record phenotypic ratios in this generation.

Exercise 8.3

You will grow green plants and variegated plants, cross these parent plants, and then grow the offspring from this cross and record the phenotypes.

You will begin planning for this lab and germinating all seeds (unless otherwise noted) *6 to 8 weeks* before the laboratory period designated for making final observations. Before coming to lab on this preliminary date, plan your schedule of activities carefully. Notice that several activities will take place outside of scheduled lab time. You are personally responsible for the success of this laboratory, and you will be responsible for maintaining plants and carrying out activities.

General Procedures for All Exercises

Complete and detailed procedures are provided with Exercises 8.1–8.3. The following procedures for planting seeds, pollinating flowers, and harvesting and germinating seeds of the next generation are common to all exercises.

Planting Seeds

Plant the appropriate seeds in quads (Figure 8.5).

1. Add one wick to each cell in a quad; wicks should extend from the base.
2. Add potting soil to each cell until half full.
3. Add three fertilizer pellets to each cell.
4. Add more soil and press to make a depression.

5. Add two or three seeds to each cell and barely cover with potting mix.
6. Using a dropper, water each cell until water drips from the wick.
7. Place quad on watering tray under fluorescent light bank (Figure 8.6). The lights should be maintained 2–3 inches above the growing plants and should be on 24 hours a day. Be sure that the wicks in the quad make good contact with the mat on the watering tray. Add 4–5 drops of 20% $CuSO_4$ or 1 anti-algae square to the water in the tray to prevent algal growth. Check the watering system regularly throughout the experiment.
8. Use tape to label your quad with your name, the date, and the plant type(s).
9. After 4 or 5 days, record phenotypes of plants in the appropriate table and thin plants, leaving only the most vigorous single plant in each cell (Figure 8.5d).

Pollinating Flowers

After two or three flowers open on most plants (approximately day 14), pollinate as follows (Figure 8.7):

1. Using a "bee-stick" (Williams, 1980), made by gluing a dry honeybee thorax to the top of a toothpick, transfer pollen from one plant to another. (You can also use a small, soft paintbrush.)
2. Save the stick by inserting it into one cell of the quad. Use it to pollinate again 2 and 4 days later.
3. After the third pollination, pinch off all unopened buds.
4. Remove and discard all new buds and shoots for the next 2 weeks.

Harvesting Seeds

Seeds are ready to harvest approximately 21 days after pollination.

1. Remove quad with plants from the watering tray and dry for 5 days.
2. Remove dry seed pods and roll them between your hands over a collecting pan to free the seeds from the pod.
3. Store seeds in an envelope labeled with your name, the date, and the seed type.

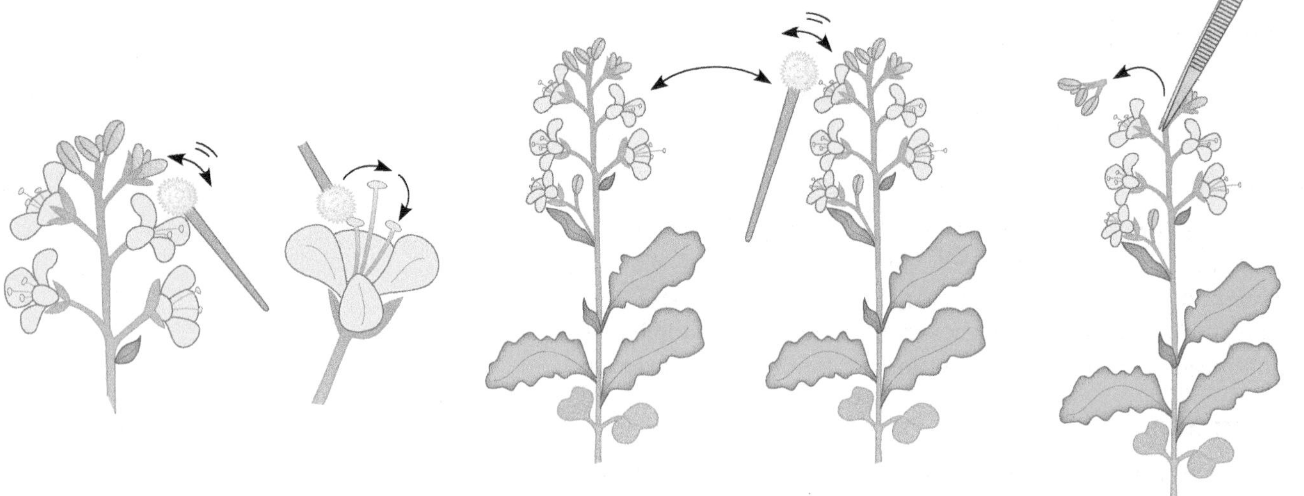

a. Pollen transfer to bee-stick   **b.** Cross-pollination   **c.** Removal of unopened buds

FIGURE 8.7
Pollination of Fast Plants. (a, b) Using a bee-stick, transfer pollen from one plant to another. Plants are self-incompatible and must be cross-pollinated. (c) After pollinating on 3 days, remove all remaining flowers and buds.

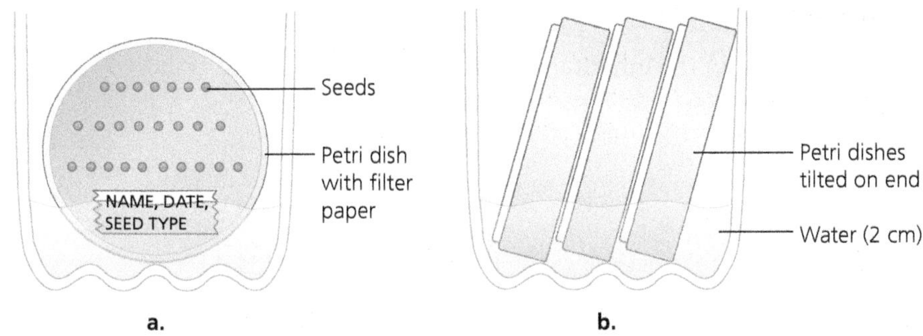

FIGURE 8.8
Germination of seeds in petri dishes.
In Exercise 8.1, seeds are germinated on moist filter paper in petri dishes.

Germinating Seeds

Seeds in Exercises 8.2 and 8.3 will be germinated in quads by planting as described above for initial seeds (pp. 197–198). Seedlings can be scored as soon as the first leaves are produced, and the phenotype of interest can be observed.

Seeds of offspring in Exercise 8.1 will be germinated in petri dishes as described below (Figure 8.8).

1. Moisten a piece of filter paper in a petri dish, labeling the paper with your name, the date, and the seed type. Pour off excess water.
2. Place 25 of the harvested seeds in neat rows in the upper two-thirds of the filter paper.
3. Place the petri dish tilted on end in a water reservoir. Add about 2 cm water.
4. Place the dish and reservoir under the light bank.
5. In approximately 48–96 hours, observe seedlings and record phenotypes in the appropriate table.
6. Tabulate results, and perform a chi-square test on class results.

EXERCISE 8.1
Inheritance of Anthocyanin Gene

Materials

12 F_1 seeds resulting from a cross previously made between a homozygous plant that has anthocyanin present and a true-breeding plant that has no anthocyanin
1 quad and related planting supplies (refer to Materials for All Exercises section)
2 mature plants with parental phenotypes

Introduction

Wild-type *Brassica rapa* has a red-purple pigment present, called *anthocyanin.* Plants with even a faint red or purple coloration in the stems or leaf stalks are considered to show the phenotype for anthocyanin. Mutant plants have no anthocyanin and are completely green. In this exercise, you will investigate the inheritance of this trait. Formulate a hypothesis for this trait that describes a mode of inheritance based on the Mendelian model. Will these traits conform to Mendel's laws? Is one trait dominant or recessive?

Hypothesis

Hypothesize about the inheritance of these traits.

Prediction

Predict the results based on your hypothesis (if/then). Outline the cross(es) for your predictions in the margin of your lab manual, and provide expected results for each in the form of a Punnett square if appropriate. Refer to Figure 8.1 if needed.

Procedure

Refer to the General Procedures section (pp. 197–199) and appropriate figures. All days are approximate. Monitor plants for initiation of flowering and seed set.

1. Refer to the checklist of activities in Table 8.2, and record dates, procedures, and any observations as you follow the procedure for this exercise. Initial the final column when you have completed each step.
2. *Day 1.* Plant F_1 hybrid seeds resulting from crosses previously made between homozygous purple plants and homozygous green plants. See Planting Seeds, (pp. 197–198).

TABLE 8.2 Checklist of Activities for Exercise 8.1. Record the date and initial the final column indicating that you have completed each step.

Approximate Day	Date	Activity	Initials
Day 1		Plant F_1 hybrid seeds. See Planting Seeds, (pp. 197–198). Regularly check water.	
Day 4 or 5		Observe seedlings and record numbers of each phenotype in Table 8.3. Thin plants to one per cell.	
Days 14, 16, 18		Pollinate on 3 days; pollinate at least 6–8 flowers. See Pollinating Flowers, (p. 198).	
Days 20 to 39		Remove buds and shoots.	
Day 39		Harvest F_2 seeds and germinate in petri dishes. See Harvesting Seeds, (p. 198), and Germinating Seeds, (p. 199).	
Day 42		Count and record numbers of each phenotype in Table 8.4.	

3. *Day 4 or 5.* Observe germinating seedlings and compare phenotypes with parental plants on demonstration. Record numbers of each phenotype in Table 8.3 in the Results section and on the class data sheet. Thin plants to one per cell. Remember to keep plants watered.
4. *Days 14, 16, 18.* Cross the F_1 plants. Pollinate plants three times, with 1 day between pollinations. Pollinate at least six to eight flowers. See Pollinating Flowers, (p. 198).
5. *Days 20 to 39.* Remove new buds and shoots. Keep plants watered.
6. *Day 39.* (It should be a minimum of 21 days after the last pollination.) Harvest the F_2 seeds and germinate them in the petri dish as directed in the General Procedures section, Harvesting Seeds, (p. 198), and Germinating Seeds, (p. 199).
7. *Day 42 (or next lab period).* Count the numbers of each phenotype, and record these data in Table 8.4 in the Results section.

You and your lab partner are responsible for the care and maintenance of your plants. Remember to check the reservoir for water, and be sure that the wicks make good contact with the watering tray, ensuring moisture flow to quads. A scientist has to remember the organism!

Results

1. Record the phenotypes of the F_1 seeds in Table 8.3 and the phenotypes of the F_2 offspring in Table 8.4. Compile results from all teams.
2. Note any changes in procedure or problems with the experiment in the margin of your lab manual or on the schedule of activities (Table 8.2).

TABLE 8.3 Phenotypes of F_1 *B. rapa* Seedlings from Crosses Between Plants with Anthocyanin (Purple) and Without Anthocyanin (Green)

Download an Excel version of this table from www.masteringbiology.com in the Study Area under Lab Media.

Number of Plants					
Team No.	Purple	Green	Team No.	Purple	Green
1			7		
2			8		
3			9		
4			10		
5			11		
6			12		
Class total	Purple:		Green:		

TABLE 8.4 Phenotypes of F_2 Seedlings from Crosses Between F_1 *B. rapa* Plants.

Download an Excel version of this table from www.masteringbiology.com in the Study Area under Lab Media.

Number of Plants					
Team No.	**Purple**	**Green**	**Team No.**	**Purple**	**Green**
1			7		
2			8		
3			9		
4			10		
5			11		
6			12		
Class total	**Purple:**		**Green:**		

3. Compare your observed and expected results by calculating the chi-square test in Table 8.5. Refer to your hypothesis and predictions and use the class data from Table 8.4. Refer to Appendix D for an explanation of the chi-square test.

Discussion

1. What do the results suggest is the inheritance pattern of this trait? Which is the dominant trait, anthocyanin present (purple) or anthocyanin absent (green)? Assign symbols and write genotypes and phenotypes of all F_1 and F_2 offspring.

TABLE 8.5 Chi-Square Calculations

Download an Excel version of this table from www.masteringbiology.com in the Study Area under Lab Media.

	Anthocyanin Present (Purple)	**Anthocyanin Absent (Green)**
Observed value (*o*)		
Expected value (*e*)		
Deviation ($o - e$) or *d*		
Deviation² (d^2)		
d^2/e		
Chi-square, $\chi^2 = \Sigma d^2/e$		

What evidence supports your conclusion?

2. Did the results support your hypothesis? Explain, describing the results of the chi-square test.

3. Explain the significance of the p value (see Appendix D).

4. If the results do not support expected Mendelian ratios, what are other possible explanations for these results? What other patterns of inheritance could explain the results?

5. What additional experiments can you suggest to further test your original or modified hypothesis? Outline your crosses.

EXERCISE 8.2

Inheritance of Plant Color: Green, Yellow-Green, and Purple (Anthocyanin Present)

Materials

12 F_1 seeds resulting from a cross previously made between a true-breeding homozygous normal bright green parent (no purple) and a true-breeding homozygous yellow-green with purple parent
1 quad and related planting supplies (see Materials for All Exercises section)
2 mature plants with the paternal phenotypes

Introduction

The wild-type plant of *Brassica rapa* is green with purple; anthocyanin is present in the stems and/or leaf stalks. However, variants exist that are bright green (no purple), yellow-green with purple, and yellow-green (no purple) (Figure 8.3). In this exercise, you will investigate the inheritance of these phenotypes. You will determine if one or two genes are involved. You will determine if the inheritance pattern suggests simple dominance and recessiveness, incomplete dominance or codominance, or other modes of inheritance. Formulate a hypothesis for these traits that describes a mode of inheritance based on the Mendelian model. Will these traits conform to Mendel's laws? Is one trait dominant or recessive?

Hypothesis

Hypothesize about the inheritance of these phenotypes.

Prediction

Predict the results of the experiment based on your hypothesis. Outline the cross(es) for your predictions in the margin of your lab manual. Provide expected offspring for each cross in a Punnett square if appropriate. Refer to Figures 8.1 and 8.2 if necessary.

Procedure

Refer to the General Procedures section (pp. 197–199) and appropriate figures. All days are approximate. Monitor plants for initiation of flowering and seed set.

1. Refer to the checklist of activities in Table 8.6, and record dates, procedures, and your observations as you follow the procedure for this exercise. Initial the final column when you have completed each step.
2. *Day 1.* Plant F_1 hybrid seeds resulting from previously made crosses between homozygous bright green (no purple) and homozygous yellow-green (with purple) plants. See Planting Seeds, pp. 197–198.
3. *Day 4 or 5.* Observe germinating seedlings and compare phenotypes with the adult parental plants on demonstration (Figure 8.3). Record the numbers of each phenotype in Table 8.7 and on the class data sheet. Modify the table to accommodate your results. Write in the headings for the phenotypes and draw in columns. Thin plants to one per cell. Remember to keep plants watered.
4. *Days 14, 16, 18.* Cross the F_1 plants. Pollinate plants three times, with 1 day between pollinations. Pollinate at least six to eight flowers. See Pollinating Flowers, p. 198.
5. *Days 20 to 39.* Remove and discard new buds and shoots. Keep plants watered.

TABLE 8.6 Checklist of Activities for Exercise 8.2. Record the date and initial the final column indicating that you have completed each step.

Approximate Day	Date	Activity	Initials
Day 1		Plant F_1 hybrid seeds. See Planting Seeds, pp. 197–198. Regularly check water.	
Day 4 or 5		Observe seedlings and record numbers of each phenotype in Table 8.7. Thin plants to one per cell.	
Days 14, 16, 18		Pollinate on 3 days; pollinate at least 6–8 flowers. See Pollinating Flowers, p. 198.	
Days 20 to 39		Remove buds and shoots.	
Day 39		Harvest F_2 seeds and germinate in quads. See Harvesting Seeds, p. 198 and Planting Seeds, pp. 197–198.	
Days 47 to 52		Count and record numbers of each phenotype in Table 8.8.	

6. *Day 39.* (It should be a minimum of 21 days after the last pollination.) Harvest the F_2 seeds and begin the next generation. Germinate seeds in a quad. See Harvesting Seeds, p. 198 and Planting Seeds, pp. 197–198.
7. *Days 47 to 52.* (Plants should be large enough to score phenotypes.) Count the numbers of each phenotype, and record these data in Table 8.8. Modify the table to accommodate your results. Write in the headings for the phenotypes and draw in columns.

You and your lab partner are responsible for the care and maintenance of your plants. Remember to check the reservoir for water, and be sure that the wicks make good contact with the watering tray, ensuring moisture flow to quads. The organism must survive to reproduce!

Results

1. Record the F_1 offspring in Table 8.7 and the F_2 offspring in Table 8.8. Modify these tables to accommodate observed phenotypes. Compile results from all teams.
2. Note any modifications of procedures or problems in the experiment in the margin of your lab manual or on your schedule of activities (see Table 8.6).
3. Compare your observed and expected results by calculating the chi-square test in Table 8.9. Refer to your hypothesis and predictions and use the class data from Table 8.8. Refer to Appendix D for an explanation of chi-square analysis.

TABLE 8.7 Phenotypes of F_1 *B. rapa* Seedlings from a Cross Between a Homozygous Normal Green Parent and a Yellow-Green with Purple Parent. Write in headings for the phenotypes and draw in columns.

Download an Excel version of this table from www.masteringbiology.com in the Study Area under Lab Media.

	Phenotypes
Team No.	
1	
2	
3	
4	
5	
6	
7	
8	
9	
10	
11	
12	
Class total	

Discussion

1. What do the results suggest is the inheritance pattern of these traits? Assign symbols and write genotypes and phenotypes of all F_1 and F_2 offspring.

   What evidence supports your conclusion?

TABLE 8.8 Phenotypes of F_2 Seedlings. Write in headings for the phenotypes and draw in columns.

Download an Excel version of this table from www.masteringbiology.com in the Study Area under Lab Media.

	Phenotypes
Team No.	
1	
2	
3	
4	
5	
6	
7	
8	
9	
10	
11	
12	
Class total	

2. Did the results support your hypothesis? Explain, describing the results of the chi-square test.

3. Write a statement explaining the significance of the *p* value (see Appendix D).

4. If the results do not support expected Mendelian ratios, what are other possible explanations for these results? What other patterns of inheritance could explain the results?

TABLE 8.9 Chi-Square Calculations. Write in headings for the phenotypes and draw in columns.

Download an Excel version of this table from www.masteringbiology.com in the Study Area under Lab Media.

	Phenotypes
Observed value (*o*)	
Expected value (*e*)	
Deviation ($o - e$) or d	
Deviation^2 (d^2)	
d^2/e	
Chi-square, $\chi^2 = \Sigma d^2/e$	

5. What additional experiments can you suggest to further test your original or modified hypothesis? Outline your crosses.

EXERCISE 8.3

Inheritance of Variegated and Nonvariegated Leaf Color

Materials

6 seeds from homozygous green plants
6 seeds from homozygous variegated plants
2 quads and related planting supplies (see Materials for All Exercises section)
5-in. by 8-in. index cards
stakes (small wooden applicator sticks)

Introduction

In one strain of *Brassica rapa* plants, wild-type plants are entirely green (nonvariegated). However, other plants have white or yellow blotches or streaks on the green leaves and stems (variegated). In this exercise, you will investigate the inheritance of this trait for leaf color. Which trait is dominant, variegated or nonvariegated? Does the inheritance of this trait follow predicted Mendelian ratios? Formulate a hypothesis for these traits that describes a mode of inheritance based on the Mendelian model.

Hypothesis

Hypothesize about the inheritance of this trait.

Prediction

Predict the results of the experiment based on your hypothesis. Outline the cross(es) for your predictions in the margin of your lab manual. Provide expected offspring for each cross in a Punnett square if appropriate. Refer to Figures 8.1 and 8.2 if necessary.

Procedure

Refer to the General Procedures section (pp. 197–199) and relevant figures. All days are approximate. Monitor plants for initiation of flowering and seed set.

1. Refer to the checklist of activities in Table 8.10 and record dates, procedures, and your observations as you follow the procedure for this exercise. Initial the final column when you complete each step.
2. About 3 or 4 days *before* day 1 in the previous exercises, plant three seeds from variegated plants in two cells of the quad. Leave the other two cells empty. (Your instructor may have already done this.) Variegated plants grow more slowly than totally green plants and must be planted earlier. See Planting Seeds, pp. 197–198.
3. *Day 1.* Plant three seeds from green plants in each of the two empty cells of your quad. See Planting Seeds, pp. 197–198.
4. *Day 4 or 5.* Thin the green seedlings to one per cell. Thin variegated seedlings, leaving one plant that is definitely variegated in each cell.
5. *Days 10 to 14.* To prevent "like" plants from pollinating each other, place a 5-in. by 8-in. card supported by two stakes across the quad, separating green from green and variegated from variegated plants. One variegated and one green plant should be on each side of the card. Be sure flowers do not extend above the card. If they do, substitute a larger card. Keep plants watered.
6. *Days 16, 18, 20.* Using a bee-stick, transfer pollen from a flower on a white or variegated stem to flowers on the totally green plant. Using a different bee-stick, carry out the **reciprocal pollination**; that is, transfer pollen from a flower of a green plant to a flower of a white or variegated plant. Pollinate plants three times, with three consecutive pollinations, with 1 day between pollinations. See Pollinating Flowers, p. 198.

TABLE 8.10 Checklist of Activities for Exercise 8.3. Record the date and initial the final column indicating that you have completed each step.

Approximate Day	Date	Activity	Initials
Before day 1		Plant seeds from variegated parents in two cells. See Planting Seeds, pp. 197–198.	
Day 1		Plant seeds from green parents in remaining two cells. See Planting Seeds, pp. 197–198.	
Day 4 or 5		Thin seedlings to one per cell. Leave clearly variegated seedlings in two cells.	
Days 10 to 14		Place card in quad so each side has a green and a variegated plant. Prevent pollination between like plants.	
Days 16, 18, 20		Cross-pollinate green and variegated flowers. Use a different bee-stick for reciprocal pollinations. Pollinate on 3 days. See Pollinating Flowers, p. 198.	
Days 20 to 39		Remove buds and shoots.	
Day 39		Harvest F_1 seeds and germinate in quads. See Harvesting Seeds, p. 198, and Planting Seeds, pp. 197–198.	
Days 47 to 52		Count and record numbers of each phenotype in Table 8.11.	

7. *Days 20 to 39.* Remove and discard new buds and shoots. Keep plants watered.
8. *Day 39.* (It should be a minimum of 21 days after the last pollination. This may be performed during the next lab period.) Harvest seeds from green plants. Next, *keeping seeds separate,* harvest seeds from variegated plants. See Harvesting Seeds, p. 198.

Collect and store seeds in separate envelopes labeled with the date and seed type. Throw away any seeds if you are unsure of the parents.

9. Begin the second generation. Germinate seeds in a quad. Plant five seeds from the green plants in each of two cells and five seeds from the variegated plants in each of the other two cells. Label quads to indicate the parents: "green" or "variegated." See Planting Seeds, pp. 197–198.
10. *Days 47 to 52.* Record phenotypes of plants from green parents and phenotypes of plants from variegated parents in Table 8.11. Modify the table to accommodate the phenotypes for each cross.

You and your lab partner are responsible for the care and maintenance of your plants. Remember to check the reservoir for water, and be sure that the wicks make good contact with the watering tray, ensuring moisture flow to quads.

Results

1. Record the offspring phenotypes in Table 8.11. Record your results separately for the reciprocal crosses: (A) seeds from green plants pollinated with pollen from variegated plants and (B) seeds from variegated plants pollinated with pollen from green plants. Modify the table to accommodate the observed phenotypes for each cross. Compile results from all teams.
2. Note any modifications of procedures or problems in the experiment in the margin of your lab manual or on your schedule of activities (see Table 8.10).

TABLE 8.11 Phenotypes of Germinated Seeds Taken from Green Plants Pollinated with Variegated Pollen (A) and Variegated Plants Pollinated with Green Pollen (B). Write in headings for the phenotypes and draw in columns.

Download an Excel version of this table from www.masteringbiology.com in the Study Area under Lab Media.

	Phenotypes A	Phenotypes B
Team No.		
1		
2		
3		
4		
5		
6		
7		
8		
9		
10		
11		
12		
Class total		

3. Compare your observed and expected results for cross **A** and cross **B.** Refer to your hypothesis and predictions and use the class data from Table 8.11.

Discussion

1. Did the results support your hypothesis? Explain, comparing your predictions and results.

2. If the results do not support expected Mendelian ratios, what are other possible patterns of inheritance that could explain the results?

3. Why would it be inappropriate to use the chi-square test for these data?

4. What additional experiments can you suggest to further test your original or modified hypothesis? Outline your crosses.

REVIEWING YOUR KNOWLEDGE

1. Review all terminology used in this laboratory. Define and provide examples of all terms indicated in bold type.
2. The fruit fly, *Drosophila,* is an extremely useful organism in the study of genetics today. *Brassica rapa,* a Fast Plant, has become a popular organism for study. Based on your knowledge and experience, what are some important characteristics of fruit flies and Fast Plants that make them ideal for genetic experiments?

3. In general terms, explain why a scientist would use the chi-square test as part of his or her data analysis. What important information does it provide?

4. Cyndi grows thousands of *B. rapa* plants from seed and carefully watches for any unusual plants that might be a new mutation. She has been watching a small seedling whose leaves continue to develop, but the stem

is not elongating. The result is a cluster, or rosette, of leaves sitting just above the soil surface. After 15 days the plant is beginning to produce flowers but is still a rosette. Cyndi is encouraged and hypothesizes that rosette is a recessive mutation. Describe the first cross she should make and the predicted results.

. Because *B. rapa* is an annual plant, the one rosette individual dies. Cyndi would like to continue her study of this mutation, but all of the offspring from her original cross were normal. There are no more plants with the rosette mutation. What can Cyndi do to continue her investigation of rosette? What would be the results of these "next steps"?

What results would you expect if Cyndi crosses her new generation of rosette plants with a known heterozygous plant of normal height?

APPLYING YOUR KNOWLEDGE

1. If two traits, dwarfism and stem color, were both on the same pair of chromosomes, would you expect an F_2 generation of 9:3:3:1 for a dihybrid cross? Use Mendel's laws and your understanding of meiosis to explain your answer. Refer to Figure 8.2, if needed.

2. A budding plant breeder discovers a rare mutant of *Brassica* that is almost entirely white. The plastids in the leaves have a mutation that results in white plastids. If the breeder uses the pollen from this mutant to pollinate a plant with normal green leaves, what kind of leaves would you expect to see in the offspring? Explain.

   If the normal green parent is used as the pollen parent, would you expect the results to be the same? Explain.

3. Stefanie is a young farmer growing new and heritage pumpkins. In her rows of pumpkins she finds a dark orange pumpkin with warts and a green pumpkin with smooth skin. She knows that orange skin color is dominant in pumpkins, but she cannot find any information on the inheritance of warty skin in pumpkins. She decides to cross the pumpkins to determine if warty skin is a dominant or recessive trait. State a hypothesis for the inheritance of warty skin in pumpkins.

Hypothesis:

Then, based on your hypothesis, predict the results of the F_1 and F_2 dihybrid crosses between the *orange warty* pumpkin and the *green smooth* pumpkin. (Assume these pumpkins are homozygous for both traits.)

The results of Stefanie's crosses were not revealed until the following summer. All of the pumpkins grown from her crosses were orange warty pumpkins. Do these results support your hypothesis?

4. Molecular geneticists interested in the evolutionary history of the human race have concentrated their research on samples of DNA from women representing all races and continents. Why might the DNA of women—and not men—be of interest?

INVESTIGATIVE EXTENSIONS

1. Wisconsin Fast Plants, like the fruit fly and Mendel's peas, offer a model system for investigating inheritance. Other interesting traits in *B. rapa* are available for extended study. Select from the following list of traits.

   - *Rosette* is a mutation that results in a small plant in which the leaves remain at soil level, because the stem does not elongate.

- *Petite/Astro* plants were developed to meet the "5-10-15" NASA specifications of 5 seed pods, no taller than 10 cm, and flowers by 15 days, so that they could be studied in space!
- *Tall* plants have an elongated internode (stem length between leaves) and the average plant height at 15 days is 28 cm.
- In addition, there are plants that have been developed to investigate two traits in dihybrid crosses: *purple stem, hairy* (the standard trait) and *non-purple stem, hairless* (two mutations). The presence of anthocyanins produces the purple stem color, and the hairless trait is for the absence of hairs on leaves and petioles.

Follow the procedures and planning schedules provided in the General Procedures for All Exercises and in Exercises 8.1, 8.2, and 8.3.

2. Genetic variation and its relationship to the environment can be investigated using the *hairy/hairless* trait. You may choose to begin your investigation by simply growing the seeds and scoring the traits to determine the range and distribution of genetic variation in your sample of plants. These plants can also be used to investigate selection and genetic variation through the selective breeding for either hairy or hairless plants and recording your results over several generations. Working with these interesting organisms will raise additional questions and hypotheses as you carefully observe, measure, and analyze the parents and offspring. Seed stocks, additional suggestions, and resources are available at www.fastplants.org and at Carolina Biological Supply.
3. What did Thomas Hunt Morgan do when his student Calvin Bridges discovered a mutant fly with white eyes in his *Drosophila* cultures? How would *you* determine the mode of inheritance for unknown traits? Using a genetics simulation program, you will be faced with the same dilemma. Unlike most genetics problems you are expected to solve, for your simulation there is no answer key! You will have to do the crosses, analyze the data, and make a logical and convincing argument based on your evidence to convince your colleagues and instructor that you have uncovered the genetics of these traits. The computer will generate populations of "flies" with known phenotypes. As the geneticists, your team must consider the possible modes of inheritance (the hypotheses), then design and complete crosses (testing your hypothesis and analyzing the results). The simulation program can be used to investigate dominant, recessive, and incomplete dominant traits, sex-linked traits, and linked autosomal traits. The instructor can set the range of traits and the level of difficulty. Two simulation programs are available for your use. The BioQUEST program "Genetics Construction Kit" (GCK) can be downloaded from the BioQUEST Library Online (http://www.bioquest.org/BQLibrary/library_result.php). A Web-based genetics simulation program, "Classical Genetics Simulator," has been developed recently by the University of Wisconsin-Madison (http://cgslab.com/). You can investigate the genetics of *Arabidopsis* (another plant in the Brassica family) or the fruit fly, *Drosophila*.

STUDENT MEDIA: BioFlix, Activities, Investigations, Videos, and Data Tables

www.masteringbiology.com (select Study Area)

Activities—Ch. 14: Monohybrid Cross; Dihybrid Cross; Gregor's Garden

Ch. 30: Angiosperm Life Cycle

Videos—Ch. 30: Time Lapse Flowering Plant Life Cycle; Flower Blooming Time Lapse; Bee Pollinating

Data Tables—Tables 8.3, 8.4, 8.5, 8.7, 8.8, 8.9, and 8.11 can be downloaded in Excel format. Look in the Study Area under Lab Media.

REFERENCES

Heitz, J. and C. Giffen. *Practicing Biology: A Student Workbook,* 6th ed. San Francisco, CA: Pearson, 2017. Chapters 14 and 15 provide a guide to solving genetics problems and practice problems.

Jungck, J. and J. Calley. "Genetics Construction Kit," *BioQUEST Library Online.* http://www.bioquest.org/BQLibrary/library_result.php (software available to download)

Klug, W. S., M. R. Cummings, C. Spencer, and M. A. Palladino. *Essentials of Genetics,* 9th ed. San Francisco, CA: Pearson, 2015.

Williams, Paul H. "Bee-Sticks, an Aid in Pollinating Cruciferae." *HortScience,* 1980, vol. 115(6) pp. 802–803.

WEBSITES

BioQUEST home: http://www.bioquest.org/

Carolina Biological Supply Co. information on Fast Plants seed stock descriptions:
http://www.fastplants.org/resources/kinds_of_plants.php

Classical Genetics Simulator developed at the University of Wisconsin–Madison with free Web access, instructor site, and class/student access: http://cgslab.com/

Genetic problems and tutorials:
http://www.biology.arizona.edu/

Human genes and diseases, an online book that can be searched and read at the website:
http://www.ncbi.nlm.nih.gov/books/NBK22183/

Wisconsin Fast Plants: http://www.fastplants.org/

Wisconsin Fast Plants Life Cycle Time-Lapse, YouTube video: http://youtu.be/JumEfAbjBjk

LAB TOPIC 9

Mendelian Genetics II: *Drosophila*

Laboratory Objectives

After completing this lab topic, you should be able to:

1. Discuss why *Drosophila* is one of the most important organisms used in eukaryotic genetics.
2. Explain how a biochemical assay can be used as an indication of biochemical phenotypes.
3. Describe the inheritance pattern of the gene for aldehyde oxidase.
4. Name genes using the convention recommended in *Drosophila* genetics.
5. Determine parental genotypes by investigating offspring.
6. Use the chi-square test to evaluate experimental results.
7. Describe gene mapping.

Introduction

As you perform the exercises in this lab topic, you will use fruit flies to explore some of the earliest investigations in genetics that led to an understanding of the principles governing the inheritance of specific traits. Early experiments in genetics were concerned with the transmission of hereditary factors from generation to generation and led to the discovery of Mendel's laws, which define the pattern of inheritance of individual genes. Later experiments identified chromosomes as the physical structures wherein the units of heredity reside and provided firm cytological evidence for the theorems of Mendelian genetics. More recent investigations have addressed the biochemical and molecular basis of gene expression.

The fruit fly, *Drosophila melanogaster*, has played an important role in the development of our knowledge of heredity since the famous scientist Thomas Hunt Morgan, an experimental embryologist at Columbia University, began using fruit flies for genetics studies in 1907. Though initially skeptical of earlier work that proposed that genes (Mendel's "hereditary particles") were located on chromosomes, Morgan's work with fruit flies convinced him that specific genes are located on specific chromosomes. Morgan's choice of the fruit fly as his experimental organism was fortunate, as this organism has many characteristics that make it easy to study. First, fruit flies have a very low chromosome number. The haploid (n) number of chromosomes is 4, and the chromosomes are designated X(1), 2, 3, and 4 (Figure 9.1). The 2, 3, and 4 chromosomes are the same in both sexes and are referred to as **autosomes** to distinguish them from the X and Y **sex chromosomes**. *Drosophila* females are characterized by two X chromosomes, whereas *Drosophila* males have an X and a Y chromosome. Chromosome 4 and the Y chromosome contain a limited number of genes, with almost the entire genetic content of the *Drosophila* genome residing on only three chromosomes: X, 2, and 3.

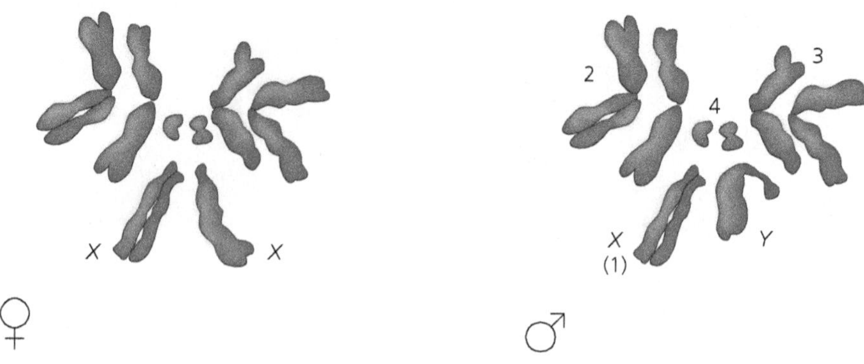

FIGURE 9.1
Metaphase chromosomes from a dividing cell in *Drosophila melanogaster*. Light microscopy has revealed four pairs of chromosomes. Females have two X chromosomes, whereas males have an X and a Y chromosome.

Another characteristic of *Drosophila* that makes it an excellent genetic research tool is its short generation time. At 25°C, a *Drosophila* culture will produce a new generation in 10 days: 1 day in the egg (embryo) stage, 5 days in the larval stage, and 4 days in the pupal stage (Figure 9.2).

You will use *Drosophila melanogaster* in each of the following exercises. You will be asked to investigate the inheritance of a gene called *aldox*, and you will determine the position of this gene on its chromosome; that is, you will map the gene.

EXERCISE 9.1

Establishing the Enzyme Reaction Controls

Materials

stereoscopic microscope
fly vials 1a and 1b
ether dropper bottles or FlyNap
re-etherizer
2 spot assay plates
large and small white index cards
toothpicks
pestle
Kimwipes®
assay mixture dropper bottles
water bottle

The *Drosophila* stocks used in this lab topic are available from Carolina Biological Supply. See the Preparation Guide for ordering instructions.

Introduction

The trait to be studied in each exercise of this lab topic is the presence or absence of the enzyme **aldehyde oxidase (AO)**, which catalyzes the oxidation of a number of aldehydes, including acetaldehyde and benzaldehyde. AO activity is controlled by one gene, the **aldox** gene. Although *Drosophila* flies possess AO activity, its physiological importance to the organism is not well understood. Mutant strains that exhibit no AO activity are available, and their viability and fertility are normal. This latter observation indicates that AO activity is not a vital enzyme activity for a fly that is reared in a laboratory setting.

To test for AO activity, you will use an **enzyme spot test**, or **spot assay**. This test works on the following principle: In the presence of AO, the substrate

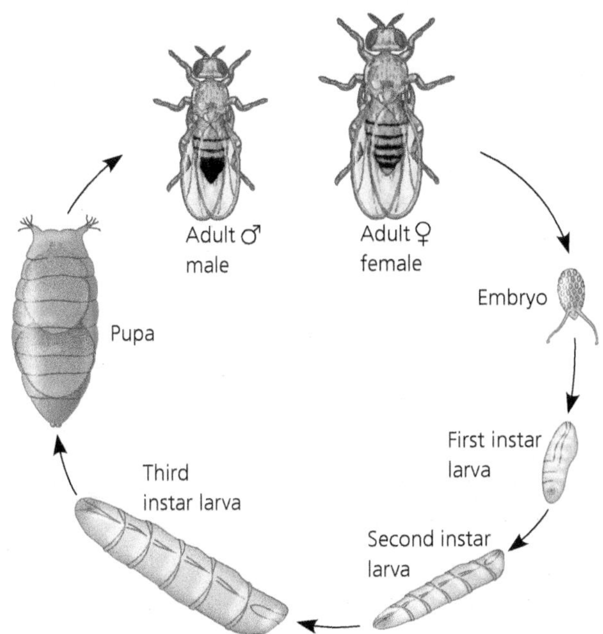

FIGURE 9.2
Developmental stages of *Drosophila melanogaster*. An embryo hatches to a larva, which undergoes two molts and then pupates. The pupae develop into adult winged flies.

benzaldehyde, when mixed with the color indicator nitroblue tetrazolium (NBT)–phenazine methosulfate (PMS), will oxidize to form benzoic acid and a blue color. The blue color indicates that the enzyme is present and active.

This reaction can be diagrammed as follows:

Substrate	**Enzyme AO**	**Product**
benzaldehyde + NBT + PMS (assay mixture)	⟶	benzoic acid + blue products

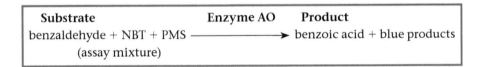

Without AO, the reaction will not proceed, and no blue color will be produced.

This first exercise will demonstrate the positive and negative enzyme reactions as seen in the spot assay. Vial 1a contains flies that have the enzyme present. What will be the results if flies from this vial are homogenized in the assay mixture? Will they demonstrate AO activity? (Remember that AO activity produces a blue color with the assay.)

Vial 1b contains flies that do not have the enzyme present. What will be the results if these flies are homogenized in the assay mixture? Will they demonstrate AO activity?

Hypothesis

Hypothesize about AO activity in flies from vial 1a and flies from vial 1b.

Prediction

Predict the results of the experiment (test) based on your hypothesis (if/then).

Procedure

1. Anesthetize the flies in vials 1a and 1b as follows:
   a. From your ether dropper bottle, place 2 or 3 drops of ether on the cotton plug of your vial. Be sure to recap the ether bottle tightly.

> Remember that ethyl ether fumes are explosive. Use in a well-ventilated room. No flames or sparks! If a spill occurs, call an instructor.

   b. Invert your vial so that the adult flies will fall asleep on the cotton plug rather than on the culture medium.
   c. When flies have become immobilized on the cotton plug, tap them onto a small white index card for examination.
   d. Using the stereoscopic microscope, examine the flies using a toothpick to turn them.
   e. A fly "re-etherizer" petri dish with a gauze pad taped in the lid is provided in case the adults begin to awaken before phenotype classification is concluded. Add a couple of drops of ether to the gauze pad in the lid and place the lid/gauze pad saturated with ether over the flies when needed.
2. From each vial, identify two or three females and two or three males, and return the rest to their appropriate vial, keeping the vial on its side until the flies wake up. (Vial 1a will be used again in later experiments.) Use the following criteria to distinguish adult males and females (Figure 9.3).
   a. *Size.* The female is generally larger than the male.
   b. *Shape of abdomen.* The female abdomen is larger and more pointed than the male abdomen.
   c. *Abdominal pigmentation.* In dorsal view, the alternating dark and light bands on the entire rear portion of the female abdomen are visible; the last few segments of the male abdomen are uniformly pigmented.
   d. *Sex comb.* On males, there is a tiny brushlike tuft of hairs on the basal tarsal segment of each foreleg. This is the most reliable characteristic for sexing males accurately.
   e. *External genitalia.* On the ventral portion of the abdomen, the female has anal plates and lightly pigmented ovipositor plates. The male has anal plates and a darkly pigmented genital arch and penis.
3. Keeping track of the sexes, place the flies from vial 1a in one row of a spot assay plate, one fly per well.
4. Place the flies from vial 1b in a different row.

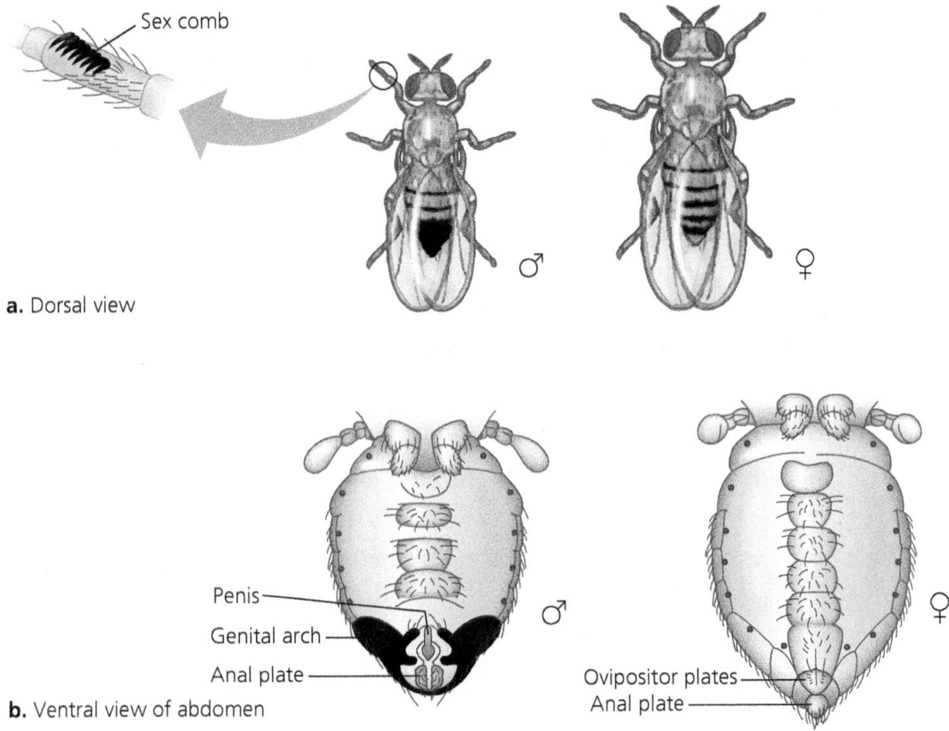

FIGURE 9.3
Characteristics of male and female *Drosophila melanogaster*. The female is larger and has alternating dark and light dorsal bands on the abdomen. (a) The dorsal abdomen of the male is uniformly pigmented. Conspicuous sex combs are visible on each foreleg of the male. (b) In ventral view, the male's darkly pigmented genital arch and penis are visible.

> The next two steps need to be done quickly, because the assay mixture is light sensitive. Cover the spot plate with an index card if there is any delay between steps.

5. Add 1 drop of assay mixture to each well.

> The assay mixture contains carcinogens. Do not allow it to contact skin. Wash hands thoroughly after performing the tests. Disposable gloves may be provided for your use. Notify an instructor if a spill occurs. If an instructor is unavailable, wipe up the spill wearing disposable gloves and using dry paper towels. Follow dry towels with towels soaked in soap and water. Dispose of all towels and gloves in a plastic bag in the trash.

6. Homogenize the flies with a pestle. Wipe off the pestle after each fly to avoid contamination. Why is it necessary to homogenize the flies?

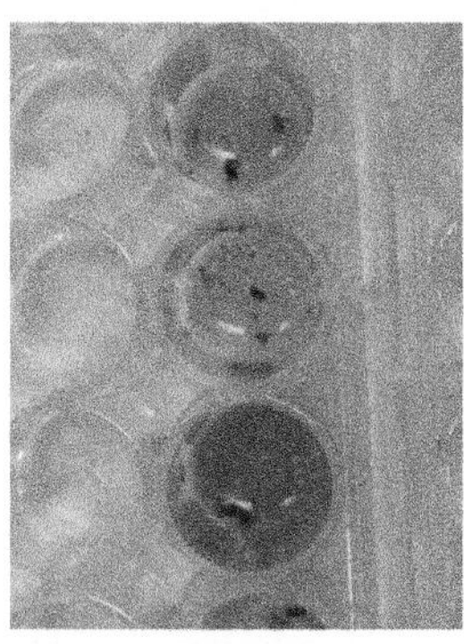

a.

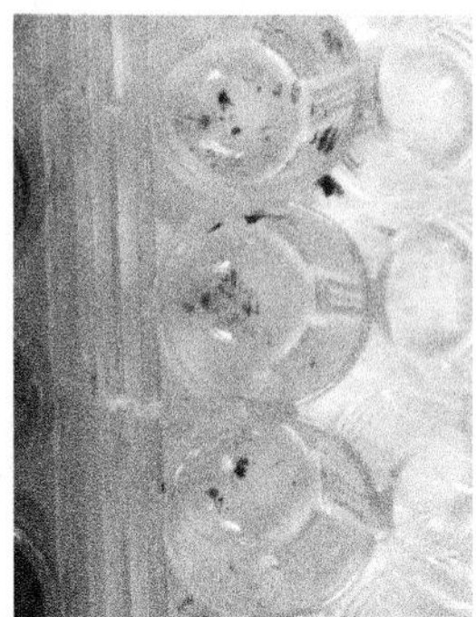

b.

FIGURE 9.4
Results of assay test. (a) If aldehyde oxidase (AO) is present in the fly, the solution will turn blue. (b) If the AO enzyme is not present in the fly, the solution will remain pale yellow, the color of the assay mixture.

7. Place the spot plate in a place away from light, such as a desk drawer.
8. After 5 minutes, check the reactions. Determine which flies demonstrate AO activity. A positive test for the presence of AO will appear as a blue color in the homogenized fly in the assay mixture (Figure 9.4a). Record the results in Table 9.1.
9. Determine which flies do *not* demonstrate AO activity (a negative reaction). The homogenized fly in the assay mixture will remain the color of the assay mixture, a pale yellow (Figure 9.4b). Record the results for each fly in Table 9.1 in the Results section.
10. Thoroughly rinse your spot plate and shake off the excess water.

Results

Complete Table 9.1 as the results of the assay tests are determined.

TABLE 9.1 AO Activity in Male and Female Flies from Vials 1a and 1b

Vial 1a (AO Present, blue color)			Vial 1b (AO Absent, no color change)		
Fly No.	Sex (F, M)	AO Activity (+/−)	Fly No.	Sex (F, M)	AO Activity (+/−)
1			1		
2			2		
3			3		
4			4		
5			5		
6			6		

Discussion

1. Do your results match your predictions?

2. Does the sex of the fly appear to have an impact on the results of the assay test?

3. Which two characteristics are most useful to your group in determining the sex of flies?

EXERCISE 9.2

Determining the Pattern of Inheritance of the Aldox Gene

Materials

vial 1a
vial 2
remaining materials from Exercise 9.1

Introduction

This exercise initiates our study of the pattern of inheritance of the aldox gene, which determines the activity of aldehyde oxidase (AO).

> The gene is called the *aldox gene;* the enzyme produced by this gene is *aldehyde oxidase* and is abbreviated AO.

In this exercise, you will determine which **allele** (form of the gene) is dominant: enzyme-present or enzyme-absent. In addition, you will determine if the gene is **autosomal** or **sex-linked**. A sex-linked gene is located on one sex chromosome but not on the other. An autosomal gene is located on any chromosome *except* a sex chromosome. Finally, you will use this information to name the gene according to conventional naming procedures used by *Drosophila* geneticists.

The organisms you will use are in vial 2, which contains the offspring (called F_1 *progeny*) from a mating between a female that *lacks the enzyme* and a male that *has the enzyme*. For now, name the enzyme-present allele *AO-present* and the enzyme-absent allele *AO-absent*. The female is from a stock of flies that, when inbred, *consistently* lacks the enzyme from generation to generation. We say that this stock **breeds true** for AO-absent. The male is from a stock that *consistently* has the enzyme from generation to generation; that is, it breeds true for AO-present. This stock, which has the enzyme, is genetically like flies most commonly found in nature and is called the **wild type**.

It is necessary to understand this background information to be able to predict the outcome of this cross. To help you in your predictions, answer the following questions.

1. What is the genotype of the female (maternal parent—lacks the enzyme) in our cross?

2. What would be the genotype of the male (paternal parent—has the enzyme) in our cross if the gene is *not* sex-linked?

3. What would be the genotype of the male if the gene *is* sex-linked?

4. What would be the genotypes of the F_1 progeny if the gene is *not* sex-linked?

5. What would be the genotype if the gene *is* sex-linked?

Hypothesis

Hypothesize about the inheritance of this gene. Is it sex-linked? According to your hypothesis, what will be the genotypes of the parents?

Prediction

Predict the offspring from the hypothesized parents. (*Hint:* Set up a Punnett square in the margin.)

Procedure

1. Anesthetize the flies in vial 2.
2. Count out 10 males and 10 females. Keep track of the sexes of flies as you place them in the spot plate. Record the sex of each fly in Table 9.2.

> It is easier to keep accurate records if you put all males in one row of the spot plate and all females in another row. Do not write on the spot plates!

3. Perform the spot assay as described in Exercise 9.1 with the following addition: From vial 1a, select a single fly and assay it with this and all following experiments. This will provide you with a *positive control* to compare with your unknown assays.

> The assay mixture is light sensitive. Cover the spot plate with an index card between steps.

4. When you see that the positive control has turned blue, indicating that the assay is working, record the results in Table 9.2.
5. Rinse out the spot plate carefully.

Results

Record the sex of each fly and the results of the spot assay tests in Table 9.2.

TABLE 9.2 Data Sheet for Exercise 9.2: Results of Spot Assay Tests on F_1 Progeny from a Female That Lacks AO Activity and a Male That Has AO Activity

Fly No.	Sex (F, M)	AO Activity (+/−)	Fly No.	Sex (F, M)	AO Activity (+/−)
1			11		
2			12		
3			13		
4			14		
5			15		
6			16		
7			17		
8			18		
9			19		
10			20		
Positive control:					

Discussion

1. What allele appears to be dominant?

2. What evidence of your results in Table 9.2 supports your answer?

3. Determine the conventional way to name this gene. The following information will assist you. For any gene with two alleles, there exists the *wild type* and the *mutation*. Recall that the wild type is the allele most commonly found in nature. In this case, the wild type is AO-present. AO-absent is considered a **mutation**, a change in the DNA of the gene. In naming genes and alleles, give the gene an appropriate name using a word or letter derived from the phenotype affected by the gene. (In this case, the gene is named *aldox* because the enzyme that it produces catalyzes the oxidation of aldehydes.) If the mutation is dominant, capitalize the first letter in the name. If the mutation is recessive, do not capitalize. Based on the results of your experiment (see Question 1), should the aldox gene be written beginning with a capital or a lowercase letter?

4. Correctly write the names of the wild-type and mutant alleles. By convention, wild-type alleles are designated by a superscript + after the name.

5. Did you hypothesize that the gene was sex-linked (on a sex chromosome) or autosomal (not on a sex chromosome)?

6. Describe how your results either support or falsify your hypothesis.

7. Write a statement describing your conclusions from this experiment.

EXERCISE 9.3

Determining Parental Genotypes Using Evidence from Progeny

Materials

vial 1a
vial 3
remaining materials from Exercise 9.1

Introduction

Once the pattern of inheritance has been determined, this information can be used to predict genotypes and phenotypes of individuals by observing parents, siblings, and offspring. Vial 3 contains the first-generation (F_1) progeny from a mating between unknown parents. The objective of this exercise is to determine the genotypes of the parents. At this stage in the investigation, you have no data from observations on which to base a hypothesis. The following procedure will allow you to collect preliminary data, then, using this data, to propose and test a hypothesis about the genotypes of the parents.

Procedure, Preliminary Observations

1. Anesthetize the flies in vial 3.
2. Count out 24 flies and place them in individual spot plate wells.
3. Perform the spot assay, including a positive control, on a fly from vial 1a.
4. Record the results on *two* data sheets, Table 9.3, and for your team on Table 9.4, the class data sheet.

Results, Preliminary Observations

1. Record results of assay tests on progeny of the unknown parents in Table 9.3.

TABLE 9.3 Data Sheet for Exercise 9.3: Results of Assay Tests on Progeny of Unknown Parents

Fly No.	AO Activity (+/−)	Fly No.	AO Activity (+/−)
1		13	
2		14	
3		15	
4		16	
5		17	
6		18	
7		19	
8		20	
9		21	
10		22	
11		23	
12		24	
Positive control:			

2. Total the number of offspring in each phenotype category.

   Your Totals: AO Present ____________ **AO Absent** ____________

3. Record your results and those of all other teams in Table 9.4. Calculate totals for the entire class.

TABLE 9.4 Class Data Sheet for Exercise 9.3: Each Team Records Its Results

Download an Excel version from www.masteringbiology.com in the Study Area under Lab Media

Team #	AO Present	AO Absent	Team #	AO Present	AO Absent
1			7		
2			8		
3			9		
4			10		
5			11		
6			12		
Total			Total		

Class Totals: AO Present ____________ **AO Absent** ____________

4. Review observations made in Exercise 9.2.
   a. Is the trait sex-linked?

   b. Which allele is dominant?

Hypothesis

Using all observations, hypothesize the genotypes of the parent flies, making sure to name the alleles correctly.

Prediction

Predict the results of the experiment (if/then). Set up a Punnett square in the margin to help you make your prediction.

Procedure—Testing Your Hypothesis

Using the class total from Table 9.4, refer to Appendix D and use Table 9.5 to perform the chi-square test to determine if the results of the exercise support or falsify your hypothesis. Calculate the expected values for each trait based on the total number of flies counted in your class. (For a discussion of the chi-square test, see Appendix D.)

Results

Complete the chi-square calculations in Table 9.5.

TABLE 9.5 Chi-Square Calculations to Evaluate Results of Exercise 9.3 (Observed Value Represents Total Class Data.)

Download an Excel version from www.masteringbiology.com in the Study Area under Lab Media

	AO Activity (+)	AO Activity (–)
Observed value (*o*)		
Expected value (*e*)		
Deviation ($o - e$) *or d*		
Deviation^2 (d^2)		
d^2/e		
Chi-square $\chi^2 = \Sigma d^2/e$		
Degrees of freedom (*df*)		
Probability (*p*) (see Appendix D)		

Discussion

1. Do the class results support or falsify your hypothesis?

2. Does this experiment support or contradict your conclusions concerning the pattern of inheritance derived from Experiment 9.2?

EXERCISE 9.4

Mapping Genes

Materials

vial 1a
vial 4
remaining materials from Exercise 9.1

Introduction

In this exercise, you will investigate the inheritance of two genes in *Drosophila*: the aldox gene and a gene that influences eye color named **sepia**. In the wild-type fly (sepia$^+$), eye color is red. In the mutant fly, eye color is dark brown. The wild-type allele is dominant over the mutant. The flies you will be studying are the offspring from a cross in which the parents differ in these two genes: a **dihybrid cross**. You will ask if these two genes are transmitted from parent to offspring *linked* together or if each gene is inherited independently of the other. If genes are transmitted from parent to offspring linked together, this means that they are located on the same chromosome. Then if the location of one gene is known, the pattern of inheritance can provide evidence that will allow you to determine the location of the second gene. With this information, you can construct a map showing gene locations.

Vial 4 contains the F_1 progeny from a mating between parents having the following genotypes:

$$\begin{array}{ccc} \textbf{Parent1} & & \textbf{Parent2} \\ \dfrac{\text{sepia}^+\ \text{aldox}^+}{\text{sepia aldox}} & \times & \dfrac{\text{sepia aldox}}{\text{sepia aldox}} \end{array}$$

where sepia$^+$ represents the dominant, wild-type eye color allele that produces red eye color and sepia represents the mutant, recessive allele that produces a dark-brown eye color (Figure 9.5).

Hypothesis

Hypothesize about the inheritance of these two genes. Are they inherited linked together or independently of each other?

Prediction

Predict the ratios of phenotypic classes of offspring resulting from the mating described above.

Use a Punnett square to illustrate your prediction.

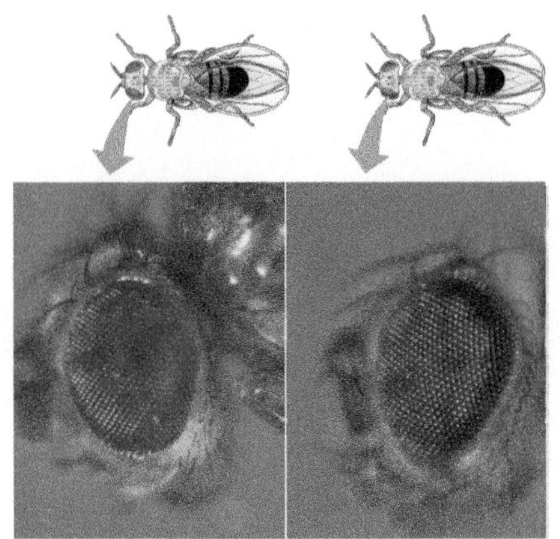

FIGURE 9.5
Eye color in *Drosophila*. In the wild-type fly on the left the eyecolor is red. The mutant fly on the right has a recessive allele that produces a dark-brown eye color.

Procedure

1. Anesthetize the flies in vial 4.
2. Count out 50 flies, and classify and separate them on the basis of eye color (red or sepia).
3. Keeping the eye colors separate, perform the spot assay, including a positive control from vial 1a.
4. Record the results in Table 9.6.
5. Rinse out the spot plate.

TABLE 9.6 Data Sheet for Exercise 9.4: Mapping Genes, Recording Eye Color and AO Activity for 50 Flies

Fly No.	Eye Color	AO (+/−)	Fly No.	Eye Color	AO (+/−)
1			26		
2			27		
3			28		
4			29		
5			30		
6			31		
7			32		
8			33		
9			34		
10			35		
11			36		
12			37		
13			38		
14			39		
15			40		
16			41		
17			42		
18			43		
19			44		
20			45		
21			46		
22			47		
23			48		
24			49		
25			50		
Positive control:					

Results

1. Record the results of the eye color classification and the spot test for each fly in Table 9.6. Then total your results.

	Your Totals
red eyes, AO-present	________
red eyes, AO-absent	________
sepia eyes, AO-present	________
sepia eyes, AO-absent	________

2. Record your results and those of all other teams in Table 9.7. Calculate totals for the entire class.

TABLE 9.7 Class Data Sheet for Exercise 9.4: Each Team Records Its Results

Download an Excel version from www.masteringbiology.com in the Study Area under Lab Media

Team #	Red Eyes, AO Present	Red Eyes, AO Absent	Sepia Eyes, AO Present	Sepia Eyes, AO Absent	Team #	Red Eyes, AO Present	Red Eyes, AO Absent	Sepia Eyes, AO Present	Sepia Eyes, AO Absent
1					7				
2					8				
3					9				
4					10				
5					11				
6					12				
Total					Total				

	Class Totals
Red eyes, AO present	________
Red eyes, AO absent	________
Sepia eyes, AO present	________
Sepia eyes, AO absent	________

3. What phenotypic classes were observed in the total class data, and in what approximate ratio?

4. On separate paper, using class totals, perform the chi-square test to determine if the results support or falsify your hypothesis (see Appendix D).

Discussion

1. Do the data support your predicted results?

2. If the results differ from what was expected, can you suggest an explanation for these differences?

3. Suppose that the aldox gene and the sepia gene are located on the same chromosome. This means that when meiosis takes place, the two genes will not assort independently but will be linked together, moving into the same gamete *unless crossing over has taken place* (refer to Lab Topic 7 to review independent assortment in meiosis). Only if crossing over takes place will **recombinant classes** of phenotypes be observed. A **recombinant chromosome** is one emerging from meiosis with a combination of alleles not present on the chromosomes entering meiosis. What are the recombinant classes of phenotypes for this cross?

4. The distance between two genes is related to the frequency of recombinants produced by crossing over during meiosis. The closer two genes are, the fewer recombinants. Geneticists use an arbitrary measure, or **map unit**, to represent the distance between two genes. A 1% recombinant frequency equals one map unit. Calculate the relationship between map units and recombinants as follows:

$$\textbf{Map units} = \frac{\textbf{number of recombinants}}{\textbf{total}} \times 100$$

5. In your experiment, did recombinant classes exist? If they did, how frequent were they?

6. Do your data suggest that aldox and sepia are on the same chromosome? If so, how far is the aldox locus from the sepia locus?

7. What would be the exact map position of the aldox locus if the sepia locus is 26.0? Are you sure?

REVIEWING YOUR KNOWLEDGE

1. List the most obvious characteristics used to determine the sex of a fruit fly.

2. A fruit fly geneticist discovered a genetic mutation that resulted in pupae and young flies with dark pigment granules in the nuclei and cytoplasm of their fat cells. After studying the inheritance of the mutation, the geneticist named the gene **Frd** (for **Freckled**). What does this name tell you about the inheritance of this gene?

   In another strain of fruit flies, geneticists discovered a mutation producing flies with no gut muscles. They named the gene controlling this phenotype **jeb** (for **jelly belly**). What does this tell about the inheritance of this gene?

3. Is the genetic map or **linkage map** produced by determining recombination frequencies an actual physical map of the chromosome? What additional techniques do geneticists use to more precisely locate relative positions of genes on chromosomes? Refer to your textbook.

APPLYING YOUR KNOWLEDGE

1. In *Drosophila* there is a gene where a mutant named ebony produces flies with a shiny black body color, darker than wild type. A gene for eye color has a scarlet mutant with eyes that are much brighter red than in the wild type.
   A cross between wild-type flies that breed true for both traits and ebony body, scarlet-eyed flies produces an F_1 generation with the wild-type phenotype for both traits. Predict the ratio of offspring phenotypes in the next generation of a cross between these F_1 flies and ebony, scarlet-eyed flies. Assume the genes are not linked.

The results of the cross between the F_1 and ebony body, scarlet-eyed flies are as follows:

288 wild-type body, wild-type eye color
102 wild-type body, scarlet eye color
110 ebony body, wild-type eye color
300 ebony body, scarlet eye color

Are the genes linked? If so, how many map units apart are they located? Show your work.

If the ebony gene is located at map unit position 70.7 on chromosome 3, where is the gene for scarlet eye color located? (Give all possible positions.)

2. Explain why we cannot use a testcross to detect linkage between two genes located on the same chromosome 50 map units apart.

INVESTIGATIVE EXTENSIONS

1. Students may investigate the developmental profile of the aldox gene. All genes are not active in all cells, nor are genes in a cell lineage active at all times in the developmental cycle of the organism. Cells constantly turn genes on and off, bringing about the patterns of development and differentiation in the organism. There are four distinct developmental stages in the fruit fly: the embryo, larva, pupa, and adult (Figure 9.2).

   Using procedures and spot assay tests described in Exercise 9.1, design an experiment that will investigate the expression of the aldox gene during the developmental cycle of the fruit fly.

2. Genetics Construction Kit" is a computer-based, problem-solving program based on fly genetics. Students perform investigative studies in classical genetics, including linkage. This program provides sets of organisms with unknown patterns of inheritance. Students may propose hypotheses, make predictions, and then cross unknown organisms and analyze crosses to discover these inheritance patterns.

STUDENT MEDIA: BioFlix, Activities, Investigations, Videos, and Data Tables

www.masteringbiology.com (select Study Area)

Activities—Ch. 15: Linked Genes and Crossing Over; Sex-Linked Genes

Investigations—Ch. 15: What Can Fruit Flies Reveal About Inheritance?

Data Tables—Tables 9.4, 9.5, and 9.7 can be downloaded in Excel format. Look in the Study Area under Lab Media.

REFERENCES

This lab topic was first published as J. G. Morgan and V. Finnerty, "Inheritance of Aldehyde Oxidase in *Drosophila melanogaster*," in *Tested Studies for Laboratory Teaching* (Volume 12), Proceedings of the 12th Workshop/Conference of the Association for Biology Laboratory Education (ABLE), Corey A. Goldman, Editor. Used by permission.

Heitz, J. and C. Giffen. *Practicing Biology: A Student Workbook*, 6th ed. San Francisco, CA: Pearson, 2017. Activities 14.1–14.4 are excellent assignments to help students prepare for this lab topic.

Jungck, J. and J. Calley, "Genetics Construction Kit," *BioQUEST Library Online*. http://www.bioquest.org/BQLibrary/library_result.php (software available to download).

Klug, W. S., M. R. Cummings, C. Spencer, and M. A. Palladino. *Essentials of Genetics*, 9th ed. San Francisco, CA: Pearson, 2015.

WEBSITES

A directory of Internet resources for research on *Drosophila* is available at the WWW Virtual Library: *Drosophila*:
http://www.ceolas.org/fly

Access this lab topic as first presented at the ABLE conference. Look under ABLE publication, previous issues, volume 12. Many excellent labs are available at the ABLE website. www.ableweb.org/

Interactive fruit fly genetics lab, requires registration: http://biologylab.awlonline.com. Select Flylab.

Classical Genetics Simulator developed at the University of Wisconsin–Madison with free Web access, instructor site, and class/student access:
http://cgslab.com

LAB TOPIC 10

Molecular Biology

To successfully complete this lab, the restriction enzyme digestions should begin immediately and incubate for 1 hour. To begin the lab, students should practice pipetting, prepare the digestions, and pour the gel (see Procedure, pp. 252–255, steps 1–3). Once the digestions are incubating and gels are poured, return to the Introduction and a discussion of the Preliminary Questions. pp. 250–251.

Laboratory Objectives

After completing this lab topic, you should be able to:

1. Describe the function of restriction enzymes.
2. Discuss the basic principles of electrophoresis in general and for DNA specifically.
3. Use gel results to estimate DNA fragment sizes.
4. Construct a tentative map of DNA molecules based on restriction fragments.
5. Explain the use of restriction enzymes and DNA fragments to map DNA molecules, and discuss the importance of mapping.
6. Discuss the universality of the genetic code.
7. Describe ways in which the technology of molecular biology is being used in industry, medicine, criminal justice, agriculture, and basic research.

Introduction

"Golden rice" with β carotene, herbicide-resistant corn, insect-resistant cotton, zebrafish that glow in the dark, goats that produce large quantities of human antithrombin, potatoes with edible vaccines—all of these examples represent organisms that have had their DNA modified through **recombinant DNA technology**, a process of introducing specific genes from one organism into another (Figure 10.1). This process creates a **transgenic organism**, or an organism with a gene from another species. This technology was made possible by the discovery in the 1970s of **restriction enzymes** in bacteria. These enzymes function to cut DNA molecules at specific nucleotide sites and are one of the basic tools of molecular biology. The DNA fragments with a gene of interest can be inserted into a bacterial DNA, and this recombinant DNA can then be **cloned** (copied). Because the genetic code is universal, genes of one organism can be expressed in another organism, for example, human clotting factor VIII is

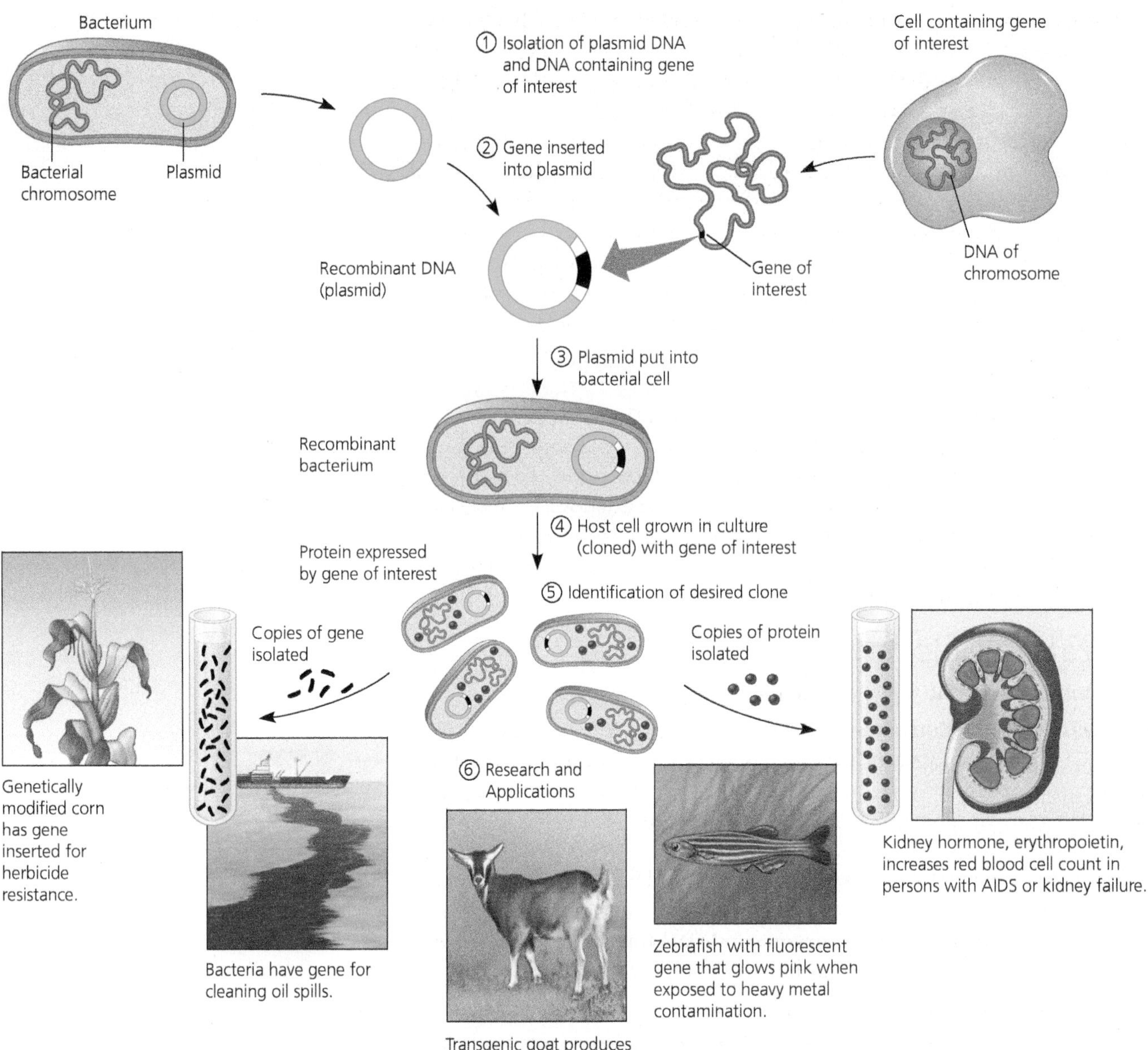
Bacterium
Bacterial chromosome
Plasmid
① Isolation of plasmid DNA and DNA containing gene of interest
Cell containing gene of interest
② Gene inserted into plasmid
Recombinant DNA (plasmid)
Gene of interest
DNA of chromosome
③ Plasmid put into bacterial cell
Recombinant bacterium
④ Host cell grown in culture (cloned) with gene of interest
Protein expressed by gene of interest
⑤ Identification of desired clone
Copies of gene isolated
Copies of protein isolated
⑥ Research and Applications
Genetically modified corn has gene inserted for herbicide resistance.
Bacteria have gene for cleaning oil spills.
Transgenic goat produces human clotting factor in milk.
Zebrafish with fluorescent gene that glows pink when exposed to heavy metal contamination.
Kidney hormone, erythropoietin, increases red blood cell count in persons with AIDS or kidney failure.

FIGURE 10.1

Overview of recombinant DNA technology. 1. The scientist isolates plasmid DNA from bacteria and purifies DNA containing the gene of interest from another cell. **2.** A piece of DNA containing the gene is inserted into the plasmid, producing recombinant DNA. **3.** The plasmid is put back into a bacterial cell. **4.** This genetically engineered bacterium is then grown in culture. The bacterial culture now contains many copies of the gene (one per cell in this example). **5.** The cell clone carrying the gene of interest is isolated. **6.** The bottom of the figure illustrates a few of the current applications of genetically engineered (transgenic) organisms.

expressed in transgenic sheep and the protein can be isolated and purified from the milk. Biotechnology has been used to produce therapeutic products and new varieties of organisms with enhanced traits as ways to address problems in agriculture, medicine, industry, conservation, and forensic science.

Essentially, the age of molecular genetics was launched in 1953, when James Watson and Francis Crick revealed the structure of the DNA double helix, and in the following decade the essence of the genetic code was solidly established. In 1978, Hamilton Smith, Daniel Nathans, and Werner Arber were awarded the Nobel Prize for developing the techniques for recombinant DNA, which have revolutionized genetics and molecular biology. Scientists can now analyze the nucleotide sequence of many genes, and in June 2000, the announcement was made that the major portions of the **human genome** had been mapped. By 2003, all 3 billion nucleotide base pairs of the human genome had been sequenced. The National Institutes of Health officially launched the first stages of the Cancer Genome Atlas, an effort to catalog genomic changes in cancer cells in humans. More recently, six federal agencies and private research laboratories initiated the 1000 Genome Project to sequence the genomes from a large number of individuals to catalog human genetic variation. In 2008, scientists reported the successful use of gene therapy (inserting a functional gene into a patient with a defective gene) to treat a form of congenital blindness (LCA); however, long-term results have been mixed. The emerging field of personalized medicine uses the molecular characteristics of humans and disease to customize drug therapy, for example, cancer treatments based on the molecular characteristics of the tumor.

As the human genome was being sequenced, laboratories all over the world began sequencing genomes of other animals and plants. The National Center for Biotechnology Information has cataloged genome sequences for over 16,000 organisms, either completed or in progress. Many of these organisms are important in biological research, for example *Drosophila, E. coli, C. elegans* (a roundworm), *Zea mays* (corn), *Danio rerio* (zebrafish), and the flowering plant *Arabidopsis.* Exciting questions related to evolution are being investigated using DNA sequencing. The chimpanzee genome has been sequenced, and recent work on sequencing genes in Neanderthal will allow comparisons of human, chimp, and Neanderthal DNA. You will learn in later lab topics that molecular biology has been used successfully to explore the evolutionary relationships among many living organisms of Earth.

Forensic science uses molecular techniques to identify individuals based on very small and sometimes degraded DNA samples that are collected after disasters and from crime scenes. Once fragments of DNA have been isolated, the process of **polymerase chain reaction**, or **PCR**, enables scientists to make many clones of the isolated DNA for comparison. The criminal justice system has come to rely heavily on DNA technology, as **DNA fingerprinting** is used to identify criminals in cases of rape and murder as well as to exonerate those convicted of crimes before this technology was available. DNA fingerprinting was used to identify victims of the 2004 tsunami and to convict traffickers in endangered species by identifying the source of elephant ivory and turtle meat.

In the following exercise you will employ some of the basic molecular techniques used in laboratories worldwide to study everything from the virus that causes AIDS to DNA extracted from mummies. Following the procedures outlined in Exercise 10.1, you will use restriction enzymes to cut bacterial DNA and separate the fragments using gel electrophoresis. Then you will be asked to answer a series of questions that will lead you through a process of analyzing your results and constructing a restriction map. The approaches to DNA recombinant technology, including using restriction enzymes and gel electrophoresis that you will use in these exercises, provide the foundation for new strategies and techniques that are rapidly emerging in molecular biology.

Student Media Video—Ch. 20: Biotechnology Lab

EXERCISE 10.1

Mapping DNA Using Restriction Enzymes and Electrophoresis

Materials

for practice pipetting:
- petri dish with sample 1% agar gel with wells
- empty microtubes
- practice pipette solution

gel electrophoresis apparatus:
- gel plates
- comb to make wells
- chamber for electrophoresis
- chamber cover

pUC 19 (plasmid) DNA
ice
restriction enzymes: *Ava* II, *Pvu* II
restriction buffers
molecular weight markers (λ DNA)
gel loading dye (bromophenol blue)
disposable gloves
15 1.5-mL microtubes (various colors)
thermometer
metric rulers
power supply with electrodes
TAE 1× buffer (Tris base; acetate; ethylenediaminetetraacetic acid, or EDTA; NaOH)
microcentrifuge (helpful, not necessary)
micropipettors and tips or microcapillary pipettes and plunger (various sizes from 1 to 100 μL)
37°C incubator or water bath
55°C water bath for agarose
portable freezer box or ice chest
digital camera and USB drive
deionized water
flask with 100 mL of 0.8% agarose containing SYBR® Safe stain in 55°C water bath

SYBR® Safe stain:
- blue light transilluminator or imaging system
- printer (optional)
- trays to transport gels

alternative methylene blue stain:
- 0.025% methylene blue
- staining trays
- light box
- deionized water

Introduction

The first step in any refined DNA analysis, such as DNA sequencing or expressing a gene in another organism, is to construct a map of the molecule. Scientists use naturally occurring enzymes to cut large DNA molecules into smaller pieces. These fragments are sorted and separated by size using a technique called **electrophoresis**. The results are then used to reconstruct the DNA molecule. This initial process of determining the size and order of DNA is called **mapping**. Recently developed strategies for mapping DNA segments rely on the advanced techniques of gene sequencing and bioinformatics that you will investigate in Lab Topic 16 Bioinformatics.

Molecular biologists use some of nature's own tools to do this analysis. In this experiment, you will use restriction enzymes called **restriction endonucleases** to help manipulate DNA molecules. In nature, bacteria use restriction enzymes as a defense against virus infections by recognizing and cutting the foreign DNA. Restriction enzymes consistently recognize a specific DNA sequence wherever it occurs in a DNA molecule (Figure 10.2) and cut the DNA at or near that site—thus, the name. *Restriction* refers to cutting, *endo* to inside a molecule (as opposed to *exo*, which refers to the ends of a molecule), and *nuclease* to the digestion of a nucleic acid such as DNA. Each restriction enzyme is named for the species of bacteria from which it is isolated. For example, ***Eco*RI** was discovered in ***Escherichia coli***, strain R, Roman numeral "I."

Cutting (also called **restricting**, or **digesting**) requires energy in the form of adenosine triphosphate (ATP) and involves a physical cleaving of chemical bonds. The specific recognition sites where the cuts occur are often **palindromic**; that is, the sequences of the complementary strands read the same on both strands from 5' to 3' (Figure 10.2). (The phrase ***race car*** is an example of a palindrome. Can you think of others?)

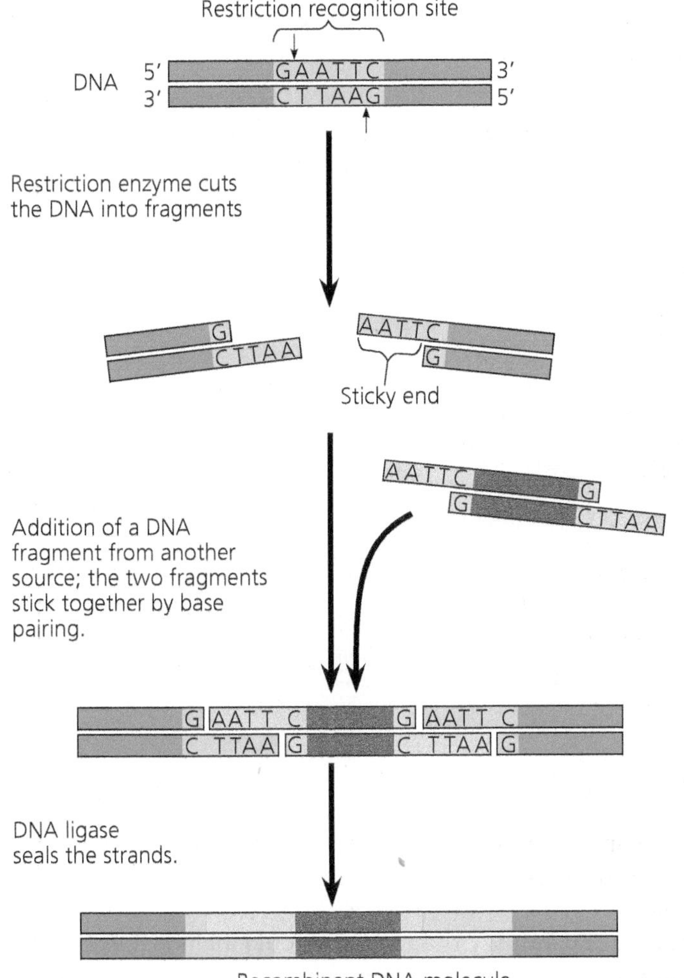

FIGURE 10.2
Using a restriction enzyme and DNA ligase to make recombinant DNA.
The restriction enzyme (*Eco*RI) recognizes a six-base-pair sequence and makes staggered cuts in the sugar-phosphate backbone within this sequence. Notice that the recognition sequence along one DNA strand is the exact reverse of the sequence along the complementary strand (that is, they are palindromic). Complementary ends will stick to each other by hydrogen bonding, rejoining fragments in their original combinations or in new recombinant combinations. The enzyme DNA ligase can then catalyze the formation of bonds joining the fragment ends. If the fragments are from two different sources, the result is *recombinant DNA*.

In this exercise, you will use restriction endonucleases in conjunction with gel electrophoresis to map the 2,686-**base-pair** (bp) pUC 19 **plasmid**. A plasmid is a relatively small extrachromosomal and circular molecule of DNA found in bacteria and yeasts. pUC 19, a plasmid found in E. coli, is one of the most significant cloning tools used in molecular biology labs. Because of its size and DNA sequence, it is an excellent system to study.

Once you have cut pUC 19 into discrete fragments, you will need a method of detecting and separating the digested products. Agarose **gel electrophoresis** is commonly used to separate these fragments (Figure 10.3). The DNA fragments are placed in the gel, and an electric current runs through the matrix of the gel-like agarose. The fragments will move through the gel at different rates, depending on their charge and size. How does the charge of a DNA molecule vary with its size? How is this different from what you see with proteins?

After running the DNA samples in a gel to which the SYBR® Safe stain has been added, the gel can be visualized using a blue light transilluminator or imaging system. SYBR® Safe is not visible on the gel while it is running, but the stain attaches to the DNA, which will be visible when exposed to blue light (300 nm wavelength). Alternatively, the gel can be stained with methylene blue, which can be viewed on a light box.

Preliminary Questions

Based on your knowledge of the structure and function of DNA and the technique of gel electrophoresis, answer the following questions before beginning your investigation.

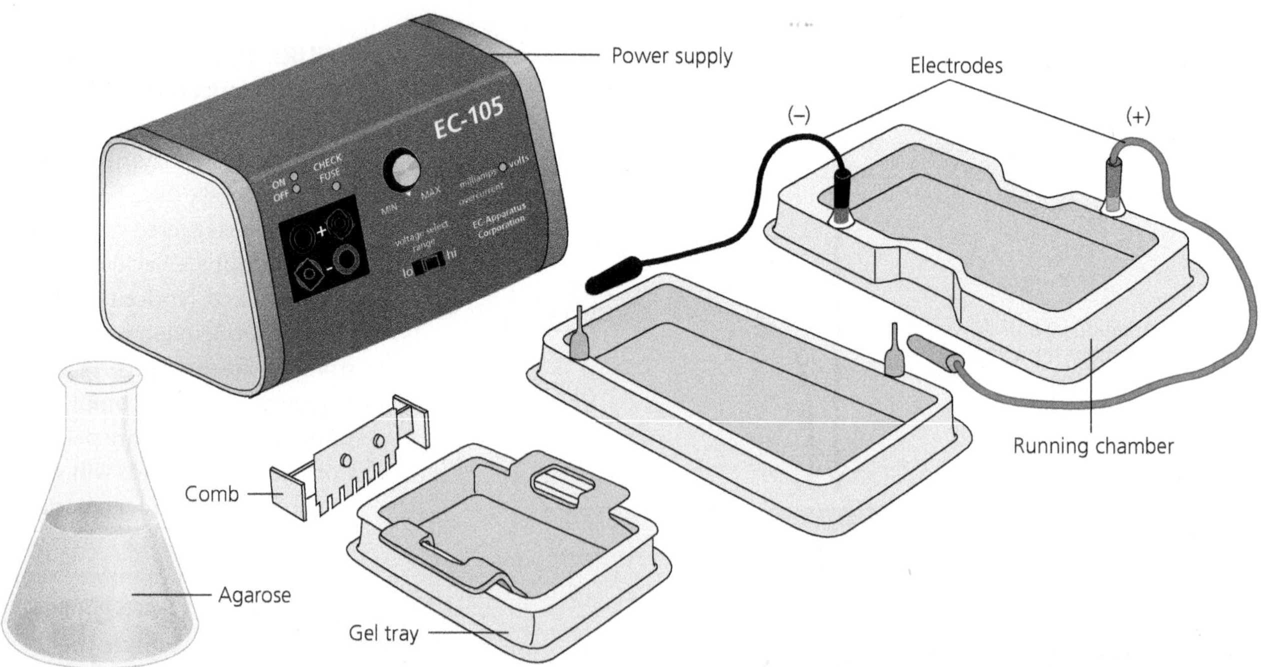

FIGURE 10.3
Electrophoresis apparatus. The apparatus should include the gel tray, comb, agarose, running chamber with electrodes, and power supply.

1. DNA will move through the gel along the electric current, its direction depending in part on the molecule's charge. In which direction, positive to negative or vice versa, will DNA move in the electric field? Why?

2. Given that all DNA molecules have similar charge-to-mass ratios, what property do you think is most important in determining migration within the gel? Why?

3. To map the plasmid, you will need to know the actual size of the fragments on the gel. Can you suggest a way to determine the size of these fragments?

4. Given that the distance a fragment of DNA moves into the gel is inversely proportional to the log of the size of the fragment in base pairs, what will a graph of fragments 0.5, 1, 5, 10, and 20 kilobases look like? (The size of DNA fragments is measured in nucleotide bases, and 1,000 bp equals 1 kilobase, Kb.) What do the axes represent?

Hypothesis

Based on your understanding of restriction enzymes and electrophoresis, state a hypothesis concerning the migration of DNA fragments.

Prediction

Predict the banding patterns you will observe in the gel based on your hypothesis.

Procedure

1. Practice pipetting and loading a gel

If you have not used a micropipettor or microcapillary pipette before, practice the technique. You may want to have your instructor check your technique because errors as small as one-half microliter can result in unsuccessful digests. Although a microliter (μL) seems very small, it can be an amazing amount in molecular biology. For example, 1 μL might contain the amount of DNA in all the genes in a human cell!

Micropipettors are designed to measure small volumes of liquid, and each pipettor can be adjusted for a range of volumes, for example, from 2 to 20 μL. *Never use a pipettor without first attaching the plastic disposable tip.*

a. Examine the micropipettor. What is the maximum volume for the pipettor (should be numbered and color coded)? Locate the window that displays the volume currently set for the pipettor. Set the micropipettor to 20 μL by rotating the dial until the correct volume appears in the display window. Sketch this window in the margin of your lab manual and show the setting to your instructor before continuing. (See Appendix C and Figures C.4a and C.4b.)

b. Locate the disposable tips that correspond to your pipettor. Attach the pipette tip and check that it is securely in place.

c. Using the practice solution of water, glycerin, and blue dye, practice drawing the liquid into the pipette tip. *Before placing the tip into the liquid,* depress the pipette plunger with your thumb to the *first stop* to eject any air.

d. Place the tip into the practice solution and *slowly* release the plunger. Remove the tip from the liquid.

e. Locate an empty microtube from your materials. Insert the pipette tip into the microtube so that the tip is close to the bottom of the tube. Touch the tip to the side of the tube. Slowly press the plunger down to the first stop and then continue to *press all the way down to the second stop* to release all the liquid from the tip. Remove the pipettor from the microtube. *Do not release the plunger yet!*

f. When the tip is completely out of the tube, *slowly* release the plunger to the starting position. *Never release the plunger inside the tube* or you could withdraw an unknown amount of solution back into the tip.

g. Discard the tip, using the release button on the pipettor. *Use a new tip each time you use the pipettor,* even if you are pipetting the same liquids. Allow other members of your team to practice pipetting before continuing to the next step.

h. Reset the pipettor to 2 μL. Sketch the window for this volume in the margin of your lab manual. Before pipetting any liquid, depress the plunger to locate the first and second stops. Note the short distance to the first stop for this extremely small volume. Note that there is very little distance between stop 1 and stop 2.

i. Repeat steps c–g, being careful to depress the plunger only to stop 1 before withdrawing the liquid. Observe the volume of liquid in the pipette tip before releasing the liquid into the microtube.

j. *At this point your instructor may direct you to continue preparing your digestions (Procedure step 2 below) or to complete this section by practicing loading the gel as directed in the next step.*

k. **Practice loading a gel.** Your instructor has prepared practice gels in a petri dish. The gel should be covered with water. *Reset the pipette volume to 20 μL.* (Compare the volume window to your original sketch.) Attach a new tip to the pipettor. Following steps c–g, draw 20 μL of the blue practice mixture into the pipette tip and hold the micropipettor with two hands—one hand to deliver the sample and the other to stabilize the end. Be sure that the sample is all the way down in the tip of the pipette and that there is no air between the sample and the tip. Carefully place the tip just inside the well but not piercing the side or bottom of the well. Slowly release the liquid into the well. All students should practice this procedure several times.

In all cases, a fresh, unused pipette tip or microcapillary pipette should be used every time a pipetting task is performed. This is true even if you are pipetting repeatedly from the same solution, especially in the case of enzymes, which are easily contaminated.

2. *Prepare the digestions*

a. Your instructor has already prepared three color-coded microtubes containing deionized (DI) water and an appropriate buffer. Obtain these three microtubes and label them **A** (for *Ava* II), **P** (for *Pvu* II), and **AP** (for both). Be sure that your labels correspond to the color key provided by the instructor. Write that color in the appropriate cell in the first column of Table 10.1.

TABLE 10.1 Contents of the Three Tubes. Write the appropriate color for each tube, A, P, and AP. These tubes will be incubated in a 37°C water bath for at least 30 minutes. Some items in the tube have already been added by the instructor. Note that each tube should have a total of 30 μL of solution.

Tube Label/Color	Instructor Adds	Student Adds Before Incubating Tube
A/ ______	24 μL DI water 3 μL *Ava* II buffer	2 μL pUC 19 DNA 1 μL *Ava* II enzyme
P/ ______	24 μL DI water 3 μL *Pvu* II buffer	2 μL pUC 19 DNA 1 μL *Pvu* II enzyme
AP/ ______	23 μL DI water 3 μL *Ava* II* buffer	2 μL pUC 19 DNA 1 μL *Ava* II enzyme 1 μL *Pvu* II enzyme

*Use the *Ava* II buffer only.

When prompted, you will add the amounts of each substance indicated in Table 10.1 to these microtubes. Read and understand steps b–d before you prepare your microtubes. *Do not begin to prepare your microtubes until step e.*

b. The buffers already *added by your instructor* are endonuclease buffers for the respective restriction enzymes. They contain a Tris buffer to maintain the pH and salts such as NaCl to maintain optimal ionic strength and $MgCl_2$ required by the enzymes for catalytic activity.

c. The concentration of pUC 19 DNA solution in Table 10.1 has been prepared by your instructor to contain 1 μg/2 μL of DNA (or 0.5 μg/μL). This DNA will be cut into smaller pieces by the restriction enzymes.

d. You will use restriction enzymes *Ava* II (found in *Anabena variabilis*) and *Pvu* II (found in *Proteus vulgaris*). Enzymes are assigned units based on the amount required to digest 1 mg of DNA in 1 hour. Thus, 5 μg of DNA requires 5 units of enzyme to completely digest the DNA in 1 hour. Given that we will be digesting 1 μg of DNA, at the appropriate time, you will add 1 μL of the appropriate enzyme to each microtube.

The restriction endonucleases must be added last to the mixture. Adding the enzyme early will result in immediate and inappropriate digestion of the DNA!

Once you understand each component added to the microtubes, proceed with step e. Accuracy in pipetting is critical for the following steps.

e. Add 2 μL of pUC 19 DNA to each tube. Tap the bottom of the tube on the lab bench to move the DNA to the bottom of the tube. Return the stock supply of DNA to the cold box.

f. Add the appropriate enzyme(s) to each tube. Tap the bottom of the tube on the lab bench. Be sure all 20 μL of the reaction mix is together at the bottom of the tube. If it is not, gently tap the bottom of the tube on the lab bench until all the liquid accumulates at the bottom. Alternatively, you may use a microcentrifuge to briefly spin and mix the solution. Return stock supplies of enzymes to the cold box.

g. Label the top of the microtube with a symbol to designate your team. Place the tubes at 37°C in a water bath or incubator for at least 45 minutes but preferably 1 hour.

3. Pour the agarose gel

While the DNA digestions are incubating, pour your gel (refer to Figure 10.4).

a. Set the comb over the gel plate so that the teeth rest just above the plate (Figure 10.4a).

b. Wearing disposable gloves, remove the agarose gel solution from the 55°C water bath. Gently swirl the flask to be sure the solution is uniformly melted.

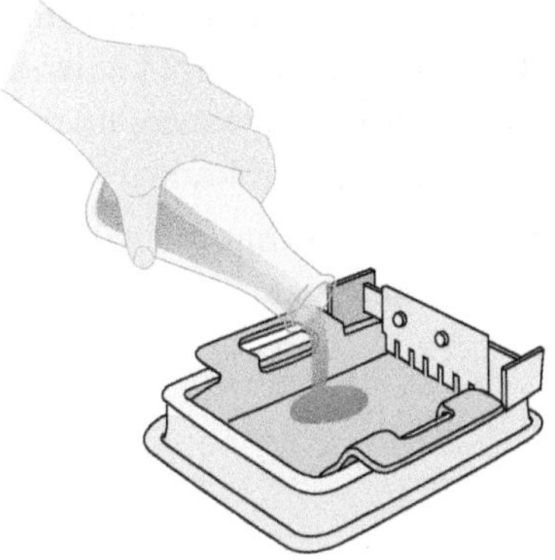

a. Pour gel with comb in position

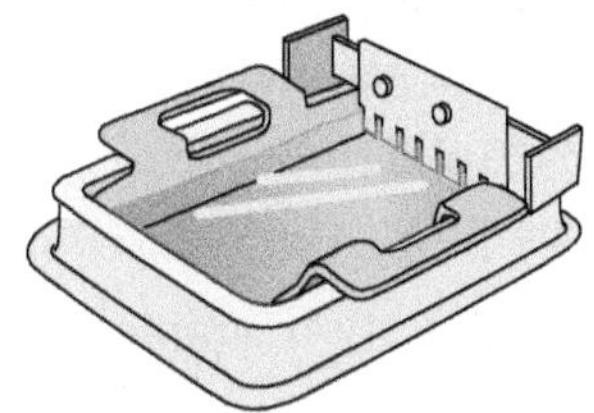
b. Gel covers half of comb teeth

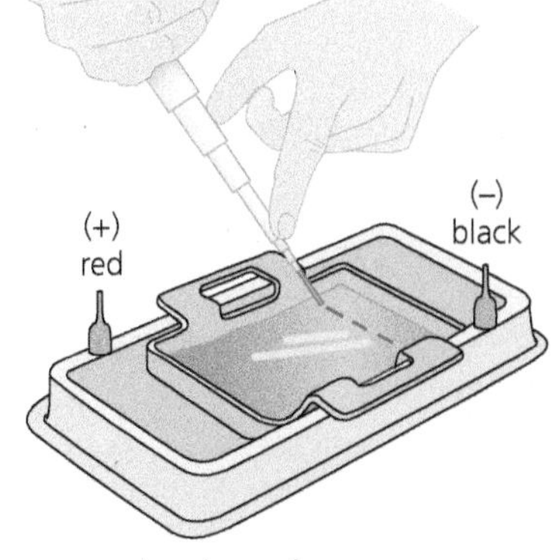

c. Load samples

FIGURE 10.4

Preparing, loading, and running the gel. (a) Pouring the gel into the gel tray. Note the position of the comb. (b) The gel is poured so that it covers approximately half the height of the comb teeth. The comb will be removed before the samples are loaded. (c) Loading samples and dye using a micropipettor. (d) Gel running with electrodes attached. Note that the dye bands are moving through the gel.

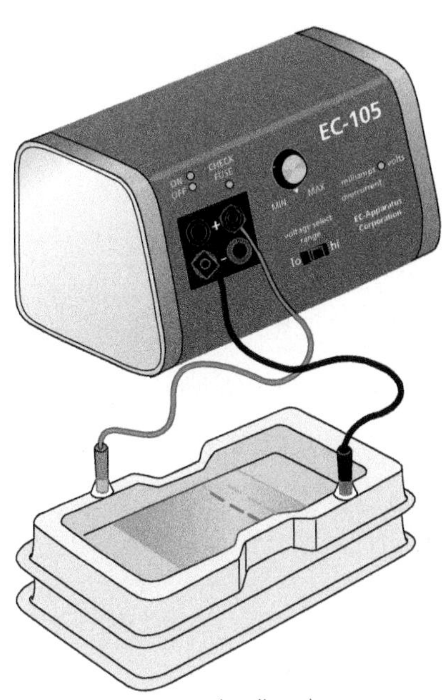

d. Gel running—note loading dye moving on gels

c. Slowly pour the gel onto the plate until the solution covers about one-half the height of the comb's teeth (Figures 10.4a and b). If the gel is too thin, it will fall apart; if it is too thick, it will take too long to run. Avoid creating bubbles. If any form, quickly and carefully pop them with a sharp object, such as a micropipette tip. (Why would bubbles be a problem?)

d. Allow the gel to solidify. It will become opaque.

e. While the gel cools and hardens, return to the Introduction and complete the Preliminary Questions (pp. 250–251).

4. Load the gel and perform electrophoresis of samples

a. When the gel has hardened, squirt some water around the comb and then pull the comb teeth from the gel slowly, carefully, and evenly. If you pull too fast, the suction will break out the bottom of the wells created by the comb.

b. Place your gel in the running chamber with the wells nearest the negative (black) electrode, and completely cover the gel with 1× TAE. Orient the running chamber so you can see in the wells. *Do not move the running chamber from this point on.*

c. Before you load the gel, you must assemble three additional samples. Obtain these tubes from the instructor. One sample (labeled 1) is for loading dye only. A second sample (labeled 2) contains **molecular weight markers**, and the third (labeled 6) is for uncut pUC 19 DNA. Your instructor will give you the key for tube colors. Write the color for each of these tubes below. These tubes contain the following, already added by your instructor:

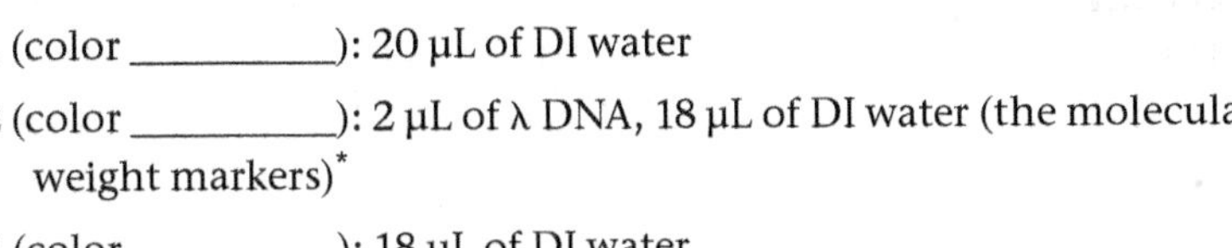
1 (color __________): 20 μL of DI water

2 (color __________): 2 μL of λ DNA, 18 μL of DI water (the molecular weight markers)*

6 (color __________): 18 μL of DI water

*The molecular weight markers (tube 2) are used to determine the size of a DNA fragment. These markers, purchased commercially, are of known sizes with which you can compare your results. You will analyze these markers simultaneously in the same gel with your fragments of unknown size.

d. Assemble the six microtubes with respective solutions that you will load on your gel and arrange in this order in a microtube rack:
   (1) Tube 1 from step c (DI water)
   (2) Tube 2 from step c (molecular weight markers)
   (3) Digestion sample A (*Ava* II digest)
   (4) Digestion sample P (*Pvu* II digest)
   (5) Digestion sample AP (*Ava* II and *Pvu* II digest)
   (6) Tube 6 from step c (DI water; you will add DNA in the next step)

e. Add 2 µL of pUC 19 DNA that has not been digested (uncut DNA) to tube 6. Tap the tube on the table. What is the purpose of the uncut or undigested DNA in tube 6?

f. Add 2 µL of loading dye to each of the six tubes. Tap the end of each tube on the table to mix.

g. Carefully load each of the six samples into their corresponding wells in your gel, as shown in Figure 10.5. Tube 1 will be in well 1, tube 2 in well 2, the *Ava* II digest in well 3, the *Pvu* II digest in well 4, the *Ava* II/*Pvu* II digest in well 5, and the uncut DNA in well 6. *Set your micropipettor to 20 µL.* Load 20 µL of each sample. Hold the micropipettor with two hands, one hand to deliver the sample and the other to stabilize the end. Be sure that the sample is all the way down in the tip of the pipette and that no air is between the sample and the tip. Place the tip of the pipette below the surface of the buffer and just into the well. Be careful not to puncture the well bottom. Slowly release the sample into the well. The density of the dye will help it sink into the well (Figure 10.4c).

h. Carefully attach positive (red) and negative (black) electrodes to the corresponding terminals (red into red, black into black) on the power supply and on the gel box (Figure 10.4d).

i. Turn the power on to about 130 volts, and make sure that small bubbles arise from the electrodes in the gel buffer, verifying current flow. Check the power source to ensure it is set to volts. Check the loading dye to make sure that the samples are running in the correct direction, toward the positive electrode (see Figure 10.4d).

Turn off the power to the gel before making any adjustments to the electrophoresis setup.

j. Run the gel until the loading dye (bromophenol blue) moves down the gel approximately 6–8 cm (this will take 30–35 minutes). Watch the gel carefully. After 10 minutes, first turn off the power, then check to make sure

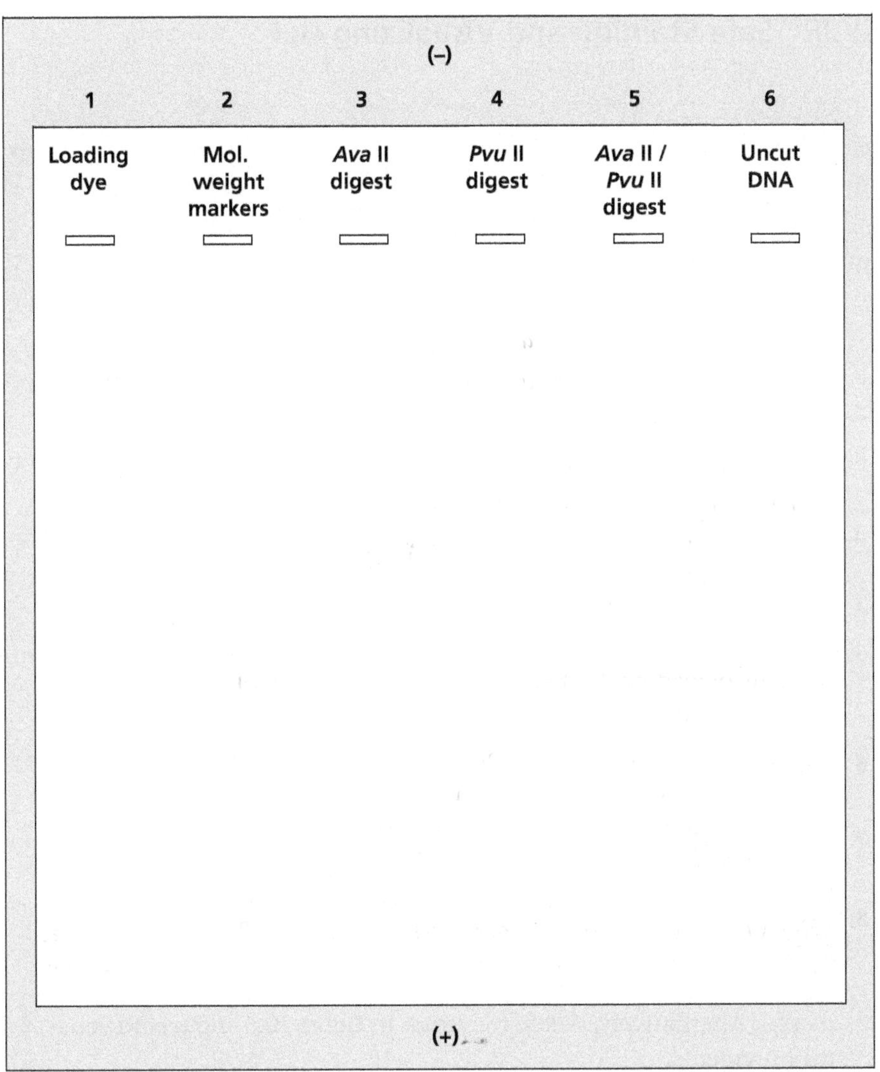

FIGURE 10.5

Wells for loading the gel. Refer to this figure to ensure that the correct sample is loaded into the appropriate well. Write the appropriate microtube color above each well.

that the gel is not hot. Accidentally making the gel or gel buffers from the wrong concentration of TAE can result in gel and buffer overheating. If this occurs, your experiment will fail.

5. Practice mapping

While the gel is running, work the practice mapping problem (Exercise 10.2) at the end of the lab topic.

6. Visualize the DNA in the gel

The loading dye that is visible on the gel does not stain the DNA but is added so that the movement of the current can be verified. The DNA must be stained by another means, by adding either SYBR® Safe or methylene blue. To visualize the DNA, proceed as follows.

SYBR® Safe Staining and Visualizing Gel

SYBR® Safe is not classified as hazardous and is visible with blue light. Follow common laboratory safety precautions. Wear gloves when handling the gel. Ask your instructor for procedures to clean up and properly dispose of your gel and gloves.

1. Turn off the power.
2. Wearing disposable gloves, remove the gel tray from the running chamber and place in a carrying tray.
3. Locate the transilluminator imaging system at your bench or in the laboratory. Slide the gel onto the tray of the imaging system.
4. Follow the directions for your imaging system to visualize the gel.
5. Use the camera or printer features to record the gel. If your imaging system has a save feature or if the image can be exported to a computer, save a copy of your gel image as a jpeg file.
6. Remove the gel from the imaging system. Follow the directions for cleaning the imaging system tray.
7. Ask your instructor for directions for proper disposal of your gel and gloves.
8. Using a printed copy or photograph, measure the distance traveled for each band on the gel and write the measurement next to the corresponding bands on the photograph with a fine point permanent marker. Alternatively, sketch the bands in Figure 10.6 and record your measurements.

Alternative Staining Procedure: Methylene Blue

1. Turn off the power. Wearing gloves, remove the gel tray from the running chamber.
2. Slide the gel out of the tray into a 0.025% solution of methylene blue in a staining tray for 30 minutes.
3. Transfer the gel to a destaining tray and destain for several hours or overnight in enough water to just cover the gel. DNA bands will become visible when the gel is viewed over a visible light box.
4. Photograph the gel using a digital camera.
5. Using the photograph, measure the distance traveled for each band on the gel and write the measurement next to the corresponding bands on the photograph with a fine point permanent marker. Alternatively, sketch the bands in Figure 10.6, and record your measurements.
6. Ask your instructor for instructions for proper disposal of your gel and gloves.

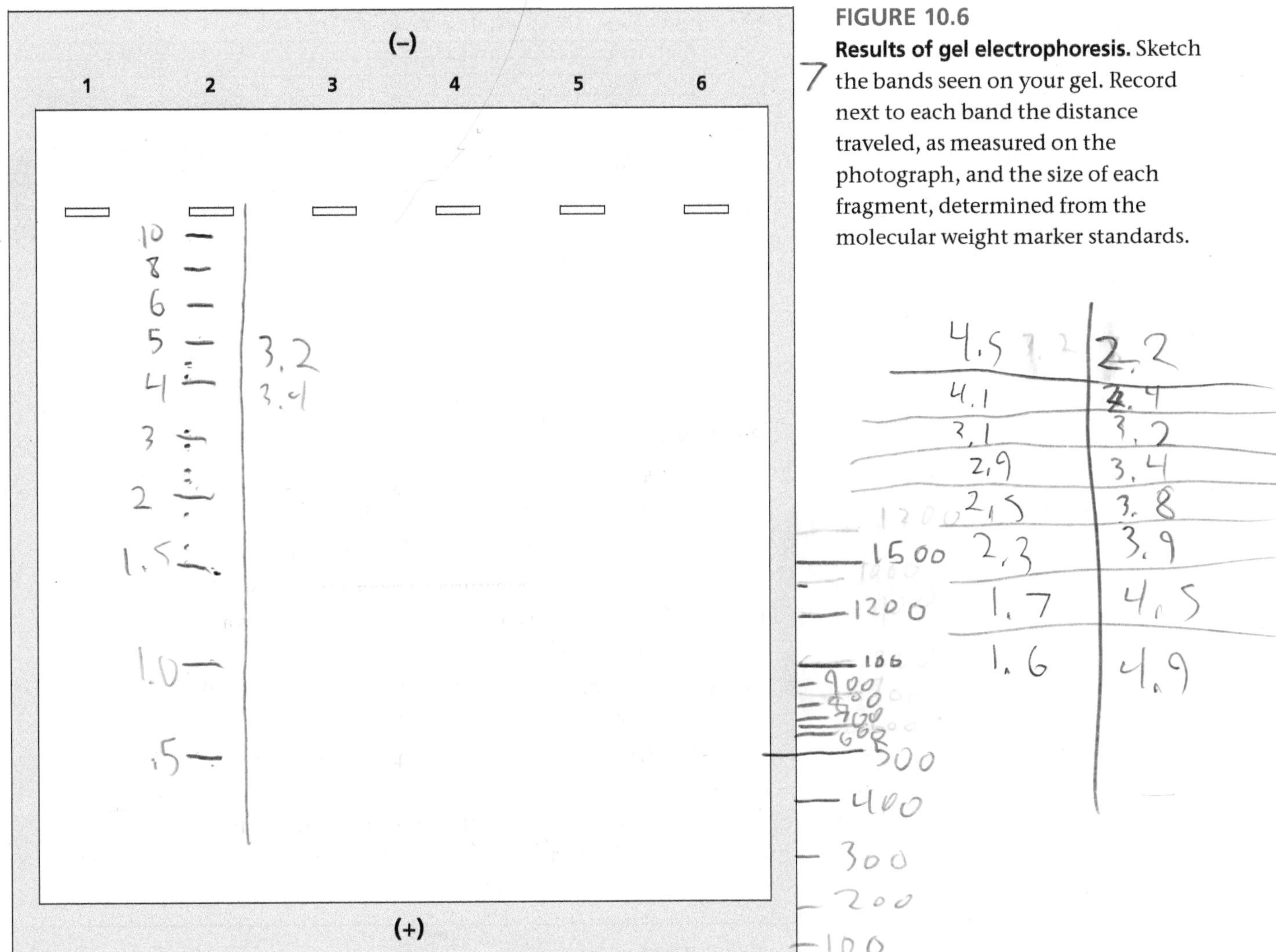

FIGURE 10.6
Results of gel electrophoresis. Sketch the bands seen on your gel. Record next to each band the distance traveled, as measured on the photograph, and the size of each fragment, determined from the molecular weight marker standards.

Results

1. What are the "controls" in the gel you ran?

Tube 6: uncut pUC19 DNA

2. In the wells where uncut DNA is run, you will see more than one band. Hypothesize: What types of DNA molecules might be represented in the different bands? How could you test your ideas? (*Hint:* Remember that the agarose gel is a matrix and consider if your uncut pUC 19 is linear or circular.)

The different DNA would be linear, circular, or super coiled. Its possible to test by doing a gel experiment

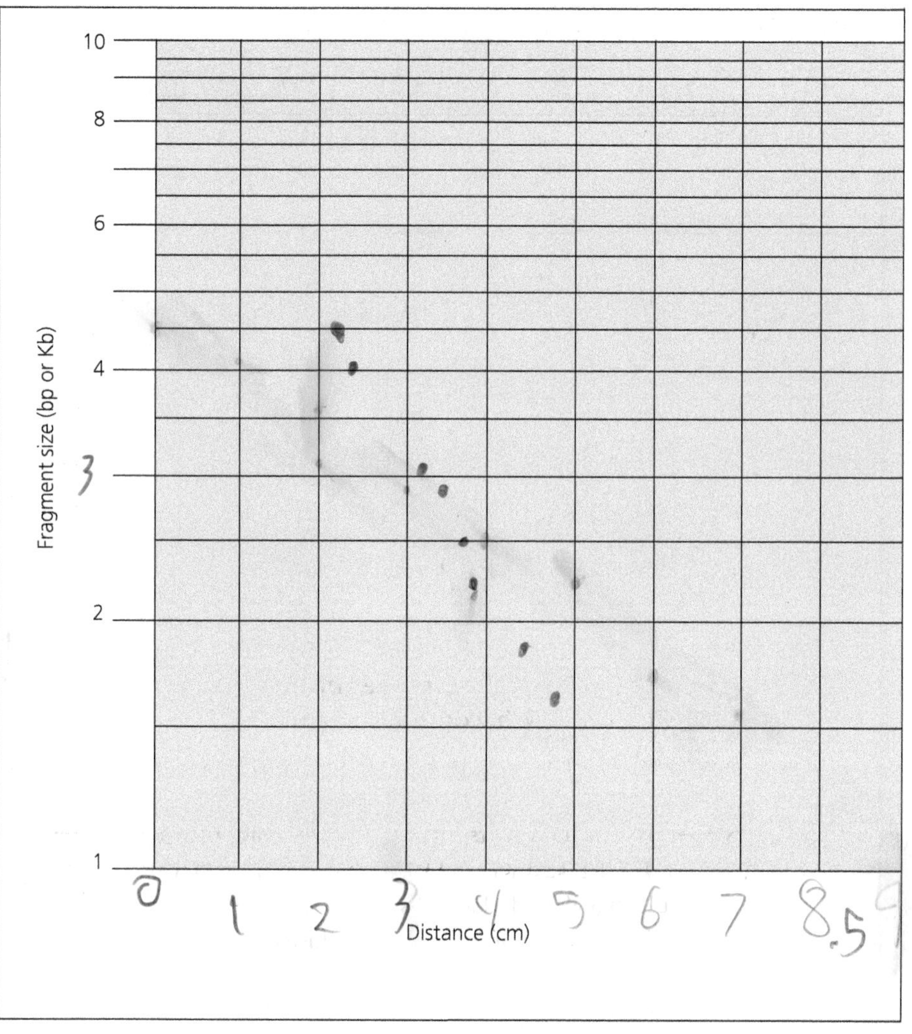

FIGURE 10.7
Relationship of distance traveled (cm) and the size of the restriction fragment (bp or Kb) for known DNA molecular weight markers. Note the distances traveled for each marker, and choose and add an appropriate scale for the x-axis. Obtain the known marker sizes from your instructor, and graph them with the corresponding distances from your gel. Note this is graphed on semi-log paper.

Kb	Distance
4.5	2.2
4.1	2.4
3.1	3.2
2.9	3.4
2.5	3.8
2.3	3.9
1.7	4.5
1.6	4.9

3. Using the photograph or Figure 10.6, locate the bands for the molecular weight markers. Which lane contained the molecular weight markers? You should have recorded the measurements for the distance traveled for each molecular weight marker fragment on your photograph or sketch.
4. Obtain the known sizes of the DNA molecular weight marker fragments from your instructor. Write these sizes (in bp or Kb) beside the molecular weight marker bands in Figure 10.6 or on your photograph. In Figure 10.7, graph the distance traveled by the marker fragments (bands) on your gel (in cm on the *x*-axis) versus the size of the corresponding fragments (in bp or Kb on the *y*-axis). Draw a best-fit line through these points on the graph.
5. Using the graph you constructed in Figure 10.7, determine the size of every other band on the gel. To do this, on the line drawn in Figure 10.7, locate the distance traveled by each restriction fragment (measured and recorded in step 3). Follow this point over to the *y*-axis to determine the number of base pairs in the restriction fragment. Record your results next to the corresponding band in Figure 10.6.

Use the following Discussion section as a guide to analyzing your data. If you have not already worked the practice problem, Exercise 10.2, do it now, before continuing to the Discussion section. Your final analysis should include a complete restriction map of the pUC 19 plasmid.

Discussion

Do not begin the discussion questions until you understand the practice problem. Then refer to your results as you answer the following questions about the map of pUC 19.

1. How many Kb is pUC 19?

   2.5

2. How many DNA fragments were produced by the enzyme digestions?

   By *Ava* II?

   1

   By *Pvu* II?

   1

   By the double digest with *Ava* II and *Pvu* II?

   3

3. Note that both the small fragments generated in the single digest are still present in the double digest. What does this mean?

   This means its linear

4. Draw the two restriction fragments for *Ava* II in the space provided. Label the two ends "*Ava* II." Indicate the size of the fragments that were produced in the double digest, and place the *Pvu* II sites on the large *Ava* II fragment. Refer to the practice problem for assistance.

AVAII

AVAII

PvuII

5. Join the fragments you have drawn above to re-create the original plasmid.

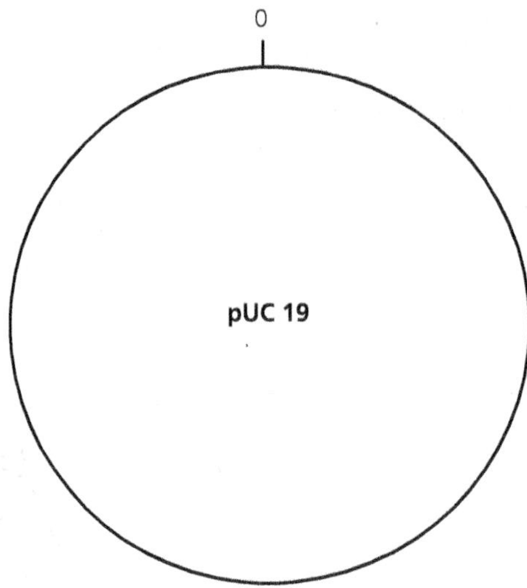

6. Now that you have mapped some DNA fragments, what could you do to further characterize the DNA you have isolated? Refer back to the Introduction to this lab topic.

EXERCISE 10.2

Practice Problem for Mapping DNA

Mapping and sequencing genomes of organisms has led to a new field of genetics called **genomics**. One approach to determining the gene sequence is mapping the DNA fragments produced by restriction enzymes. These restriction sites can serve as markers to align the DNA. An example of a DNA map is shown in Figure 10.8 for a small plasmid, LITMUS 28i, similar to pUC 19. The numerous restriction sites are indicated by the abbreviations around the perimeter of the circle. Can you find a restriction site for *Pvu* II? The location of each enzyme's restriction site was determined using a process similar to the one you will be using today. Fortunately, you are working with only two enzymes!

The construction of a DNA map is really a logic puzzle. You must compare results for the single digests with those of the double digests (those cut with two enzymes) to correctly orient the sites on the DNA molecule. As with anything else, the process becomes easier with practice. The experience you will gain as you solve the following problem will help prepare you to analyze your exercise results.

You are investigating the pathogenic organism *Bacillus anthrax*, which has recently been isolated from infected individuals. It is known that this organism contains a plasmid, and because these genetic elements often

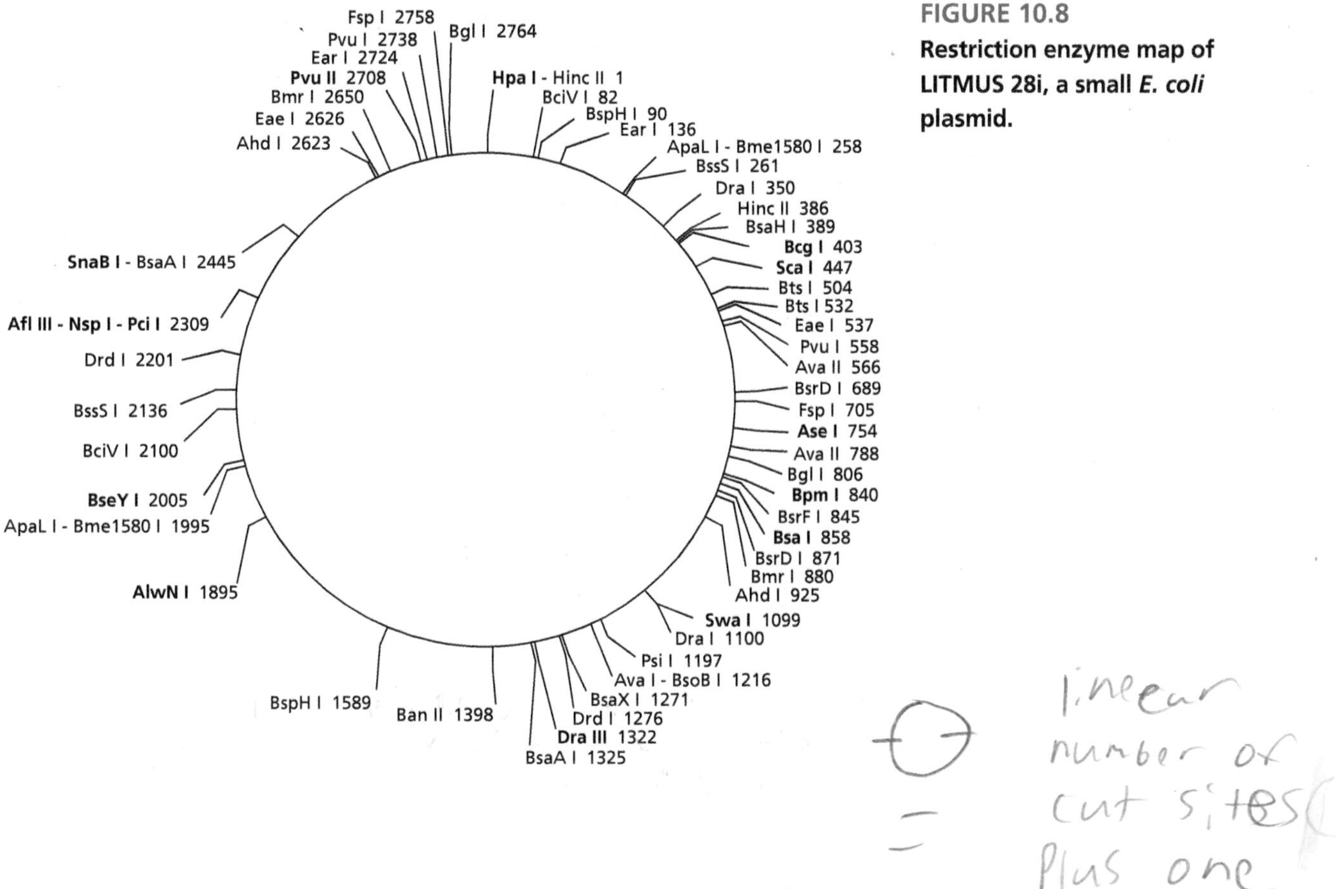

FIGURE 10.8
Restriction enzyme map of LITMUS 28i, a small *E. coli* plasmid.

linear
number of
cut sites
plus one

contain antibiotic resistance genes, you decide to begin your investigation with the plasmid. Initial studies indicate that the DNA molecule is 4,000 bp (4 Kb) long.

Procedure

1. Inspect the gel diagrammed in Figure 10.9. The restriction pattern shown in the gel diagram was produced by digestion of the bacterial plasmid with two restriction enzymes, *A* and *B*.
2. Determine the number of fragments produced by each enzyme, and determine the number of fragments produced by the double digest.

   A B AB
   2 2 4

   How does the number of fragments correlate with the number of restriction sites in the DNA molecule? Remember that plasmids are circular.

   The number of fragments is the same amount of restriction sites

3. Note that the size of the fragments has to be determined by comparison to the molecular weight markers. In this problem, the size of the fragments has already been determined and is recorded next to each band.

FIGURE 10.9
Restriction pattern generated by restriction digest of the *Bacillus anthrax* plasmid. Note the position of the origin and the molecular weight markers.

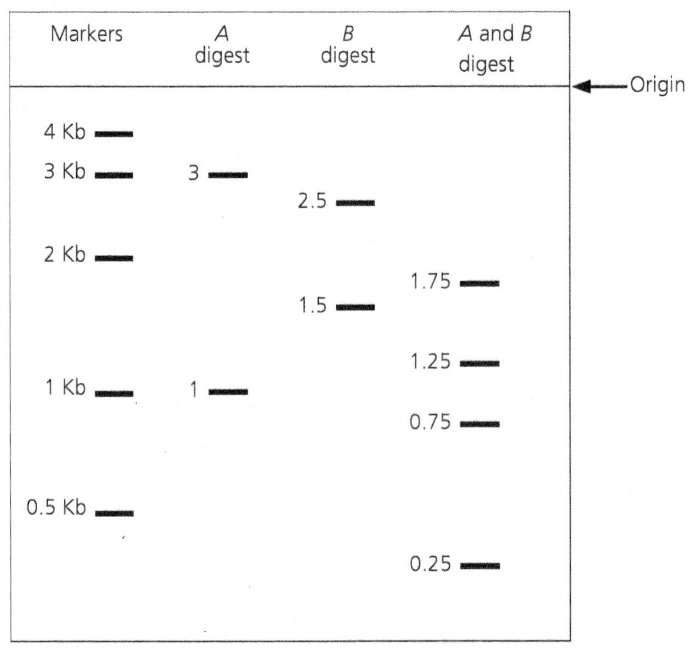

4. Make a restriction map of the isolated plasmid by determining the relative positions of the restriction sites. The best approach is to realize that the double digest is really an extension of the single digests. For example, the double digest is analogous to taking the bands you see in the *A* restriction digest and digesting these bands with the *B* enzyme. The smaller bands you see in the double digest should add up to the bands seen in the *A* digest alone.

   a. Determine which bands of the double digest come together to form the small *A* restriction fragment.

Fragments .75 and .25

   b. Determine which bands of the double digest make up the large *A* fragment.

Fragments 1.75 and 1.25

   c. Draw the two *A* restriction fragments in the space provided, and indicate where the *B* restriction sites fall on the fragments. Indicate the size of the fragments that are produced in the double digest.

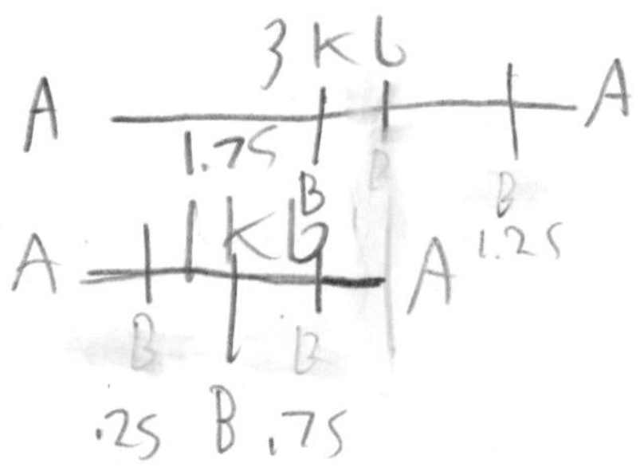

B 2.5 kb B
.75 A 1.75
B 1.5 B
.25 1.25

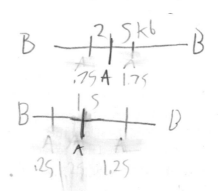

d. Do the same analysis on the *B* fragments. Compare the double digest with the single *B* digest.

e. Draw the *B* restriction fragments in the space provided, and indicate where the *A* restriction sites fall on the fragments. Indicate the size of the fragments that are produced in the double digest.

f. Align the fragments from the preceding analysis. You will notice from your drawings that there are double digest fragments of the same size. These fragments are the key to solving our puzzle because they represent overlap. If we align the large *B* fragment and the large *A* fragment, we find that we can match up restriction sites (Figure 10.10).

g. Continue to align fragments. As shown in Figure 10.10, the alignment is used to draw out one longer fragment. By continuing to align fragments, it is possible to re-create the entire restriction map. Which fragments would you try to line up next?

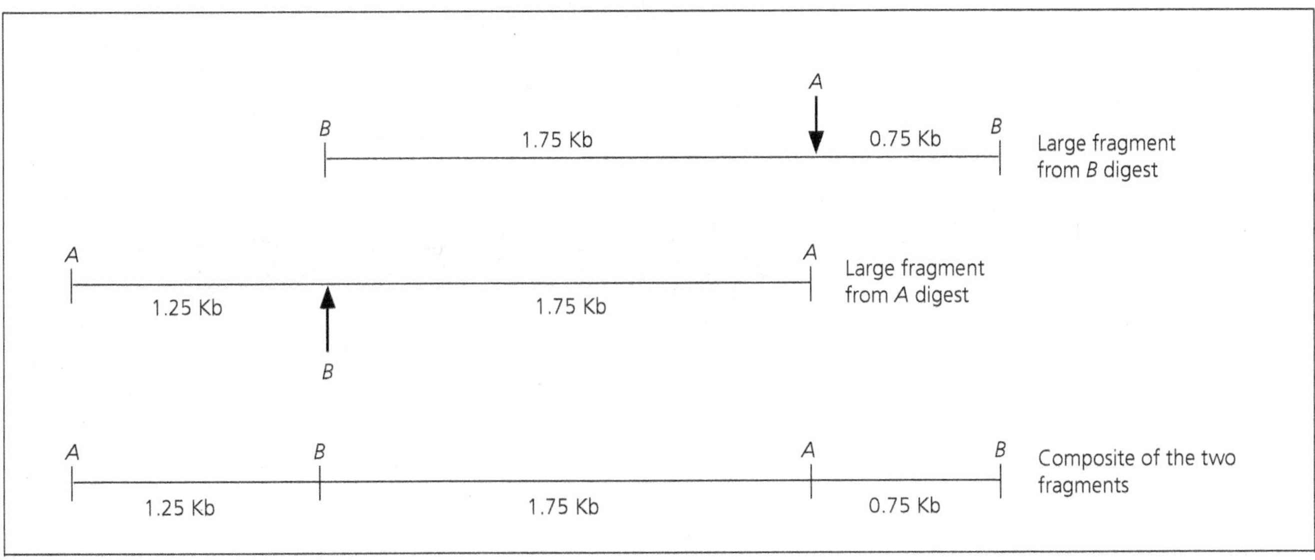

FIGURE 10.10
Alignment of the large fragments from single digests. The large fragments from both single digests can be aligned so that one longer fragment is drawn.

Because we are working with a circular DNA molecule, would you expect to eventually repeat yourself?

h. Continue to align fragments until you have finished your complete map.

i. Draw the map in the space provided. *Hint*: Arbitrarily place one of the *A* sites at position 0 (the very top of your circle) and use that as your reference site.

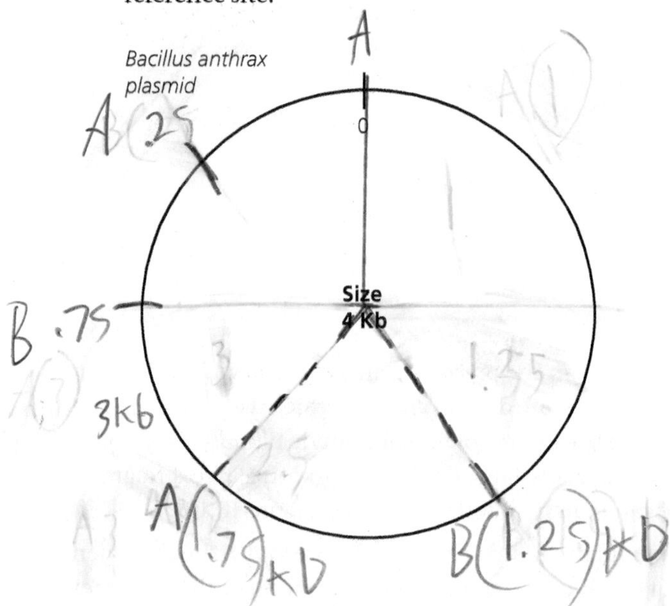

After completing the practice problem, resume your work to complete Exercise 10.1.

REVIEWING YOUR KNOWLEDGE

1. **a.** State the role of each of the following in molecular biology and provide an example if appropriate: *restriction enzyme, plasmid DNA, molecular weight markers, micropipettors.*

   b. State the role of each of the following in gel electrophoresis: *gel buffer, comb, power pack, agarose, SYBR® Safe* (or alternatively, *methylene blue*), *loading dye.*

2. The treatment of diabetes was revolutionized when transgenic bacteria were utilized to produce human insulin. What are transgenic organisms? What feature of DNA is responsible for a human gene in bacteria effectively producing the insulin protein?

3. Hypothesize about why restriction enzyme digestions are performed at 37°C.

4. Restriction endonucleases are enzymes. What type of macromolecule are endonucleases? Think of a general way that these enzymes might go about the business of cutting DNA.

5. *Bam*HI and *Eco*RI are two restriction enzymes used to digest a plasmid DNA in an experiment similar to the one completed in the laboratory. The results of the gel electrophoresis are shown in the diagram below. The number next to each band represents the size of the DNA fragment.

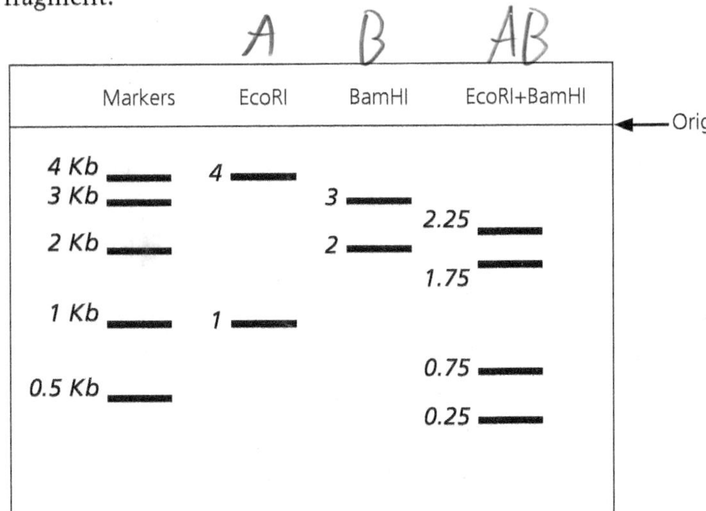

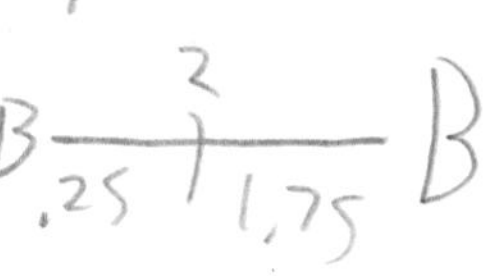

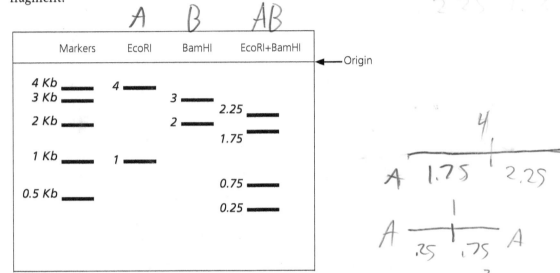

Using this information, answer the following questions.

a. What is the total length (in kilobases, Kb) of the plasmid DNA used in this experiment?

4

b. How many *Bam*HI restriction sites are present in this DNA?

2

How many *Eco*RI restriction sites are present?

2

c. How many fragments are produced by the double digest with *Bam*HI and *Eco*RI?

4

d. Using this information, construct the map for the plasmid DNA showing the location of all the restriction sites. Follow the procedure in the practice mapping problem, Exercise 10.2.

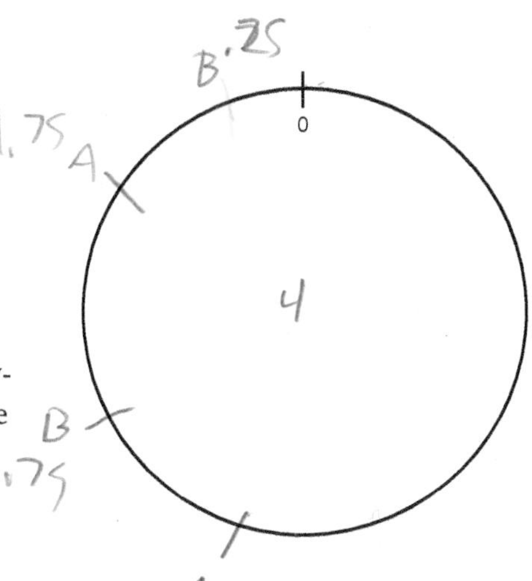

APPLYING YOUR KNOWLEDGE

1. Scientists are developing genetically engineered foods that are insect-resistant, are more nutritious than traditional foods, and can grow in stressful environments. Many scientists think that genetically modified (GM) foods will become an essential part of global strategies to provide food for the growing human population. However, the development of GM foods has raised ethical, legal, economic, and environmental questions. Using resources from the Web, the library, and your text, choose one of the following topics and prepare an essay discussing the issues and scientific evidence involving GM foods or scientists' concerns for potential effects of GM foods.
   a. Intellectual property rights: *Monsanto Canada Inc. v. Schmeiser*
   b. Environmental concerns—populations and biodiversity:
      - Will GM plants that are toxic to agricultural insect pests have a negative impact on other insects such as monarch butterflies?
      - Will the decline in insect pests have an effect on other animals, such as birds?
      - What could be the impact if crops that are herbicide-tolerant are able to cross-pollinate with weeds that are closely related?
      - Genes have been inserted into chromosomes of a fish species to make the animal grow larger. Would this give these fish a reproductive advantage over natural populations if these fish are more successful at attracting mates? What if the offspring of the genetically modified fish are less able to survive in natural environments?
   c. Human health:

      What are the safety assessment standards of the U.S. Food and Drug Administration (FDA) and what is the approval process for genetically engineered plants for human consumption? What are the assessment standards and approval process for agricultural plants that are bred using traditional breeding techniques for crossing and artificial selection?
2. More and more, in cases of rape or murder, forensic scientists are using DNA fingerprinting as evidence for the guilt or innocence of a suspect. In this process, a few selected portions of DNA from small amounts of blood or semen from the crime scene can be analyzed and compared with DNA from a suspect. If gel electrophoresis shows that two samples match, the probability that the two samples are ***not*** from the same person can be anywhere from one chance in 100,000 to one in 1 billion, depending on how the test was performed. And yet, even with compelling DNA evidence to the contrary, suspects are sometimes found "not guilty." What problems with this type of evidence can arise to create doubt about the guilt of the person?
3. The map for pUC 19 is available online at New England Biolabs, http://www.neb.com; in the search window type "pUC 19 map." Compare your map using *Pvu* II and *Ava* II with that published on the Web.

Locate the restriction sites for *Pvu* I and *Acl* I. How many restriction sites are present for each enzyme?

How many fragments would be produced from digestions of pUC 19 using these two enzymes?

INVESTIGATIVE EXTENSIONS

1. A revolutionary new molecular tool called CRISPR (clustered regulatory interspersed short palindromic repeats) relies on RNA to target specific gene sequences and then either inhibit the expression of the gene or insert a gene into the DNA. CRISPR, along with associated enzymes called Cas (CRISPR/Cas9 system), was discovered in 2012 while investigating *Streptococcus pyogenes,* bacteria that cause disease. Bacteria use this gene editing system to defend against viruses called bacteriophages. Once infected by the bacteriophage, a small bit of the phage DNA is used to develop a system to recognize and destroy similar DNA in future infections for the cell and its descendants. CRISPR/Cas9 can precisely recognize specific sequences in the DNA and can be utilized by molecular biologists to turn off the gene or insert DNA sequences directly into the organism of interest. The system is inexpensive, widely available, and easy to master.
   a. Research the CRISPR/Cas9 system. Draw and label a diagram that outlines the procedure.
   b. What components of the CRISPR/Cas9 system are equivalent to restriction enzymes that you used in the laboratory? What are the advantages of CRISPR? Are there limitations for CRISPR that can be overcome with restriction enzymes and recombinant DNA?
   c. Describe four examples of scientific research in medicine, agriculture, and the environment that are changing rapidly as a result of this new tool in molecular biology.
   d. CRISPR kits may soon be available on the internet. What are the issues with their widespread availability—both pro and con?
   e. In 2015, CRISPR was selected as the top scientific "Breakthrough of the Year" by *Science*. In the same year, the National Academy of Science convened a meeting of world-renowned scientists, who agreed to extend a moratorium on the use of CRISPR in human embryos (sometimes called editing the human germ line). Describe the published research that prompted this moratorium and discuss the issues.
2. Francis Collins, the Director of the National Institutes of Health, recently described the revolutionary changes that have occurred over the last decade as a result of genome science, and the potential for medical applications and understanding the human condition. Perhaps the most revolutionary changes will come in the area of personalized medicine. Therapeutic treatments will be optimized based on the molecular

characteristics of the disease and the patient. The consequences are challenging and encompass medicine, the pharmaceutical industry, legal and ethical issues, and economic considerations. Read Dr. Collins's article and answer the following questions.

Collins, F. "Has the Revolution Arrived?" Nature, *2010, vol. 4464, pp. 674–675.*

a. What are the specific genomic techniques that are being used in personalized medicine?

b. What are the possibilities for improving medical treatment of disease using genomics and personalized medicine? Are there examples of successful treatments?

c. What are the ethical issues surrounding personalized medicine, and which of these issues do you consider to be the most important?

d. What are the legal and economic issues that will have to be addressed before personalized medicine becomes common practice?

STUDENT MEDIA: BioFlix, Activities, Investigations, and Videos

www.masteringbiology.com (select Study Area)

BioFlix—Ch. 16: DNA Replication

Activities—Ch. 20: Applications of DNA Technology; Restriction Enzymes; Cloning a Gene in Bacteria; Gel Electrophoresis of DNA; Analyzing DNA Fragments Using Gel Electrophoresis; DNA Fingerprinting; Making Decisions About DNA Technology: Golden Rice

Investigations—Ch. 20: How Can Gel Electrophoresis Be Used to Analyze DNA? LabBench: Molecular Biology

Videos—Ch. 20: Biotechnology Lab

REFERENCES

Alberts, B., R. Beachy, D. Baulcombe, G. Blobel, S. Datta, Nina Fedoroff, D. Kennedy, F. Khush, J. Peacock, M. Rees, and P. Sharp. "Standing Up for GMOs." *Science*, 2013, vol. 342 (6152), p. 1320

Collins, F. "Has the Revolution Arrived?" *Nature*, 2010, vol. 4464, pp. 674–675.

Collins, F. and A. Barker. "Mapping the Cancer Genome." *Scientific American*, 2007, vol. 296(3), pp. 50–57.

Fahlgren, N., R. Bart, L. Herrera-Estrella, R. Rellan-Alvarez, D. Chitwood, and J. Dinneny. "Plant Scientists: GM Technology Is Safe." *Science*, 2016, vol. 351(6275), p. 824.

Feuillet, C. and K. Eversole. "Solving the Maze." *Science*, 2009, vol. 326, pp. 1071–1072.

Klug, W. S., M. R. Cummings, C. A. Spencer, and M. A. Palladino. *Essentials of Genetics*, 9th ed. San Francisco: Pearson, 2016.

Miklos, D. A. and G. A. Freyer. *DNA Science: A First Course*, 2nd ed. Cold Spring Harbor, NY: Cold Spring Harbor Press, 2010.

Phillips, T. "Genetically Modified Organisms. (GMOs): Transgenic Crops and Recombinant DNA Technology." *Nature Education*, 2008, vol. 1(1), published online at http://www.nature.com/scitable/topicpage/Genetically-Modified-Organisms-GMOs-Transgenic-Crops-and-732.

Pray, L. "Recombinant DNA Technology and Transgenic Animals." *Nature Education*, 2008, vol. 1(1), published online at http://www.nature.com/scitable/topicpage/Recombinant-DNA-Technology-and-Transgenic-Animals-34513.

Raney, T. and P. Pingali. "Sowing a Gene Revolution." *Scientific American*, 2007, vol. 297, pp. 104–111.

Stoneking, M. and J. Krause. "Learning About Human Population History from Ancient and Modern Genomes." *Nature Reviews: Genetics*, 2011, vol. 12, pp. 603–614.

Wheelwright, J. "The Revolution Will Be Edited." *Discover*, 2016, June, pp. 40–49.

Whittall, H. 2008. "The Forensic Use of DNA: Scientific Success Story, Ethical Minefield." *Biotechnology Journal*, 2008, vol. 3, pp. 303–305.

WEBSITES

The Cancer Genome Atlas (TCGA), National Cancer Institute; studying cancer through biotechnology, including sequencing genomes: http://cancergenome.nih.gov

DNA Learning Center at Cold Spring Harbor Laboratories; animations for restriction enzymes and gel elctrophoresis: https://www.dnalc.org/resources/animations/

Issues in Biotechnology, American Institutes of Biological Sciences; provides resources, links, and interviews: http://www.actionbioscience.org

Nature Education; "Genetically Modified Organisms (GMOs);" includes videos, resources, data and discussion of issues: http://www.nature.com/scitable/spotlight/gmos-6978241

National Eye Institute, NIH, provides information on the biotechnology and success of gene therapy for Leber Congenital Amaurosis (LCA), a form of congenital blindness: http://www.nei.nih.gov/lca

Nature News. "CRISPR: Gene Editing Is Just the Beginning," by Heidi Ledford, May 7, 2016. Includes information, images and videos: http://www.nature.com/news/crispr-gene-editing-is-just-the-beginning-1.19510

1000 Genomes: A Deep Catalog of Human Genetic Variation: http://www.1000genomes.org

NOTES

NOTES

LAB TOPIC 11

Population Genetics: The Hardy-Weinberg Equilibrium

Laboratory Objectives

After completing this lab topic, you should be able to:

1. Explain Hardy-Weinberg equilibrium in terms of allelic and genotypic frequencies and relate these to the expression $(p + q)^2 = p^2 + 2pq + q^2 = 1$.
2. Describe the conditions necessary to maintain Hardy-Weinberg equilibrium.
3. Use the bead model to demonstrate conditions for evolution.
4. Test hypotheses concerning the effects of evolutionary change (migration, mutation, genetic drift by either bottleneck or founder effect, and natural selection) using a computer model.

Introduction

Charles Darwin's unique contribution to biology was not that he "discovered evolution" but, rather, that he proposed a mechanism for evolutionary change—**natural selection**, the differential survival and reproduction of individuals in a population. In *On the Origin of Species*, published in 1859, Darwin described natural selection and provided convincing evidence in support of **evolution**, the change in populations over time. Evolution was accepted as a **theory** with great explanatory power supported by a large and diverse body of evidence. Population genetics, the fusion of evolutionary biology with genetics of populations, investigates how evolutionary change can arise due to natural selection, as well as migration, genetic drift, and mutation. The latest techniques of molecular biology have provided scientists with new ways to measure genetic variation and changes in gene frequencies in populations.

Ayala (1982) defines *evolution* as "changes in the genetic constitution of populations." A **population** is a group of organisms of the same species that occur in the same area and have the opportunity to interbreed or share a common **gene pool**, all the alleles at all gene loci of all individuals in the population. The population is considered the basic unit of evolution. The small-scale changes in the genetic structure of populations from generation to generation are referred to as **microevolution**. *Populations evolve; individuals do not.* Can you explain this statement in terms of the process of natural selection?

In 1908, English mathematician G. H. Hardy and German physician W. Weinberg independently developed models of population genetics that showed that the process of heredity by itself did not affect the genetic structure of a population. The **Hardy-Weinberg equilibrium** states that the frequency of alleles in an ideal population will remain the same from generation to generation.

Furthermore, the equilibrium genotypic frequencies will be established after one generation of random mating. This equilibrium occurs only if certain conditions are met:

1. The population is very large.
2. Matings are random.
3. There are no net changes in the gene pool due to mutation; that is, mutation from A to a must be equal to mutation from a to A.
4. There is no migration of individuals (gene flow) into and out of the population.
5. There is no selection; all genotypes are equal in survival and reproductive success.

One classic example of microevolution and the Hardy-Weinberg equilibrium is the case of the peppered moth. During the 19th century, collections of peppered moths in Great Britain revealed that almost all moths were light colored. Only a few of the dark forms were seen in collections before mid-century. The light moths rested during the day on the lichen-covered bark of trees, where they were camouflaged and protected from predators. Under Hardy-Weinberg equilibrium, the frequencies of the color forms would be maintained in each generation for large, random-breeding populations with no change in the mutation rate and migration rate, as long as the environment was relatively stable and no selection occurred. The process of heredity would not change the frequency of the two forms of the moth. However, during the Industrial Revolution in the early 1900s, pollution increased and light-colored lichens on trees disappeared. The frequency of dark moths increased as light moths declined. Which of the conditions just described was responsible for the change in frequencies of light and dark moths?

Basically, the Hardy-Weinberg equilibrium provides a **testable** model in which gene frequencies do not change and *evolution does not occur*. By testing the fundamental hypothesis of the Hardy-Weinberg equilibrium, biologists have investigated the roles of mutation, migration, population size, nonrandom mating, and natural selection in effecting evolutionary change in natural and human populations. Although some populations maintain genetic equilibrium, the exceptions are intriguing to scientists as they investigate the forces at work in populations.

The latest techniques of molecular biology have provided scientists with new ways to measure genetic variation and changes in gene frequencies in populations. Currently, scientists are using the tools of populations genetics to trace the migration of humans across Europe, understand the consequences of conservation practices for critically endangered species, develop models for innovative protocols to reduce antibiotic resistance, and study the epidemiology of human disease.

Use of the Hardy-Weinberg Equation

Fundamental to population genetics is determining frequencies of alleles and genotypes in populations, and then measuring the change in those frequencies over generations. The Hardy-Weinberg equation provides a mathematical formula for calculating allelic and genotypic frequencies. If we begin with a population with two alleles at a single gene locus—a dominant allele, A, and a recessive allele, a—then the frequency of the dominant allele is p, and the frequency of the recessive allele is q. Therefore, $p + q = 1$. If the frequency of one allele, p, is known for a population, the frequency of the other allele, q, can be determined by using the formula $q = 1 - p$.

During sexual reproduction, the frequency of each type of gamete produced is equal to the frequency of the alleles in the population. If the gametes combine at random, then the probability of AA in the next generation is p^2. For example, if the frequency of A is $0.5 = p$, then the probability of AA in the next generation is $p^2 (p \times p) = 0.25$ (Figure 11.1). Then the probability of aa in the next generation ($q \times q$) is q^2. The heterozygote can be obtained two ways, with either parent providing a dominant allele, so the probability would be $2pq$ (Figure 11.1).

To summarize:

$$p^2 = \text{frequency of } AA$$

$$2pq = \text{frequency of } Aa$$

$$q^2 = \text{frequency of } aa$$

The general equation then becomes

$$p^2 + 2pq + q^2 = 1$$

Follow the steps in this example.

1. If alternate alleles of a gene, A and a, occur at equal frequencies, p and q, then during sexual reproduction, 0.5 of all gametes will carry A and 0.5 will carry a.
2. Then $p = q = 0.5$.
3. Once allelic frequencies are known for a population, the genotypic makeup of the next generation can be predicted from the general equation. In this case,

$$0.25AA + 0.5Aa + 0.25aa = 1$$
$$p^2 \quad + \quad 2pq \quad + \quad q^2 \quad = 1$$

   This represents the results of random mating as shown in Figure 11.1.
4. The genotypic frequencies in the population are specifically

$$p^2 = \text{frequency of } AA = 0.25$$

$$2pq = \text{frequency of } Aa = 0.50$$

$$q^2 = \text{frequency of } aa = 0.25$$

5. The allelic frequencies remain $p = q = 0.5$.

		Eggs	
		A egg $p = 0.5$	*a* egg $q = 0.5$
Sperm	*A* sperm $p = 0.5$	*AA* $p^2 = 0.25$	*Aa* $pq = 0.25$
	a sperm $q = 0.5$	*aA* $pq = 0.25$	*aa* $q^2 = 0.25$

FIGURE 11.1
Random mating in a population at Hardy-Weinberg equilibrium. The combination of alleles in randomly mating gametes maintains the allelic and genotypic frequency generation after generation. The gene pool of the population remains constant, and the populations do not evolve.

Calculating frequencies—sample problems

1. In actual populations, the frequencies of alleles are not usually equal. For example, *in a large random mating population* of jimsonweed, 4% of the population might be white (a recessive trait), and the frequency of the white allele could be calculated as the square root of 0.04.
   a. White individuals $= q^2 = 0.04$ (genotypic frequency); therefore,

$$q = \sqrt{0.04} = 0.2 \text{ (allelic frequency)}.$$

   b. Because $p + q = 1$, the frequency of p is $(1 - q)$, or 0.8. So 4% of the population is white, and 20% of the alleles in the gene pool is for white flowers and the other 80% is for purple flowers. (Note that you could not

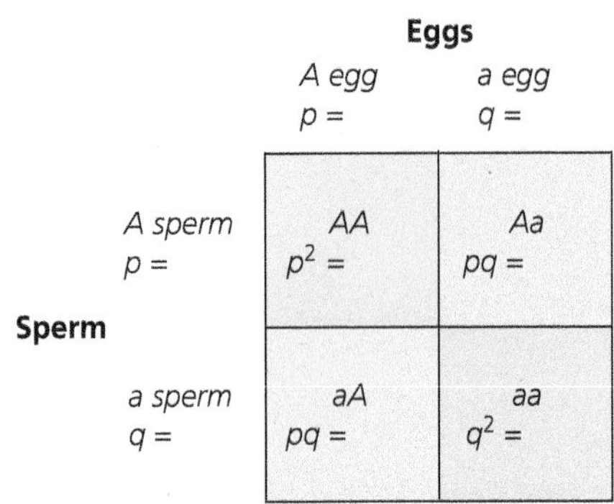

FIGURE 11.2

Random mating for a population at Hardy-Weinberg equilibrium. Complete the mating combinations for purple and white flowers.

determine the frequency of A by taking the square root of the frequency of all individuals with purple flowers because you cannot distinguish the heterozygote and the homozygote for this trait.)

c. The genotypic frequencies of the next generation now can be predicted from the general Hardy-Weinberg equation. First determine the results of random mating by completing Figure 11.2 (refer to Figure 11.1).

d. What will be the genotypic frequencies from generation to generation, provided that alleles p and q remain in genetic equilibrium?

$AA =$

$Aa =$

$aa =$

Note: *Using the square root to obtain the allelic frequency is only valid if the population meets the conditions for a population in Hardy-Weinberg equilibrium.*

2. *In populations when all three phenotypes can be distinguished, the observed genotype and allele frequencies can be calculated from the phenotype numbers.* In human populations, blood antigens include the proteins coded for by the *MN* gene. Individuals may be homozygous *MM*, heterozygous *MN*, or homozygous *NN*. In a population of 100 individuals, the following phenotypes were recorded:

36 *MM*; 48 *MN*; 16 *NN*

a. To determine the allelic frequencies of M, add all the M alleles and divide by the total number of alleles in the gene pool. Note: These are diploid organisms, so each individual has 2 alleles.

Allelic frequency of M (p):

$$72(\text{from } MM) + 48(\text{from } MN) = 120/200(\text{total alleles}) = 0.6$$

Allelic frequency of N (q):

$$32(\text{from } NN) + 48(\text{from } MN) = 80/200(\text{total alleles}) = 0.4$$

b. To determine the *observed* genotypic frequencies, divide the number of each genotype by the total number of individuals:

Observed genotypic frequencies:

$$MM = 36/100 = 0.36$$
$$MN = 48/100 = 0.48$$
$$NN = 16/100 = 0.16$$

Note: If the population is in Hardy-Weinberg equilibrium, then the observed numbers should be very close to the expected. Expected genotypic frequencies are calculated by using the allelic frequencies (p, q) in the equilibrium equation ($p^2 + 2pq + q^2 = 1$).

In the following exercises, you will use both of these approaches (counting and the square root of q^2) in your investigations and simulations of population genetics.

The genetic equilibrium will continue indefinitely if the conditions of Hardy-Weinberg are met. How often in nature do you think these conditions are met? Although natural populations may seldom meet all the conditions, Hardy-Weinberg equilibrium serves as a valuable model from which we can predict genetic changes in populations as a result of natural selection or other factors. This allows us to understand quantitatively and in genetic language how evolution operates at the population level.

EXERCISE 11.1

Testing Hardy-Weinberg Equilibrium Using a Bead Model

Materials

plastic or paper bag containing 100 beads of two colors

Introduction

Working in pairs, you will test Hardy-Weinberg equilibrium by simulating a population using colored beads. The bag of beads represents the gene pool for the population. Each bead should be regarded as a single gamete, the two colors representing different alleles of a single gene. Each bag should contain 100 beads of the two colors in the proportions specified by the instructor. Record in the spaces provided below the color of the beads and the initial frequencies for your gene pool.

A = ____________ color ____________ allelic frequency

a = ____________ color ____________ allelic frequency

1. How many diploid individuals are represented in this population?

2. What would be the color of the beads for a homozygous dominant individual?
3. What would be the color of the beads for a homozygous recessive individual?
4. What would be the color of the beads for a heterozygous individual?

Hypothesis

State the Hardy-Weinberg equilibrium in the space provided. This will be your hypothesis.

Predictions

Predict the genotypic frequencies of the population in future generations (if/then).

Procedure

1. Without looking, randomly remove two beads from the bag. These two beads represent one diploid individual in the next generation. Record in the margin of your lab manual the diploid genotype (*AA*, *Aa*, or *aa*) of the individual formed from these two gametes.
2. *Return the beads to the bag* and shake the bag to reinstate the gene pool. By replacing the beads each time, the size of the gene pool remains constant, and the probability of selecting any allele should remain equal to its frequency. This procedure is called **sampling with replacement**.

3. Repeat steps 1 and 2 (select two beads, record the genotype of the new individual, and return the beads to the bag) until you have recorded the genotypes for 50 individuals who will form the next generation of the population.
4. Total the number of *AA*, *Aa*, and *aa* genotypes removed and record the numbers in *Table 11.2* for later use in calculating *observed* frequencies.

Results

1. *Before calculating the results of your experiment*, determine the *expected* frequencies of genotypes and alleles for the population. To do this, use the original allelic frequencies for the population provided by the instructor. (Recall that the frequency of $A = p$, and the frequency of $a = q$.) Calculate the expected genotypic frequencies using the Hardy-Weinberg equation ($p^2 + 2pq + q^2 = 1$). The number of individuals expected for each genotype can be calculated by multiplying 50 (total population size) by the expected frequencies. Record these results in Table 11.1.

TABLE 11.1 *Expected* **Genotypic and Allelic Frequencies for the Next Generation Produced by the Bead Model**

Parent Population		Expected Next Generation				
Allelic Frequency		Genotypic Number (and Frequency)			Allelic Frequency	
A	*a*	*AA*	*Aa*	*aa*	*A*	*a*

2. Next, *using the results of your experiment*, calculate the *observed* frequencies in the new population created as you removed beads from the bag. Using the number of diploid individuals for each genotype in Table 11.2, calculate the frequencies for the three genotypes (*AA, Aa, aa*). Add the numbers of each allele, and calculate the allelic frequencies for *A* and *a*. Hint: How many total alleles are in the population? These values are the observed frequencies in the new population. Genotypic frequencies and allelic frequencies should each equal 1.

TABLE 11.2 *Observed* **Genotypic and Allelic Frequencies for the Next Generation Produced by the Bead Model**

Parent Population		Observed Population				
Allelic Frequency		Genotypic Number (and Frequency)			Allelic Frequency	
A	*a*	*AA*	*Aa*	*aa*	*A*	*a*

3. To compare your observed results with those expected, you can use the statistical test, chi-square. Table 11.3 will assist in the calculation of the chi-square test. *Note: To calculate chi-square you must use the actual number of individuals for each genotype, not the frequencies*. See Appendix D for an explanation of this statistical test.

   Degrees of freedom = 2

   Level of significance, $p < 0.05$

TABLE 11.3 Chi-Square of Results from the Bead Model

Download an Excel version from www.masteringbiology.com in the Study Area under Lab Media

	AA	*Aa*	*aa*
Observed value (*o*)			
Expected value (*e*)			
Deviation (*o* − *e*) = *d*			
d^2			
d^2/e			
Chi-square (χ^2) = $\Sigma d^2/e$			

4. Is your calculated χ^2 value greater or smaller than the given χ^2 value (Appendix D, Table D.2) for the degrees of freedom and p value for this problem?

Discussion

1. What proportion of the population was homozygous dominant?

   Homozygous recessive?

   Heterozygous?

2. Were your observed results consistent with the expected results based on your statistical analysis? If not, can you suggest an explanation?

3. Compare your results with those of other students. How variable are the results for each team?

4. Do your results match your predictions for a population at Hardy-Weinberg equilibrium?

What would you expect to happen to the frequencies if you continued this simulation for 25 generations?

Is this population evolving?

Explain your response.

5. Consider each of the conditions for the Hardy-Weinberg model. Does this model meet each of those conditions?

EXERCISE 11.2

Simulation of Evolutionary Change Using the Bead Model

Under the conditions specified by the Hardy-Weinberg model (random mating in a large population, no mutation, no migration, and no selection), the genetic frequencies should not change, and evolution should not occur. In this exercise, the class will modify each of the conditions and determine the effect on genetic frequencies in subsequent generations. *You will simulate the evolutionary changes that occur when these conditions are not met.*

Working in teams of two or three students, you will simulate *two* of the experimental scenarios presented and, using the bead model, determine the changes in genetic frequencies over several generations. The scenarios include the migration of individuals between two populations, also called **gene flow**; the effects of small population size, called **genetic drift**; and examples of **natural selection**.

Simulation 1. *All teams will begin by simulating the effect of genetic drift, specifically, the bottleneck effect* (Experiment A.1).

Simulation 2. *For the second simulation, you can choose to investigate migration, one of two examples of natural selection, or founder effect—another example of genetic drift.*

The effects of mutation take longer to simulate with the bead model, so you will use computer simulation to consider these in Exercise 11.3.

The procedure for investigating each of the conditions will follow the general procedures described as follows. Before beginning one of the simulation experiments, be sure you understand the following procedures to be used.

Procedure

1. Sampling with Replacement

Unless otherwise instructed, the gene pool size will be 100 beads. Each new generation will be formed by randomly choosing 50 diploid individuals represented by pairs of beads. After removing each pair of beads (representing the genotype of one individual) from the bag, replace the pair before removing the next set, *sampling with replacement*. Continue your simulations for several generations. For example, if the starting population has 50 beads each of *A* and *a* (allelic frequency of 0.5), then in the next generation you might produce the following results:

Number of individuals: 14 *AA*, 24 *Aa*, 12 *aa*

Number of alleles (beads): 28 *A* + 24 *A*, 24 *a* + 24 *a*

Total number of alleles: 100

$$\text{Frequency of } A\text{: } 28 + 24 = \frac{52}{100} = 0.52$$

$$\text{Frequency of } a\text{: } 24 + 24 = \frac{48}{100} = 0.48$$

In this example, the frequency should continue to approximate 0.5 for *A* and *a*.

2. Reestablishing a Population with New Allelic Frequencies

In some cases, the number of individuals will decrease as a result of the simulation. In those cases, return the population to 100 but reestablish the population with new allelic frequencies. For example, if you eliminate by selecting against all homozygous recessive (*aa*) individuals in your simulation, then the resulting frequencies would be:

Number of individuals: 14 *AA*, 24 *Aa*, 0 *aa*

Number of alleles (beads): 28 *A* + 24 *A*, 24 *a*

Total number of alleles: 76

$$\text{Frequency of } A\text{: } 28 + 24 = \frac{52}{76} = 0.68$$

$$\text{Frequency of } a\text{: } 24 = \frac{24}{76} = 0.32$$

To reestablish a population of 100, then, the number of beads should reflect these new frequencies. *Adjust the number of beads* so that *A* is now 68/100 and *a* is 32/100. Then continue the next round of the simulation.

If this information is not clear to you, ask for assistance before beginning your simulations.

Experiment A. Simulation of Genetic Drift

Materials

plastic or paper bag containing 100 beads, 50 each of two colors
additional beads as needed

Introduction

Genetic drift is the change in allelic frequencies in small populations as a result of chance alone. In a small population, combinations of gametes may not be random, owing to sampling error. (If you toss a coin 500 times, you expect about a 50:50 ratio of heads to tails; but if you toss the coin only 10 times, the ratio may deviate greatly in a small sample owing to chance alone.) As the size of a population decreases, the effects of genetic drift increase. **Genetic fixation** occurs when all members of the population have the same allele for a gene; all other alleles have been eliminated. Fixation of one allele is a common result of genetic drift in small natural populations. Genetic drift is a significant evolutionary force in situations known as the **bottleneck effect** and the **founder effect**. *All teams will investigate the bottleneck effect. You may choose the founder effect* (p. 287) *for your second simulation.*

1. Bottleneck effect

A **bottleneck** occurs when a population undergoes a drastic reduction in size as a result of chance events (not differential selection), such as a volcanic eruption or overharvesting by humans. For example, northern elephant seals experienced a bottleneck when they were hunted almost to extinction with only 30 surviving animals in the 1890s. In this case, the gene pool and the amount of genetic variation has been reduced to these 30 animals. In Figure 11.3, the beads pass through a bottleneck, which results in an unpredictable combination of beads that pass to the other side. Note that the frequency of the alleles in the population has changed due to chance alone. These beads would constitute the beginning of the next generation.

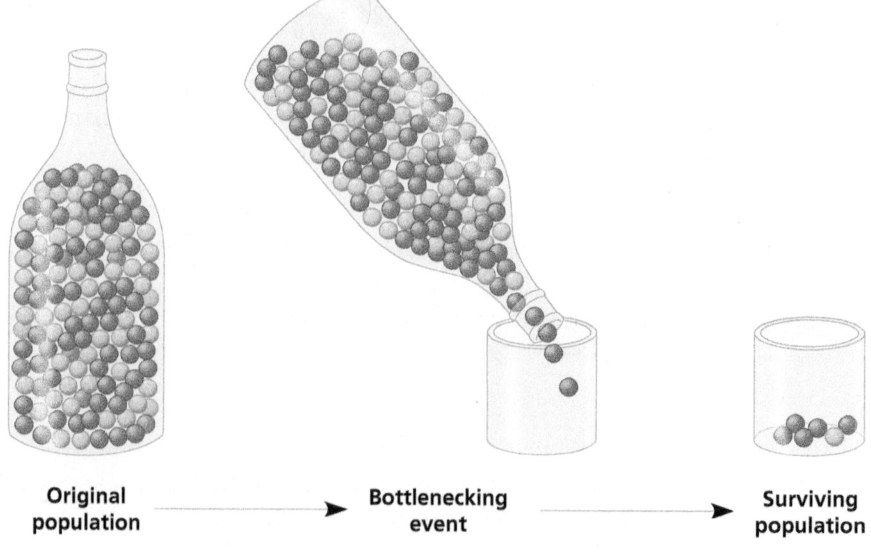

FIGURE 11.3

The bottleneck effect. The gene pool can drift by chance when the population is drastically reduced by factors that act unselectively. Bad luck, not bad genes! The resulting population will have unpredictable combinations of genes. What has happened to the amount of variation?

Hypothesis

As your hypothesis, either propose a hypothesis that addresses the bottleneck effect specifically or state the Hardy-Weinberg equilibrium.

Prediction

Either predict equilibrium values as a result of Hardy-Weinberg or predict the type of change that you expect to occur in a small population (if/then).

Procedure

1. To investigate the bottleneck effect, establish a starting population containing 50 individuals (how many beads?) with a frequency of 0.5 for each allele (Generation 0).
2. *Without replacement*, randomly select 5 individuals (only 10% of the population survives), two alleles at a time. This represents a drastic reduction in population size. On a separate sheet of paper, record the genotypes and the number of *A* and *a* alleles for the new population.
3. Count the numbers of each genotype and the numbers of each allele. Using these numbers, determine the genotypic frequencies for *AA*, *Aa*, and *aa* and the new allelic frequencies for *A* (p) and *a* (q) for the surviving 5 individuals. These are your *observed* frequencies. Enter these frequencies in Table 11.4, Generation 1.
4. Using the new observed allelic frequencies, calculate the *expected* genotypic frequencies (p^2, $2pq$, q^2). Record these frequencies in Table 11.4, Generation 1.
5. Reestablish the population to 50 individuals using the new allelic frequencies (refer to the example in the Procedure on p. 283). Repeat steps 2, 3, and 4. Record your results in the appropriate generation in Table 11.4.
6. Reestablish the gene pool with new frequencies after each generation until one of the alleles becomes fixed in the population for several generations.
7. Summarize your results in the Discussion section.

Results

1. How many generations did you simulate?
2. In the margin of your lab manual, sketch a graph of the change in p and q over time. You should have two lines, one for each allele.

TABLE 11.4 Changes in Allelic and Genotypic Frequencies for Simulations of the Bottleneck Effect, an Example of Genetic Drift. First, record frequencies based on the observed numbers in your experiment. Then, using the observed allelic frequencies, calculate the expected genotypic frequencies.

	Genotypic Frequency Observed			Allelic Frequency Observed		Genotypic Frequency Expected		
Generation	*AA*	*Aa*	*aa*	*A* (*p*)	*a* (*q*)	p^2	$2pq$	q^2
0	—	—	—	0.5	0.5	0.25	0.50	0.25
1								
2								
3								
4								
5								
6								
7								
8								
9								
10								

3. Did one allele go to fixation in that time period? Which allele?

   Remember, genetic fixation occurs when the gene pool is composed of only one allele. The others have been eliminated. Did the other allele ever appear to be going to fixation?

4. Did any of the expected genotypic frequencies go to fixation? If none did, why not?

5. Compare results with other teams. Did the same allele go to fixation for all teams? If not, how many became fixed for *A* and how many for *a*? The probability of fixation for any allele is equal to the starting allelic frequency. Are the class data consistent with this prediction?

Discussion

1. Compare the pattern of change for p and q. Is there a consistent trend or do the changes suggest chance events? Look at the graphs of other teams before deciding.

2. Explain your observations of genetic fixation for the replicate simulations completed by the class. What would you expect if you simulated the bottleneck effect 100 times?

3. How might your results have differed if you had started with different allelic frequencies, for example, $p = 0.2$ and $q = 0.8$?

4. Because only chance events—that is, the effect of small population size—are responsible for the change in gene frequencies, would you say that evolution has occurred? Explain.

5. Since the 1890s when northern elephant seals were hunted almost to extinction, the populations have increased to 30,000 animals. In a 1974 study of genetic variation, scientists found that all elephant seals are fixed for the same allele at 24 loci. Explain.

On completion of the bottleneck simulation, choose one or two of the remaining scenarios to investigate. Confirm your selection with your instructor. All scenarios should be completed by at least one team in the laboratory.

2. Founder effect

When a small group of individuals becomes separated from the larger parent population, the allelic frequencies in this small gene pool may be different from those of the original population as a result of chance alone. This occurs when a group of migrants becomes established in a new area not currently inhabited by the species—for instance, the colonization of an island—and is therefore referred to as the **founder effect**. For example, in 1975, 20 desert bighorn sheep colonized Tiburon Island off the coast of Mexico; by 1999, the population had increased to 650 sheep. A comparison with the other populations of the species in Arizona showed that the island populations had fewer alleles per gene and overall lower genetic variation due to the small size of the colonizing population and inbreeding.

Hypothesis

As your hypothesis, either propose a hypothesis that addresses founder effect specifically or state the Hardy-Weinberg equilibrium.

Prediction

Either predict equilibrium values as a result of Hardy-Weinberg or predict the type of change that you expect to occur in a small population (if/then).

Procedure

1. To investigate the founder effect, establish a starting population with 50 individuals with starting allelic frequencies of your choice. Record the frequencies you have selected for Generation 0 in Table 11.5 in the Results section at the end of this exercise.
2. *Without replacement*, randomly select 5 individuals, two alleles at a time, to establish a new founder population. On a separate sheet of paper, record the genotypes and the number of *A* and *a* alleles for the new population.
3. Calculate the new frequencies for *A* (p) and *a* (q) and the genotypic frequencies for *AA* (p^2), *Aa* ($2pq$), and *aa* (q^2) for the *founder population*, and record this information as Generation 1 in Table 11.5.
4. Reestablish the population to 50 diploid individuals using the new allelic frequencies. (Refer to the example in the Procedure on p. 283.)
5. Follow the founder population through several generations in the new population. From this point forward, each new generation will be produced by *sampling 50 individuals with replacement*. After each generation, reestablish the new population based on the new allelic frequencies. Continue until you have sufficient evidence to discuss your results with the class.
6. Summarize your results in the Discussion section at the end of this exercise. You will want to compare the founder population with the original population and compare equilibrium frequencies if appropriate. Each group will present its results to the class at the end of the laboratory.

Experiment B. Simulation of Migration: Gene Flow

Materials

2 plastic or paper bags, each containing 100 beads of two colors
additional beads as needed

Introduction

The migration of individuals between populations results in **gene flow**. In a natural population, gene flow can be the result of the immigration and emigration of individuals or gametes (for example, pollen movement). The rate and direction of migration and the starting allelic frequencies for the two populations can affect the rate of genetic change. In the following experiment, the migration rate is equal in the two populations, and the starting allelic frequencies differ for the two. Work in teams of four students.

Hypothesis

As your hypothesis, either propose a hypothesis that addresses migration specifically or state the Hardy-Weinberg equilibrium.

Prediction

Either predict equilibrium values as a result of Hardy-Weinberg or predict the type of change that you expect to observe as a result of migration (if/then).

Procedure

1. To investigate the effects of gene flow on population genetics, establish *two populations of 100 beads each.* Choose different starting allelic frequencies for *A* and *a* for each population. For example, in Figure 11.4, population 1 might start with 90 *A* and 10 *a* while population 2 might have 50 of each allele initially. For comparison, you might choose one population with the same starting frequencies as your population in Exercise 11.1. Record the allelic frequencies for each starting population, Generation 0, in Table 11.5 in the Results section at the end of this exercise.
2. Select 10 individuals (how many alleles?) from each population, and allow them to migrate to the other population by exchanging beads. Do not sample with replacement in this step.
3. Select 50 individuals from each of these new populations, *sampling with replacement.* On a separate sheet of paper, record the genotypes and the number of *A* and *a* alleles for population 1 and population 2.

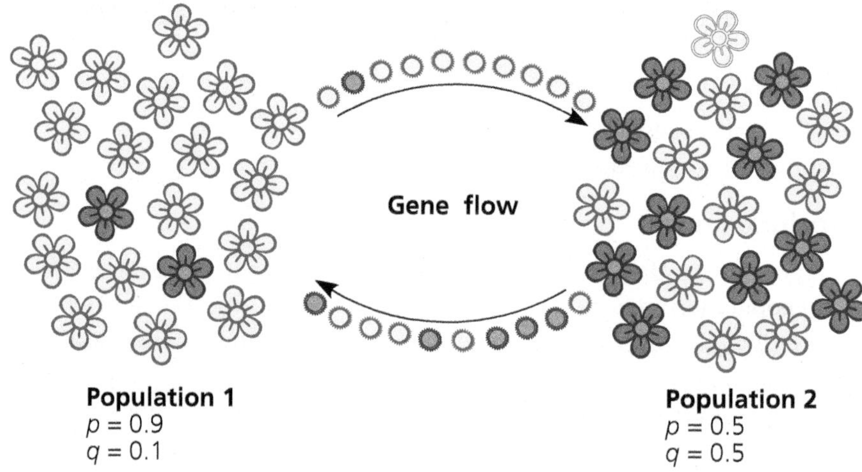

FIGURE 11.4
Migration: Gene Flow. Migration rates are constant between two populations that differ for starting allelic frequencies.

4. Calculate the new allelic and genotypic frequencies in the two populations following migration. Record your results in Table 11.5 in the Results section. Reestablish each bag based on the new allelic frequencies.
5. Repeat this procedure (steps 2 to 4) for several generations. In doing this, you are allowing migration to take place with each generation. Continue until you have a sufficient number to allow you to discuss your results in class successfully.
6. Summarize your results in the Discussion section at the end of this exercise. Each group will present its results to the class at the end of the laboratory.

Experiment C. Simulation of Natural Selection

Materials

plastic or paper bag containing 100 beads of two colors
additional beads as needed

Introduction

Natural selection, the differential survival and reproduction of individuals, was first proposed by Darwin as the primary mechanism for evolution. Although other factors have since been found to be involved in evolution, selection is still considered a very important mechanism. Natural selection is based on the observation that individuals with certain heritable traits are more likely to survive and reproduce than those lacking these advantageous traits. Therefore, the proportion of offspring with advantageous traits will increase in the next generation. As a result, the genotypic frequencies will change in the population. Whether traits are advantageous in a population depends on the environment and the selective agents (which can include physical and biological factors). Choose one of the following evolutionary scenarios to model natural selection in population genetics.

1. Rock pocket mice—adaptive melanism

Rock pocket mice, *Chaetodipus intermedius*, inhabit the desert southwest, living among the rocky outcrops and desert floor. In the Pinacate Desert in Arizona, the light-colored sandstone rocks are adjacent to areas of dark rocks produced from lava flows. The rock pocket mice exhibit dark and light forms, an example of melanism (Figure 11.5). Coat color in mammals is affected by several genes that control the amount of two kinds of melanin found in melanocytes. In rock pocket mice, the Mc1r gene codes for a protein called MC1R (melanocortin 1 receptor), which signals the production of enzymes in the pathway. When the normal Mc1r gene is present, eumelanin, the dark-colored pigment, decreases, and phaeomelanin, the light-colored pigment, increases. The mutated version of the Mc1r gene results in increased amounts of eumelanin, producing the dark coat color. Michael Nachman and associates at the University of Arizona studied the population genetics of rock pocket mice in the Pinacate Desert. They found that the frequency of the mice with dark coat color is greater in populations captured on the dark outcrops, and the light-colored mice are in higher frequency on the sandstone rocks of the desert floor. Their results suggest that the dark mice on the black rock have a selective advantage relative to the light mice, because they are protected from predators by their cryptic coloration. Consider the light mice with the normal Mc1r gene. In what habitat do they have a selective advantage relative to the dark mice?

In rock pocket mice, coat color is controlled by a single gene with two allelic forms, dark and light. The dark coat color is dominant (*A*), and the light coat color is recessive (*a*). The light mice would be *aa*, but the dark form could be either *AA* or *Aa*.

Hypothesis

As your hypothesis, either propose a hypothesis that addresses natural selection occurring in either dark or light rock environments specifically or state the Hardy-Weinberg equilibrium.

a.

b.

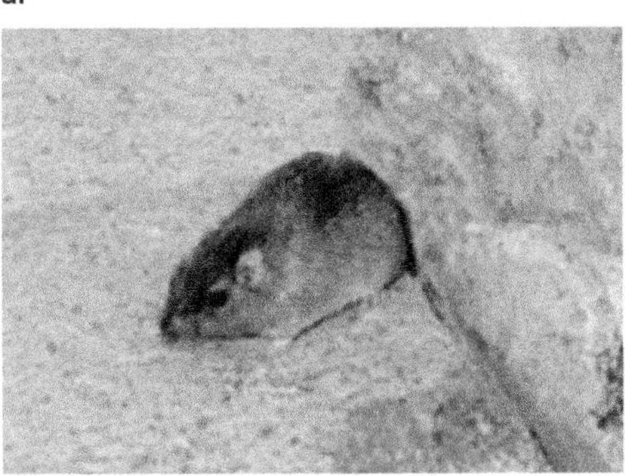

c.

d.

FIGURE 11.5
Two color forms of the rock pocket mouse. The dark and light forms are shown on the desert sandstone (a and c) and on the dark rock outcrop (b and d). In which situation would the light pocket mouse have an advantage?

Prediction

Either predict equilibrium values as a result of the Hardy-Weinberg equilibrium or predict the type of change that you expect to observe as a result of natural selection in the polluted environment (if/then).

Procedure

1. To investigate the effect of natural selection on the frequency of light and dark mice, establish a population of 50 individuals with allelic frequencies of light (*a*) + 0.9 and dark (*A*) + 0.1. Record the frequencies of the starting population, Generation 0, in Table 11.5 in the Results section at the end of this exercise.
2. Determine the genotypes and phenotypes of the population by selecting 50 individuals, two alleles at a time, by sampling with replacement (see Procedure, p. 283). On a separate sheet of paper, record the genotypes and phenotypes of each individual.

   Now, assume that the population inhabits the dark rock outcrops and that in this new population predators eat 50% of the light mice but only 10% of the dark mice. How many light mice must you eliminate from your starting population of 50 individuals? How many dark mice? Can selection distinguish dark mice with *AA* and *Aa* genotypes? How will you decide which dark mice to remove? Remove the appropriate number of individuals of each phenotype.
3. Calculate new allelic frequencies for the remaining population. Reestablish the population with these new allelic frequencies for the 100 alleles. (Refer to the Procedure on p. 283.) Record the new frequencies in Table 11.5 in the Results section.
4. Continue the selection procedure, recording the frequencies, for several generations until you have sufficient evidence to discuss your results with the class.
5. Summarize your results in the Discussion section at the end of this exercise. Each group will present its results to the class at the end of the laboratory.

2. Sickle-cell disease

Sickle-cell disease is caused by a mutant allele (Hb^-), which, in the homozygous condition, in the past was often fatal to people at quite young ages. The mutation causes the formation of abnormal, sickle-shaped red blood cells that clog vessels, cause organ damage, and are inefficient transporters of oxygen (Figure 11.6). Individuals who are heterozygous (Hb^+Hb^-) have the sickle-cell trait (a mild form of the disease), which is not fatal. Scientists were surprised to find a high frequency of the Hb^- allele in populations in Africa until they determined that heterozygous individuals have a *selective advantage* in resisting malaria. Although the homozygous condition may be lethal, the heterozygotes are under both a positive and a negative selection force. In malarial countries in tropical Africa, the heterozygotes are at an advantage compared to either homozygote.

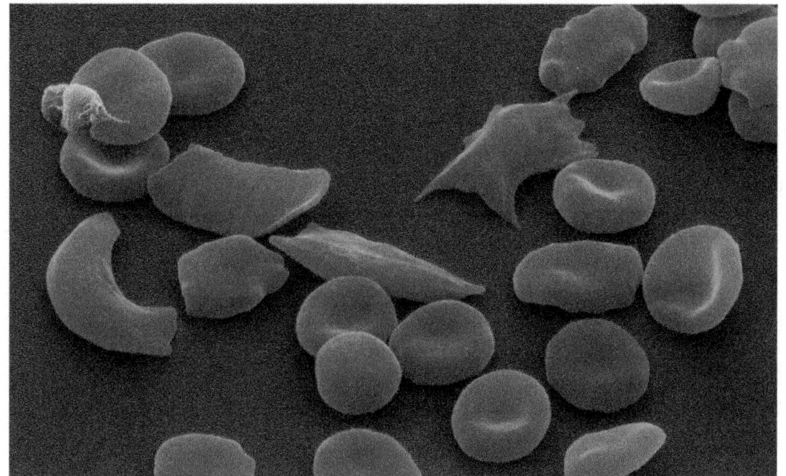

a.

Homozygous recessive individuals

Abnormal hemoglobin

Sickling of red blood cells

Rapid destruction of sickle cells

Clumping of cells and interference with blood circulation

Collection of sickle cells in the spleen

Anemia

Local failures in blood supply

Overactivity of bone marrow

Heart damage

Muscle and joint damage

Gastrointestinal tract damage

Increase in amount of bone marrow

Dilation of heart

Lung damage

Brain damage

Kidney damage

Weakness and fatigue

Poor physical development

Pneumonia

Paralysis

Kidney failure

Skull deformation

Impaired mental function

Heart failure

Rheumatism

Abdominal pain

Enlargement, then fibrosis of spleen

b.

FIGURE 11.6

Effects of the sickle-cell allele.
(a) Normal red blood cells are disk-shaped. The jagged shape of sickled cells causes them to pile up and clog small blood vessels. (b) The results include damage to a large number of organs.

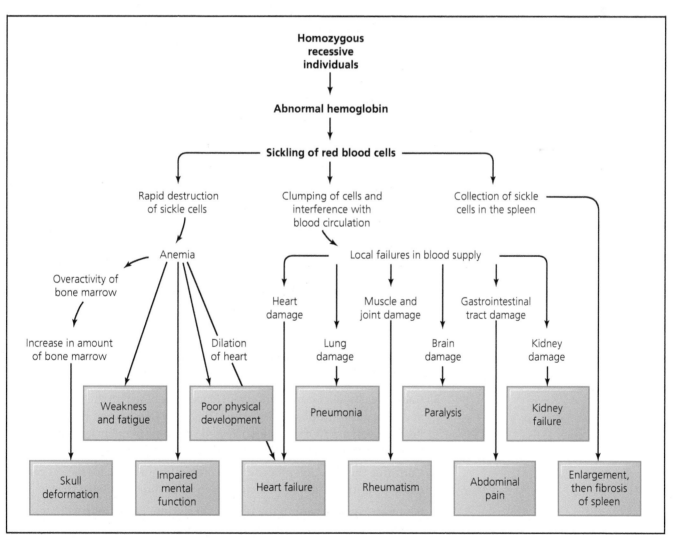

Hypothesis

As your hypothesis, either propose a hypothesis that addresses natural selection in tropical Africa specifically or state the Hardy-Weinberg equilibrium.

Prediction

Either predict equilibrium values as a result of the Hardy-Weinberg equilibrium or predict the type of change that you expect to observe as a result of natural selection in tropical Africa (if/then).

Procedure

1. To investigate the role of selection under conditions of heterozygote advantage, establish a population of 50 individuals with allelic frequencies of 0.5 for both alleles. Record the frequencies of the starting population, Generation 0, in Table 11.5 in the Results section at the end of this exercise.

   For our model, assume that the selection force on each genotype is

   Hb^+Hb^+: 30% die of malaria

   Hb^+Hb^-: 10% die of malaria

   Hb^-Hb^-: 100% die of sickle-cell disease

2. Determine the genotypes and phenotypes of the population by selecting 50 individuals, two alleles at a time, by *sampling with replacement* (see Procedure, p. 283). On a separate sheet of paper, record the 50 individuals. Select against each genotype by eliminating individuals according to the percent mortality shown above. How many Hb^+Hb^+ must be removed? How many Hb^+Hb^-? How many Hb^-Hb^-? Remove the appropriate number of beads for each color.
3. Calculate new allelic and genotypic frequencies based on the survivors, and reestablish the population using these new frequencies. (Refer to the Procedureon p. 283 if necessary.) Record your new frequencies in Table 11.5 in the Results section.
4. Continue the selection procedure, recording frequencies, for several generations, until you have sufficient evidence for your discussion in class.
5. Summarize your results in the Discussion section. Each group will present its results to the class at the end of the laboratory.

Results and Discussion for Selected Simulation

Results

Each team will record the results of its chosen experiment in Table 11.5. Modify the table to match the information you need to record for your simulation. If using the Excel version of the table, customize the table for your simulation. Make copies to share with other teams.

Discussion

In preparation for presenting your results to the class, answer the following questions about your chosen simulation. Choose one team member to make the presentation, which should be organized in the format of a scientific paper.

1. Which of the conditions that are necessary for Hardy-Weinberg equilibrium were met?

TABLE 11.5 Changes in Allelic and Genotypic Frequencies for Simulations Selected in Exercise 11.2. Calculate expected frequencies based on the actual observed numbers in your experiment.

Download an Excel version from www.masteringbiology.com in the Study Area under Lab Media

Experiment ______________________ Simulation of ______________________

Generation	Allelic Frequency		Genotypic Frequency		
	p	q	p^2	$2pq$	q^2
0					
1					
2					
3					
4					
5					
6					
7					
8					
9					
10					

2. Which condition was changed?

3. Briefly describe the scenario that your team simulated.

4. What were your predicted results?

5. How many generations did you simulate?

6. Sketch a graph of the change in p and q over time. You should have two lines, one for each frequency.

7. Describe the changes in allelic frequencies p and q over time. Did your results match your predictions? Explain.

8. Describe the changes in the genotypic frequencies.

9. Compare your final allelic and genotypic frequencies with those of the starting population.

10. If you can, formulate a general summary statement or conclusion.

11. Would you expect your results to be the same if you had chosen different starting allelic frequencies? Explain.

12. Critique your experimental design and outline your next simulation.

Be prepared to take notes, sketch graphs, and ask questions during the student presentations. You are responsible for understanding and answering questions for all conditions simulated in the laboratory.

EXERCISE 11.3

Open-Inquiry Computer Simulation of Evolution

Simulations involving only a few generations are fairly easy with the bead model, but this model is too cumbersome to obtain information about long-term changes or to combine two or more factors. Computer simulations will allow you to model 50 or 100 generations quickly, to change the size of the population, or to do a series of simulations changing one factor and comparing the results. As with any scientific investigation, design your experiment to test a hypothesis and make predictions before you begin.

Several computer simulation programs are available; the instructor will demonstrate the software. From the following suggestions, including mutation described below, or your own ideas, pursue one of the factors in more detail or several in combination. Consult with your instructor before beginning your simulations. Be prepared to present your results in the next lab period in either oral or written form. Follow the procedure provided with the simulation program.

Mutation

Evolution can occur only when there is variation in a population, and the ultimate source for that variation is mutation. The mutation rate at most gene loci is actually very small (1×10^{-5} per gamete per generation), and the forward and backward rates of mutation are seldom at equilibrium. Evolutionary change as a result of mutation alone would occur very slowly, thus requiring many generations of simulation. Changes in this condition are best considered using a computer simulation that easily handles the lengthy process. *Do not simulate this condition using the bead model.*

Other Suggestions for Computer Simulations

1. The migration between natural populations seldom occurs at the same rate in both directions. Devise a model to simulate different migration rates between two populations. Compare your results with those of the bead model.
2. How does the probability of genetic fixation differ for populations that have different starting allelic frequencies? Simulate genetic drift using a variety of allelic frequencies.
3. How large must a population be to avoid genetic drift? How many simulations do you need to run at each population size?

4. Devise a model to simulate selection for recessive and lethal traits. Then, using the same starting allelic frequencies, select against the dominant lethals. Compare rate of change of allelic frequency as a result of lethality for dominant and recessive traits.
5. Mutation alone has only small effects on evolution over long periods of time. Using realistic rates of mutation, compare the evolutionary change for mutation alone with simulations that combine mutation and natural selection.
6. Combine factors (usually two at a time). For example, combine mutation and natural selection, using realistic estimates of mutation rates. Combine genetic drift and migration. Compare your results to simulations of each condition alone.

REVIEWING YOUR KNOWLEDGE

1. Define and provide examples of the following terms: *evolution, population, gene pool, gene flow, genetic drift, bottleneck effect, founder effect, natural selection, genetic fixation, genotypic frequency, allelic frequency*, and *model*.
2. State the five conditions necessary for Hardy-Weinberg equilibrium.
3. Describe how gene flow, genetic drift, and natural selection bring about evolution.
4. Chemokines are receptor proteins found in cell membranes of immune cells. The C-C chemokine receptor-5 protein is a receptor that allows the HIV-1 virus to enter immune cells and is encoded by the CCR5 gene. A mutant allele has a 32 base pair deletion that makes the receptor protein nonfunctional and is coded for by the mutant allele $\Delta 32$. The normal allele is CCR5-1 (w^+). Individuals who are $\Delta 32/\Delta 32$ are resistant to HIV-1 infection, those who are heterozygous $w^+/\Delta 32$ are still affected by HIV-1 but the progression to AIDS is slower, and w^+/w^+ individuals are susceptible to HIV-1.
   A study of 548 individuals in Italy found the following genotypes present in the population:

   $w^+/w^+ = 491$ $w^+/\Delta 32 = 51$ $\Delta 32/\Delta 32 = 6$

   a. Calculate the observed genotypic frequencies from the numbers above.

   b. What are the allelic frequencies of w^+ (p) and $\Delta 32$ (q)?

   c. What are the expected genotypic frequencies if this population is at Hardy-Weinberg equilibrium?

5. Migration occurs at a constant rate between two populations of field mice. In one population, 65% of the population is white; in the other population, only 15% is white. What would you expect to happen to the allelic frequencies of these two populations over time?

6. Explain the difference between evolution and natural selection. Use an example to illustrate your answer.

7. Genetic drift is the result of random events and has a greater effect on small populations. Based on your simulations of genetic drift, describe the effect of drift on genetic variation.

APPLYING YOUR KNOWLEDGE

1. Pingelap is a tiny island in the Pacific Ocean that is part of the Federated States of Micronesia. In 1774, a typhoon devastated the island and of the 500 people who inhabited the island, only 20 survived. One male survivor was heterozygous for a rare type of congenital color blindness. After four generations, some children were born with complete color blindness and hypersensitivity to light due to defective cones (achromatopsia). The islands remained isolated for over 150 years. Now there are about 3,000 Pingelap descendants, including those on a nearby island and in a small town on a third island. The frequency of individuals who are homozygous recessive and have the disease is 5%, and another 30% is heterozygous carriers. This compares to the average of 1% carriers for the human population at large. Explain the conditions that are responsible for these differences in genetic frequencies for the Pingelap population and the general human population.

   Three daughters of the male survivor on Pingelap, who were carriers for color blindness, moved to the nearby island of Mokil. This new island was even smaller and had fewer people. How would this event change the frequency of color blindess in the Mokil population? Is this an example of founder effect or bottleneck? Explain.

Some Pingelap citizens were transplanted to a small town on yet a third larger island, the capital of this island group. What factors do you think affected the frequency of color blindness on this island? Would you expect the frequency of the color-blind allele to be the same on this island as it is on Pingelap? Explain.

2. Cystic fibrosis (CF) is caused by a genetic mutation resulting in defective proteins in secretory cells, mainly in the epithelial lining of the respiratory tract. The one in every 2,000 Caucasian babies who has the disease is homozygous for the recessive mutant. Although medical treatment is becoming more effective, in the past, most children with CF died before their teens. About 20 Caucasians in 2,000 are carriers of the trait, having one mutant and one normal allele, but they do not develop the disease. According to rules of population genetics, the frequency of the homozygous recessive genotype should be rarer than it is. What is one possible explanation for the unusually high frequency of this allele in Caucasian populations?

3. In 1990, scientists investigated the genetic variation in the current populations of Northern Elephant Seals, whose numbers have reached 150,000 compared to only 30 individuals remaining in 1890 after they were overhunted. They sampled a stretch of mitochondrial DNA known to be variable, with as many as 30 polymorphic (more than one allele) loci. For the Northern Elephant Seal, they found only 2 loci were polymorphic. However, in a similar sample of the Southern Elephant Seal, they detected 23 polymorphic loci in the same region of the DNA. Given what you know about the population genetics of the Northern Elephant Seal, describe at least one scenario that would explain the differences in these two species (Hoelzel et al., 1993).

INVESTIGATIVE EXTENSIONS

Evolutionary biologists are interested in documenting the amount of genetic variation in populations. One way to estimate genetic variation is to determine the proportion of the population that is heterozygous. You can continue investigations of variation in populations by exploring the lab topic "Population Genetics: Determining Genetic Variation." This lab was published in previous editions of the laboratory manual.

STUDENT MEDIA: BioFlix, Activities, Investigations, Videos, and Data Tables

www.masteringbiology.com (select Study Area)

BioFlix—Ch. 23: Mechanisms of Evolution (Animation), Allele Frequencies (Animation)

Activities—Ch. 23: Causes of Evolutionary Change

Investigations—Ch. 22: How Do Environmental Changes Affect a Population? What Are the Patterns of Antibiotic Resistance?

Ch. 23: How Can Frequency of Alleles Be Calculated?
Biology Labs OnLine: Population Genetics Lab: Background, Assignments 1, 2, 4, 5, 6, 7; Evolution Lab

Videos—Ch. 22: Biointeractive Videos: The Making of the Fittest: Natural Selection and Adaptation (Pocket Mouse); Ch. 23: The Origin of Species: The Beak of the Finch, The Making of the Fittest: Natural Selection in Humans (Sickle Cell Anemia)

Data Tables—Tables 11.3, 11.4, and 11.5 can be downloaded in Excel format. Look in the Study Area under Lab Media.

REFERENCES

Some parts of this lab topic were modified from J. C. Glase, 1993, "A Laboratory on Population Genetics and Evolution: A Physical Model and Computer Simulation," pages 29–41, in *Tested Studies for Laboratory Teaching*, Volume 7/8 (C. A. Goldman and P. L. Hauta, Editors). Proceedings of the 7th and 8th Workshop/Conference of the Association for Biology Laboratory Education (ABLE), 187 pages. Used by permission.

Ayala, F. J. *Population and Evolutionary Genetics: A Primer*. Menlo Park, CA: Benjamin Cummings, 1982.

Halliburton, R. *Introduction to Population Genetics*. San Francisco, CA: Benjamin Cummings, 2004.

Hardy, G. H. "Mendelian Proportions in a Mixed Population." *Science* (July 10, 1908), pp. 49–50. In this interesting early application of the "Hardy-Weinberg principle," this letter written by Hardy to the editor of *Science* refutes a report that dominant characters increase in frequency over time.

Hedrick, P. W., G. Gutierrez-Espeleta, and R. Lee. "Founder Effect in an Island Population of Bighorn Sheep." *Molecular Ecology*, 2001, vol. 10(4), pp. 851–857.

Heitz 6th, J. and C. Giffen. *Practicing Biology: A Student Workbook*, 6th ed. San Francisco, CA: Pearson, 2017. Chapter 23 provides a guide to solving population genetics problems and practice problems.

Hoelzel, A., J. Halley, S. O'Brien, C. Campagna, T. Arnbom, B. LeBoeuf, K. Ralls, and G. Dover. "Elephant seal genetic variation and the use of simulation models to investigate historical population bottlenecks." *J. Heredity*, 1993, vol. 84(6), pp. 443–449.

Nachman, M. W., H. E. Hoekstra, and L. D'Agostino. "The Genetic Basis of Adaptive Melanism in Pocket Mice," *Proc. Natl. Acad. Sci.* 2003, vol. 100, pp. 5268–5273.

Price, F. and V. Vaughn. "Evolve," *BioQuest Library Online*. http://www.bioquest.org/BQLibrary/library_-result.php (software available to download).

Sidoti, A., R. D'Angelo, C. Rinaldi, G. De Luca, F. Pino, C. Salpietro, D. E. Giunta, F. Saltamacchia, and A. Amato. "Distribution of the Mutated 32 Allele of the CCR5 Gene in a Sicilian Population." *International Journal of Immunogenetics*, 2005, vol. 32, pp. 193–196.

Sundin, O. H., J. Yang, Y. Li, D. Zhu, J. Hurd, T. Mitchell, E. Silva, and I. Maumenee. "Genetic Basis of Total Colourblindness Among the Pingelapese Islanders." *Nature Genetics*, 2000, vol. 25, pp. 289–293.

Zimmer, C. and D. Emlen. *Evolution: Making Sense of Life*, 2nd ed. Greenwood Village, CA: Roberts and Company, 2015

WEBSITES

For activities related to the rock pocket mice and other research:
http://www.hhmi.org/biointeractive/activities/pocketmouse.html

Howard Hughes Medical Institute Biointeractive; includes an Evolution page with resources for students and teachers:
http://www.hhmi.org/biointeractive/evolution-collection

Issues in Evolution, American Institutes of Biological Sciences; provides resources, interviews, and issues:
http://www.actionbioscience.org/evolution/

Library of resources at PBS Evolution website:
http://www.pbs.org/wgbh/evolution/library/index.html

Malaria movie showing sickle cell and malaria; importance of mutation:
http://www.pbs.org/wgbh/evolution/library/01/2/l_012_02.html

Short films including natural selection and adaptation:
http://www.hhmi.org/biointeractive/short-films-collection

Simulations for population genetics are available at several sites:
http://www.cbs.umn.edu/populus/
http://scit.us/redlynx/
http://evolution.gs.washington.edu/popgen/popg.html

Video of toxic newt and garter snake—evolutionary arms race:
http://www.pbs.org/wgbh/evolution/library/01/3/l_013_07.html

NOTES

LAB TOPIC 12

Bacteriology

Laboratory Objectives

After completing this lab topic, you should be able to:

1. Describe bacterial structure: colony morphology, cell shape, growth patterns.
2. Describe the results of Gram staining and discuss the implications to cell wall chemistry.
3. Describe a scenario for succession of bacterial and fungal communities in aging milk, relating this to changes in environmental conditions such as pH and nutrient availability.
4. Practice aseptic techniques in producing bacterial streaks, smears, and lawns.
5. Describe the ecology and control of bacteria, applying these concepts to real-life situations.

Introduction

Humans have named and categorized organisms for hundreds—perhaps even thousands—of years. *Taxonomy* is an important branch of biology that deals with naming and classifying organisms into distinct groups or categories. Much of the work of early taxonomists included recording characteristics of organisms and grouping them based on appearance, habitat, or perhaps medicinal value. As scientists began to understand the processes of genetics and evolution by natural selection, they realized the value of classifying organisms based on phylogeny, or evolutionary history. Early in the study of *systematics*, the scientific discipline that classifies organisms based on their evolutionary relationships, scientists obtained information about phylogeny from studies of development or homologous features—common features resulting from common genes. In recent years, the discipline of *molecular systematics,* also called *phylogenetics,* has become important, where scientists use biochemical evidence—studies of nucleic acids and proteins—to investigate relationships among organisms, leading to revisions in the taxonomic scheme.

Systematists continue to grapple with the complex challenge of organizing the diversity of life into categories. In the 1960s, the early system of classifying all living organisms as either plants or animals was replaced by a five-kingdom scheme that placed all prokaryotic organisms (organisms whose cells lack a distinct membrane-bound nucleus and membrane-bound organelles) in the kingdom Monera and eukaryotic organisms (organisms with membrane-bound nuclei and membrane-bound organelles in their cells) in the kingdoms Protista, Fungi, Plantae, and Animalia. In the late 1970s, the microbiologist Carl Woese and his colleagues at the University of Illinois were studying DNA and ribosomal RNA sequences in prokaryotes when they discovered a group of organisms that were dramatically different from other prokaryotes. Because of their vast differences, Woese proposed a three-domain system of classification that has been validated over decades of research involving many studies. One recent study that supported the three-domain proposal involved the analysis of nearly 100 completely sequenced genomes. In the three-domain system, the three domains—Bacteria, Archaea, and Eukarya—are a taxonomic level higher and include the kingdoms, historically the broadest taxonomic category. In the three-domain system, prokaryotes are placed in one of the two domains Bacteria or Archaea. Most of the familiar organisms commonly called bacteria are placed in the domain Bacteria. Prokaryotic organisms placed in the domain Archaea share many traits with common bacteria, but they have many unique traits. The three-domain classification renders the kingdom Monera obsolete because its members are in two domains. All eukaryotic organisms (eukaryotes) are categorized in the domain Eukarya. Three of the kingdoms of the five-kingdom scheme now placed in Eukarya—Fungi, Plantae, and Animalia—are multicellular organisms. Members of the former kingdom Protista are now in the domain Eukarya, but studies based on molecular genetics have dramatically reorganized classification schemes for the protists. You will study protistan classification further in Lab Topic 13 Protists. The reorganization of organisms within the domains has led some biologists to abandon the term "kingdom" and to propose placing organisms into "supergroups."

All of the organisms studied in this lab topic are common bacteria, small, relatively simple, **prokaryotic**, single-celled organisms. **Prokaryotes**, from the Greek for "prenucleus," have existed on Earth longer and are more widely distributed than any other organismal group. They are found in almost every imaginable habitat: air, soil, and water, in extreme temperatures and harsh chemical environments. They can be photosynthetic (cyanobacteria, formerly called blue-green algae), using light as the source of energy, or chemosynthetic, using inorganic chemicals as the source of energy, but most are heterotrophic, absorbing nutrients from the surrounding environment. Many bacteria are beneficial to other organisms and the environment. For example, some species play an important role in decomposing dead organisms and waste products. Many bacteria are harmful to other organisms. For example, *pathogenic* bacteria cause diseases in humans and other animals.

Most bacteria have a cell wall, a complex layer outside the cell membrane. The most common component found in the cell wall of organisms in the domain Bacteria is peptidoglycan, a complex protein-carbohydrate polymer. There are no membrane-bound organelles in bacteria and the genetic material is not bound by a nuclear envelope. Bacteria do not have chromosomes as described in Lab Topic 7; their genetic material is a single circular molecule of DNA. In addition, bacteria may have smaller rings of DNA called **plasmids** (see page 250), consisting of only a few genes. They reproduce by a process called **binary fission**, in which the cell duplicates its components and divides into two cells. These cells usually become independent, but they may remain attached in linear chains or grapelike clusters. In favorable environments, individual bacterial cells rapidly proliferate, forming colonies consisting of millions of cells.

Differences in colony morphology and the shape of individual bacterial cells are important distinguishing characteristics of bacterial species. In Exercise 12.1, working independently, you will observe and describe the morphology of colonies and individual cells of several bacterial species. You will examine and describe characteristics of bacteria growing in plaque on your teeth. You and your lab partner will compare the results of all lab studies.

EXERCISE 12.1

Investigating Characteristics of Bacteria

Because of the small size and similarity of cell structure in bacteria, techniques used to identify bacteria are different from those used to identify macroscopic organisms. Staining reactions and properties of growth, nutrition, and physiology are usually used to make final identifications of species. The structure and arrangement of cells and the morphology of colonies contribute preliminary information that can help us determine the appropriate test necessary to make final identification. In this exercise, you will use the tools at hand, microscopes and unaided visual observations, to learn some characteristics of bacterial cells and colonies.

When you are working with bacteria, it is very important to practice certain **aseptic techniques** to ensure that the cultures being studied are not contaminated by organisms from the environment and that organisms are not released into the environment.

1. Wear a lab coat, a lab apron, or a clean old shirt over your clothes to lessen chances of staining or contamination accidents.
2. Wipe the lab bench with disinfectant before and after the lab activities.
3. Wash your hands before and after performing an experiment. If directed by your instructor, use disposable gloves.
4. Using the alcohol lamp or Bunsen burner, flame all nonflammable instruments used to manipulate bacteria or fungi before and after use.
5. Place swabs and toothpicks in the disposal container immediately after use. *Never place one of these used items on the lab bench!*

The bacteria used in these exercises are not pathogenic (disease-producing); nevertheless, use appropriate aseptic techniques and work with care! If a spill occurs, notify the instructor. If no instructor is available, wear disposable gloves, and wipe up the spill with paper towels. Follow this by washing the affected area with soap and water and a disinfectant. Dispose of the gloves and soiled towels in the autoclavable plastic bag provided.

Lab Study A. Colony Morphology

Materials

disinfectant
stereoscopic microscope
metric ruler
agar plate cultures with bacterial colonies

Introduction

A **bacterial colony** grows from a single bacterium and is composed of millions of cells. Each colony has a characteristic size, shape, consistency, texture, and color (colony morphology), all of which may be useful in preliminary species identification. Bacteriologists use specific terms to describe colony characteristics. Figure 12.1 illustrates some of the terminology that may be used to

FIGURE 12.1
Terminology used in describing bacterial colonies. (a) Common shapes, (b) margins, and (c) surface characteristics are illustrated.

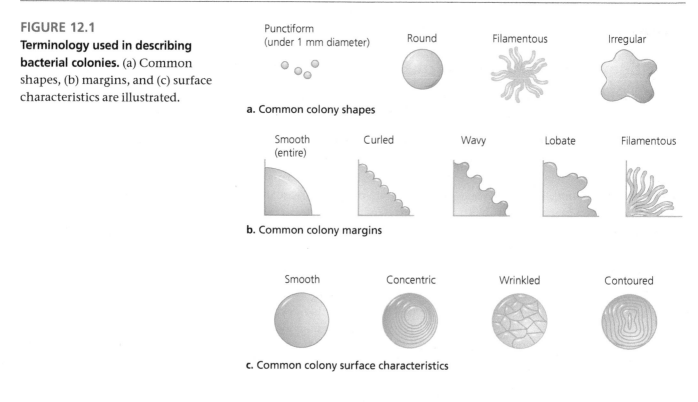

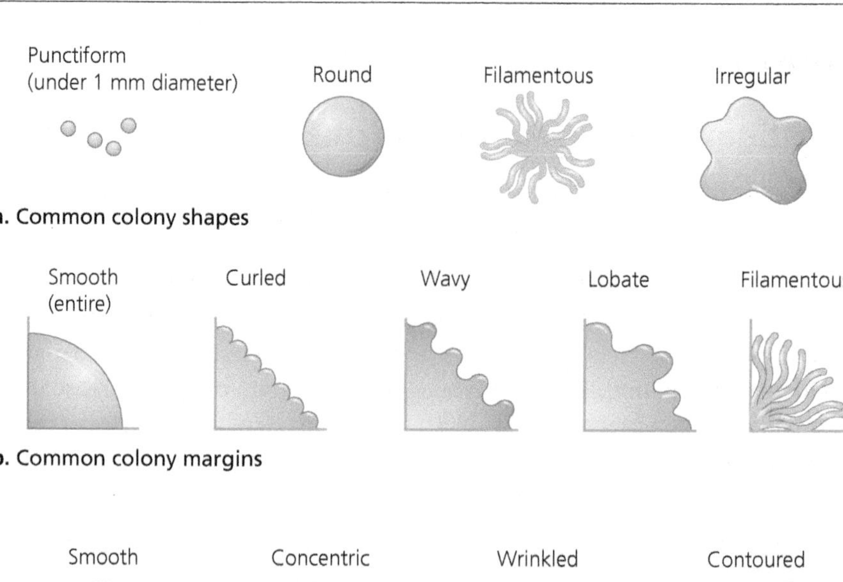

describe colony morphology. Use the figure to become familiar with this terminology and describe the bacterial species provided. Occasionally, one or more **fungal colonies** will contaminate the bacterial plates. Fungi may be distinguished from bacteria by the *fuzzy* appearance of the colony (Figure 12.2). The body of a fungus is a mass of filaments called **hyphae** in a network called a **mycelium**. Learn to distinguish fungi from bacteria.

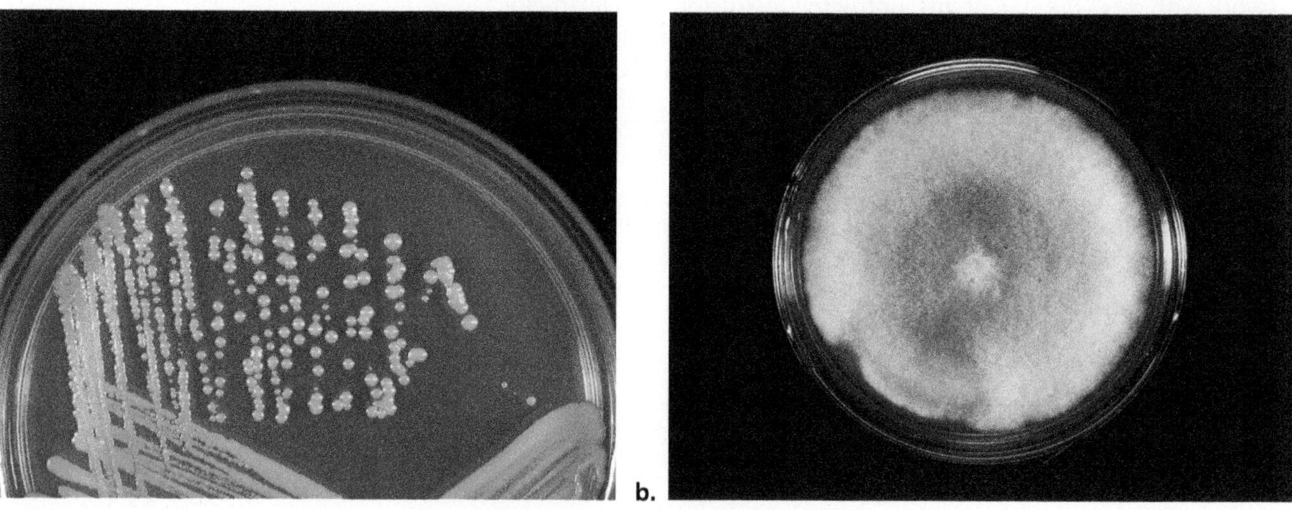

FIGURE 12.2
Distinguishing bacteria and fungi. (a) Bacterial colonies growing on a nutrient agar plate. Bacteria have been isolated using the streak technique (see Exercise 12.3, Procedure step 3). (b) Fungi growing on an agar plate. Note the filamentous fungal body, the mycelium, consisting of hyphae that give fungal colonies a fuzzy appearance.

Procedure

1. Wipe the work area with disinfectant and wash your hands.
2. Set up your stereoscopic microscope.
3. Obtain one of the bacterial plates provided. Leaving the plate closed (unless otherwise instructed), place it on the stage of the microscope.
4. Examine a typical individual, separate colony. Measure the size and note the pigmentation (none or color) of the colony, and record this information in Table 12.1 in the Results section.
5. Using the diagrams in Figure 12.1, select appropriate terms that describe the colony.
6. Record your observations in Table 12.1.
7. Sketch one colony in the margin of your lab manual, illustrating the characteristics observed.
8. Repeat steps 2 to 6 with two additional species. Your lab partner should examine three different species.
9. Observe Figure 12.2a. Describe the shape, margin, surface, and pigmentation (color) of colonies of this bacterial species.

Results

1. Complete Table 12.1 using terms from Figure 12.1 to describe the three bacterial cultures you observed.
2. Compare your observations with those of your lab partner.

Discussion

1. What are the most common colony shapes, colony margins, and colony surface characteristics found in the species observed by you and your lab partner?

2. Based on your observations, comment on the reliability of colony morphology in the identification of a given bacterial species.

TABLE 12.1 Characteristics of Bacterial Colonies

Name of Bacteria	Size	Shape	Margin	Surface	Pigmentation (none or specific color)
1.					
2.					
3.					

Lab Study B. Morphology of Individual Cells

Materials

compound microscope
prepared slides of bacillus, coccus, and spirillum bacteria
blank slide
clean toothpick
clothespin
timer or clock with second hand
dropper bottle of deionized (DI) water
dropper bottle of crystal violet stain
squirt bottle of DI water
alcohol lamp or Bunsen burner
staining pan
discard beaker with weak bleach solution for used toothpicks

Introduction

Microscopic examination of bacterial cells reveals that most bacteria can be classified according to three basic shapes: **bacilli** (rods), **cocci** (spheres), and **spirilla** (spirals, or corkscrews). In many species, cells tend to adhere to each other and form clusters or chains of cells. In some environments, different species may associate in a complex polysaccharide matrix, creating a **community**, or assemblage of species of bacteria that adheres to a surface. These communities, called **biofilms**, are found in moist environments where nutrients are plentiful. Examples of environments that support biofilms are soils, water pipes, medical devices such as the tubes used in kidney dialysis, and the plaque found on your teeth. Anthropologists use the study of plaque on fossil teeth to determine the diets and habitats of early hominins. For example, plaque found on the teeth of a 2-million-year-old fossil of *Australopithecus sediba* disclosed that this early hominin ate fruit, leaves, wood, and bark, indicating it lived in woodland habitats. In this lab study, you will examine prepared slides of bacteria that illustrate the three basic cell shapes, and then you will examine and describe bacteria growing in your mouth.

Procedure

1. To become familiar with the basic shapes of bacterial cells, using the compound microscope, examine prepared slides of the three types of bacteria, and make a sketch of each shape in the space provided.

2. Your mouth is a type of ecosystem with at least 600 species of bacteria living in this warm, moist environment. Protein and carbohydrate materials from food particles accumulate at the gum line in your mouth and create an ideal environment for bacteria to grow. This mixture of materials and bacteria is a biofilm called **plaque**. To investigate the forms of bacteria found on your teeth, prepare a stained slide of plaque.
   a. Set out a clean slide.
   b. Place a drop of water on the slide. This must air-dry, so make the drop of water *small*.
   c. Using a fresh toothpick, scrape your teeth near the gum line and mix the scraping in the drop of water.
   d. Use the toothpick to spread this plaque–water mixture into a thin film and allow it to air-dry. Place your used toothpick in the discard container.

e. When the smear is dry, hold the slide with a clothespin and pass it quickly over the flame of an alcohol lamp or Bunsen burner several times at a 45° angle. This should warm the slide but not cook the bacteria. Briefly touch the warm slide to the back of your hand. If it is too hot to touch, you are allowing it to get too warm.

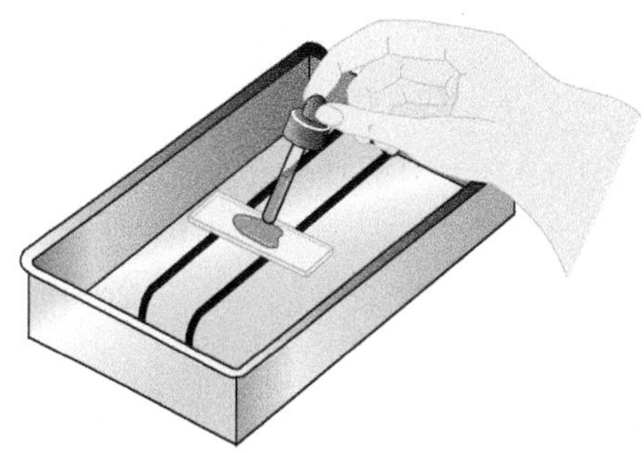

FIGURE 12.3
Apply several drops of crystal violet stain to the slide supported in a staining pan or tray.

Keep long hair and loose clothing away from the flame. Extinguish the flame immediately after use.

f. Place the slide on the support of a staining pan or tray and apply three or four drops of crystal violet stain to the smear (Figure 12.3).

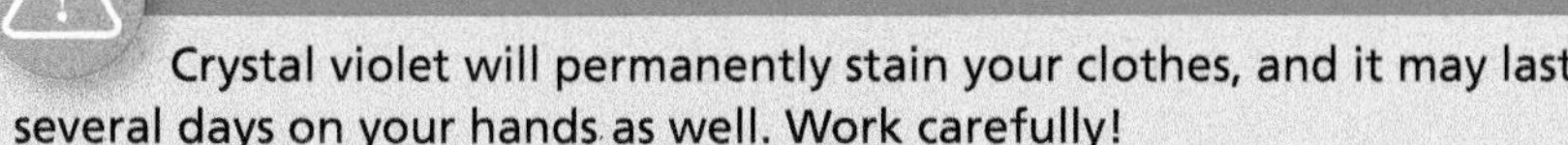

Crystal violet will permanently stain your clothes, and it may last several days on your hands as well. Work carefully!

g. Leave the stain on the smear for 1 minute.

h. Wash the stain off with a gentle stream of water from a squirt bottle so that the stain goes into the staining pan (Figure 12.4).

i. Blot the stained slide gently with a paper towel. Do not rub hard or you will remove the bacteria.

3. Examine the bacteria growing in the plaque on your teeth and determine bacterial forms. Use the highest magnification on your compound microscope. If you have an oil immersion lens, after focusing on the high-dry power, without changing the focus knobs, rotate the high-dry objective to the side, add a drop of immersion oil directly to the bacterial smear, and carefully rotate the oil immersion objective into place. Focus with the fine adjustment only. After observing the slide, rotate the oil immersion objective away from the slide and wipe the objective carefully, using lens paper to remove all traces of oil.

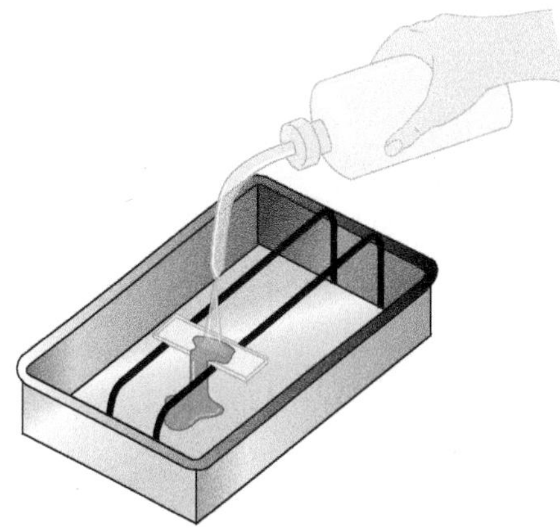

FIGURE 12.4
Gently rinse the stain into the staining pan.

Results

1. Record the individual cell shapes of bacteria present in plaque.

2. What shapes are absent?

3. Estimate the relative abundance of each shape.

Discussion

1. Discuss with your lab partner information you have learned from your dentist or health class about the relationship among plaque, dental caries (cavities), and gum disease.

2. Suggest an explanation for differences in the proportion of each type of bacteria in the bacterial community of plaque.

Lab Study C. Identifying Bacteria by the Gram Stain Procedure

Materials

compound microscope
blank slides
alcohol lamp or Bunsen burner
clean toothpicks
staining pan
cultures of *Micrococcus, Bacillus, Serratia*, and *E. coli*
dropper bottles of Gram iodine, crystal violet, safranin, DI water, 95% ethyl alcohol/acetone mixture
squirt bottle of DI water

Introduction

Gram stain is named for the Danish scientist, Hans Christian Gram, who developed and published procedures for the technique in 1884. A German physician, Paul Ehrlich (1854–1915), who is best-known for his Nobel Prize–winning research in immunology and chemotherapy, actually invented the precursor technique to Gram staining of bacteria. Ehrlich's staining procedures were revolutionary in that they made it possible to use microscopy to detect bacteria in sections of organ tissue. The significance of this discovery was portrayed in the 1940s movie, *Dr. Ehrlich's Magic Bullet.*

Gram stain separates bacteria into groups depending on their reaction to this stain. Bacteria react by testing either **gram-positive**, **gram-negative**, or **gram-variable**, with the first two groups being the most common. Although the exact mechanisms are not completely understood, scientists know that the response of cells to the stain is due to differences in the complexity and chemistry of the bacterial cell wall. Recall that bacterial cell walls contain a complex polymer, **peptidoglycan**. The cell walls of gram-negative bacteria contain less peptidoglycan than the cell walls of gram-positive bacteria. In addition, the cell walls of gram-negative bacteria are more complex, containing various polysaccharides, proteins, and lipids not found in gram-positive bacteria. Gram-staining properties play an important role in bacterial classification.

Gram stain relies on the use of three stains: crystal violet (purple), Gram iodine, and safranin (pink/red). *Gram-positive* bacteria (with the thicker peptidoglycan layer) retain the crystal violet/iodine stain and appear blue/purple. *Gram-negative* bacteria lose the blue/purple stain but retain the safranin and appear pink/red (Figure 12.5).

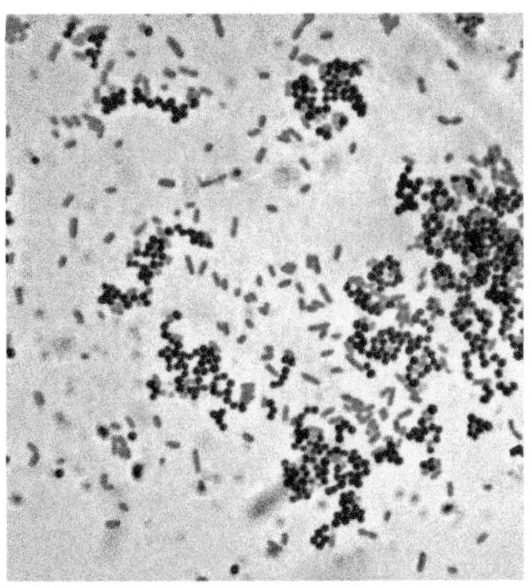

FIGURE 12.5
Gram-positive bacteria appear purple (coccus bacteria in this photomicrograph). **Gram-negative bacteria appear pink** (the rod shaped bacteria seen here).

In summary:

Gram-Negative Bacteria	Gram-Positive Bacteria
more complex cell wall	simple cell wall
thin peptidoglycan cell wall layer	thick peptidoglycan cell wall layer
outer lipopolysaccharide wall layer	no outer lipopolysaccharide wall layer
retain safranin	retain crystal violet/iodine
appear pink/red	appear blue/purple

In this lab study, you will prepare and stain slides of two different bacterial species. One member of the lab team should stain *Micrococcus* and *Bacillus*. The other member should stain *Serratia* and *E. coli*.

Procedure

1. Using a *clean* slide, prepare smears as directed for the plaque slide (Lab Study B, steps 2a to 2e), substituting the bacterial species for the plaque. If you are using a liquid bacterial culture, do *not* add water to your slide (step 2b). Label the slide with your initials and the name of the bacterial species being investigated.
2. Support the slide on the staining tray and cover the smear with three or four drops of crystal violet. Wait 1 minute.
3. Rinse the stain gently into the staining pan with water from the squirt bottle.
4. Cover the smear with Gram iodine for 1 minute, setting the stain.
5. Rinse it again with water.
6. Destain (remove the stain) by dropping the 95% alcohol/acetone mixture down the slanted slide one drop at a time. At first a lot of violet color will rinse away. Continue adding drops until only a faint violet color is seen in the alcohol rinse. Do not overdo this step (Figure 12.6). You should be able to see some color in the smear on the slide. If not, you have destained too much. The alcohol/acetone removes the crystal violet stain from the gram-negative bacteria. The gram-positive bacteria will not be destained.
7. Using the water squirt bottle, rinse immediately to prevent further destaining.
8. Cover the smear with safranin for 30 to 60 seconds. This will stain the destained gram-negative bacteria a pink/red color. The gram-positive bacteria will be unaffected by the safranin (Figure 12.7).

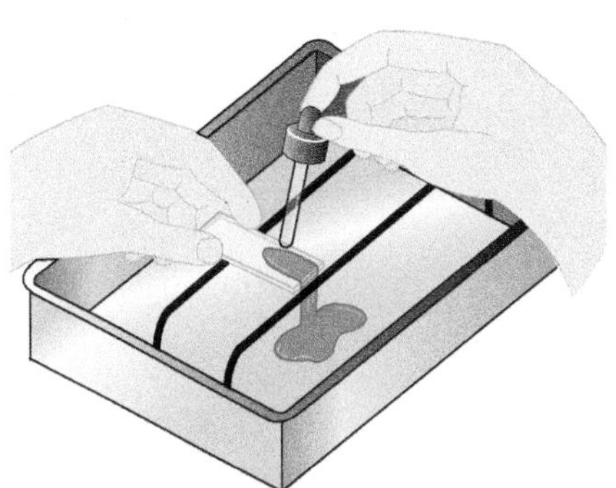

FIGURE 12.6
Destain by dropping 95% ethyl alcohol/acetone down the slanted slide until only a faint violet color is seen in the solution.

	Step 1 Crystal violet	Step 2 Gram iodine	Step 3 Alcohol/acetone wash	Step 4 Safranin
Gram-positive cell:	Purple	Blue/purple	Remains blue/purple	Remains blue/purple
Gram-negative cell:	Purple	Blue/purple	Loses stain	Pink/red

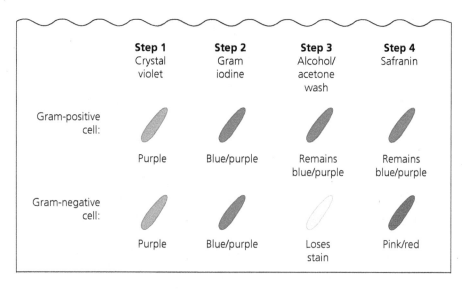

FIGURE 12.7

The Gram stain. Crystal violet and Gram iodine stain all cells blue/purple. Alcohol/acetone destains gram-negative cells. Safranin stains gram-negative cells pink/red.

9. Briefly rinse the smear with water as above. Blot it lightly with a paper towel and let it dry at room temperature.
10. Examine each slide using the highest magnification on your microscope.

If you use oil immersion, remove all traces of oil from the objective after observing the slide.

Results

Record your observations of the results of the Gram stain in Table 12.2.

TABLE 12.2 Bacteria Observed and Results of Gram Stain

Name of Bacteria	Results of Gram Stain
1.	
2.	

Discussion

1. Which of the bacteria observed are probably more closely related taxonomically?

2. What factors can modify the expected results of this staining procedure?

EXERCISE 12.2

Ecological Succession of Bacteria in Milk

Materials

pH paper
flasks of plain and chocolate whole milk, aged 1, 4, and 8 days
TGY agar plates of each of the milk types
supplies from Exercise 12.1 for Gram stains

Introduction

Milk is a highly nutritious food containing carbohydrates (lactose, or milk sugar), proteins (casein, or curd), and lipids (butterfat). This high level of nutrition makes milk an excellent medium for the growth of bacteria. Pasteurizing milk does not sterilize it (sterilizing kills *all* bacteria) but merely destroys pathogenic bacteria, leaving many bacteria that will multiply very slowly at refrigerated temperatures; but at room temperature, these bacteria will begin to grow and bring about milk spoilage. Biologists have discovered that as milk ages, changing conditions in the milk bring about a predictable, orderly succession of microorganism communities (associations of species).

Community succession is a phenomenon observed in the organizational hierarchy of all living organisms, from bacterial communities in milk to animal and plant communities in a maturing deciduous forest. In each example, as one community grows, it modifies the environment, and a different community develops as a result.

In this laboratory exercise, you will work in pairs and observe successional patterns in two types of milk, plain whole milk and milk with sucrose and chocolate added. You will record changes in the environmental conditions of the two types of milk as they age. Consider the following information about milk bacteria and their environments as you continue this exercise.

1. *Lactobacillus* (gram-positive rod) and *Streptococcus* (gram-positive coccus) survive pasteurization.
2. *Lactobacillus* and *Streptococcus* ferment lactose to lactate and acetic acid.
3. An acidic environment causes casein to solidify, or curd.
4. Two bacteria commonly found in soil and water, *Pseudomonas* and *Achromobacter* (both gram-negative rods), digest butterfats and give milk a putrid smell.
5. Yeasts and molds (both fungi) grow well in acidic environments.

Scenario

Propose a scenario (the hypothesis) for bacterial succession in each type of milk.

On each lab bench are four flasks of plain whole milk and four flasks of chocolate milk. One flask of each has been kept under refrigeration. One flask of each has been at room temperature for 24 hours, one for 4 days, and one for 8 days. On each bench there are also TGY (tryptone, glucose, yeast) agar plate cultures of each of the types of milk.

One team of two students should work with plain milk, another with chocolate milk. Teams will then exchange observations and results.

Procedure

1. Using the pH paper provided, take the pH of each flask. Record your results in Table 12.3 in the Results section.
2. Record the odor (sour, putrid), color, and consistency (coagulation slight, moderate, chunky) for the milk in each flask.
3. Using the TGY agar plates, observe and describe bacterial/fungal colonies in each age and type of milk. Use the vocabulary you developed while doing Exercise 12.1.
4. Prepare Gram stains of each different bacterial type on each plate using the staining instructions in Exercise 12.1, Lab Study C.
5. Record the results of the Gram stains in Table 12.3 in the Results section.

Results

Complete Table 12.3, describing the characteristics of each milk culture and the bacteria present in each.

TABLE 12.3 Physical Features and Bacterial/Fungal Communities of Aged Plain and Chocolate Milk

Age/Type of Milk	Environmental Characteristics (pH, Consistency, Odor, Color)	Organisms Present (Bacteria: Gram +/–, Shapes; Yeasts or Fungi)
Refrigerated plain		
24-hr plain		
4-day plain		
8-day plain		
Refrigerated chocolate		
24-hr chocolate		
4-day chocolate		
8-day chocolate		

Discussion

1. Describe the changing sequence of organisms and corresponding environmental changes during succession in plain milk. Do the results of your investigation match your hypothesis?

2. Describe the changing sequence of organisms and corresponding environmental changes during succession in chocolate milk. Do the results of your investigation match your hypothesis?

3. Compare succession in plain and chocolate milk. Propose reasons for differences.

4. Propose an experiment to test the environmental factors and/or organisms changing in your proposed scenario for milk succession.

5. Perform an experiment similar to the one in this Exercise to investigate community succession in soy milk, lactose-free milk, or almond milk.

EXERCISE 12.3

Bacteria in the Environment

Bacteria thrive in a wide variety of natural environments. They are present in all environments where higher forms of life exist, and many flourish in extreme environments where no higher life forms exist—boiling hot springs, extremely salty bodies of water, and waters with extreme pH.

In the following experiments, you will sample different environments, testing for the presence of bacteria and fungi. In Experiment A, pairs of students will investigate one of six different environments. Each pair will report results to the entire class. In Experiment B, your team will investigate an environment of your choice.

Experiment A. Investigating Specific Environments

Materials

sterile agar plates
wax pencil
2 cotton-tipped sterile swabs
bacterial inoculating loop
alcohol lamp or Bunsen burner
piece of raw chicken in a petri dish
disposable gloves for Team 1
soil samples
samples of pond or stream water
plant leaves or other plant parts
hand soap
Parafilm strips
discard receptacle
forceps

Introduction

The instructor will assign team numbers to each pair of students. Each pair (team) of students will sample bacteria and fungi from one of six environments: food supply, soil, air, pond water, a plant structure, and hands. Read the instructions for all investigations. Think about the following questions, and before you begin your investigation, hypothesize about the relative growth of bacteria and fungi in the different environments.

Where in the environment would bacteria be more common, and where would fungi be more common? Would any of these environments be free of bacteria or fungi?

Seal all plates with Parafilm after preparation! Discard all used swabs in the designated receptacle!

Hypothesis

Hypothesize about the presence of bacteria and fungi in the different environments.

Prediction

Predict the results of the experiment based on your hypothesis (if/then).

Procedure

Team 1—bacteria and fungi on food

1. Wearing disposable gloves, obtain a sterile agar plate and label the bottom "chicken." Then obtain the petri dish containing a piece of raw chicken.
2. Open the dish containing the piece of chicken, and swab the chicken surface using a sterile cotton swab.

Avoid touching the chicken. Use disposable gloves and the swab. Always wash your hands thoroughly after touching raw chicken, owing to the potential presence of *Salmonella*, bacteria that cause diarrhea.

3. Isolate bacteria by the **streak plate** method.
   a. Carefully lift the lid of the agar plate to 45° and lightly streak the swab back and forth across the top quarter of the agar (Figure 12.8a). Close the lid and *discard the swab in the receptacle provided*. Minimize exposure of the agar plate to the air.
   b. Flame the bacterial inoculating loop using the alcohol lamp or Bunsen burner. Allow the loop to cool; starting at one end of the swab streak, lightly streak the microorganism in the pattern shown in Figure 12.8b. Do not gouge the medium.
   c. Reflame the loop and continue to streak as shown in Figure 12.8c and described in the figure legend.

   By the end of the last streak, the bacteria should be separated and reduced in density so that only isolated bacteria remain. These should grow into isolated, characteristic colonies.
4. Discard the disposable gloves as directed by your instructor.
5. Write the initials of your team members, the lab room, and the date on the bottom of the agar plate.
6. Seal the plate with Parafilm and place it in the area indicated by the instructor.
7. Incubate the culture for 1 week and observe the results during the next laboratory period.

The following week, to avoid exposure to potentially pathogenic bacteria, do not open the petri dish containing the chicken bacteria. Wash hands after handling cultures.

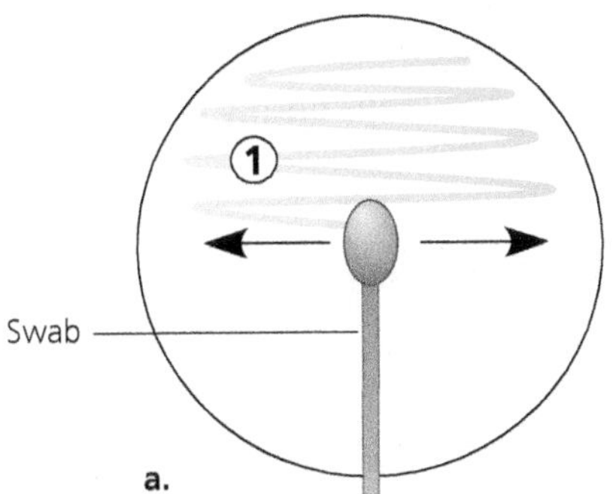

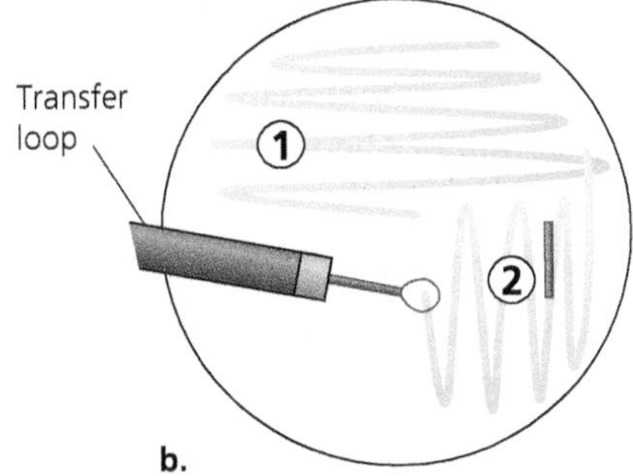

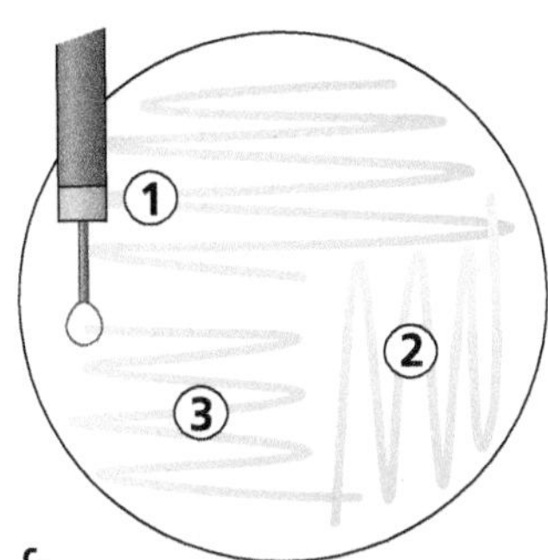

FIGURE 12.8
Isolating bacterial colonies using the streak technique. (a) Streak the swab over the top quarter of the agar plate, region 1. (b) Using the newly flamed, cooled loop, pick up organisms from region 1 and streak them into region 2.
(c) Reflame the loop, allow it to cool, and pick up organisms from region 2 and streak them into region 3.

Team 2—bacteria and fungi in soil

1. Holding the lid in place, invert an agar plate and label the bottom "soil."
2. Using a cotton swab, pick up a small amount of soil from the sample.
3. Prepare a streak culture by following step 3 in the procedure for Team 1.
4. Write the initials of your team members, the lab room, and the date on the bottom of the agar plate.
5. Seal the plate with Parafilm and place it in the area indicated by the instructor.
6. Incubate the culture for 1 week and observe the results during the next laboratory period.

Team 3—bacteria and fungi in air

1. Holding the lid in place, invert an agar plate and label the bottom "air."
2. Collect a sample of bacteria by leaving the agar plate exposed (lid removed) to the air in some interesting area of the room for 10 to 15 minutes. Possible areas might be near a heat duct or an animal storage bin. Write the sample site in your lab manual margin or on the bottom of your agar plate.
3. If additional agar plates are available, you may choose to sample several sites.
4. Write the initials of your team members, the lab room, and the date on the bottom of the agar plate.
5. Seal the plate with Parafilm and place it in the area indicated by the instructor.
6. Incubate the culture(s) for 1 week and observe the results during the next laboratory period.

Team 4—bacteria and fungi in water supplies

1. Holding the lid in place, invert an agar plate and label the bottom "pond (or stream) water."
2. Using a sterile cotton swab, take a sample from the pond or stream water.
3. Prepare a streak culture by following step 3 in the procedure for Team 1.
4. Write the initials of your team members, the lab room, and the date on the bottom of the agar plate.
5. Seal the plate with Parafilm and place it in the area indicated by the instructor.
6. Incubate the culture for 1 week and observe the results during the next laboratory period.

Team 5—bacteria and fungi on plant structures

1. Obtain a sample from a plant. This might be a leaf, a part of a flower, or another plant structure.
2. Holding the lid in place, invert an agar plate and label the bottom "plant ______," naming the part you are testing. If you are testing a leaf, indicate if it is the top or bottom of the leaf.
3. Open the plate and place the plant part flat on the agar surface. You may need to use forceps. Be sure there is good contact. Remove and discard the plant part. Close the plate.
4. Write the initials of your team members, the lab room, and the date on the bottom of the agar plate.

5. Seal the plate with Parafilm and place it in the area indicated by the instructor.
6. Incubate the culture for 1 week and observe the results during the next laboratory period.

Team 6—bacteria and fungi on human hands

1. Draw a line across the center of the bottom of an agar plate. Write "unwashed" on the dish bottom on one side of the line and "washed" on the other side of the line.
2. Select one person who has not recently washed his or her hands to be the test subject. The subject should open the agar plate and *lightly* press three fingers on the agar surface in the half of the dish marked "unwashed." Do not break the agar. Close the plate.
3. The subject should wash his or her hands for 1 minute and repeat the procedure, touching the agar with the same three fingers on the side of the dish marked "washed."
4. Write the initials of your team members, the lab room, and the date on the bottom of the agar plate.
5. Seal the plate with Parafilm and place it in the area indicated by the instructor.
6. Incubate the culture for 1 week and observe the results during the next laboratory period.

Results

Include results from the entire class.

1. During the following laboratory period, observe your agar cultures of bacteria and fungi from the environment and record your observations in Table 12.4.
2. Place your agar culture on the demonstration table and make a label of the environment being investigated. All students should observe every culture.
3. Observe the agar plates prepared by your classmates. Record observations in Table 12.4.

TABLE 12.4 Abundance and Types of Colonies Associated with Food (Raw Chicken), Soil, Air, Water, Plant Structures, and Hands

Environment	Colony Type(s) and Abundance
Chicken	
Soil	
Air	
Pond or stream water	
Plant structure	
Hands before washing	
Hands after washing	

Discussion

1. How did the plates differ in the number and diversity of bacterial and fungal colonies?

2. Did your predictions match your observations? Describe any discrepancies.

3. What factors might be responsible for your results?

4. Based on the results of your experiments, suggest health guidelines for workers in the food industry, as well as for schoolchildren or others who might be concerned with sanitary conditions.

Experiment B. Investigating the Environment of Your Choice

Materials

agar plates
sterile swab
capped test tube of sterile water
Parafilm strips
discard receptacle

Introduction

In the previous experiment, you tested specific environments for the presence of bacteria and fungi. In this lab study, you will study an environment of your choice. If extra agar plates are available, you may choose to investigate bacteria in an environment before and after some treatment, such as bacteria on the water fountain before and after cleaning.

Seal all plates with Parafilm after preparation!

Hypothesis

Hypothesize about the growth of bacteria in an environment of your choice.

Prediction

Predict the results of the experiment based on your hypothesis.

Procedure

1. Decide what environment you will investigate. It might be some environment in the lab room or somewhere in the biology building. For example, think about technology items you share. Carry the sterile cotton swab and agar plate to the environment, and use the swab to collect the sample. If you are collecting from a dry surface, you should first dip the cotton swab in the sterile water and then swab the surface. If you apply any treatment to the surface, describe the treatment in the margin of your lab manual.

Do not do throat or ear swabs! Pathogenic bacteria may be present.

2. Open the agar plate and lightly streak the swab back and forth across the agar. Discard the swab in the receptacle provided.
3. Label the bottom of the agar plate to indicate the environment tested. Record the environment tested in the Results section.
4. Write the initials of your team members, the lab room, and the date on the bottom of the agar plate.
5. Seal the plate with Parafilm and place it in the area indicated by the instructor.
6. Incubate the culture for 1 week and then observe and describe the results during the next laboratory period.

Results

1. What environment did you investigate? Indicate any treatment you applied.
2. Characterize the bacterial and fungal colonies from your experiment.

Discussion

1. Do your results match your predictions for the presence of bacteria and fungi in this environment?
2. What factors might be responsible for your results?

EXERCISE 12.4

Controlling the Growth of Bacteria

Bacteria are found almost everywhere on Earth, and most species are directly or indirectly beneficial to other organisms. Bacteria are necessary to maintain optimum environments in animal and plant bodies and in environmental systems. However, even beneficial species, if they are reproducing at an uncontrolled rate, are potentially harmful or destructive to their environment. In addition, several species of bacteria and fungi are known to be pathogenic, that is, to cause disease in animals and plants. Their growth must be controlled. Agents have been developed that control bacterial and fungal growth. In this exercise, you will investigate the efficacy of three of these growth-controlling agents: antibiotics, antiseptics, and disinfectants.

Lab Study A. Using Antibiotics to Control Bacterial Growth

Materials

agar plate
metric ruler
sterile swab
wax pencil
Parafilm strips
broth cultures of *Micrococcus*, *Bacillus*, *Serratia*, and *E. coli*
antibiotic dispenser with antibiotic disks

Introduction

An **antibiotic** is a chemical produced by a bacterium or fungus that has the potential to control the growth of another bacterium or fungus. Many antibiotics are selective, however, having their inhibiting effect on only certain species of bacteria or fungi. In this lab study, you will apply an assortment of antibiotics to a lawn culture of a bacterial species. Working in pairs, you will determine which antibiotics are able to control the growth of the bacteria by measuring zones of inhibition. Each pair of students in a group of eight should culture a different bacterium. All four species should be cultured.

A lawn of bacteria is like a lawn of grass—a uniform, even layer of organisms covering an entire surface. Prepare the lawn of bacteria carefully. The success of this experiment will largely depend on the quality of your lawn.

Hypothesis

Hypothesize about the effect of different antibiotics on the growth of bacteria.

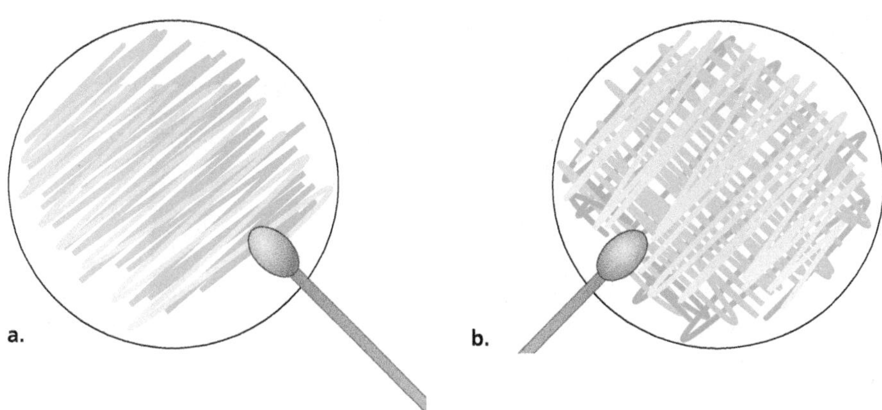

FIGURE 12.9
Preparation of a bacterial lawn.
(a) Apply the bacteria evenly over the entire agar surface. (b) Rotate the plate and swab at right angles to the first application.

Prediction

Predict the results of the experiment based on your hypothesis.

Procedure

1. Label the bottom of an agar plate with your initials, the lab room, the date, and a word to indicate the experiment (such as "antibiotic").
2. Prepare a bacterial lawn.
   a. Insert a sterile swab into the bacterial culture in liquid nutrient broth.
   b. Allow the swab to drip for a moment before taking it out of the culture tube, but do *not* squeeze out the tip. The swab should be soaked but not dripping.
   c. Carefully lift the lid of the agar plate to about 45° and swab the *entire* surface of the agar, taking care to swab the bacteria to the edges of the dish (Figure 12.9a).
   d. Rotate the plate 45° and swab the agar again at right angles to the first swab (Figure 12.9b). Close the lid.
3. Carry the agar plate swabbed with bacteria to the demonstration table.
4. Remove the plate lid, place the antibiotic disk dispenser over the plate, and dispense the disks (Figure 12.10). (Each disk has been saturated with a particular antibiotic. The symbol on the disk indicates the antibiotic name. Your instructor will provide a key to the symbols.)
5. Replace the lid, seal the plate with Parafilm, and place the plate in the area indicated by the instructor. Incubate the dishes at 37°C for 24–48 hours and then refrigerate them.
6. Next week, examine the cultures to determine bacterial sensitivity to antibiotics. Measure the diameter of the **zone of inhibition** (area around disk where bacteria growth has been inhibited) for each antibiotic (Figure 12.11).
7. Record the measurement for your bacterial species and each antibiotic in Table 12.5. If the antibiotic had no effect on bacterial growth, record the size of the zone as 0.

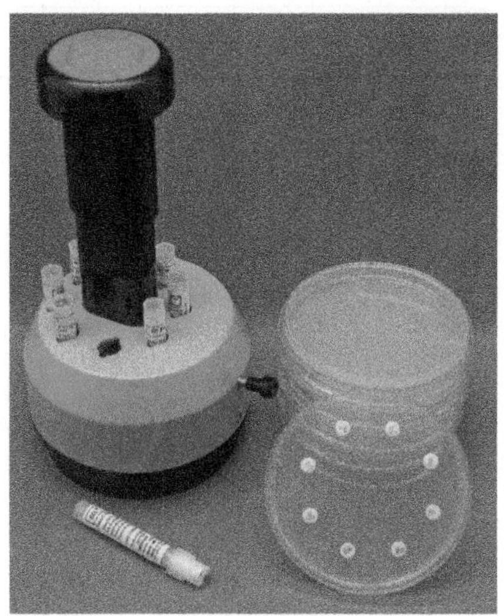

FIGURE 12.10
A typical antibiotic dispenser.

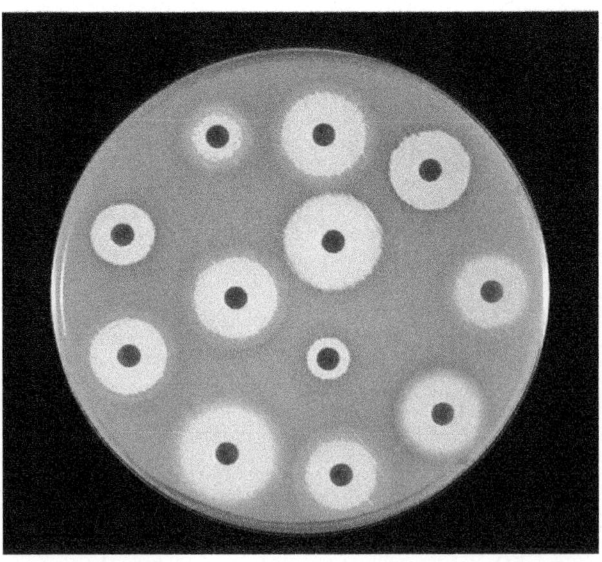

FIGURE 12.11
Zones of inhibition. Clear areas around antibiotic disks are zones where bacterial growth has been inhibited.

Results

1. Write the symbol for each antibiotic you have used in the appropriate row of Table 12.5. Using your results and the results from other teams, complete Table 12.5. Record the size of the zone of inhibition for all species of bacteria and all antibiotics.

TABLE 12.5 Results of Antibiotic Sensitivity Tests (size of inhibition zone for each antibiotic is given in centimeters.)

Bacteria (Name, Gram + or –)	Antibiotic							
1.								
2.								
3.								
4.								

2. Use the following arbitrary criteria to rank relative bacterial sensitivity to antibiotics:

   NS = not sensitive = no zone of inhibition
   S = sensitive = zone size above 0 but less than 1 cm
   VS = very sensitive = zone size greater than 1 cm
   Write the designation in each blank in the table.

Discussion

1. Did your results support your hypothesis? Was the zone of inhibition the same for all bacteria?

2. Were any bacteria very sensitive (greater than 1 cm) to all antibiotics? If so, which bacteria?

3. Based on your results, which antibiotic would you prescribe for each microorganism?

4. Were the results different for gram-positive and gram-negative bacteria?

5. Can you think of alternative explanations for the differential effectiveness of some antibiotics?

Lab Study B. Using Antiseptics and Disinfectants to Control Bacterial Growth

Materials

agar plate
forceps
metric ruler
sterile swab
wax pencil
Parafilm strips
broth cultures of *Micrococcus, Bacillus, Serratia*, and *E. coli*
paper disks soaking in disinfectants
paper disks soaking in antiseptics
paper disks soaking in sterile water

Introduction

Other agents besides antibiotics are often used to control bacterial growth. Those used to control bacteria on living tissues such as skin are called **antiseptics**. Those used on inanimate objects are called **disinfectants**. Antiseptics and disinfectants do not kill all bacteria, as would occur in sterilization, but they reduce the number of bacteria on surfaces.

Hypothesis

Hypothesize about the effect of antiseptics and disinfectants on the growth of bacteria.

Prediction

Predict the results of the experiment based on your hypothesis.

Procedure

1. Holding the lid in place, invert a sterile agar plate and label the bottom with your initials, the lab room, and the date. Draw four circles on the bottom. Number the circles.
2. Using the same bacterial culture as you used in Lab Study A, prepare a lawn culture as instructed in Lab Study A.
3. Carry the closed agar plate swabbed with bacteria to the demonstration table.
4. Open the agar plate; using forceps soaking in alcohol, pick up a disk soaked in one of the antiseptics or disinfectants, shake off the excess liquid, and place the disk on the agar above one of the circles. Repeat this procedure with two more antiseptics and/or disinfectants. Place a disk soaked in sterile water above the fourth circle to serve as a control.
5. Record the name of the agent placed above each numbered circle in Table 12.6. (Example: 1 = Lysol, 2 = Listerine, and so on.) Seal the plate with Parafilm.
6. Place the agar plate in the area indicated by the instructor. Incubate the agar plates at 37°C for 24–48 hours and then refrigerate them.
7. Next week, examine the cultures to determine the bacterial sensitivity to disinfectants and antiseptics. Measure the diameter of the zone of inhibition for each agent.
8. Record the measurement for your bacterial species and each inhibiting agent in Table 12.6. If the agent had no effect on bacterial growth, record the size of the zone as 0.

Results

1. Using your results and the results from other teams, complete Table 12.6. Record the size of the zone of inhibition for all species of bacteria and all antiseptics and disinfectants.
2. Use the following arbitrary criteria to rank relative bacterial sensitivity to antiseptics and disinfectants:

   NS = not sensitive = no zone of inhibition
   S = sensitive = zone size above 0 but less than 1 cm
   VS = very sensitive = zone size greater than 1 cm
   Write the designation in each blank in the table.

TABLE 12.6 Results of Sensitivity Tests of Antiseptics and Disinfectants (size of inhibition zones given in centimeters)

	Antiseptic/Disinfectant/Control			
Bacteria	**1.**	**2.**	**3.**	**4. Control**
1.				
2.				
3.				
4.				
5.				

Discussion

1. Did your results support your hypothesis? Explain.

2. Based on your results, which disinfectant is most effective in controlling the growth of bacteria?

3. Which antiseptic is most effective?

4. In which situations is it appropriate to use a disinfectant?

5. An antiseptic?

REVIEWING YOUR KNOWLEDGE

1. Once you have completed this lab topic, you should be able to define and use the following terms, providing examples if appropriate: *sterilize, pasteurize, nutrient broth* and *agar, coccus, bacillus, spirillum, biofilm, antibiotic, antibiotic resistance, antiseptic, disinfectant, peptidoglycan, aseptic technique*.
2. Compare the techniques used to prepare a lawn culture and a streak culture.
3. Return to Figure 12.11 and observe the zones of inhibition for each antibiotic disk.
   a. What is the significance of the clear zone around the disks?

   b. Is the bacterium growing on this plate resistant to any of the antibiotics being tested here?

4. A liquid that has been sterilized may be considered pasteurized, but one that has been pasteurized may not be considered sterilized. Why not?

APPLYING YOUR KNOWLEDGE

1. Would you expect the community of bacteria in plaque sampled 1 week *after* you have your teeth cleaned to differ from the community of bacteria found 1 week *before* you have your teeth cleaned? Explain. In your answer, consider the results of the milk succession experiment.

2. When you were younger and were sick with a respiratory infection, it is likely that your parents used a portable home humidifier in your room. Room humidifiers are good environments for the growth of biofilms. Using information you learned in Exercise 12.4, what would you suggest as an effective way to clean a humidifier?

3. Bacterial species that are harmful, as well as others that are beneficial, are found living in the human body. To slow the rate of developing antibiotic resistance in bacteria, physicians are being encouraged to use "narrow spectrum" antibiotics—those that target only a few bacterial types. How can the information learned by antibiotic sensitivity testing be used by physicians who must choose antibiotics that inhibit the growth of bacteria causing disease but that do not interfere with beneficial bacteria?

4. Search the Web for information about milk seen in boxes on grocery store shelves. How is this milk prepared? How would you expect bacterial succession in milk prepared in this fashion to differ from succession in milk as investigated in Exercise 12.2?

5. Death rates due to infectious diseases declined steadily in the United States throughout most of the 20th century. However, since the 1980s, infectious-disease–related deaths in the United States have been increasing significantly. Speculate about possible factors that may be contributing to this increase.

6. Mary, a patient in a nursing home, has been diagnosed with a urinary tract infection (UTI). Her physician prescribes Cipro, a popular antibiotic that is very effective in controlling UTIs and has been effective previously when treating these infections in Mary. However, this time, the antibiotic has no effect and the infection persists.
   a. Propose an explanation for the difference in Mary's response to the antibiotic from previous treatments.

   b. If you were Mary's physician, how would you proceed with her treatment?

7. In 1929, Alexander Fleming, a Scottish physician, discovered the first antibiotic when he noticed that colonies of certain staphylococcus bacteria growing in culture plates appeared to die when the plates became contaminated with the fungus *Penicillium*. Fleming concluded that a substance diffusing from the fungus into the growth medium was causing the bacteria to lyse (break down), and he called this substance penicillin.
   In its natural environment, what would be the adaptive advantage of a fungus producing and secreting a bacterial inhibitor?

8. Two potentially fatal bacterial infections that are becoming more common in hospitals, doctors' offices, and nursing homes are MRSA (methicillin-resistant *Staphylococcus aureus*) and C. diff (*Clostridium difficile*), the latter being one of the most important causes of infectious diarrhea in the U.S.

   Using Web resources, describe causes, symptoms, treatments, and prevention of these two serious bacterial diseases.

INVESTIGATIVE EXTENSIONS

Topics related to bacteriology are popular studies because they apply to many areas of ecology and human health. Diseases caused by waterborne bacteria cause almost 2 million deaths annually worldwide. In recent years, illnesses caused by contamination of foods such as ground beef, leafy greens, prepackaged cookie dough, and shellfish have been reported, leading to a heightened awareness of potential food contamination and changes in agricultural and food-processing practices.

Recent outbreaks of new flu strains in the United States have made the public more aware of the spread of bacteria and viruses by contact with infected people. In May 2009, the Centers for Disease Control and Prevention (CDC) issued hand-washing recommendations to reduce disease transmission. This agency recommended providing more hand-washing facilities in public settings and the use of alcohol-based hand sanitizer dispensers if soap and water are not available.

These ideas represent a few of the current popular issues that relate to bacteriology. The opportunity to perform an investigative extension will allow you to ask any number of interesting questions relating to these topics and to design an experiment to investigate one or more of these questions.

Investigating Soil Bacteria

The study of bacteria is important not only to health professionals, but also to ecologists. Microbial ecologists know the importance of having bacterial diversity in soils in all ecosystems. Certain bacteria in soil serve as decomposers that consume simple carbon compounds releasing inorganic materials useful to other organisms in the soil food web. Other bacteria—for example, nitrogen-fixing bacteria—form mutualistic relationships with plants, benefiting both bacteria and plant. Some species of bacteria are important in nitrogen cycling and degradation of pollution, and some soil bacteria are pathogenic and cause plant diseases.

Using ideas and protocols in this laboratory topic, design an experiment to investigate soil microbial diversity. First survey your campus or community and propose possible questions about soil bacteria in your area. For example, you might compare samples of soil taken from the athletic field that is regularly treated with pesticides to those from a natural area, a flower garden, a wetland, an organic garden, and so forth.

Before you begin your experiment, review safety notes and aseptic techniques (Exercise 12.1). Wear gloves when handling plates and bacteria and wash your hands frequently. Remember to keep careful records of procedures and results for each component of your experiment.

After preparing and incubating plates (room temperature is usually adequate), record the numbers of fungal and bacterial colonies on each plate and the morphology—shape and color—of bacterial colonies. Then, using techniques for streaking plates, prepare plates of individual colonies. After the plates have incubated, use laboratory protocols for determining cell shapes and Gram stain characteristics. Look for motility in freshly prepared slides.

With your partners, decide how you will present your results. Design tables and graphs to collect and present data. If digital and/or microscope photographic equipment is available, photograph your plates and the bacteria on your slides. Photographs may be used in a written or oral presentation of your experiment.

Investigating Bacteria and Human Health

Using techniques provided in this lab topic and following aseptic techniques and protocols (Exercise 12.1), design an experiment to investigate one of the following:

1. Investigate the efficacy of hand washing by varying the type of soap (liquid, bar, antibacterial, deodorant) or the manner of washing (scrubbing time, use of a brush). (Exercise 12.3)
2. Investigate the efficacy of waterless hand sanitizers in killing bacteria after various activities—for example, using the restroom, touching raw chicken, shaking hands, or cracking a raw egg. Wear disposable gloves when performing this experiment. (Exercise 12.3)

3. The chicken industry and the U.S. Food and Drug Administration (FDA) have recently been criticized for having low health and safety standards. Pursue this topic by a survey of brands or handling techniques. Health officials now recommend that all eggs be cooked before eating to avoid *Salmonella*. Determine the extent of contamination in store-bought eggs and in eggs from local sources. (Exercise 12.3)
4. Design an experiment to test bacterial succession in plaque. (Exercise 12.2)
5. Onions, garlic, green tea, cinnamon, honey, wasabi, and grapefruit seeds have all been suggested as having medicinal and antiseptic properties. Design an experiment to test this. (Exercise 12.4)

STUDENT MEDIA: BioFlix, Activities, Investigations, and Videos

www.masteringbiology.com (select Study Area)

Activities—Ch. 26: Classification Schemes

Ch. 27: Prokaryotic Cell Structure and Function; Classification of Prokaryotes

REFERENCES

Ackerman, J. "The Ultimate Social Network." *Scientific American*, 2012, vol. 306(6), pp. 36–43. A discussion of the effects of beneficial bacteria on human health.

Bell, L. "Down in the Mouth. Microbes from the Gums May Be Causing a Variety of Diseases." *Science News*, 2016, vol. 189, pp. 18–21.

Dill, B. and H. Merilles. "Microbial Ecology of the Oral Cavity," in *Tested Studies for Laboratory Teaching (Volume 10). Proceedings of the 10th Workshop/Conference of the Association for Biology Laboratory Education (ABLE)*, Corey Goldman, editor, 1989.

Gillen, A. L. and R. P. Williams. "Pasteurized Milk as an Ecological System for Bacteria." *The American Biology Teacher*, 1988, vol. 50, pp. 279–282.

Jacob, N. P. "Investigating Arabia Mountain: A Molecular Approach." *Science*, 2012, vol. 335, pp. 1588–1589.

Levy, S. B. "The Challenge of Antibiotic Resistance." *Scientific American*, 1998, vol. 278, pp. 46–53.

Nester, E. W., D. G. Anderson, and C. E. Roberts. *Microbiology: A Human Perspective*, 7th ed. Dubuque, IA: WCB/McGraw-Hill, 2012. Good discussion of ecological succession of microbes.

Pommerville, J. C. *Alcamo's Fundamentals of Microbiology*, 10th ed. Boston, MA: Jones and Bartlett Publishers, 2013. Good discussion of the role of bacteria in dental disease.

Urry, L., M. Cain, S. Wasserman, P. Minorsky, and J. Reece. *Campbell Biology*, 11th ed. San Francisco, CA: Pearson, 2017.

Spellberg, B., R. Guidos, D. Gilbert, J. Bradley, H. Boucher, W. Scheld, J. Bartlett, J. Edwards, and the Infectious Disease Society of America. "The Epidemic of Antibiotic-Resistant Infections: A Call to Action for the Medical Community from the Infectious Diseases Society of America," *Clinical Infectious Diseases*, 2008, vol. 46, Issue 2, pp. 155–164.

Zimmer, C. "What Is a Species?" *Scientific American*, 2008, vol. 298, pp. 72–79. Discusses changing criteria used in taxonomy.

WEBSITES

For information on a variety of topics relating to bacteria, visit this site maintained by the American Society for Microbiology:
http://www.microbeworld.org

Information about the domains Archaea and Bacteria and links to additional websites:
http://www.ucmp.berkeley.edu/archaea/archaea.html
http://www.ucmp.berkeley.edu/bacteria/bacteria.html

Information about responsible antibiotic use:
http://www.cdc.gov/getsmart/antibiotic-use/fast-facts.html

LAB TOPIC 13

Protists

This lab topic gives you another opportunity to practice the scientific process introduced in Lab Topic 1. Before going to lab, review scientific investigation in Lab Topic 1 and carefully read Lab Topic 13. Be prepared to use this information to design an experiment with protists.

Laboratory Objectives

After completing this lab topic, you should be able to:

1. Discuss the diversity of protists, and the current interest in their phylogenetic relationships.
2. Describe a current hypothesis for organizing clades of protists based on recent molecular evidence of relationships.
3. Identify and describe representative organisms in several major protistan clades.
4. Discuss the ecological role and economic importance of protists.
5. Describe the characteristics and representative organisms of the green algae and their relationship to plants.
6. Design and perform an independent investigation of a protist.

Introduction

Unicellular eukaryotic organisms originated over 2 billion years ago, and today they are found in every habitable region of Earth. The enormous diversity of organisms, their numerous adaptations, and their cellular complexity reflect the long evolutionary history of eukaryotes. For almost 30 years, scientists placed these diverse groups of unicellular organisms into the kingdom Protista. The Protista usually included all organisms not placed in the other eukaryotic kingdoms of Plants, Animals, and Fungi. This catchall kingdom included not only the unicellular eukaryotes, but also their multicellular relatives, like the giant kelps and seaweeds. However, scientists now agree that the designation kingdom Protista should be abandoned and these eukaryotic organisms should be placed in the domain Eukarya along with fungi, plants, and animals. (Recall from Lab Topic 12 that prokaryotes are placed in domains Bacteria or Archaea.) In this lab topic we will refer to this diverse group as *protists*, meaning a general term rather than a taxonomic category.

Most protists are unicellular, although there are colonial and multicellular species (see Lab Topic 2 for definitions of these terms). Historically protists were classified based on characteristics such as cellular structure, reproductive strategies, life cycles, and modes of nutrition. For example, one system of classification separated protists into **algae** and **protozoa**. Algae included autotrophic (photosynthetic) species that convert the sun's energy to organic compounds. The energy stored by autotrophs is called **primary production**. Protozoa was the designation for heterotrophic protists that obtain their food by absorbing large organic molecules or by **phagocytosis**—the uptake of large particles or whole organisms by the pinching inward of the cell membrane. Some protozoa, euglenoids for example, are **mixotrophic**, capable of photosynthesis and phagocytosis. However, these criteria were found to be unreliable when attempting to classifying protists based on their phylogeny, or evolutionary history, because each of these modes of nutrition are found in many different protist lineages. The emergence of molecular and biochemical research, particularly the ability to sequence genes and even whole genomes, has provided strong evidence for a dramatic reconstruction of phylogenetic relationships of protists.

Scientists performing phylogenetic studies have suggested that grouping organisms into **clades** can be meaningful for indicating evolutionary relationships. A clade is a group of species, all of which are descended from one ancestral species, representing one phylogenetic group. One current hypothesis for classifying organisms in the domain Eukarya places protists, along with plants, fungi, and animals, into four "**supergroups**." The first supergroup, **Excavata**, gets its name from protists included in this group that have an "excavated" feeding groove on one side of the cell body. Some of the protists in this group have unique mitochondria, and others have unique flagella.

A second supergroup proposed in this hypothesis has been given the informal name "**SAR**." The protists in this supergroup are classified together based on DNA sequences in their genomes, and include **Stramenopiles**, **Alveolates**, and **Rhizarians**.

The third supergroup, **Unikonta**, includes the protistan clade **Amoebozoans** and another clade, the Opisthokonts (studied in other lab topics). Animals and fungi are Opisthokonts.

The fourth supergroup, **Archaeplastida**, includes the protistan groups red algae and green algae, and also the plants that will be studied in Lab Topics 14 and 15. Including plants in this supergroup reflects the amount of evidence supporting the origin of plants from the green algae.

As more information from a variety of sources becomes available, major groupings or clades will surely be modified. These investigations into the nature of eukaryotic diversity demonstrate the process of scientific inquiry. New technologies, new ideas, and novel experiments are used to test hypotheses, and the resulting evidence must be consistent with the existing body of knowledge and classification scheme. The results lead to modification of our hypotheses and further research. No matter how many groups or clades are proposed, remember that this is a reflection of the evolution of eukaryotes over the rich history of the Earth. It is not surprising that the diversity of life does not easily fit into our constructed categories.

In this lab topic, we will study diverse examples of protists. These protists represent some of the most common clades. As you investigate the diversity of protists and their evolutionary relationships in this exercise, ask questions about the nutritive mode of each. Note morphological characteristics of examples studied. Ask which characteristics are found in organisms in the same clade and those shared with organisms in other clades or groups. Many of these characteristics are examples of evolutionary convergence. Ask questions about the ecology of the organisms. What means of locomotion do they possess, if any? What role do they play in an ecosystem? Do they have any economic value? Where do they live? (Protists live in a diversity of habitats, but most are aquatic. A great variety of protists may be found in **plankton**, the community of organisms found floating in the ocean or in bodies of fresh water.)

If you complete all of the lab topics in this laboratory manual, you will have studied examples of all the major groups of living organisms with the exception of those in domain Archaea. Bacteria are investigated in Lab Topic 12, and several additional protists are introduced in Lab Topic 2. Fungi are studied in Lab Topic 17, and you will investigate plant evolution and animal evolution in subsequent lab topics. Table 13.1 *is an overview of major clades and examples of each that will be investigated in this lab topic.*

At the end of this lab topic, you will be asked to design a simple experiment to further your investigation of the behavior, ecology, or physiology of one of the organisms studied. As you proceed through the exercises, ask questions about your observations and consider an experiment that you might design to answer one of your questions.

TABLE 13.1 Major Clades of Protists and Examples Studied in This Lab Topic

Supergroup	Clade	Lab study	Examples
Excavata (Exercise 13.1)	Euglenozoans	13.1A	Kinetoplastids—-*Trypanosoma*
		13.1B	Euglenids—*Euglena* sp.
"SAR" (Exercise 13.2)	Stramenopiles	13.2A	Diatoms Brown algae Water mold—*Saprolegnia*
	Alveolates	13.2B	Ciliates—Paramecia Dinoflagellates Apicomplexan—*Plasmodium sp.*
	Rhizarians	13.2C	Foraminiferans Radiolarians
Unikonta (Exercise 13.3)	Amoebozoans	13.3A	Tubulinids—*Amoeba* sp.
		13.3B	Plasmodial slime molds—*Physarum* Cellular slime molds—*Dictyostelium*
Archaeplastida (Exercise 13.4)	Red algae	13.4A	Rhodophytes—*Porphyra*
	Green algae	13.4B	Chlorophytes—*Spirogyra,Ulva, Chlamydomonas,* Charophytes—*Chara*

EXERCISE 13.1

Supergroup Excavata

The supergroup **Excavata** includes *diplomonads*, *parabasalids*, and *euglenozoans*. Many organisms in this supergroup have an "excavated" feeding groove, some have unique mitochondria, and some have unusual flagella. Only examples of **euglenozoans** will be studied here.

Lab Study A. Euglenozoans—Example: *Trypanosoma levisi*

Materials

compound microscope
prepared slides of *Trypanosoma levisi*

Introduction

Organisms in the clade **Euglenozoa** are grouped together based on the ultrastructure (structure that can be seen only with an electron microscope) of their **flagella** and their mitochondria. Included in this group are some heterotrophs, some autotrophs, and some parasitic species. The many diverse single-celled and colonial flagellates have been a particular challenge to taxonomists. Under the old two-kingdom system of classification, the heterotrophic flagellates were classified as animals, and the autotrophic flagellates (with chloroplasts) were classified as plants. However, euglenozoans include members of each type. The common flagellated, mixotrophic *Euglena* belongs in this clade.

Two groups included in the clade Euglenozoa are the **kinetoplastids** and the **euglenids**. The organism that you will investigate in this lab study, *Trypanosoma levisi*, is an example of a kinetoplastid. DNA within the mitochondria of these protists is organized into a mass called a *kinetoplast*. Organisms in the genus *Trypanosoma* are parasites that alternate between a vertebrate and an invertebrate host. *Trypanosoma levisi* lives in the blood of rats and is transmitted by fleas. Its flagellum originates near the posterior end but passes to the front end as a marginal thread of a long undulating membrane. Another organism in this same genus, *T. brucei*, causes African sleeping sickness in humans. Its invertebrate host is the tsetse fly.

Procedure

1. Obtain a prepared slide of *Trypanosoma levisi* (Figure 13.1) and observe it using low, intermediate, and high powers in the compound microscope.
2. Locate the organisms among the blood cells of the parasite's host.
3. Identify the **flagellum**, the **undulating membrane**, and the **nucleus** in several organisms.

Results

1. In the margin of your lab manual, draw several representative examples of *T. levisi* and several blood cells to show relative cell sizes.
2. Turn to Table 13.5 near the end of this lab topic and list the characteristics, ecological roles, and economic importance of *T. levisi*.

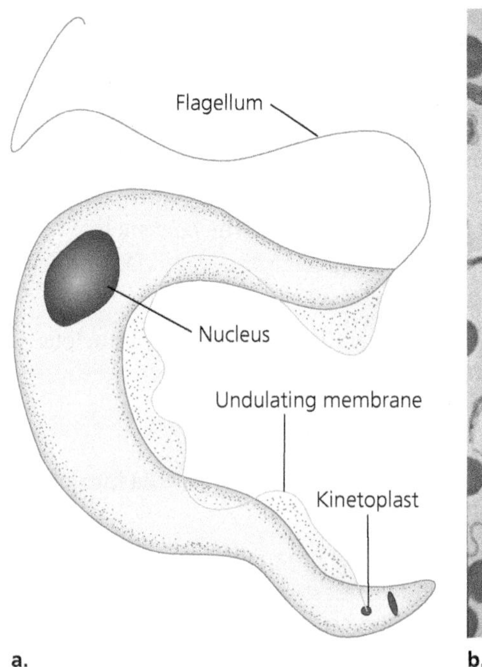

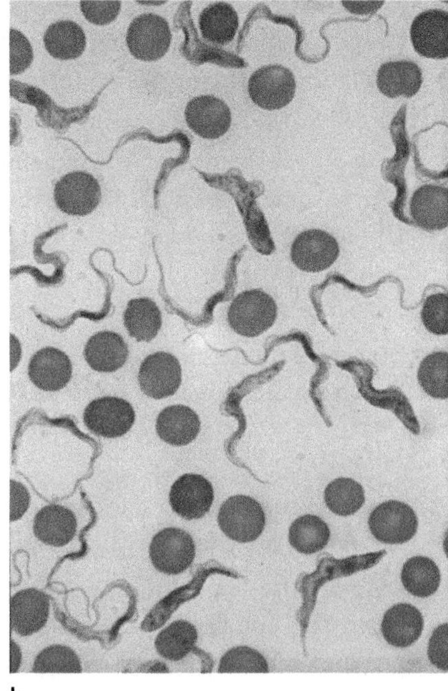

FIGURE 13.1
***Trypanosoma*, a euglenozoan.**
(a) *Trypanosma* is a flagellated parasite that lives in the blood of its mammalian host. The flagellum originates near the posterior end, but passes along an undulating membrane to the anterior end. (b) The round cells in this photograph are red blood cells (erythrocytes) in the infected human. Trypanosomes are seen in the plasma surrounding the cells.

Lab Study B. Euglenozoans—Example: *Euglena*

Materials

compound microscope
microscope slides and coverslips
living cultures of *Euglena* sp.
prepared microscope slides of *Euglena* sp.
transfer pipettes
Protoslo, 10% methyl cellulose, or other quieting agent

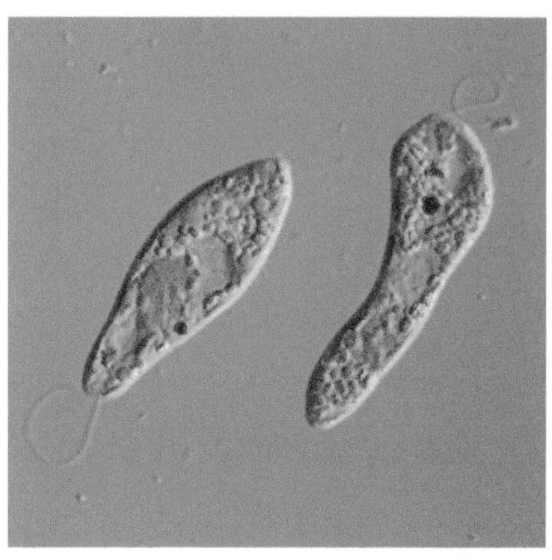

FIGURE 13.2
***Euglena* sp.** A pigmented eyespot is seen near the base of the flagellum in each cell. The nucleus appears as the clear area near the center of the cell.

Introduction

Euglena (Figure 13.2) is an example of a euglenid in the clade Euglenozoa. Most euglenids are unicellular and move using flagella. They may be *autotrophic*, containing chloroplasts and performing photosynthesis, but some may be *heterotrophic* or even *mixotrophic*, absorbing or engulfing nutrients from the environment. Experiments have shown that some species that are normally autotrophic, if grown in darkness, will lose their chloroplasts and become heterotrophic. One distinguishing characteristic of euglenids is the presence of a **pellicle**, made of strips of protein, lying just below the cell membrane. Species of *Euglena* are common in the surface scum of freshwater ponds and slowly moving waters. Some species are elongate and narrow, and others are more rounded. You may see examples of both shapes today in lab.

Procedure

1. Place a drop of liquid from the living *Euglena* culture on a clean microscope slide. Add a coverslip and examine on low, intermediate, and high powers using phase contrast if possible. Observe the method and direction of locomotion. To slow cell movement, add a *small* drop of Protoslo or methyl cellulose to the edge of the coverslip and allow it to diffuse under the coverslip.

2. Note the shape of the cells and striations in the cell surface. Where is the **flagellum** located? Describe its movement—whiplash or spiraling? Does it *pull* or *push* the organism through the water?
3. Identify other cellular structures. Look for a pigmented **eyespot** positioned near the base of the flagellum. The eyespot shields directional light from the nearby **light detector** (not visible at this magnification). This arrangement of eyespot and light detector allows the *Euglena* to determine the direction of light as it moves through the water. Do you see **chloroplasts** and a centrally located **nucleus**?
4. Observe a prepared slide of *Euglena*. In this stained preparation a nucleus with a **nucleolus** is easier to identify in each cell.

Results

1. Describe the shape and movement of *Euglena*. Where are flagella located and where are they positioned as the cells move?
2. Describe the shape, numbers, and locations of organelles seen in living organisms and then in the prepared slide. Describe the surface striations. Are they linear? Irregular? What makes these striations?
3. Speculate about the adaptive role of the eyespot and light detector in *Euglena*.
4. Turn to Table 13.5 and list the characteristics, ecological roles, and economic importance of euglenids.

EXERCISE 13.2

Supergroup "SAR"

Three clades are included in the SAR supergroup: **Stramenopiles**, **Alveolates**, and **Rhizarians**. These diverse protists were grouped together based on DNA sequence similarities. You will investigate examples of each of these groups in this exercise.

Lab Study A. Stramenopiles—Examples: Diatoms, Brown Algae, and Water Molds

Materials

compound microscope
stereoscopic microscope
slides and coverslips
living cultures of diatoms
transfer pipettes
prepared slides of diatomaceous earth (demonstration only)
demonstration materials of brown algae
agar plate cultures of *Saprolegnia*
prepared slides of *Saprolegnia*
dropper bottles of water
forceps

Introduction

The clade Stramenopila includes diatoms (phylum Bacillariophyta), golden algae (phylum Chrysophyta), brown algae (phylum Phaeophyta), and water molds (phylum Oomycetes). These organisms are grouped in this clade based on biochemical evidence and the structure of their flagella (when present). The flagellum has many hairlike lateral projections.

In this lab study you will investigate three examples: diatoms, brown algae, and a water mold, *Saprolegnia*. Diatoms and brown algae are autotrophic organisms that play an important role in primary production in oceans. *Saprolegnia* is a heterotrophic organism that lives in freshwater environments such as quiet ponds and lakes, and is often found growing on dead fish in aquaria.

Diatoms (Bacillariophyta)

Diatoms are important autotrophic organisms in plankton. In fact, they are the most important photosynthesizers in cold marine waters. They can be unicellular, or they can aggregate into chains or starlike groups. Protoplasts of these organisms are enclosed by a cell wall made of silica that persists after the death of the cell. These cell wall deposits are mined as **diatomaceous earth** and have numerous economic uses (for example, in swimming pool filters and as an abrasive in toothpaste and silver polish). Perhaps the greatest value of diatoms, however, is the carbohydrate and oxygen they produce that can be utilized by other organisms. Ecologists are concerned about the effects of acid rain and changing climatic conditions on populations of diatoms and their rate of primary productivity.

Diatom cells are either elongated, boat-shaped, bilaterally symmetrical **pennate** forms or radially symmetrical **centric** forms (Figure 13.3). The cell wall consists of two valves, one fitting inside the other, in the manner of the lid and bottom of a petri dish.

Procedure

1. Prepare a wet mount of diatoms from marine plankton samples or other living cultures.
2. Observe the organisms on low, intermediate, and high powers.

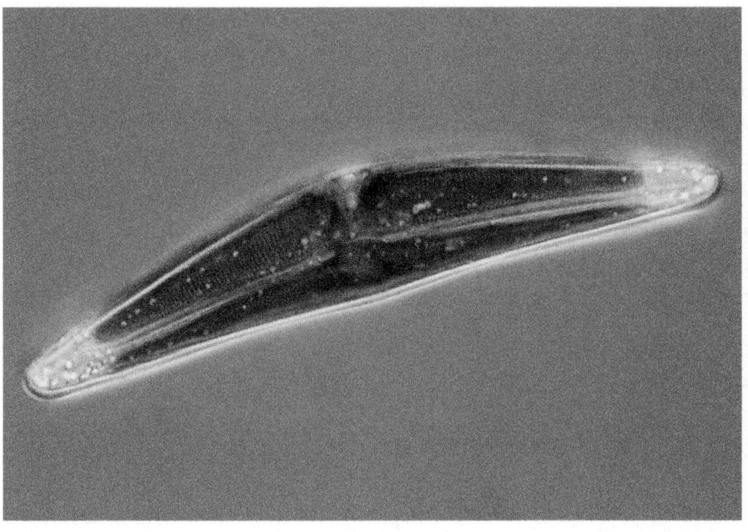

a.

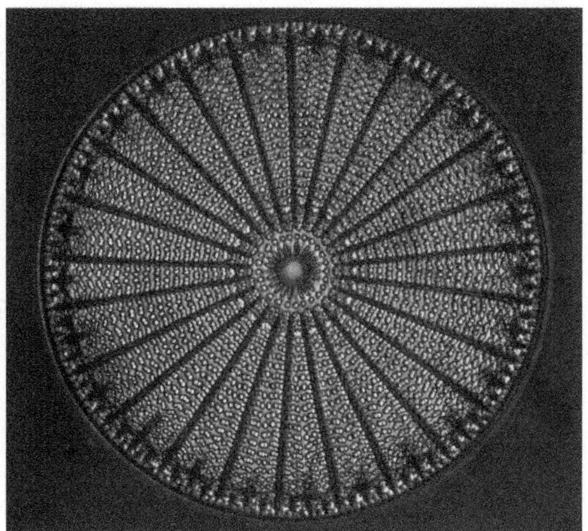

b.

FIGURE 13.3

Diatoms are important autotrophs found in plankton. Many different species and forms exist. All have cell walls made of silica. (a) A bilaterally symmetrical pennate form. (b) A radially symmetrical centric form.

3. Describe the form of the diatoms in your sample. Are they centric, pennate, or both?
4. If you are studying living cells, you may be able to detect locomotion. The method of movement is uncertain, but it is thought that contractile fibers just inside the cell membrane produce waves of motion on the cytoplasmic surface that extends through a groove in the cell wall. What is the body form of motile diatoms?
5. Observe a single centric form on high power and note the intricate geometric pattern of the cell wall. Can you detect the two valves?
6. Look for chloroplasts in living forms.
7. Observe diatomaceous earth on demonstration and identify pennate and centric forms.

Results

1. Sketch several different shapes of diatoms in the margin of your lab manual.
2. Turn to Table 13.5 and list the characteristics, ecological roles, and economic importance of diatoms.

Student Media Videos—Ch. 28: Diatoms Moving; Various Diatoms

Brown Algae (Phaeophyta)

Some of the largest algae, the **kelps**, are brown algae. The Sargasso Sea is named after the large, free-floating brown algae *Sargassum*. These algae appear brown because of the presence of the brown pigment **fucoxanthin** in addition to chlorophyll *a*. Brown algae are perhaps best known for their commercial value. Many health benefits have been claimed for edible brown algae. Extracts of **algin**, a polysaccharide in the cell wall of some brown algae, are used commercially as thickening or emulsifying agents in paint, toothpaste, ice cream, pudding, and in many other commercial food products. *Laminaria*, known as *kombu* in Japan, is added to soups, used to brew a beverage, and covered with icing as a dessert. This brown algae is extensively cultivated in the seas of Japan and Korea.

Procedure

Observe examples of brown algae on demonstration (Figure 13.4).

Results

1. In Table 13.2, list the names and distinguishing characteristics of each brown algal species on demonstration. Compare the examples with those illustrated in Figure 13.4.
2. Turn to Table 13.5 and list the key characteristics, ecological roles, and economic importance of brown algae.

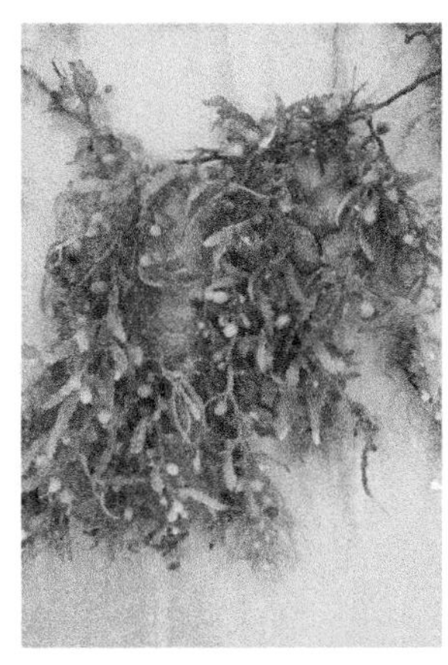

a. b. c.

FIGURE 13.4
Examples of multicellular brown algae. The body of a brown alga consists of broad blades, a stemlike stipe, and a holdfast for attachment. These body parts are found in the kelps (a) sea palm (*Postelsia*) and (b) *Nereocystis*. Rounded air bladders for flotation are seen in (c) *Sargassum* and other species of brown algae.

TABLE 13.2 Representative Brown Algae

Name	Body Form (single-celled, filamentous, colonial, leaflike; broad or linear blades)	Characteristics (pigments, reproductive structures, structures for attachment and flotation)

Water Mold *Saprolegnia* (Oomycetes)

Saprolegnia sp. (Figure 13.5) is an example of a water mold, an Oomycete. Water molds were formerly considered to be fungi because the body of these organisms consists of fungus-like hyphae organized into mycelia.(You will learn more about hyphae and mycelia in Lab Topic 17 Fungi.) However, there are key differences between water molds and fungi. For example, cell walls in water molds are made of cellulose, not chitin as in fungi. The most convincing evidence separating water molds from fungi is the molecular and genetic differences. *Saprolegnia* is common in aquaria and quiet ponds, where it may be found growing on dying or dead animals such as insects, fish, and even amphibians. This mode of growth reflects the **saprophytic** nutrition (deriving nutrients by absorption from decaying matter) of most water molds. Other species in this group are **parasites**, living on and harming other organisms (a host). One destructive species of water mold causes *potato late blight*, the plant disease responsible for the Irish potato famine of the mid-1800s, when a million people in Ireland died and a million more migrated to other countries. In

FIGURE 13.5
Saprolegnia. A saprophytic water mold on a decaying goldfish.

the year 2000, a species of water mold (*Phytophthora ramorum*) was identified as the cause of *sudden oak death*, a disease killing a wide range of oak species in California and Oregon. Some species of water molds continue to plague farmers today and, if not controlled, can be devastating to a variety of crops.

Procedure

1. Obtain a petri dish containing a living culture of *Saprolegnia* and return to your lab bench to study the organism. Keep the dish closed.
2. With the aid of your stereoscopic microscope, examine the culture. Look for hyphae and reproductive structures that may contain spores in either the asexual or sexual stages of the life cycle. Describe the appearance of the culture in the Results section.
3. If directed by your instructor, add a drop of water to a microscope slide and use forceps to remove a portion of the culture and add to the water drop. Add a coverslip. Observe the slide using a compound microscope, first on low and then intermediate and high powers. Describe the appearance of filaments and any reproductive structures. Are reproductive cells present? Make sketches in the Results section.
4. Observe a prepared slide of *Saprolegnia*. Large round reproductive structures may be visible.

Results

1. From your observations, describe the vegetative (nonreproductive) body of *Saprolegnia*. Do you see filaments? This observation may help you understand why water molds were initially incorrectly classified as fungi.
2. Sketch and describe any reproductive structures observed. Did they appear to contain spore-like structures?
3. In Table 13.5, list the characteristics, ecological role, and economic importance of *Saprolegnia*.

Lab Study B. Alveolates—Examples: Paramecia, Dinoflagellates, and *Plasmodium* sp.

Materials

compound microscope
slides and coverslips
cultures of living *Paramecium caudata*
Protoslo or other quieting agent
solution of yeast stained with Congo red
cultures of *Paramecium caudata* that have been fed yeast stained with Congo red (optional)
dropper bottle of 1% acetic acid
transfer pipettes
living cultures or prepared slides of dinoflagellates
microscope slides of *Plasmodium* sp. life cycle
wall charts diagramming the life cycle of *Plasmodium* sp.

Introduction

Alveolates are single-celled organisms; some are heterotrophic, others autotrophic, and still others are parasitic. The common characteristic of all alveolates is the presence of membrane-bound saclike structures (**alveoli**) packed into a continuous layer just inside and supporting the cell membrane. New groupings of protistans into clades place ciliates, dinoflagellates, and apicomplexans (e.g., *Plasmodium* sp.) in the Alveolates.

A Ciliate—*Paramecium caudatum*

The first example you will investigate in this lab study is *Paramecium caudatum*, a heterotrophic organism that moves about using cilia (short projections from the cell surface). Cilia are generally shorter and more numerous than flagella. Internally both structures are similar in their microtubular arrangement.

Procedure

1. Using the compound microscope, examine a living paramecium (Figure 13.6). Place a drop of water from the bottom of the culture on a clean microscope slide. Add a *small* drop of Protoslo or some other quieting solution to the water drop, then add the coverslip.
2. Observe paramecia on the compound microscope using low, then intermediate powers.
3. Describe the movement of a single paramecium. Does movement appear to be directional or is it random? Does the organism reverse direction only when it encounters an object, or does it appear to reverse direction even with no obstruction?
4. Locate a large, slowly moving organism, switch to high power, and identify the following organelles:

   Oral groove: depression in the side of the cell that runs obliquely back to the mouth that opens into a **gullet**.
   Food vacuole: forms at the end of the gullet. Food vacuoles may appear as dark vesicles throughout the cell.
   Macronucleus: large, grayish body in the center of the cell. The macronucleus has many copies of the genome and controls most cellular activities, including feeding, water balance, and the asexual reproductive process of binary fission (the cell divides, giving rise to two cells).
   Micronucleus: often difficult to see in living organisms, this small round body may be lying close to the macronucleus. Micronuclei are involved

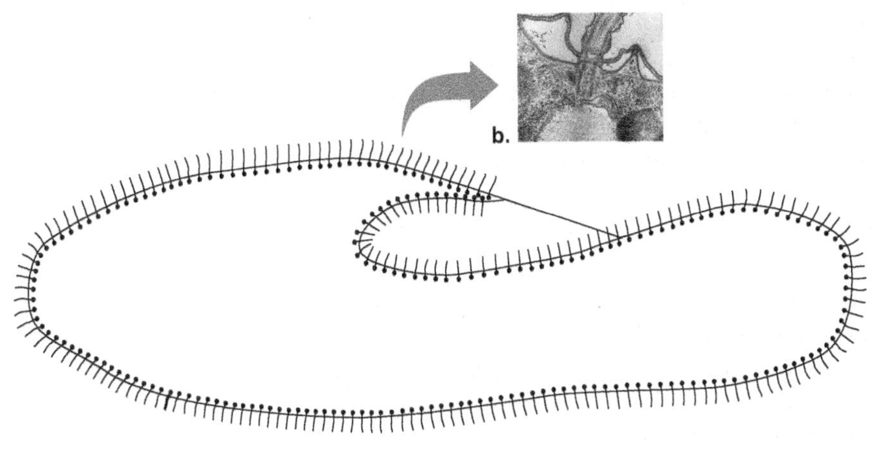

FIGURE 13.6
Paramecium. (a) Complete the drawing of a paramecium, labeling organelles and structures. (b) Electron micrograph of the region just under the cell membrane in an alveolate. Sac-like alveoli are visible here.

in sexual reproduction. Many species of paramecia have more than one micronucleus.

Contractile vacuole: used for water balance, two of these form, one at each end of the cell. Each contractile vacuole is made up of a ring of radiating tubules and a central spherical vacuole. Your organism may be under osmotic stress because of the Protoslo, and the contractile vacuoles may be filling and collapsing as they expel water from the cell.

5. Observe feeding in a paramecium. Add a drop of yeast stained with Congo red to the edge of the coverslip and watch as it diffuses around the paramecium. Study the movement of food particles from the oral groove to the gullet to the formation of a food vacuole that will subsequently move through the cell as the food is digested in the vacuole. You may be able to observe the discharge of undigested food from the food vacuole at a specific site on the cell surface.
6. Observe the discharge of **trichocysts**, structures that lie just under the outer surface of the paramecium. When irritated by a chemical or attacked by a predator, the paramecium discharges these long thin threads that may serve as a defense mechanism, as an anchoring device, or to capture prey. Make a new slide of paramecia. Add a drop of 1% acetic acid to the edge of the coverslip and carefully watch a paramecium. Describe the appearance of trichocysts in this species.

Results

1. Complete the drawing of a paramecium (Figure 13.6), labeling all the organelles and structures shown in bold in the text.
2. Turn to Table 13.5 and list the characteristics, ecological roles, and economic importance of paramecia.

MB Student Media Videos—Ch. 28: *Paramecium* Vacuole; *Paramecium* Cilia

Dinoflagellates

Swirl your hand through tropical ocean waters at night and you may notice a burst of tiny lights. Visit a warm, stagnant inlet and you might notice that the water appears reddish and dead fish are floating on the surface. Both of these phenomena may be due to activities of dinoflagellates—single-celled organisms that are generally photosynthetic. Some dinoflagellates are able to bioluminesce, or produce light. They sometimes can *bloom* (reproduce very rapidly) and cause the water to appear red from pigments in their bodies. If the organisms in this "red tide" are a species of dinoflagellate that releases toxins, fish and other marine animals can be poisoned. Red tides in the Chesapeake Bay are thought to be caused by *Pfiesteria*, a dinoflagellate that produces deadly toxins resulting in invertebrate and fish kills, and that also may be implicated in human illness and death. Dinoflagellates have a cellulose cell wall often in the form of an armor of numerous plates with two perpendicular grooves, each containing a flagellum. Most of these organisms are autotrophic and play an

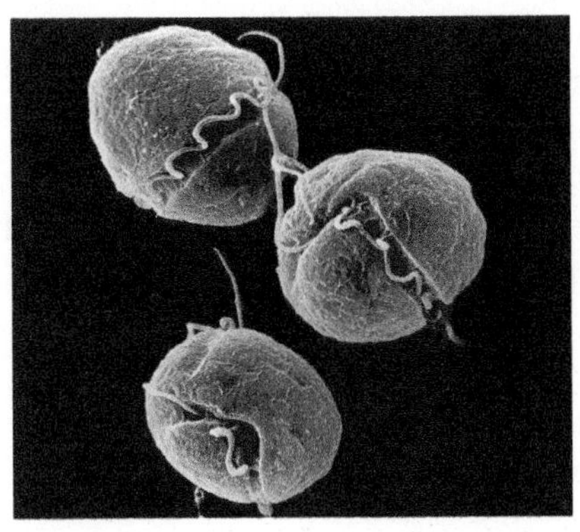
a.

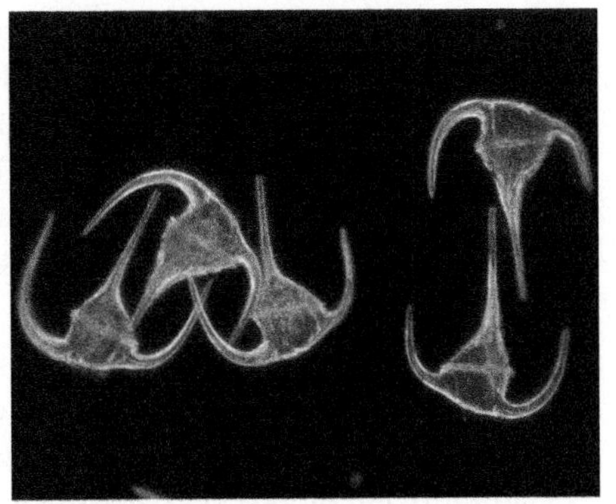
b.

FIGURE 13.7
Dinoflagellates. (a) In this SEM the cellulose plates of the cell wall are visible. Note also the two perpendicular grooves, each containing a flagellum. (b) Note the elongated spines of the cell wall plates in this species.

important role in **primary production** in oceans—photosynthesis that ultimately provides food for all marine organisms.

Dinoflagellates have traditionally been considered algae, but they are now thought to share a common ancestor with ciliates, as evidenced by the presence of alveoli.

Procedure

1. Obtain a prepared slide or make a wet mount of dinoflagellates (Figure 13.7).
2. Focus the slide on low power and attempt to locate the cells. You may have to switch to intermediate power to see them.
3. Switch to high power.
4. Identify the perpendicular **grooves** and the **cellulose plates** making up the cell wall. Are the plates in your species elongated into spines? **Flagella** may be visible in living specimens.

Results

1. Draw several examples of cell shapes in the margin of your lab manual. Note differences between the species on your slide and those in Figure 13.7.
2. Turn to Table 13.5 and list the characteristics, ecological roles, and economic importance of dinoflagellates.

Student Media Video—Ch. 28: Dinoflagellate

Apicomplexan—*Plasmodium* sp.

Other protists included in the clade Alveolates are apicomplexans, named from the complex of organelles in the apex of the cell. Most apicomplexans are parasitic, using this complex of organelles to penetrate their host's cells. They have no means of locomotion, and their reproductive structures are called spores. Included in this group are organisms in the genus *Plasmodium*. This protist has played an important role in human health worldwide. At least five species in this genus can infect humans, causing the devastating disease malaria. These organisms have a complex life cycle that alternates between female *Anopheles* mosquitos and humans. When an infected mosquito bites a person, the parasite is injected into the human, first entering the bloodstream, then passing to the

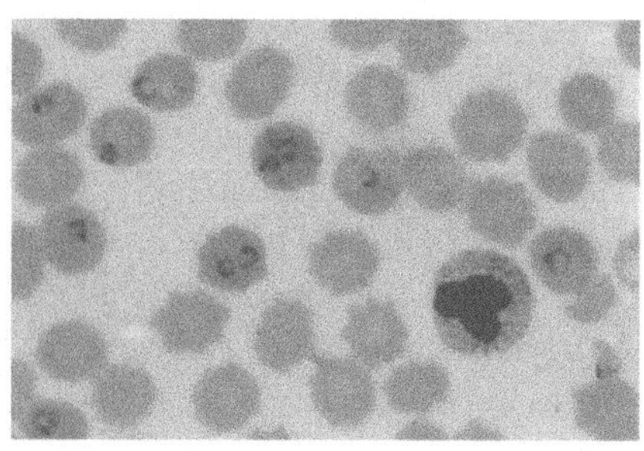

FIGURE 13.8
***Plasmodium* sp.** This parasite is seen in red blood cells in this photomicrograph of a human blood smear.

liver, and then moving back to the bloodstream, ultimately infecting the red blood cells. Sexual stages in the life cycle are transmitted back to the mosquito when it bites the infected human, continuing the life cycle. You will observe a prepared slide of *Plasmodium* sp. in the blood of the human host. Depending on the stage of the life cycle on your slide, the parasites may be seen among the red blood cells or within the red blood cells, appearing as a ring-like structure.

Procedure

1. Using the compound microscope, examine a prepared slide of *Plasmodium* sp. (Figure 13.8). Begin with the lowest power on your microscope and identify red blood cells (erythrocytes).
2. Increase the magnification, focusing and scanning the slide with each magnification increase. Examine carefully, looking first outside the red blood cells, and then look for cells with darker granules inside.
3. Consult diagrams of the *Plasmodium* life cycle posted in your lab room or in your text to help identify the stage observed on your slide.

Results

1. In the margin of your lab manual, sketch a view of your slide showing several red blood cells and parasites.
2. Turn to Table 13.5 and list the characteristics, ecological roles, and economic importance of *Plasmodium* sp.

Lab Study C. Rhizarians—Examples: Foraminiferans and Radiolarians

Materials

compound microscope
prepared slides of foraminiferans
prepared slides of radiolarian skeletons (demonstration only)

Introduction

Rhizarians include foraminiferans, radiolarians, and cercozoans. These organisms are grouped together based mainly on molecular evidence, although they do share several morphological characteristics. Many are amoeboid, heterotrophic organisms with **threadlike pseudopodia** or cellular extensions used in feeding and, in some species, locomotion. You will study examples of foraminiferans and radiolarians.

Foraminiferans

Foraminiferans, commonly called **forams**, are another example of organisms that move and feed using pseudopodia. Forams are marine planktonic (freely floating) or benthic (bottom-dwelling) organisms that secrete a calcium carbonate shell-like **test** (a hard outer covering) made up of chambers. In many species, the test chambers are secreted in a spiral pattern, and the organism resembles a microscopic snail. Although most forams are microscopic, some species, called *living sands*, may grow to the size of several centimeters, an astounding size for a single-celled protist. Threadlike pseudopodia extend through special pores in the calcium carbonate test. The test can persist after the organism dies, becoming part of marine sand. Remains of tests can form vast limestone deposits.

Procedure

1. Obtain a prepared slide of representative forams (Figure 13.9).
2. Observe the organisms first on the lowest power of the compound microscope and then on intermediate and high powers.
3. Note the arrangement and attempt to count the number of chambers in the test. In most species, the number of chambers indicates the relative age of the organism, with older organisms having more chambers. Which are more abundant on your slide, older or younger organisms?
4. Chambers can be arranged in a single row, in multiple rows, or wound into a spiral. Protozoologists determine the foram species based on the appearance of the test. Are different species present?

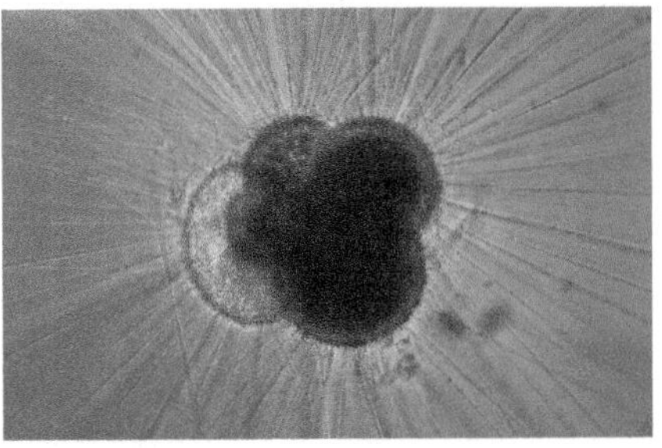

FIGURE 13.9
Forams. These heterotrophic organisms move using threadlike pseudopodia. Their shell-like tests are made of calcium carbonate.

Results

1. Sketch several different forams in the margin of your lab manual. Note differences in the organisms on your slide and those depicted in Figure 13.9.
2. Turn to Table 13.5 and list the characteristics, ecological roles, and economic importance of forams.

Radiolarians

The **radiolarians** studied here are common in marine plankton. They secrete skeletons of silicon dioxide that can, as with the forams, collect in vast deposits on the ocean floor. Their threadlike pseudopodia, called **axopodia**, extend outward through pores in the skeleton in all directions from the central spherical cell body.

Procedure

1. Observe slides of radiolarians on demonstration (Figure 13.10).
2. Observe the size and shape of the skeletons and compare your observations with Figure 13.10.

Results

1. Sketch several different radiolarians skeletons in the margin of your lab manual, noting any differences between the organisms on demonstration and those in the figure.
2. Turn to Table 13.5 and list the characteristics, ecological roles, and economic importance of radiolarians.

FIGURE 13.10
Radiolarians. These organisms are supported by a skeleton of silicon dioxide. They use threadlike pseudopodia to obtain food.

EXERCISE 13.3

Supergroup Unikonta

The supergroup **Unikonta** includes many diverse organisms with their phylogenetic relationships supported by molecular systematics. Fungi and animals are included in this supergroup and will be studied in subsequent lab topics. Examples studied in this exercise are protists in the clade Amebozoans. These organisms have pseudopodia as do foraminifera and radiolarians, but the structure is different. Rather than threadlike pseudopodia, amebozoans' pseudopodia are *lobe-shaped*. Based on their amoeboid characteristics, their phagocytic mode of obtaining nutrition, and molecular systematics, both amoeba and slime molds are included in the clade **Amoebozoa**.

Lab Study A. Amoebozoan Tubulinids—Example: *Amoeba*

Materials

cultures of *Amoeba proteus* (if amoeba were not studied in Lab Topic 2)
compound microscope
slides and coverslips
stereoscopic microscope

Amoeba

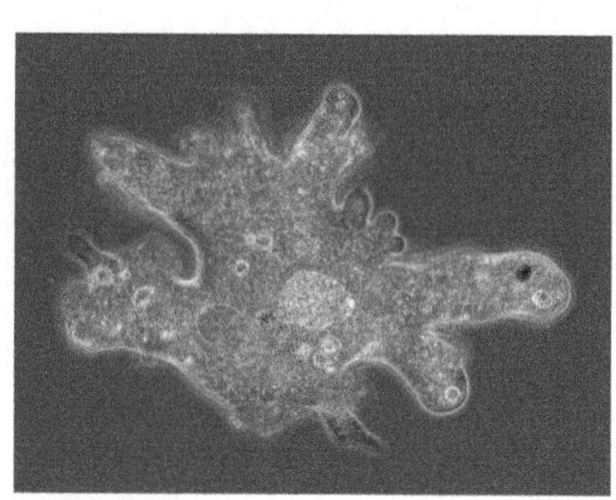

FIGURE 13.11
Amoeba proteus. These organisms use lobe-shaped pseudopodia to move and ingest food.

In Lab Topic 2, you studied *Amoeba proteus*, a protozoan species of organisms that move using lobe-shaped pseudopodia (Figures 2.6 and 13.11). Amoeba have no fixed body shape and they are naked; that is, they do not have a shell. Different species may be found in a variety of habitats, including soil, freshwater, and marine habitats. Recall that pseudopodia are cellular extensions. As the pseudopod extends, endoplasm flows into the extension. By extending several pseudopodia in sequence and flowing into first one and then the next, the amoeba proceeds along in an irregular, slow fashion. Pseudopodia are also used to capture and ingest food. When a suitable food particle such as a bacterium, another protist, or a piece of detritus (fragmented remains of dead organisms) contacts an amoeba, a pseudopod will flow completely around the particle and take it into the cell by phagocytosis.

If you did not observe *Amoeba proteus* or some other naked amoeba in Lab Topic 2, use the following procedure to observe these organisms.

Procedure

1. Obtain a drop of water containing living *Amoeba* (Figure 13.11) from the class culture.
2. To transfer a specimen to your slide, follow these procedures:
   a. Use the dissecting microscope to focus on the bottom of the dish. The amoeba will appear as a whitish, irregularly shaped organism attached to the bottom.
   b. Transfer a drop with several amoebas to your microscope slide.
   c. Cover your preparation with a clean coverslip.
3. Observe the amoeba using your compound microscope.
   a. Scan the slide at low power to locate an amoeba. Center the specimen in your field of view; then switch to higher powers.
   b. The following structures may be seen in the amoeba: **Endoplasm** is the granular cytoplasm containing the cell organelles. **Contractile vacuoles** are clear, spherical vesicles of varying sizes that gradually enlarge as they fill with excess water. These vacuoles serve an excretory function for the amoeba. **Food vacuoles** are small, dark, irregularly shaped vesicles within the endoplasm. They contain undigested food particles. You may observe movement by **pseudopodia** and perhaps **phagocytosis**.

Student Media Videos—Ch. 28: Amoeba; Amoeba Pseudopodia

Lab Study B. Amoebozoan Slime Molds—Examples: *Physarum* and *Dictyostelium*

Materials

stereoscopic microscope
Physarum growing on agar plates
Dictyostelium growing on agar plates

Introduction

William Crowder, in a classic *National Geographic* article (April 1926) describes his search for strange creatures in a swamp on the north shore of Long Island. This is his description of his findings: "Behold! Seldom ever before had such a gorgeous sight startled my unexpectant gaze. Spreading out over the bark [of a dead tree] was a rich red coverlet . . . consisting of thousands of small, closely crowded, funguslike growths. . . . A colony of these tiny organisms extended in an irregular patch . . . covering an area nearly a yard in length and slightly less in breadth. . . . Each unit, although actually less than a quarter of an inch in height, resembled . . . a small mushroom, though more marvelous than any I have ever seen."

The creatures described by Crowder are heterotrophic organisms called **slime molds**. They have been called plants, fungi, animals, fungus animals, protozoa, Protoctista, Protista, Mycetozoa, and probably many more names. Classifying slime molds as fungi (as in previous classification schemes) causes difficulties because whereas slime molds are phagocytic like protozoa, fungi are never phagocytic but obtain their nutrition by absorption. Characteristics other than feeding mode, including cellular ultrastructure, cell wall chemistry, and other molecular characteristics, indicate that slime molds fit better with the amoeboid protists than with the fungi. These studies suggest that slime molds descended from unicellular amoeba-like organisms.

There are two types of slime molds, plasmodial slime molds and cellular slime molds. In this lab study, you will observe the plasmodial slime mold *Physarum* and the cellular slime mold *Dictyostelium*. In a plasmodial slime mold the vegetative stage is called a **plasmodium**, and it consists of a multinucleate mass of protoplasm totally devoid of cell walls. This mass feeds on bacteria as it creeps along the surface of moist logs or dead leaves. When conditions are right, it is converted into one or more reproductive structures, called **fruiting bodies**, that produce spores (Figure 13.12).

FIGURE 13.12
Slime mold fruiting bodies.

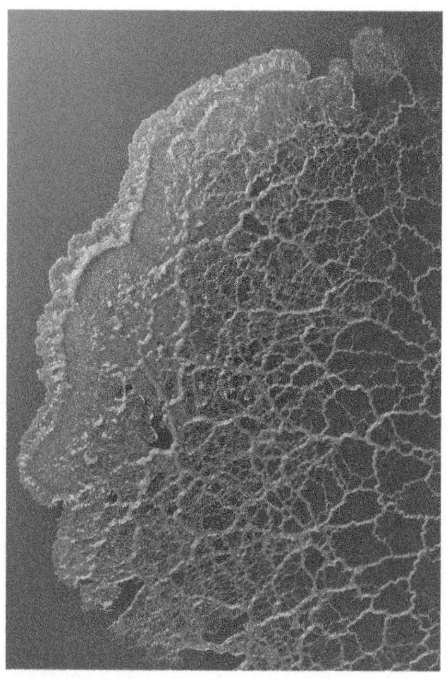

FIGURE 13.13
Plasmodial slime mold. Slime molds are protists that share some characteristics with both protozoa and fungi. The vegetative stage of a plasmodial slime mold includes an amoeboid phase consisting of a multinucleate mass known as a plasmodium.

Physarum—A Plasmodial Slime Mold

Procedure

1. Obtain a petri dish containing *Physarum* and return to your lab bench to study the organism. Keep the dish closed.
2. With the aid of your stereoscopic microscope, examine the plasmodium (Figure 13.13). Describe characteristics such as color, size, and shape. Look for a system of branching veins. Do you see any movement? Speculate about the source of the movement. Is the movement unidirectional or bidirectional—that is, flows first in one direction and then in the other? Your instructor may have placed oat flakes or another food source on the agar. How does the appearance of the plasmodium change as it contacts a food source?
3. Examine the entire culture for evidence of forming or mature fruiting bodies. Are the fruiting bodies stalked or are they sessile, that is, without a stalk? If a stalk is present, describe it.

Results

1. Sketch the plasmodium and fruiting bodies in the margin of your lab manual. Label structures where appropriate.
2. Turn to Table 13.5 and list the characteristics, ecological roles, and economic importance of slime molds.

Media Videos—Ch. 28: Plasmodial Slime Mold Streaming; Plasmodial Slime Mold

Dictyostelium—A Cellular Slime Mold

Like the plasmodial slime molds, the natural habitat of cellular slime molds is the forest floor, living on bacteria on decaying logs or dead leaves. Unlike plasmodial slime molds in which the feeding stage is a *multinucleate* plasmodium, cellular slime molds consist of *individual amoeboid cells* that function independently under optimum conditions. However, when food is scarce, individual cells will begin to migrate toward a central cell, directed by a chemical secretion. Cells continue to aggregate, forming a "pseudoplasmodium"—an aggregation that migrates as a unit until a suitable food supply is located. Once this is found, the aggregation dramatically develops into a fruiting body that produces asexual spores. When these spores are released from the fruiting body they develop into individual amoeboid cells. To see amazing videos of this process, go to youtube.com and search for "cellular slime mold."

Procedure

1. Obtain an agar plate containing *Dictyostelium* (Figure 13.14) and return to your lab bench to examine the organism. Keep the dish closed.
2. With the aid of your stereoscopic microscope, examine the aggregation of cells. Describe characteristics of the aggregate such as color, size, and shape.
3. Examine the entire culture for evidence of migrating cells—trails through the mass of cells that may be pathways.
4. Look for evidence of forming or mature fruiting bodies. Describe the color, size, and shape of these. Where are they located in the aggregate? Do the fruiting bodies have a stalk or are they sessile? If a stalk is present, describe it.

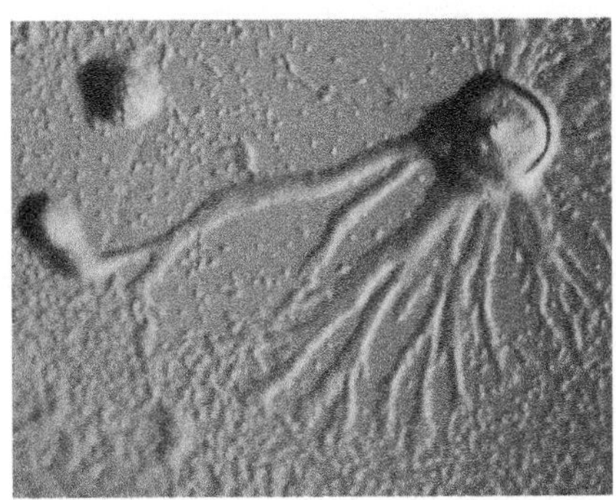

FIGURE 13.14
Cellular slime mold *Dictyostelium*. The feeding stage is a multinucleate plasmodium consisting of individual amoeboid cells.

Results

1. Sketch an aggregate and any fruiting bodies in the margin of your lab manual. Label structures observed.
2. Turn to Table 13.5 and list the characteristics, ecological roles, and economic importance of cellular slime molds.

EXERCISE 13.4

Supergroup Archaeplastida

The three main clades of photosynthetic organisms—red algae (**Rhodophyta**), green algae (**Chlorophyta** and **Charophyta**), and plants—are included in the supergroup **Archaeplastida**. Some species of red algae and green algae are unicellular, but many are large, multicellular species, informally called "seaweeds." In this exercise you will investigate examples of red algae and two groups of green algae, chlorophytes and charophytes. You will investigate plant diversity in Lab Topics 14 and 15.

Lab Study A: Red Algae (Rhodophyta)

Materials

examples of red algae on demonstration

Introduction

The simplest red algae are single celled, but most species have a macroscopic, multicellular body form. The red algae, unlike all the other algae, do not have flagella at any stage in their life cycle. Red algae are photosynthetic with plastids containing chlorophyll *a* and the accessory pigments **phycocyanin** and **phycoerythrin** that often mask the chlorophyll, making the algae appear red. These pigments absorb green and blue wavelengths of light that penetrate deep into ocean waters. Many red algae also appear green or black or even blue, depending on the depth at which they are growing. Because of this, color is not always a good characteristic to use when determining the classification of algae. Recall that in Lab Topic 12 you grew bacteria and fungi on plates of agar. This substance, **agar**, is a polysaccharide extracted from cell walls of red algae. Another extract of red algae cell walls, **carrageenan**, is used to give the texture of thickness and richness to foods such as dairy drinks and soups. In Asia and elsewhere, the red algae *Porphyra* (known as *nori*) are used as seaweed wrappers for sushi. The cultivation and production of *Porphyra* constitute a billion-dollar industry.

Procedure

Observe the examples of red algae that are on demonstration (Figure 13.15).

Results

1. In Table 13.3, list the names and distinguishing characteristics of the red algae on demonstration. Compare the demonstration examples with those illustrated in Figure 13.15.
2. Turn to Table 13.5 and list the key characteristics, ecological roles, and economic importance of red algae.

a.

b.

c.

FIGURE 13.15

Examples of multicellular red algae (phylum Rhodophyta). (a) Some red algae have deposits of carbonates of calcium and magnesium in their cell walls and are important components of coral reefs. (b) Most red algae have delicate, finely dissected blades. (c) *Porphyra* (or *nori*) is used to make sushi.

TABLE 13.3 Representative Red Algae

Name	Body Form (single-celled, filamentous, colonial, leaflike)	Characteristics (reproductive structures, structures for attachment or flotation, pigments)

Lab Study B. Green Algae (Chlorophyta and Charophyta)—The Protist–Plant Connection

Materials

compound microscope
microscope slides and coverslips
transfer pipettes
living cultures and prepared slides of *Chlamydomonas* sp.
cultures or prepared slides of *Spirogyra* sp.
living or preserved *Ulva lactuca*
living or preserved *Chara* sp.

Introduction

The green algae are photosynthetic organisms divided into two main groups, chlorophytes and charophytes. These groups include the largest numbers of algal species and perhaps the most important from both ecological and evolutionary perspectives. You can find examples of unicellular, motile and nonmotile, colonial, filamentous, and multicellular species that inhabit both freshwater and marine environments. Molecular systematics, including nuclear and chloroplast DNA sequencing, has supported the close relationship between green algae and plants. Other characteristics support this hypothesis as well. For example, both green algae and plants store the starch amylase, and chlorophylls *a* and *b* are found in both. However, it is the charophytes that are considered to be most closely related to plants because of several unique similarities. The method of cell wall formation during cell division, the organization of proteins that synthesize cellulose, and the structure of sperm, when present, are all similarities of charophytes and plants.

Green algae play an important ecological role as primary producers, providing food for animals in aquatic ecosystems; and organisms in one genus, *Ulva*, commonly called sea lettuce, are eaten by humans in salads and cooked in soups. Because of the rapid reproductive rate of some green algae, particularly those in the genus *Chlorella*, some scientists have proposed the use of these green algae for biofuels and for human and livestock food. *Chlorella* has been shown to be a good source of proteins, fats, and carbohydrates, and small chlorella "farms" have been established in various countries. The discovery of a "*Chlorella* growth factor" that is reported to stimulate growth and wound repair in animals has led to the formation of a multibillion-dollar segment of the health food industry.

In this lab study you will view several body forms of green algae: single-celled, filamentous, colonial, and multicellular. Finally, you will observe the multicellular, branched charophyte *Chara* (the stonewort), believed to be most similar to the green algae that gave rise to plants over 475 million years ago.

If you completed Lab Topic 2 Microscopes and Cells, you may remember observing aggregates of single-celled algae, *Protococcus*, and the spherical green algae *Volvox*. Your instructor may ask you to review your notes and drawings from Lab Topic 2. In this lab study you will observe the single-celled green algae *Chlamydomonas*, the filamentous algae *Spirogyra*, and the multicellular algae *Ulva* (chlorophyte) and *Chara* (charophyte) (Figures 13.16 and 13.17).

Chlamydomonas—A Unicellular Chlorophyte

Chlamydomonas in the group Chlorophyta is a unicellular green algae that swims with two flagella (Figure 13.16a). There are many species in this genus that are distributed in moist soil, ponds, other freshwater habitats, and in marine environments.

Procedure

1. Place a drop of liquid from the *Chlamydomonas* class culture on a clean microscope slide and add a coverslip.
2. Observe the organisms using the compound microscope on low, intermediate, and high powers using phase contrast if possible.
3. Describe the shape of individual cells and the swimming pattern. Look for the two flagella. Where are they attached to the cell? Do they appear to pull or push the cell through the water?
4. Identify organelles in the cytoplasm. The clear **nucleus** may be visible with phase contrast. Each cell has a large cup-shaped **chloroplast** containing a proteinaceous body surrounded by starch granules called a **pyrenoid**. A pigmented **eyespot** may be visible inside the chloroplast.
5. Examine a prepared slide of *Chlamydomonas*, and identify the **flagella**, **chloroplast**, **pyrenoid**, and **nucleus** in each cell if possible.

Spirogyra—A Filamentous Chlorophyte; *Ulva*—A Multicellular Chlorophyte

Procedure

1. Using your compound microscope, observe living materials or prepared slides of the filamentous algae *Spirogyra* (Figure 13.16b). This organism is common in small freshwater ponds. The most obvious structure in the cells of the filament is a long chloroplast. Can you determine how the algae got its name? Describe the appearance of the chloroplast.

   Can you see a nucleus in each cell of the filament?

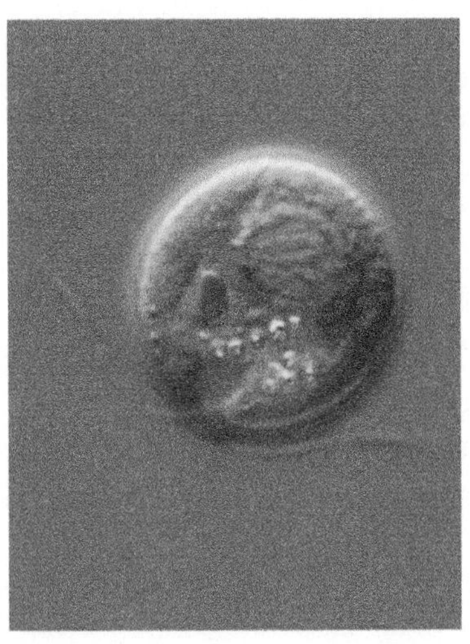

a.

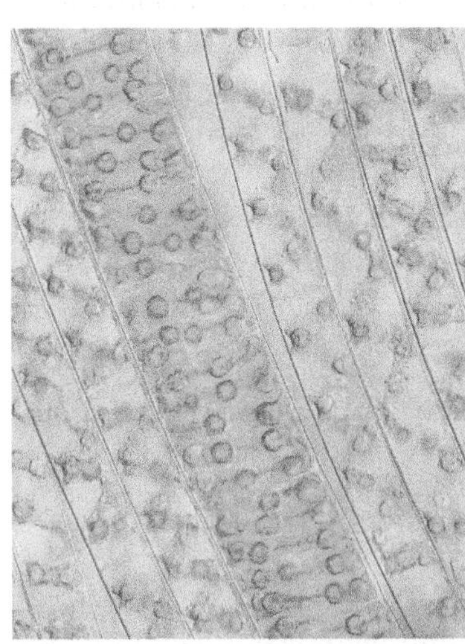

b.

c.

FIGURE 13.16

Examples of chlorophytes. (a) *Chlamydomonas*, a unicellular chlorophyte. (b) A filamentous green algae, *Spirogyra*. (c) Some green algae are multicellular, as in *Ulva*, sea lettuce.

2. Observe the living or preserved specimen of *Ulva* sp., commonly called sea lettuce (Figure 13.16c). These multicellular algae are commonly found on rocks or docks in marine and brackish water and is used as food in some cultures.

   a. Describe the appearance and body form of *Ulva*.

   b. Are structures present that would serve to attach *Ulva* to its substrate (dock or rock)? If so, describe them.

   c. Compare your specimen of *Ulva* with that shown in the figure.

Chara—A Multicellular Charophyte

Procedure

1. Examine the living or preserved specimen of the multicellular green algae *Chara* (Figure 13.17). These algae grow in muddy or sandy bottoms of clear lakes or ponds. Its body form is so complex that it is often mistaken for a plant, but careful study of its structure and reproduction confirms its classification as green algae.
2. Note the cylindrical branches attached to nodes. Compare your specimen to Figure 13.17. Sketch the appearance of your specimen in the margin of your lab manual.

Results

1. In Table 13.4, list the names and distinguishing characteristics of each green algal species studied. Compare these examples with those illustrated in Figures 13.16 and 13.17.
2. Turn to Table 13.5 and list the key characteristics, ecological roles, and economic importance of green algae.

FIGURE 13.17
Chara is a multicellular charophyte. Algae in this group are believed to be similar to the green algae that gave rise to plants.

TABLE 13.4 Representative Green Algae

Name	Body Form (single-celled, filamentous, colonial, leaflike)	Characteristics (pigments, specialized structures, flagella, structures for attachment)
Chlamydomonas		
Spirogyra		
Ulva		
Chara		

Discussion

1. Describe the mechanism for feeding in amoeboid, flagellated, and ciliated protozoans.

2. How do you think amoeboid organisms with skeletons, such as the radiolarians, move food to their cell bodies?

3. Compare the appearance and rate of locomotion in amoeboid, flagellated, and ciliated organisms observed in this exercise.

4. Describe mechanisms for defense in the organisms studied.

5. Single-celled protists use several strategies to protect or maintain the shape of their cells. This is illustrated in *Euglena*, diatoms, *Paramecium*, and dinoflagellates. How is cell shape maintained in these organisms?

6. What is one characteristic that you could observe under the microscope to distinguish diatoms and dinoflagellates?

7. Slime molds and water molds were once placed in the kingdom Fungi. What characteristics suggest that these organisms are protistan? Why do these examples illustrate the need for multiple kinds of evidence to determine phylogenetic relationships?

8. What important ecological role is shared by the macroscopic algae (green, red, and brown)?

9. Based on your observations in the laboratory, what two characteristics might you use to distinguish brown and red algae?

10. What characteristics of green algae have led scientists to conclude that this group includes the ancestors of plants, most likely the charophytes?

EXERCISE 13.5

Designing and Performing an Open-Inquiry Investigation

Materials

protozoa and algae cultures
cultures of slime molds *Physarum, Didymium, Dictyostelium*
cultures of *Saprolegnia*
sterile agar plates
sterile agar with oat flakes
sterile agar with sugar
sterile agar with albumin
sterile agar with pH 6, 7, or 8
pH solutions, pH 6, 7, or 8
aluminum foil
small lamps
aquarium mold retardants from pet store
culture media for *Paramecium*
ice
Chrysanthemum and *Yucca* plants
commercial chemicals that may control water mold

Introduction

In this exercise, you will choose one of the organisms observed in this lab topic and design a simple experiment answering a question about its behavior, growth patterns, or interactions with other species.

Use Lab Topic 1 as a reference for designing and performing a scientific investigation. Be ready to assign tasks to members of your lab team. Be sure that everyone understands the techniques that will be used. Your experiment will be successful only if you plan carefully, cooperate with your team members, perform lab techniques accurately and systematically, and record and report data accurately. If your experiment requires materials other than those provided, ask your laboratory instructor about their availability.

Procedure

1. **Develop a research question**. Collaborating with your research partner(s), read the following potential questions, and choose a question to investigate using this list or an original question proposed by your team. You may want to check your text and other sources for supporting information. You should be able to explain the rationale behind your choice of question. *Write your question in the margin of your lab manual.*

   a. Investigate responses of protists to light and darkness.

      Choose one of these organisms and ask if they are negatively or positively phototrophic (swim, move, or grow *away from* or *toward* a light source):

      Amoeba, Euglena, Paramecium, dinoflagellates, diatoms, *Saprolegnia, Physarum, Dictyostelium, Chlamydomonas*.

      You should be able to relate this question to some relevant situation in the environment or behavior of the organism—for example, seasonal or diurnal movements in aquatic habitats; mode of obtaining nutrition; preferred environmental conditions.

      What changes do you see in *Euglena* and other photosynthetic protists when grown in the dark? Does the culture remain alive?

   b. Investigate slime mold behavior.

      (1) Do plasmodia of the same species of slime mold unite when growing on the same agar plate? What about different species of slime mold?

      (2) Do slime mold plasmodia or "pseudoplasmodia" demonstrate chemotaxis (response to chemical stimuli such as food molecules)? Do the two behave differently?

      (3) What happens to slime molds when nutrients are scarce or abundant? What would happen if you put a drop of bacteria on an agar plate where a slime mold is growing?

      (4) What happens to slime molds when grown in different temperatures or pH?

      (5) Valerian root, hops, passion flower, and wild lettuce are all used as sleep aids because of their purported sedative properties. Adamatzky (2011) reported an experiment that showed that plasmodia of the slime mold *Physarum polycephalum* were more attracted to sleep aid tablets than to their usual diet of oats and honey. Compare the response of a plasmodial slime mold to herbal extracts of each of these plants.

   c. Investigate paramecia.

      (1) Will varying the molarity of the culture medium change the rate of contractile vacuole formation in paramecia?

      (2) In lab you fed paramecia yeast stained with Congo red. How do paramecia respond when fed something other than potential

food? Will they engulf and make food vacuoles with inanimate particles such as fine silt or sediment?

(3) What potential environmental pollutants will cause trichocyst discharge in paramecia?

d. Investigate the water mold *Saprolegnia*.

Do chemical agents purchased at a pet store effectively control *Saprolegnia* growth? Are extracts from plants with known antimicrobial activity, such as *Chrysanthemum*, *Yucca*, or others, just as effective as commercial chemicals in controlling *Saprolegnia* growth?

2. **Formulate a testable hypothesis.** (Refer to Lab Topic 1, Exercise 1.1, Lab Study B, Developing Hypotheses.)
   Hypothesis:

3. **Summarize the essential elements of the experiment.** (Use separate paper.)

4. **Predict the results of your experiment based on your hypothesis.** (Refer to Lab Topic 1, Exercise 1.2, Lab Study C, Making Predictions.)
   Prediction: (if/then)

5. **Outline the procedures used in the experiment.** (Refer to Lab Topic 1, Exercise 1.2, Lab Study B, Choosing or Designing a Procedure.)
   a. On a separate sheet of paper, list in numerical order each exact step of your procedure.
   b. Remember to include the number of replicates (usually a minimum of five), levels of treatment, appropriate time intervals, and controls for each procedure.
   c. If you have an idea for an experiment that requires materials other than those provided, ask your laboratory instructor about availability. If possible, additional supplies will be provided.
   d. When carrying out an experiment, remember to quantify your measurements when possible.

6. **Perform the experiment,** making observations and collecting data for analysis.

7. **Record results, including observations and data,** on a separate sheet of paper. Design tables and graphs, at least one of each. Be thorough when collecting data. Do not just write down numbers, but record what they mean as well. Do not rely on your memory for information that you will need when reporting your results.

8. **Prepare your discussion.** Discuss your results in light of your hypothesis.
   a. Review your hypothesis. Review your results (tables and graphs). Do your results support or falsify your hypothesis? Explain your answer, using data for support.

b. Review your prediction. Did your results correspond to the prediction you made? If not, explain how your results are different from your predictions, and why this might have occurred.

c. If you had problems with the procedure or questionable results, explain how they might have influenced your conclusion.

d. If you had an opportunity to repeat and expand this experiment to make your results more convincing, what would you do?

e. Summarize the conclusion you have drawn from your results.

9. **Be prepared to report your results to the class.** Prepare to persuade your fellow scientists that your experimental design is sound and that your results support your conclusions.

10. If your instructor requires it, **submit a written laboratory report** in the form of a scientific paper (see Appendix A). Keep in mind that although you have performed the experiments as a team, you must turn in a lab report of *your original writing*. Your tables and figures may be similar to those of your team members, but your paper must be the product of your own literature search and creative thinking.

REVIEWING YOUR KNOWLEDGE

1. Complete Table 13.5 comparing characteristics of all protists investigated in this lab topic.
2. Using observations of pigments present, body form, and distinguishing characteristics of the three groups of macroscopic green, brown, and red algae, speculate about where they might be most commonly found in ocean waters.

TABLE 13.5 Comparison of Protists Studied in This Lab Topic

Group (Clade)	Example(s)	Characteristics	Ecological Role	Economic Importance
Euglenozoans	***Trypanosoma levisi***			
	Euglena			
Stramenopiles	**Diatoms**			
	Brown algae			
	Water molds			
Alveolates	**Paramecia**			
	Dinoflagellates			
	Plasmodium			
Rhizarians	**Foraminiferans**			
	Radiolarians			

(continued)

TABLE 13.5 Comparison of Protists Studied in This Lab Topic (*continued*)

Group (Clade)	Example(s)	Characteristics	Ecological Role	Economic Importance
Amoebozoans	Amoeba			
	Physarum			
	Dictyostelium			
Rhodophyta	Red algae			
Chlorophyta and Charophyta	Green algae: *Chlamydomonas, Spirogyra, Ulva, Chara*			

APPLYING YOUR KNOWLEDGE

1. In 1950, the living world was classified simply into two kingdoms: plants and animals. More recently, scientists developed the five-kingdom system of classification: plants, animals, monerans, protists, and fungi. In 2000, there was a general consensus among scientists that three domains with more than five kingdoms was a better system for classifying the diversity of life on Earth. With the growth of molecular genetics and phylogenetics, the phylogeny of protists, in the domain Eukarya, has undergone the most radical reconstruction of all groups of organisms, leading to the suggestion of creating "supergroups" rather than kingdoms. Using the protists studied in this lab topic, explain why the classification of this diverse group in particular is problematic. How is solving the problem of organizing protistan diversity a model for understanding the process of science?

2. Scientists are concerned that the depletion of the ozone layer will result in a reduction of populations of marine algae such as diatoms and dinoflagellates. Recall the ecological role of these organisms and comment on the validity of this concern.

STUDENT MEDIA: BioFlix, Activities, Investigations, and Videos

www.masteringbiology.com (select Study Area)

Activities—Ch. 26: Classification Schemes; Ch. 28: Tentative Phylogeny of Eukaryotes

Investigations—Ch. 28: What Kinds of Protists Do Various Habitats Support?

Videos—Ch. 28: Euglena; Euglena Motion; *Paramecium* Vacuole; *Paramecium* Cilia; Dinoflagellate; Diatoms Moving; Various Diatoms; Amoeba; Amoeba Pseudopodia; Plasmodial Slime Mold Streaming; Plasmodial Slime Mold; *Chlamydomonas*; Life Cycle of a Malaria Parasite; HHMI Seeing the Invisible: VanLeeuwenhoek's first glimpses of the microbial world

REFERENCES

Adamatzky, A. "On Attraction of Slime Mould *Physarum polycphalum* to Plants with Sedative Properties." *Nature Precedings*, 2011, May 31.

Adams, M. "Teaching with Chlamydomonas." *Tested Studies for Laboratory Teaching*, 2011, vol. 32, pp. 237–244.

Anderson, R. "What to Do with Protists?" *Australian Systematic Botany*, 1998, vol. 11, p. 185.

Burki, F. "The Eukaryotic Tree of Life from a Global Phylogenomic Perspective." *Cold Spring Harbor Perspectives in Biology*, 2014, vol. 6(5), : doi: 10.1101/cshperspect.a016147

Crowder, W. "Marvels of Mycetozoa." *National Geographic Magazine*, 1926, vol. 49, pp. 421–443.

Doolittle, W. F. "Uprooting the Tree of Life." *Scientific American*, 2000, vol. 282, pp. 90–95.

Milius, Susan. "Reinventing the Treetop of Life." *Science News*, 2015, vol.188, no.13, p. 23.

Urry, L., M. Cain, S. Wasserman, P. Minorsky, and J. Reece. *Campbell Biology*, 11th ed. San Francisco, CA: Pearson, 2017.

WEBSITES

Chlamydomonas Teaching Website:
www.chlamycollection.org/info.html

Links to pictures of red, brown, and green algae:
https://www.sonoma.edu/users/c/cannon/MarineAlgae.html

Protist image data, with excellent page links:
http://megasun.bch.umontreal.ca/protists/protists.html

Seaweeds:
http://www.seaweed.ie/

The Tree of Life web project—a collaborative effort of biologists from around the world to present information on diversity and phylogeny of organisms:
http://tolweb.org/tree/

Videos of amoeba and other protozoa in motion and slime mold migration and behavior:
http://www.youtube.com. Search for "amoeba in motion" and "cellular slime mold."

LAB TOPIC 14

Plant Diversity I: Bryophytes (Nonvascular Plants) and Seedless Vascular Plants

Laboratory Objectives

After completing this lab topic, you should be able to:

1. Describe the distinguishing characteristics of bryophytes and seedless vascular plants.
2. Discuss the ancestral and derived features of bryophytes and seedless vascular plants relative to their adaptations to the land environment.
3. Recognize and identify representative members of each phylum of bryophytes and seedless vascular plants.
4. Describe the general life cycle and alternation of generations in the bryophytes (moss) and the seedless vascular plants (fern), and discuss the differences between the life cycles of the two groups of plants using examples.
5. Identify fossil members and their extant counterparts in the seedless vascular plants.
6. Describe homospory and heterospory, including the differences in spores and gametophytes.
7. Discuss the ecological role and economic importance of these groups of plants.

Introduction

In the history of life on Earth, one of the most revolutionary events was the colonization of land by plants, then by animals. Comparison of the morphology, life cycles, biochemistry, and genetics of extant (living) plants and phyla of algae indicate that the first plants shared a common ancestor with the green algae, specifically the charophytes. Plants and charophytes share a number of distinctive traits not shared with other groups of algae, including the structure of the flagellated sperm, the formation of the cell plate in mitosis, and gene sequences in the nuclear and chloroplast DNA. The first colonists of land are thought to be most similar to the living, multicellular green algae *Zygnema, Chlorocaete,* and *Chara* (studied in Lab Topic 13 Protists). Once simple ancestral plants arrived on land over 470 million years ago, they faced new and extreme challenges in their physical environment. Only individuals that were able to survive the variations in temperature, moisture, gravitational forces, UV radiation, and substrate would thrive. Out of this enormous selective regime would come new and different adaptations and new and different life-forms: the plants.

Scientists continue to investigate the boundaries for the Kingdom Plantae given the evidence connecting green algae, charophytes, and plants. Plants generally have complex, multicellular bodies that are specialized for a variety of functions. In this lab manual, members of the Kingdom **Plantae** have

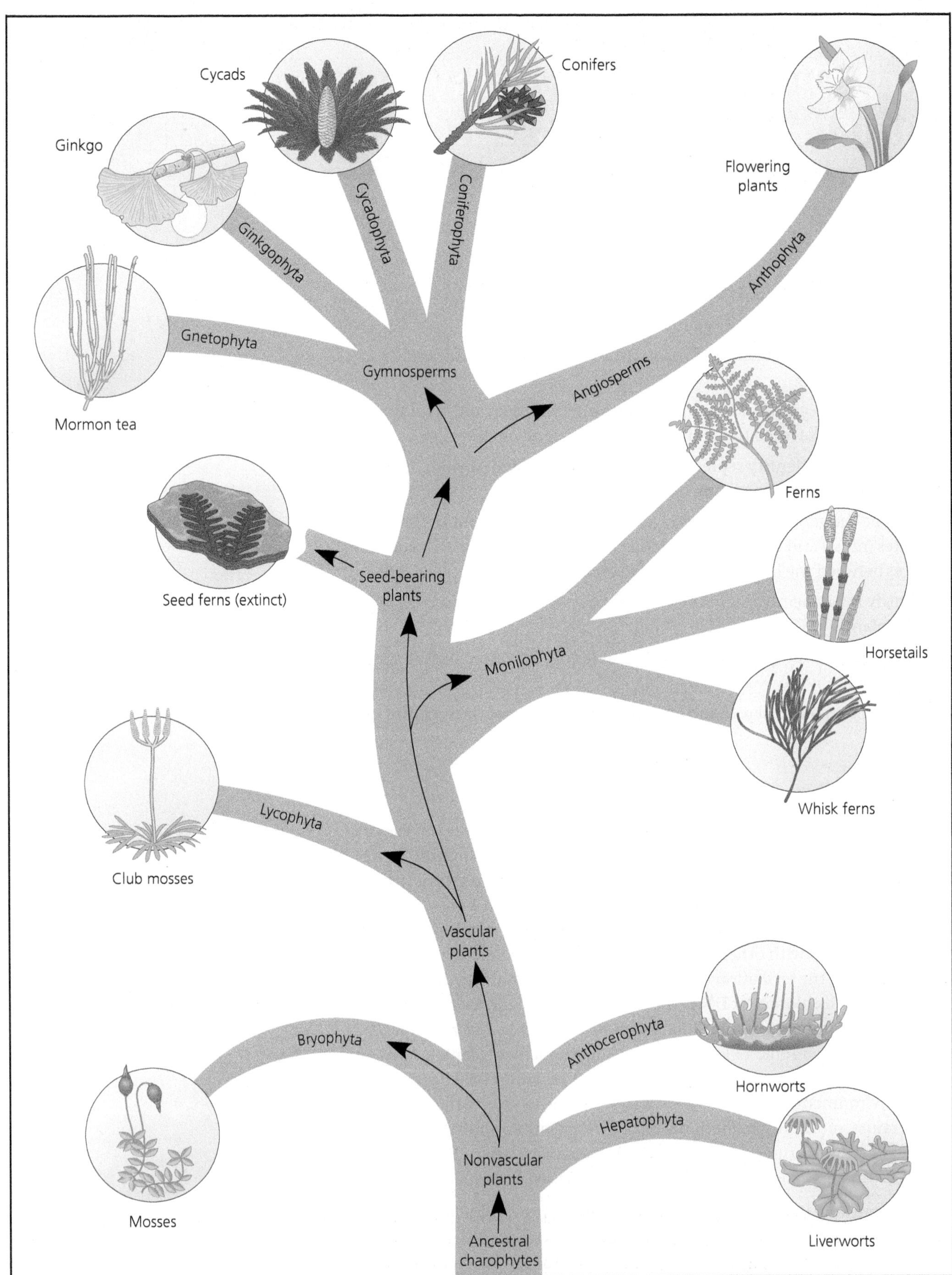
Cycads
Conifers
Ginkgo
Flowering plants
Cycadophyta
Coniferophyta
Ginkgophyta
Anthophyta
Gnetophyta
Gymnosperms
Angiosperms
Mormon tea
Ferns
Seed-bearing plants
Seed ferns (extinct)
Horsetails
Monilophyta
Whisk ferns
Lycophyta
Club mosses
Vascular plants
Bryophyta
Anthocerophyta
Hornworts
Hepatophyta
Nonvascular plants
Mosses
Ancestral charophytes
Liverworts

TABLE 14.1 Classification of Plants

Classification	Examples and Common Name	
Bryophytes (Nonvascular Plants)		
Phylum Bryophyta	Mosses	
Phylum Hepatophyta	Liverworts	
Phylum Anthocerophyta	Hornworts	
Vascular Plants		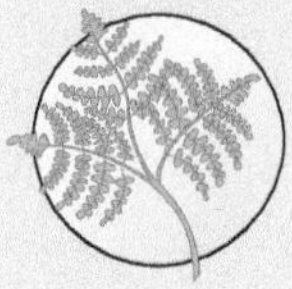
Seedless Plants		
Phylum Lycophyta	Club mosses	
Phylum Monilophyta	Ferns, horsetails, whisk ferns	
Seed Plants		
Gymnosperms		
Phylum Coniferophyta	Conifers	
Phylum Cycadophyta	Cycads	
Phylum Ginkgophyta	Ginkgo	
Phylum Gnetophyta	Mormon tea	
Angiosperms		
Phylum Anthophyta	Flowering plants	

similar alternation-of-generations life cycles, produce multicellular embryos, and have evolved specialized multicellular structures for protection of the vulnerable stages of sexual reproduction. The plant body is usually covered with a waxy cuticle that prevents desiccation. However, the waxy covering also prevents gas exchange, a problem solved by the presence of openings called **stomata** (sing., **stoma**) in most plants. Some plants have developed vascular tissue for efficient movement of materials throughout these complex bodies, which are no longer bathed in water. As described in the following section, the reproductive cycles and reproductive structures of these plants are also adapted to the land environment.

In the two plant diversity labs, you will be investigating the diversity of plants (Table 14.1 and Figure 14.1), some of which will be familiar to you (flowering plants, pine trees, and ferns) and some of which you may never have seen before (whisk ferns, horsetails, and liverworts). You will study the bryophytes and seedless vascular plants in this lab topic, Plant Diversity I, and seed plants in Lab Topic 15 Plant Diversity II.

FIGURE 14.1 (AT LEFT)

Evolution of plants. The bryophytes probably evolved from ancestral charophytes over 470 million years ago. Seedless vascular plants dominated Earth 325 million years ago, and representatives of two phyla have survived until the present. Seed plants replaced the seedless plants as the dominant plants, and today flowering plants are the most diverse and successful group in an amazing variety of habitats. The representatives studied in Plant Diversity I and Plant Diversity II are indicated.

To maintain your perspective in the face of all this diversity—and to reinforce the major themes of these labs—bear in mind the following questions.

1. What are the special adaptations of these plants to the land environment?
2. How are specialized plant structures related to functions in the land environment?
3. What are the major trends in the plant kingdom as plant life evolved over the past 500 million years?
4. In particular, how has the fundamental reproductive cycle of alternation of generations been modified in successive groups of plants?

Plant Life Cycles

All plants have a common sexual reproductive life cycle called **alternation of generations**, in which plants alternate between a haploid **gametophyte** generation and a diploid **sporophyte** generation (Figure 14.2). In living plants, these two generations differ in their morphology, but they are still the same species. In all plants except the bryophytes (mosses and liverworts), the diploid sporophyte generation is the dominant (more conspicuous) generation.

The essential features in the alternation-of-generations life cycle, beginning with the sporophyte, are:

- The diploid sporophyte undergoes meiosis to produce haploid **spores** in a protective, nonreproductive multicellular structure called the **sporangium**.
- Dividing by mitosis, the spores germinate to produce the haploid gametophyte.

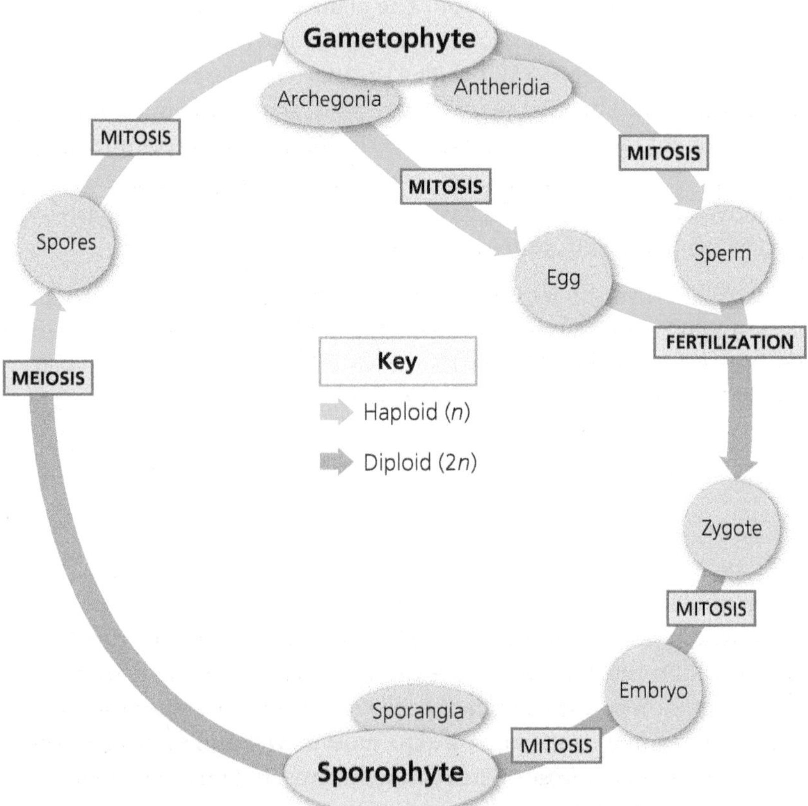

FIGURE 14.2
Alternation of generations. In this life cycle, a diploid sporophyte plant alternates with a haploid gametophyte plant. Note that haploid spores are produced on the sporophyte by meiosis, and haploid gametes are produced in the gametophyte by mitosis. *Using a colored pencil, indicate the structures that are haploid, and with another color, note the structures that are diploid.*

- The gametophyte produces **gametes** inside a multicellular structure made of nonreproductive cells, forming **gametangia** (sing., **gametangium**).
- There are two types of gametangia: **Eggs** are produced by mitosis in **archegonia** (sing., **archegonium**), and **sperm** are produced in **antheridia** (sing., **antheridium**).
- The gametes fuse (**fertilization**), usually by entrance of the sperm into the archegonium, forming a diploid **zygote**. The zygote divides by mitosis to produce the multicellular embryo and then the mature sporophyte of the next diploid generation.

Note that both gametes and spores are haploid in this life cycle. Unlike the animal life cycle, *the plant life cycle produces gametes by mitosis; spores are produced by meiosis*. The difference between these two cells is that gametes fuse with other gametes to form the zygote and restore the diploid number, whereas haploid spores germinate to form a new haploid gametophyte plant.

Review the generalized diagram of this life cycle in Figure 14.2. *Using colored pencils, distinguish the structures that are diploid and those that are haploid.* As you become familiar with variations of this life cycle through specific examples, you will want to continue referring to this general model for review.

Major trends in the evolution of this life cycle include:

> the increased importance of the sporophyte as the photosynthetic and persistent plant that dominates the life cycle; the reduction and protection of the gametophyte within the body of the sporophyte; the development of vascular tissue for transport; transition from homospory to heterospory; and the evolution of seeds and then flowers.

Bryophytes (Nonvascular Plants) and Seedless Vascular Plants

In this lab topic, terrestrial plants will be used to illustrate how life has undergone dramatic changes during the past 500 million years. Not long after the transition to land, plants diverged into at least two lineages, one to the nonvascular bryophytes and the other to vascular plants (see Figure 14.1). Nonvascular bryophytes first appear in the fossil record dating over 450 million years ago and remain unchanged, whereas the vascular plants appear around 425 million years ago and have undergone enormous diversification. As you review the evolution of plants, refer to the geological time chart for an overview of the history of life on Earth (Figure 14.3).

EXERCISE 14.1

Bryophytes (Nonvascular Plants)

The nonvascular plants, informally called bryophytes, are composed of three phyla of related plants that share some key characteristics and include mosses (Bryophyta) and liverworts (Hepatophyta). The third phylum, hornworts (Anthocerophyta), will not be seen in lab. (See again Figure 14.1 and Table 14.1.) The term *bryophytes* does not refer to a taxonomic category; rather, bryophytes are a group of plants that share a number of traits in common, appear to have evolved

Years Ago (millions)	Era / Period / Epoch	Life on Earth
	CENOZOIC Quaternary — Holocene, Pleistocene	• Origin of agriculture; ice ages; origin of *Homo* genus
2.6	Neogene — Pliocene	• Large carnivores; bipedal human ancestor
5.3	Paleogene — Miocene	• Forests dwindle; grassland spreads; early humans
23	Oligocene	• Anthropoid apes
34	Eocene	• Diversification of mammals and flowering plants
56	Paleocene	• Specialized flowers, pollinators, and seed distributors
66	**MESOZOIC** Cretaceous	• Flowering plants established and diversified; many modern families present; extinction of most dinosaurs
145	Jurassic	• Origin of birds; reptiles dominant; cycads and ferns abundant; first modern conifers and immediate ancestors of flowering plants
200	Triassic	• First dinosaurs and mammals; forests of gymnosperms and ferns; cycads
252	**PALEOZOIC** Permian	• Diversification of gymnosperms and reptiles; extinction of many organisms at end of Permian
299	Carboniferous	• First treelike plants; giant woody lycopods and sphenopsids form extensive forests in swampy areas; evolution of early seeds (seed ferns); amphibians dominant
359	Devonian	• Sharks and fishes dominant in the oceans
419	Silurian	• Diversification of vascular plants
443	Ordovician	• Diversification of algae and fungi; plants and animals colonize land
485	Cambrian	• Diversification of major animal phyla
541	**PRECAMBRIAN**	• Origin of bacteria, archaea, and eukaryotes
Earth is about 4.6 billion years old		

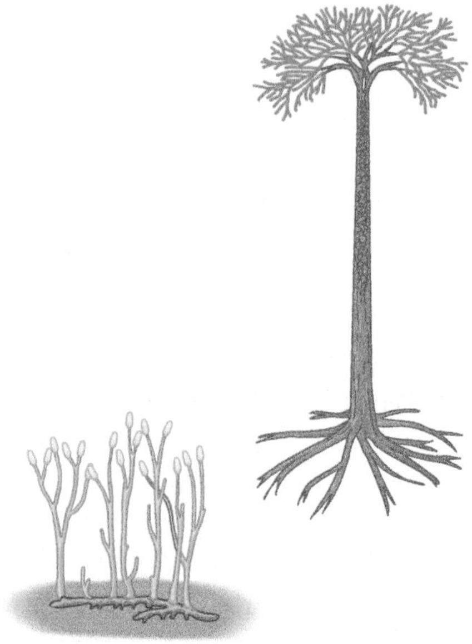

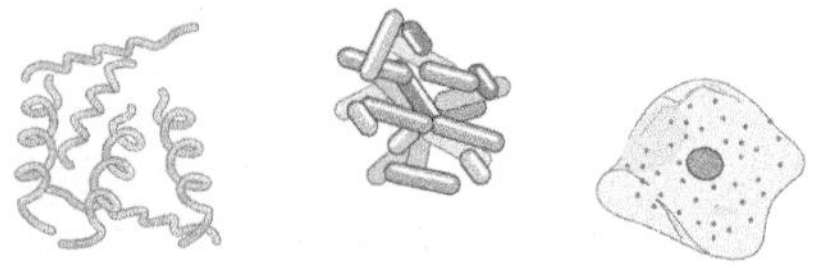

into several different groups independently, and do not form a single clade. They are small plants generally lacking vascular tissue (specialized cells for the transport of material), although simple water-conducting cells appear to be present in some mosses. The life cycle for the bryophytes differs from that of all other plants because the gametophyte is the dominant and conspicuous plant. Because bryophytes are nonvascular, they are restricted to moist habitats for their reproductive cycle and have never attained the size and importance of other groups of plants. The gametophyte plants remain close to the ground, enabling the motile sperm to swim from the antheridium to the archegonium and fertilize the egg. They have a cuticle but lack stomata on the surface of the gametophyte **thallus** (plant body that lacks vascular tissue), which is not organized into roots, stems, and leaves. Stomata are present on the sporophyte in some mosses and hornworts.

Bryophytes are not important economically, with the exception of sphagnum moss, which in its harvested and dried form is known as *peat moss*. Peat moss is absorbent, is an antibacterial agent, and was reportedly once used as bandages and diapers. Today peat moss is used in the horticultural industry, and dried peat is burned as fuel in some parts of the world. Peatlands cover more than 3% of the Earth's surface and store 400 billion metric tons of organic carbon. Harvesting and burning peat releases CO_2 to the atmosphere, thus contributing to changes in the global carbon cycle and climate change. Mosses are also ecologically important as pioneer species that colonize bare rock during primary succession.

Lab Study A. Bryophyta: Mosses

Materials

living examples of mosses
prepared slides of *Mnium* archegonia and antheridia
colored pencils

Introduction

The mosses are the most common group of bryophytes occurring primarily in moist environments, but are also found in dry habitats that are periodically wet (Figure 14.4). Refer to Figure 14.5 as you investigate the moss life cycle, which is representative of the bryophytes.

Procedure

1. Examine living colonies of mosses on demonstration. Usually you will find the two generations, gametophyte and sporophyte, growing together.
2. Identify the leafy **gametophytes** and the dependent **sporophytes**, which appear as elongated structures growing above them. Note that mosses do not have true leaves, because they lack specialized vascular tissue. Tug gently at the sporophyte and notice that it is attached to the gametophyte. Recall that the sporophyte develops and matures while attached to the gametophyte and receives its moisture and nutrients from the gametophyte.

FIGURE 14.3 (AT LEFT)
Geological time chart. The history of life can be organized into time periods that reflect changes in the physical and biological environment. Refer to this table as you review the evolution of plants in Plant Diversity I and Plant Diversity II.

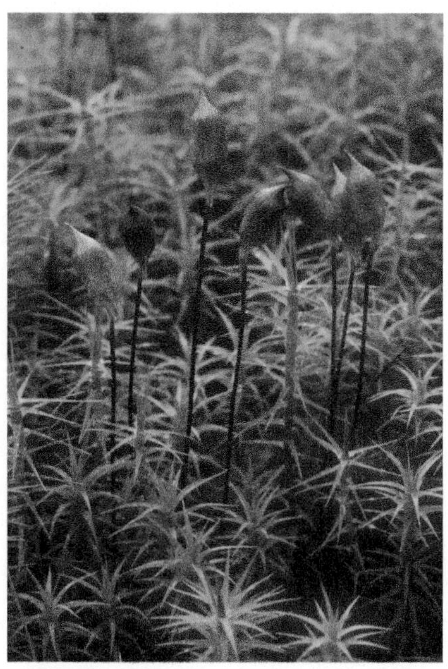

FIGURE 14.4
Moss with sporophytes growing out of leafy gametophytes. Spores develop by meiosis in the sporangia of the sporophytes.

3. The gametes are produced by the gametophyte in **gametangia** *by mitosis.* Gametangia protect the gametes but are not readily visible without a microscope. Observe under the microscope's low-power lens prepared slides containing long sections of heads of the unisex moss *Mnium*, which contain the gametangia. One slide has been selected to show the **antheridia** (male); the other is a rosette of **archegonia** (female). Sperm-forming tissue will be visible inside the antheridia. On the archegonial slide, look for an archegonium. The moss archegonium has a very long neck and rounded base. It will be difficult to find an entire archegonium in any one section. Search for a single-celled **egg** in the base of the archegonium.
4. Refer to Figure 14.5 as you follow the steps of fertilization through formation of the gametophyte in the next generation. The sperm swim through a film of water to the archegonium and swim down the neck to the egg, where fertilization takes place. The diploid **zygote** divides by mitosis and develops into an embryonic sporophyte within the archegonium. As the sporophyte matures, it grows out of the gametophyte but remains attached, deriving water and nutrients from the gametophyte body. **Spores** develop *by meiosis* in the **sporangium** at the end of the sporophyte. The haploid spores are discharged from the sporangium and in a favorable environment develop into new haploid gametophytes.

Results

1. Review the structures and processes observed and then label the diagram of the moss life cycle in Figure 14.5.
2. Using colored pencils, indicate if structures are haploid or diploid and circle the processes of mitosis and meiosis.

Discussion

Refer to Plant Life Cycles in the Introduction and Figure 14.2, the generalized diagram of the plant life cycle.

1. Are the spores produced by the moss sporophyte formed by meiosis or mitosis? Are they haploid or diploid? Do the spores belong to the gametophyte or sporophyte generation?
2. Are the gametes haploid or diploid? Are they produced by meiosis or mitosis?
3. Is the dominant generation for the mosses the gametophyte or the sporophyte?
4. Describe the structure of the archegonium and the sporangium. Relate these structures to their functions for moss survival in the terrestrial environment.

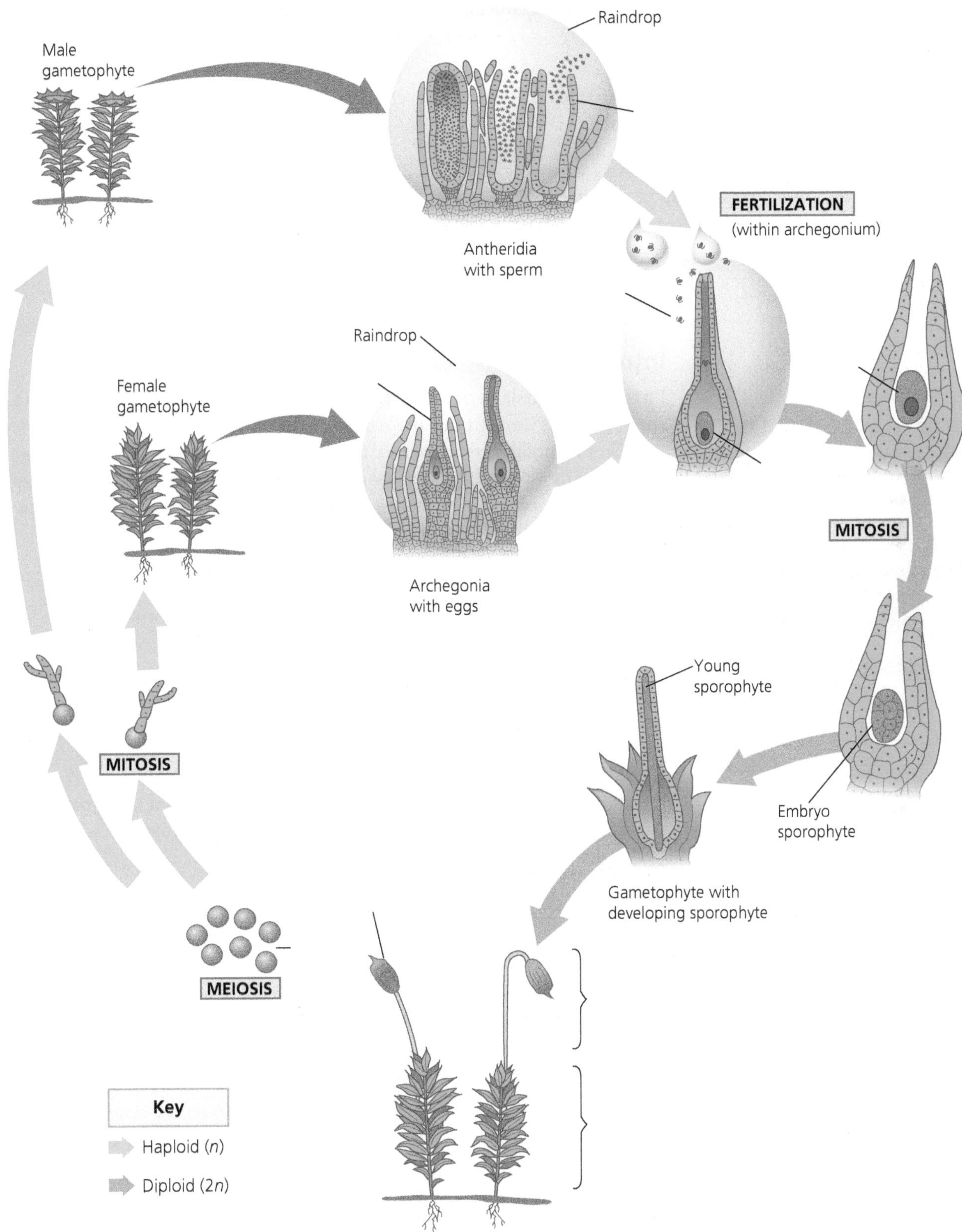

FIGURE 14.5

Moss life cycle. The leafy moss plant is the gametophyte, and the sporophyte is dependent on it, deriving its water and nutrients from the body of the gametophyte. Review this variation of alternation of generations and label the structures described in Lab Study A. *Using colored pencils, highlight the haploid and diploid structures in different colors. Circle the processes of mitosis and meiosis.*

5. Can you suggest any ecological role for mosses?

6. What feature of the life cycle differs for bryophytes compared with all other plants?

Lab Study B. Hepatophyta: Liverworts

Materials

living liverworts

Introduction

Liverworts are so named because their bodies are flattened and lobed (Figure 14.6). Early herbalists thought that these plants were beneficial in the treatment of liver disorders. Although less common than mosses, liverworts can be found along streams on moist rocks, but because of their small size, you must look closely to locate them. Recent fossil evidence suggests that liverworts may have been the first group of plants to diverge from the common ancestor of all plants.

FIGURE 14.6
Liverworts. Gemmae cups on the surface of this bryophyte function in asexual reproduction.

Procedure

Examine examples of liverworts on demonstration. Liverworts have a flat **thallus** (undifferentiated plant body). Note the **rhizoids**, rootlike filaments on the lower surface that primarily anchor plants and do not absorb or transport materials. Observe the **pores** on the surface of the leaflike thallus. These openings function in gas exchange; however, unlike stomata they are always open because they lack guard cells. On the upper surface of the thallus you should see circular cups called **gemmae cups**, which contain flat disks of green tissue called **gemmae**. The gemmae are washed out of the cups when it rains, and they grow into new, genetically identical liverworts.

Results

Sketch the overall structure of the liverwort in the margin of your laboratory manual. Label structures where appropriate.

Discussion

1. Is the plant you observed the gametophyte or sporophyte?

2. Are the gemmae responsible for asexual or sexual reproduction? Explain.

3. Why are these plants, like most bryophytes, restricted to moist habitats, and why are they always small?

4. In this lab topic, as in Plant Diversity II (Lab Topic 15) and Plant Anatomy (Lab Topic 20), you are asked to complete tables that summarize features advantageous to the adaptation of plant groups to the land environment. You may be asked to compare **derived traits** (features that have newly originated in a group of plants) with **ancestral traits** (features that have changed little and are shared with an ancestral group). Note these terms are used for relative comparisons. For example, for bryophytes, motile sperm might be considered an ancestral feature, whereas the cuticle would be considered derived.

   Complete Table 14.2, relating the features of bryophytes to their success in the land environment. Refer to the lab topic introduction for assistance.

TABLE 14.2 Ancestral and Derived Features of Bryophytes as They Relate to Adaptation to Land

Ancestral Features	Derived Features

EXERCISE 14.2

Seedless Vascular Plants

Seedless vascular plants are analogous to the first terrestrial vertebrate animals, the amphibians, in their dependence on water for external fertilization and development of the unprotected, free-living embryo. Both groups were important in the Paleozoic era but have undergone a steady decline in importance since that time. Seedless vascular plants were well suited for life in the vast swampy areas that covered large areas of the Earth in the Carboniferous period but were not suited for the drier areas of the Earth at that time or for later climatic changes that caused the vast swamps to decline and disappear. The fossilized remains of these swamp forests are the coal deposits of today (Figures 14.3 and 14.14).

Although living representatives of the seedless vascular plants have survived for millions of years, their limited adaptations to the land environment have restricted their range. All seedless vascular plants have vascular tissue, which is specialized for conducting water, nutrients, and photosynthetic products. Their life cycle is a variation of alternation of generations, in which the sporophyte is the dominant plant; the gametophyte is usually independent of the sporophyte. These plants generally have well-developed stems, leaves, and roots, as well as stomata and structural support tissue. However, because they still retain the ancestral feature of motile sperm that require water for fertilization, the gametophyte is small and restricted to moist habitats.

Economically, the only important members of this group are the ferns, a significant horticultural resource.

The phyla included in the seedless vascular plants are Lycophyta and Monilophyta (see again Table 14.1 and Figure 14.1).

The living examples of lycophytes are small club mosses, spike mosses, and quillworts. (Although named "mosses," these plants have vascular tissue and therefore are not true mosses.) The monilophytes include ferns, horsetails, and whisk ferns, which are remarkably different in overall appearance. Current evidence from analysis of morphology and both chloroplast and nuclear genes indicates that these diverse plants share a common ancestor and should all be included in the phylum Monilophyta. This evidence also suggests that monilophytes are more closely related to seed plants than they are to lycophytes.

Lab Study A. Lycophyta: Club Mosses

Materials

living *Selaginella* and *Lycopodium*
preserved *Selaginella* with microsporangia and megasporangia
prepared slide of *Selaginella strobilus*, l.s.

Introduction

Living members of Lycophyta are usually found in moist habitats, including bogs and streamsides (Figure 14.7a and b). However, one species of *Selaginella*, the resurrection plant, inhabits deserts. It remains dormant throughout periods of low rainfall, but then comes to life—resurrects—when it rains. During the Carboniferous period, lycophytes were conspicuous parts of the flora and formed the forest canopy; they were the ecological equivalent of today's oaks, hickories, and pines (Figure 14.14).

Bryophytes and most seedless vascular plants produce one type of spore (**homospory**), which gives rise to the gametophyte by mitosis. One advanced feature occasionally seen in seedless vascular plants is the production of two kinds of spores (**heterospory**). Large spores called **megaspores** divide by mitosis to produce the female gametophyte. The numerous small spores, **microspores**, produce the male gametophytes by mitosis. Heterospory and separate male and female gametophytes, as seen in *Selaginella*, are unusual in seedless vascular plants, but characteristic of all seed-producing vascular plants.

Procedure

1. Examine living club mosses, *Selaginella* and *Lycopodium*. Are they dichotomously branched? (The branches would split in two, appearing to form a Y.) Locate sporangia, which may be present either clustered at the end of the leafy stem tips, forming **strobili**, or **cones**, or dispersed

FIGURE 14.7
Lycophytes. (a) *Lycopodium.* (b) *Selaginella* often lives in moist habitats, but this species may survive in the desert.

along the leafy stems. Note that these plants have small leaves, or bracts, along the stem.

2. Examine preserved strobili of *Selaginella*. Observe the round sporangia clustered in sporophylls (leaflike structures) at the tip of the stem (Figure 14.8a). These sporangia contain either four megaspores in the **megasporangia** or numerous microspores in the **microsporangia**. Can you observe any differences in the sporangia or spores?

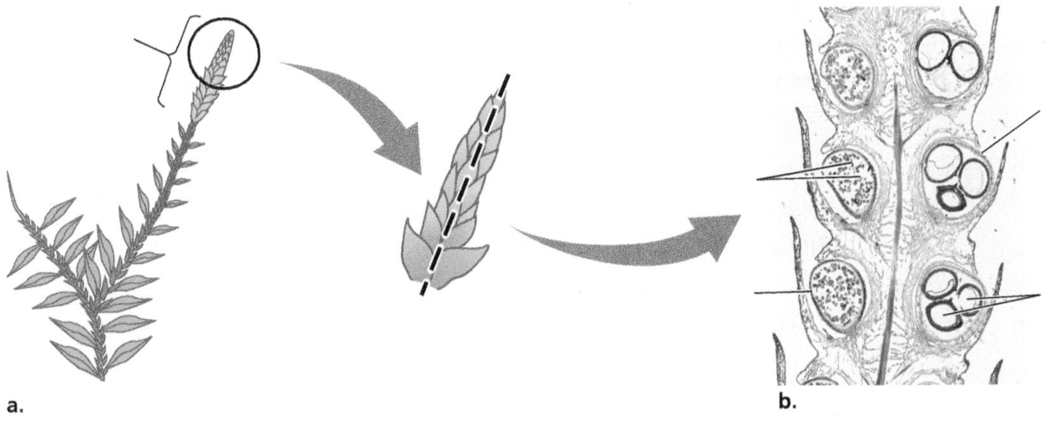

FIGURE 14.8
Selaginella. (a) The leafy plant is the sporophyte. The sporangia are clustered at the tips in strobili. (b) Photomicrograph of a longitudinal section through the strobilus of *Selaginella*.

3. Observe the prepared slide of a long section through the strobilus of *Selaginella*. Begin your observations at low power. Are both microspores and megaspores visible on this slide?

   How can you distinguish these spores?

4. Identify the **strobilus**, **microsporangium**, **microspores**, **megasporangium**, and **megaspores** and label Figure 14.8a and b.

Results

1. Sketch the overall structure of the club mosses in the margin of your lab manual. Label structures where appropriate.
2. Review Figure 14.8 of *Selaginella*. Using a colored pencil, highlight the structures that are haploid and part of the gametophyte generation.

Discussion

1. Are these leafy plants part of the sporophyte or the gametophyte generation? Do you have any evidence to support your answer?

2. What features did you observe in *Selaginella* that convinced you that it is a seedless vascular plant?

3. Are microspores and megaspores produced by mitosis or meiosis? (Review the life cycle in Figure 14.2.)

4. Will megaspores divide to form the female gametophyte or the sporophyte?

Having trouble with life cycles? Return to the Introduction and review the generalized life cycle in Figure 14.2. Reread the introduction to the study of seedless vascular plants. The key to success is to determine where meiosis occurs and to remember the ploidal level for the gametophyte and the sporophyte.

Lab Study B. Monilophyta: Ferns, Horsetails, and Whisk Ferns

Materials

living and/or preserved whisk ferns (*Psilotum*)
living and/or preserved horsetails (*Equisetum*)
living ferns

Introduction

If a time machine could take us back 400 million years to the Silurian period, we would find that vertebrate animals were confined to the seas, and early vascular plants had begun to diversify on land (Figure 14.3). By the Carboniferous period, ferns, horsetails, and whisk ferns grew alongside the lycophytes. Previously, these three groups of seedless vascular plants were placed in separate phyla: Pterophyta (ferns), Sphenophyta (horsetails), and Psilophyta (whisk ferns). Strong evidence from molecular biology now reveals a close relationship among these three groups, supporting a common ancestor for the group and their placement in one phylum, Monilophyta.

Scientists continue to actively investigate the relationships of this complex and diverse group of plants.

FIGURE 14.9
Whisk fern, *Psilotum*. The small spherical structures on the stems are sporangia.

Psilophytes (**whisk ferns**) are diminutive, dichotomously branched (repeated Y branches), photosynthetic stems that reproduce sexually by aerial spores. Today, whisk ferns can be found in some areas of Florida and in the tropics (Figure 14.9). Sphenophytes (**horsetails**) have green jointed stems with occasional clusters of leaves or branches. Their cell walls contain silica that gives the stem a rough texture. These plants were used by pioneers to scrub dishes—thus their name, scouring rushes. In cooler regions of North America, horsetails grow as weeds along roadsides (Figure 14.10). **Ferns** are the most successful group of seedless vascular plants, occupying habitats from the desert to tropical rain forests. Most ferns are small plants that lack woody tissue (Figure 14.11). An exception is the tree ferns found in tropical regions. Many cultivated ferns are available for home gardeners.

In this lab study you will investigate the diversity of monilophytes, including whisk ferns, horsetails, and a variety of ferns. The plants on demonstration are sporophytes, the dominant generation in seedless vascular plants. You will investigate the life cycle of a fern in Lab Study C, Fern Life Cycle.

FIGURE 14.10
Horsetail, *Equisetum*. Horsetail stems with strobili containing sporangia.

Procedure

1. Examine a living **whisk fern** (*Psilotum nudum*) on demonstration. This is one of only two extant genera of psilophytes.
2. Observe the spherical structures on the stem. If possible, cut one open and determine the function of these structures. Note the dichotomous branching, typical of the earliest plants.
3. Examine the **horsetails** (*Equisetum* sp.) on demonstration. Note the ribs and ridges in the stem. Also examine the nodes or joints along the stem where branches and leaves may occur in some species. Locate the **strobili** in the living or preserved specimens on demonstration. These are clusters of **sporangia**, which produce **spores**.

FIGURE 14.11
Ferns. Fern fronds are the sporophyte stage of the life cycle. Fiddleheads in the inset are young fronds unfurling.

FIGURE 14.12
Fern frond with sori. Clusters of sporangia located on the underside of the leaf are called sori. In this fern, each sorus has an indusium covering the sporangia.

4. Examine the living **ferns** on demonstration. Note the deeply dissected leaves, which arise from an underground stem called a **rhizome**, which functions like a root to anchor the plant. Roots arise from the rhizome. Observe the dark spots, or **sori** (sing., **sorus**), which are clusters of sporangia, on the underside of some leaves, called **sporophylls** (Figure 14.12).

Results

1. Sketch the overall structure of the whisk fern, horsetail, and fern in the margin of your lab manual. Label structures where appropriate.
2. Are there any leaves on the whisk fern? On the horsetails?
3. Are sporangia present on the whisk fern? On the horsetails? On the ferns?

Discussion

1. Are the spores in the sporangia produced by mitosis or meiosis?
2. Are the sporangia haploid or diploid? Think about which generation produces them.
3. Once dispersed, will these spores produce the gametophyte or sporophyte generation?

Lab Study C. Fern Life Cycle

Materials

living ferns
living fern gametophytes with archegonia and antheridia
living fern gametophytes with young sporophytes attached
stereoscopic microscope
compound microscope
prepared slide of fern gametophytes with archegonia, c.s.
colored pencils
Protoslo®
glycerol in dropping bottle

Introduction

In the previous lab study you examined the features of the fern sporophyte. In this lab study you will examine the fern life cycle in more detail, beginning with the diploid sporophyte.

Procedure

1. Examine the sporophyte leaf with sori (sporophyll) at your lab bench (Figure 14.12). Make a wet mount of a sorus, using a drop of glycerol, and do not add a coverslip. Examine the sporangia using a dissecting microscope. You will find the stalked **sporangia** in various stages of development. Find a sporangium still filled with **spores** and observe carefully for a few minutes, watching for movement. The sporangia will open and fling the spores into the glycerol.
2. Refer to Figure 14.13 as you observe the events and important structures in the life cycle of the fern. The haploid spores of ferns fall to the ground and grow into heart-shaped **gametophyte** plants. All seedless terrestrial plants depend on an external source of water for a sperm to swim to an egg to effect fertilization and for growth of the resulting sporophyte plant. The gametangia, which bear male and female gametes, are borne on the underside of the gametophyte. Egg cells are produced by mitosis in urnlike structures called **archegonia**, and sperm cells are produced by mitosis in globular structures called **antheridia**. Archegonia are usually found around the notch of the heart-shaped gametophyte, whereas antheridia occur over most of the undersurface.
3. To study whole gametophytes, make a slide of living gametophytes. View them using the stereoscopic microscope or the scanning lens on the compound microscope. Note their shape and color and the presence of **rhizoids,** rootlike multicellular structures. Locate archegonia and antheridia. Which surface will you need to examine? Sketch in the margin of your lab manual any details not included in Figure 14.13.
4. If you have seen antheridia on a gametophyte, remove the slide from the microscope. Gently but firmly press on the coverslip with a pencil eraser. View using the compound microscope first on intermediate power and then on high power. Look for motile **sperm** swimming with a spiral motion. Each sperm has two flagella. Add a drop of Protoslo to slow down the movement of the sperm.

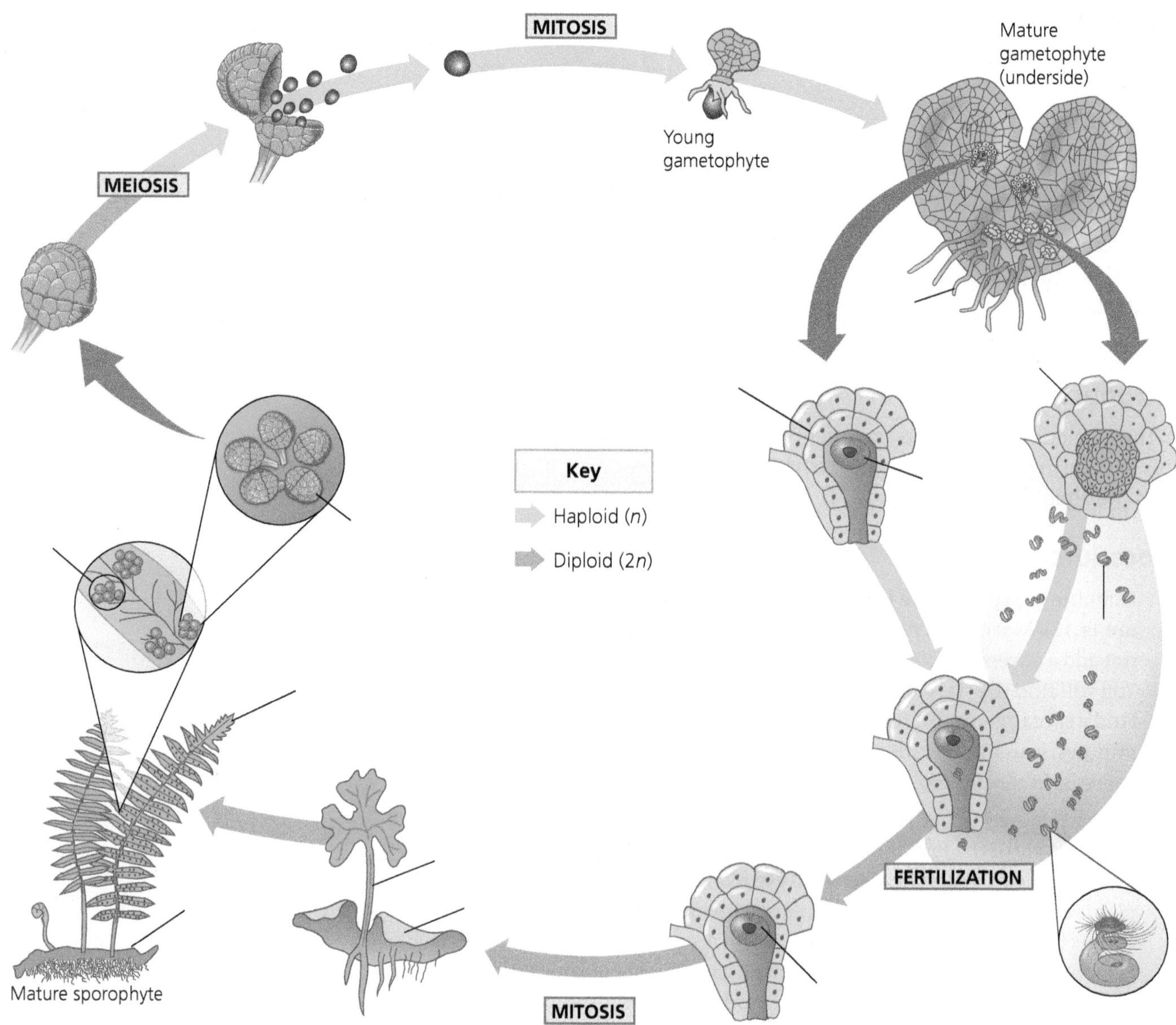

FIGURE 14.13

Fern life cycle. The familiar leafy fern plant is the sporophyte, which alternates with a small, heart-shaped gametophyte. Review this life cycle, a variation of alternation of generations, and label the structures and processes described in Lab Study C. *Using colored pencils, highlight the haploid and diploid structures in different colors.*

5. Observe the cross section of a fern gametophyte with archegonia. Each archegonium encloses an **egg**, which may be visible on your slide.
6. Make a wet mount of a fern gametophyte with a **young sporophyte** attached. Look for a young **leaf** and **root** on each sporophyte.
7. Share slides of living gametophytes with archegonia, antheridia and sperm, and sporophytes until everyone has observed each structure.

Results

1. Review the structures and processes observed, and then label the stages of fern sexual reproduction outlined in Figure 14.13.
2. Using colored pencils, circle those parts of the life cycle that are sporophytic (diploid). Use another color to encircle the gametophytic (haploid) stages of the life cycle. Highlight the processes of meiosis and mitosis.

Discussion

Refer to Figure 14.2, the generalized diagram of the plant life cycle, and Figure 14.13, a representation of the fern life cycle.

1. Are the spores produced by the fern sporophyte formed by meiosis or mitosis? Do the spores belong to the gametophyte or the sporophyte generation?
2. Are the gametes produced by mitosis or meiosis?
3. Are the archegonia and antheridia haploid or diploid? Think about which generation produces them.
4. Is the dominant generation for the fern the gametophyte or the sporophyte?
5. Describe one derived feature of ferns compared to bryophytes and explain why this feature is important to success in the terrestrial environment.
6. Can you suggest any ecological role for ferns?

Lab Study D. Fossils of Seedless Vascular Plants

Materials

fossils of extinct lycophytes (*Lepidodendron, Sigillaria*)
fossils of extinct sphenophytes (*Calamites*)
fossils of extinct ferns

Introduction

If we went back in time 300 million years to the Carboniferous period, we would encounter a wide variety of vertebrate amphibians moving about vast swamps dominated by spore-bearing forest trees. Imagine a forest of horsetails and lycophytes the size of trees, amphibians as large as alligators, and enormous dragonflies and roaches! Seedless vascular plants were at their peak during this period and were so prolific that their carbonized remains form the bulk of Earth's coal deposits. Among the most spectacular components of the coal-swamp forest were 100-foot-tall lycophyte trees belonging to the fossil genera *Lepidodendron* and *Sigillaria*, tree ferns, 60-foot-tall horsetails assigned to the fossil genus *Calamites*, and seed ferns (now extinct) (Figures 14.3 and 14.14). In 2004, geologists and paleobotanists recognized a 1,000-hectare fossilized forest from 300 million years ago in Illinois. The ancient forest, which lies above a coal seam, was discovered when tree trunks of lycyophytes and sphenophytes were discovered in the roof of the coal mine. The entire forest had been preserved when an ancient earthquake collapsed the site into a lake bottom. This rare preserved forest has provided a look into the complex relationships within the Carboniferous swamp forest that once dominated the Earth.

Procedure

Examine flattened fossil stems of *Lepidodendron, Sigillaria, Calamites*, and fossil fern foliage, all of which were recovered from coal mine tailings. Compare these with their living relatives, the lycophytes (club mosses), sphenophytes (horsetails), and ferns, which today are diminutive plants found in restricted habitats.

Results

1. For each phylum of seedless vascular plants, describe those characteristics that are similar for both living specimens and fossils. For example, do you observe dichotomous branching and similar shape and form of leaves, stems, or sporangia? Refer to the living specimens or your sketches.

   Lycophytes:

   Sphenophytes:

   Ferns:

2. Sketch the overall structure of the fossils in the margin of your lab manual. How would you recognize these fossils at a later date? Label structures where appropriate.

Discussion

The lycophytes, sphenophytes, and ferns were once the giants of the plant kingdom and dominated the landscape. Explain why they are presently restricted to certain habitats and are relatively small in stature.

a.

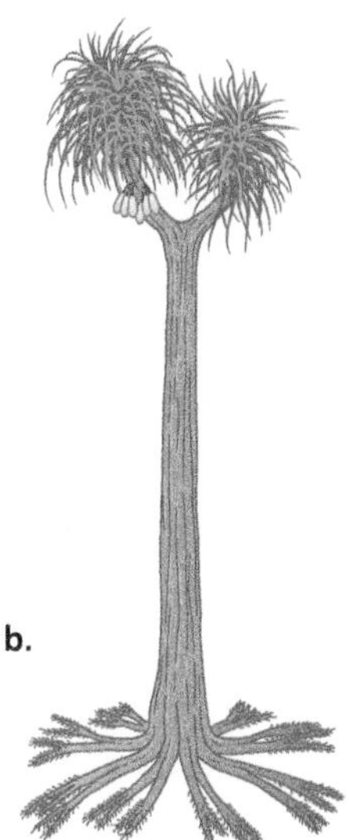

b.

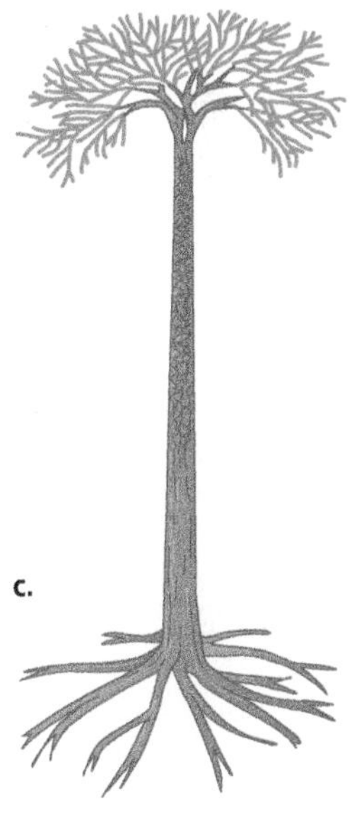

c.

d.

FIGURE 14.14

Seedless vascular plants of the Carboniferous period. (a) Reconstruction of a swamp forest dominated by lycophytes (b) *Sigillaria* and (c) *Lepidodendron.* (d) *Calamites* was a relative of horsetails.

REVIEWING YOUR KNOWLEDGE

1. Complete Table 14.3, indicating the ancestral and derived features of seedless vascular plants relative to success in land environments. Recall that in this context, the term *ancestral* refers to a trait shared with a group of common ancestors, for example, the bryophytes. The term *derived* indicates a trait that has arisen more recently, for example, in the seedless vascular plants. Traits shared with the bryophytes (such as sperm requiring water for fertilization) are ancestral, whereas the presence of vascular tissue is derived.

TABLE 14.3 Ancestral and Derived Features of Seedless Vascular Plants as They Relate to Adaptation to Land

Ancestral Features	Derived Features

2. For each of the listed features, describe its contribution, if any, to the success of plants.

   gametangium

   cuticle

   rhizoid

   motile sperm

   vascular tissue

   gemma

3. Complete Table 14.4. Identify the function of the structures listed. Indicate whether they are part of the gametophyte or sporophyte generation, and provide an example of a plant that has this structure.

TABLE 14.4 Structures and Functions of the Bryophytes and Seedless Vascular Plants

Structure	Function	Sporophyte/ Gametophyte	Example
Antheridium			
Archegonium			
Spore			
Gamete			
Rhizome			
Gemma			
Sporangium			
Strobilus			
Sorus			

4. What is the major difference between the alternation of generations in the life cycles of bryophytes and seedless vascular plants?

APPLYING YOUR KNOWLEDGE

1. The fossil record provides little information about ancient mosses. Do you think that bryophytes could ever have been large, tree-sized plants? Provide evidence from your investigations to support your answer.

2. On a walk through a botanical garden, you notice a small leafy plant that is growing along the edge of a small stream in a shady nook. You hypothesize that this plant is a lycophyte. What information can you gather to test your hypothesis?

3. Fern antheridia release sperm that then swim toward archegonia in a watery film. The archegonia release a fluid containing chemicals that attract the sperm. This is an example of chemotaxis, the movement of cells or organisms in response to a chemical. What is the significance of chemotaxis to fern (and moss) reproduction?

4. Scientists investigating the evolutionary history (phylogeny) of plants represent their hypotheses as phylogenetic trees or branching diagrams. Groups of plants that share a common ancestor are called clades, which are represented as lines connected at a branching point (the common ancestor). See the example below.

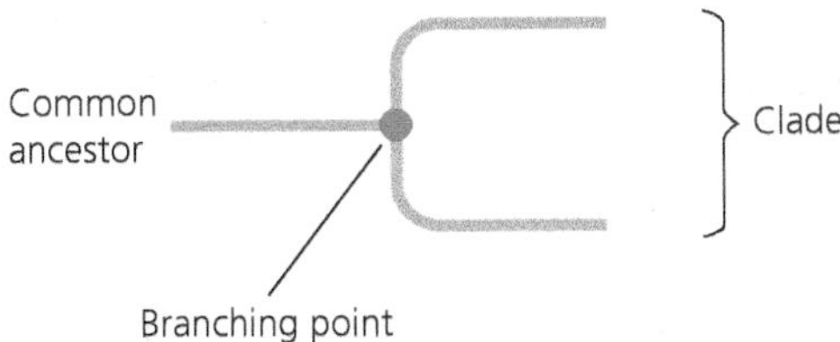

Which two of the phylogenetic trees below best represents the evolutionary history of plants? Explain your choice of tree using your results from this laboratory topic.

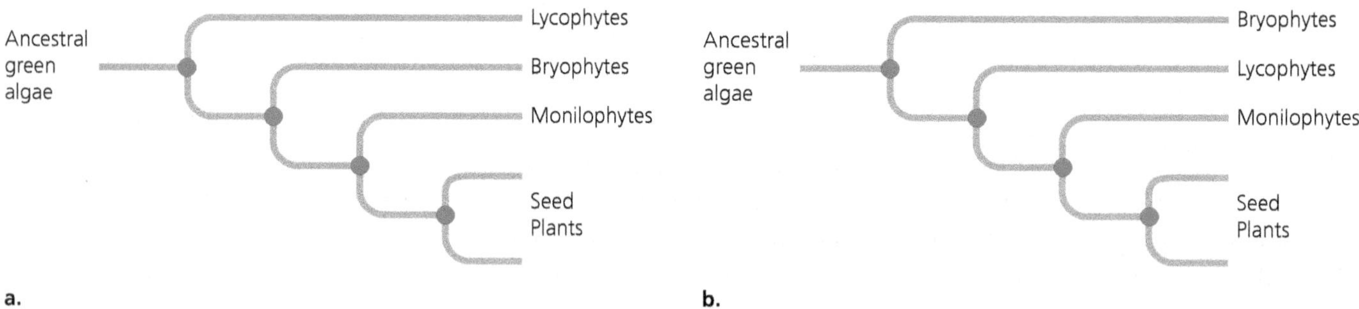

5. Heterospory occasionally occurs in lycophytes and ferns, and in all seed plants. Botanists are convinced that heterospory must have originated more than once in the evolution of plants. Can you suggest one or more advantages that heterospory might provide to plants?

INVESTIGATIVE EXTENSIONS

C-Ferns, Ceratopteris, are excellent experimental organisms for investigating the alternation-of-generations life cycle, comparisons of seedless vascular plants with seed plants, and the physiology of ferns. These small ferns have a short life cycle of 12–14 days from spore to mature gametophyte, and they can be grown successfully in small laboratory spaces. *C-Fern Express*, a new strain of *C-Fern*, has an even faster life cycle. Amazingly, the motile sperm, easily visible under the microscope, can swim for 2 hours in buffer! The following are questions that you might investigate.

1. Fern archegonia secrete pheromones to attract swimming sperm for fertilization, an example of chemotaxis (see Question 3 in Applying Your Knowledge). Are sperm attracted to other compounds as well, including organic acids that might be present in the fluid secreted by the archegonia? Are there other compounds that might also be attractants? What are the common characteristics of the compounds that are attractants, for example, type of compound or chemical structure?

2. *C-Ferns* produce two types of gametophytes, hermaphrodites (both archegonia and antheridia present) and males (antheridia only). What factors affect the proportion of hermaphrodite and male gametophytes in the population—for example, temperature, light, or population density?

3. Are sperm from male gametophytes more or less attracted by pheromones compared with sperm from the hermaphrodites? Do sperm from both types of gametophytes respond to the same concentration of pheromones or other organic attractants?
Resources, laboratory procedures, additional student research questions, and preparation instructions are provided at the *C-Fern* website, http://cfern.bio.utk.edu/index.html. Growing materials are available at Carolina Biological Supply, including *C-Fern* Chemotaxis Kit and *C-Fern* Life Cycle Kit.

STUDENT MEDIA: BioFlix, Activities, Investigations, and Videos

www.masteringbiology.com (select Study Area)

Activities—Ch. 29: Highlights of Plant Phylogeny; Moss Life Cycle; Fern Life Cycle

Investigations—Ch. 26: How Is Phylogeny Discovered by Comparing Proteins?
Ch. 29: What Are the Different Stages of a Fern Life Cycle?

REFERENCES

Armstrong, J. E. *How the Earth Turned Green: A Brief 3.8-Billion-Year History of Plants*, Chicago: University of Chicago Press, 2014.

C-Fern Web Manual, 2009. Available for download at http://www.c-fern.org/.

Evert, R. F. and S. E. Eichhorn. *Raven Biology of Plants*, 8th ed. New York: W.H. Freeman, 2013.

Hickock, L. G. and T. R. Warne. *C-Fern Manual*. Burlington, NC: Carolina Biological Supply, 2000.

Kendrick, P. and P. Davis. *Fossil Plants*. Washington, DC: Smithsonian, 2004.

Mauseth, J. D. *Botany: An Introduction to Plant Biology*, 5th ed. Sudbury, MA: Jones and Bartlett Publishers, 2012.

Niklas, K. J. *Plant Evolution: An Introduction to the History of Life*, Chicago: University of Chicago Press, 2016.

Tidwell, W. D. *Common Fossil Plants of Western North America*, 2nd ed. Washington, DC: Smithsonian Books, 2010.

WEBSITES

C-Fern website with information, suggestions, and support:
http://www.c-fern.org/

Fern basics, images, and current research:
http://amerfernsoc.org/

Illinois State Geological Survey, "A 300 Million Year Old Pennsylvanian Age Mire Forest." Includes photos and reconstruction illustrations:
http://isgs.illinois.edu/research/coal/pennsylvanian-age-mire-forest

An introduction to plants, including morphology, evolution, and fossils. Images and resources provided:
http://www.ucmp.berkeley.edu/plants/plantae.html

Links to images of plants:
http://botit.botany.wisc.edu/Resources/

The Tree of Life Web Project is a collaborative project of biologists worldwide. Information is provided on all major groups of living organisms, including plants:
http://tolweb.org/Embryophytes

LAB TOPIC 15

Plant Diversity II: Seed Plants

> Before lab, read the introductory material on gymnosperms and angiosperms and complete Table 15.1 by listing (and comparing) the traits of each.

Laboratory Objectives

After completing this lab topic, you should be able to:

1. Identify characteristics and examples of the phyla of seed plants.
2. Describe the life cycle of a gymnosperm (pine tree) and an angiosperm.
3. Describe features of flowers that ensure pollination by insects, birds, bats, and wind.
4. Describe factors influencing pollen germination.
5. Identify types of fruits, recognize examples, and describe dispersal mechanisms.
6. Relate the structures of seed plants to their functions in the land environment.
7. Compare the significant features of life cycles for various plants and state their evolutionary importance.
8. Summarize major trends in the evolution of plants and provide evidence from your laboratory investigations.

Gymnosperms

For over 470 million years, plants have been adapting to the rigors of the land environment. The nonvascular bryophytes, with their small and simple bodies, survived in habitats that remained moist for at least part of their life cycle. The earliest seed plant fossils are from 360 million years ago, and during the cool Carboniferous period, vascular seedless plants dominated the landscape of the swamp forests that covered much of Earth. Although these plants were more complex and better adapted to the challenges of the land environment, they still were dependent on water for sperm to swim to the egg. During the Mesozoic era, 250 million years ago, Earth became warmer and drier and the swamp forests declined, presenting another challenge to terrestrial plants and animals. Earth at that time was a world dominated by reptilian vertebrates, including the flying, running, and climbing dinosaurs. The landscape was dominated by a great variety of seed-bearing plants called **gymnosperms** (literally, "naked seeds"). Gymnosperms appeared in the Carboniferous along with other seed plants (now extinct); however, the ancestral origin of gymnosperms remains uncertain. During the Mesozoic, a number of distinct gymnosperm groups diversified, and a few of the spore-bearing plants survived (see Figure 14.1 and Table 14.1 in Lab Topic 14). As you review the evolution of plants, refer to the geological time chart in Figure 14.3 for an overview of the history of life on Earth.

Vertebrate animals became fully terrestrial during the Mesozoic with the emergence of reptiles, which were free from a dependence on water for sexual reproduction and development. The development of the amniotic egg along with an internal method of fertilization made this major transition possible. The amniotic egg carries its own water supply and nutrients, permitting early embryonic development to occur on dry land, a great distance from external water. In an analogous manner, the gymnosperms became free from dependence on water through the development of a process of internal fertilization via the **pollen grain** and development of a **seed**, which contains a dormant embryo with a protective seed coat and a supply of food.

Several features of the gymnosperms have been responsible for their success. They are heterosporous (a trait shared with some lycophtes) and have reduced (microscopic) gametophytes. The male gametophyte is contained in a multinucleated pollen grain, and the female gametophyte is retained within the sporangium in the ovule of the sporophyte generation. The pollen grain is desiccation resistant and adapted primarily for wind pollination, removing the necessity for fertilization in a watery medium. The pollen tube conveys the sperm nucleus to an egg cell, and the embryonic sporophyte develops within the gametophyte tissues, which are protected by the previous sporophyte generation. The resulting seed is not only protected from environmental extremes, but also is packed with nutritive materials and can be dispersed away from the parent plant. In addition, gymnosperms have advanced vascular tissues: xylem for transporting water and nutrients and phloem for transporting photosynthetic products. The xylem cells are called *tracheids* and are more efficient for transport than those of the seedless vascular plants.

Angiosperms

A visit to Earth 65 million years ago, during the late Cretaceous period, would reveal a great diversity of mammals and birds and a landscape dominated by **flowering plants**, or **angiosperms** (phylum **Anthophyta**). Ultimately, these plants would diversify (to more than 250,000 species) and become the most numerous, widespread, and important plants on Earth. Angiosperms now occupy well over 90% of the vegetated surface of Earth and contribute virtually 100% of our agricultural food plants.

The evolution of the **flower** resulted in enormous advances in the efficient transfer and reception of pollen. Whereas gymnosperms (with a couple of interesting exceptions) are wind-pollinated, producing enormous amounts of pollen that reach the appropriate species by chance, the process of flower pollination is mediated by specific agents—insects, birds, and bats—in addition to water and wind. Pollination agents such as the insect are attracted to the flower with its rewards of nectar and pollen. Animal movements provide precise placement of pollen on the receptive portion of the female structures, increasing the probability of fertilization. The process also enhances the opportunity for cross-fertilization among distant plants and therefore the possibility of increased genetic variation.

Angiosperm reproduction follows the trend for reduction in the size of the gametophyte. The pollen grain contains the male gametophyte, and the multinucleated **embryo sac** is all that remains of the female gametophyte. This generation continues to be protected and dependent on the adult sporophyte plant. The female gametophyte provides nutrients for the developing sporophyte embryo through a unique triploid **endosperm** tissue. Another unique feature of angiosperms is the **fruit**. The seeds of the angiosperm develop within the flower ovary, which matures into the fruit. This structure provides protection and enhances dispersal of the young sporophyte into new habitats.

In addition to advances in reproductive biology, the angiosperms evolved other advantageous traits. All gymnosperms are trees or shrubs, with a large investment in woody, persistent tissue; and their life cycles are long (five or more years before they begin to reproduce and two to three years to produce a seed). Flowering plants, on the other hand, can be woody, but many are herbaceous, with soft tissues that survive up to a few years. It is possible for angiosperms to go from seed to seed in less than one year. As you perform the exercises in this lab, think about the significance of this fact in terms of the evolution of this group. How might generation length affect the rate of evolution? Angiosperms also have superior conducting tissues. Xylem tissue is composed of *tracheids* (as in gymnosperms), but also contains large-diameter, open-ended *vessels.* The phloem cells, called *sieve-tube elements,* provide more efficient transport of the products of photosynthesis. The cell structure and organization of plants will be investigated in Lab Topic 20 Plant Anatomy.

Review the characteristics of gymnosperms and angiosperms described in this introduction, and *summarize in Table 15.1 the advantageous traits of these groups relative to their success on land.* You should be able to list several characteristics for each. At the end of the lab, you will be asked to modify and complete the table, based on your investigations.

You will want to return to this table after the laboratory to be sure that the table is complete and that you are familiar with all of these important features.

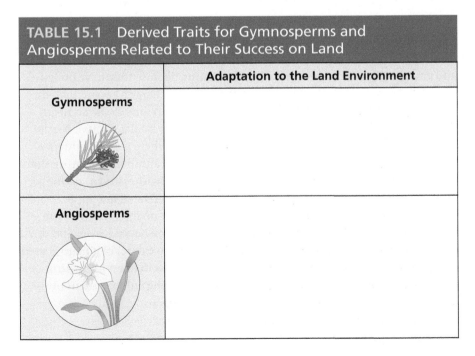

TABLE 15.1 Derived Traits for Gymnosperms and Angiosperms Related to Their Success on Land

	Adaptation to the Land Environment
Gymnosperms	
Angiosperms	

EXERCISE 15.1

Gymnosperms

The term *gymnosperms* refers to a diverse group of seed plants that do not produce flowers. Although they share many characteristics, including the production of pollen, they represent four distinct groups, or phyla. In this exercise, you will observe members of these phyla and investigate the life cycle of a pine, one of the most common gymnosperms.

Lab Study A. Phyla of Gymnosperms

Materials

living or pressed examples of conifers, ginkgos, cycads, and Mormon tea

Introduction

Gymnosperms are economically and ecologically important plants that have diversified into four phyla, but include only 1,000 species (see Figure 14.1 and Table 14.2 in Lab Topic 14). The name gymnosperm ("naked seeds") refers to the fact that the seeds are unprotected on the surface of a bract, usually in cones. The largest and best-known phylum is **Coniferophya**, which includes pines and other cone-bearing trees and shrubs. In addition to the gymnosperm innovations of seeds and pollen grains, conifers have thick cuticles, needle-like leaves, and resin ducts that reduce water loss and contribute to their success in dry environments (both cold and hot). Important conifers include spruces, hemlocks, junipers, yews, and pines (Figure 15.1a). Conifers are the oldest non-clonal organisms on Earth, with the bristlecone pine in the White Mountains of Nevada holding the record at 5,067 years old at this time and still growing. The next oldest organisms are a cypress in Iran and a yew in Wales. The largest and tallest trees on Earth are also conifers. Conifers are culturally and economically important as building materials, resins, fuel, and in making paper. Taxol, one of the leading chemotherapy drugs for breast cancer, was first isolated from yew trees.

Cycads (**Cycadophyta**), which have a palmlike appearance, are found primarily in tropical regions scattered around the world (Figure 15.1b). Cycads and ginkgos (Figure 15.1 b, c, and d) are intriguing because they have pollen grains that develop a pollen tube, but rather than discharge a sperm nucleus, some have flagellated sperm that swim from the pollen tube to the archegonium and fertilize the egg (a transition from seedless vascular plants). Beetles and weevils are often found around cycad cones, and there is some evidence that these insects may carry pollen from one cycad to another, indicating that at least a few cycads are insect pollinated. Ginkgos (**Ginkgophyta**) no longer grow in the wild, but this native of China has been cultivated in temple grounds in China and Japan for centuries and appears unchanged from the 150-mya fossils (Figure 15.1c and d). With their flat fan-shaped leaves that turn golden in the fall, ginkgos are prized as urban trees. An extract of ginkgo is used as an herbal medicine, and there is evidence for its effectiveness in treating some cases of dementia and altitude sickness and for the enhancement of

a.

b.

c.

d.

e.

FIGURE 15.1
Phyla of gymnosperms.
(a) Coniferophyta. First-, second-, and third-year cones of bristlecone pine. Some bristlecone pines are more than 4,000 years old. (b) Cycadophyta. Pollen is produced in male cones. (c) Ginkgophyta. Ginkgo with fleshy seeds. (d) Ginkgophyta. Male stroboli produce pollen. (e) Gnetophyta. Mormon tea occurs in the deserts of North and Central America. Inset, enlarged male cones.

memory in healthy patients. **Gnetophyta** is composed of three distinct and unusual groups of plants: gnetums, which are primarily vines of Asia, Africa, and South America; *Welwitschia,* a rare desert plant with two leathery leaves; and Mormon tea (*Ephedra*), desert shrubs of North and Central America (Figure 15.1e). Recent studies indicate that *Gnetum* and *Ephedra* may have some angiosperm-like characteristics. Compounds from *Ephedra,* ephedrines, used in diet aids and decongestants, have raised serious concerns due to side effects that include cardiac arrest.

Procedure

1. Observe demonstration examples of all phyla of gymnosperms and be able to recognize their representatives. Note any significant ecological and economic role for these plants.
2. In the margin of your manual, sketch the overall structure of the plants. Label structures where appropriate.

Results

1. Record your observations in Table 15.2.
2. Are there any reproductive structures present for these plants? If so, make notes in the margin of your lab manual.

TABLE 15.2 Phyla of Gymnosperms

Phyla	Examples	Characteristics/Comments
Coniferophyta		
Ginkgophyta		
Cycadophyta		
Gnetophyta		

Discussion

1. What are the key characteristics shared by all gymnosperms?

2. What is the ecological role of conifers in forest systems?

3. What economically important products are provided by conifers?

4. What economically important products are provided by other gymnosperms?

Lab Study B. Pine Life Cycle

Materials

living or preserved pine branch, male and female cones (one, two, and three years old)
fresh or dried pine pollen or prepared slide of pine pollen
coverslips
prepared slides of male and female pinecones
colored pencils
slides

Review the pine life cycle (Figure 15.2) before you begin. Follow along as you complete the exercise.

Introduction

Gymnosperms are almost exclusively **wind-pollinated** trees or shrubs, most bearing unisexual, male, and female reproductive structures on different parts of the same plant. Gymnosperms are **heterosporous**, producing two kinds of spores: male **microspores**, which develop into **pollen** that contains the male gametophyte, and female **megaspores**. The megaspore develops into the female gametophyte, which is not free-living as with ferns, but is retained within the **megasporangium** and nourished by the sporophyte parent plant. The female gametophye develops **archegonia**, each containing an egg. Numerous pollen grains (containing the male gametophytes) are produced in each **microsporangium**, and when they are mature they are released into the air and conveyed by wind currents to the female cone. **Pollen tubes** grow through the tissue of the megasporangium, and the **sperm nucleus** is released to fertilize the egg. After fertilization, development results in the formation of an **embryo**. A **seed** is a dormant embryo embedded in the nutritive tissue of the female gametophyte and surrounded by the hardened **seed coat**.

Having trouble with life cycles? Return to Lab Topic 14 (Plant Diversity I) and review the generalized life cycle (Figure 14.2). The key to success is to determine where meiosis occurs and to remember the ploidal level for the gametophyte and sporophyte.

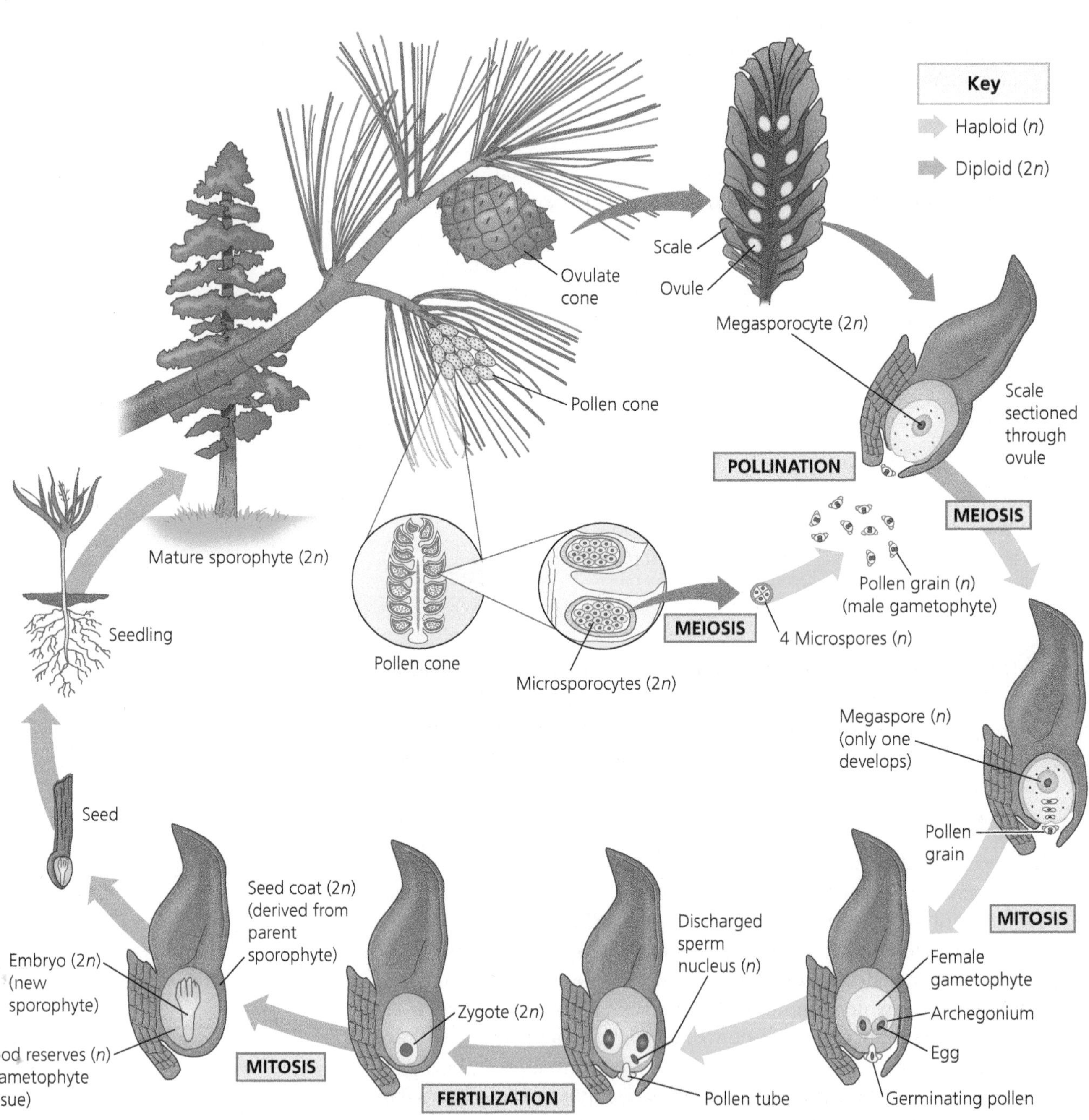

FIGURE 15.2

Pine life cycle. Observe the structures and processes as described in Exercise 15.1. *Using colored pencils, indicate the structures that are haploid or diploid. Circle the terms* mitosis, meiosis, *and* fertilization.

Procedure

1. Pine sporophyte:
   a. Examine the pine branch and notice the arrangement of leaves in a bundle. A new twig at the end of the branch is in the process of producing new clusters of leaves. Is this plant haploid or diploid?

   b. Examine the small **cones** produced at the end of the pine branch on this specimen or others in lab. Recall that cones contain clusters of sporangia. What important process occurs in the sporangia?

   c. Locate an ovulate cone and a pollen cone. Elongated male **pollen cones** are present only in the spring, producing pollen within overlapping bracts, or scales. The small, more rounded female cones (which look like miniature pinecones) are produced on stem tips in the spring and are called **ovulate cones**. Female cones persist for several years. Observe the overlapping scales, which contain the sporangia.
   d. In the margin of your lab manual, sketch your observations for future reference.
2. Male gametophyte—development in pollen cones:
   a. Examine a longitudinal section of the pollen cone on a prepared slide and identify its parts. Observe that pollen cones are composed of radiating scales, each of which carries two elongated sacs on its lower surface. The sacs are the **microsporangia**. **Microsporocytes**, diploid cells within microsporangia, divide by meiosis. Each produces four haploid **microspores**, which develop into **pollen grains**.
   b. Observe a slide of pine pollen. If pollen is available, you can make a wet mount. Note the wings on either side of the grain. The pollen grain contains the greatly reduced male gametophyte surrounded by the pollen coat that prevents desiccation. Once mature, pollen will be wind dispersed, sifting down into the scales of the female cones.
   c. Sketch, in the margin of your lab manual, your observations for future reference.
3. Female gametophyte—development in ovulate cones:
   a. Examine a longitudinal section of a young ovulate cone on a prepared slide. Note the **ovule** (containing the megasporangium) on the upper surface of the scales. Diploid **megasporocytes** contained inside will divide by meiosis to produce haploid **megaspores**, the first cells of the gametophyte generation. In the first year of ovulate cone development, pollen sifts into the soft bracts (pollination) and the pollen tube begins to grow, digesting the tissues of the ovule.
   b. Observe a second-year cone at your lab bench. During the second year, the ovule develops a multicellular female gametophyte with two **archegonia** in which an **egg** will form. Fertilization will not occur until the second year, when the pollen tube releases a sperm nucleus

into the archegonium, where it unites with the egg to form the **zygote**. In each ovule only one of the archegonia and its zygote develops into a **seed**.

c. Observe a mature cone at your lab bench. The development of the embryo sporophyte usually takes another year. The female gametophyte will provide nutritive materials stored in the seed for the early stages of growth. The outer sporophyte tissues of the ovule will harden to form the **seed coat**.

d. In the margin of your lab manual, sketch your observations for future reference.

Results

1. Review the structures and processes observed.
2. Using colored pencils, indicate the structures of the pine life cycle in Figure 15.2 that are haploid or diploid, and circle the processes of mitosis, meiosis, and fertilization.

Discussion

1. What is the function of the wings on the pollen grain?

2. Why is wind-dispersed pollen an important phenomenon in the evolution of plants?

3. Are microspores and megaspores produced by mitosis or meiosis?

4. Describe the structure and function of a seed.

5. Can you think of at least two ways in which pine seeds are dispersed?

6. One of the major trends in plant evolution is the reduction in size of the gametophytes. Describe the male and female gametophyte in terms of size and location.

EXERCISE 15.2

Angiosperms

All flowering plants (angiosperms) are classified in the phylum **Anthophyta** (Gk. *anthos,* "flower"). A unique characteristic of angiosperms is the **carpel**, a structure in which **ovules** are enclosed. After fertilization, the ovule develops into a seed (as in the gymnosperms), whereas the carpel matures into a **fruit** (unique to angiosperms). Other important aspects of angiosperm reproduction include additional reduction of the gametophyte, double fertilization, and an increase in the rapidity of the reproduction process.

The **flowers** of angiosperms are composed of male and female reproductive structures, which are frequently surrounded by attractive or protective leaflike structures collectively known as the **perianth** (Figure 15.3). The flower functions both to protect the developing gametes and to ensure **pollination** and **fertilization**. Although many angiosperm plants are self-fertilized, cross-fertilization is important in maintaining genetic diversity. Plants, rooted and stationary, often require transfer agents to complete fertilization. Various insects, birds, and mammals transfer pollen from flower to flower. The pollen then germinates into a pollen tube and grows through the female carpel to deliver the sperm to the egg.

Plants must attract pollinators to the flower. What are some features of flowers that attract pollinators? Color and scent are important, as is the shape of the flower. Nectar and pollen provide nutritive rewards for the pollinators as well. The shape and form of some of the flowers are structured to accommodate pollinators of specific size and structure, providing landing platforms, guidelines, and even special mechanisms for the placement of pollen on body parts. While the flower is encouraging the visitation by one type of pollinator, it also may be excluding visitation by others. The more specific the relationship between flower and pollinator, the more probable that the pollen of that species will be successfully transferred. But many successful flowers have no specific adaptations for particular pollinators and are visited by a wide variety of pollinators. Some plants do not have colorful, showy flowers and are rather inconspicuous, often dull in color, and lack a perianth. These plants are usually wind pollinated, producing enormous quantities of pollen and adapted to catch pollen in the wind (Figure 15.4a).

The origin and diversification of angiosperms cannot be understood apart from the coevolutionary role of animals in the reproductive process. Colorful petals, strong scents, nectars, food bodies, and unusual perianth shapes all relate to pollinator visitation. Major trends in the evolution of angiosperms involve the development of mechanisms to exploit a wide variety of pollinators (Figure 15.4).

In Lab Study A, you will investigate a variety of flowers, observing their shape, structure, and traits that might attract pollinators of various kinds. Following this, in Lab Study B, you will use a key to identify the probable pollinators for some of these flowers. You will follow the life cycle of the lily in Lab Study C and complete the lab by using another key to identify types of fruits and their dispersal mechanisms.

Note: Before beginning Lab Study A, turn to Lab Study C, Procedure step 3, to set up the pollination experiments, as the pollen must incubate for 30 to 60 minutes. Then return to Lab Study A.

Lab Study A. Flower Morphology

Materials

living flowers provided by the instructor and/or students
stereoscopic microscope

Introduction

Working in teams of two students, you will investigate the structure of the flower (Figure 15.3). The instructor will provide a variety of flowers, and you may have brought some with you to lab. You will need to take apart each flower carefully to determine its structure, as it is unlikely that all your flowers will follow the simple diagram used to illustrate the structures. Your observations will be the basis for predicting probable pollinators in Lab Study B.

Procedure

1. Examine fresh flowers of four different species, preferably with different floral characteristics.
2. Identify the parts of each flower using Figure 15.3 and the list provided following the heading Floral Parts. You may be able to determine the floral traits for large, open flowers by simply observing. However, to really understand the flower structure, you should remove the flower parts stepwise, beginning with the outer sepals and continuing toward the center of the flower. To observe the details of smaller flowers and structures, use the stereoscopic microscope. For example, the ovary is positively identified by the presence of tiny crystal-like ovules, and these are best seen with the stereoscopic microscope.
3. In the margin of your lab manual, sketch any flower shapes or structures that you might need to refer to in the future.

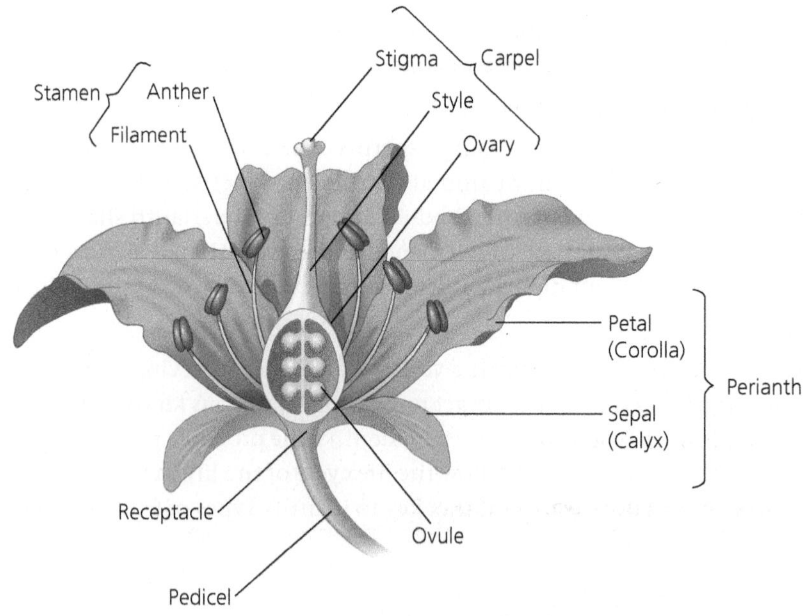

FIGURE 15.3
Flower structures. Determine the structures of flowers in the laboratory by reviewing this general diagram.

4. Record the results of your observations in Table 15.3. You will determine pollinators in Lab Study B.

Floral Parts

Pedicel: stalk that supports the flower.
Receptacle: tip of the pedicel where the flower parts attach.
Sepal: outer whorl of bracts, which may be green, brown, or colored like the petals; may appear as small scales or be petal-like.
Calyx: all the sepals, collectively.
Petal: colored, white, or even greenish whorl of bracts located just inside the sepals.
Corolla: all the petals, collectively.
Perianth: the corolla (petals) and calyx (sepals) all together.
Stamen: pollen-bearing structure, composed of filament and anther.
Filament: stalk that supports the anther.
Anther: pollen-producing structure that terminates the stamen.
Carpel: female reproductive structure, composed of the stigma, style, and ovary, often pear-shaped and located in the center of the flower.
Stigma: receptive tip of the carpel, often sticky or hairy, where pollen is placed; important to pollen germination.
Style: tissue connecting stigma to ovary, often long and narrow, but may be short or absent; pollen must grow through this tissue to fertilize the egg.
Ovary: base of carpel; protects ovules inside, matures to form the fruit.

Results

Summarize your observations of flower structure in Table 15.3.

TABLE 15.3 Flower Morphology and Pollinators

Features	Plant Names			
	1	2	3	4
Number of petals				
Number of sepals				
Parts absent (petals, stamens, etc.)				
Color				
Scent (+/–)				
Nectar (+/–)				
Shape (including corolla shape: tubular, star, etc.)				
Special features (landing platform, guidelines, nectar spur, etc.)				
Predicted pollinator (see Lab Study B)				

Discussion

What structures or characteristics did you observe in your (or other teams') investigations that you predict are important to pollination?

MB Student Media Videos—Ch. 30: Flower Blooming (Time Lapse); Flowering Plant Life Cycle

Lab Study B. Pollinators

Materials

living flowers provided by the instructor and/or students
stereoscopic microscope

Introduction

Flowers with inconspicuous sepals and petals are usually pollinated by wind (Figure 15.4a). Most showy flowers are pollinated by animals. Some pollinators tend to be attracted to particular floral traits, and, in turn, some groups of plants have coevolved with a particular pollination agent that ensures successful reproduction. Other flowers are generalists, pollinated by a variety of organisms, and still others may be visited by only one specific pollinator (Figure 15.4b, c, and d). Based on the floral traits that attract common pollinators (bees, flies, butterflies, and hummingbirds), you will predict the probable pollinator for some of your flowers using a **dichotomous key**. (Remember, *dichotomous* refers to the branching pattern and means "divided into two parts.")

In biology, we use a key to systematically separate groups of organisms based on sets of characteristics. Most keys are based on couplets, or pairs of characteristics, from which you must choose one or the other—thus the term *dichotomous.* For example, the first choice of characteristics in a couplet might be *plants with showy flowers and a scent,* and the other choice in the pair might be *plants with tiny, inconspicuous flowers and no scent.* You must choose one or the other statement. In the next step, you would choose from a second pair of statements listed directly below your first choice. With each choice, you would narrow the group more and more until, as in this case, the pollinator is identified. *Each couplet or pair of statements from which you must choose will be identified by the same letter or number.*

a.

FIGURE 15.4
Flower pollination. (a) Wind-pollinated flowers are inconspicuous and produce enormous quantities of pollen. (b) Flowers pollinated by bees may be irregular in shape and have a landing platform. (c) Hummingbirds pollinate red tubular flowers. (d) Bats pollinate night-blooming flowers with pale sepals and petals.

b.

c.

d.

Key to Pollination

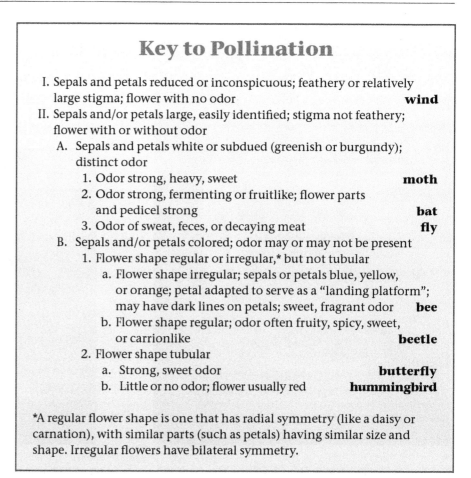

I. Sepals and petals reduced or inconspicuous; feathery or relatively large stigma; flower with no odor **wind**

II. Sepals and/or petals large, easily identified; stigma not feathery; flower with or without odor

  A. Sepals and petals white or subdued (greenish or burgundy); distinct odor

    1. Odor strong, heavy, sweet **moth**
    2. Odor strong, fermenting or fruitlike; flower parts and pedicel strong **bat**
    3. Odor of sweat, feces, or decaying meat **fly**

  B. Sepals and/or petals colored; odor may or may not be present

    1. Flower shape regular or irregular,* but not tubular
      a. Flower shape irregular; sepals or petals blue, yellow, or orange; petal adapted to serve as a "landing platform"; may have dark lines on petals; sweet, fragrant odor **bee**
      b. Flower shape regular; odor often fruity, spicy, sweet, or carrionlike **beetle**
    2. Flower shape tubular
      a. Strong, sweet odor **butterfly**
      b. Little or no odor; flower usually red **hummingbird**

*A regular flower shape is one that has radial symmetry (like a daisy or carnation), with similar parts (such as petals) having similar size and shape. Irregular flowers have bilateral symmetry.

Procedure

Using the Key to Pollination, classify the flowers used in Lab Study A based on their floral traits and method of pollination.

Results

1. Return to Table 15.3 and compare the pollinators you predicted with your results using the key.
2. If you made sketches of any of your flowers, you may want to indicate the pollinator associated with that flower.

Discussion

1. Review the Key to Pollination and describe the characteristics of flowers that are adapted for pollination by each of the following agents:
   a. wind

   b. hummingbird

c. bat

d. fly

2. Discuss with your lab partner other ways in which keys are used in biology. Record your answers in the space provided.

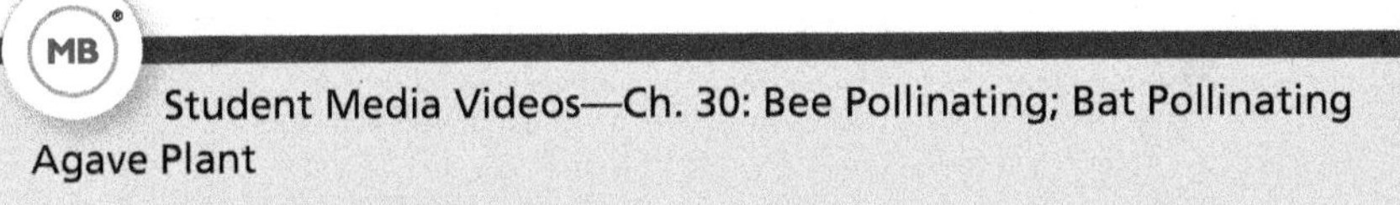

MB
Student Media Videos—Ch. 30: Bee Pollinating; Bat Pollinating Agave Plant

Lab Study C. Angiosperm Life Cycle

Materials

pollen tube growth medium in dropper bottles
dropper bottle of water
petri dish with filter paper to fit inside
prepared slides of lily anthers and ovary
dissecting probe
brush bristles
compound microscope
flowers for pollen—bridal veil, sweet clover, snapdragon and others

Introduction

In this lab study, you will study the life cycle of flowering plants, including the formation of pollen, pollination, fertilization of the egg, and formation of the seed and fruit. You will also investigate the germination of the pollen grain as it grows toward the egg cell.

i

Refer to Figures 15.3 (flower structures) and 15.5 (angiosperm life cycle) as you complete the exercise.

Procedure

1. Pollen grain—the male gametophyte:
   a. Examine a prepared slide of a cross section through the **stamens** of *Lilium*. The slide shows six anthers and may include a centrally located ovary that contains ovules.
   b. Observe a single **anther**, which is composed of four **anther sacs** (microsporangia). Note the formation of **microspores** (with a single nucleus) from diploid **microsporocytes**. You may also see mature **pollen grains** with two nuclei. One nucleus is the **tube nucleus**, and the other is the **generative nucleus**.

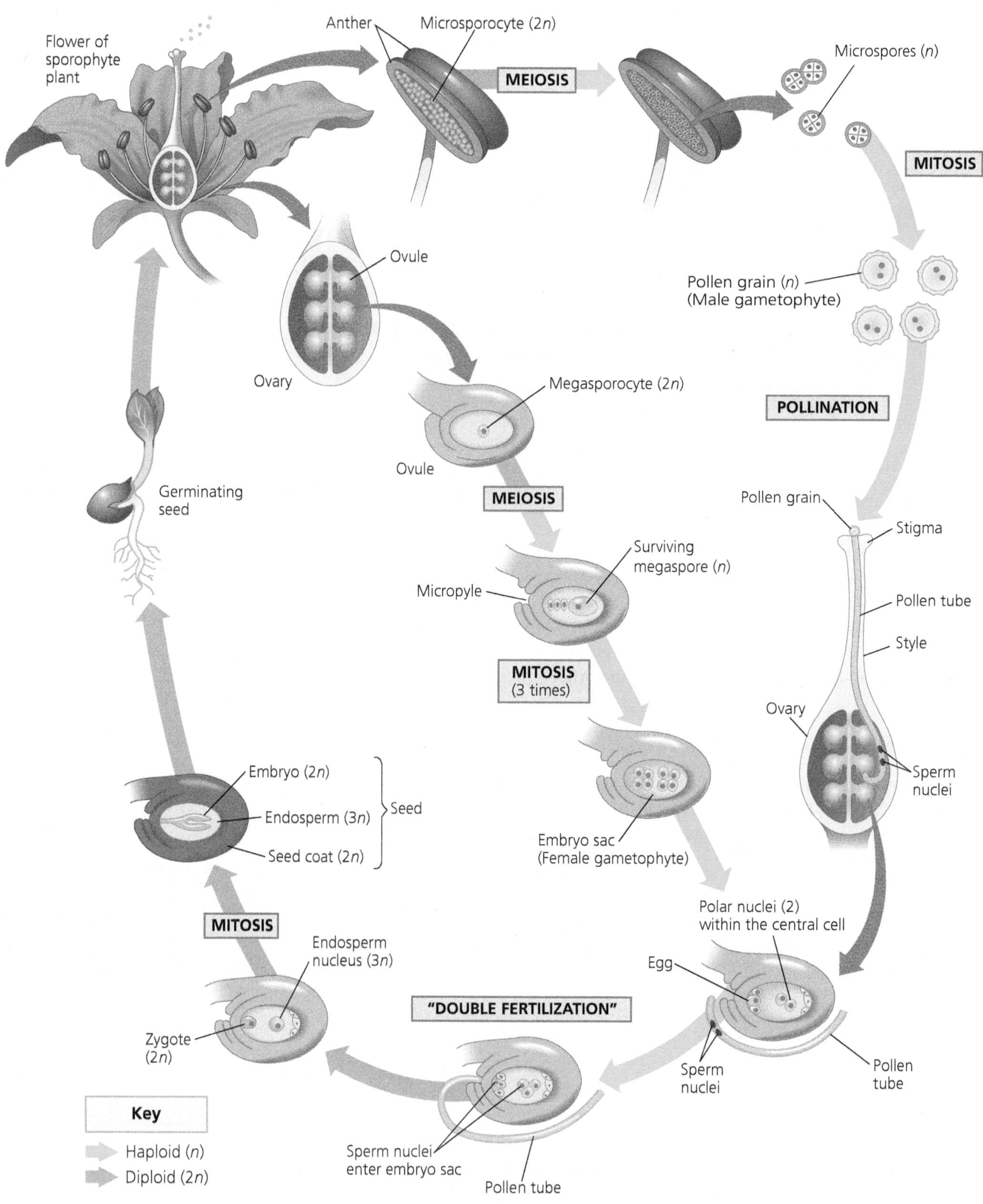

FIGURE 15.5

Angiosperm life cycle. Observe the structures and processes as described in Exercise 15.2. *Using colored pencils, indicate the structures that are haploid or diploid. Circle the terms* mitosis, meiosis, *and* double fertilization.

2. Development of the female gametophyte:

   a. Examine a prepared slide of the *Lilium* ovary and locate the developing ovules. Each **ovule**, composed of the megasporangium and other tissues, contains a diploid **megasporocyte**, which produces **megaspores** (haploid), only one of which survives. The megaspore will divide three times by mitosis to produce the eight nuclei in the **embryo sac**, which is the microscopic female gametophyte. The two nuceli in the center of the embryo sac are the **polar nuclei**. Near the opening to the ovule (**micropyle**), one of the three nuclei is the **egg cell** (Figure 15.6). Note that angiosperms do not even produce an archegonium.

   b. Your slide will not contain all stages of development, and it is almost impossible to find a section that includes all eight nuclei. Locate the three nuclei near the micropyle. One of these is the egg. The two polar nuclei in the center are part of the large **central cell**.

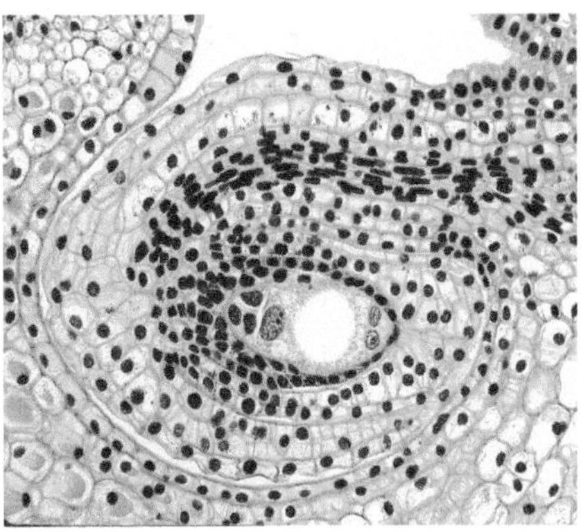

FIGURE 15.6
Ovule containing an eight-nucleate embryo sac. Not all nuclei are visible.

3. Pollination and fertilization:

   When pollen grains are mature, the anthers split and the pollen is released. When pollen reaches the stigma, it germinates to produce a **pollen tube**, which grows down the style and eventually comes into contact with the opening to the ovule. During this growth, the generative nucleus divides into two **sperm nuclei**. One sperm nucleus fuses with the egg to form the **zygote**, and the second fuses with the two polar nuclei to form the triploid **endosperm**, which will develop into a rich nutritive material for the support and development of the embryo. The fusion of the two sperm nuclei with nuclei of the embryo sac is referred to as **double fertilization**. *Formation of triploid endosperm and double fertilization are unique to angiosperms.*

   You will examine pollen tube growth by placing pollen in pollen growth medium to stimulate germination. Pollen from some plants germinates easily; for others a very specific chemical environment is required. Work with a partner, following the next steps.

   a. Using a dissecting probe, transfer some pollen from the anthers of one of the plants available in the lab to a slide on which there are 2 to 3 drops of pollen tube growth medium and a few brush bristles or grains of sand (to avoid crushing the pollen). Add a coverslip. Alternatively, touch an anther to the drop of medium, then add brush bristles and a coverslip.

   b. Examine the pollen under the compound microscope. Observe the shape and surface features of the pollen.

   c. Prepare a humidity chamber by placing moistened filter paper in a petri dish. Place the slide in the petri dish, and place it in a warm environment.

   d. Examine the pollen after 30 minutes and again after 60 minutes to observe pollen tube growth. The pollen tubes should appear as long, thin tubes extending from the surface or pores in the pollen grain (Figure 15.7).

   e. Record your results in Table 15.4 in the Results section. Indicate the plant name and the times when pollen tube germination was observed.

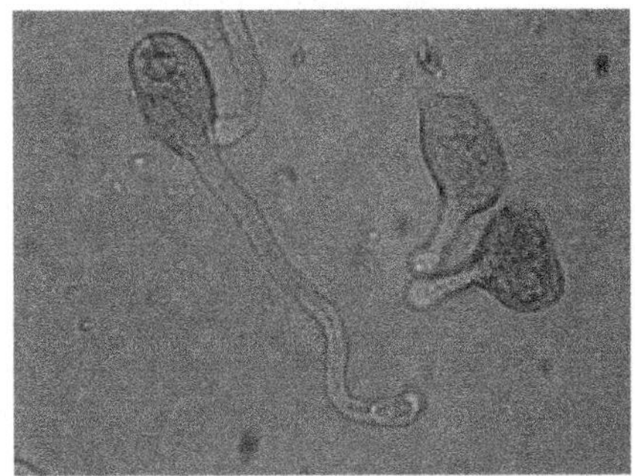

FIGURE 15.7
Pollen tubes germinating in growth medium.

4. Seed and fruit development:

   The zygote formed at fertilization undergoes rapid mitotic divisions, forming the embryo. The endosperm also divides; the mature ovule forms a seed. The **seed** is composed of the sporophyte embryo (2*n*), the nutritive endosperm (3*n*), and the seed coat (2*n*) that forms from the outer sporophyte tissue of the ovule. At the same time, the surrounding ovary and other floral tissues are forming the fruit. In Lab Study D, you will investigate the types of fruits and their function in dispersal.

Results

1. Review the structures and processes observed in the angiosperm life cycle, Figure 15.5. Indicate the haploid and diploid structures in the life cycle, using two different colored pencils.

Having trouble with life cycles? Return to Lab Topic 14 Plant Diversity I, and review the generalized life cycle in Figure 14.2. The key to success is to determine where meiosis occurs and to remember the ploidal level for the gametophyte and sporophyte.

2. Sketch your observations of the slides in the margin of your lab manual for later reference.
3. Record the results of the pollen germination studies in Table 15.4. Compare your results with those of other teams that used different plants. This is particularly important if your pollen did not germinate.

TABLE 15.4 Results of Pollen Germination Studies

Plant Name	30 min(+/–)	60 min(+/–)

Discussion

1. What part of the life cycle is contained in the mature pollen grain?

2. How does the female gametophyte in angiosperms differ from the female gametophyte in gymnosperms?

3. Do you think that all pollen germinates indiscriminately on all stigmas? How might pollen germination and growth be controlled? How could you design an experiment to test your ideas? See Investigative Extensions at the end of the lab.

Lab Study D. Fruits and Dispersal

Materials

variety of fruits provided by the instructor and/or students

Introduction

The **seed** develops from the ovule, and inside the seed coat is the embryo and its nutritive tissues. The **fruit** develops from the ovary or from other tissues in the flower. It provides protection for the seeds, and both the seed and the fruit may be involved in dispersal of the sporophyte embryo.

Procedure

1. Examine the fruits and seeds on demonstration.
2. Use the Key to Fruits on the next page to help you complete Table 15.5. Remember to include the dispersal mechanisms for fruits and their seeds in the table.

Results

1. Record in Table 15.5 the fruit type for each of the fruits keyed. Share results with other teams so that you have information for all fruits in the lab.
2. For each fruit, indicate the probable method of dispersal—for example, wind; water; gravity; ingestion by birds, mammals, or insects; or adhesion to fur and socks.
3. For some fruits, the seeds rather than the fruit are adapted for dispersal. In the milkweed, for example, the winged seeds are contained in a dry ovary. Indicate in Table 15.5 if the seeds have structures to enhance dispersal. Recall that seeds are inside fruits. The dandelion "seed" is really a fruit with a fused ovary and seed coat.

TABLE 15.5 Fruit Types and Dispersal Mechanisms

Plant Name	Fruit Type	Dispersal Method

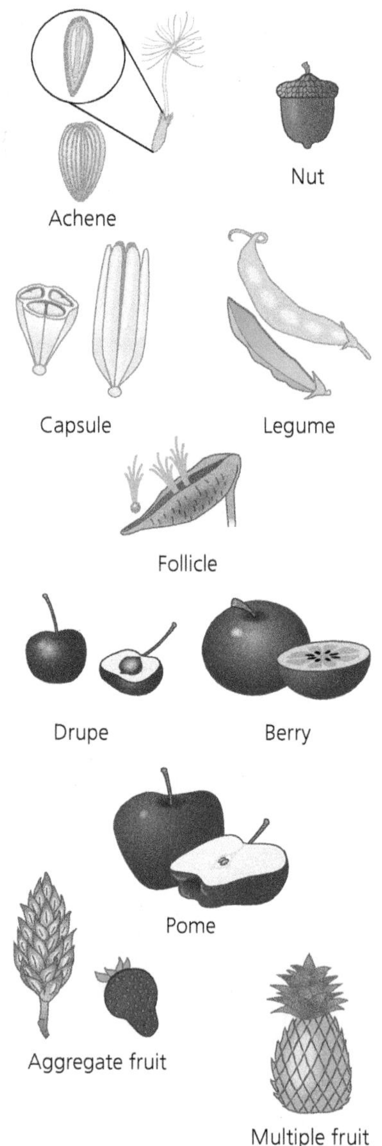

Key to Fruits

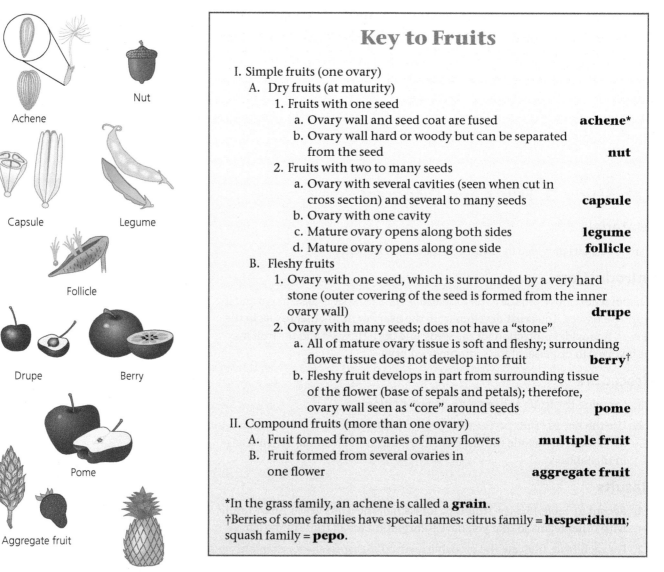

I. Simple fruits (one ovary)
- A. Dry fruits (at maturity)
  - 1. Fruits with one seed
    - a. Ovary wall and seed coat are fused — **achene***
    - b. Ovary wall hard or woody but can be separated from the seed — **nut**
  - 2. Fruits with two to many seeds
    - a. Ovary with several cavities (seen when cut in cross section) and several to many seeds — **capsule**
    - b. Ovary with one cavity
    - c. Mature ovary opens along both sides — **legume**
    - d. Mature ovary opens along one side — **follicle**
- B. Fleshy fruits
  - 1. Ovary with one seed, which is surrounded by a very hard stone (outer covering of the seed is formed from the inner ovary wall) — **drupe**
  - 2. Ovary with many seeds; does not have a "stone"
    - a. All of mature ovary tissue is soft and fleshy; surrounding flower tissue does not develop into fruit — **berry**†
    - b. Fleshy fruit develops in part from surrounding tissue of the flower (base of sepals and petals); therefore, ovary wall seen as "core" around seeds — **pome**

II. Compound fruits (more than one ovary)
- A. Fruit formed from ovaries of many flowers — **multiple fruit**
- B. Fruit formed from several ovaries in one flower — **aggregate fruit**

*In the grass family, an achene is called a **grain**.
†Berries of some families have special names: citrus family = **hesperidium**; squash family = **pepo**.

Discussion

1. How might dry fruits be dispersed? Fleshy fruits?

2. Describe the characteristics of an achene, drupe, and berry.

REVIEWING YOUR KNOWLEDGE

1. Complete Table 15.6. Compare mosses, ferns, conifers, and flowering plants relative to sexual life cycles and adaptations to the land environment. Return to Table 15.1 and modify your entries.

TABLE 15.6 Comparison of Important Characteristics of Plants

Features	Moss	Fern	Conifer	Flowering Plant
Gametophyte or sporophyte dominant				
Water required for fertilization				
Vascular tissue (+/–)				
Homosporous or heterosporous				
Seed (+/–)				
Pollen grain (+/–)				
Fruit (+/–)				
Examples				

2. All seed plants are heterosporous and produce pollen grains as well as seeds. However, there are examples of transitions between seedless vascular plants (lack seeds; generally homosporous) and seed plants. What seedless vascular plants are heterosporous?

What groups of seed plants produce pollen grains, but also have flagellated sperm that swim from the pollen tube into the ovule to fertilize the egg?

3. Identify the function of each of the following structures found in seed plants. Consider their function in the land environment.

pollen grain:

microsporangium:

flower:

carpel:

seed:

fruit:

endosperm:

4. Plants have evolved a number of characteristics that attract animals and ensure pollination, but what are the benefits to animals in this relationship?

5. Why is internal fertilization essential for true terrestrial living?

6. Review the introduction to Lab Topic 14 Plant Diversity I, and describe the major trends in the evolution of plants.

APPLYING YOUR KNOWLEDGE

1. Explain how the rise in prominence of one major group (angiosperms, for example) does not necessarily result in the total replacement of a previously dominant group (gymnosperms, for example).

2. In 1994, naturalists in Australia discovered a new genus and species of conifer, *Wollemia nobilis*, growing in a remote area not far from the city of Sydney. This was the first new conifer discovered since 1948. Wollemi pine (not really a pine) is in a family that had a global distribution 90 million years ago in the Cretaceous period. Scientists used a variety of different evidence to decide where this rare tree should be placed in the "Tree of Life." What evidence would be necessary to determine that this is in the phylum Coniferophyta?

   To support the conservation of this ecosystem and three groups of trees (less than 100 individuals), botanical gardens are propagating these trees for sale. Although seeds have been collected, the commercially available plants are from cuttings and tissue cultures. Based on your understanding of the life cycle of conifers, why is it not practical to reproduce Wollemi pines from seed or even to sell the seeds?

3. Your neighbor's rose garden is being attacked by Japanese beetles, so she dusts her roses with an insecticide. Now, to her dismay, she realizes that the beans and squash plants in her vegetable garden are flowering, but are no longer producing vegetables. She knows beetles feed on leaves of roses and squash plants. What is the problem? Explain to your neighbor the relationship among flowers, fruits (vegetables, in the gardening language), and insects.

4. In December 2012, Lindenmayer and colleagues reported in the journal *Science* on the global decline in large old trees. They describe the significant contributions that large old trees make in terms of ecosystem function and the factors responsible for the decline in these trees globally. For example, the density of large trees in Yosemite National Park declined by 24% in the 60 years leading up to 1990. In Sweden, the density of large old trees has gone from 19 per hectare to 1 per hectare. Review the article and determine the "critical ecological role" particular to large trees, as well as the threats that are responsible for decline (Lindenmayer, Laurance, & Franklin, 2012.).

5. Seed plants provide food, medicine, fibers, beverages, building materials, dyes, and psychoactive drugs. Using Web resources, your textbook, and library references, describe examples of human uses of plants in Table 15.7. Indicate whether each example is a gymnosperm or angiosperm. Based on your research, what is the relative economic importance of angiosperms and gymnosperms?

TABLE 15.7 Uses of Seed Plants: Angiosperms and Gymnosperms

Uses of Plants	Example	Angiosperm/ Gymnosperm
Food		
Beverage		
Medicine		
Fibers		
Materials		
Dyes		

INVESTIGATIVE EXTENSIONS

Pollen germinates easily in the laboratory for some species and not at all for others. In some species, a biochemical signal is required from the stigma to initiate germination. What are the advantages to the species if pollen germinates easily or if it requires a biochemical signal? What external and internal factors might affect pollen germination and pollen tube growth? Consider the flowers available in the laboratory that did not germinate using the growth medium and the basic procedure in Lab Study C, Procedure step 3. Design your own independent investigation to study factors that affect pollen germination based on your observations in this lab topic and additional research, or consider one of the following suggestions.

1. Are factors present in the stigma necessary for pollen germination? Mince a small piece of the stigma in sucrose and then add it to a slide with pollen in pollen growth medium. (The sucrose concentration in the growth medium can also be varied, as this may affect pollen germination).
2. Will pollen from closely related species germinate more effectively than pollen from distantly related species? For example, try different species of the Mustard family or even different varieties of one species of *Brassica*. Remember that Fast Plants (*B. rapa*) are mustards, as is *Arabidopsis,* another plant used in genetic studies.
3. What essential micronutrients are needed for pollen germination? Some species are sensitive to micronutrients in the growth medium, including boron and calcium. Research various growth media and test these with flowers that failed to germinate. Which of the micronutrients in the growth medium is necessary for pollen germination in your plants? Prepare growth medium omitting one of each of the components and observe pollen germination.
4. How do environmental factors, such as temperature and light, affect the rate of pollen tube growth?

STUDENT MEDIA: BioFlix, Activities, Investigations, and Videos

www.masteringbiology.com (select Study Area)

Activities—Ch. 29: Terrestrial Adaptations of Plants; Highlights of Plant Phylogeny; Ch. 30: Pine Life Cycle; Angiosperm Life Cycle; Ch. 38: Seed and Fruit Development

Investigations—Ch. 30: How Are Trees Identified by Their Leaves? Ch. 38: What Tells Desert Seeds When to Germinate?

Videos—Ch. 30: Flower Blooming Time Lapse; Time Lapse of Flowering Plant Life Cycle; Bee Pollinating; Bat Pollinating *Agave* Plant; Ch. 38: Seed and Fruit Development

REFERENCES

Armstrong, J. E. *How the Earth Turned Green: A Brief 3.8-Billion-Year History of Plants*, Chicago: University of Chicago Press, 2014.

Evert, R. F. and S. E. Eichhorn. *Raven Biology of Plants*, 8th ed. New York: W.H. Freeman, 2013.

Kessler, R. and M. Harley. "Diamonds of the Dust." *Natural History*, 2014, vol. 122, pp. 28–33.

Levetin, E. and K. McMahon. *Plants and Society,* 7th ed. New York: McGraw Hill Co., 2015.

Lindenmayer, D. B., W. F. Laurance, and J. F. Franklin. "Global Decline in Large Old Trees." *Science*, 2012, vol. 338, pp. 1305–1306.

Mauseth, J. D. *Botany: An Introduction to Plant Biology,* 6th ed. Sudbury, MA: Jones and Bartlett Publishers, 2016.

McLoughlin, S. and V. Vajda. "Ancient Wollemi Pines Resurgent." *American Scientist,* 2005, vol. 93, pp. 540–547.

Niklas, K. J. *Plant Evolution: An Introduction to the History of Life*, Chicago: University of Chicago Press, 2016.

Rui, M. *The Pollen Tube: A Cellular and Molecular Perspective.* New York: Springer, 2006.

Taylor, L. P. and P. K. Hepler. "Pollen Germination and Tube Growth." *Annual Review of Plant Physiology and Plant Molecular Biology,* 1997, vol. 48, pp. 461–491.

WEBSITES

American Institute of Biological Science website, *action-bioscience*. Original interview with Pam Soltis, "Flowering Plants: Keys to Earth's Evolution and Human Well-Being," plus extensive website links to plant resources: http://www.actionbioscience.org/genomics/soltis.html

Images, maps, and additional links for plants of North America: http://plants.usda.gov/

Plant Conservation Alliance site with projects, medicinal and other plant uses: http://www.nps.gov/plants/

Society for Economic Botany: http://www.econbot.org/

TED Talks with amazing video of flowers and pollination. "Every pollen grain has a story": https://www.ted.com/talks/jonathan_drori_every_pollen_grain_has_a_story?language=en

"The beautiful tricks of flowers": https://www.ted.com/talks/jonathan_drori_the_beautiful_tricks_of_flowers?language=en

Tree of Life Project, includes phylogeny of living organisms, movies, references, current research, and ideas for independent investigations: http://tolweb.org/tree?group=Spermatopsida&contgroup=Embryophytes

University of California Berkeley Museum of Palentology site with images and resources for both living and fossil seed plants: http://www.ucmp.berkeley.edu/plants/plantae.html

University of Michigan Dearborn site for the uses of plants by Native Americans: http://naeb.brit.org/

LAB TOPIC 16

Bioinformatics: Molecular Phylogeny of Plants

Laboratory Objectives

After completing this lab topic, you should be able to:

1. Describe a phylogenetic tree and explain the relationships depicted in a tree using appropriate terminology.
2. Describe the evolutionary history (phylogeny) of plants.
3. Construct a phylogenetic tree using morphological evidence.
4. Use the tools of bioinformatics to test hypotheses in the form of phylogenetic trees.
5. Describe the steps used in the construction of phylogenetic trees based on molecular data.
6. Analyze results, and write an illustrated report, using critical thinking skills and synthesizing data from a variety of sources.
7. Describe evolutionary relationships, including common ancestry and changes over time in lineages.
8. Discuss the importance of bioinformatics and molecular evidence in evolutionary analysis and in other areas of science.

Introduction

In Lab Topic 14 Plant Diversity I and Lab Topic 15 Plant Diversity II, you investigated the diversity of the plant kingdom, comparing representative plants from the major phyla. You studied vegetative and reproductive structures, observed fossils, and investigated moss, fern, pine, and flowering plant life cycles. During your investigations, you described the evolutionary relationships of the phyla as you determined shared ancestral and shared derived traits for these groups of plants. Plants have been classified into 10 phyla based on their similarities and differences. Morphological and biochemical features shared by all plants provide evidence that the closest living relatives to plants are complex green algae, the charophytes. With advances in biotechnology in the 1970s, the analysis and sequencing of DNA provided additional genetic evidence for this relationship. In 1994, the Green Plant Phylogeny Research Coordination Group was formed at the University of California, Berkeley, to facilitate the efforts of various research groups working on the study of genetic relationships between plants and their algal ancestors. This began the "Deep Green" project, a large-scale collaborative study of plant evolution based on examination of nuclear and chloroplast gene sequences of algae and plants. Recent advances in faster and more efficient methods of sequencing the whole genome of organisms has inspired collaborative efforts across all taxa including plants. Recent collaborations include, Assembling the Tree of Life, oneKP, and Genome 10K. The study of the evolutionary history of organisms (phylogeny) has expanded in the last decade as scientists use molecular evidence to test our understanding of these relationships for all major groups of organisms.

In this lab topic you will use your previous laboratory experience and knowledge to develop a hypothesis for the evolutionary relationships among plants and their green algal relatives. You will test your hypothesis by comparing molecular evidence from DNA sequences of a chloroplast gene using the tools of bioinformatics (the discipline connecting computer science and biology). You will construct a phylogenetic tree that represents plant evolutionary history. Before leaving the laboratory, you will prepare a final report incorporating your hypothesis, results, and conclusions.

Understanding Phylogenetics

Phylogenetics is the study of evolutionary relationships among groups of living and extinct organisms on Earth. Scientists analyze structural, reproductive, physiological, or molecular changes in specific characters. In particular, they utilize **homologous** characteristics, those features that are similar as a result of shared ancestry (common descent). Molecules, including DNA and proteins, can also reveal homology, as the DNA sequences (and the information for protein sequences) are inherited from one generation to the next.

Phylogenetic systematics is the field of biology that examines morphological characteristics, biochemical pathways, and gene sequences to establish relationships among groups of organisms. To show evolutionary relationships, a **phylogenetic tree** (a visual representation of the lineages among organisms) is constructed of pairs of branching points or dichotomies depicting two groups or lineages that share a common ancestor. *All phylogenetic trees are hypotheses based on available evidence that are to be tested, modified, and tested again.* One type of phylogenetic tree, known as a **rooted tree**, contains a root, nodes, branches, and clades (Figure 16.1). The **root** of the tree represents a common ancestor from which all the organisms in the group are derived. Within the tree are several **nodes**. A node represents a branching point for a taxonomic unit (such as group, phylum, genus, or species). **Branches**, the lines that extend from nodes, establish how closely related one group is to the other. Groups sharing a node share a common ancestor and make up a **clade**. The branching pattern of the tree reflects the number of changes or differences that have accumulated among the species or groups. For example, in

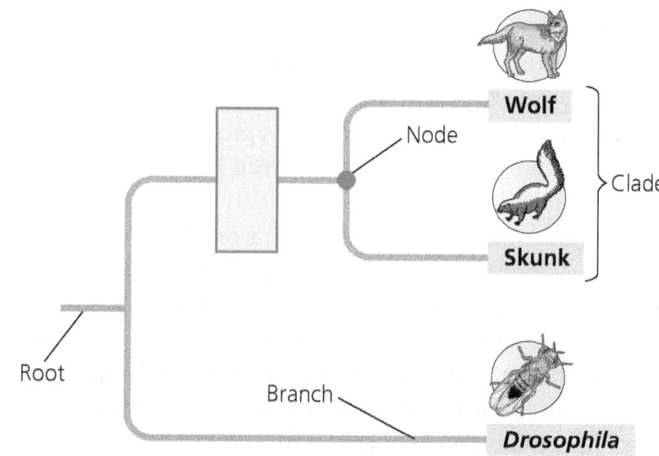

FIGURE 16.1
Terminology associated with a phylogenetic tree. In the box on the upper branch, write an example of a derived trait that all mammals in this clade share in common.

Figure 16.1, wolves, skunks, and *Drosophila* (fruit flies) are all animals, and therefore their **shared ancestral traits** are eukaryotic, multicellular organisms that consume food. But wolves and skunks share some traits that are not shared by *Drosophila.* These shared traits (for example, hair and milk production), which have arisen since they diverged from other animals, are known as **derived traits** and determine the specific clade—mammals—of the organisms that share them. *In the box in Figure 16.1, write in an example of a derived trait that all mammals in this clade share in common.* A phylogenetic tree shows the patterns of ancestry and descent, but not the actual ages or times. From this simple phylogenetic tree we can conclude that wolves and skunks are more closely related to each other and that they share a common ancestor more recently than they do with *Drosophila.* Note that these patterns of descent do not suggest that one species on a branch comes from another on the adjacent branch. For example, wolves are not descended from skunks, but rather, they are both descended from a common mammalian ancestor that is now extinct.

Phylogenetic trees can be constructed using a variety of different evidence and methods to establish the nodes, branches, and topology of the tree. Refer to your textbook (see *Campbell Biology,* Chapter 26: Phylogeny and the Tree of Life) for further information on methodologies.

Molecular Phylogenetics

Rapid technological advances in molecular biology have allowed scientists to obtain DNA sequences of genes and complete and partial genomes of a large variety of organisms. **Molecular phylogenetics** is the field that examines evolutionary relationships among groups specifically based on changes in DNA and protein structure. The number of differences in DNA sequences between different groups reflects the accumulation of mutations over the time since they shared a common ancestor. For example, because wolves and skunks share a common ancestor more recently, then they would have fewer nucleotide differences than between wolves and *Drosophila*. Phylogenetic trees can be constructed to illustrate these differences in DNA sequences for groups of organisms.

Four steps are important in the construction of phylogenetic trees based on molecular data (NCBI, 2004).

1. **Align** similar DNA sequences from different groups to detect similarities and differences in nucleotide bases. Sequences are compared nucleotide by nucleotide taking into account the potential for deletions and insertions (Figure 16.2).
2. **Establish sequence variation** by observing the level of homology or similarity of sequences among groups.
3. **Build a tree** by arranging groups based on the percentage of matching bases for sequences and other factors.
4. **Evaluate the tree,** including the analysis of the resulting tree and comparison with trees constructed with nonmolecular data.

Using DNA sequences in phylogenetics can generate very large data sets. Imagine comparing 10 different species for a gene sequence with 100 nucleotides and the possibility of 4 different nucleotides at each position. To cope with such huge amounts of data, scientists use the tools of bioinformatics to construct molecular phylogenies.

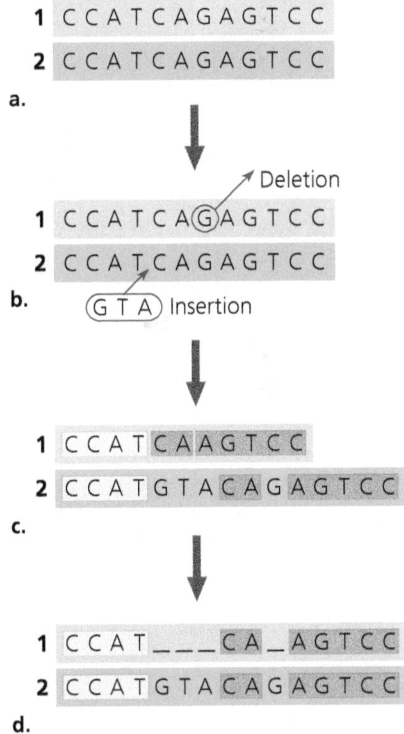

FIGURE 16.2

DNA sequence alignment. (a) Species 1 and 2 have homologous DNA sequences from a shared ancestor. (b) Over time mutations occur, including deletions and insertions. (c) In comparing the sequences some areas align and others do not due to mutations. (d) Using computational software the sequence gaps are added and the shared sequences are aligned.

Using Bioinformatics

Bioinformatics is the field of science combining the disciplines of biology, computational biology, and computer science. Bioinformatics uses computer algorithms (a clear set of instructions for solving a problem) to conduct analyses of large biological data sets. Applications include aligning DNA and protein sequences, predicting protein structure, creating restriction enzyme maps for a DNA sequence, and predicting translation products. Many of these computer algorithms are now available for public access and have become a staple analysis tool for molecular biologists. One important website is the **National Center for Biotechnology Information (NCBI)** established by the National Library of Medicine at the National Institutes of Health. NCBI manages GenBank®, an international database and collection of all publicly available DNA sequences, which number over 213 billion nucleotide bases from over 490,000 species as of June 2016. NCBI also provides links to tools for molecular data analysis such as BLAST, a program used to find DNA sequences with similarity to an unknown sequence. Another excellent resource for bioinformatics applications is **Biology WorkBench**, a Web interface that allows rapid access to biological databases, such as GenBank®, and analysis tools, such as BLAST and ClustalW, a program to align multiple sequences.

In the following laboratory exercise, you will use bioinformatics tools in Biology WorkBench to construct a phylogenetic tree or gene tree using DNA sequences of plants and green algae.

Arabidopsis

Chara

EXERCISE 16.1

Establishing Molecular Phylogenetic Relationships Among Rubisco Large Subunit Genes

Materials

computers with Internet access, one per pair of students
2 computer desktop folders: "images" and "rbcL nucleotide seq" (available at www.masteringbiology.com in the Study Area under Lab Media and then select *Investigating Biology* Lab Materials)
printer access
living examples of *Arabidopsis, Chara, Equisetum, Lilium, Marchantia, Pinus, Polypodium, Polytrichum,* and *Zamia,* if available
24 sheets of large paper (11" × 17")
botany reference books

Introduction

The ribulose bisphosphate carboxylase (Rubisco) protein is essential to carbon fixation in photosynthesis and is found in green algae and all plants. Therefore, it is an ideal choice to establish phylogenetic relationships among green algae and plants using nucleotide sequences. The gene sequence for the large subunit (rbcL) of the Rubisco protein has been isolated from a large variety of algae and plants. The DNA sequences of rbcL genes for *Arabidopsis, Chara, Equisetum, Lilium, Marchantia, Pinus, Polypodium, Polytrichum,* and *Zamia* are available in GenBank.

Working in pairs, you will use your knowledge of plant morphology and reproduction and laboratory experience from previous lab topics to develop a phylogenetic tree that depicts the evolutionary relationships for the organisms listed in the previous paragraph. *This tree will serve as your hypothesis, which you will test using molecular data for the rbcL nucleotide sequences.* You will access the supercomputer, which will use the tools of bioinformatics to compare the sequences for rbcL genes for each of the organisms, then perform an alignment of sequences and construct a phylogenetic tree reflecting the relationships among these genes and groups.

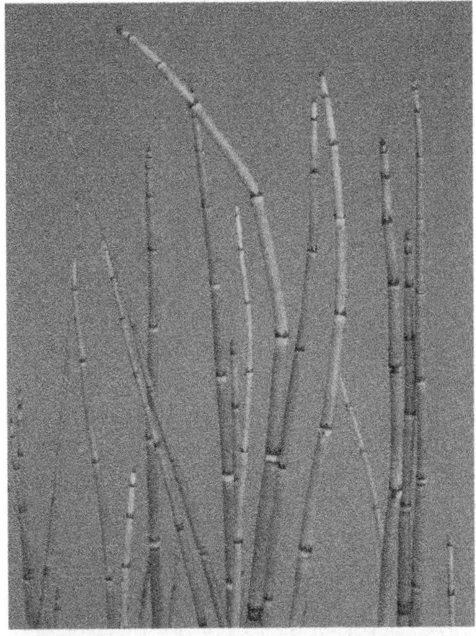

Equisetum

Lilium

Marchantia

Developing Hypothesis and Prediction

Examine the organisms selected for this exercise and determine which are charophytes, bryophytes, monilophytes, gymnosperms, or angiosperms. The plants may be on demonstration in the laboratory and images are provided in a folder on the desktop of your computer and in the margin of your lab manual. Based on your knowledge of plant diversity and the *ancestral* and *derived* characteristics of the major phyla of plants, develop a hypothesis that is consistent with the evidence. Follow these steps in developing your hypothesis for the phylogenetic relationships of the organisms in the laboratory. Refer to Table 15.6 to review morphological and life cycle characteristics of plants studied in Lab Topic 14 Plant Diversity I and Lab Topic 15 Plant Diversity II. Consult your textbook for an overview of plant morphology and diversity.

- Closely observe the plants on demonstration in the laboratory and the images provided for this lab topic. Compare these with the plant groups in Lab Topic 14 Plant Diversity I and Lab Topic 15 Plant Diversity II. How many different plant groups (taxa) are represented by these plants?
- For these major groups of plants (charophytes, bryophytes, monilophytes, gymnosperms, and angiosperms), what are the ancestral and derived traits?
- Consider how you might arrange these groups. Which groups are more closely related to each other? What derived traits do they share?
- Consider which clades might share a common ancestor and group those clades together into a larger branching pattern. What ancestral traits do these share in common?

Hypothesis—Drawing a Phylogenetic Tree

Based on the hypothesized relationships, draw a phylogenetic tree showing appropriate branching to reflect where each organism would be placed relative to the others. The tree you draw is a visual representation of your hypothesis for plant relationships, which is consistent with diverse sources of evidence: morphology, vegetative and reproductive structures, variations in life cycles, and fossils. Refer to the Introduction at the beginning of this lab topic to review terminology and concepts.

(Note: This tree will be referred to as the "morphological" tree, even though other evidence was included in its development.)

- Make preliminary sketches and a final drawing on the large sheets of paper available in the laboratory.
- Use your laboratory manual (Lab Topics 14 and 15) and your textbook (*Campbell Biology*, 11th edition, Chapters 26, 29, and 30) to guide you in constructing your tree.
- Begin by identifying two groups that are more closely related to each other. What derived traits do they share? Draw these as a clade.

Pinus

Polypodium

Polytrichum

Zamia

- Consider which of the remaining groups are closely related to the clade you just formed. Group these together into a branching pattern to make a larger clade. Review clades, nodes, and branch patterns in Figure 16.1.
- At each step, identify traits that are shared within the group as you build your tree until all the organisms have been added into your tree.
- At the node for each clade, ask yourself what characteristic(s) do these groups share in common? Write this shared derived trait at the node for each successive clade.
- What is the *root* of your tree?
- Your tree should have paired branches or sister taxa. Review your tree and make changes if you have included *three* branches extending from one node. Consider which two are most closely related and revise your drawing.
- Label the **root**, **clades**, **nodes**, and **branches**.

Prediction

You will be testing your hypothesis that the morphological tree accurately represents plant phylogeny. You will use molecular data from the nucleotide sequences of rbcL. Do you think the molecular evidence will support (be consistent with) your hypothesis or falsify it? Write your prediction below.

Using Biology WorkBench

Imagine how many comparisons are possible for nine species, comparing changes at each nucleotide position in rbcL DNA! To make this possible, you will use Biology WorkBench with access to the San Diego Supercomputer Center, which will compare and align the nucleotide sequences for all of your organisms using the multiple sequence alignment program, ClustalW. The supercomputer will then construct phylogenetic trees using a mathematical algorithm to select the one phylogeny from those possible that is best explained by the molecular data. *Remember that the more similar two sequences are, the more closely related they are (the more recent their common shared ancestor). The more nucleotide changes that occur, the more time since two groups shared a common ancestor, and therefore they are more distantly related.* There are many computer programs that use different algorithms for analyzing these large data sets. In this exercise you will use a multiple sequence analysis program.

You and your lab partner will do the evaluation and analysis of results, comparing the molecular phylogenetic tree to your hypothesized phylogenetic tree, which was based on morphology and life cycle features.

> While using Biology WorkBench, do not use the *"Back" button* on your Web browser. Use only the "Return" or "Abort" keys on the WorkBench menus. *Using the "Back" button could overload the system leading to the loss of your work.*

Procedure

1. Open a new Word document. Next, open a Web browser and type http://workbench.sdsc.edu/ to enter the Biology WorkBench website.

   Note that during this exercise you will move back and forth between these two windows.

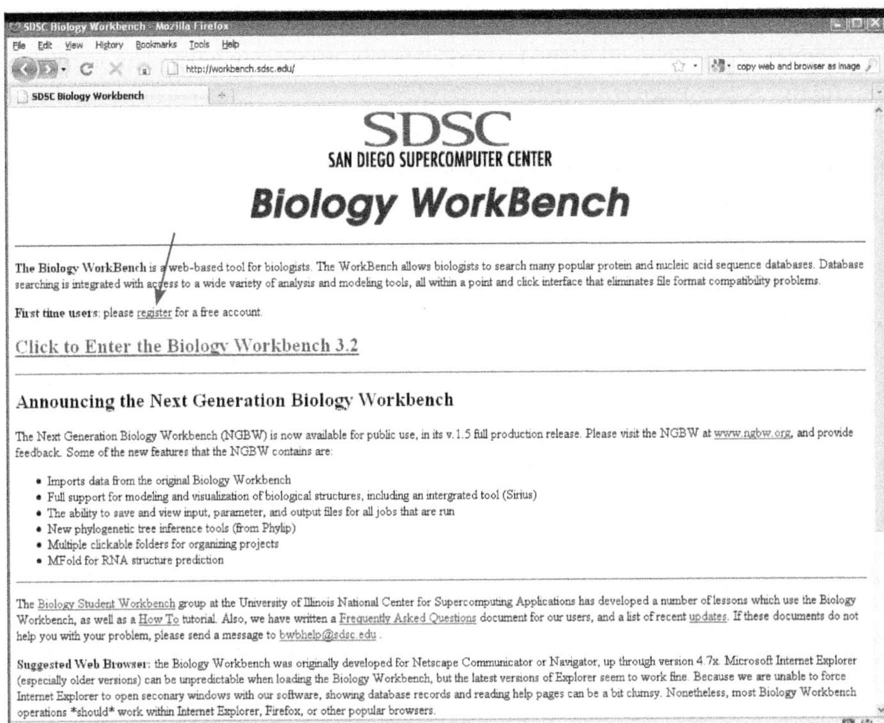

FIGURE 16.3
Biology WorkBench home page. Register and enter WorkBench.

2. To acquire access to Biology WorkBench, you must register your team. Click on the link "register" under first-time users (Figure 16.3). Supply your name (choose one member of the team) and your e-mail address. Select a user ID and a password that is easy to remember; *be sure to write it down.* Click on the "Register" button. Only one member of the team should register.
3. After you register, a new dialog box will appear on your screen asking for your user ID and password; enter the user ID and password you just selected. This will allow you to enter the WorkBench site. Alternatively, if you return to the main page, click on "Enter the Biology WorkBench 3.2" (Figure 16.3), which will prompt you for the user ID and password.
4. When you enter Biology WorkBench, the first page describes the Web tools and provides an introduction. Scroll toward the end of the page and select "Session Tools" (Figure 16.4). This will lead you to a new page entitled "Default Session."

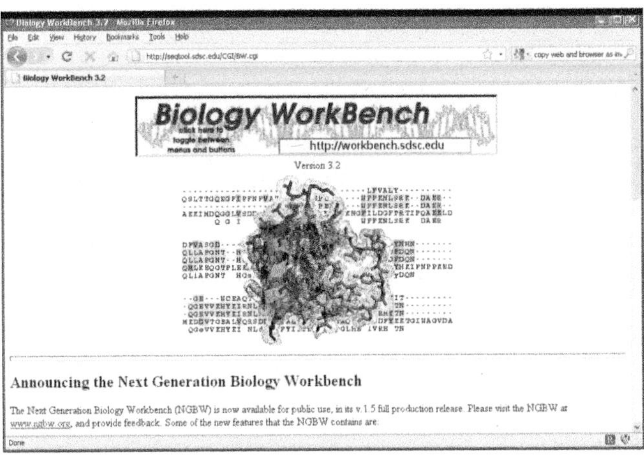

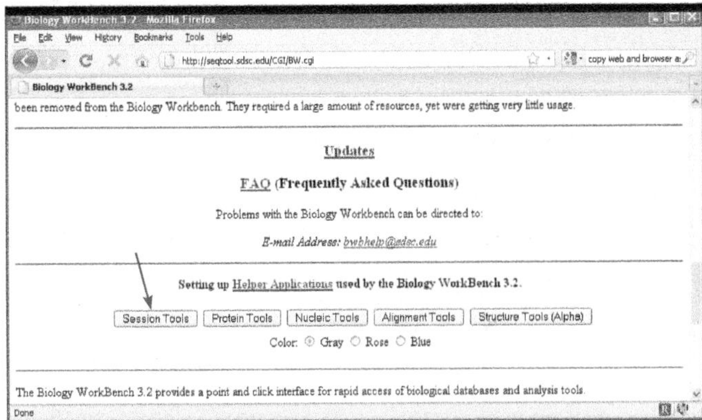

FIGURE 16.4
Biology WorkBench. Selecting "Session Tools."

5. Using the scroll menu in the middle of the page, select "Start New Session" (Figure 16.5). Then click the button "Run" immediately below the scroll menu. A new page will open.

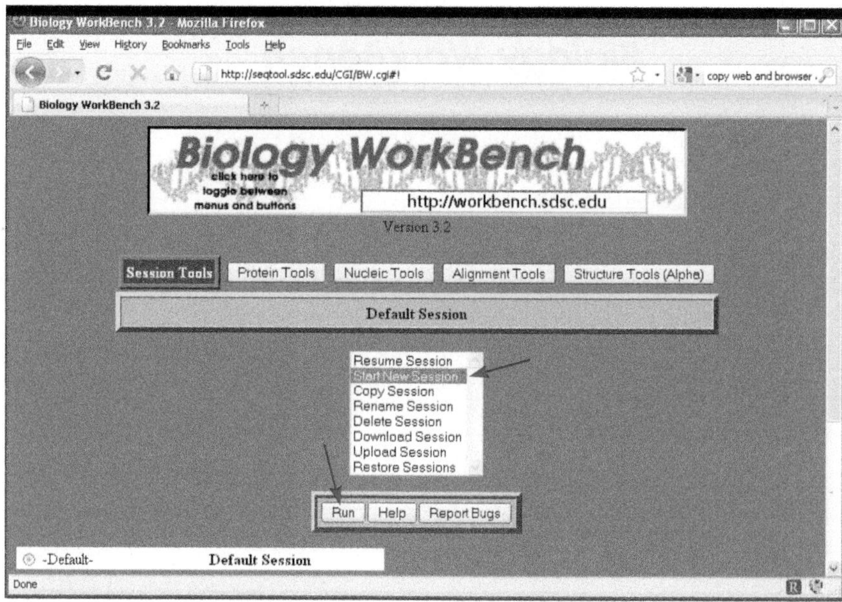

FIGURE 16.5
Biology WorkBench. Starting a new session.

Select a name for your session—such as "molecular phylogeny lab"—and type this in the box labeled "Session Description." Click the button "Start New Session" right below the session description (Figure 16.6).

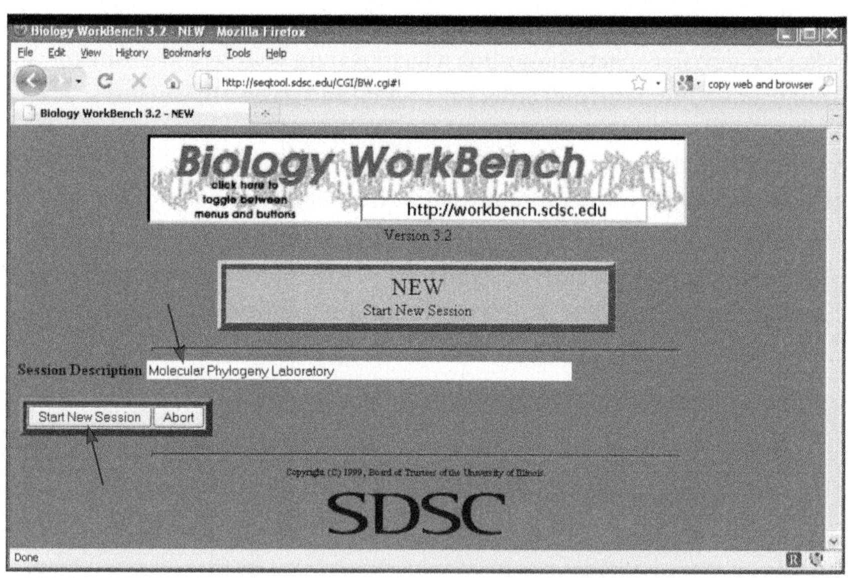

FIGURE 16.6
Biology WorkBench. Entering a session description.

6. A new page will open with your new session listed (Figure 16.7). Next, select "Nucleic Tools" from the menu at the top. After you select this option, a new page will appear with a scroll menu in the middle of the page.

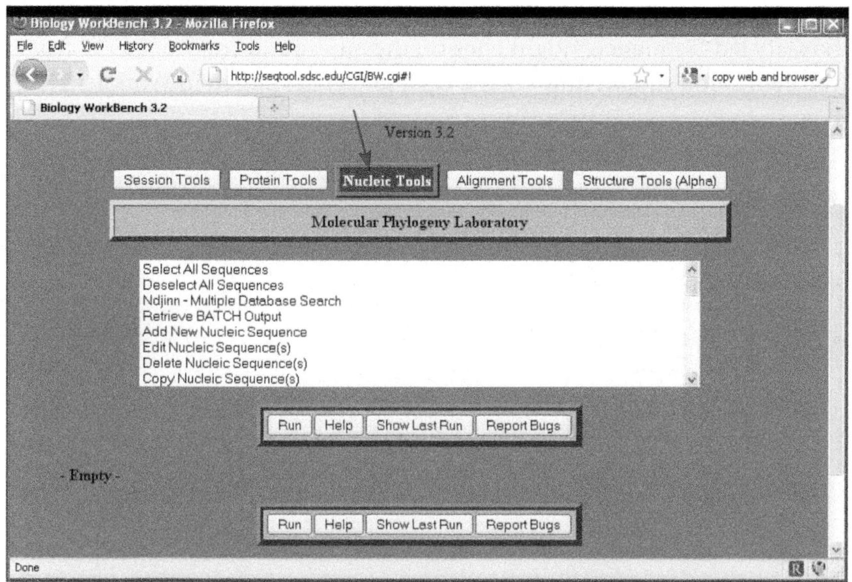

FIGURE 16.7
Biology WorkBench. Selecting "Nucleic Tools."

7. Minimize (but do not close) the Biology WorkBench window. *(If you accidentally close the WorkBench window, follow the steps to log in again.)* Locate the Word file labeled "rbcL nucleotide seq" on your computer desktop. Open the document, then go to "Edit" and "Select All" to highlight the entire text of the document. Right click on the mouse and select "Copy."
8. Minimize the Word document and return to the Biology WorkBench window. Select "Add New Nucleic Sequence" on the scroll menu in the middle of the page (Figure 16.8). Click on "Run."

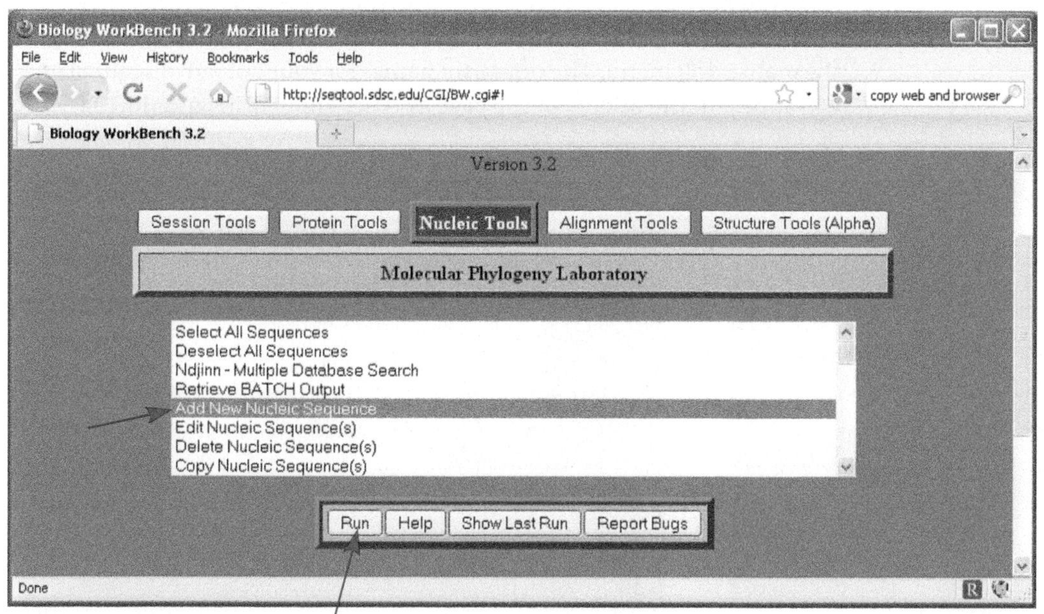

FIGURE 16.8
Biology WorkBench. Adding a new nucleic acid sequence.

A new page requesting input of new sequences will open. Click on the large box labeled "Sequence." Right click on the mouse and choose "Paste" from the menu. You should now see the DNA sequences pasted in WorkBench (Figure 16.9).

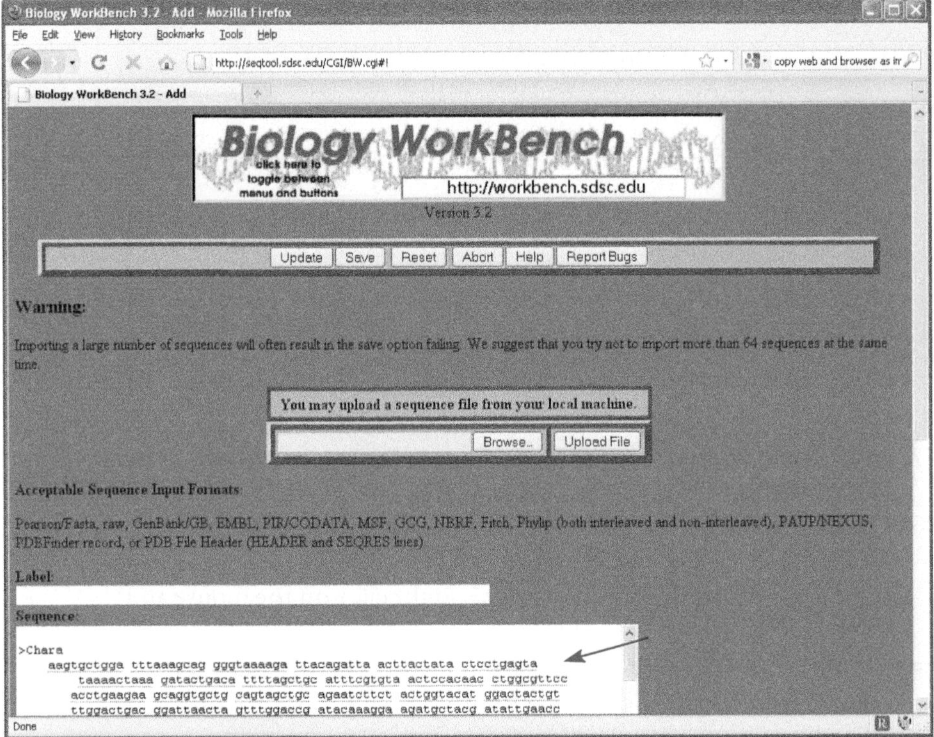

FIGURE 16.9
Biology WorkBench. Pasting the nucleotide sequences.

9. After you have pasted the rbcL sequences into the Sequence box, click "Save" at the bottom of the Web page. A new page will open, containing the main scroll menu. The organism names for your uploaded sequences will appear in the middle of the page as "User Entered" (Figure 16.10).

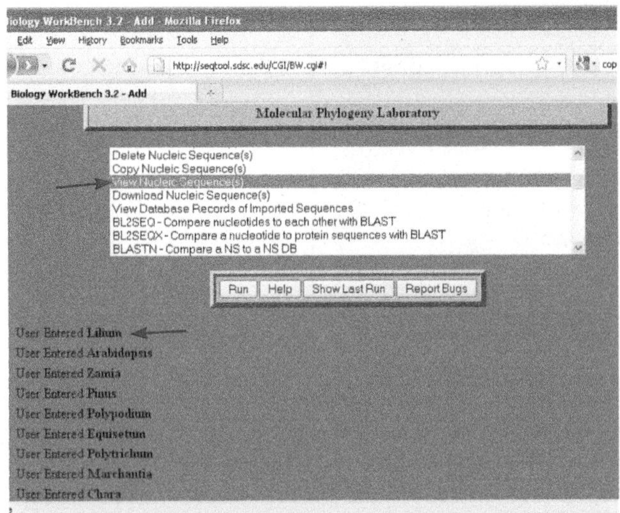

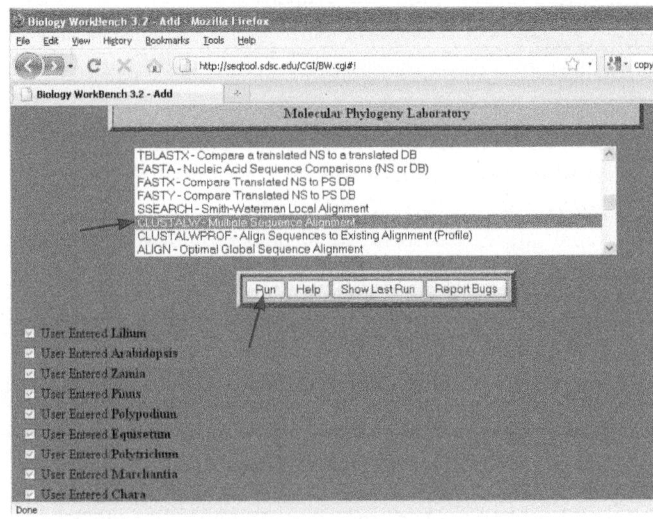

FIGURE 16.10
Biology WorkBench. Entered sequences.

Select one of the uploaded sequences by checking the box on the left of the sequence. Select "View Nucleic Sequence" from the scroll menu (scroll down to find this). Then hit "Run." This allows you to check to be sure that the sequence has been uploaded for the organism listed. (You do not have to check all the organisms.) Next hit "Return" at the bottom of the page to return to the previous page.

Do not use the "Back" button!

10. Now select *all* of the uploaded sequences under the main menu by checking the appropriate boxes. From the scroll menu in the middle of the page, choose "CLUSTALW—Multiple Sequence Alignment" and click "Run" (Figure 16.10). The program will select the sequences and open a new page.
11. In the new ClustalW page, below the selected sequence files, find "Guided Tree Display" (Figure 16.11). Change this to "Rooted Tree." Next click on "Submit" at the bottom of the page. *Be patient* as ClustalW analyzes and processes the data. Do not click on links or arrow keys while the computer is retrieving data. A new page will appear with the results.

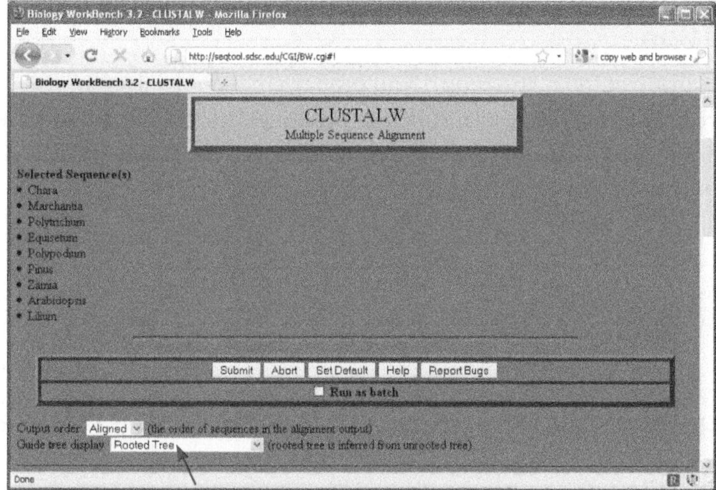

FIGURE 16.11
Biology WorkBench. Options in ClustalW.

12. Scroll to the bottom of the ClustalW results page. Find the phylogenetic tree for the entered sequences. Place the cursor on this image, right click on the mouse, and choose "Copy." *Do not close the browser window!* Return to the open Word document on your desktop and paste the image into the document (alternatively you can drag the image into the document). Save the document to the Desktop. Keep the Biology WorkBench/ClustalW browser window open to continue analyzing the results as instructed. *Closing this window will result in the loss of your alignment results.*
13. Print a copy of the molecular tree pasted in the Word document.
14. Return to the Biology WorkBench/ClustalW window. Scroll to the sequence alignments located above the phylogenetic tree. Review the sequence alignments and record any interesting observations about the sequence variability among the rbcL genes of the selected organisms.

Scroll down below the tree to find the diagnostic messages that include the number of base pairs for each species and the similarity score for pairwise comparisons (sum of identical matches between two species divided by the total number of bases). These values are only one part of the information used in the complex algorithm to create the phylogenetic tree. Continue your analysis in the following Results section.

Results

1. Record any information about the variation in rcbL sequences you observed in the ClustalW analysis.
2. Review the results. Can you determine the number of nucleotides for this gene? Note: The nucleotide sequence for some of the species may not include the entire rbcL sequence.
3. Observe the alignments. What is the significance of the nucleotides highlighted in blue? What do the dashes represent?
4. List the pairs of plants and/or algae that have the most similar alignment (greatest similarity score). This is only one piece of information that is used to determine the final tree.

Discussion

Discuss your results in the report you prepare in Exercise 16.2 In this report you will analyze your molecular phylogenetic tree and compare your results with the morphological tree that you hypothesized for these nine organisms.

EXERCISE 16.2

Analyzing Phylogenetic Trees and Reporting Results

Materials

computer with printer access
computer desktop folder with images
scissors and tape

Introduction

The supercomputer completed the enormous task of aligning the nucleotide sequences and developing the phylogenetic tree that best fits the data, looking at shared sequences (evidence of common ancestry) and sequence differences (accumulated after divergence from a common ancestor). Now you will compare your results to your hypothesized phylogenetic tree. If you predicted that the molecular data would be consistent (or not) with your hypothesis, what evidence do you have to support or falsify your hypothesis? *Once you have completed your analysis, you and your partner will use your results and analysis to construct a report to submit at the end of the laboratory period.*

Procedure

1. Along with your lab partner, examine the observations you made about the sequence alignment results provided by ClustalW. Examine the phylogenetic tree constructed based on rbcL genes of different organisms. Label the **root**, a **branch**, a **node**, and a **clade** within your phylogenetic tree. Paste the images of each organism from the folder on your computer desktop onto your molecular phylogenetic tree. (*Note:* Expand the tree image to maximize the size. You may cut and paste using the computer or print the images and use scissors and tape!)
2. Compare your molecular phylogenetic tree to your hypothesized morphological tree. Observe the similarities and differences in the branching patterns and clades. Begin by reviewing clades that are consistent, indicating that they share particular traits derived from a common ancestor.
3. For those clades that are not consistent, describe the branching pattern. Compare your results with information provided in your textbook and the websites at the end of the lab topic. In some cases the evolutionary relationships may be unresolved given the evidence that is currently available. For the groups that do not have a clear dichotomous (paired) branching pattern, what are possible explanations? Refer to your textbook, websites, and lab topics on plant diversity.
4. The two phylogenetic trees are supported by different types of evidence. What evidence was used to create the phylogenetic tree using bioinformatics in Biology WorkBench?

   What types of evidence support your hypothesized "morphological" tree?

5. Together with your partner, create and type a one-page report that includes the following components.

- *Hypothesis and prediction.* Include your hypothesized morphological tree as a figure (numbered with a title) as part of your hypothesis. Be sure that the phyla are identified on the tree and briefly describe the hypothesized relationships.
- *Results.* Using the results of the computer comparison of nucleotides sequences, in the text of your report briefly describe the relationships depicted in the molecular tree. Include the molecular phylogenetic tree as a figure.
- *Analysis and discussion.* Discuss any differences between the two phylogenetic trees using your research and understanding of phylogenies to support your explanations. Did you support or falsify your hypothesis?

Turn in your report (include the names of both partners) along with the figures before leaving the laboratory.

REVIEWING YOUR KNOWLEDGE

1. Define and describe the following terms: *phylogeny, phylogenetic tree, phylogenetic systematics, molecular phylogenetics, bioinformatics, sequence alignment, homology.*
2. Draw a simple phylogenetic tree for two sister clades with a common ancestor. Define the following terms and use them to label your diagram: *clade, node, root,* and *branch.*
3. What is rbcL, and why is it a particularly useful molecule for studying evolutionary relationships in plants and green algae?
4. What are the limitations in using rcbL to construct a phylogenetic tree?

APPLYING YOUR KNOWLEDGE

1. For the following phylogenetic tree, label the boxes on the branches with the derived trait that is shared by the members of each clade. For example, on the lower branch a box has already been labeled for the feature "cuticle," which is shared by all plants.

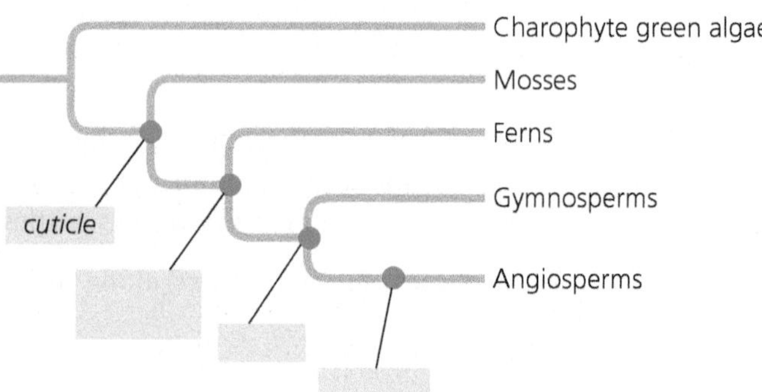

2. Can you suggest reasons why a phylogeny based on molecular evidence and a phylogeny based on morphology and other evidence might not be exactly the same?

3. Zoologists worldwide are sequencing a mitochondrial gene CO1 (cytochrome c oxidase subunit), which is found in all animals and appears to be distinctive for each species. The sequence of nucleotides can be used as a universal DNA bar code. By comparing the CO1 DNA sequence for an animal to a growing database of DNA sequences, scientists can accurately identify any animal and also discover species not previously known to science. How might DNA bar coding, which uses molecular biology and bioinformatics, be useful in enforcing international laws for banning the import of endangered species? How might these approaches stimulate the study of biodiversity in remote areas?

4. Cetaceans (whales, dolphins, and porpoises) are generally classified into two groups, toothed whales (including dolphins and porpoises) and baleen or filter-feeding whales. However, recent molecular evidence suggests two alternative phylogenetic trees: one that indicates sperm whales are closely related to baleen whales or the alternative that sperm whales are more closely related to the common ancestor of all cetaceans. In a recent study, Nikaido et al. (2007) tested these alternative phylogenetic trees (hypotheses) using additional DNA sequences. Answer the following questions about the tree below resulting from their research.

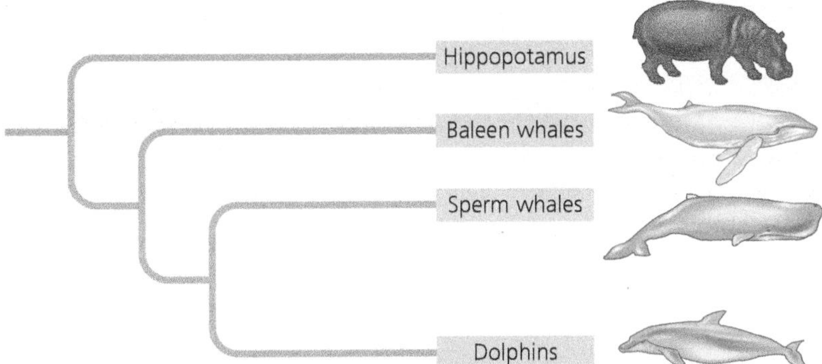

Are sperm whales more closely related to baleen whales or dolphins?

Which group (dolphins, sperm whales, or baleen whales) do you predict will have more nucleotides in common with hippos?

Why do you think hippos are included in this tree? Are they the common ancestor to all whales?

5. When scientists compare the phylogenetic trees based on molecular data with existing trees based on morphological characteristics, in over 90% of the cases the molecular data confirm the relationships previously recognized. However, there are some surprises. For example, water lilies and the water lotus were classified as close relatives, but recent molecular analyses have shown that the water lotus is more closely related to the sycamore tree than to water lilies. Scientists have to critically evaluate these conflicts. Do you think that the evidence from molecular analyses should be the deciding factor in resolving these conflicts with other kinds of evidence (morphology, reproduction, biogeography, fossils, etc.)? How do you think scientists should weigh the evidence?

INVESTIGATIVE EXTENSIONS

You may have time to modify your analysis during the laboratory period, or you may be inspired to pursue bioinformatics and molecular phylogeny as an independent investigation.

1. To modify the analysis of the species featured in this laboratory, consider doing the same comparisons (using all species), but select "Unrooted Tree" in the ClustalW screen. Then compare your results for rooted and unrooted trees. Go to the NCBI website to read about rooted and unrooted trees (http://www.ncbi.nlm.nih.gov/Class/NAWBIS/Modules/Phylogenetics/phylo9.html).
2. Do you think the results of your analysis might be different if you selected only some of the nine species used in this lab topic? Return to Figure 16.10 in the Procedure section and select only those sequences that you would like to investigate in a simplified or at least different phylogenetic analysis. What effect does removing one species have? Consider the unresolved relationships where three species formed a clade. What happens if you remove one of the three?

STUDENT MEDIA: BioFlix, Activities, Investigations, and Videos

www.masteringbiology.com (select Study Area)

Activities— Ch. 26: Phylogenetic Trees; Ch. 29: Highlights of Plant Phylogeny

Investigations—Ch. 26: How Is Phylogeny Determined by Comparing Proteins?

REFERENCES

This lab topic was coauthored with Dr. Nitya Jacob, Associate Professor of Biology, Oxford College of Emory University.

Cameron, K. M. "Bar Coding for Botany." *Natural History*, 2007, vol. 114(2), pp. 52–57.

Evert, R. F. and S. E. Eichhorn. *Raven Biology of Plants*, 8th ed. New York: W.H. Freeman, 2013.

Klug, W. S., M. R. Cummings, C. A. Spencer, and M. A. Palladino. *Essentials of Genetics*, 9th ed. San Francisco: Pearson, 2016. See "Exploring Genomics—

Manipulating Recombinant DNA: Restriction Mapping and Designing PCR Primers" for an excellent exercise using Webcutter to create a restriction map.

Mauseth, J. D. *Botany: An Introduction to Plant Biology*, 6th ed. Sudbury, MA: Jones and Bartlett Publishers, 2016.

National Center for Biotechnology Information. Revised April 2004. "Just the Facts: A Basic Introduction to the Science Underlying NCBI Resources—Systematics and Molecular Phylogenetics," published online at http://www.ncbi.nlm.nih.gov/About/primer/phylo.html.

Nikaido, M., O. Piskurek, and N. Okada. "Toothed Whale Monophyly Reassessed by SINE Insertion Analysis: The Absence of Lineage Sorting Effects Suggests a Small Population of a Common Ancestral Species." *Molecular Phylogenetics and Evolution*, 2007, vol. 43(1), pp. 216–224.

Urry, L., M. Cain , S. Wasserman, P. Minorsky, and J. Reece. *Campbell Biology*, 11th ed. San Francisco, CA: Pearson, 2017.

Valentini, A., F. Pompanon, and P. Taberlet. "DNA Barcoding for Ecologists." *Trends in Ecology and Evolution*, 2009, vol. 24, pp. 110–117.

Zimmer, C. and D. Emlen. *Evolution: Making Sense of Life*, 2nd ed. New York: W.H. Freeman, 2015.

WEBSITES

Barcode for Life sponsored by the Smithsonian Institution and includes international barcoding initiatives CBOL and iBOL. Learn more about bar coding and access the data bases:
http://www.barcodeoflife.org/

Bedrock: Bioinformatics Education Dissemination: Reaching Out, Connecting and Knitting-Together. Integrating bioinformatics in undergraduate biology curriculum:
http://www.bioquest.org/bedrock/

Biology WorkBench home page:
http://workbench.sdsc.edu

CyVerse (formerly iPlant), NSF funded project provides data management for life sciences:
http://www.iplantcollaborative.org/learning-center

DNA Subway, educational resources for teaching and open inquiry investigations:
http://www.iplantcollaborative.org/learning-center/dna-subway

Genome 10K project to sequence 10,000 vertebrates, one from every genus as a genomic zoo:
https://genome10k.soe.ucsc.edu/

National Center for Biotechnology Information (NCBI) home page and information on rooted and unrooted phylogenies:
http://www.ncbi.nlm.nih.gov/
http://www.ncbi.nlm.nih.gov/Class/NAWBIS/Modules/Phylogenetics/phylo9.html

OneKP (1000 Plants) international initiative to sequence 1000 plants:
www.onekp.com/

Tree of Life Web Project site for exploring diversity and phylogeny:
http://www.tolweb.org/tree/home.pages/abouttol.html

University of California Berkley, understanding evolution website with excellent pages on phylogenetics:
http://evolution.berkeley.edu/evolibrary/home.php

NOTES

LAB TOPIC 17

The Kingdom Fungi

This lab topic gives you another opportunity to practice the scientific process introduced in Lab Topic 1. Before going to lab, review scientific investigation in Lab Topic 1 and carefully read this lab topic and review fungi in your text. Be prepared to use this information to design an experiment with fungi.

Laboratory Objectives

1. Describe the three most common phyla of the kingdom Fungi, identifying and describing representative organisms in each.
2. Describe differences in reproduction in fungal phyla.
3. Discuss the ecological role and economic importance of fungi.
4. Design and perform an independent investigation of an organism in the kingdom Fungi.

Introduction

In 1992, scientists studying a northern Michigan hardwood forest reported the first discovery of a massive fungus, an individual clone of the species *Armillaria bulbosa* that occupied an area greater than 15 hectares (37 acres) and weighed over 10,000 kg (11 tons). At the time it was discovered, it was reported to be among the oldest living organisms on Earth, with its age estimated to be more than 1,500 years. Since that discovery, several more huge fungi have been reported. In 2008, the United States Department of Agriculture reported the discovery of "the world's largest living organism" in the Blue Mountains of eastern Oregon. *Armillaria ostoyae,* commonly called the "shoe-string" fungus or "honey mushroom," lives as a parasite on several species of conifers. DNA testing has confirmed that the largest known individual of this fungus covers 2,385 acres and is estimated to weigh between 7,000 and 35,000 tons. Using estimates of the growth rate of this fungus per year, scientists have concluded that this fungus could be between 2,000 and 8,000 years old (Schmitt and Tatum, 2008). Fungi are remarkable organisms with unusual characteristics. In the past they were included in three different kingdoms—animal, plant, or protista. Now they are classified in their own kingdom, Fungi, in the super-group Unikonta. Recent DNA analysis supports the hypothesis that fungi are in the opisthokont clade, which includes animals and certain protistan relatives.

The kingdom Fungi includes a diverse group of organisms found in a wide range of environments. Approximately 100,000 species of fungi have been described and assigned scientific names, but it is estimated that this kingdom may include over 1.5 million species. Most species are terrestrial, distributed from polar regions to deserts, but other species are found in freshwater habitats, hot springs, and one species, called the kerosene fungus, is found growing in fuel tanks of aircraft.

Fungi may be unicellular (yeasts) or multicellular. The vegetative body of multicellular fungi is made up of threadlike filaments called *hyphae* organized into a mass called a *mycelium*. Hyphae grow from the tip, and, as demonstrated by the huge fungus mentioned above, can grow almost indefinitely in favorable conditions. Hyphal cells have cell walls made of *chitin* rather than cellulose as found in plant cell walls. You may recall that chitin is the main component of insect exoskeletons. Another difference in fungal cells and plant cells is the method of storing carbohydrates. Whereas plants store carbohydrates in the form of starch (amylose), fungi store carbohydrates in the form of glycogen. All fungi are heterotrophic organisms that obtain their nutrients by *absorption*, digesting their food outside their bodies and absorbing the digestion products into their cells. Most are *saprophytic* (feed on dead organic matter) but many are *parasitic* (feed on living hosts). They often have complex life cycles with alternating sexual and asexual (vegetative) reproduction. They may produce spores either asexually by mitosis or sexually by meiosis. When reproducing sexually, fungi produce reproductive structures called *fruiting bodies* made of closely interwoven hyphae. Sexual spores form in these fruiting bodies.

Fungi are beneficial to humans in many ways. The fungus *Penicillium* is used to produce antibiotics, and cyclosporine, a drug used to suppress the immune system after organ transplants, is derived from the fungus *Tolypocladium inflatum*. Yeast, a single-celled fungus, is used in the production of wine, beer, and leavened bread. Fungi are also a source of food in many cultures, with truffles being the most expensive food in the world. White truffles are edible subterranean fungi that can sell for as much as $3,600 per pound. The dark green areas seen in Roquefort cheese are patches of the mold *Penicillium roquefortii* growing in cow's milk.

In ecosystems, fungi share with bacteria the essential role of decomposition, returning to the ecosystem the nutrients trapped in dead organisms. One extremely important ecological role played by fungi is their mutualistic association with roots of most plants, forming "mycorrhizae." Mycorrhizal fungi increase the plant's ability to capture water and provide the plant with minerals and essential elements. In turn, the plants supply the fungi with nutrients such as carbohydrates. This association greatly enhances plant growth, and may have played a role in plant colonization of land.

Although many fungi are beneficial, others play destructive roles in nature. Scientists have estimated that fungal diseases are responsible for more deaths worldwide than malaria and tuberculosis, causing hundreds of thousands of deaths annually from meningitis, lung disease, and pneumonia. Athlete's foot, jock-itch, and ringworm are fungal diseases commonly known to humans. Histoplasmosis is a respiratory disease in humans caused by a fungus found in soil and in bat and bird droppings. Annually fungi destroy about a third of all food crops leading to billions of pounds of food wastage worldwide. Wheat rust, soybean root and stem rot, and corn smut are examples of plant diseases caused by fungi. The ergot fungus that parasitizes rye causes convulsive ergotism in humans who eat bread made with infested grains. The bizarre behavior of young women who were later convicted of witchcraft in Salem Village, Massachusetts, in 1692, has been attributed to convulsive ergotism. In September, 2015, the British newspaper, *The Guardian*, reported that over 40 cases of mushroom poisonings had been diagnosed in Syrian refugees. Toxicologists (scientists who study poisons) suspected that families foraging in woods near refugee shelters in Germany found mushrooms that appeared to be edible types common in Syria, but were actually the poisonous "death cap" species. Several of these refugees died from liver failure due to the fungus.

In this lab topic you will survey examples of major fungal groups. You will also have the opportunity to design and perform an open-inquiry investigation with an organism of your choice.

EXERCISE 17.1

A Survey of Major Fungal Groups

In this exercise you will learn about the structure of typical fungi and the characteristics of three important phyla of fungi: Zygomycota, Ascomycota, and Basidiomycota. You will learn details of life cycles and investigate example fungi in each phylum. You will see examples of lichens that are associations between fungi and algae or cyanobacteria. As you observe these examples, consider interesting questions that you might ask about fungi behavior, diversity, or ecology. You can choose one of these questions to develop an independent investigation and design your own experiment in Exercise 17.2.

Lab Study A. Zygote Fungi—Phylum Zygomycota

Materials

compound microscope
stereoscopic microscope
cultures of *Rhizopus stolonifer* with sporangia
microscope slide of *Rhizopus* with sporangia and zygosporangia
cultures of *Pilobolus crystallinus* on demonstration
forceps
ethyl alcohol in a small beaker
alcohol lamp and matches
dropper bottles of water
slides and coverslips

Introduction

In this lab study you will investigate two examples of zygomycetes (fungi in the phylum Zygomycota): the common bread mold, *Rhizopus stolonifera*, and the "shotgun fungus," *Pilbolus crystallinus*. As you study these organisms, note general characteristics of fungi, including **hyphae** organized into the body of the fungus, the **mycelium**. The life cycle of these fungi differs from the life cycles you have studied for plants and animals. In zygomycetes, cells of the hyphae are haploid. In this life cycle, haploid cells may fuse to form diploid nuclei that undergo meiosis. Meiosis does not produce *gametes*, but produces haploid spores that divide by mitosis to produce hyphae. Follow the life cycles of these organisms, noting sexual and asexual stages and the events that result in alternations between haploid and diploid stages.

Rhizopus Stolonifer

Rhizopus is commonly called bread mold, but it may be found growing on many different foods as they spoil. This species reproduces both sexually and asexually. Haploid hyphae grow over a substrate, for example, a slice of bread, giving the bread a fuzzy appearance. In *asexual reproduction*, certain hyphae grow upright and develop **sporangia**, round black structures, on their tips. Haploid spores develop in the sporangia following mitosis, and when they are mature, they are dispersed through the air. If they fall on a suitable substrate, they will absorb water and germinate, growing a new mycelium (Figure 17.1).

Rhizopus also reproduces sexually in certain environmental conditions. In *sexual reproduction,* adjacent hyphae (with their haploid nuclei) of different but compatible mating types (designated as + and –) produce extensions called **gametangia**. The gametangia from the two different mating types then fuse in a process called **plasmogamy**, a term that describes the fusion of the *cytoplasm* of two cells, but not the nuclei. Two fused gametangia with their haploid nuclei develop into a **zygosporangium**. As a zygosporangium matures, the haploid nuclei then fuse (**karyogamy**), producing diploid nuclei. Meiosis follows in these nuclei, and haploid spores are produced in sporangia borne on filaments that emerge from the zygosporangium (Figure 17.1).

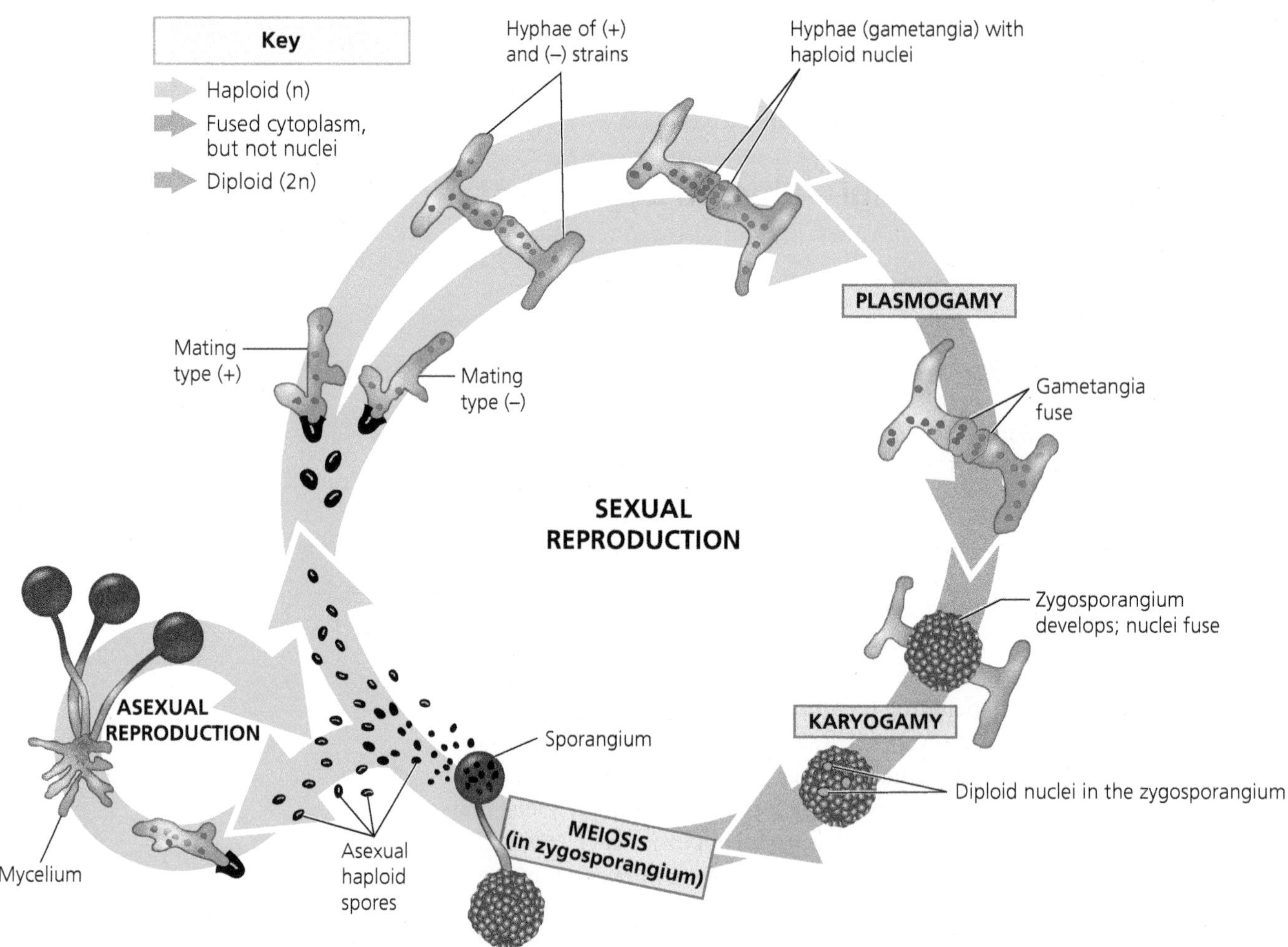

FIGURE 17.1

***Rhizopus stolonifer* life cycle.** *Rhizopus* reproduces asexually by sporangia, producing genetically identical haploid spores. In sexual reproduction, (+) and (–) mating types fuse, ultimately forming a zygosporangium with multiple diploid nuclei. Meiosis follows in these nuclei, producing haploid spores.

Pilobolus Crystallinus

Pilobolus crystallinus (also called the *fungus gun*, or *shotgun fungus*) is another member of the phylum Zygomycota. This fungus is called a **coprophilous** fungus because it grows on dung. It displays many unusual behaviors, one of which is that it is positively phototropic. Perhaps you can investigate this behavior in Exercise 17.2. Bold et al. (1980) describe asexual reproduction in *Pilobolus*. This species has sporangia as does *Rhizopus*, but rather than dispersing single spores, in *Pilobolus* the sporangium is forcibly discharged as a unit; the dispersion is tied to moisture and diurnal cycles. In nature, in the early evening the sporangia form; shortly after midnight, a swelling appears below the sporangium (Figure 17.2). Late the following morning, turgor pressure continues to increase until the swelling explodes, rocketing the sporangium as far as 2 meters. The sticky sporangium will adhere to grass leaves and subsequently may be eaten by an animal—horse, cow, or rabbit. The intact sporangia pass through the animal's digestive tract and are excreted, and the spores germinate in the fresh dung. See the Websites section at the end of this lab topic for an exciting video of *Pilobolus* sporangium discharge.

FIGURE 17.2
***Pilobolus crystallinus*, the shotgun fungus.**

In this lab study you will investigate *Rhizopus* and observe *Pilobolus* on demonstration.

Procedure

Rhizopus

1. Obtain a culture of *Rhizopus* and carry it to your lab station.
2. Examine it using the stereoscopic microscope.
3. Identify the **mycelia**, **hyphae**, and **sporangia**.
4. Review the life cycle of *Rhizopus* (Figure 17.1). Locate the structures in this figure that are visible in your culture. Circle the structures involved in asexual reproduction.
5. Using forceps and aseptic technique (see Appendix C), remove a small portion of the mycelium with several sporangia and make a wet mount.
6. Examine the hyphae and sporangia using the compound microscope.

   Are spores visible? How have the spores been produced? Is this sexual or asexual reproduction?

   How do the spores compare with the hyphal cells genetically?

   How would spores produced by sexual reproduction differ from spores produced asexually?

7. Observe the prepared microscope slide of *Rhizopus* and identify hyphae, sporangia, zygosporangia, and spores.

Pilobolus

1. Using the stereoscopic microscope, observe cultures of *Pilobolus* (Figure 17.2) growing on rabbit dung agar on demonstration. Do not open the petri dish.
2. Identify the **sporangia**, **mycelia**, and **hyphae**. What color are the sporangia and spores? Can you see the clear swelling below the sporangia? What is the function of this structure?

Results

1. Review the life cycle of *Rhizopus* and the structures observed in the living culture, on the microscope slide you prepared, and on the prepared microscope slide, comparing your observations with Figure 17.1. What asexual structures did you observe in the living culture?

   Were asexual structures visible on the microscope slides? Draw these structures in the margin of your lab manual, pointing out differences from those seen in the figure.

   Did you observe sexual structures (zygosporangia)? How could you distinguish the asexual sporangium from the zygosporangium? Draw details of these structures.

2. Review the structures observed in *Pilobolus* and compare with Figure 17.2. Describe the appearance of vegetative and reproductive structures in the cultures seen in lab.
3. Turn to Table 17.1 in the Reviewing Your Knowledge section and describe characteristics of members of the phylum Zygomycota.

Discussion

1. The body form of most fungi, including *Rhizopus*, is a mycelium composed of filamentous hyphae. Using your observations as a basis for your thinking, state why this body form is well adapted to the fungus mode of nutrition.

2. Refer back to the description of *Pilobolus*. Speculate about the adaptive advantage of having a system to propel sporangia, as seen in *Pilobolus*.

3. What advantage can you suggest for the positive phototropic response seen in many coprophilous fungi such as *Pilobolus*?

Lab Study B. Sac Fungi—Phylum Ascomycota

Materials

compound microscope
stereoscopic microscope
dried or preserved *Peziza* specimen
prepared microscope slide of *Peziza* ascocarp
prepared microscope slide of *Penicillium*
demonstrations: cultures of *Penicillium,* Roquefort cheese, preserved or fresh morels, plastic mounts of ergot ascocarp in rye or wheat

Introduction

Fungi in the phylum Ascomycota are called cup fungi or sac fungi (Figure 17.3). All fungi in this phylum share the characteristic of producing sexual spores called ascospores in sac-like cells called asci (sing. ascus). This largest group of fungi includes edible fungi, morels, and truffles, whose fruiting bodies form underground and are dispersed by deer, pigs, and dogs attracted by their strong odor. Also included in this phylum are several deadly plant and animal parasites. For example, chestnut blight and Dutch elm disease have devastated native populations of chestnut and American elm trees. The fungi causing these diseases were introduced into the United States from Asia and Europe. Powdery mildew is a common disease of plants caused by different species of ascomycetes, each species attacking one plant species. Many food crops, including grapes, soybeans, melons, apples, pears, and strawberries, may be attacked by powdery mildew. Several species of the sac fungus *Aspergillus* are toxic to humans. One species produces one of the most potent known natural carcinogens. Other species may cause allergic reactions and lung infections, especially in people with weakened immune systems. You have already examined one example of the phylum Ascomycota in Lab Topic 7, when you studied meiosis and crossing over in *Sordaria fimicola.*

When present, *sexual reproduction* in the ascomycota fungi produces either four or eight haploid **ascospores** after meiosis in an **ascus**. Asci form within a fruiting body called an **ascocarp**. In *Sordaria,* the ascocarp (called a perithecium) is a closed, spherical structure that develops a pore at the top for spore dispersal. In *Peziza* that you will examine in this lab study, the asci are borne on open cup-shaped ascocarps.

In *asexual reproduction*, spores are produced by mitosis, but rather than being enclosed within a sporangium as in zygote fungi, the spores, called **conidia**, are produced on the *surface* of special reproductive hyphae called **conidiophores**. Other features of sac fungi also vary. For example, yeasts are ascomycetes, yet they are single-celled organisms. Yeasts most frequently reproduce asexually by *budding*, a process in which small cells form by pinching off the parent cell. When they reproduce sexually, however, they produce asci, each of which produces four or eight ascospores. Although observations of sexual reproduction in *Penicillium* are incomplete, recent molecular studies and DNA sequencing and the formation of conidia in this important fungus indicate that it should be classified as an ascomycete.

a.

b.

FIGURE 17.3
Examples of sac fungi, phylum Ascomycota. (a) Note the cup-shaped ascocarp in this cup fungus, the orange peel fungus. (b) Morels are cup fungi that resemble mushrooms.

In this lab study you will examine an ascomycete life cycle and the structure of *Peziza* with asci borne on open cup-shaped fruiting bodies. You will observe demonstrations of additional examples of Ascomycota, including *Penicillium* growing in lab cultures and in Roquefort cheese.

Procedure

Life Cycle of an Ascomycete

1. Examine Figure 17.4 illustrating asexual and sexual stages in the life cycle of an ascomycete with an open, cup-shaped ascocarp. As you follow the diagram and the description of the life cycle that follows, *circle on the diagram the items in bold* in the description.
2. In the figure, locate the *sexual reproduction* stage of the life cycle. At this point in the life cycle, haploid **ascospores** have germinated to produce haploid hyphae that form the mycelium. The hyphae may be of two different mating types (+) and (–). When specialized cells in hyphae of two mating types come into contact, they begin to enlarge and **plasmogamy** follows (cytoplasm fuses, but not nuclei). New hyphae with cells containing haploid nuclei *from each mating type* begin to grow from the united hyphae. These hyphae are called **dikaryotic hyphae** (meaning "two nuclei").
3. Dikaryotic hyphae and hyphae with nuclei from only one mating type grow into a hyphal mass, forming a cup-shaped fruiting body called an **ascocarp** that emerges from the soil. As the ascocarp grows, just inside the rim the tips of the dikaryotic hyphae begin to develop, initially forming a dikaryotic **ascus. Karyogamy** follows as the two haploid nuclei in the forming ascus fuse, creating *one diploid nucleus in each ascus*. The diploid nucleus, called the **zygote**, immediately undergoes meiosis producing *four haploid nuclei*, and each of these nuclei now divide by mitosis, producing *eight haploid* **ascospores**. Ascospores contain new gene combinations because of karyogamy and meiosis. As the ascocarp matures, ascospores are released from the asci and, in a favorable environment (soil or rotting wood), germinate into hyphae, creating mycelia.
4. In the figure, locate the *asexual reproduction* stage of the life cycle. Asexual reproduction takes place in the mycelium as cells at the tips of some

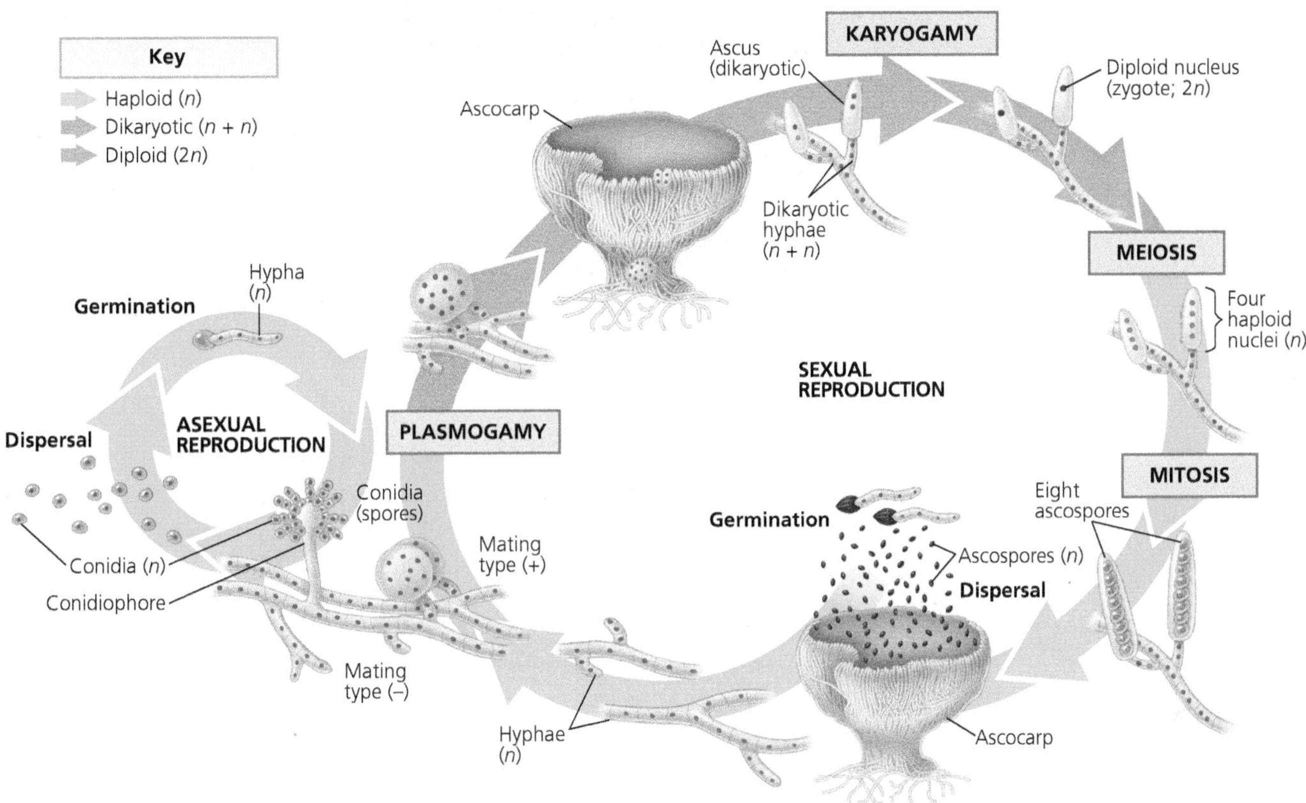

FIGURE 17.4

Life cycle of an ascomycete. In sexual reproduction, dikaryotic hyphae in the ascocarp grow asci in which karyogamy is followed by meiosis, producing four haploid nuclei. The four nuclei divide by mitosis resulting in eight haploid ascospores in each ascus. The ascospores are then dispersed and germinate. In asexual reproduction, mitosis produces genetically identical haploid spores called conidia.

hyphae called **conidiophores** produce haploid spores called **conidia**. These spores are genetically identical and germinate to form new hyphae. Both (+) and (–) mating strains can produce conidia.

Peziza

1. Obtain a dried or preserved **ascocarp** (fruiting body) of *Peziza*. Notice the shape of the **ascocarp** that bears **asci** within the cup (not visible with the naked eye). Fungi with ascocarps shaped in this fashion are called **cup fungi**. The cup may be supported by a stalk.
2. Examine a prepared slide of a section through an ascocarp of *Peziza* using low and intermediate magnifications on the compound microscope. Identify **asci**. How many spores are present per ascus? Are they diploid or haploid?

3. Complete the sketch of the ascocarp section that follows, labeling **asci**, **ascospores**, **hyphae**, and **mycelium**.

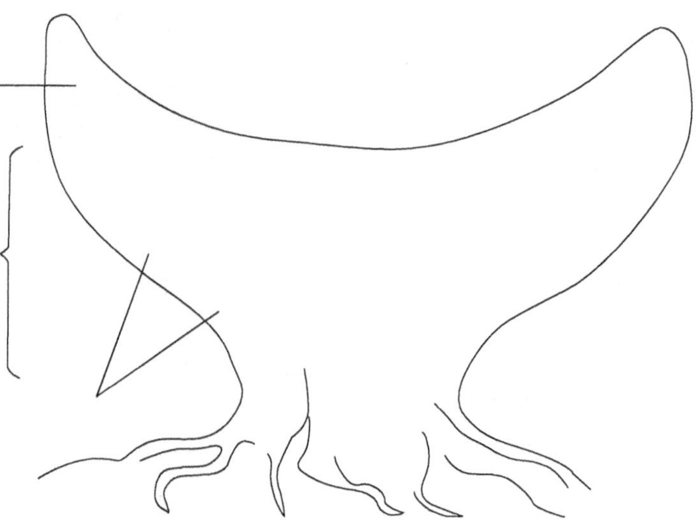

Penicillium

FIGURE 17.5
***Penicillium* growing on an orange.**

1. Observe the *Penicillium* culture and Roquefort cheese on demonstration. You may have observed *Penicillium* growing on oranges or other foods left too long in your refrigerator (Figure 17.5). The green veins seen in the Roquefort cheese (a kind of blue cheese) are growths of *Penicillium* sp. in sheep milk cultures, supposedly grown only in France.
2. Describe the texture and the color of the mycelium in the culture dish.
3. Examine a prepared slide of *Penicillium* and focus on the hyphae. Using intermediate and then high power, locate branched, broom-like structures, **conidiophores**. Asexual spores (**conidia**) are formed here by mitosis.

Ascomycetes Diversity

1. Observe demonstrations of preserved or dried **morels** (called "molly moochers" or "hickory chickens" in the southern Appalachians) (Figure 17.3b). These fungi resemble mushrooms, but the "cap" has a convoluted appearance. Asci are located inside the ridges. Morels are highly prized edible fungi. (Morels should never be eaten raw, however, due to small amounts of toxins in their tissues that are rendered harmless by cooking.)
2. Observe demonstrations of the mature inflorescence of wheat or rye grass infected with the ascomycete *Claviceps purpurea*, the **ergot** fungus. The large black structures seen among the grains are the ergot.

Results

1. Review the life cycle of an ascomycete (Figure 17.4) and the structures observed in *Peziza*, morels, and ergot. Modify Figure 17.3 to reflect features of your examples not included in this figure.
2. Sketch and describe ergot examples in the margin of your lab manual.
3. Sketch your observations of *Penicillium* as seen in the prepared slide in the margin of your lab manual. Include a conidiophore with conidia in your drawing. Note any features that may be important in distinguishing this organism.
4. Turn to Table 17.1 in the Reviewing Your Knowledge section and describe characteristics of ascomycetes.

Discussion

1. What characteristics are common to all sac fungi that reproduce sexually?

2. Compare the appearance of *Penicillium* with that of *Rhizopus*.

3. Traditionally, the classification of fungi has been based on the nature of sexual stages of the life cycle. For *Penicillium*, however, no sexual stages of the life cycle have been observed. Without evidence from sexual stages, speculate about other possible sources of evidence that scientists may use in classification.

Lab Study C. Club Fungi—Phylum Basidiomycota

Materials

compound microscope
stereoscopic microscope
fresh, ripe mushroom basidiocarps
a variety of mushrooms collected from nature (optional)
prepared slides of *Coprinus* pileus sections
demonstrations of fresh or preserved examples of club fungi: bird's-nest fungi, puffballs, shelf fungi, and corn smut (ears of corn infected with the fungus *Ustilago maydis,* or commercially prepared cans of "corn mushroom" cuitlacoche)

Introduction

Fungi classified in the phylum Basidiomycota (club fungi) produce sexual spores (**basidiospores**) on club-shaped cells, called **basidia**, usually born on large, fleshy fruiting bodies called **basidiocarps**. This phylum includes the fungi that cause the plant diseases wheat rust and corn smut, as well as the more familiar puffballs, shelf fungi, earth star, and edible and nonedible mushrooms (the latter often called *toadstools*). In this lab study, you will study features of the life cycle of free-living mushrooms, and then you will survey a diversity of club fungi.

Although most mushrooms are relatively small when mature, in 2011, a basidiocarp weighing over 900 pounds and measuring 33 feet long was discovered on an island in China. Basidiocarps grow from a mycelia mass. When they grow upward around the rim of an underground mass, a "fairy ring" of mushrooms appears. In asexual reproduction, **conidia** form by budding from hyphal tips by mitosis.

Procedure

Life Cycle of a Basidiomycete

1. Examine Figure 17.6, sexual reproduction in the life cycle of a club fungus. Begin by noting the **basidiospores** produced by **meiosis** that germinate to haploid hyphae of different mating types (–) and (+). The hyphae grow to form haploid mycelia.
2. Observe **plasmogamy**; the two different mating types come into contact and the cells fuse. From the fusion point new hyphae begin to grow into a mycelium, but the *nuclei do not fuse*, resulting in a **dikaryotic** mycelium.
3. The dikaryotic mycelium can grow almost indefinitely, but when certain environmental stimuli are present a closely woven mycelial mass will begin to emerge as a fruiting body or mushroom, a **basidiocarp**. Eventually, in the gills of the basidiocarp, the ends of hyphae will grow microscopic club-shaped structures called **basidia**, each containing two haploid nuclei. The two nuclei in each basidium fuse (**karyogamy**) forming a diploid nucleus that undergoes meiosis resulting in four haploid nuclei that develop into basidiospores.
4. As you observe a mushroom in this lab study, you will identify the gills and other parts of a typical basidiocarp and you will examine a slide to see basidia with basidiospores. Continue to refer to Figure 17.6 as you investigate the structure of a mushroom and basidiospore development.

Basidiocarp (Mushroom) Structure

1. Obtain a fresh mushroom, a **basidiocarp**, and identify its parts. The stalk is the **stipe**; the cap is the **pileus**. Look under the pileus and identify **gills**. Haploid spores form on the surface of the gills (described in the life cycle above and in step 6). Examine the gills with the stereoscopic microscope. Do you see spores?
2. Prepare a *spore print*. (Optional activity if additional mushrooms are available in the laboratory. Your instructor may ask you to bring your own examples of mushrooms to lab for this activity.)

   When identifying mushrooms, the color of spores is often used as a distinguishing characteristic, and a scientist may use a spore print to help identify a mushroom species. A spore print is made by cutting the stalk of a ripe mushroom as close to the pileus as possible and then placing the pileus with the gill side down on a piece of white paper for at least several hours, allowing the spores to drop to the paper. If directed by your instructor, prepare a spore print to be observed after several hours or in next week's lab. Place the pileus in a protected area indicated by your instructor, and cover it with a dish or plastic wrap to prevent excessive drying.

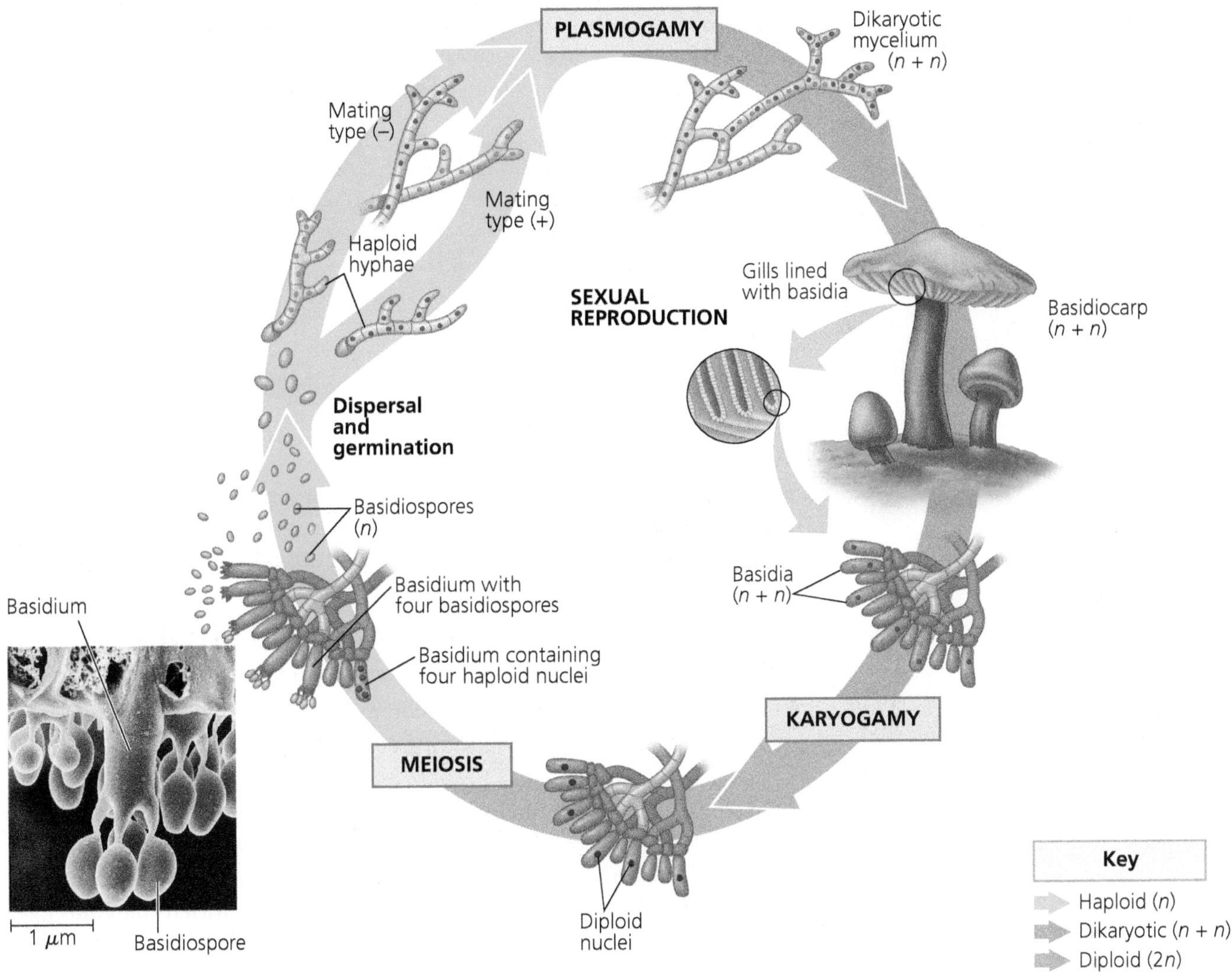

FIGURE 17.6

Sexual reproduction in the life cycle of a basidiomycete. Hyphae of mycelia of different mating types fuse, and eventually form fruiting bodies called basidiocarps (mushrooms). Dikaryotic hyphae on the surface of the gills in the basidiocarp develop into cells called basidia. In these structures the nuclei fuse, undergo meiosis, and produce haploid basidiospores.

3. Label the parts of the basidiocarp in Figure 17.7a.
4. Obtain a prepared slide of a section through the pileus of *Coprinus* or another mushroom. Observe it on the compound microscope using low and then intermediate powers. Is your slide a *cross section* or a *longitudinal section* through the pileus?

   Make a sketch in the lab manual margin indicating the plane of your section through the basidiocarp. Compare your section with the fresh mushroom you have just studied and with Figure 17.7b.

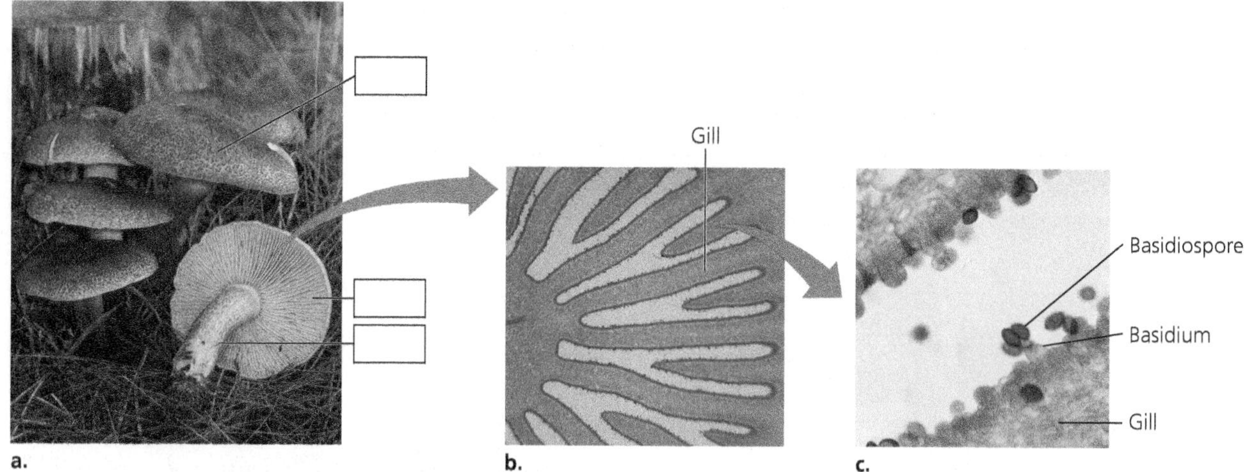

FIGURE 17.7

Details of mushroom structure. (a) A whole mushroom, or basidiocarp. Label the **pileus** (cap), **stipe** (stalk), and **gills**. (b) A cross section through the pileus showing basidia on the gills. (c) Enlargement of the gill surface showing basidia and basidiospores.

5. On the prepared slide, observe the surface of several gills using high power. Look for small club-shaped **basidia** on the gill surface.
6. Focus carefully on the end of a basidium. Do you see four small extensions or knob-like protuberances growing out from the end (see Figure 17.7c)?

   Before these protuberances begin to grow, a young basidium contains two haploid nuclei. These nuclei then fuse to form *one diploid nucleus*. Meiosis takes place in this nucleus, forming *four haploid nuclei*. As meiosis is taking place, the four protuberances grow and the haploid nuclei migrate, one into each of these protuberances. Each protuberance with its haploid nucleus becomes a **basidiospore**. When the spores are mature, they are released from the basidium and are dispersed by the wind.
7. Review again the life cycle of a basidiomycete in Figure 17.6. Circle the items in this diagram that you observed on the mushroom and slide that you studied. Modify the diagram to reflect any differences that you noted.

Basidiomycetes Diversity

Observe other examples of basidiomycetes on demonstration. These may include the bird's-nest fungus; earthstar and other puff balls; shelf fungi; and a parasitic club fungus, corn smut.

1. The small, hard, lentil-shaped structures seen in the mature, hollow fruiting bodies of **bird's-nest fungi** (Figure 17.8a) contain the basidiospores that may be dispersed by animals or rain.
2. The group of basidiomycetes called puffballs includes **common puffballs** and one unusual puffball called **earthstar** (Figure 17.8b). The fruiting bodies of all puffballs are edible and are best to eat when pure white inside. The basidiospores in a common puffball develop in the center of the ball-shaped fruiting body and are surrounded by a distinct outer wall. Spores inside turn yellow as they mature and in most species are discharged from a pore on the surface of the fruiting body. In the earthstar, the outer layer of the fruiting body wall splits along radial fissures. When wet, this layer will open out into a structure that resembles a star. This exposes the basidiospore-filled sac inside and spores are dispersed through a pore at the top.

3. **Shelf fungi**, sometimes called bracket fungi (Figure 17.8c), are basidiomycetes that include some of the most common and familiar fungi. The young fruiting bodies of these fungi may be soft and pliable but become tough, leathery, corky, or woody when mature. These fungi are commonly seen in a forest on living or dead trees, but they also grow on lumber and are responsible for wood rot. Treating lumber with wood preservatives to prevent this damage costs the industry large sums of money.
4. **Corn smut—a parasitic club fungus**
   Corn smut, *Ustilago maydis,* is most obviously found in the ears and tassels of corn plants where it causes tumor-like galls that form when the fungus stimulates the corn cells to increase in size and number (Figure 17.8d).

FIGURE 17.8

Examples of club fungi, phylum Basidiomycota. (a) **Bird's-nest fungi**—basidiospores form in small fruiting bodies in the "nest." (b) **Earthstar**—a puffball. Spores develop in the ball-shaped, centrally located fruiting body. (c) **Shelf fungi**—common fungi found growing on living or dead trees. (d) **Corn smut**—a parasitic club fungus that causes tumor-like galls in ears of corn.

As the galls grow, the interiors turn from silvery-white to black as spores develop in structures that are analogous to mushrooms. The term "smut" refers to the black and dusty masses of spores within these tissues. Although destructive to corn crops, in Mexico immature galls, called huitlacoche or cuitlacoche, are eaten in tortilla-based foods and soups.

Observe examples of galls from corn ears infected with corn smut. These may be fresh materials, or they may be from commercially preserved cuitlacoche.

Results

1. Review the structures observed in the living mushroom and on the prepared slide of a section taken through a mushroom pileus. Review any changes or additions that you made to the diagram of the basidiomycete life cycle (Figure 17.6) and Figure 17.7.
2. Turn to Table 17.1 in the Reviewing Your Knowledge section and describe characteristics of basidiomycetes.
3. If you prepared a spore print, return to lab to observe your spore print later in the week or next week during lab. Do you see spores? What color are they? If you would like to preserve your spore print, you can spray it with a light coating of art fixative or hair spray. Hold the can at least 12 inches away from the print.

Discussion

1. State the characteristics shared by all basidiomycetes.

2. Basidia within a basidiocarp are obvious for mushrooms, but this may not be as obvious for other examples that you have observed. For each of the examples of basidiomycetes that you have observed, speculate about the location of structures analogous to the basidia observed in mushrooms. Think about the formation of spores by meiosis.

   bird's-nest fungi

   common puffballs

   earthstar puffball

   shelf fungi

   corn smut

Lab Study D. Lichens

Materials

compound microscope
stereoscopic microscope
blank slides and coverslips
dissecting needles
water in dropper bottle
prepared slides of sections through lichens
examples of foliose, crustose, and fruticose lichens on demonstration

Introduction

Lichens are symbiotic associations between fungi and algae or photosynthetic bacteria forming a body that can be consistently recognized. The fungal component usually gives the lichen its form, and is usually a sac fungus or a club fungus. The lichen body, called a **thallus**, varies in shape and colors, depending on the species of the components. Reproductive structures can be bright red or pink or green. Photosynthesis in the algae provides nutrients for the fungus, and the fungus provides a moist environment for the algae or bacteria. Because lichens can survive in extremely harsh environments, they are often the first organisms to colonize a newly exposed environment such as volcanic flow or rock outcrops, and they play a role in soil formation. The fungi in many lichens may have antibiotic properties, and in the past, lichens were often used to dress wounds. In fact, it is estimated that as many as 50% of all lichens may secrete antibiotic substances.

Procedure

1. Observe the demonstrations of different lichen types: those with a leafy thallus (**foliose**—Figure 17.9a), a crust-like thallus (**crustose**—Figure 17.9b), or a branching, cylindrical thallus (**fruticose**—Figure 17.9c). Look for cup-shaped or club-like reproductive structures produced by the fungal component of the lichen.
2. Make a wet mount of a living lichen designated by your instructor: place a small piece of lichen in a drop of water on a microscope slide. Use dissecting needles to tease apart the tissues. Add a coverslip and view using low, intermediate, and high powers on the microscope. Look for hyphae and small photosynthetic cells dispersed among the hyphae.
3. Observe a prepared slide of a section through a lichen thallus. Identify hyphae and photosynthetic cells.

Results

1. Sketch the lichens on demonstration in the margin of your lab manual.
2. Label any visible reproductive structures, and, if possible, indicate whether the fungal component is a sac fungus or a club fungus.
3. Identify and label each according to lichen type.

Discussion

1. Imagine that you are the first scientist to observe lichen microscopically. What observations would lead you to conclude that the lichen is composed of a fungus and an algal species?

a.

b.

c.

FIGURE 17.9
Lichen types. Lichens may have (a) a leafy thallus (foliose), (b) a crust-like thallus (crustose), or (c) a cylindrical thallus (fruticose).

2. Lichens are a symbiotic relationship between fungi and either algae or photosynthetic bacteria (cyanobacteria). How could you determine if the photosynthetic component in the lichen you are studying is algae or cyanobacteria?

3. It is reported that many lichens possess antibiotic properties. What adaptive advantage can you suggest for this characteristic of lichens?

EXERCISE 17.2

Designing and Performing an Open-Inquiry Investigation

Introduction

In this exercise, you will choose one of the organisms observed in this lab topic and design a simple experiment answering a question about its behavior, growth patterns, or interactions with other species. Use Lab Topic 1 as a reference for designing and performing a scientific investigation. Be ready to assign tasks to members of your lab team. Be sure that everyone understands the techniques that will be used. Your experiment will be successful only if you plan carefully, cooperate with your team members, perform lab techniques accurately and systematically, and record and report data accurately.

Materials

living cultures of *Pilobolus crystallinus, Rhizopus, Penicillium*
a variety of living lichens
sterile agar plates to grow each species
sterile potato dextrose agar plates
sterile agar with sugar
sterile agar with albumin
sterile agar prepared with pH 6, 7, or 8 buffers
aluminum foil (for light directional studies)
various breads from the health food store—wheat, rye, corn, potato, rice
bread with and without preservatives
sterilized dung from various animals
mycorrhizae inoculate
sterile forest soil (microwave to kill bacteria and fungi)
natural forest soil
cultures of living bacteria, e.g., *E. coli*
filter paper disks to use with fungal or lichen extracts

Procedure

1. **Collaborating with your research partners, read the following potential questions or research projects and choose an investigation using this list or an original question proposed by your team.** You may want to check your text and other sources for supporting information. You should be able to explain the rationale behind your choice of question. *Write your question in the margin of your lab manual.*
   a. Do the same fungi grow on different varieties of bread?
   b. How effective are preservatives in preventing fungal growth on foods?
   c. Is *Pilobolus* phototaxic? What about other fungi?
   d. Does succession take place in dung cultures of fungi? Refer to the milk bacteria succession study in Lab Topic 12 Bacteriology and design a similar experiment to investigate this phenomenon in fungi growing on dung.
   e. Is there a difference in the growth of plants growing with and without mycorrhizae (grow seeds in sterile and natural forest soil or use a mycorrhizae inoculate)?
   f. Can the growth of fungi be altered by supplying different nutrients (e.g., sugar or albumin) in the agar culture?
   g. Investigate the common lichens in your area. Collect local lichens from various habitats. You may compare collected lichens by making temporary slides (squash tissue on blank microscope slide) and identifying the fungal and photosynthetic portions of each.
   h. How does the form of lichen differ among species growing in different habitats? For example, collect lichens from tree bark and another habitat (e.g. growing on rocks), and determine differences in structure between them.
   i. Test the antibiotic potential of various lichens. Are reports that lichens secrete antibiotics supported by your results? Make an extract of the lichen thallus and refer to Lab Topic 12 Bacteriology for suggestions for how to design an experiment to test this.

j. What organisms other than fungi and algae live in lichens? Look for microscopic organisms by soaking the lichen overnight in fresh water that has been aged to remove chlorine. The next day, pour the water through a sieve to strain out larger particles, and then look at samples of the water using the compound microscope.

k. Compare the diversity of organisms found on lichens collected from different habitats.

l. Collect mushrooms from wooded areas on your campus. Make spore prints of each and determine if you could use differences in spore prints to identify these mushrooms.

m. Follow the succession of fungal species on samples of dung. Identify at least the phylum of each newly appearing species.

2. **Formulate a testable hypothesis.** (Refer to Exercise 1.1, Lab Study B, Developing Hypotheses.)

   Hypothesis:

3. **Summarize the essential elements of the experiment.** (Use separate paper.)

4. **Predict the results of your experiment based on your hypothesis.** (Refer to Lab Topic 1, Exercise 1.2, Lab Study C, Making Predictions.)

   Prediction: (if/then)

5. **Outline the procedures used in the experiment.** (Refer to Exercise 1.2, Lab Study B, Choosing or Designing a Procedure.)

   a. On a separate sheet of paper, list in numerical order each exact step of your procedure.

   b. Remember to include the number of replicates (usually a minimum of five), levels of treatment, appropriate time intervals, and controls for each procedure.

   c. If you have an idea for an experiment that requires materials other than those provided, ask your laboratory instructor about availability. If possible, additional supplies will be provided.

   d. When carrying out an experiment, remember to quantify your measurements when possible.

6. **Perform the experiment**, making observations and collecting data for analysis.

7. **Record results, including observations and data** on a separate sheet of paper. Design tables and graphs, at least one of each. Be thorough when collecting data. Do not just write down numbers, but record what they mean as well. Do not rely on your memory for information that you will need when reporting your results.

8. **Prepare your discussion.** Discuss your results in light of your hypothesis.
   a. Review your hypothesis. Review your results (tables and graphs). Do your results support or falsify your hypothesis? Explain your answer, using data for support.
   b. Review your prediction. Did your results correspond to the prediction you made? If not, explain how your results are different from your predictions, and why this might have occurred.
   c. If you had problems with the procedure or questionable results, explain how they might have influenced your conclusion.
   d. If you had an opportunity to repeat and expand this experiment to make your results more convincing, what would you do?
   e. Summarize the conclusion you have drawn from your results.
9. **Be prepared to report your results to the class.** Prepare to persuade your fellow scientists that your experimental design is sound and that your results support your conclusions.
10. **If required by your instructor, submit a written laboratory report** in the form of a scientific paper (see Appendix A). Keep in mind that although you have performed the experiments as a team, you must submit a lab report of *your original writing*. Your tables and figures may be similar to those of your team members, but your paper must be the product of your own literature search and creative thinking.

REVIEWING YOUR KNOWLEDGE

1. Complete Table 17.1 comparing characteristics of fungal phyla.
2. Compare spore formation in the sexual stages of the life cycles of sac fungi and club fungi.

TABLE 17.1 Comparison of Fungi by Major Features

Phylum	Example(s)	Sexual Reproductive Structures	Asexual Reproductive Structures
Zygomycota (Zygote Fungi)			
Ascomycota (Sac or Cup Fungi)			
Basidiomycota (Club Fungi)			

APPLYING YOUR KNOWLEDGE

1. Imagine an ecosystem with no fungi. How would it be modified?

2. Speculate about a possible evolutionary advantage to the *fungus* for the following:

   a. *Penicillium* makes and secretes an antibiotic.

   b. *Ergot* fungus (parasitizes rye grain) produces a chemical that is toxic to animals.

STUDENT MEDIA: BioFlix, Activities, Investigations, and Videos

www.masteringbiology.com (select Study Area)

Activities—Ch. 26: Classification Schemes; Ch. 28: Tentative Phylogeny of Eukaryotes; Ch. 31: Fungal Reproduction and Nutrition; Fungal Life Cycles

Investigations—Ch. 31: How Does the Fungus *Pilobolus* Succeed as a Decomposer?

REFERENCES

Ahmadjian, V. "Lichens Are More Important Than You Think." *BioScience*, 1995, vol. 45, p. 124.

Alexopoulos, C., C. Mims, and M. Blackwell. *Introductory Mycology*, 4th ed. New York: John Wiley and Sons, Inc., 1996.

Barron, G. *Mushrooms of Northeast North America: Midwest to New England*. Auburn, WA: Lone Pine Publishing, 1999.

Bessett, W., W. C. Roody, and A. R. Bessette. *Mushrooms of the Southeastern United States*. Syracuse, NY: Syracuse University Press, 2007.

Bold, H., C. J. Alexopoulos, and T. Delevoryas. *Morphology of Plants and Fungi*. New York: Harper & Row, 1980, p. 654.

Brodo, I., S. D. Sharnoff, S. Sharnoff. *Lichens of North America*. New Haven, CT: Yale University Press, 2001.

Doolittle, W. F. "Uprooting the Tree of Life." *Scientific American*, 2000, vol. 282, pp. 90–95.

Light, K. H. "Take a Likin' to Lichens!" *The Tennessee Conservationist*, 2012, vol. 78, pp. 28–31.

Litten, W. "The Most Poisonous Mushrooms." *Scientific American*, 1975, vol. 232.

Milius, S. "A Partnership Apart: Lichen Mutualism." *Science News*, 2009, vol. 176(10), pp. 17–20.

Miller, K. "Mushroom Manifesto." *Discover*, July 2013, pp. 38–47.

Petersen, Jens H. *The Kingdom of Fungi*. Princeton and Oxford: Princeton University Press, 2012.

Schmitt, C., and M. Tatum. "Location of the World's Largest Living Organism (The Humongous Fungus)." *USDA*, 2008.

Smith, M., J. Bruhn, and J. Anderson. "The Fungus *Armillaria bulbosa* Is Among the Largest and Oldest Living Organisms." *Nature*, 1992, vol. 356, pp. 428–431.

Stephenson, S. *The Kingdom Fungi*. Portland, OR: Timber Press, 2010.

Urry, L., M. Cain, S. Wasserman, P. Minorsky, and J. Reece. *Campbell Biology*, 11th ed. San Francisco, CA: Pearson, 2017.

WEBSITES

Fungus reported to be the largest organism on Earth:
http://www.scientificamerican.com/article.cfm?id=strange-but-true-largest-organism-is-fungus (October 4, 2007)

Leaf-cutter ants/fungus symbiotic relationship:
Leafcutter Ants—The First Agriculture:
https://www.youtube.com/watch?v=RH3KYBMpxOU

Lichens of North America photos and links to other lichen web sites:
http://www.lichen.com

North American Mycological Association:
http://namyco.org

Photos and life cycles of fungi:
http://botit.botany.wisc.edu/toms_fungi/

Report on deadly human fungal diseases:
www.dailymail.co.uk/sciencetech/article-2912033/

Ted Talks on the importance of fungi, including amazing images:
http://www.ted.com/talks/paul_stamets_on_6_ways_mushrooms_can_save_the_world.html

Ustilago maydis, corn smut:
http://www.metapathogen.com/ustilago

Videos of *Pilobolus*:
http://www.bbc.co.uk/nature/life/Pilobolus_crystallinus
http://www.youtube.com/watch?v=TrKJAojmB1Y (excellent)
http://www.plantpath.cornell.edu/PhotoLab/TimeLapse2/Pilobolus1_credit_FC.html

LAB TOPIC 18

Animal Diversity I: Porifera, Cnidaria, Platyhelminthes, Mollusca, and Annelida

Laboratory Objectives

After completing this lab topic, you should be able to:

1. Discuss characteristics of representative animals and phylogenetic relationships of selected animal phyla in the clade Metazoa.
2. Compare the anatomy of the chosen animals, describing similarities and differences in organs and body form that allow each animal to carry out body functions.
3. Discuss the relationship between body form and the lifestyle or niche of the organism.

Introduction

Animals are classified in the domain **Eukarya**, and based on molecular evidence in the diverse supergroup **Opisthokonta**, that also includes fungi and several groups of protists. All animals share a common ancestor forming a clade **Metazoa**. They are **multicellular** organisms and are **heterotrophic**, meaning that they obtain food by ingesting other organisms or their by-products. The ancestor of all animals is thought to have lived somewhere between 675 and 800 million years ago, but most animal fossils range in age from 565 to 550 million years old, with most body forms appearing by the end of the Cambrian period (see Figure 14.3).

Since the beginning of the scientific study of animals, scientists have attempted to sort and group closely related organisms, and categories used in classification schemes have changed over time. Based on the latest molecular evidence, comparative anatomy, and patterns of development, current phylogenetic trees divide the metazoa into two major groups: the sponges in the phylum **Porifera**, and the clade **Eumetazoa**, which includes all other animals. This division is made because the body form of sponges is very different from that of other animals, leading most biologists to conclude that sponges are not closely related to any other animal groups.

Animals in Eumetazoa differ in basic physical characteristics, such as symmetry and body form. Animals may be **radially symmetrical** (parts arranged around a central axis) or **bilaterally symmetrical** (right and left halves are mirror images). Some animals have a saclike body form with only one opening into a digestive cavity. Others have two outer openings, a mouth and an anus, and the digestive tract forms essentially a "tube within a tube." Differences in early developmental patterns have led to other important criteria for animal classification. You will learn more about animal development in Lab Topic 25. It will be helpful to read the "Overview of Stages of Early Development" in that lab topic before proceeding with this study of animal characteristics. Some differences in animal groups are based on the number of embryonic germ layers (layers of tissue that form early in development from which all the other tissues in the body arise). Most animals have three germ layers, **ectoderm**, **mesoderm**, and **endoderm**. Animals may also differ in the type of body cavity (coelom) that forms in the embryo, and the embryonic development of the digestive tract. The origin of the mouth and anus has led scientists to describe two developmental modes in bilaterally symmetrical animals, **protostome development** and **deuterostome development**. These two types of development differ in several basic characteristics, including the type of cleavage (early divisions of a fertilized egg), the manner of coelom formation, and the origin of the mouth and anus. An embryonic structure, the blastopore, develops into a mouth in the protostomes and into an anus in the deuterostomes.

In the study of the protists in Lab Topic 13, you learned that "traditional" phylogenetic trees have been challenged by the results of molecular studies that initially investigated the gene coding for the small sub-unit of ribosomal RNA. Now hundreds of genes from many organisms are used in phylogenetic studies, not only for protists, but also for animals. Particularly in the protostomes, molecular studies have led to a regrouping of many traditionally established phylogenetic relationships. For example, for over 200 years zoology publications have assumed that annelids (segmented worms in the phylum Annelida) and arthropods (e.g., insects) are closely related based on their segmented bodies. Zoologists also noted, however, that annelids have developmental patterns similar to several groups that are not segmented. For example, annelids are like molluscs (e.g., clams) in having a developmental stage called the "trochophore larva." Recent molecular evidence helps to clarify this puzzle as it supports the hypothesis that annelids and molluscs are closely related, and separate from arthropods.

Taxonomists now separate animals in Eumetazoa into two groups: those with radial symmetry and those with bilateral symmetry (clade **Bilateria**), and most phyla of animals are classified in one of the three clades of bilateral animals—**Deuterostomia**, **Lophotrochozoa**, and **Ecdysozoa**. Annelids (phylum Annelida), molluscs (phylum Mollusca), and several more phyla not studied in this lab topic are placed in the clade Lophotrochozoa. The name reflects the trochophore larvae found in annelids and molluscs. Also included in this clade are flatworms (phylum Platyhelminthes). Although flatworms lack such characteristics as a body cavity, the "tube-within-a-tube" body plan with mouth and anus, and elaborate internal organs that characterize the annelids and molluscs, recent molecular evidence indicates that they should be grouped with annelids and molluscs in the Lophotrochozoa clade. Evidence from nucleotide sequences in several genes indicates that roundworms or nematodes (phylum Nematoda), arthropods (phylum Arthropoda), and several other phyla belong in the clade Ecdysozoa. Animals in this clade undergo molting (ecdysis), or the shedding of an outer body cover. In nematodes this covering is called the **cuticle**. In arthropods the covering is the **exoskeleton**.

Another surprising result of rRNA and other molecular evidence is that the nature of the body cavity may not be a characteristic that indicates major phylogenetic branching. A true coelom or pseudocoelom may have been independently gained or lost many times in evolutionary history. In traditional phylogenetic groupings, flatworms and nematodes were considered primitive, neither group having a "true" coelom. Molecular evidence has now moved nematodes to a different position with arthropods in the metazoan tree.

Figure 18.1 is a tree diagram representing the phylogeny of several major animal phyla based on the most recent molecular evidence. This phylogenetic tree represents a hierarchy of clades nested within larger clades. Much work remains to resolve the branching order within the lophotrochozoan and ecdysozoan clades. Scientists are collecting evidence from studies based on mitochondrial DNA sequencing, ribosomal genes, *Hox* genes, and genes coding for various proteins.

In addition to phylogenetic relationships among groups of organisms, there are two more unifying themes to be considered as you study the animals reviewed in lab topics 18 and 19. One theme is the relationship between *form and function*—how does the form (anatomy) of a structure relate to its function? A second theme is the relationship between *form and environment*—how does the form of a structure enable an organism to survive in its environment? The questions at the end of the lab topics will assist with this.

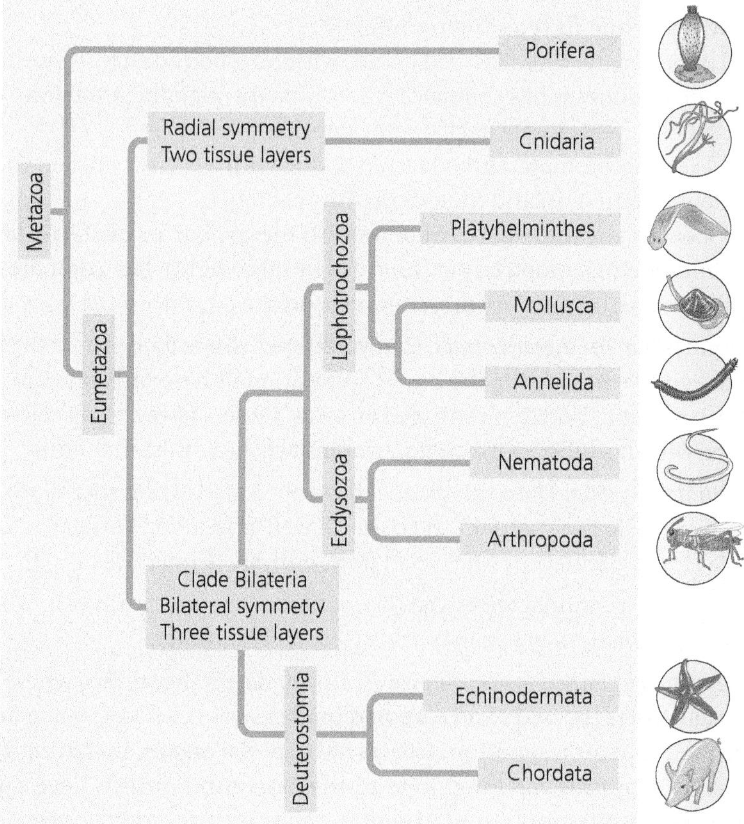

FIGURE 18.1
Phylogenetic organization of animals studied in this lab topic and Lab Topic 19 based on new molecular evidence. Animals with protostome development are assigned to one of two clades, Lophotrochozoa or Ecdysozoa.

In this and the following lab topic, you will investigate body form and function in examples of nine major groups of animals. You will use these investigations to ask and answer questions comparing general features of morphology and relating these features to the lifestyle of each animal. The animals you will study in this lab topic are the sponge, the radially symmetrical hydra, and four examples of animals in the clade Lophotrochozoa: planaria, clamworm, earthworm, and clam (mussel). In Lab Topic 19 you will study three examples of Ecdysozoa (roundworm, crayfish, and grasshopper) and three examples of Deuterostomia (sea star, lancelet, and pig). All of the animals studied in these lab topics except the pig are **invertebrates**, animals that do not have a vertebral column (backbone). Only the pig is a **vertebrate** (has a backbone).

In your comparative study of these organisms, you will investigate 13 characteristics and you will be asked to describe these characteristics for each animal.

Before you begin the dissections, become familiar with the following characteristics and their descriptions:

1. *Symmetry.* Is the animal (a) radially symmetrical (parts arranged around a central axis), (b) bilaterally symmetrical (right and left halves are mirror images), or (c) asymmetrical (no apparent symmetry)?
2. *Tissue organization.* Are cells organized into well-defined tissue layers (structural and functional units)? How many distinctive layers are present?
3. *Body cavity.* Is a body cavity present? A body cavity—the space between the gut and body wall—is present only in three-layered organisms, that is, in organisms with the embryonic germ layers ectoderm, mesoderm, and endoderm. There are three types of body forms related to the presence of a body cavity and its type (Figure 18.2).
   a. Acoelomate, three-layered bodies without a body cavity. Tissue from the mesoderm fills the space where a cavity might be; therefore, the tissue layers closely pack on one another.
   b. Pseudocoelomate, three-layered bodies with a cavity between the endoderm (gut) and mesoderm (muscle).
   c. Coelomate, three-layered bodies with the coelom, or cavity, *within* the mesoderm (completely surrounded by mesoderm). In coelomate organisms, mesodermal membranes suspend the gut within the body cavity.
4. *Openings into the digestive tract.* Can you detect where food enters the body and digestive waste exits the body? Some animals have only one opening, which serves as both a mouth and an anus. Others have a body called a "tube within a tube," with an anterior mouth and a posterior anus.
5. *Circulatory system.* Does this animal have open circulation (the blood flows through coelomic spaces in the tissue as well as in blood vessels), or does it have closed circulation (the blood flows entirely through vessels)?
6. *Habitat.* Is the animal terrestrial (lives on land) or aquatic (lives in water)? Aquatic animals may live in marine (sea) or fresh water.
7. *Organs for respiration (gas exchange).* Can you detect the surface where oxygen enters the body and carbon dioxide leaves the body? Many animals use their skin for respiration. Others have special organs, including gills in aquatic organisms and lungs in terrestrial organisms. Insects have a unique system for respiration, using structures called *spiracles* and *tracheae.*
8. *Organs for excretion.* How does the animal rid its body of nitrogenous waste? In many animals, these wastes pass out of the body through the skin by

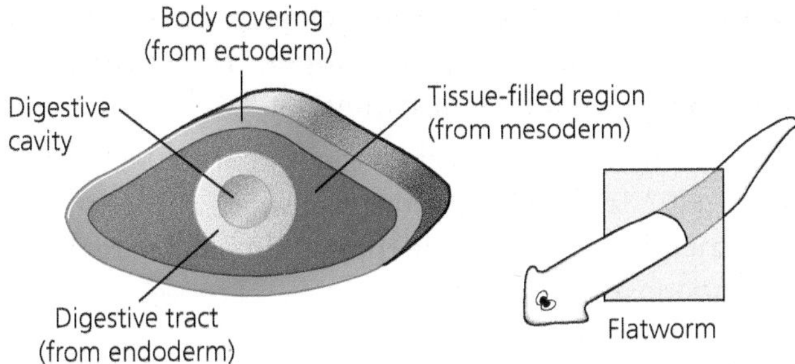

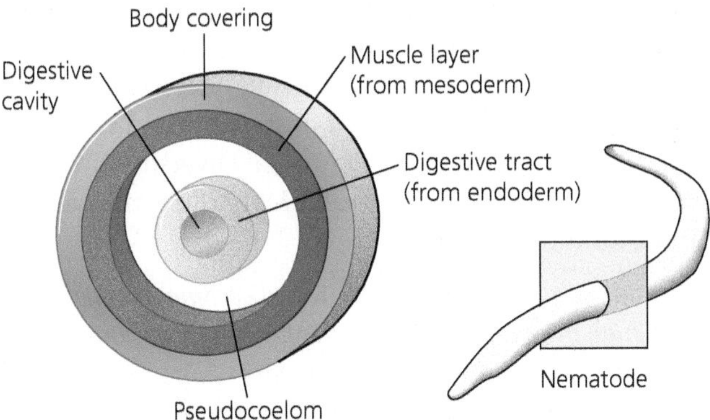

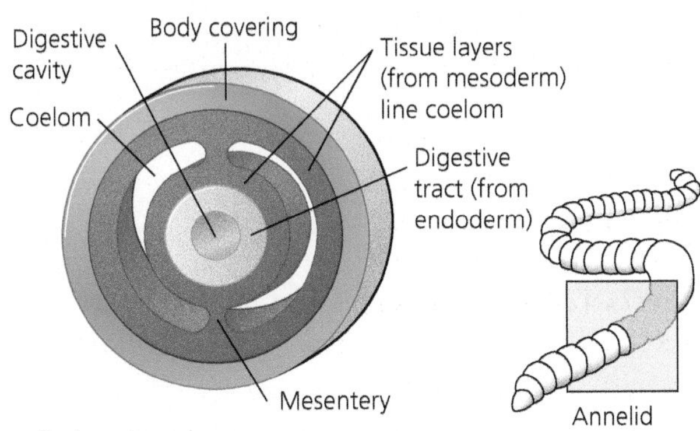

FIGURE 18.2

Three types of body cavities. (a) In acoelomate animals, there is no body cavity and mesoderm fills the space where a cavity might be. (b) In pseudocoelomate animals, the body cavity lies between tissues derived from endoderm and mesoderm. (c) In coelomate animals, the body cavity is lined with mesoderm.

diffusion. In others, there are specialized structures, such as Malpighian tubules, lateral excretory canals, lateral canals with flame cells, structures called *nephridia,* and kidneys.

9. *Type of locomotion.* Does the organism swim, crawl on its belly, walk on legs, burrow in the substrate, or fly? Does it use cellular structures, such as cilia, to glide its body over the substrate?
10. *Support systems.* Is there a skeleton present? Is it an endoskeleton (inside the epidermis or skin of the animal), or is it an exoskeleton (outside the body wall)? Animals with no true skeleton can be supported by water: Fluid within and between cells and in body chambers such as a gastrovascular cavity or coelom provides a "hydrostatic skeleton."
11. *Segmentation.* Can you observe linear repetition of similar body parts? The repetition of similar units, or segments, is called *segmentation.* Segments

can be more similar (as in the earthworm) or less similar (as in a lobster). Can you observe any degree of segmentation? Have various segments become modified for different functions?

12. *Appendages.* Are there appendages (organs or parts attached to a trunk or outer body wall)? Are these appendages all along the length of the body, or are they restricted to one area? Are they all similar, or are they modified for different functions?
13. *Type of nervous system.* Do you see a brain and nerve cord? Is there more than one nerve cord? What is the location of the nerve cord(s)? Are sensory organs or structures present? Where and how many? Do you see signs of cephalization (the concentration of sensory equipment at the anterior end)? What purpose do such structures serve (for example, eyes for light detection)?

As you carefully study or dissect each organism, refer to these 13 characteristics, observe the animal, and record your observations in the summary table, Table 19.1, near the end of Lab Topic 19 Animal Diversity II. You may find it helpful to make sketches of difficult structures or dissections in the margin of your lab manual for future reference.

Before you begin this study, read Appendix E *and become thoroughly familiar with dissection techniques, orientation terms, and planes and sections of the body. Be able to use the terms associated with bilateral symmetry—anterior, posterior, dorsal, ventral, proximal, and distal—as you dissect and describe your animals.*

Wear gloves while dissecting preserved animals.

EXERCISE 18.1

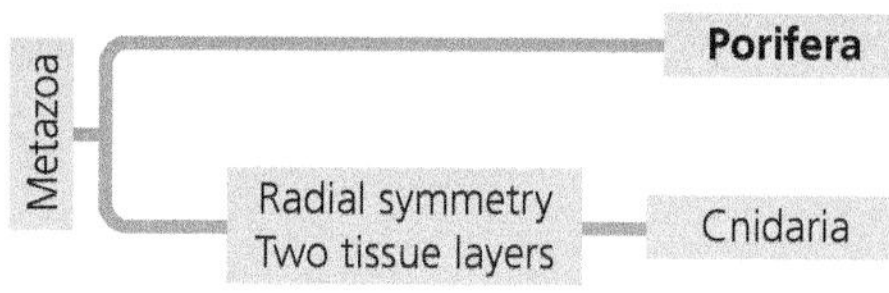

Phylum Porifera—Sponges (*Scypha*)

Materials

dissecting needle
compound microscope
stereoscopic microscope
preserved and dry bath sponges
prepared slide of *Scypha* in longitudinal section
preserved *Scypha* or *Grantia* in watch glass
microscope slides
depression slide
coverslips
forceps
dropper bottle with household bleach

Introduction

Within the animal kingdom, sponges are separated from all other animals because of their unique body form. You will observe the unique sponge structure by observing first a preserved specimen and then a prepared slide of a section taken through the longitudinal axis of the marine sponge *Scypha* (Figure 18.3). You will observe other more complex and diverse sponges on demonstration.

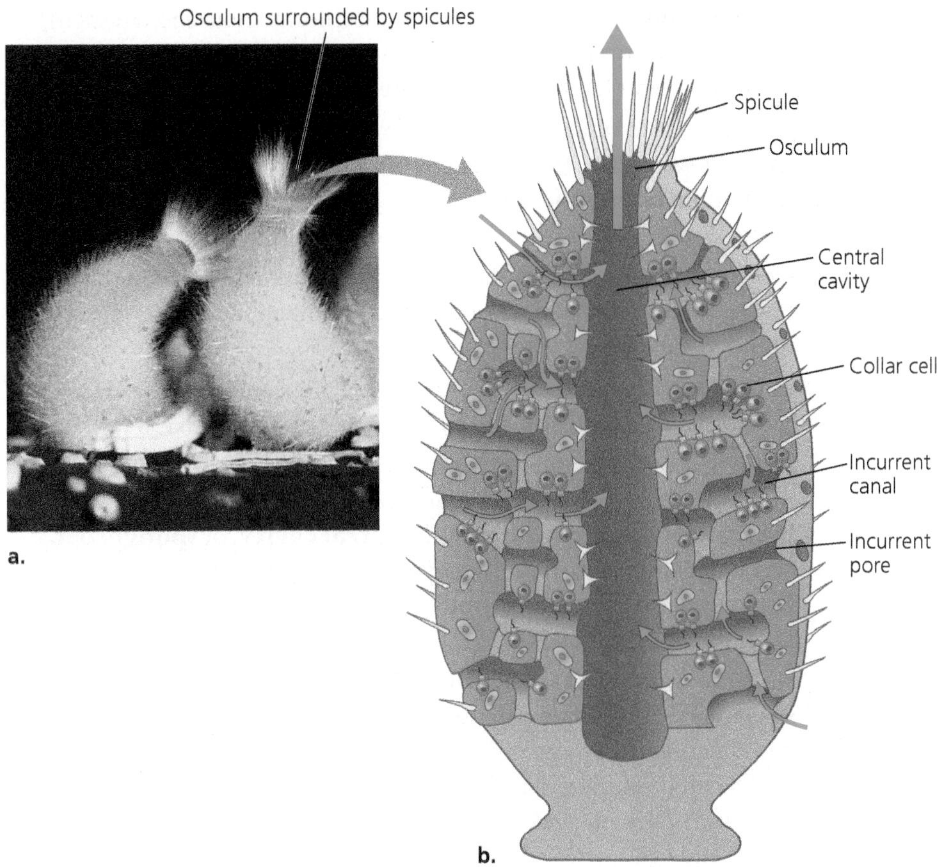

FIGURE 18.3
The sponge *Scypha*. (a) The entire sponge; (b) a longitudinal section through the sponge. Water passing through incurrent pores and canals passes into a central cavity, the spongocoel.

Procedure

1. Obtain the preserved sponge *Scypha* or *Grantia* and observe its external characteristics using the stereoscopic microscope, comparing your observations with Figure 18.3a.
   a. Note the vaselike shape of the sponge and the **osculum**, a large opening to the body at one end. The end opposite the osculum attaches the animal to the substrate.
   b. Note the invaginations in the body wall, which form numerous folds and channels. You may be able to observe needlelike **spicules** around the osculum and protruding from the surface of the body. These spicules are made of calcium carbonate; they give support and protection to the sponge body and prevent small animals from entering the sponge's internal cavity.
2. Isolate and observe the needlelike spicules from the sponge body using this procedure:
   a. Using forceps, remove a small piece of the preserved sponge and place it in the well of a *depression slide.*

b. Carefully add 2–3 drops of sodium hypochlorite (household bleach), and allow this to stand for approximately 15 minutes. The bleach will dissolve the soft tissue, leaving isolated spicules.

c. Use a pipette to carefully transfer a drop of liquid and debris from the depression slide to a clean microscope slide. Add a coverslip. Dry the bottom of the slide if needed and place it on the compound microscope.

d. Observe the slide, first on the lowest, then intermediate, and then on high power. Switch to phase-contrast, if available. Sketch several spicules in the margin of your lab manual.

3. Using the compound microscope, examine a prepared slide of a sponge body in longitudinal section and compare it with Figure 18.3b.

   a. Again, locate the osculum. This structure is not a mouth, as its name implies, but an opening used as an outlet for the current of water passing through the body wall and the **central cavity**, or **spongocoel**. The water enters the central cavity from channels and pores in the body. The central cavity is not a digestive tube or body cavity, but is only a channel for water.

   b. Note the structure of the body wall. Are cells organized into definite tissue layers (well-defined structural and functional units), or are they best described as a loose organization of various cell types? Various cells in the body wall carry out the functions of digestion, contractility, secretion of the spicules, and reproduction (some cells develop into sperm and eggs). One cell type unique to sponges is the **choanocyte**, or **collar cell**. These cells line the central cavity and the channels leading into it. Each collar cell has a flagellum extending from its surface. The collective beating of all flagella moves water through the sponge body. Small food particles taken up and digested by collar cells are one major source of nutrition for the sponge. How would you hypothesize about the movement of oxygen and waste throughout the sponge body and into and out of cells?

FIGURE 18.4
Bath sponges have a body form consisting of large and small canals and chambers.

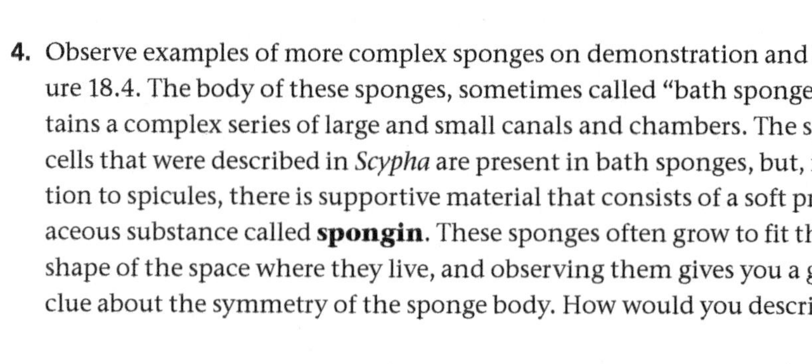

4. Observe examples of more complex sponges on demonstration and in Figure 18.4. The body of these sponges, sometimes called "bath sponges," contains a complex series of large and small canals and chambers. The same cells that were described in *Scypha* are present in bath sponges, but, in addition to spicules, there is supportive material that consists of a soft proteinaceous substance called **spongin**. These sponges often grow to fit the shape of the space where they live, and observing them gives you a good clue about the symmetry of the sponge body. How would you describe it?

Results

Complete the summary table, Table 19.1, near the end of Lab Topic 19, filling in all information for sponge characteristics in the appropriate row. This information will be used to answer questions in the Applying Your Knowledge section at the end of Lab Topic 19 Animal Diversity II.

EXERCISE 18.2

Phylum Cnidaria—Hydras (*Hydra*)

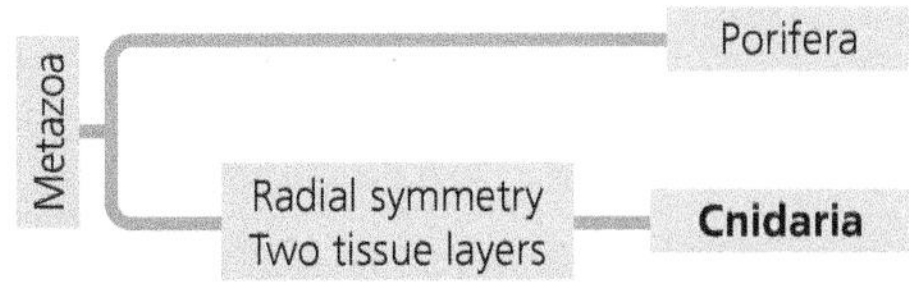

Materials

stereoscopic microscope
compound microscope
living *Hydra* culture
water flea culture
dropper bottles of water, 1% acetic acid, and methylene blue
prepared slide of *Hydra* sections
watch glass
depression slide
pipettes and bulbs
microscope slide and coverslip

Introduction

Cnidarians are a diverse group of organisms, all of which have a **tissue grade** of organization, meaning that tissues, but no complex organs, are present. Included in this group are corals, jellies, sea anemones, and Portuguese men-of-war. Most species are marine; however, there are a few freshwater species. Two body forms are present in the life cycles of many of these animals—an umbrella-like, free-swimming stage called a **medusa**, and a cylindrical, attached or stationary form called a **polyp**. The polyp forms often grow into colonies of individuals. Life cycles of most cnidarians include both polyp and medusa stages, but some species exist only as polyps or medusae. In this exercise you will observe some of the unique features of this group by observing the solitary freshwater organism *Hydra* (Figure 18.5).

Procedure

1. Place several drops of freshwater pond or culture water in a watch glass or depression slide. Use a dropper to obtain a living hydra from the class culture, and place the hydra in the drop of water. Hydra usually exist as single polyps that reproduce by budding, so occasionally you might see a smaller polyp growing from the side of a larger polyp. Using a stereoscopic microscope, observe the hydra structure and compare it with Figure 18.5a. Note any movement, the symmetry, and any body structures present. Note the **tentacles** that surround the "mouth," the only opening into the central cavity. Tentacles are used in capturing food and in performing a certain type of locomotion, much like a "handspring." To accomplish this motion, the hydra attaches its tentacles to the substrate and flips the basal portion of its body completely over, reattaching the base to a new position. If water fleas (*Daphnia*) are available, place one or two near the tentacles of the hydra and note the hydra's behavior. Set aside the hydra in the depression slide and return to it in a few moments.
2. Study a prepared slide of *Hydra* sections using the compound microscope and compare your observations with Figure 18.5b and c.

   Are definite tissue layers present? If so, how many?

   Given what you know of embryology, what embryonic layers would you guess give rise to the tissue layers of this animal's body?

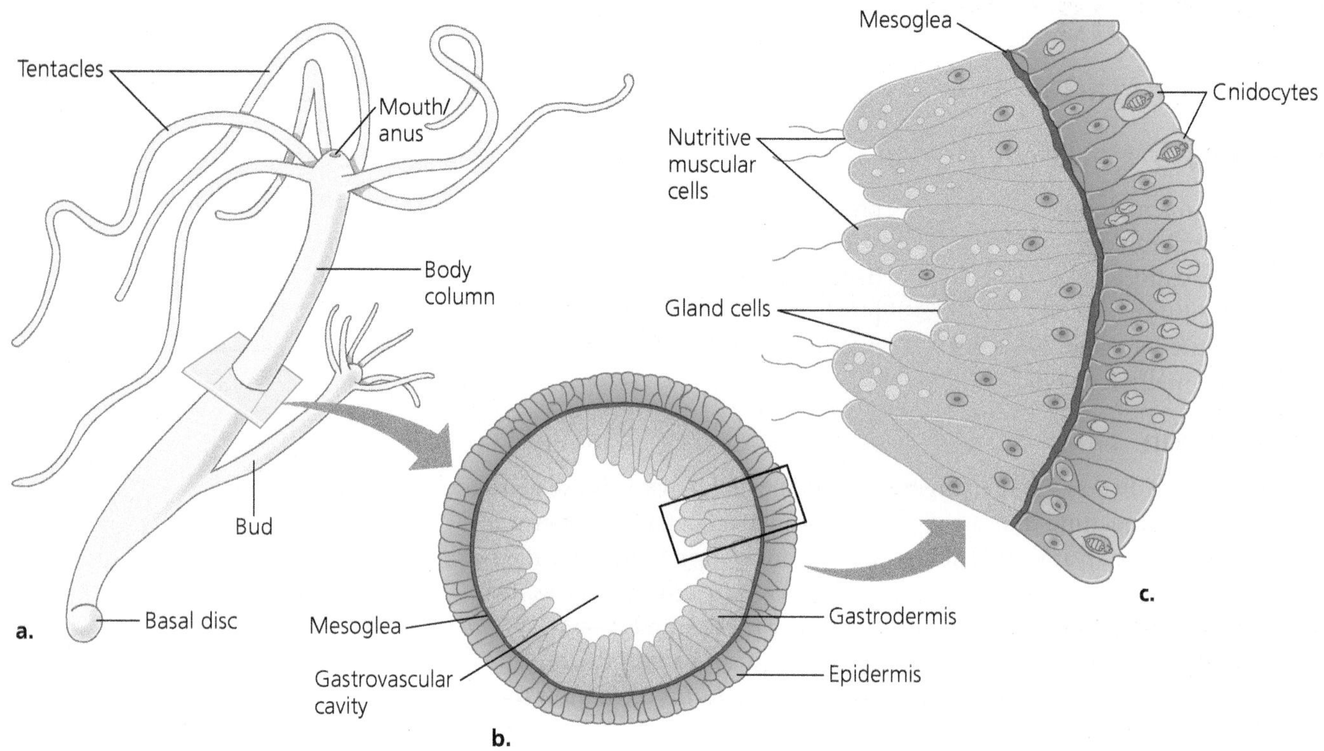

FIGURE 18.5

Hydra. (a) *Hydra* growing a new individual (a bud); (b) enlargement showing a cross section through the body wall, revealing two tissue layers; and (c) further enlargement showing details of specialized cells in the body wall, including cnidocytes.

3. Not visible with the microscope is a network of nerve cells in the body wall, which serves as the nervous system. There is no concentration of nerve cells into any kind of brain or nerve cord.
4. Observe the central cavity, called a **gastrovascular cavity**. Digestion begins in this water-filled cavity (**extracellular digestion**), but many food particles are drawn into cells in the **gastrodermis** lining the cavity, where **intracellular digestion** occurs.
5. Do you see signs of a skeleton or supportive system? How do you think the body is supported? Are appendages present?

6. Recalling the whole organism and observing this cross section, are organs for gas exchange present? How is gas exchange accomplished?

7. Do you see any organs for excretion?

8. Are specialized cell types seen in the layers of tissues?

Cnidarians have a unique cell type called **cnidocytes**, which contain a stinging organelle called a **nematocyst**. When stimulated, the nematocyst will evert from the cnidocyte with explosive force, trapping food or stinging predators. Look for these cells.

9. Return to your living hydra and water fleas in the depression slide. Has the hydra captured a flea? Can you tell how the flea was captured? Where is the flea now—do you see a bulge in the hydra body?

10. To better observe cnidocytes and nematocysts in your living hydra, follow this procedure:
    a. Using a pipette, transfer the hydra to a drop of water on a microscope slide and carefully add a coverslip.
    b. Use your microscope to examine the hydra, first on low, then intermediate, and finally on high power, focusing primarily on the tentacles. The cnidocytes will appear as swellings. If your microscope is equipped with phase contrast, switch to phase. Alternatively, add a drop of methylene blue to the edge of the coverslip. Locate several cnidocytes with nematocysts coiled inside.
    c. Add a drop of 1% acetic acid to the edge of the coverslip and, watching carefully using intermediate power, observe the rapid discharge of the nematocyst from the cnidocyte.
    d. Using high power, study the discharged nematocysts that will appear as long threads, often with large spines, or barbs, at the base of the thread.

Results

Complete the summary table, Table 19.1, recording all information for *Hydra* characteristics in the appropriate row. You will use this information to complete Table 19.2 and to answer questions in the Applying Your Knowledge section at the end of Lab Topic 19 Animal Diversity II.

Discussion

What major differences have you detected between *Scypha* and *Hydra* body forms? List and describe them.

MB

Student Media Videos—Ch. 33: *Hydra* Building; *Hydra* Eating *Daphnia*; Jelly Swimming; Thimble Jellies

EXERCISE 18.3

Phylum Platyhelminthes—Planarians (*Dugesia*)

Lophotrochozoa
Platyhelminthes
Mollusca
Annelida

Materials

stereoscopic microscope
compound microscope
living planarian
watch glass
transfer pipette
prepared slide of whole mount of planarian
prepared slide of planarian cross sections

Introduction

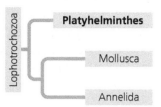

FIGURE 18.6
Dugesia, a freshwater planarian with two pigmented eyespots between the two auricles on its anterior end.

The phylum Platyhelminthes (clade Lophotrochozoa) includes both parasitic and free-living flatworms with thin bodies that are *dorsoventrally* flattened. Tapeworms and flukes are well-known parasites of humans and other animals. In this exercise you will investigate an example of a free-living flatworm, a planarian, *Dugesia*. They are found under rocks, leaves, and debris in freshwater ponds and creeks. They move over these surfaces using a combination of muscles in the body wall and cilia on their ventral sides (Figure 18.6).

Procedure

1. Add a dropperful of pond or culture water to a watch glass. Using a transfer pipette, *carefully* obtain a living planarian from the class culture and add it to the watch glass. Using your stereoscopic microscope, observe the planarian. Describe its locomotion. Is it directional? What is the position of its head? Does its body appear to contract?

   As you observe the living planarian, you will see two striking new features with regard to symmetry that you did not see in the two phyla previously studied. What are they?

2. Add a *small* piece of fresh liver to the water near the planarian. The planarian may approach the liver and begin to feed by extending a long tubular **pharynx** out of the **mouth**, a circular opening on the ventral side of the body. If the planarian feeds, it will curve its body over the liver and extend the pharynx, which may be visible in the stereoscopic microscope.

   After observing the planarian's feeding behavior, return it to the culture dish, if possible without the liver.

3. Using the lowest power on the compound microscope, observe the prepared slide of a whole planarian and compare it with Figure 18.7.

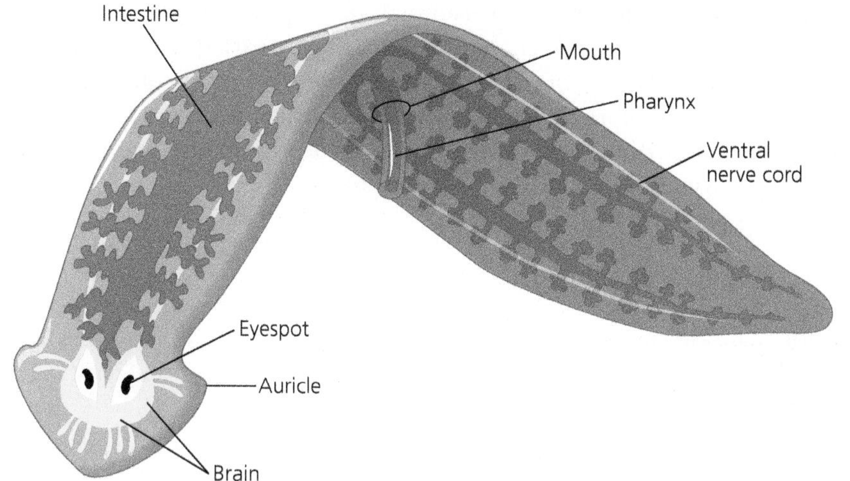

FIGURE 18.7
A planarian. The digestive system consists of a mouth, a pharynx, and a branched intestine. A brain and two ventral nerve cords (plus transverse nerves connecting them, not shown) make up the nervous system.

Do not observe these slides using high power! The high power objective may crack the coverslip, resulting in damage to the lens.

Examine the body for possible digestive tract openings. How many openings to the digestive tract are present?

Observe again the pharynx and the mouth. The pharynx lies in a **pharyngeal chamber** inside the mouth. The proximal end of the pharynx opens into a dark-colored, branched intestine. If the intestine has been stained on your slide, you will see the branching more easily.

4. Continue your study of the whole planarian. The anterior blunt end of the animal is the head end. At each side of the head is a projecting **auricle**. It contains a variety of sensory cells, chiefly of touch and chemical sense. Between the two auricles on the dorsal surface are two pigmented **eyespots**. These are pigment cups into which retinal cells extend from the brain, with the photosensitive end of the cells inside the cup. Eyespots are sensitive to light intensities and the direction of a light source but can form no images. Beneath the eyespots are two cerebral ganglia that serve as the **brain**. Two ventral nerve cords extend posteriorly from the brain. These are connected by transverse nerves to form a ladderlike **nervous system**.
5. Study the prepared slide of cross sections of a planarian. You will have several sections on one slide. One section should have been taken at the level of the pharynx and pharyngeal chamber. Do you see a body cavity in any of the sections? (The pharyngeal chamber and spaces in the gut are not a body cavity.) What word describes this body cavity condition (see Figure 18.2a)?

   a. How many tissue layers can be detected? Speculate about their embryonic origin.

Flatworms are the first group of animals to have three well-defined embryonic tissue layers, enabling them to have a variety of tissues and organs. Reproductive organs and simple excretory organs consisting of two lateral excretory canals and "flame cells" that move fluid through the canals are derived from the embryonic mesoderm. Respiratory, circulatory, and skeletal systems are lacking.

b. How do you think the body is supported?

c. How does gas exchange take place?

Results

1. In the space on the following page, diagram the flatworm as seen in a cross section at the level of the pharynx. Label the **epidermis**, **muscle** derived from **mesoderm**, the lining of the digestive tract derived from **endoderm**, the **pharynx**, and the **pharyngeal chamber**.

2. Complete the summary table, Table 19.1, recording all information for planarian characteristics in the appropriate row. You will use this information to complete Table 19.2 and answer questions in the Applying Your Knowledge section at the end of Lab Topic 19 Animal Diversity II.

Discussion

One of the major differences between Cnidaria and Platyhelminthes is radial versus bilateral symmetry. Discuss the advantage of radial symmetry for sessile (attached) animals and bilateral symmetry for motile animals.

EXERCISE 18.4

Phylum Mollusca—Clams

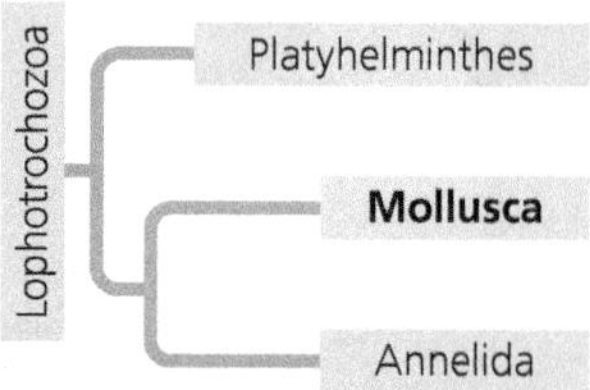

Materials

dissecting instruments
dissecting pan
preserved clam or mussel
disposable gloves

Introduction

Second only to the phylum Arthropoda in numbers of species, the phylum Mollusca (clade Lophotrochozoa) includes thousands of species living in many diverse habitats. Most species are marine. Others live in fresh water or on land. Many molluscs are of economic importance, being favorite human foods. Molluscs include such diverse animals as snails, slugs, clams, squids, and octopuses. Although appearing diverse, most of these animals share four characteristic features: (1) a hard external **shell** for protection; (2) a thin structure called the **mantle**, which secretes the shell; (3) a **visceral mass** in which most organs are located; and (4) a muscular **foot** used for locomotion.

In this exercise, you will dissect a clam, a molluscan species in the clade Bivalvia with a shell made of two parts called **valves**. Most clams are marine, although many genera live in freshwater lakes and ponds. You may have eaten molluscs, including clams, oysters, and scallops, which are all bivalves.

Wear gloves while dissecting preserved animals.

Procedure

1. Observe the external anatomy of the preserved clam. Certain characteristics will become obvious immediately. Can you determine symmetry, support systems, and the presence or absence of appendages? Are there external signs of segmentation? Record observations in Table 19.1.
2. Before you continue making observations, determine the dorsal, ventral, anterior, posterior, right, and left regions of the animal. Identify the two valves. The valves are held together by a **hinge** near the **umbo**, a hump on the valves. The hinge and the umbo are located **dorsally**, and the valves open **ventrally**. The umbo is displaced **anteriorly**. Hold the clam vertically with the umbo away from your body, and cup one of your hands over each valve. The valve in your right hand is the right valve; the valve in your left hand is the left valve. The two valves are held together by two strong **adductor** muscles inside the shell. Compare your observations with Figure 18.8.

Be cautious as you open the clam! Hold the clam in the dissecting pan in such a way that the scalpel will be directed toward the bottom of the pan.

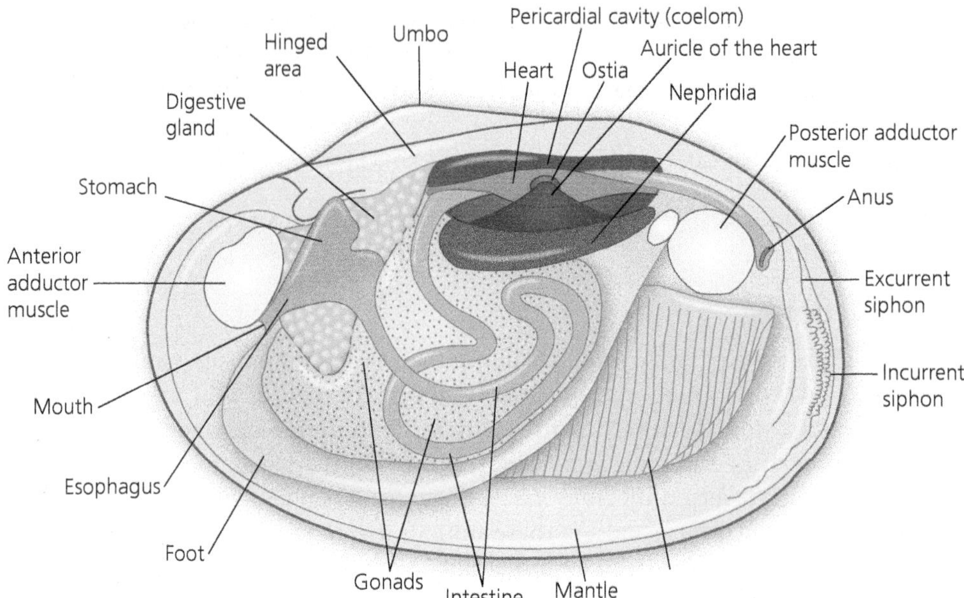

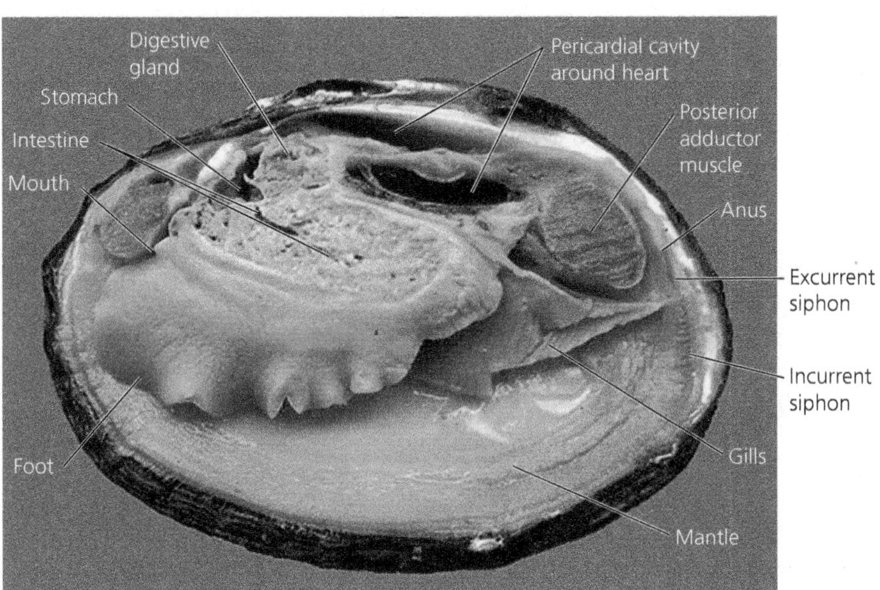

FIGURE 18.8
Anatomy of a clam. The soft body parts are protected by the shell valves. Two adductor muscles hold the valves closed. Most major organs are located in the visceral mass. Note in this figure the left mantle, the left pair of gills, and half of the visceral mass have been removed.

3. To study the internal anatomy of the clam, you must open it by prying open the valves. (A wooden peg may have been inserted between the two valves.) Insert the handle of your forceps or scalpel between the valves and twist it to pry the valves farther open. Place your clam in the dissecting pan with the clam's dorsal side supported on the pan bottom. This will allow you to make your cuts with the scalpel blade directed toward the pan bottom and not your hand. Carefully insert a scalpel blade, directed toward the dorsal side of the animal, into the space between the left valve and a flap of tissue lining the valve. The blade edge should be just ventral to (that is, below) the anterior adductor muscle (see Figure 18.8). The flap of tissue is the left **mantle**. Keeping the scalpel blade pressed flat against the left valve, carefully loosen the mantle from the valve and press the blade dorsally. You will feel the tough **anterior adductor muscle**. Cut through this muscle near the valve.

4. Repeat the procedure at the posterior end and cut the posterior adductor muscle. Lay the clam on its right valve and carefully lift the left valve. As you do this, use your scalpel to loosen the mantle from the valve. If you have been successful, you should have the body of the clam lying in the right valve. It should be covered by the mantle. Look for pearls between the mantle and the shell. How do you think pearls are formed?

5. Look at the posterior end of the animal where the left and right mantle come together. Hold the two mantle flaps together and note the two gaps formed. These gaps are called **incurrent** (ventral) and **excurrent** (dorsal) **siphons**. Speculate about the function of these siphons (see Figure 18.9).

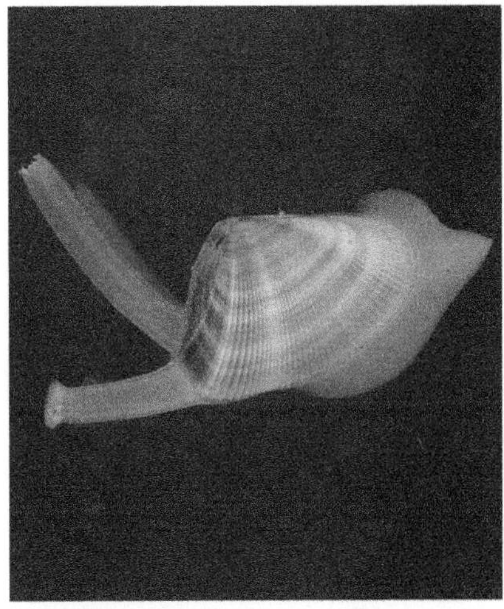

FIGURE 18.9
Donax, a small bivalve often seen in the lower intertidal zone of beaches. Two tubular siphons are seen extending out of the shells. These siphons are more easily identified in *Donax* than in the clam that you are dissecting, but they have the same function. Notice the extended foot used for burrowing in sand.

6. Lift the mantle and identify the **visceral mass** and the **muscular foot**.
7. Locate the **gills**, which have a pleated appearance. One function of these structures is obvious, but they have a second function as well. As water comes into the body (how would it get in?), it passes through the gills, and food particles are trapped on the gill surface. The food is then moved anteriorly (toward the mouth) by coordinated ciliary movements.
8. Locate the **mouth** between two flaps of tissue just ventral to the anterior adductor muscle. Look just above the posterior adductor muscle and locate the **anus**. How is it oriented in relation to the excurrent siphon?

9. Imagine that this is the first time you have seen a clam. From the observations you have made, what evidence would indicate whether this animal is aquatic or terrestrial?

10. The **heart** of the clam is located in a sinus, or cavity, just inside the hinge, dorsal to the visceral mass (see Figure 18.8). This cavity, called the *pericardial cavity,* is a reduced **true coelom**. The single ventricle of the heart actually surrounds the **intestine** passing through this cavity. Thin auricles, usually torn away during the dissection, empty into the heart via openings called **ostia**. Blood passes from **sinuses** in the body into the auricles. What type of circulatory system is this?

11. Ventral to the heart and embedded in mantle tissue you will find a pair of greenish-brown tissue masses, the **nephridia**, or kidneys. The kidneys remove waste from the pericardial cavity.
12. Open the visceral mass by making an incision with the scalpel, dividing the mass into right and left halves. Begin this incision just above the foot and cut dorsally. You should be able to open the flap produced by this cut and see organs such as the **gonads**, **digestive gland**, **intestine**, and **stomach**. Clam chowder is made by chopping up the visceral mass.

13. It is difficult to observe the nervous system in the clam. It consists of three ganglia, one near the mouth, one in the foot, and one below the posterior adductor muscle. These ganglia are connected by nerves.

    Now that you have dissected the clam, you should have concluded that there is no sign of true segmentation. Also, appendages (attached to a trunk or body wall) are absent.

Results

Complete the summary table, Table 19.1, recording all information for clam characteristics in the appropriate row. Use this information to complete Table 19.2 and to answer questions in the Applying Your Knowledge section at the end of Lab Topic 19 Animal Diversity II.

Discussion

List several features of clam anatomy that enable it to survive in a marine environment.

EXERCISE 18.5

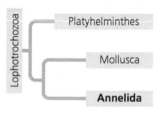

Phylum Annelida—Clamworms (*Nereis*) and Earthworms (*Lumbricus terrestris*)

Lophotrochozoa
Platyhelminthes
Mollusca
Annelida

The phylum Annelida (clade Lophotrochozoa) includes a diverse group of organisms that have adapted to a variety of environments. Examples range in size from microscopic to several meters in length. Most species are marine, living free in the open ocean or burrowing in ocean bottoms. Others live in fresh water or in soils. Members of one group of annelids, the leeches, are parasitic and live on the blood or tissues of their hosts. The ability of annelids to inhabit such a wide range of environments is due in part to their segmented body design. In annelids, segmentation (the linear repetition of body segments) includes both external and internal structures of several systems. This is not true for two other phyla of animals that demonstrate segmentation, Arthropods and Chordates, studied in Lab Topic 19. In this exercise, you will study the clamworm, a marine annelid, and the earthworm, a terrestrial species. Keep in mind features that are adaptations to marine and terrestrial habitats as you study these organisms.

Lab Study A. Clamworms (*Nereis*)

Materials

dissecting tools
dissecting pan
preserved clamworm
disposable gloves
dissecting pins

Introduction

Species of *Nereis* (clamworms) are commonly found in mudflats and on the ocean floor. People who live near the ocean often use these worms for fishing bait much like earthworms are used inland. These animals burrow in sediments during the day and emerge to feed at night. As you observe the clamworm, note features that are characteristic of all annelids, as well as features that are special adaptations to the marine environment (Figure 18.10).

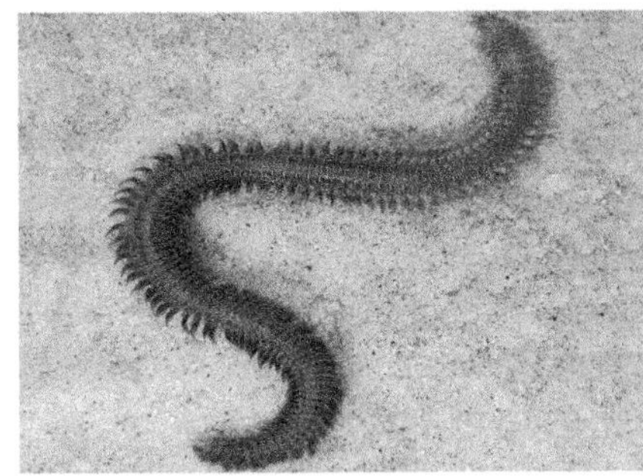

FIGURE 18.10
Nereis, a marine worm commonly found in mudflats and burrowing in the ocean floor.

Procedure

1. Observe the preserved, undissected clamworm and compare it with Figure 18.11. How would you describe the symmetry of this organism?
2. Determine the anterior and posterior ends. At the anterior end, the well-differentiated head bears **sensory appendages**. Locate the mouth, which leads into the digestive tract.
3. A conspicuous new feature of these organisms is the presence of **segmentation**, the division of the body along its length into segments. In annelids, segmentation includes both external and internal structures of several systems. Posterior to the head region, the segments bear fleshy outgrowths called **parapodia**. Each parapodium contains several terminal bristles called **chaetae**. In Lab Study B, you will see that the earthworm has chaetae but does not have parapodia. Suggest functions for parapodia and chaetae in the marine clamworm.
4. Holding the animal in your hand and using sharp-pointed scissors, make a middorsal incision the full length of the body. Carefully insert the tip of the scissors and lift up with the tips as you cut. Pin the opened body in the dissecting pan but do not put pins through the head region.
5. Locate the **intestine**. Do you see the "tube-within-a-tube" body plan?

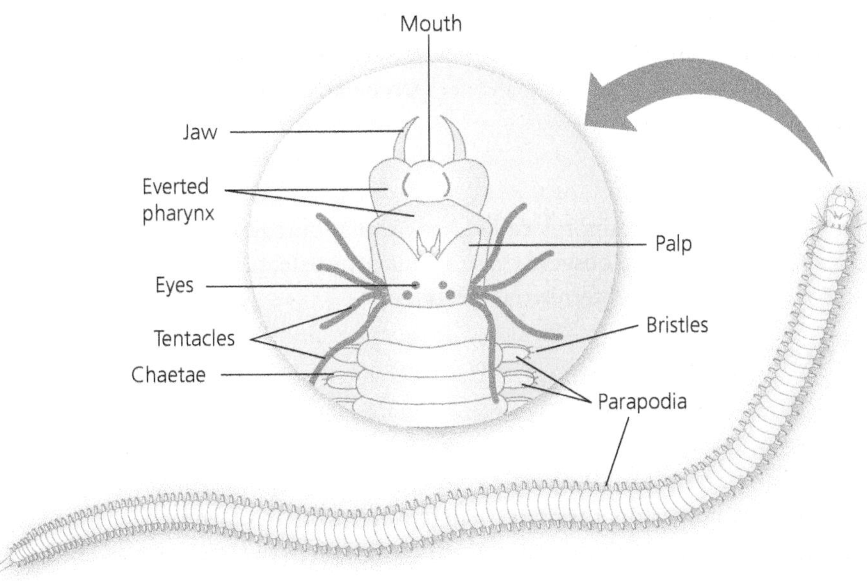

FIGURE 18.11
The clamworm, *Nereis.* The head has sensory appendages, and each segment of the body bears two parapodia with chaetae.

6. Two **muscle layers**, one inside the skin and a second lying on the surface of the intestine, may be visible with the stereoscopic microscope. With muscle in these two positions, what kind of coelom does this animal have (see Figure 18.2c)?

7. Continuing your observations with the unaided eye and the stereoscopic microscope, look for **blood vessels**, particularly a large vessel lying on the dorsal wall of the digestive tract. This vessel is contractile and propels the blood throughout the body. You should be able to observe smaller lateral blood vessels connecting the dorsal blood vessel with another on the ventral side of the intestine. As you will see, in the earthworm these connecting vessels are slightly enlarged as "hearts" around the anterior portion of the digestive tract (around the esophagus). This is not as obvious in *Nereis*. What is this type of circulatory system, with blood circulating through continuous closed vessels?

8. Gas exchange must take place across wet, thin surfaces. Do you see any organs for gas exchange (gills or lungs, for example)? How do you suspect that gas exchange takes place?

9. Do you see any signs of a skeleton? What would serve as support for the body?

10. Clamworms and earthworms have a small bilobed brain (a pair of ganglia) lying on the surface of the digestive tract at the anterior end of the worm. You can see this more easily in an earthworm.

Lab Study B. Earthworms (*Lumbricus terrestris*)

Materials

dissecting instruments
compound microscope
stereoscopic microscope
preserved earthworm
prepared slide of cross section of earthworm

Introduction

Lumbricus species, commonly called *earthworms,* burrow through soils rich in organic matter. As you observe these animals, note features that are adaptations to the burrowing, terrestrial lifestyle.

Procedure

1. Obtain a preserved earthworm and identify its anterior end by locating the mouth, which is overhung by a fleshy dorsal protuberance called the **prostomium**. The anus at the posterior end has no such protuberance. Also, a swollen glandular band, the **clitellum** (a structure that secretes a cocoon that holds eggs), is located closer to the mouth than to the anus (Figure 18.12).

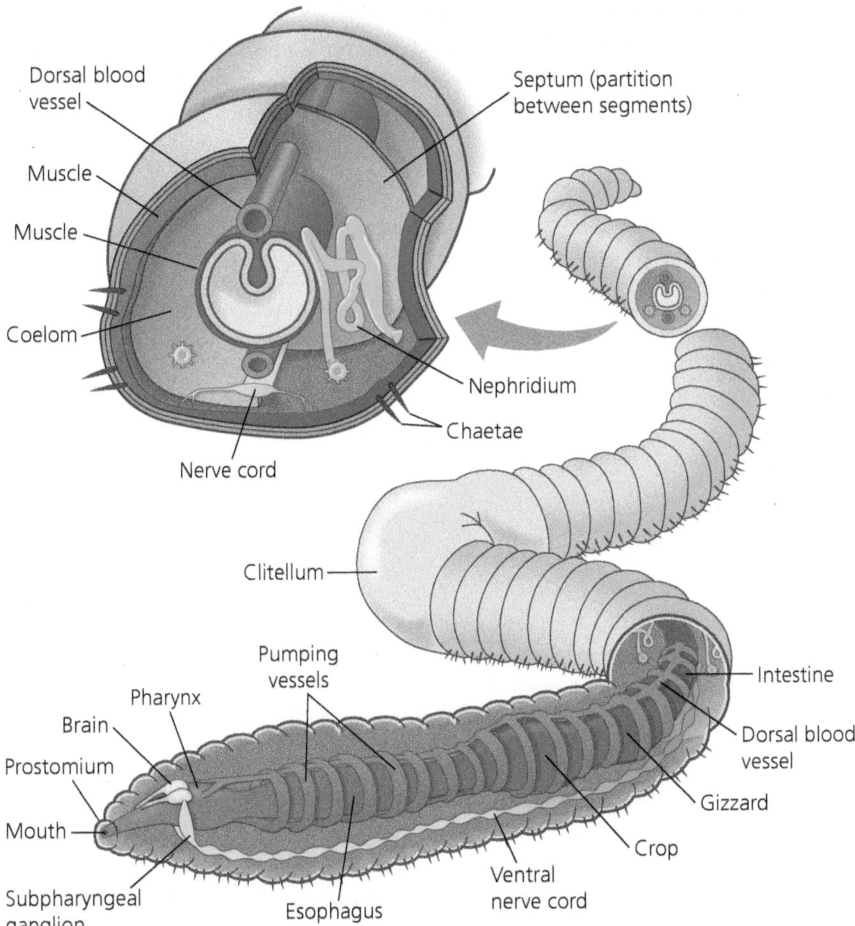

FIGURE 18.12

The earthworm. The small brain leads to a ventral nerve cord. A pair of nephridia lies in each segment.

a. Using scissors, make a middorsal incision along the anterior third of the animal, as you did for *Nereis*. You can identify the dorsal surface in a couple of ways. The prostomium is dorsal, and the ventral surface of the worm is usually flattened, especially in the region of the clitellum. Cut to the prostomium. Pin the body open in a dissecting pan near the edge. You may need to cut through the septa that divide the body cavity into segments.

b. Using a stereoscopic microscope or hand lens, look for the small **brain** just behind the prostomium on the surface of the digestive tract. Note the two nerves that pass from the brain around the pharynx and meet ventrally. These nerve tracts continue posteriorly as a **ventral nerve cord** lying in the floor of the coelom.

c. Look for the large **blood vessel** on the dorsal wall of the digestive tract. You may be able to see the enlarged lateral blood vessels (**hearts**) around the anterior portion of the digestive tract.

d. Identify (from anterior to posterior) the **pharynx**, **esophagus**, **crop** (a soft, swollen region of the digestive tract), **gizzard** (smaller and more rigid than the crop), and **intestine**.

e. Excretion in the clamworm and earthworm is carried out by organs called **nephridia**. A pair of these minute, white, coiled tubes is located in each segment of the worm body. Nephridia are more easily observed in the earthworm than in *Nereis* and should be studied here. To view these organs, cut out an approximately 2-cm-long piece of the worm posterior to the clitellum and cut it open along its dorsal surface. Cut through the septa and pin the piece to the dissecting pan near the edge to facilitate observation with the stereoscopic microscope. The coiled tubules of the nephridia are located in the coelomic cavity, where waste is collected and discharged to the outside through a small pore.

2. Using the compound microscope, observe the prepared slide of a cross section of the earthworm.

   a. Locate the **thin cuticle** lying outside of and secreted by the **epidermis**. Recall the habitat of this organism and speculate about the function of the cuticle.

   b. Confirm your decision about the type of coelom by locating **muscle layers** inside the epidermis and also lying on the surface of the **intestine** near the body cavity.

   c. Locate the **ventral nerve cord**, lying in the floor of the coelom, just inside the muscle layer.

Results

Complete the summary table, Table 19.1, recording all information for clamworm and earthworm characteristics in the appropriate row. You will use this information to complete Table 19.2 and answer questions in the Applying Your Knowledge section at the end of Lab Topic 19 Animal Diversity II.

Student Media Video—Ch. 33: Earthworm Locomotion

Discussion

A major new feature observed in the phylum Annelida is the segmented body. Speculate about possible adaptive advantages provided by segmentation.

By the end of today's laboratory period, you should have completed observations of all animals described in Animal Diversity I. The next lab topic, Animal Diversity II, is a continuation of this investigation and will present similar laboratory objectives. In Animal Diversity II, you will continue asking questions and making comparisons as you did in Animal Diversity I. By the end of the two lab topics, you should be able to use what you have learned about the animals to discuss and answer questions about the unifying themes of these laboratory topics.

REVIEWING YOUR KNOWLEDGE

1. What are the characteristics of sponges that have led scientists to classify them in a group separate from the Eumetazoa?

2. Zoologists have described two modes of development in animals in the clade Bilateria: (1) protostome development seen in animals in the clades Ecdysozoa and Lophotrochozoa, and (2) deuterostome development as in the phyla Echinodermata and Chordata. What criteria are used to distinguish between these two modes of development?

3. Molecular evidence has led taxonomists to group bilaterally symmetrical animals in one of three clades, Deuterostomia, Lophotrochozoa, or Ecdysozoa. Give one distinguishing characteristic of each of these groups other than molecular differences.

APPLYING YOUR KNOWLEDGE

1. A hydra (*Chlorophyra viridissima*) is bright green, and yet it does not synthesize chlorophyll. Think about the structure of the hydra and its feeding and digestive habits. What do you think is the origin of the green pigment in this species?

INVESTIGATIVE EXTENSIONS

1. Earthworms are among the most familiar inhabitants of soil. They play an important role in improving the texture of and adding organic matter to soil. You may have read Darwin's estimate that over 50,000 earthworms may inhabit one acre of British farmland. Earthworms are readily available from biological supply houses, or you may collect your own to use in experiments. Following are questions that you might investigate.
   a. Why do earthworms come out of their burrows when it rains? Is it because they may drown in the water in their burrows? Does rain

stimulate mating behavior, and the worms are coming to the surface to mate? Does the pH of the soil change as it rains, and is the burrow becoming too acidic or alkaline? What *is* the optimum pH range for earthworms? Does rain create conditions more favorable for migration to new habitats?

   b. What effects do chemicals used in agriculture have on earthworm populations? Compare numbers and health of earthworms in containers of soil to which varying amounts of fertilizers, pesticides, or herbicides have been added.

   c. Do earthworms in the soil stimulate plant growth? Compare the biomass of plants grown in containers with and without earthworms present.

2. An amazing diversity of organisms has evolved from the foot–mantle–visceral mass body plan of molluscs. Living terrestrial, freshwater, and marine snails and bivalves are available from biological supply houses, aquarium supply stores, or from ponds or terrestrial sites in your area. Consider the following questions that you might investigate. (A Web search for "snail experiments" yields over 1,140,000 entries, including experiments being performed in the International Space Station.)

   a. What effect does sedimentation have on aquatic snail populations? Consider changes in water chemistry and/or substrate.

   b. What effect does temperature have on the growth and/or reproduction of aquatic or terrestrial snails or slugs? Why would this question be of interest?

   c. Invasive aquatic plants have become a major concern of scientists worldwide. For example, water hyacinth, introduced into ponds in the southern United States, chokes ponds and waterways, in some cases hindering human and fish navigation. Ponds may become so choked that they dry up, destroying habitat for alligators, turtles, fish, and other native species. Design a greenhouse experiment to test the efficacy of aquatic snails in controlling the growth of invasive aquatic plants.

3. Review regeneration in planaria in the lab activity found at http://www.hhmi.org/biointeractive/activities/planaria/planaria_regen_activity.pdf. Use this as the basis for designing an independent investigation of regeneration in planaria.

STUDENT MEDIA: BioFlix, Activities, Investigations, and Videos

www.masteringbiology.com (select Study Area)

Activities—Ch. 32: Animal Phylogenetic Tree; Ch. 33: Characteristics of Invertebrates

Investigations—Ch. 32: How Do Molecular Data Fit Traditional Phylogenies?

Videos—Ch. 33: *Hydra* Budding; *Hydra* Eating *Daphnia*; *Hydra* Releasing Sperm; Jelly Swimming; Thimble Jellies; Earthworm Locomotion

REFERENCES

Adoutte, A., G. Balavoine, N. Lartillot, O. Lespinet, B. Prud'homme, and R. de Rosa. "The New Animal Phylogeny: Reliability and Implications." *Proc. Natl. Acad. Sci. USA*, 2000, vol. 97(9), pp. 4453–4456.

Balavoine, G. "Are Platyhelminthes Coelomates Without a Coelom? An Argument Based on the Evolution of *Hox* Genes." *American Zoologist*, 1989, vol. 38, pp. 843–858.

DeSalle, R. and B. Schierwater. "An Even 'Newer' Animal Phylogeny." *BioEssays*, 2008, 30, pp. 1043–1047.

Erwin, D., J. Valentine, and D. Jablonski. "The Origin of Animal Body Plans." *American Scientist*, 1997, vol. 85, pp. 126–137.

Halanych, K. M. "The New View of Animal Phylogeny." *Annual Review of Ecology, Evolution, and Systematics*, 2004, vol. 35, pp. 229–256.

Mallatt, J. and C. Winchell. "Testing the New Animal Phylogeny: First Use of Combined Large-Subunit and Small-Subunit rRNA Sequences to Classify Protostomes." *Molecular Biology and Evolution*, 2002, vol. 19, pp. 289–301.

WEBSITES

Encyclopedia of Life—an online reference source and database for all known species:
http://www.eol.org

Planaria as a model organism for understanding stem cell biology:
http://www.hhmi.org/biointeractive/stemcells/planarian_regen.html

The Tree of Life Web project provides information on all major groups of organisms, including invertebrates:
http://tolweb.org/Animals/2374

University of Michigan Museum of Zoology, Animal Diversity Web. Includes descriptions of many invertebrates and vertebrates, links to insect keys, and references:
http://animaldiversity.ummz.umich.edu/site/index.html

NOTES

LAB TOPIC 19

Animal Diversity II: Nematoda, Arthropoda, Echinodermata, and Chordata

This lab is a continuation of observations of organisms in the clade Metazoa as discussed in Animal Diversity I. Return to Lab Topic 18 and review the objectives of the lab on page 473. Review the descriptions of the 13 characteristics you are investigating in the study and dissection of these animals (pp. 476–478).

In this lab topic you will study examples of two phyla included in the clade **Ecdysozoa**, Nematoda (Exercise 19.1), and Arthropoda (Exercise 19.2). These organisms have coverings on their body surfaces (exoskeletons) that they shed as they grow, a process called *ecdysis* or *molting*. In Exercises 19.3 and 19.4, you will study two phyla in the clade **Deuterostomia**, Echinodermata and Chordata.

As you continue your study of representative organisms, continue to record your observations in Table 19.1 near the end of this lab topic. Keep in mind the big themes you are investigating:

1. What clues do similarities and differences among organisms provide about phylogenetic relationships?
2. How is body form related to function?
3. How is body form related to environment and lifestyle?
4. What characteristics can be the criteria for major branching points in producing a phylogenetic tree (representing animal classification)?

EXERCISE 19.1

Phylum Nematoda—Roundworms (*Ascaris*)

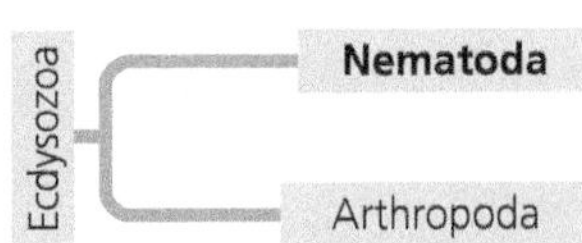

Materials

dissecting instruments
dissecting pan
dissecting pins
compound microscope
disposable gloves
preserved *Ascaris*
prepared slide of cross section of *Ascaris*
hand lens (optional)

Introduction

Roundworms, or nematodes (clade Ecdysozoa), are among the most abundant and widely distributed organisms on earth. One species of roundworm, *Caenorhabditis elegans*, has become a model research organism in biology. NASA has used *C. elegans* in experiments to test the effects of space flight on human physiology. The resilience of these organisms became apparent when *C. elegans* specimens survived the Space Shuttle *Columbia* disaster in February, 2003. In 2006, the Nobel Prize in Medicine was awarded to two scientists for their studies of gene regulation using this soil worm as their research organism, and *C. elegans* continues to be important in many disciplines of biological research around the world. *Caenorhabditis elegans* is a small roundworm—only 1 millimeter in length. *Ascaris*, the roundworm you will study in this exercise, is considerably larger.

Recall that ecdysozoans secrete exoskeletons that must be shed as the animal grows. Nematodes are covered with a proteinaceous **cuticle** that sheds periodically. *Ascaris* lives as a parasite in the intestines of mammals such as horses, pigs, and humans. Most often these parasites are introduced into the mammalian body when food contaminated with nematode eggs is eaten. Keep in mind the problem of adaptation to a parasitic lifestyle as you study the structure of this animal.

Wear disposable gloves while dissecting preserved animals.

Procedure

1. Wearing disposable gloves, obtain a preserved *Ascaris* and determine its sex. Females are generally larger than males. The posterior end of the male is sharply curved.
2. Use a hand lens or a stereoscopic microscope to look at the ends of the worm. A mouth is present at the anterior end. Three "lips" border this opening. A small slitlike **anus** is located ventrally near the posterior end of the animal.
3. Open the animal by making a middorsal incision along the length of the body with a sharp-pointed probe or sharp scissors. Remember that the anus is slightly to the ventral side (Figure 19.1). Be careful not to go too deep. Once the animal is open, pin the free edges of the body wall to the dissecting pan, spreading open the body. Pinning the animal near the edge of the pan will allow you to view it using the stereoscopic microscope. As you study the internal organs, you will note that there is a **body cavity**. This is not a so-called "true" coelom, however, as you will see shortly when you study microscopic sections. From your observations, you should readily identify such characteristics as symmetry, tissue organization, and digestive tract openings.
   a. The most obvious organs you will see in the dissected worm are **reproductive organs**, which appear as masses of coiled tubules of varying diameters.
   b. Identify the flattened **digestive tract**, or intestine, extending from mouth to anus. This tract has been described as a "tube within a tube," the outer tube being the body wall.
   c. Locate two pale lines running laterally along the length of the body in the body wall. The excretory system consists of two longitudinal tubes lying in these two **lateral lines**.

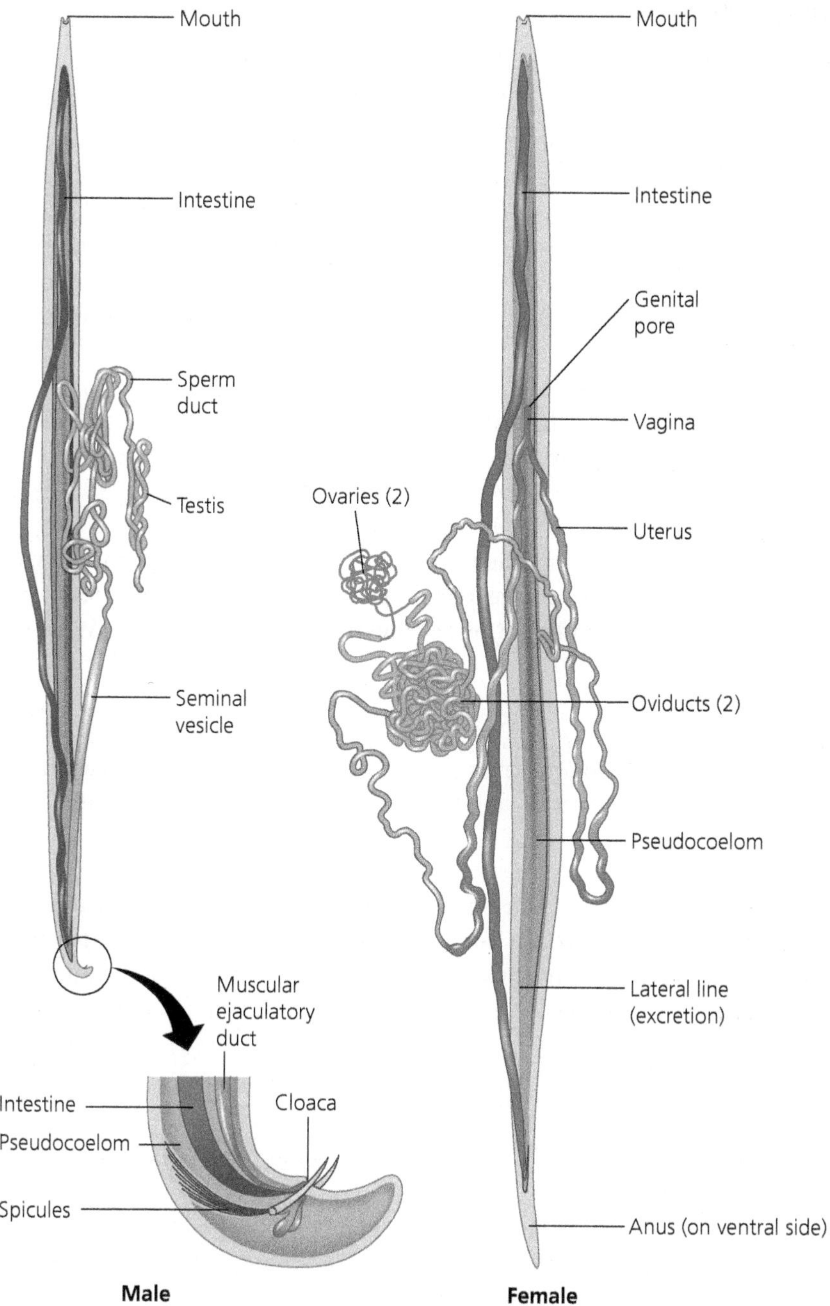

FIGURE 19.1
Male and female *Ascaris*. The digestive tract originates at the mouth and terminates in the anus. Reproductive structures fill the body cavity.

d. There are no organs for gas exchange or circulation. Most parasitic roundworms are essentially anaerobic (require no oxygen).

e. How would nourishment be taken into the body and be circulated?

f. The nervous system consists of a ring of nervous tissue around the anterior end of the worm, with one dorsal and one ventral nerve cord. These structures will be more easily observed in the prepared slide.

g. Do you see signs of segmentation in the body wall or in the digestive, reproductive, or excretory systems?

h. Do you see signs of a support system? What do you think supports the body?

4. Using the compound microscope, observe a prepared slide of a cross section through the body of a female worm. Note that the body wall is made up of (from outside inward) **cuticle** (noncellular), **epidermis** (cellular), and **muscle fibers**. The muscle (derived from mesoderm) lies at the outer boundary of the body cavity. Locate the **intestine** (derived from endoderm). Can you detect muscle tissue adjacent to the endodermal layer?

   What do we call a coelom that is lined by mesoderm (outside) and endoderm (inside) (see Figure 18.2b)?

5. Most of the body cavity is filled with reproductive organs. You should see cross sections of the two large **uteri**, sections of the coiled **oviducts** with small lumens, and many sections of the **ovaries** with no lumen. What do you see inside the uteri?

6. By carefully observing the cross section, you should be able to locate the **lateral lines** for excretion and the dorsal and ventral **nerve cords**.

Results

1. Sketch the cross section of a female *Ascaris*. Label the **cuticle**, **epidermis**, **muscle fibers**, **intestine**, **body cavity** (give specific name), **reproductive organs** (**uterus**, **oviduct**, **ovary**), **lateral lines**, and **dorsal** and **ventral nerve cords**.

2. List some features of *Ascaris* that are possible adaptations to parasitic life.

3. Complete the summary table, Table 19.1, recording all information for roundworm characteristics in the appropriate row. You will use this information to complete Table 19.2 and answer questions in the Applying Your Knowledge section at the end of this lab topic.

Student Media Video—Ch. 33: *C. elegans* Crawling

Discussion

1. Discuss the significance of an animal's having two separate openings to the digestive tract, as seen in *Ascaris*.

2. What are the advantages of a body cavity being present in an animal?

EXERCISE 19.2

Phylum Arthropoda

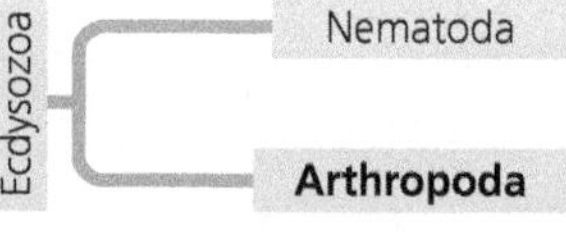

With astronomical numbers of individuals and more than a million species identified, the phylum Arthropoda (clade Ecdysozoa) is the largest and most diverse phylum of animals. Evidence indicates that arthropods first lived on Earth over half a billion years ago. They can be found in almost every imaginable habitat: marine waters, fresh water, and almost every terrestrial niche. Many species are directly beneficial to humans, serving as a source of food. Others make humans miserable by eating their homes, infesting their domestic animals, eating their food, and biting their bodies. These organisms have an exoskeleton that periodically sheds as they grow. In this exercise, you will observe the morphology of two arthropods: the crayfish (an aquatic arthropod) and the grasshopper (a terrestrial arthropod).

Lab Study A. Crayfish (*Cambarus*)

Materials

dissecting instruments
dissecting pan
preserved crayfish
disposable gloves

Introduction

Crayfish live in streams, ponds, and swamps, usually protected under rocks and vegetation. They may walk slowly over the substrate of their habitat, but they can also swim rapidly using their tails. The segmentation seen in annelids is seen also in crayfish and all arthropods; however, you will see that the segments are grouped into functional units.

Procedure

1. Obtain a preserved crayfish, study its external anatomy, and compare your observations with Figure 19.2. Describe the body symmetry, supportive structures, appendages, and segmentation, and state the adaptive advantages of each characteristic.
   a. body symmetry

   b. supportive structures

   c. appendages

   d. segmentation

2. Identify the three regions of the crayfish body: the **head**, **thorax** (fused with the head), and **abdomen**. Note the appendages associated with each region. Speculate about the functions of each of these groups of appendages.

a.

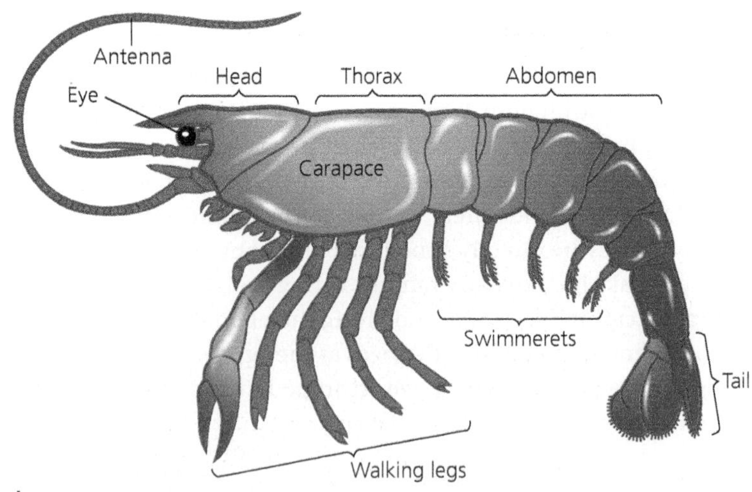

b.

FIGURE 19.2

An aquatic arthropod. (a) *Cambarus,* a freshwater crayfish. (b) External anatomy of a crayfish. The body is divided into head, thorax, and abdominal regions. Appendages grouped in a region perform specific functions.

a. head appendages

b. thoracic appendages

c. abdominal appendages

3. Feathery **gills** lie under the lateral extensions of a large, expanded exoskeletal plate called the **carapace** (see Figure 19.2). To expose the gills, use scissors to cut away a portion of the plate on the left side of the animal. What is the function of the gills? Speculate about how this function is performed.

4. Remove the dorsal portion of the carapace to observe other organs in the head and thorax. Compare your observations with Figure 19.3.
   a. Start on each side of the body at the posterior lateral edge of the carapace and make two lateral cuts extending along each side of the thorax and forward over the head, meeting just behind the eyes. This should create a dorsal flap in the carapace.
   b. Carefully insert a needle under this flap and separate the underlying tissues as you lift the flap.
   c. Observe the **heart**, a small, angular structure located just under the carapace near the posterior portion of the thorax. (If you were not successful in leaving the tissues behind as you removed the carapace, you may have removed the heart with the carapace.) Thin threads leading out from the heart are **arteries**. Look for holes in the heart wall. When blood collects in **sinuses** around the heart, the heart relaxes, and these

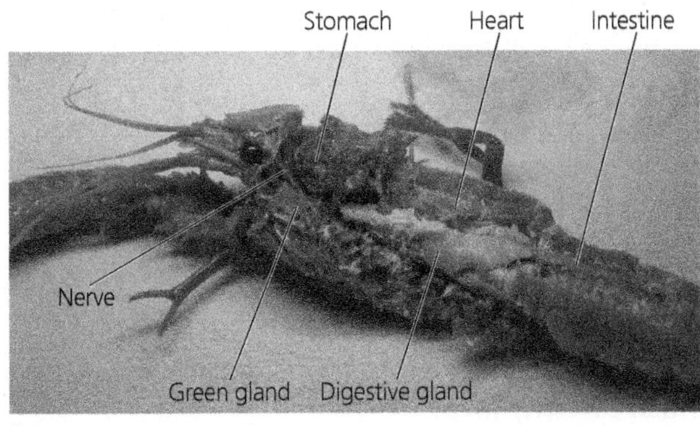

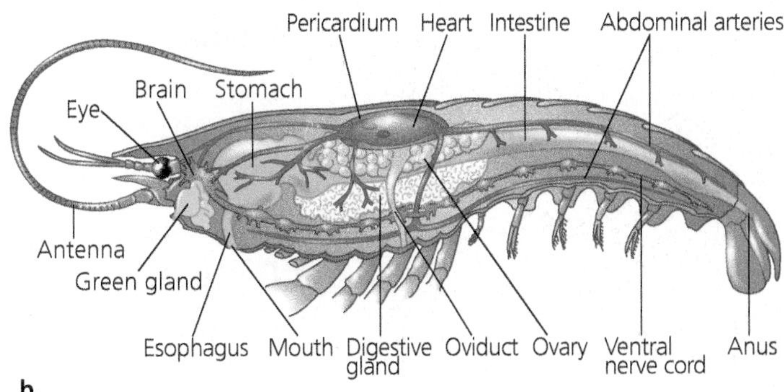

FIGURE 19.3
Internal anatomy of the crayfish. Large digestive glands fill much of the body cavity. The intestine extends from the stomach through the tail to the anus. The green glands lie near the brain in the head.

holes open to allow the heart to fill with blood. The holes then close, and the blood is pumped through the arteries, which distribute it around the body. Blood seeps back to the heart, since no veins are present. What is the name given to this kind of circulation?

d. Locate the **stomach** in the head region. It is a large, saclike structure. It may be obscured by the large, white **digestive glands** that fill the body cavity inside the body wall. Leading posteriorly from the stomach is the **intestine**. Make longitudinal cuts through the exoskeleton on either side of the dorsal midline of the abdomen. Lift the dorsal portion of the exoskeleton and trace the intestine to the anus. (When shrimp are "deveined" in preparation for eating, the intestine is removed.) Given all of the organs and tissues around the digestive tract and inside the body wall in the body cavity, what kind of coelom do you think this animal has?

e. Turn your attention to the anterior end of the specimen again. Pull the stomach posteriorly (this will tear the esophagus) and look inside the most anterior portion of the head. Two **green glands** (they do not look green), the animal's excretory organs, are located in this region. These are actually long tubular structures that resemble nephridia but are compacted into a glandular mass. Waste and excess water pass from these glands to the outside of the body through pores at the base of the antennae on the head.

f. Observe the **brain** just anterior to the green glands. It lies in the midline with nerves extending posteriorly, fusing to form a **ventral nerve cord**.

Student Media Video—Ch. 33: Lobster Mouth Parts

Results

Complete Table 19.1, recording all information for crayfish characteristics in the appropriate row. You will use this information to complete Table 19.2 and answer questions in the Applying Your Knowledge section at the end of this lab topic.

Discussion

How does the pattern of segmentation differ in the crayfish and the earthworm studied in Lab Topic 18 Animal Diversity I?

Lab Study B. Grasshoppers (*Romalea*)

Materials

dissecting instruments
dissecting pan
preserved grasshopper
disposable gloves

Introduction

The grasshopper, an insect, is an example of a terrestrial arthropod (Figure 19.4a). Insects are the most successful and abundant of all land animals. They are the principal invertebrates in dry environments, and they can survive extreme temperatures. They are the only invertebrates that can fly. As you study the grasshopper, compare the anatomy of this terrestrial animal with that of the aquatic crayfish, just studied. This comparison should suggest ways that terrestrial animals have solved the problems of life out of water.

Procedure

1. Observe the external anatomy of the grasshopper. Compare your observations with Figure 19.4a and b.
   a. Note the symmetry, supportive structures, appendages, and segmentation of the grasshopper.
   b. Observe the body parts. The body is divided into three regions: the **head**, the **thorax** (to which the legs and wings are attached), and the **abdomen**. Examine the appendages on the head, speculate about their functions, and locate the mouth opening into the digestive tract.
   c. Turning your attention to the abdomen, locate small dots along each side. These dots are **spiracles**, small openings into elastic air tubes, or

a.

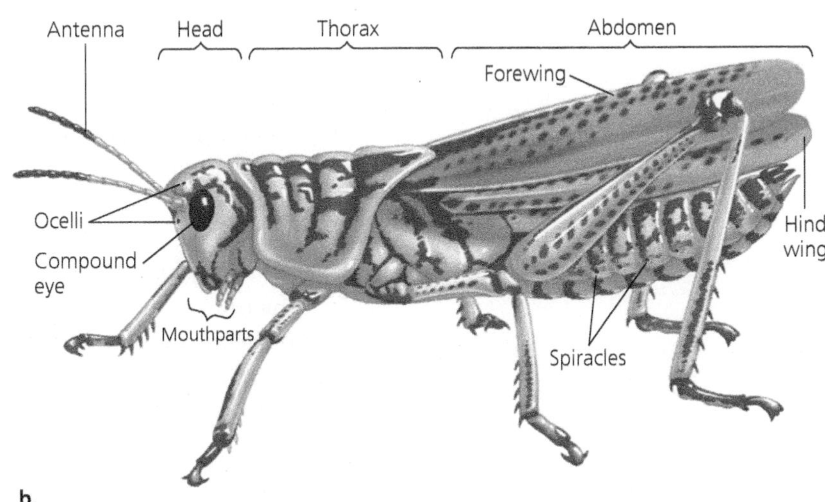

b.

FIGURE 19.4
A terrestrial arthropod. (a) A grasshopper with segmented body and jointed appendages. (b) External anatomy of the grasshopper. The body is divided into head, thorax, and abdominal regions. Wings and large legs are present. Small openings, called *spiracles,* lead to internal tracheae, allowing air to pass into the body.

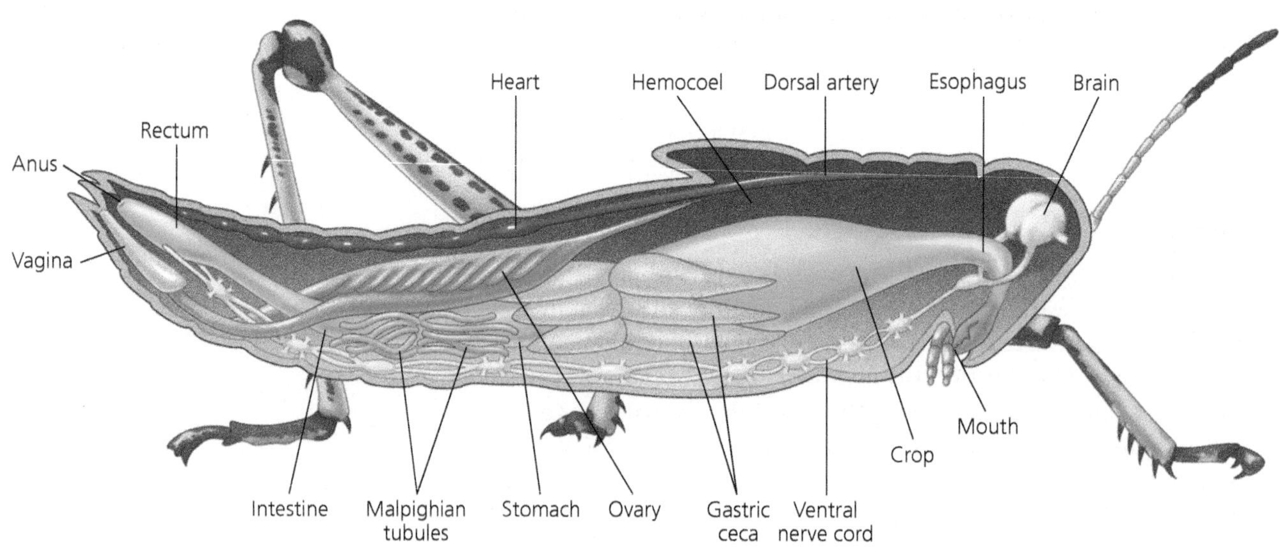

FIGURE 19.5

Internal anatomy of the grasshopper. The digestive tract, extending from mouth to anus, is divided into specialized regions: the esophagus, crop, stomach, intestine, and rectum. Gastric ceca attach at the junction of the crop and the stomach. Malpighian tubules empty excretory waste into the anterior end of the intestine.

tracheae, that branch to all parts of the body and constitute the respiratory system of the grasshopper. This system of tubes brings oxygen directly to the cells of the body.

2. Remove the exoskeleton. First take off the wings and, starting at the posterior end, use scissors to make two lateral cuts toward the head. Remove the dorsal wall of the exoskeleton and note the segmented pattern in the muscles inside the body wall. Compare your observations with Figure 19.5 as you work.

   a. A space between the body wall and the digestive tract, the **hemocoel** (a true coelom), in life is filled with colorless blood. What type of circulation does the grasshopper have?

   The heart of a grasshopper is an elongate tubular structure lying just inside the middorsal body wall. This probably will not be visible.

   b. Locate the digestive tract and again note the mouth. Along the length of the tract are regions specialized for specific functions. A narrow **esophagus** leading from the mouth expands into a large **crop** used for food storage. The crop empties into the **stomach**, where digestion takes place. Six pairs of fingerlike extensions called **gastric pouches** or **ceca** connect to the digestive tract where the crop and the stomach meet. These pouches secrete digestive enzymes and aid in food absorption. Food passes from the stomach into the **intestine**, then into the **rectum**, and out the **anus**. Distinguish these regions by observing constrictions and swellings along the tube. There is usually a constriction between the stomach and the intestine where the Malpighian tubules (discussed below) attach. The intestine is shorter and usually smaller in diameter than the stomach. *The intestine expands into an enlarged rectum that absorbs excess water from any undigested food, and relatively dry excrement passes out the anus.*

c. The excretory system is made up of numerous tiny tubules, the **Malpighian tubules,** which empty their products into the anterior end of the intestine. These tubules remove wastes and salts from the blood. Locate these tubules.

d. Push aside the digestive tract and locate the **ventral nerve cord** lying medially inside the ventral body wall. Ganglia are expanded regions of the ventral nerve cord found in each body segment. Following the nerve cord anteriorly, note that branches from the nerve cord pass around the digestive tract and meet, forming a brain in the head.

Results

Complete Table 19.1, recording all information for grasshopper characteristics in the appropriate row. You will use this information to complete Table 19.2 and answer questions in the Applying Your Knowledge section at the end of this lab topic.

Discussion

1. Describe how each of the following external structures helps the grasshopper live successfully in terrestrial environments.

   a. Exoskeleton

   b. Wings

   c. Large, jointed legs

   d. Spiracles

2. Describe how each of the following internal structures helps the grasshopper live successfully in terrestrial environments.

   a. Tracheae

   b. Malpighian tubules

   c. Rectum

EXERCISE 19.3

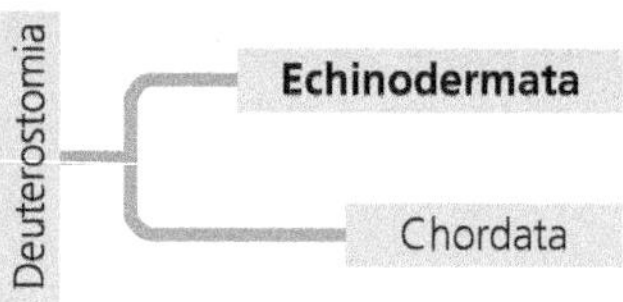

Phylum Echinodermata—Sea Star

Echinodermata is one of three phyla in the clade Deuterostomia. You will study another deuterostome phylum in Exercise 19.4, phylum Chordata. The third phylum, Hemichordata, will not be studied. Examples of echinoderms include the sea star, sea urchin, sea cucumber, and sea lily. Some of the most familiar animals in the animal kingdom are in the phylum Chordata—fish, reptiles, amphibians, and mammals. Take a look at a sea star (starfish) in the saltwater aquarium in your lab or in a tidal pool on a rocky shore. What are the most obvious characteristics of this animal? Then imagine a chordate—a fish, dog, or even yourself. You might question why these two phyla are considered closely related phylogenetically. The most obvious difference is a very basic characteristic—the sea star has radial symmetry and most chordates that you imagine have bilateral symmetry. The sea star has no head or other obvious chordate features and it crawls around using hundreds of small suction cups called tube feet. Most chordates show strong cephalization and move using appendages. Your conclusion from the superficial observations might be that these two phyla are not closely related. Your observations are a good example of the difficulty faced by taxonomists when comparing animals based only on the morphology of adults. Taxonomists must collect data from studies of developmental and molecular similarities before coming to final conclusions.

In this and the following exercise, you will examine an echinoderm, the adult sea star, and two chordates, asking questions about their morphology and adaptation to their habitats. You may not be convinced of their phylogentic relationships, however, until you complete Lab Topic 25 Animal Development, when you will study early development in sea urchins and sea stars. In that lab topic, you will see that chordates and echinoderms have similar early embryonic developmental patterns, including the formation of the mouth and anus and the type of cleavage.

Materials

preserved sea stars
fine-tipped scissors and other dissecting instruments
dissecting pan
disposable gloves

Introduction

The sea star (also called starfish) is classified in the phylum Echinodermata. They are marine animals with an endoskeleton of small, spiny calcareous plates bound together by connective tissue. Their symmetry is radial pentamerous (five-parted). They have no head or brain and few sensory structures. All animals in this phylum have a unique **water-vascular system** that develops from mesoderm and consists of a series of canals carrying water that enters the body through an outer opening, the **madreporite**. The canals are located inside the body and include a ring around the central disk of the body and tubes or canals that extend out into each arm. The canals then terminate in many small structures called **tube feet**, visible along the groove on the oral side of each arm. Each tube foot ends blindly in an attachment disk (podium) that

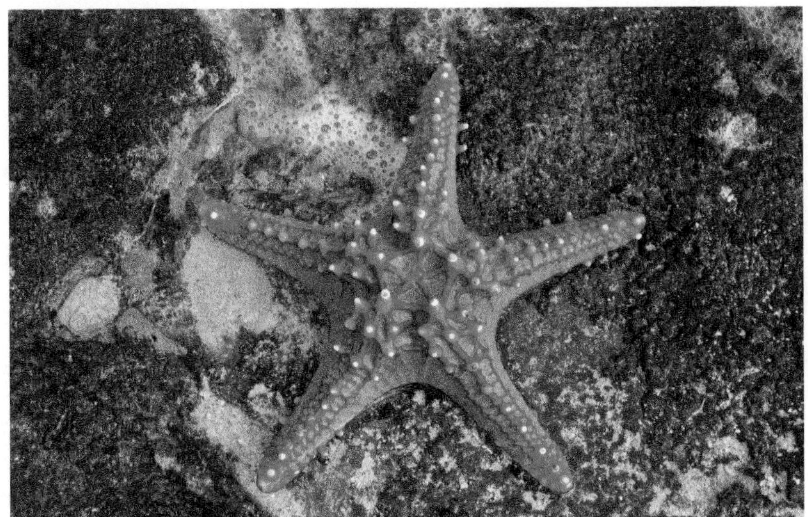

FIGURE 19.6
Sea star. The radial symmetry of echinoderm adults is evident in this deuterostome.

extends to the outside of the body. A small canal connects each podium to a small spherical sac (ampulla) inside the body. Using a combination of muscle contraction and adhesive chemicals, a sea star can extend and attach the feet to hard surfaces such as the surface of a clamshell, or rocks on the ocean shore (Figure 19.6).

Procedure

1. Observe the external anatomy of a sea star. Compare your observations with Figure 19.7a and b. Locate the **aboral** surface—the "upper" surface away from the mouth (Figure 19.7a). The downside is the **oral** surface where the mouth is located (Figure 19.7b).
2. Count the number of arms that extend out from the **central disk**. Echinoderms are usually pentamerous, meaning that their arms are in multiples of five. Occasionally a sea star with six arms will be found. Arms that are damaged or lost can be regenerated, and an extra arm may regenerate.
3. Observe the animal's aboral surface (Figure 19.7a). Locate the **madreporite**, a small porous plate displaced to one side of the central disk that serves to take water into the water vascular system.

   Notice that the surface of the animal's body is spiny. The spines project from calcareous plates of the **endoskeleton**. The endoskeleton is derived from the embryonic germ layer mesoderm. In life, the entire surface of the body is covered with an **epidermis** derived from ectoderm that may not be visible with the naked eye.
4. Observe the animal's oral surface (Figure 19.7b). The **mouth**, surrounded by spines, is located in the central disk with grooves extending from the mouth out into each arm. **Tube feet** lie along these grooves.
5. Open the body of the sea star to observe the internal structures (Figure 19.7c). Use fine-tipped scissors to cut off and discard about 1/2 inch from the tip of the arm opposite the madreporite (see Figure 19.7a). Then carefully cut along each side of this arm to where it joins the central disk. Cut across the top of the arm at the margin of the disk and remove this portion of the body wall. Observe the inside surface of the removed piece to see the endoskeleton and its calcareous plates.

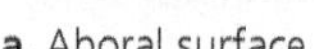

a. Aboral surface

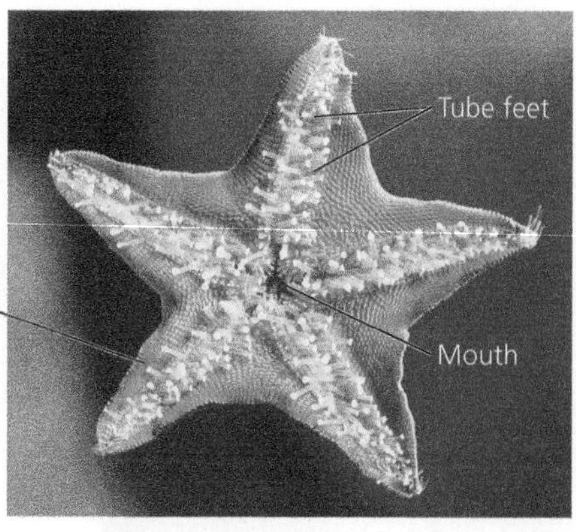

b. Oral surface

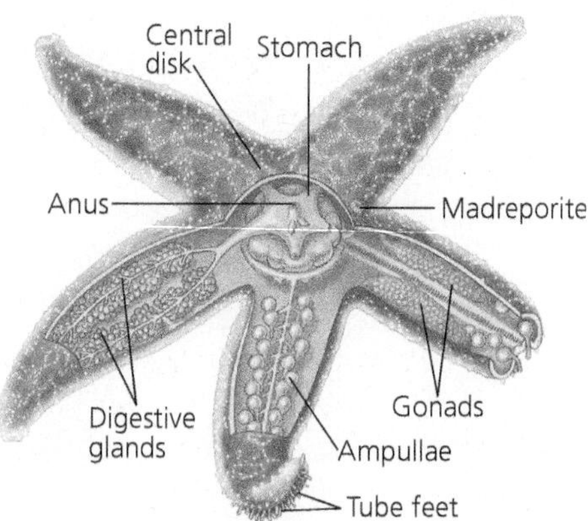

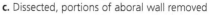

c. Dissected, portions of aboral wall removed

FIGURE 19.7
Anatomy of a sea star. (a) aboral surface (b) oral surface (c) dissected sea star with portions of the aboral wall removed

6. Continue using the scissors to carefully cut through the body wall along the side of each arm, and then around the central disk, but leave the madreporite in place by carefully cutting around it. This will allow you to remove the entire aboral surface of the body (excluding the madreporite).
7. Inside the body the organs are located in a **true body cavity** that in life is filled with coelomic fluid that carries oxygen and absorbed food to various parts of the body. Small delicate projections of the body cavity protrude between the plates of the endoskeleton to the outside of the body. These projections, covered with epidermis, are called **skin gills** or dermal branchiae, and function in the exchange of oxygen and carbon dioxide with the water bathing the animal's body. In addition, nitrogenous waste passes through these skin gills into the surrounding water; these structures thus have both respiratory and excretory functions.
8. Refer to Figure 19.7c as you locate the following structures in your dissected sea star. The central disk contains the folded **stomach**, a portion of which can be everted through the mouth on the oral side of the animal. A small anus is located on the aboral body surface (now removed), although very little fecal material is ejected here. Most digestion takes place in the stomach. If the stomach is everted into the body of a clam, the digested clam broth is then sucked up into the sea star body. After feeding, the sea star draws in its stomach by contracting its stomach muscles.
9. Conspicuous organs in the coelom of the animal's arms are **digestive glands** lying on top of **gonads**. Push aside the digestive glands to observe the gonads. Then look along the inside of the arm below the gonads to observe small fleshy **ampullae** lying along the grooves extending from the central disk into each arm. Each ampulla connects with a podium of a tube foot, as described in the Introduction above. Other systems cannot be easily observed in this preparation. A reduced circulatory system (hemal system) exists, but its function is not well defined. It consists of tissue strands and unlined sinuses. The nervous system includes a nerve ring around the mouth and radial nerves with epidermal nerve networks. There is no central nervous system.

Results

Complete Table 19.1, recording in the appropriate row all information you have been able to observe. You will use this information to complete Table 19.2 and answer questions in the Applying Your Knowledge section at the end of this lab topic.

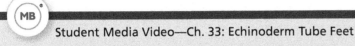

Student Media Video—Ch. 33: Echinoderm Tube Feet

Discussion

1. Imagine that you are a zoologist studying sea stars for the first time. What characteristics would you note from the dissection of an adult animal that might give a clue to its phylogenetic relationships—that it belongs in the clade Deuterostomia?

2. What structures have you observed that appear to be unique to echinoderms?

3. How would you continue your study to obtain more information that might help in classifying these animals?

4. Given the fact that other deuterostomes are bilaterally symmetrical, what is one explanation for the radial symmetry of most adult echinoderms?

EXERCISE 19.4

Phylum Chordata

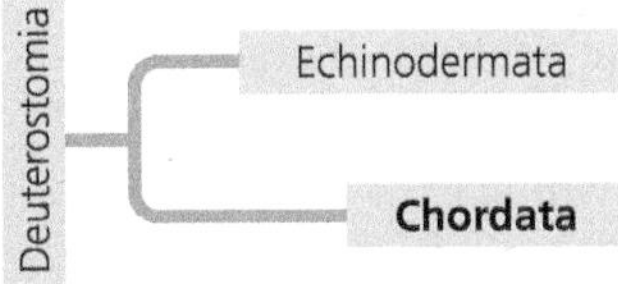

Up to this point, all the animals you have studied in Lab Topics 18 and 19 are commonly called **invertebrates**, a somewhat artificial designation based on the absence of a backbone. The phylum Chordata studied in this exercise includes two subphyla of invertebrates and a third subphylum that includes those animals that have a backbone, called **vertebrates**. Vertebrates have an endoskeleton of cartilage or bone and a head and skull. Chordates inhabit terrestrial and aquatic (freshwater and marine) environments. One group has developed the ability to fly. The body plan of chordates is unique in that these animals demonstrate a complex of four important characteristics at some stage in their development. In this exercise, you will discover these characteristics.

You will study two chordate species: the lancelet, an invertebrate in the subphylum Cephalochordata, and the pig, a vertebrate. The third subphylum, Urochordata, will not be studied.

Lab Study A. Lancelets (*Branchiostoma*)

Materials

compound microscope
stereoscopic microscope
preserved lancelet in watch glass
prepared slide of whole mount of lancelet
prepared slide of cross section of lancelet

Introduction

Lancelets (also known as amphioxi) are marine animals that burrow in sand in tidal flats. They feed with their head end extended from their burrow. They resemble fish superficially, but their head is poorly developed, and they have unique features not found in fish or other vertebrates. They retain the four unique characteristics of chordates throughout their life cycle and are excellent animals to use to demonstrate these features. In this lab study, you will observe preserved lancelets, prepared slides of whole mounts, and cross sections through the body of a lancelet (Figure 19.8).

FIGURE 19.8
Branchiostoma. The lancelet is a small chordate that lives in coastal waters.

Procedure

1. Place a *preserved lancelet* in water in a watch glass and observe it using the stereoscopic microscope. Handle the specimen with care and *do not dissect it.* Note the fishlike shape of the slender, elongate body. Locate the anterior end by the presence at that end of a noselike **rostrum** extending over the mouth region, surrounded by small tentacles. Notice the lack of a well-defined head. Look for the segmented muscles that surround much of the animal's body. Can you see signs of a tail? If the animal you are studying is mature, you will be able to see two rows of 20 to 25 white gonads on the ventral surface of the body.
2. Return the specimen to the correct container.
3. Observe the *whole mount slide* of the lancelet and compare your observations with Figure 19.9.

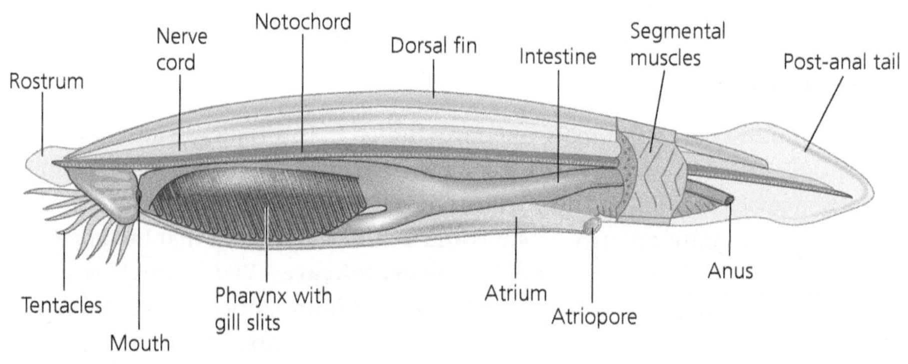

FIGURE 19.9
The lancelet, whole mount. The rostrum extends over the mouth region. The pharynx, including the pharyngeal gill slits, leads to the intestine, which exits the body at the anus. Note that a tail extends beyond the anus. Structures positioned from the dorsal surface of the body inward include a dorsal fin, the nerve cord, and the notochord.

Use only the lowest power on the compound microscope to study this slide.

a. Scan the entire length of the body wall. Do you see evidence of segmentation in the muscles or in other organs or structures?

b. Look at the anterior end of the animal. Do you see evidence of a sensory system? Describe what you see.

c. Locate the mouth of the animal at the anterior end. See if you can follow a tube from just under the rostrum into a large sac with numerous gill slits. This sac is the **pharynx with gill slits**, a uniquely chordate structure. Water and food pass into the pharynx from the mouth. Food passes posteriorly from the pharynx into the intestine, which ends at the anus on the ventral side of the animal, several millimeters before the end. The extension of the body beyond the anus is called a **post-anal tail**. Think of the worms you studied in Lab Topic 18 Animal Diversity I. Where was the anus located in these animals? Was a post-anal region present? Explain.

d. Water entering the mouth passes through the gill slits and collects in a chamber, the **atrium**, just inside the body wall. The water ultimately passes out of the body at a ventral pore, the **atripore**. Surprisingly, the gill slits are not the major gas exchange surface in the lancelet body. Because of the great activity of ciliated cells in this region, it is even possible that blood leaving the gill region has less oxygen than that entering the region. The function of gill slits is simply to strain food from the water. The major site for gas exchange is the body surface.

e. Now turn your attention to the dorsal side of the animal. Beginning at the surface of the body and moving inward, identify the listed structures and speculate about the function of each one.

dorsal fin:

nerve cord:

notochord:

The nerve cord is in a dorsal position. Have you seen only a dorsal nerve cord in any of the animals previously studied?

The notochord is a cartilage-like rod that lies ventral to the nerve cord and extends the length of the body. Have you seen a notochord in any of the previous animals?

The lancelet circulatory system is not visible in these preparations, but the animal has **closed circulation** with dorsal and ventral aortae, capillaries, and veins. Excretory organs, or nephridia (not visible here), are located near the true coelom, which surrounds the pharynx.

4. Observe the *slide of cross sections* taken through the lancelet body. There may be several sections on this slide, taken at several positions along the length of the body. Find the section through the pharynx and compare it with Figure 19.10.

Study this slide on the lowest power.

In cross section, it is much easier to see the structural relationships among the various organs of the lancelet. Identify the following structures and label them on Figure 19.10.

a. **Segmental muscles**. They are located on each side of the body, under the skin.

b. **Dorsal fin**. This projects upward from the most dorsal surface of the body.

c. **Nerve cord**. You may be able to see that the nerve cord contains a small central canal, thus making it hollow. The nerve cord is located in the dorsal region of the body, ventral to the dorsal fin between the lateral bundles of muscle.

d. **Notochord**. This is a large oval structure located just ventral to the nerve cord.

e. **Pharynx with gill slits**. This structure appears as a series of dark triangles arranged in an oval. The triangles are cross sections of **gill bars**. The spaces between the triangles are **gill slits**, through which water passes into the surrounding chamber.

Results

1. Complete the diagram of the lancelet cross section in Figure 19.10. Label all the structures listed in step 4 of the Procedure section.
2. Complete Table 19.1, recording all information for lancelet characteristics in the appropriate row. You will use this information to complete Table 19.2 and answer questions in the Applying Your Knowledge section at the end of this lab topic.

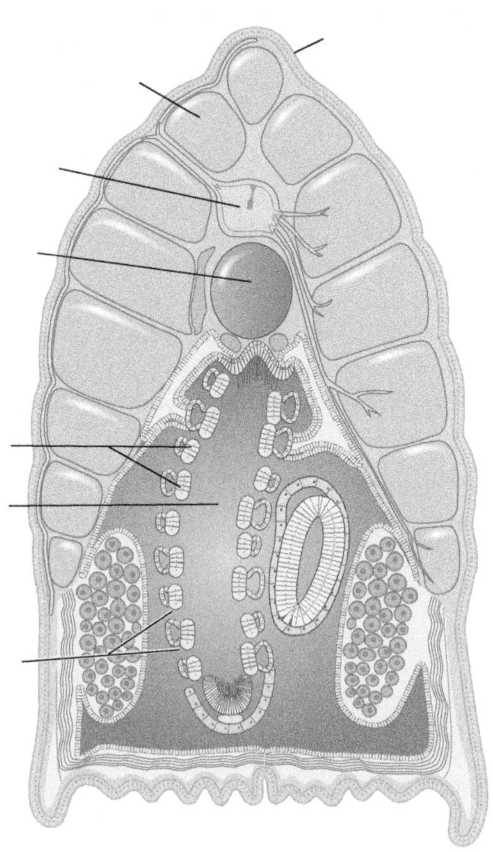

FIGURE 19.10
Cross section through the pharyngeal region of the lancelet.

Discussion

Describe the uniquely chordate features that you have detected in the lancelet that were not present in the animals previously studied.

Lab Study B. Fetal Pigs (*Sus scrofa*)

Materials

preserved fetal pig
dissecting pan
disposable gloves

Introduction

The pig is a terrestrial vertebrate. You will study its anatomy in detail in Lab Topics 22, 23, and 24. In this lab study, working with your lab partner, you will observe external features only, observing those characteristics studied in other animals in previous exercises. Compare the organization of the vertebrate body with the animals previously studied. As you dissect the pig in subsequent labs, come back to these questions and answer the ones that cannot be answered in today's lab study.

Procedure

1. Obtain a preserved fetal pig from the class supply and carry it to your desk in a dissecting pan.

Use disposable gloves to handle preserved animals.

2. With your lab partner, read each of the following questions. Drawing on observations you have made of other animals in the animal diversity lab studies, predict the answer to each question about the fetal pig. Then examine the fetal pig and determine the answer, if possible. Give evidence for your answer based on your observations of the pig, your knowledge of vertebrate anatomy, or your understanding of animal phylogeny.
   a. What type of symmetry does the pig body have?
      Prediction:

      Evidence:

   b. How many layers of embryonic tissue are present?
      Prediction:

      Evidence:

   c. Are cells organized into distinct tissues?
      Prediction:

      Evidence:

   d. How many digestive tract openings are present? Would you describe this as a "tube within a tube"?
      Prediction:

      Evidence:

   e. Is the circulatory system open or closed?
      Prediction:

      Evidence:

f. What is the habitat of the animal?

Prediction:

Evidence:

g. What are the organs for respiration?

Prediction:

Evidence:

h. What are the organs for excretion?

Prediction:

Evidence:

i. What is the method of locomotion?

Prediction:

Evidence:

j. Are support systems internal or external?

Prediction:

Evidence:

k. Is the body segmented?

Prediction:

Evidence:

l. Are appendages present?

Prediction:

Evidence:

m. What is the position and complexity of the nervous system?

Prediction:

Evidence:

Results

Complete Table 19.1, recording all information for pig characteristics in the appropriate row. You will use this information to complete Table 19.2 and answer questions in the Applying Your Knowledge section that follows.

REVIEWING YOUR KNOWLEDGE—LAB TOPICS 18 AND 19

1. Complete the summary table, Table 19.1, recording in the appropriate row information about characteristics of all animals studied.
2. Using Table 19.1, complete Table 19.2. Categorize all animals studied based on the 13 basic characteristics discussed in Lab Topic 18. Use this information to answer questions in the Applying Your Knowledge section that follows.

APPLYING YOUR KNOWLEDGE

1. Using specific examples from the animals you have studied in Lab Topics 18 and 19, describe ways that organisms have adapted to specific environments.
   a. Compare organisms adapted to aquatic environments with those from terrestrial environments.
   b. Compare adaptations of the parasitic *Ascaris* with the earthworm or clamworm, free-living organisms.
2. In Lab Topic 18 you studied the free-living flatworm, *Planaria*. The phylum Platyhelminthes also includes many examples of parasitic flatworms, for example, tapeworms and trematodes (flukes). Using Web resources, choose an example of a parasitic flatworm and compare morphological differences between this organism and the planarian that reflect specific lifestyle adaptations.
3. In your studies of animal phyla, you observed segmentation in widely diverse clades, for example, annelids (Lophotrochozoa), arthropods (Ecdysozoa), and chordates (Deuterostomia). How can you explain this in terms of their evolutionary history?

4. Upon superficial examination, the body form of certain present-day animals might be described as simple, yet these animals may have developed specialized structures, perhaps unique to their particular phylum. Illustrate this point using examples from some of the simpler organisms you have dissected.

5. From Lab Topics 18 and 19, one might conclude that certain trends can be detected, trends from ancestral features (those that arose early in the evolution of animals) to more derived traits (those that arose later). However, animals with ancestral characteristics still successfully exist on Earth today. Why is this so? Why have the animals with derived traits not completely replaced the ones with ancestral traits? Use examples from the lab to illustrate your answer.

6. A major theme in biology is the relationship between form and function in organisms. Select one of the major characteristics from Table 19.1, and illustrate the relationship of form and function for this characteristic using examples from the organisms studied.

7. Will, a student in the Coastal Biology course, has found an animal he cannot identify in the lower beach sand of Jekyll Island. What are some questions that he should ask to help determine the phylum of the animal?

8. As you review the information you have recorded in Tables 19.1 and 19.2, do you see a relationship between *symmetry* and (1) the organization of the nervous system and (2) the number of tissue layers?

TABLE 19.1 Summary Table of Animal Characteristics

Animal	Symmetry	Tissue Organization	Type of Body Cavity	Digestive Openings	Circulatory System	Habitat	Respiratory Organs
Sponge							
Hydra							
Planarian							
Clam							
Clamworm/ earthworm							
Roundworm							
Crayfish							
Grasshopper							
Sea star							
Lancelet							
Pig							

TABLE 19.1 Summary Table of Animal Characteristics (*continued*)

Animal	Excretory System	Locomotion	Support System	Segmentation	Appendages	Nervous System Organization
Sponge						
Hydra						
Planarian						
Clam						
Clamworm/ earthworm						
Roundworm						
Crayfish						
Grasshopper						
Sea star						
Lancelet						
Pig						

TABLE 19.2 Comparison of Organisms by Major Features

1. Tissue Organization a. distinct tissues absent: b. distinct tissues present:	5. Circulatory System a. none: b. open: c. closed:
2. Symmetry a. radial: b. bilateral:	6. Habitat a. aquatic: b. terrestrial: c. parasitic:
3. Body Cavity a. acoelomate: b. pseudocoelomate: c. coelomate:	7. Organs for Gas Exchange a. skin: b. gills: c. lungs: d. spiracles/tracheae:
4. Openings to Digestive Tract a. one: b. two:	

TABLE 19.2 Comparison of Organisms by Major Features (*continued*)

8. Organs for Excretion (list organ and animals)	11. Segmented Body a. no: b. yes:
	12. Appendages a. yes: b. no:
9. Type of Locomotion (list type and animals)	13. Nervous System a. ventral nerve cord: b. dorsal nerve cord: c. other:
10. Support System a. external: b. internal: c. hydrostatic:	

INVESTIGATIVE EXTENSIONS

1. Scientists worldwide are concerned about reports of climate change. Crayfish are common inhabitants of freshwater streams, ponds, and swamps, all of which may be affected by a warming earth. Design an experiment to test the thermal limits that can be tolerated by crayfish.

   Design similar experiments to test the effects of pesticides, fertilizers, herbicides, petroleum products, or human wastes—all of which may be present in runoff into crayfish habitat from farmlands or urban development.

2. Arthropods are the dominant animals on Earth, both in number of species and number of individuals. Now that you are familiar with the characteristics of insects (terrestrial arthropods), using an entomology (study of insects) text or insect identification key, determine the diversity of arthropods, or specifically insects, found in various habitats. You might sample a given amount of soil taken from several different environments. For example, you might compare arthropod diversity in a deciduous forest with that of a cultivated field with that of a manicured lawn.

   You might also investigate arthropod diversity in habitats that differ in moisture. Compare well-drained soil (along a ridge) with saturated soil (along a creek, in a marsh, or in a bog).

3. N. A. Cobb, famous nematologist, writes: "If all the matter in the universe except the nematodes were swept away, our world would still be dimly recognizable . . . We should find its mountains, hills, vales, rivers, lakes, and oceans represented by a film of nematodes" (Cobb, 1915). Nematodes are everywhere and many are readily available for study.

   Design an experiment to investigate the diversity of nematodes present in several sources; for example, fresh and rotting fruits, soils collected from different sources or treated with different chemicals, roots of plants—trees or vegetables grown for human consumption. (Many nematodes are plant parasites, and many have been imported on foods or nursery stock.)

   Are there nematodes in drinking water? How could you investigate this question? How could you collect the nematodes?

STUDENT MEDIA: BioFlix, Activities, Investigations, and Videos

www.masteringbiology.com (select Study Area)

Activities—Ch. 32: Animal Phylogenetic Tree; Ch. 33: Characteristics of Invertebrates; Ch. 34: Characteristics of Chordates; Ch. 41: Feeding Mechanisms of Animals

Investigations—Ch. 32: How Do Molecular Data Fit Traditional Phylogenies? Ch. 33: How Are Insect Species Identified?

Videos—Ch. 33: *C. elegans* Crawling; Butterfly Emerging; Bee Pollinating; Lobster Mouthparts; Echinoderm Tube Feet.

REFERENCES

Aguinaldo, A. M., J. M. Turbeville, L. S. Linford, M. C. Rivera, J. R. Garey, R. A. Raff, and J. A. Lake. "Evidence for a Clade of Nematodes, Arthropods, and Other Moulting Animals." *Nature*, 29 May 1997 (387), pp. 489–493.

Cobb, N. A. "Nematodes and Their Relationships." *Year Book Dept. Agric. 1914*, pp. 457–490. Washington, DC: Dept. Agric., 1915.

DeSalle, R. and B. Schierwater. "An Even 'Newer' Animal Phylogeny." *BioEssays*, 2008, 30, pp. 1043–1047.

Halanych, K. M. "The New View of Animal Phylogeny." *Annual Review of Ecology, Evolution, and Systematics*, 2004, vol. 35, pp. 229–256.

Hickman, C. P., L. S. Roberts, S. Keen, D. J. Eisenhour, A. Larson, and H. I'Anson. *Integrated Principles of Zoology*, 16th ed. Boston: McGraw Hill, 2014.

WEBSITES

Descriptions of many invertebrates and vertebrates, links to insect keys, references:
http://animaldiversity.ummz.umich.edu/

Search this website for nematodes to find experiments using these animals in space:
http://www.nasa.gov/

The Tree of Life Web project provides information on all major groups of organisms, including invertebrates:
http://tolweb.org/Bilateria

NOTES

LAB TOPIC 20

Plant Anatomy

Laboratory Objectives

After completing this lab topic, you should be able to:

1. Identify and describe the structure and function of each plant cell type and tissue type.
2. Describe the organization of tissues and cells in each plant organ.
3. Relate the function of an organ (root, stem, and leaf) to its structure.
4. Describe primary and secondary growth and identify the location of each in the plant.
5. Relate primary and secondary growth to the growth habit (woody or herbaceous).
6. Discuss adaptation of plants to the terrestrial environment as illustrated by the structure and function of plant anatomy.
7. Apply your knowledge of plants to the kinds of produce you find in the grocery store.

Introduction

Vascular plants have been successful on land for over 425 million years, and their success is related to their adaptations to the land environment. Consider that aquatic algae live most often in a continuously homogeneous environment. The requirements for life surround algae, so relatively minor structural adaptations have evolved for functions such as reproduction and attachment. In contrast, the terrestrial habitat, with its extreme environmental conditions, presents numerous challenges for the survival of plants. Consequently, plants have evolved structural adaptations for functions such as the absorption of underground water and nutrients, the anchoring of the plant in the substrate, the elevation and support of aerial parts of the plant, and the transport of materials throughout the relatively large plant body. In vascular plants, the structural adaptations required for these and other functions are divided among three vegetative plant organs: stems, roots, and leaves. Unlike animal organs, which are often composed of unique cell types (for example, cardiac muscle fibers are found only in the heart, osteocytes only in bone), plant organs have many tissues and cell types in common, but they are organized in different ways. The structural organization of basic tissues and cell types in different plant organs is directly related to their different functions. For example, leaves function as the primary photosynthetic organ and generally have thin, flat blades that maximize light absorption and gas exchange. Specialized cells of the root epidermis are long extensions that promote one of the root functions, absorption. The interrelationship of structure and function at different levels of organization is a major theme in biology that you will continue to explore in this lab topic.

Use the figures in this lab topic for orientation and as a study aid. Be certain that you can identify all items by examining the living specimens and microscope slides. These, and not the diagrams, will be used in the laboratory evaluations.

Summary of Basic Plant Tissue Systems and Cell Types

The plant body is organized into **tissue systems** based on their shared structural and functional features. There are three tissue systems—**dermal**, **ground**, and **vascular**—that are continuous throughout the organs of roots, stems, and leaves (Figure 20.1). The plant tissues that actively and continuously divide by mitosis are called **meristematic tissues**. These are located in specific regions—for example the root tip. Following is a review of plant tissue systems and the most common types of cells seen in plant organs, as well as their functions. Other specialized cells will be described as they are discussed in lab. Refer back to this summary as you work through the exercises.

Dermal Tissue System: Epidermis

The **epidermis** forms the outermost layer of cells, usually one cell thick, covering the entire plant body. The epidermal cells are often flattened and rectangular in shape (Figure 20.2a and b). Specialized epidermal cells include the **guard**

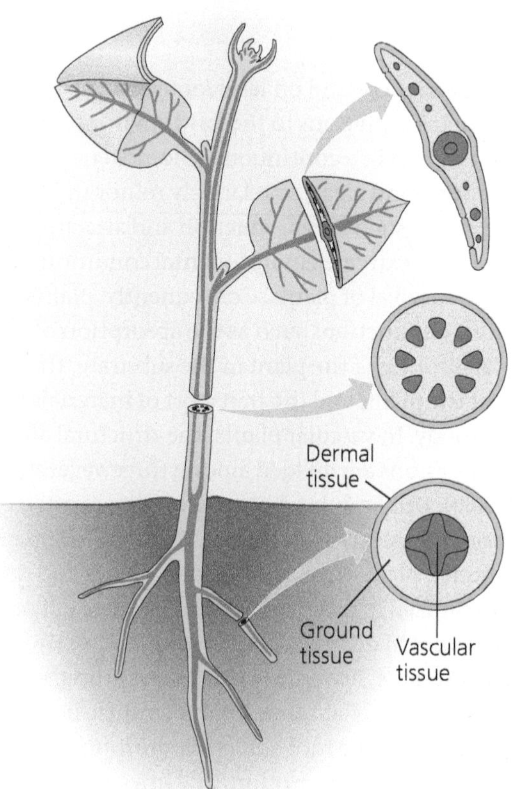

FIGURE 20.1
Three-tissue system. Roots, stems, and leaves are constructed of three basic tissue systems: dermal (blue), ground (yellow), and vascular (purple).

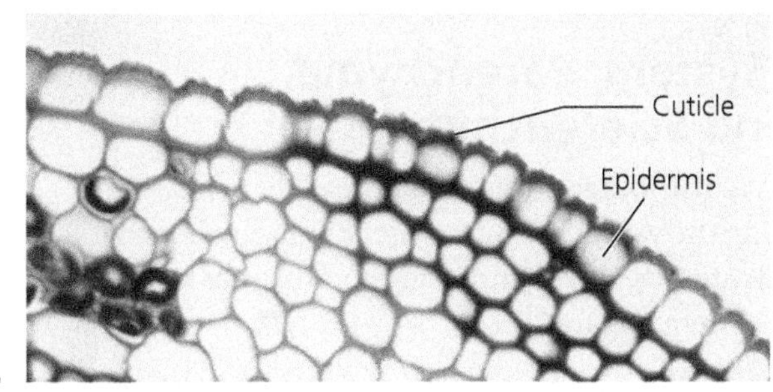

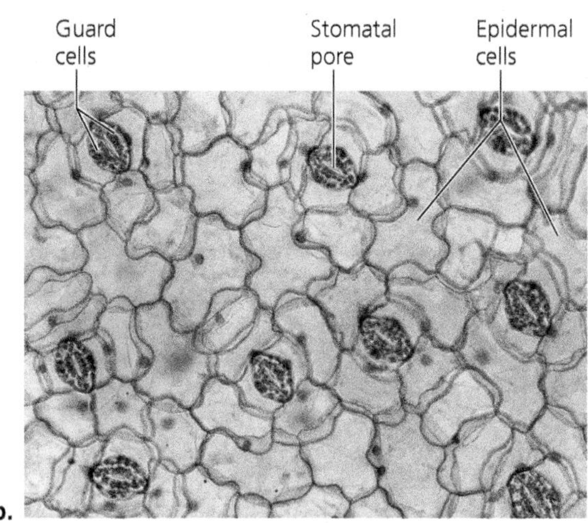

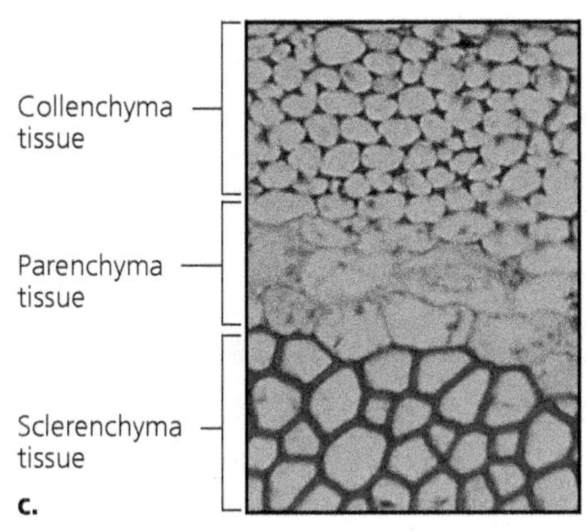

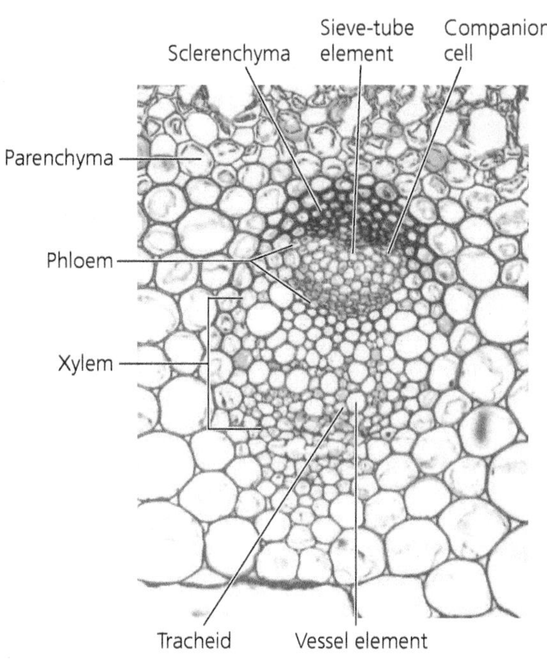

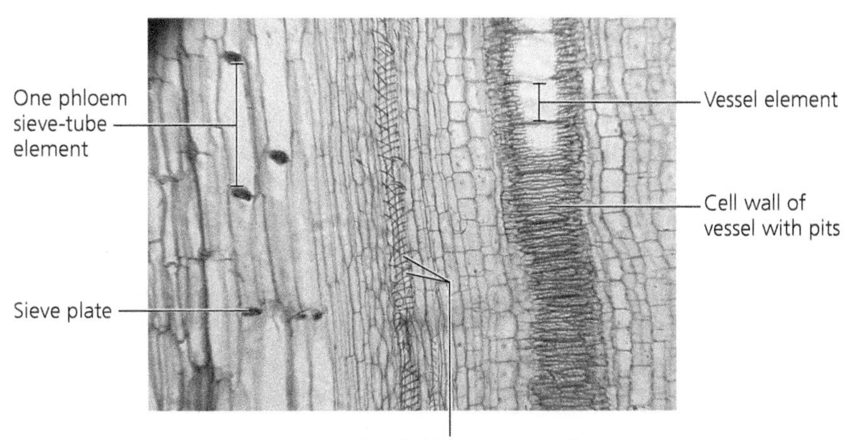

FIGURE 20.2

Plant tissue systems and cell types. (a) Dermal tissue—a single layer of epidermis covered by waxy cuticle. (b) Leaf surface showing epidermis with stomata and guard cells. (c) Ground tissue—cross section of pumpkin stem. (d) Vascular tissue—cross section of a vascular bundle in a buttercup stem. (e) Long section through xylem and phloem of pumpkin stem.

cells of the stomata, hairs called **trichomes**, and unicellular **root hairs**. Most epidermal cells on aboveground structures are covered by a waxy **cuticle**, which prevents water loss. The epidermis provides protection and regulates gas exchange and transpiration (water evaporation).

Ground Tissue System: Parenchyma, Collenchyma, and Sclerenchyma

The ground tissue system is distributed throughout the plant beneath the epidermis and surrounding the vascular tissues. Parenchyma, collenchyma, and/or sclerenchyma cells are typically found in ground tissue as seen in the cross section of a pumpkin stem (Figure 20.2c).

Parenchyma cells are the most common cell in plants and are characteristically thin-walled with large vacuoles. These cells may function in photosynthesis, support, storage of materials, and lateral transport.

Collenchyma cells are usually found near the surface of the stem, leaf petioles, and veins. These living cells are similar to parenchyma cells but are characterized by an uneven thickening of cell walls. They provide flexible support to young plant organs.

Sclerenchyma cells have thickened cell walls that may contain a strong polymer, lignin. They provide strength and support to mature plant structures and may be dead at functional maturity. The most common type of sclerenchyma cells are long, thin **fibers**.

Vascular Tissue System: Xylem and Phloem

The vascular tissue system functions in the transport of materials throughout the plant body. Xylem tissue and phloem tissue are complex tissues (composed of several cell types) seen in the cross section of a buttercup stem (Figure 20.2d) and a long section of a pumpkin stem (Figure 20.2e).

Xylem cells form a complex vascular tissue that functions in the transport of water and minerals throughout the plant and provides support. **Tracheids** and **vessel elements** are the primary water-conducting cells. Tracheids are long, thin cells with perforated tapered ends. Vessel elements are larger in diameter, open-ended, and joined end to end, forming continuous transport systems referred to as **vessels**. Parenchyma cells are present in the xylem and function in storage and lateral transport. Sclerenchyma fibers in the xylem provide additional support.

Phloem tissue transports the products of photosynthesis throughout the plant as part of the vascular tissue system. This complex tissue is composed of living, conducting cells called **sieve-tube elements** or sieve-tube members, which lack a nucleus and have perforated sieve **plates** for end walls. The cells are joined end to end throughout the plant. Each sieve-tube element is associated with one or more adjacent **companion cells**, which are thought to regulate sieve-tube member function. Phloem parenchyma cells function in storage and lateral transport, and phloem fibers provide additional support.

Meristematic Tissue: Primary Meristem, Cambium, and Pericycle

Primary meristems consist of small, actively dividing undifferentiated cells located in buds of the shoot and in root tips of plants. These cells produce the primary tissues along the plant axis throughout the life of the plant. You will study primary meristems in the apical bud in Exercise 20.2 (Figure 20.4).

Pericycle is a layer of meristematic cells just outside the vascular cylinder in the root. These cells divide to produce lateral branch roots (Exercise 20.3, Lab Study B, Figure 20.7c).

Vascular cambium is a lateral meristem also composed of small, actively dividing cells that are located between the xylem and phloem vascular tissue. These cells divide to produce secondary growth, which results in an increase in circumference (Exercise 20.4, Figure 20.11).

Cork cambium is a lateral meristem located just inside the cork layer of a woody plant. These cells divide to produce secondary growth that replaces the primary epidermis as the root and stem expand (Exercise 20.4, Figure 20.11).

This lab topic begins with a study of the whole plant and then continues with investigations of the cells, tissues, and organs. You will investigate the structure and function of vascular plants in the following exercises:

- Study of the shape and form (morphology) of a herbaceous plant
- Investigation of the primary plant body derived from apical meristem
- Observation of the three organs, and the tissues and cells of the plant body: stems, roots, and leaves
- Investigation of secondary growth in stems of a woody plant
- Application of your knowledge of plant organs to plants commonly found in the grocery store

EXERCISE 20.1

Plant Morphology

Materials

living bean or geranium plant
squirt bottle of water
paper towels

Introduction

As you begin your investigation of the structure and function of plants, you need an understanding of the general shape and form (morphology) of the whole plant. In this exercise, you will study a bean or geranium plant, identifying basic features of the three vegetative organs: *roots, stems, and leaves*. In the following exercises, you will investigate the cellular structure of these organs and their tissues as viewed in cross sections. Refer to the living plant for orientation before you view your slides.

Procedure

1. Working with another student, examine a living **herbaceous** (non-woody) plant and identify the following structures in the *shoot (stems and leaves)*.
   a. **Nodes** are regions of the stem from which leaves, buds, and branches arise and that contain meristematic tissue (areas of cell division).
   b. **Internodes** are the segments of the stem located between the nodes.
   c. **Terminal buds** are located at the tips of stems and branches. They enclose the shoot apical meristem, which gives rise to leaves, buds, and all primary tissue of the stem. Only stems produce buds.
   d. **Axillary**, or **lateral**, **buds** are located in the leaf axes at nodes; they may give rise to lateral branches.
   e. Leaves consist of flattened **blades** attached at the node of a stem by a stalk, or **petiole**.
2. Observe the *root* structures by gently removing the plant from the pot and loosening the soil from the root structure. You may need to rinse a few roots with water to observe the tiny, active roots. Identify the following structures.
   a. **Primary** and **secondary roots**. The primary root is the first root produced by a plant embryo and may become a long taproot. Secondary roots arise from meristematic tissue deep within the primary root.
   b. Root tips consist of a **root apical meristem** that gives rise to a **root cap** (protective layer of cells covering the root tip) and to all the primary tissues of the root. A short distance from the root tip is a zone of **root hairs** (specialized epidermal cells), the principal site of water and mineral absorption.

Results

1. Label Figure 20.3.
2. Sketch in the margin of your lab manual any features not included in this diagram that might be needed for future reference. For example, your plant may have small green bracts (leaflike structures) at the base of the petiole. These are called stipules.

Discussion

1. Look at your plant and discuss with your partner the possible functions of each plant organ. Your discussion might include evidence observed in the lab today or prior knowledge. Describe proposed functions (more than one) for each organ.

   Stems:

   Roots:

   Leaves:

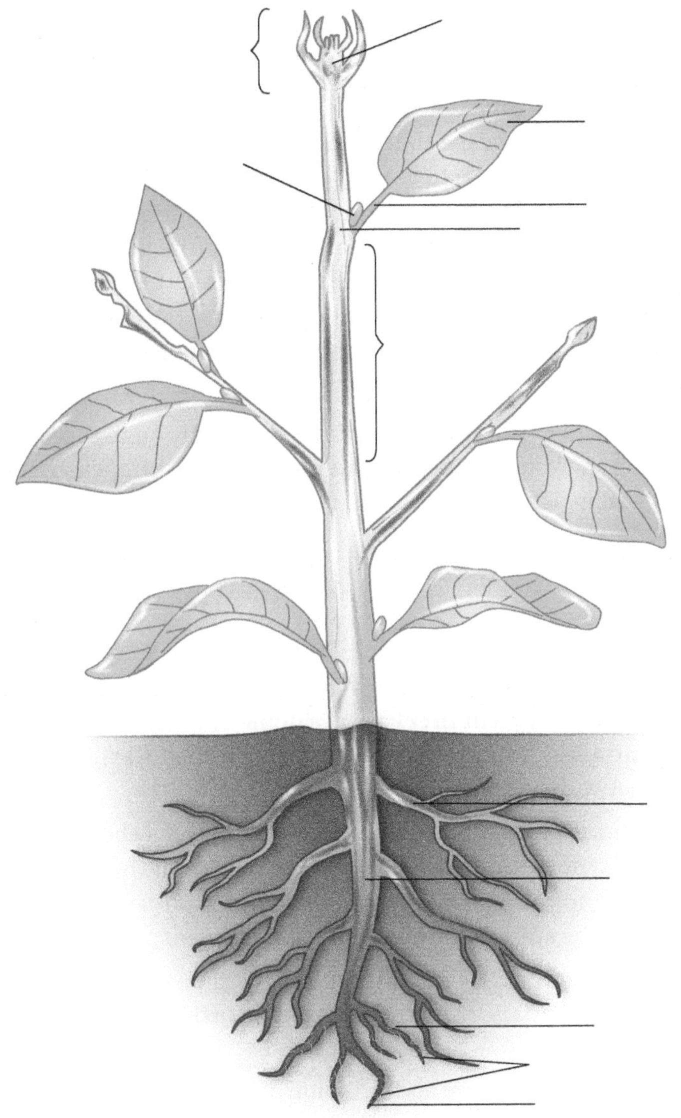

FIGURE 20.3

A herbaceous plant. The vegetative plant body consists of roots, stems, and leaves. The buds are located in the axils of the leaves and at the shoot tip. The roots also grow from meristem tissues in the root tip. Label the diagram based on your observations of a living plant and the structures named in Exercise 20.1.

2. Imagine that you have cut each organ—roots, stems, and leaves—in cross section. Sketch the overall shape of that cross section in the margin of your lab manual. Remember, you are not predicting the internal structure, just the overall shape.

EXERCISE 20.2

Plant Primary Growth and Development

Materials

prepared slides of *Coleus* stem (long section)
compound microscope

Introduction

Plants exhibit **indeterminate growth**, as they produce new cells throughout their lifetime as a result of cell divisions in meristems. Tissues produced from apical meristems are called **primary tissues**, and this growth is called

primary growth. Primary growth occurs along the plant axis at the shoot tip and the root tip. Certain meristem cells divide in such a way that one cell product becomes a new root or shoot cell and the other cell remains in the meristem and continues to divide. Beyond the zone of active cell division, new cells become enlarged and specialized (**differentiated**) for specific functions (resulting, for example, in vessels, parenchyma, and epidermis). Using the model plant *Arabidopsis,* research into the genetic and molecular basis of cell differentiation has rapidly advanced. The entire genome of *Arabidopsis* has been sequenced, and mutations can be inserted that prevent specific gene functions. Subsequently, scientists can observe the effect of these "knockout mutations" by studying the corresponding changes in growth and differentiation.

In this exercise, you will examine a longitudinal section through the tip of the stem, observing the youngest tissues and meristems at the apex, then moving down the stem, where you will observe more mature cells and tissues.

Procedure

1. Examine a prepared slide of a longitudinal section through a terminal bud of *Coleus.* Use low power to get an overview of the slide; then increase magnification. Locate the **apical meristem**, a dome of tissue nestled between the **leaf primordia**, young developing leaves. Locate the axillary **bud primordia** between the leaf and the stem.
2. Move the specimen under the microscope so that cells may be viewed at varying distances from the apex. The youngest cells are at the apex of the bud, and cells of increasing maturity and differentiation can be seen as you move away from the apex. Follow the early development of vascular tissue, which differentiates as the primordial leaves develop.
   a. Locate the narrow, dark tracks of **undifferentiated vascular tissue** in the leaf primordia.
   b. Observe changes in cell size and structure of the vascular system as you move away from the apex and end with a distinguishable vessel element of the **xylem**, with its spiral cell wall thickening in the older leaf primordia and stem. You may need to use the highest power on the microscope to locate these spiral cell walls.

Results

1. Label Figure 20.4, indicating the structures visible in the young stem tip.
2. Modify the figure or sketch details in the margin of the lab manual for future reference.

Discussion

1. Describe the changes in cell size and structure in the stem tip. Begin at the youngest cells at the apex and continue to the xylem cells.

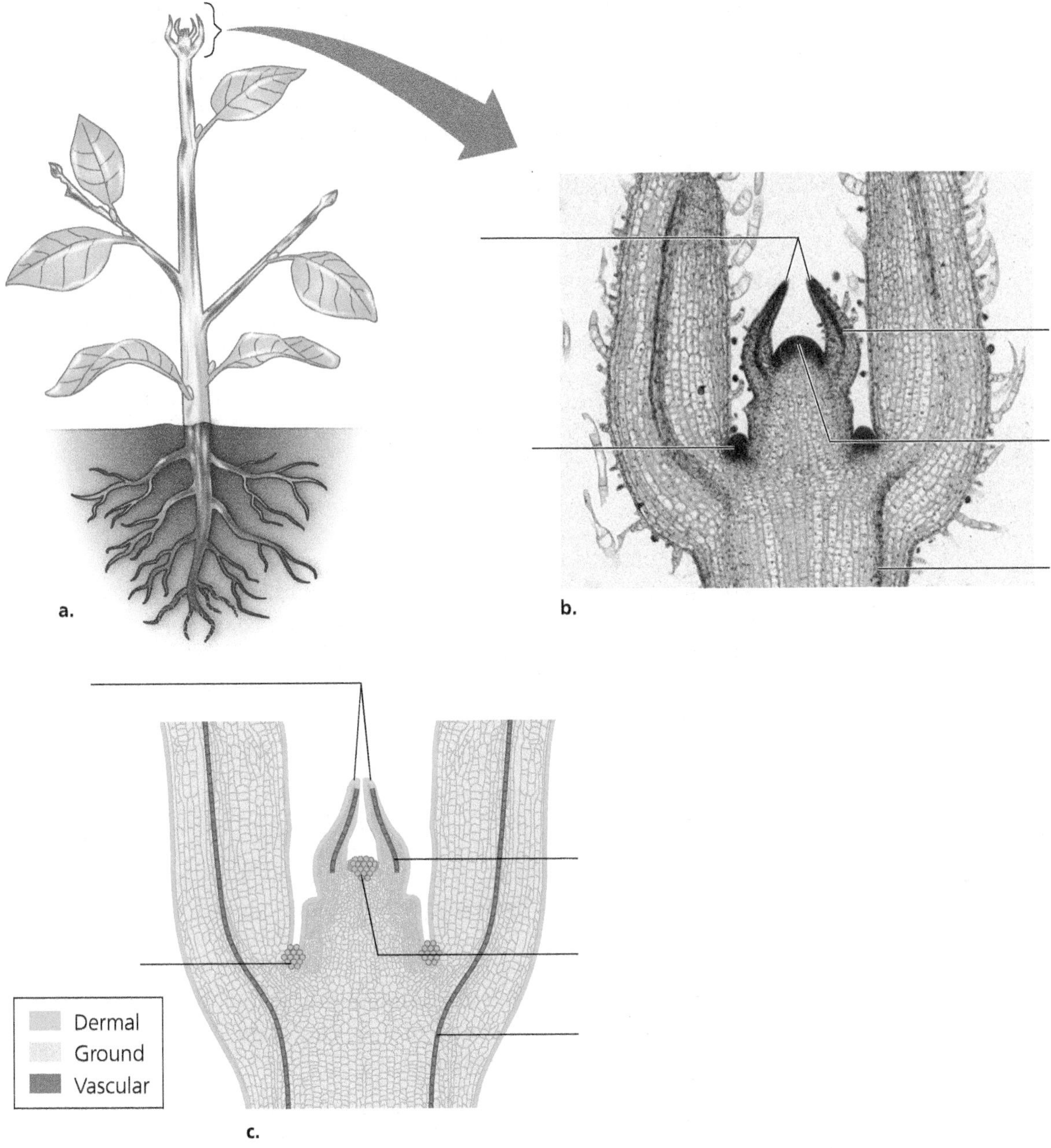

FIGURE 20.4

***Coleus* stem tip.** (a) Diagram of entire plant body. (b) Photomicrograph of a longitudinal section through the terminal bud. (c) Line diagram of the growing shoot tip with primordial leaves surrounding the actively dividing apical meristem. The most immature cells are at the tip of the shoot and increase in stages of development and differentiation farther down the stem. Label the cells and structures described in Exercise 20.2.

2. The meristems of plants continue to grow throughout their lifetime, an example of indeterminate growth. Imagine a 200-year-old oak tree, with active meristem producing new buds, leaves, and stems each year. Contrast this with the growth pattern in humans.

EXERCISE 20.3

Cell Structure of Primary Tissues

All **herbaceous** (nonwoody) flowering plants produce a complete plant body composed of primary tissue, derived from apical primary meristem. This plant body consists of *organs*—roots, stems, leaves, flowers, fruits, and seeds—and *tissue systems*—**dermal**, **ground**, and **vascular** (Figure 20.1). In this exercise, you will investigate the cellular structure and organization of plant organs and tissues by examining microscopic slides. You will make your own thin cross sections of stems, and view prepared slides of stems, roots, and leaves. Secondary growth in woody stems will be examined in Exercise 20.4.

Lab Study A. Stems

Materials

prepared slide of herbaceous dicot stem
dropper bottle of distilled water
small petri dish with 50% ethanol
dropper bottle of 50% glycerine
dropper bottle of 0.2% toluidine blue stain
nut-and-bolt microtome
warm paraffin
living plant for sections
new single-edged razor blade
forceps
microscope slides
coverslips
compound microscope
dissecting needles

Introduction

A stem is usually the main stalk, or axis, of a plant and is the only organ that produces buds and leaves. Stems support leaves and conduct water and inorganic substances from the root to the leaves and carbohydrate products of photosynthesis from the leaves to the roots. Most herbaceous stems are able to photosynthesize. Stems exhibit several interesting adaptations, including water storage in cacti, carbohydrate storage in some food plants, and thorns that reduce herbivory in a variety of plants.

You will view a prepared slide of a cross section of a stem, and, working with another student, you will use a simple microtome—an instrument used for cutting thin sections for microscopic study—to make your own slides. You will embed the stem tissue in paraffin and cut thin sections. You will stain your sections with toluidine blue, which will help you distinguish different cell types. This simple procedure is analogous to the process used to make prepared slides used in subsequent lab studies.

Read through the procedure and set up the materials before beginning.

Procedure

1. Embed the sections of the stem.
   a. Using a new single-edged razor blade, cut a 0.5-cm section of a young bean stem.
   b. Obtain a nut-and-bolt microtome. The nut should be screwed just into the first threads of the bolt. Using forceps, carefully hold the bean stem upright inside the nut.
   c. Pour the warm paraffin into the nut until full. Continue to hold the top of the stem until the paraffin begins to harden. While the paraffin completely hardens, continue the exercise by examining the prepared slide of the stem.
2. Examine a prepared slide of a cross section through the herbaceous dicot stem (Figure 20.6). As you study the stem tissues and cells, refer back to Summary of Basic Plant Tissue Systems and Cell Types and Figure 20.2.
3. Identify the **dermal tissue system**, characterized by a protective cell layer covering the plant. It is composed of the **epidermis** and the **cuticle**. Occasionally, you may also observe multicellular **trichomes** (hairs and glands) on the outer surface of the plants.
4. Locate the **ground tissue system**, background tissue that fills the spaces between epidermis and vascular tissue. Identify the **cortex region** located between the vascular bundles and the epidermis. It is composed mostly of **parenchyma**, but the outer part may contain **collenchyma** as well.
5. Next find the **pith region**, which occupies the center of the stem, inside the ring of vascular bundles; it is composed of parenchyma. In herbaceous stems, these cells provide support through turgor pressure. This region is also important in storage of water and materials.
6. Now identify the **vascular system**, a continuous system of xylem and phloem providing transport and support. In your stems and in many stems, the **vascular bundles** (clusters of xylem and phloem) occur in rings that surround the pith; however, in some groups of flowering plants, the vascular tissue is arranged in a complex network.
7. Observe that each bundle consists of *phloem tissue toward the outside and xylem tissue toward the inside*. A narrow layer of vascular cambium, which may become active in herbaceous stems, is situated between the xylem and the phloem. Take note of the following information as you make your observations.

   Phloem tissue is composed of three cell types:

   a. Dead, fibrous, thick-walled **sclerenchyma cells** that provide support for the phloem tissue and appear in a cluster as a **bundle cap**.
   b. **Sieve-tube elements**, which are large, living, elongated cells that lack a nucleus at maturity. They become vertically aligned to form sieve tubes, and their cytoplasm is interconnected through sieve plates located at the ends of the cells. Sieve plates are not usually seen in cross sections.
   c. **Companion cells**, which are small, nucleated parenchyma cells connected to sieve-tube cells by means of cytoplasmic strands.

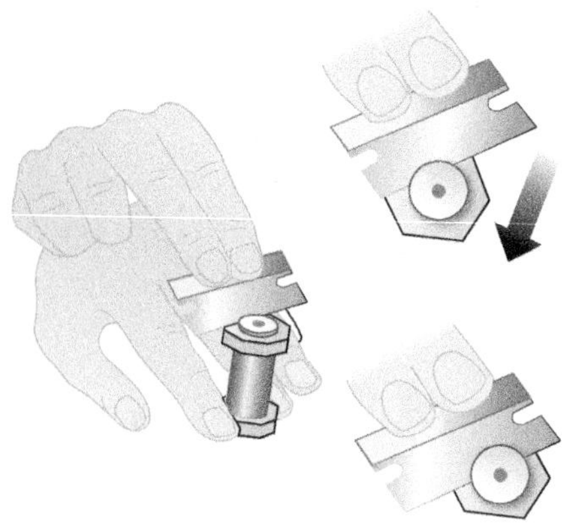

FIGURE 20.5
Using the nut-and-bolt microtome. A piece of stem is embedded in paraffin in the bolt. As you twist the bolt up, slice thin sections to be stained and viewed. Slide the entire blade through the paraffin to smoothly slice thin sections. Follow the directions in Exercise 20.3, Lab Study A, carefully.

Xylem tissue is made up of two cell types:

a. **Tracheids**, which are elongated, thick-walled cells with closed, tapered ends. They are dead at functional maturity, and their interior spaces are interconnected through pits in the cell walls.

b. **Vessel elements**, which are cylindrical cells that are large in diameter and dead at functional maturity. They become joined end to end, lose their end walls, and form long, vertical vessels.

Vascular cambium is a type of meristematic tissue that is located between the xylem and the phloem and which actively divides to give rise to secondary tissues (You will study these tissues in woody stems in Exercise 20.4).

8. *Complete the Results section on the next page for this slide, then return to step 9 to prepare and observe your own handmade sections of stem preparations.*
9. Cut the stem sections in the hardened paraffin.

   a. Support the nut-and-bolt microtome with the bolt head down and, using the razor blade, carefully slice off any excess paraffin extending above the nut. Be careful to slice in a direction away from your body and to keep your fingers away from the edge of the razor blade (Figure 20.5).

   Be careful to keep fingers and knuckles away from the razor blade. Follow directions carefully.

   b. Turn the bolt *just a little*, to extend the stem/paraffin above the edge of the nut.

   c. Produce a thin section by slicing off the extension using the full length of the razor blade, beginning at one end of the blade and slicing to the other end of the blade (see Figure 20.5). *Use the entire blade surface, not a sawing action.*

   d. Transfer each section to a small petri dish containing 50% ethanol.

   e. Continue to produce thin sections of stem in this manner. The thinnest slices may curl, but this is all right if the stem section remains in the paraffin as you make the transfer. Cell types are easier to identify in very thin sections or in the thin edges of thicker sections.

10. Stain the sections.

    a. Leave the sections in 50% ethanol in the petri dish for 5 minutes. The alcohol *fixes*, or preserves, the tissue. Using dissecting needles and forceps, carefully separate the tissue from the surrounding paraffin.

    b. Using forceps, move the stem sections, free of the paraffin, to a clean slide.

    c. Add several drops of toluidine blue to cover the sections. Allow the sections to stain for 10 to 15 seconds.

    d. Carefully draw off the stain by placing a piece of paper towel at the edge of the stain.

    e. Rinse the sections by adding several drops of distilled water to cover the sections. Draw off the excess water with a paper towel. Repeat this step until the rinse water no longer looks blue.

    f. Add a drop of 50% glycerine to the sections and cover them with a coverslip, being careful not to trap bubbles in the preparation.

g. Observe your sections using a compound microscope. Survey the sections at low or intermediate power, selecting the specimens with the clearest cell structure. You may have to study more than one specimen to see all structures. The thinnest edges of sections will provide the clearest view, particularly of the cells in vascular bundles.

11. Follow steps 3–7 above and identify *all structures and cells*. Incorporate your observations into the Results section (step 4 below).

Results

1. Label the stem section in Figure 20.6b and 20.6c.
2. Were any epidermal trichomes present in your stem?

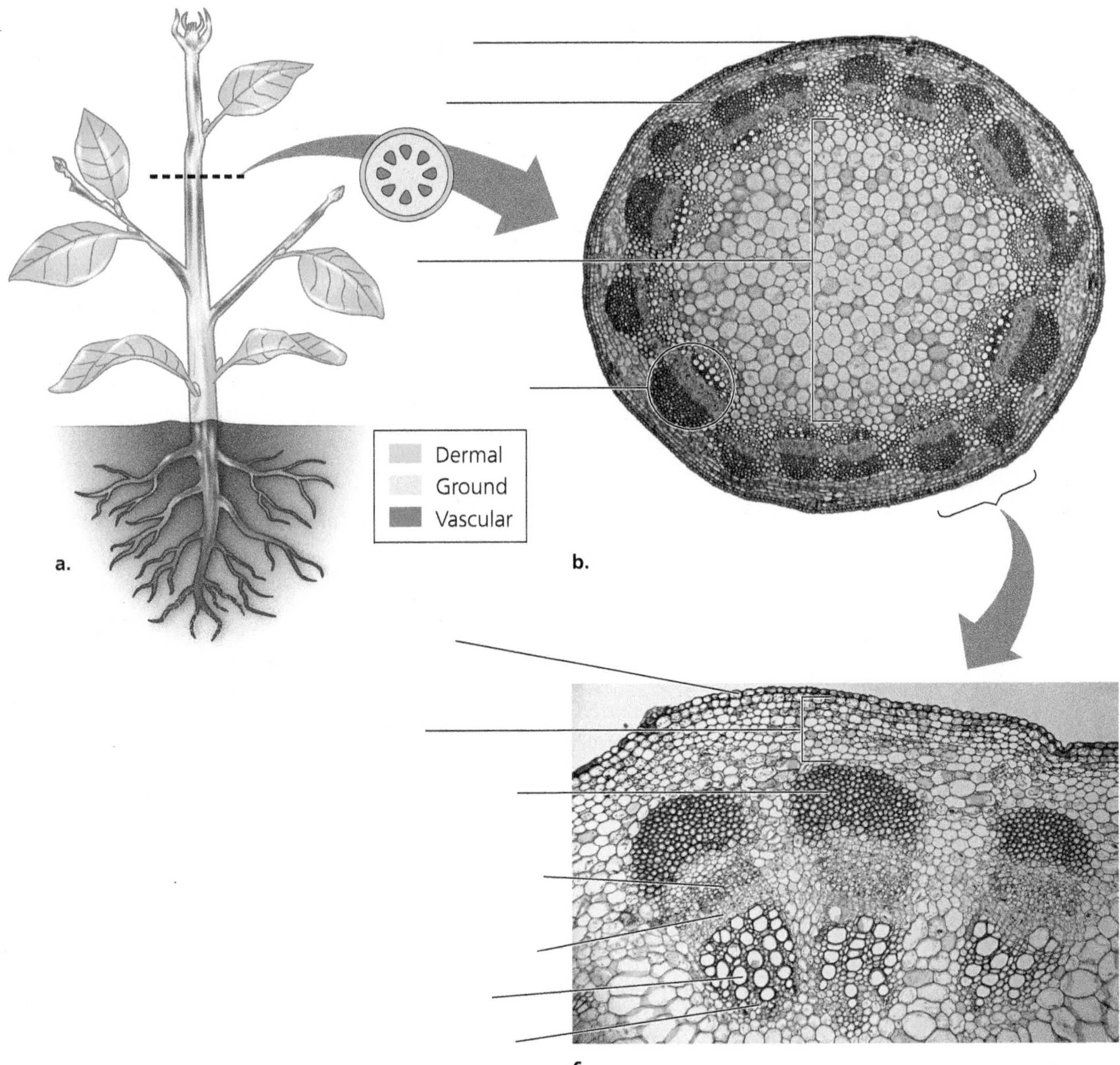

FIGURE 20.6
Stem anatomy. (a) Diagram of whole plant. (b) Photomicrograph of cross section through the stem portion of the plant. (c) Enlargement of one vascular bundle as seen in cross section of the stem.

3. Note any features not described in the procedure. Sketch these in the margin of your lab manual for future reference. Return to Procedure step 9 in this lab study and complete the preparation of hand sections of the bean stem.
4. Compare your hand sections with the prepared slide. Modify Figure 20.6 or sketch your hand sections in the margin. Is there any evidence of vascular cambium and secondary growth (Exercise 20.4)? Compare your results with those of other students.

i The functions of cells were described in the Summary of Basic Plant Tissue Systems and Cell Types, which appeared near the beginning of this lab topic (Figure 20.2).

Discussion

1. Which are larger and more distinct, xylem cells or phloem cells?

2. What types of cells provide support of the stem? Where are these cells located in the stem?

3. For the cells described in your preceding answer, how does their observed structure relate to their function, which is support?

4. What is the function of xylem? Of phloem? Which of these have living cells at maturity? Why are living cells important to the function of one type of tissue and not to the other?

5. The pith and cortex are made up of parenchyma cells. Describe the many functions of these cells. Relate parenchyma cell functions to their observed structure.

6. What differences did you observe in the prepared stem sections and your hand sections? What factors might be responsible for these differences?

Lab Study B. Roots

Materials

prepared slide of buttercup (*Ranunculus*) root (cross section)
demonstration of fibrous roots and taproots
colored pencils
compound microscope

Introduction

Roots and stems often appear to be somewhat similar, except that roots grow in the soil and stems above the ground. However, some stems (rhizomes) grow underground, and some roots (adventitious roots) grow aboveground. Roots and stems may superficially appear similar, but they differ significantly in their functions. One of the major themes of biology is that structure and function are closely related at all levels of the hierarchy of life. Therefore, we would expect that the structure of stems and roots might differ in important ways.

What are the primary functions of stems?

Roots have four primary functions:

1. Anchorage of the plant in the soil
2. Absorption of water and minerals from the soil
3. Conduction of water and minerals from the region of absorption to the base of the stem
4. Starch storage to varying degrees, depending on the plant

Hypothesis

The working hypothesis for this investigation is that the *structure* of the plant body is related to particular *functions.*

Prediction

Based on the hypothesis, make a prediction about the similarity of root and stem structures that you expect to observe (if/then).

You will now test your hypothesis and predictions by observing the external structure of roots and their internal cellular structure and organization in a prepared cross section. This investigation is an example of collecting evidence from observations rather than conducting a controlled experiment.

Procedure

1. Examine the external root structure. When a seed germinates, it sends down a **primary root**, or **radicle**, into the soil. This root sends out side branches called lateral roots, and these in turn branch out until a root system is formed.

   If the primary root continues to be the largest and most important part of the root system, the plant is said to have a **taproot** system. If many main roots are formed, the plant has a **fibrous root** system. Most grasses have a fibrous root system, as do trees with roots occurring within 1 m of the soil surface. Carrots, dandelions, and pine trees are examples of plants having taproots.

   a. Observe examples of fibrous roots and taproots on demonstration in the laboratory.

   b. Sketch the two types of roots in the margin of your lab manual.

2. Examine the internal root structure.

   a. Study a slide of a cross section through a buttercup (*Ranunculus*) root. Note that the root lacks a central pith. The vascular tissue is located in the center of the root and is called the **vascular cylinder** (Figure 20.7b).

   b. Look for a cortex located between the vascular cylinder and the epidermis. The **cortex** is primarily composed of large parenchyma cells filled with numerous purple-stained organelles. Which of the four functions of roots listed in the introduction to this lab study do you think is related to these cortical cells and their organelles?

   c. Identify the following tissues and regions and label Figure 20.7b and 20.7c accordingly: **epidermis**, parenchyma of **cortex**, **vascular cylinder**, **xylem**, **phloem**, **endodermis**, and **pericycle**. The endodermis and the pericycle are unique to roots. The endodermis is the innermost cell layer of the cortex. The walls of endodermal cells have a band called the **Casparian strip**—made of **suberin**, a waxy material—that extends completely around each cell, as shown in Figure 20.8. This strip forms a barrier to the passage of anything moving between adjacent cells of the endodermis. All water and dissolved materials absorbed by the epidermal root hairs and transported inward through the cortex must first pass through the living cytoplasm of endodermal cells before entering the vascular tissues. The pericycle is a layer of dividing cells immediately inside the endodermis; it gives rise to lateral roots. Refer to Summary of Basic Plant Tissue Systems and Cell Types and Figure 20.2.

Results

1. Review Figure 20.7 and note comparable structures in Figure 20.8.
2. Using a colored pencil, highlight the structures or cells found in the root but not seen in the stem.

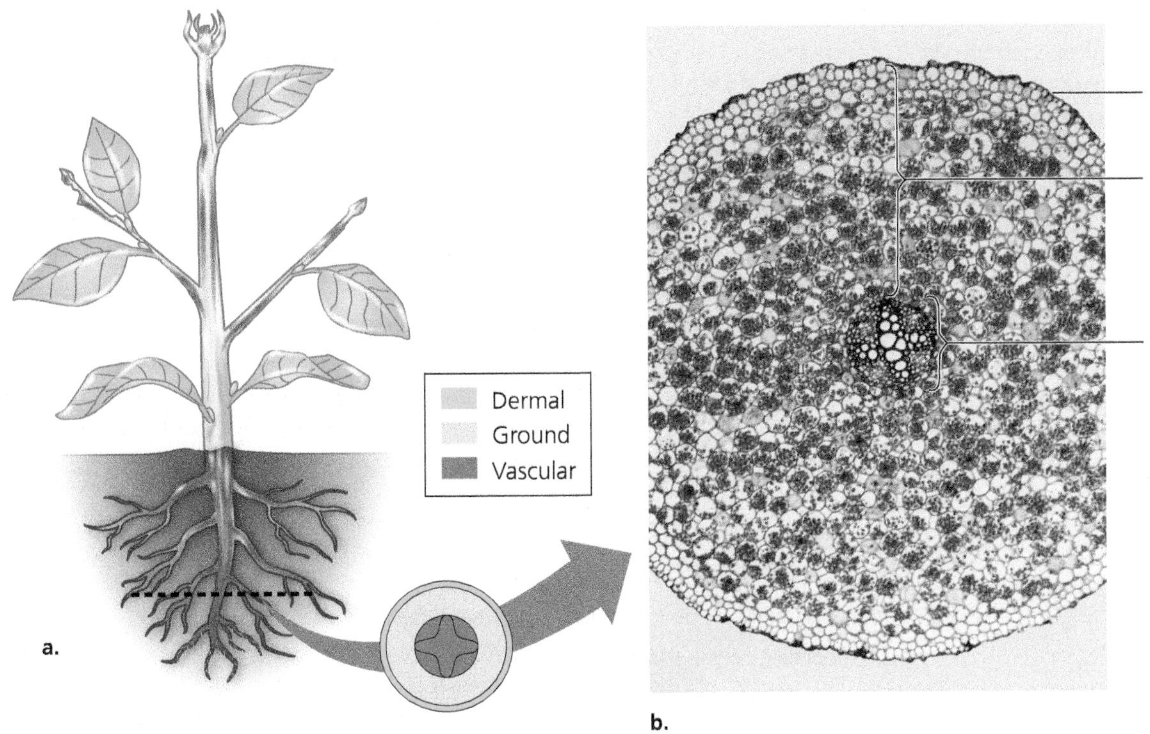

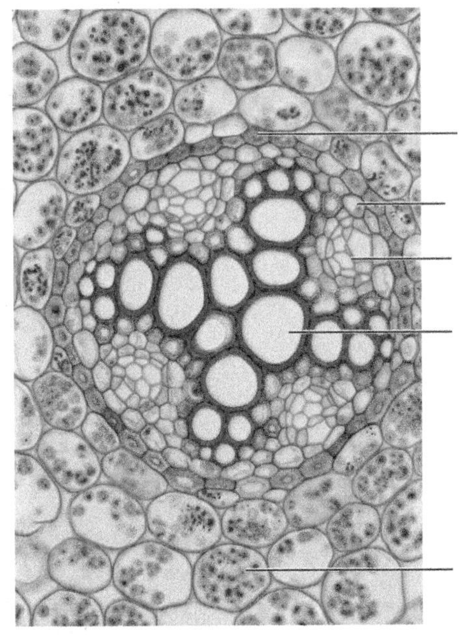

FIGURE 20.7

Cross section of the buttercup root. (a) Whole plant. (b) Photomicrograph of a cross section of a root. (c) Enlargement of the vascular cylinder. Label the root based on your observations of a prepared microscope slide.

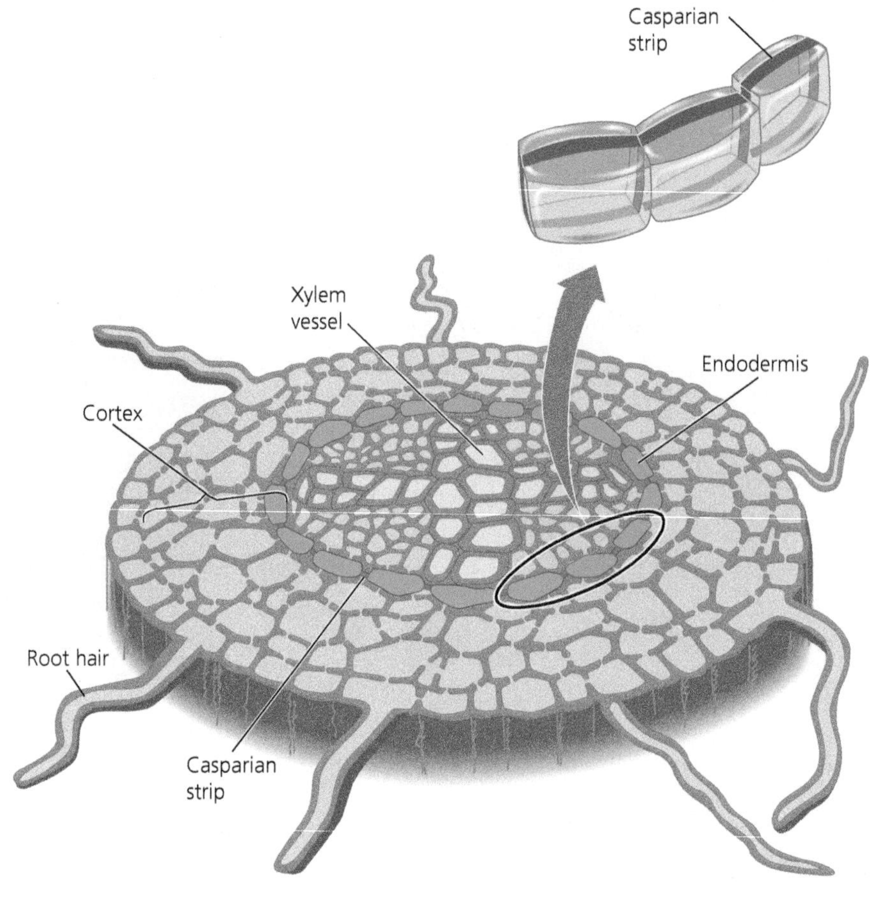

FIGURE 20.8

Root endodermis. The endodermis is composed of cells surrounded by a band containing *suberin,* called the *Casparian strip* (seen in enlargement), that prevents the movement of materials along the cells' walls and intercellular spaces into the vascular cylinder. Materials must cross the cell membrane before entering the vascular tissue.

Discussion

1. Suggest the advantage of taproots and of fibrous roots under different environmental conditions.

2. Did your observations support your hypothesis and predictions?

3. Compare the structure and organization of roots and stems. How do these two organs differ?

4. Explain the relationship of structure and function for two structures or cells found only in roots.

5. Note that the epidermis of the root lacks a cuticle. Can you explain why this might be advantageous?

6. What is the function of the endodermis? Why is the endodermis important to the success of plants in the land environment?

MB Student Media Video—Ch. 35: Root Growth in a Radish Seed (time-lapse)

Lab Study C. Leaves

Materials

prepared slide of lilac (*Syringia*) leaf
slides
compound microscope
coverslips
dropper bottles of water
leaves of purple heart (*Setcreasia*) kept in saline and DI water

Introduction

Leaves are organs especially adapted for photosynthesis. The thin blade portion provides a very large surface area for the absorption of light and the uptake of carbon dioxide through stomata. The leaf is basically a layer of parenchyma cells (the **mesophyll**) between two layers of epidermis. The loose arrangement of parenchyma cells within the leaf allows for an extensive surface area for the rapid exchange of gases. Specialized epidermal cells called **guard cells** surround the stomatal opening and allow carbon dioxide uptake and oxygen release, as well as evaporation of water at the leaf surface. Guard cells are photosynthetic (unlike other epidermal cells), and are capable of changing shape in response to complex environmental and physiological factors. Current research indicates that the opening of the stomata is the result of the active uptake of K^+ and subsequent changes in turgor pressure in the guard cells.

In this lab study, you will examine the structure of a leaf in cross section. You will observe stomata on the leaf epidermis and will study the activity of guard cells under different conditions.

Procedure

1. Before beginning your observations of the leaf cross section, compare the shape of the leaf on your slide with Figure 20.9a and b.

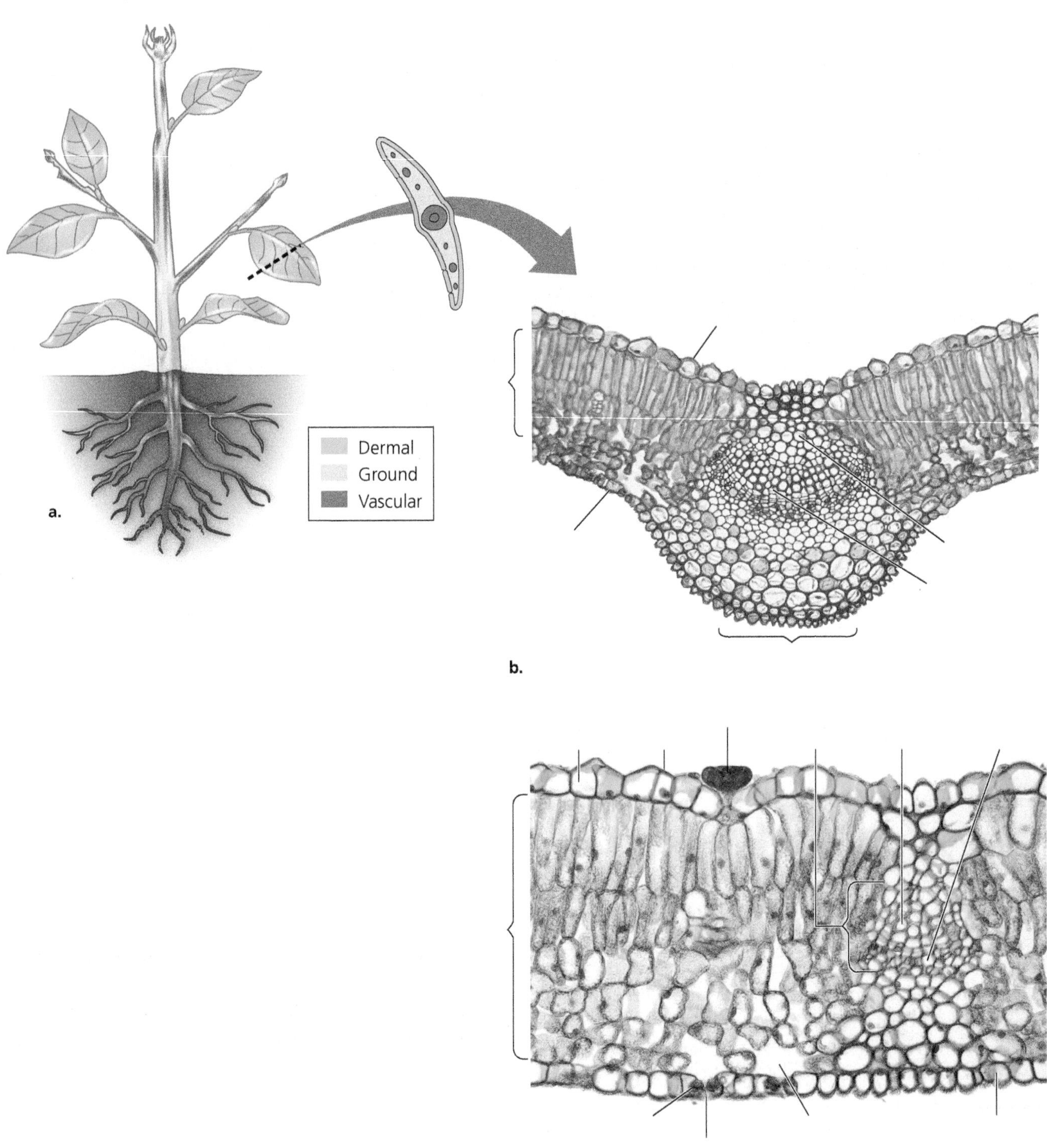

FIGURE 20.9
Leaf structure. (a) Whole plant. (b) Photomicrograph of a leaf cross section through the midvein. (c) Photomicrograph of a leaf cross section near the midvein.

2. Observe the internal leaf structure.

   a. Examine a cross section through a lilac leaf and identify the following cells or structures: **cuticle** (a waxy layer secreted by the epidermis), **epidermis** (upper and lower), parenchyma with chloroplasts (**mesophyll**), **vascular bundle** with **phloem** and **xylem,** and **stomata** with **guard cells** and **substomatal chamber. Trichomes,** hairs or glands, may be visible on the leaf surface. Refer to Summary of Basic Plant Tissue Systems and Cell Types and Figure 20.2.

   b. The vascular bundles of the leaf are often called **veins** and can be seen in both cross section and longitudinal sections of the leaf. Observe the structure of cells in the central midvein. Is xylem or phloem on top in the leaf?

   c. Observe the distribution of stomata in the upper and lower epidermis. Where are they more abundant?

   d. Label the cross section of the leaf in Figure 20.9.

3. Observe the leaf epidermis and stomata.

   a. Obtain two *Setcreasia* leaves, one placed in saline for an hour and the other placed in distilled water for an hour.

   b. Label two microscope slides, one "saline" and the other "H_2O."

   c. To remove a small piece of the lower epidermis, fold the leaf in half, with the lower epidermis to the inside. Tear the leaf, pulling one end toward the other, stripping off the lower epidermis (Figure 20.10). If you do this correctly, you will see a thin purple layer of lower epidermis at the torn edge of the leaf.

   d. Remove a small section of the epidermis from the leaf in *DI water* and mount it in water on the appropriate slide, being sure that the outside surface of the leaf is facing up. View the slide at low and high power on your microscope, and observe the structure of the stomata. Sketch your observations in the margin of your lab manual.

   e. Remove a section of the epidermis from the leaf in *saline* and mount it on the appropriate slide in a drop of the *saline.* Make sure that the outside surface of the leaf is facing up. View the slide with low power on your microscope, and observe the structure of the stomata. Sketch your observations in the margin of your lab manual.

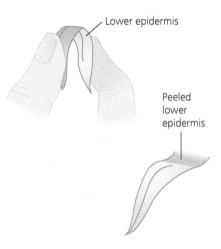

FIGURE 20.10
Preparation of leaf epidermis peel. Bend the leaf in half and peel away the lower epidermis. Remove a small section of lower epidermis and make a wet mount.

Results

1. Review the leaf cross section in Figure 20.9.
2. Describe the structure of the stomata on leaves kept in DI water.
3. Describe the structure of the stomata on leaves kept in saline.

Discussion

1. Describe the functions of leaves.

2. Provide evidence from your observations of leaf structure to support the hypothesis that structure and function are related. Be specific in your examples.

3. Explain the observation that more stomata are found on the lower surface of the leaf than on the upper.

4. Explain the differences observed, if any, between the stomata from leaves kept in DI water and those kept in saline. Utilize your knowledge of osmosis to explain the changes in the guard cells. (In this activity, you stimulated stomatal closure by changes in turgor pressure due to saline rather than K^+ transport.)

EXERCISE 20.4

Structure of Tissues Produced by Secondary Growth

Materials

prepared slides of basswood (*Tilia*) stem
compound microscope

Introduction

Secondary growth is responsible for the increase in circumference in woody plants and arises from meristematic tissue called **cambium.** Vascular cambium and cork cambium are two types of cambium. The **vascular cambium** is a single layer of meristematic cells located between the secondary phloem and secondary xylem. Dividing cambium cells produce a new cell at one time toward the xylem, at another time toward the phloem. Thus, each cambial cell produces files of cells, one toward the inside of the stem, another toward the outside, resulting in an increase in stem circumference. The secondary phloem cells become differentiated into sclerenchyma fiber cells, sieve-tube elements, and companion cells. Secondary xylem cells become differentiated into tracheids and vessel elements. Certain cambial cells produce parenchyma ray cells that can extend radially through the xylem and phloem of the stem.

The **cork cambium** is a type of meristematic tissue that divides, producing cork tissue to the outside of the stem and other cells to the inside. The cork cambium and the secondary tissues derived from it are called **periderm**. The periderm layer replaces the epidermis and cortex in stems and roots with secondary growth. These layers are continually broken and sloughed off as the woody plant grows and expands in diameter.

Procedure

1. Examine a cross section of a woody stem (Figure 20.11).
   a. Observe the cork cambium and periderm in the outer layers of the stem. The outer **cork** cells of the periderm have thick walls impregnated with a waxy material called **suberin**. These cells are dead at maturity. The thin layer of nucleated cells that may be visible next to the cork cells is the **cork cambium**. The **periderm** includes the layers of cork and associated cork cambium. The term **bark** is used to describe the periderm and phloem on the outside of woody plants.
   b. Observe the cellular nature of the listed tissues or structures, beginning at the periderm and moving inward to the central pith region. **Sclerenchyma fibers** have thick, dark-stained cell walls and are located in bands in the phloem. **Secondary phloem** cells with thin cell walls alternate with the rows of fibers. The **vascular cambium** appears as a thin line of small, actively dividing cells lying between the outer phloem tissue and the extensive secondary xylem. **Secondary xylem** consists of distinctive open cells that extend in layers to the central **pith** region. Lines of parenchyma cells one or two cells thick form **lateral rays** that radiate from the pith through the xylem and expand to a wedge shape in the phloem, forming a **phloem ray**.
2. Note the **annual rings** of xylem, which make up the **wood** of the stem surrounding the pith. Each annual ring of xylem has several rows of **early wood**, thin-walled, large-diameter cells that grew in the spring and, outside of these, a few rows of **late wood**, thick-walled, smaller-diameter cells that grew in the summer, when water is less available.
3. By counting the annual rings of xylem, determine the age of your stem. Note that the phloem region is not involved with determining the age of the tree.

Results

1. Review Figure 20.11.
2. Sketch in the margin of your lab manual any details not represented in the figure that you might need for future reference.
3. Indicate on your diagram the region where primary tissues can still be found.

Discussion

1. What has happened to the several years of phloem tissue production?

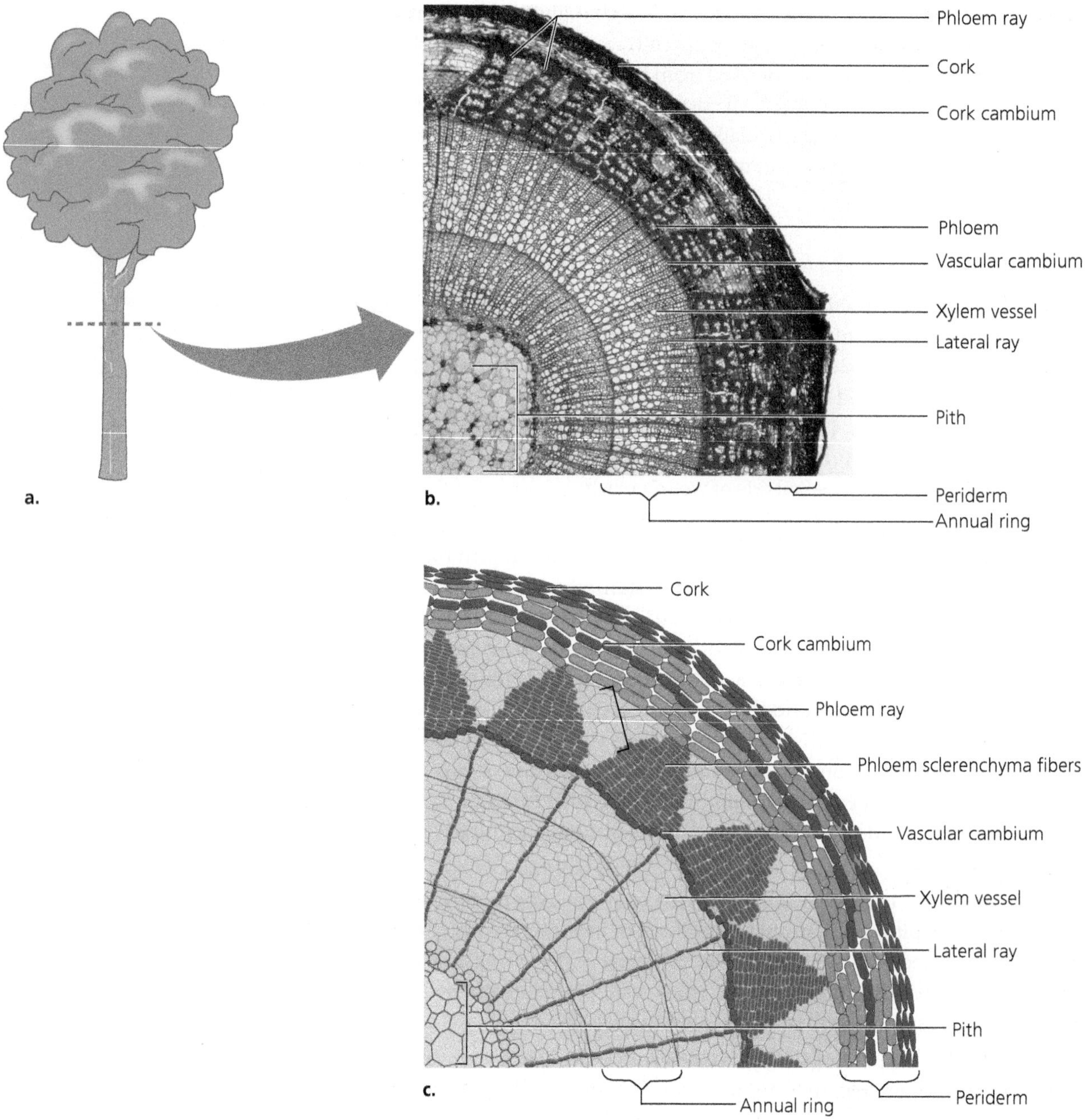

FIGURE 20.11

Secondary growth. (a) Whole woody plant. (b) Photomicrograph of a cross section of a woody stem. (c) Compare the corresponding diagram with your observations of a prepared slide. If necessary, modify the diagram to correspond to your specimen.

2. Based on your observations of the woody stem, does xylem or phloem provide structural support for trees?

3. What function might the ray parenchyma cells serve?

4. How might the structure of early wood and late wood be related to seasonal conditions and the function of the cells? Think about environmental conditions during the growing season.

EXERCISE 20.5

Grocery Store Botany: Modifications of Plant Organs

Materials

variety of produce: squash, lettuce, celery, carrot, white potato, sweet potato, asparagus, onion, broccoli, and any other produce you wish to examine

Introduction

Every day you come into contact with the plant world, particularly in the selection, preparation, and enjoyment of food. Most agricultural food plants have undergone extreme selection for specific features. For example, broccoli, cauliflower, cabbage, and brussels sprouts are all members of the same species that have been selected for different characteristics through generations of plant breeding. In this exercise, you will apply your botanical knowledge of plant organs and tissues to the laboratory of the grocery store.

Procedure

1. Working with another student, examine the numerous examples of root, stem, and leaf modifications on demonstration. (There may be some reproductive structures as well. Refer to Lab Topic 15 Plant Diversity II, if needed.)
2. For each grocery item, determine the type of plant organ, its modification, and its primary function. How will you decide what is a root, stem, or leaf? Review the characteristics of these plant organs and examine your produce carefully.

Results

Complete Table 20.1.

Discussion

1. What feature of the white potato provided key evidence in deciding the correct plant organ?

2. Based on your knowledge of the root, why do you think roots have been selected so often as food sources?

TABLE 20.1 Grocery Store Botany

Name of Item	Plant Organ (Root, Stem, Leaf, Flower, Fruit)	Function/Features (Storage, Support, Reproduction, Photosynthesis)

REVIEWING YOUR KNOWLEDGE

1. Use Table 20.2 to describe the structure and function of the cell types seen in lab today. Indicate the location of these in the various plant organs examined. Refer to Summary of Tissue Systems and Cell Types and Figures 20.2, 20.6, 20.7, 20.9, and 20.11.
2. Some tissues are composed of only one type of cell; others are more complex. List the cell types observed in xylem and in phloem.

   Xylem:

   Phloem:

3. What characteristic of sieve-tube structure provides a clue to the role of companion cells?

TABLE 20.2 Structure and Function of Plant Cells

Cell Type	Structure	Function	Plant Organ
Epidermis			
Guard cells			
Parenchyma			
Collenchyma			
Sclerenchyma			
Tracheids			
Vessels			
Sieve tubes			
Endodermis			
Primary meristems			
Vascular cambium			
Pericycle			
Periderm			
Ray parenchyma			

4. Compare primary and secondary growth. What cells divide to form primary tissue? To form secondary tissue? Can a plant have both primary growth and secondary growth? Explain, providing evidence to support your answer.

APPLYING YOUR KNOWLEDGE

1. Cells of the epidermis frequently retain a capability for cell division. Why is this important? (*Hint:* What is their function?)

2. Why is the endodermis essential in the root but not in the stem?

3. Growth patterns and anatomical features can be used to identify different tree species. Wood identification is an important tool used by forensic scientists, archaeologists, antique restorers, customs agents, physicians, and veterinarians. Choose one of these fields to locate a case study and describe how wood identification using anatomical features was significant in solving a problem.

4. The density of stomata on the lower surface of leaves is affected by genetics and the environment. As the concentration of CO_2 decreases, the density of stomata increases. Hypothesize about the trends in stomatal density that might be expected in response to climate change and increased carbon dioxide concentration.

5. The belt buckle of a standing 20-year-old man may be a foot higher than it was when he was 10, but a nail driven into a 10-year-old tree will be at the same height 10 years later. Explain.

6. The number of annual rings in trees growing in temperate climates corresponds to the age of the trees. Scientists use tree ring analysis (dendrochronology) to age trees. By comparing patterns of tree ring size in cores from living trees with patterns in older dead trees and even samples of wood

from old buildings, scientists have been able to piece together a chronology back thousands of years. Scientists in Europe are working collaboratively to develop a history of climate change for the last 10,000 years using tree ring analysis (dendroclimatology). How would regional climate changes affect the growth of trees? What features of secondary growth in trees would provide evidence for scientists to determine regional climate change? Why do you think this research is being conducted in temperate forests, but not in tropical forests?

7. The oldest living organisms on Earth are plants. Some bristlecone pines are about 4,600 years old, and a desert creosote bush is known to be 10,000 years old. What special feature of plants provides for this incredible longevity? How do plants differ from animals in their pattern of growth and development?

8. Many of the structural features studied in this laboratory evolved in response to the environmental challenges of the terrestrial habitat. Complete Table 20.3 naming the cells, tissues, and organs that have allowed vascular plants to adapt to each environmental factor.

TABLE 20.3 Adaptations of Plant Cells and Structures to the Land Environment

Environmental Factor	Adaptations to Land Environment
Desiccation	
Transport of materials between plant and environment	
Gas exchange	
Anchorage in substrate	
Transport of materials within plant body	
Structural support in response to gravity	
Sexual reproduction without water	
Dispersal of offspring from immobile parent	

INVESTIGATIVE EXTENSIONS

1. A walk across campus, through the forest, or along the path at the botanical garden will reveal an amazing diversity in the structure of roots, stems, and leaves. This outward diversity is accompanied by anatomical diversity in structure that is directly related to the functioning of these organs. Using the nut-and-bolt technique in Exercise 20.3, you can investigate questions relating to structural diversity. For other techniques for studying plant anatomy, see http://www.publicbookshelf.com/public_html/Methods_in_Plant_Histology/. Develop an investigation based on your observations of plant structures or consider one of the following suggestions.
   a. C_3 and C_4 plants use different photosynthetic pathways with corresponding differences in leaf anatomy. Using Web and other resources, identify C_3 and C_4 species that are commercially available and compare the leaf anatomy for two related species.
   b. Angiosperms are divided into several groups, including eudicots and monocots. One feature that distinguishes these two groups is the difference in the organization of stem and root tissues. Investigate differences in anatomy for monocots and eudicots.
2. Plants balance the complex problems of gas exchange, temperature regulation, and water loss through stomata opening and closing. Using techniques from this lab topic, investigate microenvironmental (at the scale of the plant) factors, such as light, temperature, wind, or soil moisture, that might affect stomata function. See Brewer (1992) for a method for making leaf casts using nail polish and then observing these surface replicas under the microscope.

STUDENT MEDIA: BioFlix, Activities, Investigations, and Videos

www.masteringbiology.com (select Study Area)

BioFlix—Ch. 35: Tour of a Plant Cell; Ch. 36: Water Transport in Plants

Activities—Ch. 35: Root, Stem, and Leaf Sections; Primary and Secondary Growth; Ch. 36: Transport of Xylem Sap; Translocation of Phloem Sap

Investigations—Ch. 35: What Are Functions of Monocot Tissues?

Videos—Ch. 35: Root Growth in a Radish Seed (time-lapse); Ch. 36: Turgid Elodea

REFERENCES

Beck, C. *An Introduction to Plant Structure and Development: Plant Anatomy for the Twenty-First Century*, 2nd ed. Cambridge: Cambridge University Press, 2010.

Brewer, C. A. "Responses by Stomata on Leaves to Microenvironmental Conditions," in *Tested Studies for Laboratory Teaching* (volume 12), Proceedings of the 13th Workshop/Conference of the Association for Biology Laboratory Education (ABLE), 1992, Corey A. Goldman, Editor.

Evert, R. F. and S. E. Eichhorn. *Raven Biology of Plants*, 8th ed. New York: W.H. Freeman, 2013.

Levetin, E. and K. McMahon. *Plants and Society*, 7th ed. New York: McGraw-Hill, 2015.

Mauseth, J. D. *Botany: An Introduction to Plant Biology*, 6th ed. Sudbury, MA: Jones and Bartlett Publishers, 2016.

Taiz, L. and E. Zeiger. *Plant Physiology*, 6th ed. Sunderland, MA: Sinauer, 2014.

WEBSITES

American Botanical Society Online Image Collection for plant anatomy:
http://cms.botany.org/media/collection/id.24.html

American Society of Plant Biologists website with links for undergraduates:
http://www.aspb.org/

Click on General Botany and browse this site for images of plant cells and tissues and plant organ anatomy:
http://botit.botany.wisc.edu

InsideWood database includes images and information on modern and fossil wood:
http://insidewood.lib.ncsu.edu/search;jsessionid=AF80C8095DF5DF1D0E30D8FED2A7027A?0

Photographic Atlas of Plant Anatomy by J. Curtis, N. Lersten, and M. Nowak, with excellent images:
http://botweb.uwsp.edu/anatomy/

LAB TOPIC 21

Plant Growth

This lab topic gives you another opportunity to practice the scientific process introduced in Lab Topic 1. Before going to lab, review scientific investigation in Lab Topic 1 and carefully read Lab Topic 21. Be prepared to use this information to design an experiment for plant growth.

Laboratory Objectives

After completing this lab topic, you should be able to:

1. Describe external and internal factors that influence the germination of angiosperm seeds.
2. Explain the effect of auxin on plant growth.
3. Explain the effect of gibberellins on the growth of dwarf corn seedlings.
4. Define and give examples of phototropism and gravitropism.
5. Design and execute an experiment testing factors that influence seed germination and plant growth.
6. Present the results of the experiment in oral or written form.

Introduction

Plant growth and development are determined by the interactions of external environmental conditions and internal cellular processes, beginning with the formation of the seed. As you studied the life cycle of angiosperms (Lab Topic 15 Plant Diversity II), you observed that fertilization of the egg in the female gametophyte results in an embryo protected in a **seed** consisting of the young embryonic sporophyte, food, and a protective seed coat. Seeds begin their development in the parent plant, but once mature, they are dispersed by a variety of mechanisms. Most seeds go through a period of dormancy, but when the environmental conditions (internal and external) are favorable, a dormant seed will begin to **germinate**; that is, it resumes its development and embryonic growth. Most plants continue to grow as long as they live, a condition known as **indeterminate growth**. Indeterminate growth is possible because of plant tissues called **meristems**, which actively divide throughout the plant's life. Seed germination and plant growth are regulated by external factors such as light, temperature, nutrients, and water availability. The seed coat may constrain dormancy because of impermeability to water and oxygen, its rigidity, and inhibitors. Plants respond to these conditions and stimuli by internal processes that are triggered by plant **hormones**. Hormones (in plants and animals) are chemical messengers that are produced in small quantities in one part of the organism and transported to another site, where they induce some specific effect. However, there are a few plant hormones that are produced and act locally. In addition to seed germination, plant hormones regulate plant growth and responses to the environment, including gravity and light. Some hormones stimulate growth, whereas others inhibit, and hormones can have different effects depending on the tissue, concentration of the hormone, and the ratios of multiple hormones. Recent research with mutant strains of the model plant *Arabidopsis* has allowed plant scientists to detect new groups of hormones. Auxin was the first hormone discovered, and the current list also includes gibberellins, cytokinins, abscisic acid, and ethylene, in addition to three newly recognized hormones—brassinosteroids, strigolactones, and jasmonates.

In this lab topic, you will work in teams to investigate external stimuli and internal mechanisms that influence the germination of seeds and the growth of plants. You will complete several brief introductory experiments (Experiment A of Exercises 21.1, 21.2, and 21.3); then your team will propose one or more testable hypotheses based on questions generated during these first experiments. You will then design and carry out an independent investigation based on your hypotheses (Exercise 21.4). You may design an experiment that can be completed in the laboratory period. However, you should plan to make observations of your plants over several days or at the beginning of the next laboratory. Your instructor will tell you if you will be able to return to the lab to make observations or if you should carry your experiment elsewhere for observations. You may be given time to finalize your results and presentation at the beginning of the next laboratory period. Your team should discuss the results and prepare an oral presentation (Appendix A Scientific Writing and Communication). One member of your team will present your team's results to the class for discussion. You should be prepared to persuade the class that your experimental design is sound and that your results support your conclusions. If assigned by the lab instructor, each of you will submit an independent laboratory report describing your experiment and results in the format of a scientific paper (see Appendix A).

The following summarizes the components of each activity in this lab topic.

- Complete Experiment A in Exercises 21.1, 21.2, and 21.3.
- Discuss possible questions that your research team might investigate.
- Choose one interesting question from your list; be certain you can develop a *testable hypothesis* and *prediction.*
- Design and initiate your open-inquiry investigation (Exercise 21.4).
- Complete the experiment and report your results during the following laboratory period.

General Procedures for Independent Investigations: Germinating Seeds and Growing Plants

The following sections provide general procedures you will use in designing your open-inquiry investigation (Exercise 21.4). Return to these procedures when planning your investigation.

Experimental Plants

You may choose plants used in the lab studies for your independent investigation. These include *Zea mays* (corn), *Phaseolus vulgaris* (pinto bean), *Phaseolus limensis* (lima bean), *Coleus blumei* (a common ornamental annual with colorful variegated leaves), and *Brassica rapa* (related to *Arabidopsis*, mustard, and cabbage). If you decide to use different plant species, check with your laboratory instructor about the availability of additional plants.

Germinating Bean and Pea Seeds

Bean and pea seeds can be germinated by first submerging them in a 10% sodium hypochlorite solution for 5 minutes to kill bacteria and fungus spores on their surfaces. Follow this with a distilled water rinse and plant the seeds 1 cm deep in flats of vermiculite, a clay mineral that looks like mica and is frequently used as a starting medium for seeds. Add water or a test solution to the flats daily or as needed.

Growing Wisconsin Fast Plants

The *Brassica rapa* seeds used in this exercise were developed by Dr. Paul Williams of the University of Wisconsin, Madison. Dr. Williams used traditional breeding techniques to produce plants, called Wisconsin Fast Plants™, that can complete an entire breeding cycle from seed to seed in 35 days (Figure 21.1). Because of the rapid growth and shortened breeding cycle of these plants, they are excellent investigative tools for use in plant growth experiments. *Brassica rapa* and *Arabidopsis* (a model plant used in molecular biology and genetics research) are in the same plant family, Brassicaceae or Mustard family. For additional information on procedures, resources, and materials for growing Fast Plants see http://www.fastplants.org/how_to_grow/.

A. Seed Germination Exercises

Brassica rapa seeds can be germinated by placing them on wet filter paper in the lid of a petri dish. Stand the dish, tilted on its end, in a water reservoir such as the bottom of a 2-L soft-drink bottle (Figure 21.2a and b). The dish and reservoir

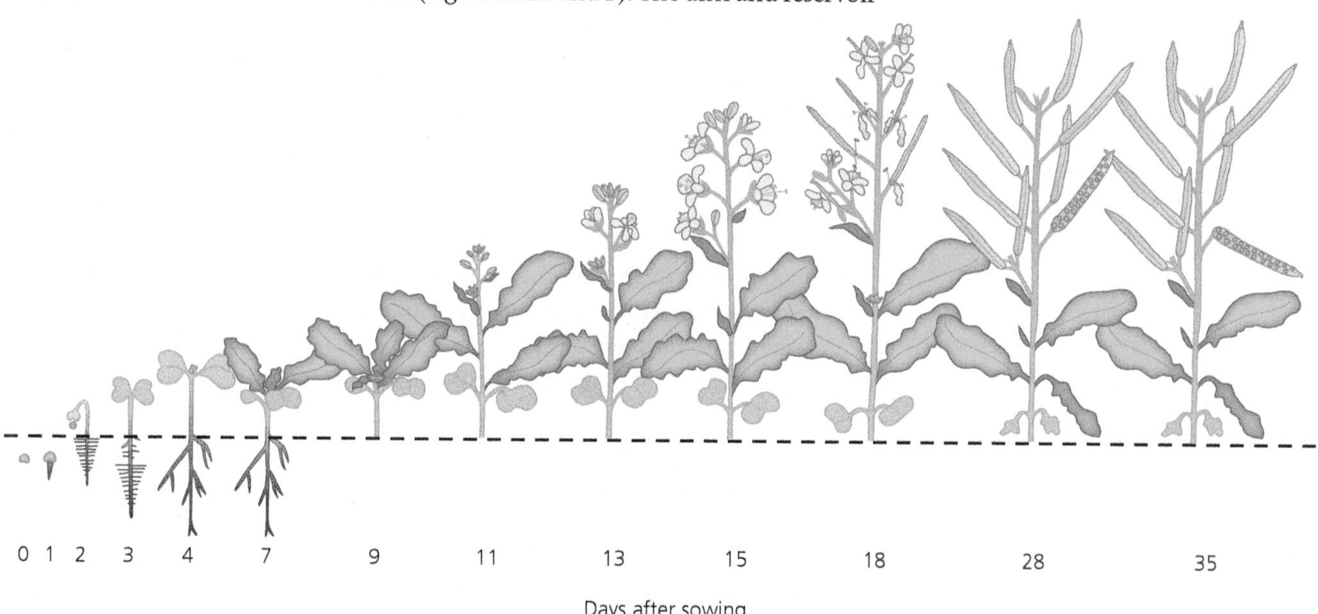

FIGURE 21.1

Life cycle of Fast Plant *Brassica rapa*. These plants can complete their entire life cycle from seed to seed in 35 days.

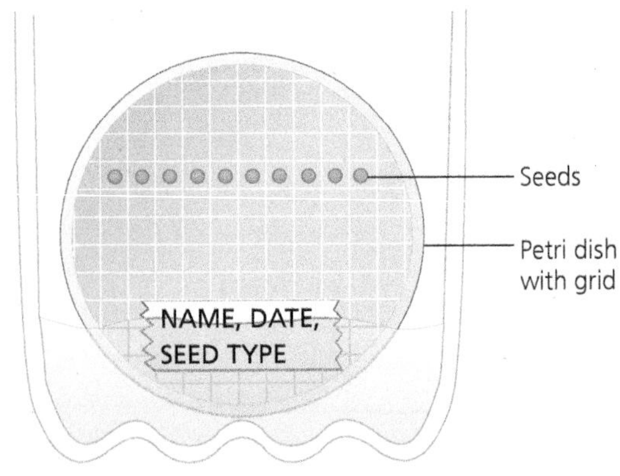

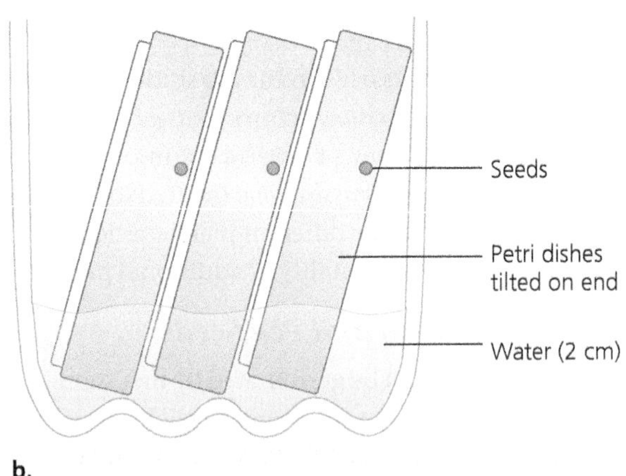

FIGURE 21.2

Germinating *Brassica rapa* seeds in a petri dish. (a) Place the seeds on wet filter paper in the lid of a petri dish. Attach a grid to the outside of the lid for easy seedling measurement. (b) Stand the dishes on end in a water reservoir.

should be placed under fluorescent lights. Germination begins within 24 hours, and observations can be made for several days. It is important to keep the filter paper moist by carefully adding water. If you wish to make quantitative measurements of seed germination, tape a transparent grid sheet marked in measured increments to the outside of the petri dish lid. Place the wet filter paper in the lid, as before, and plant the seeds at a particular position in relation to the grid. As the seeds germinate and grow, you can easily use the grid to measure their size. Use a permanent marker to indicate the starting position and growth (Figure 21.2a).

B. Tropism Studies

Environmental stimuli, such as light and gravity, can elicit a growth response in which the plant grows toward or away from the stimulus. This kind of directional response is called a **tropism**. In Exercise 21.2, you will be investigating responses to light (**phototropism**) and to gravity (**gravitropism**).

1. *Phototropism. Brassica rapa* seeds can be germinated in empty 35-mm clear or black film canisters (Figure 21.3). The canisters can be used as is or black canisters can be modified by punching holes in the sides to allow light to enter the chamber. Place small, appropriately sized squares of wet blotting paper or floral foam disks in the lid, and place two or three seeds on the blotting paper or foam disks. (Do not use filter paper; it dries out too quickly.) Invert the canister and snap it into the lid. Holes punched into the canister can be covered with different-colored filters, and the size and number of the holes can be varied to alter the quality or quantity of light hitting the plants.
2. *Gravitropism.* Seeds and seedlings may be used to demonstrate gravitropism. For investigating *gravitropism* and *seed germination*, prepare a windowless black film canister with wet blotting paper as described for phototropism. Place three seeds on the blotting paper. Place the chamber horizontally and use a permanent marker to indicate which side is up. The chamber will resemble Figure 21.3, but there will be no holes and the chamber will be placed on its side. (*Hint: Tape a film canister lid to the outside of the canister to keep it from rolling.*) Within 1 or 2 days you will be able to observe the effects of gravitropism.

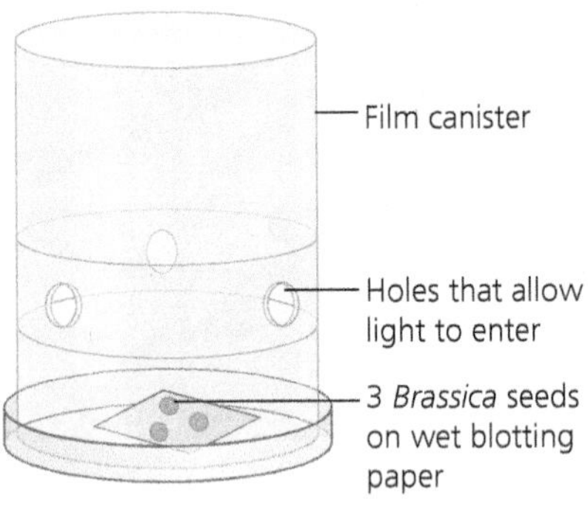

FIGURE 21.3

Growing *Brassica rapa* seeds in film canisters. Place seeds on moist blotting paper in the lid. Holes can be punched in the canister to allow light to enter the chamber.

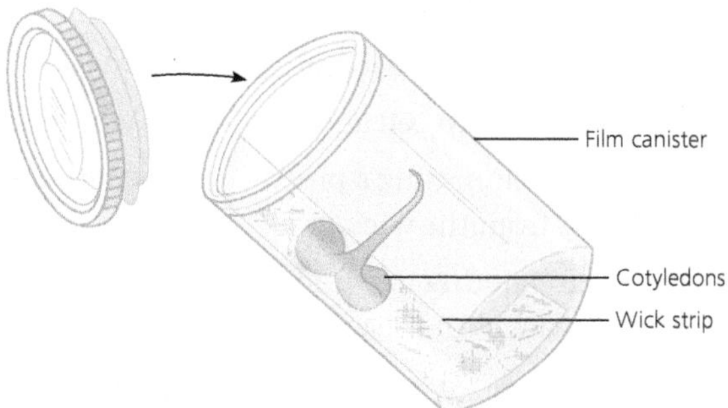

FIGURE 21.4
Growing *Brassica rapa* seedlings in film canisters. Attach the seedling to the center of the wick strip with the cotyledons against the wick as shown.

To observe gravitropism in 3-day-old seedlings, attach a wick and grid germination strip to the inside wall of a windowless black film canister, with the grid strip between the wick and the side of the canister. Add just enough water to cover the bottom of the canister, and have the wick in contact with the water. Attach a young seedling with cotyledons present to the center of the wick strip with the cotyledons against the wick and the hypocotyl pointing out, into the canister. (See Figure 21.4.)

C. Growing *Brassica rapa* Seedlings in Quads

Scientists working with Fast Plants suggest germinating seeds in small, commercially available Styrofoam™ containers called *quads,* which contain four cells, or chambers. To germinate seeds in quads (Figure 21.5):

1. Add a wick to each cell to draw water from the source into the soil.
2. Add potting mix until each cell is about half full.

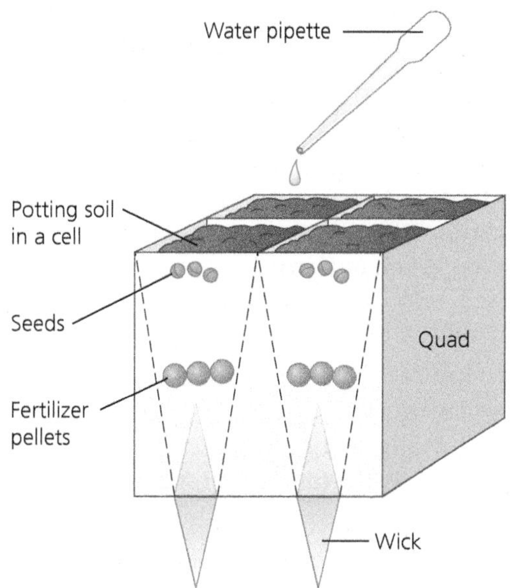

FIGURE 21.5
Growing *Brassica rapa* plants in quads. Pull a wick through the hole in each cell. Add potting soil, fertilizer, and seeds. Initially, water using a pipette.

3. Add three fertilizer pellets.
4. Add more soil and press to make a depression.
5. Add two or three seeds to each cell and cover them with potting mix.
6. Carefully water each section using a pipette until water soaks through the potting mix and drips from the wick.
7. Place the quad on the watering tray under fluorescent lights. Be sure the wicks make good contact with the wet mat on the tray.

After the seeds begin to germinate, you can manipulate the plants in many different ways to investigate plant growth.

EXERCISE 21.1

Factors Influencing Seed Germination

Seeds are the means of reproduction, dispersal, and, frequently, survival for a plant. Plants are immobile and can colonize new habitats and escape inhospitable weather only through the dispersal and dormancy of seeds. *Dormancy* is a special condition of arrested growth in which the seed cannot germinate without a specific set of environmental cues. The optimum environmental conditions for germination (breaking dormancy) are very different for plants in the desert or swamp, for tiny or large seeds, and for seeds that begin to grow in early spring or late fall. Some seeds require that the seed coat be broken, whereas others can only germinate after passing through the digestive system of a bird or mammal. Consider the range of environmental cues and the role of chemical signals (hormones) to initiate germination. In this exercise, you will observe germinating bean and *Brassica rapa* seeds. The beans have been germinating in a moist environment for several days. The *Brassica rapa* seeds are germinating on wet filter paper in the lid of a petri dish.

Experiment A. Germinating Bean and *Brassica rapa* Seeds

Materials

germinating bean seeds
petri dishes with germinating *Brassica rapa* seeds
stereoscopic microscope or hand lens

Introduction

In this lab study, you will examine seed and seedling morphology. Working individually, you will identify seed parts in two plants, a species of bean and *Brassica rapa.* As you investigate the morphology of seeds, consider the role of each structure in facilitating the function of the seed.

Procedure

1. Examine a germinating bean seed and identify the **seed coat**, **cotyledons** (seed leaves), and **embryo** consisting of the **radicle** (embryonic root), **hypocotyl** (plant axis below the cotyledons), and **epicotyl** with young leaves (plant axis above the cotyledons).

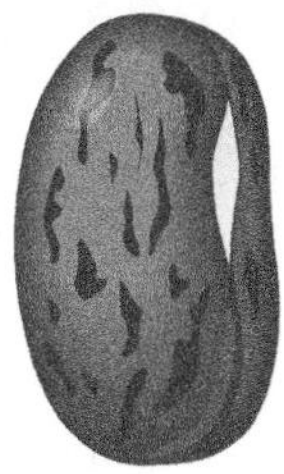

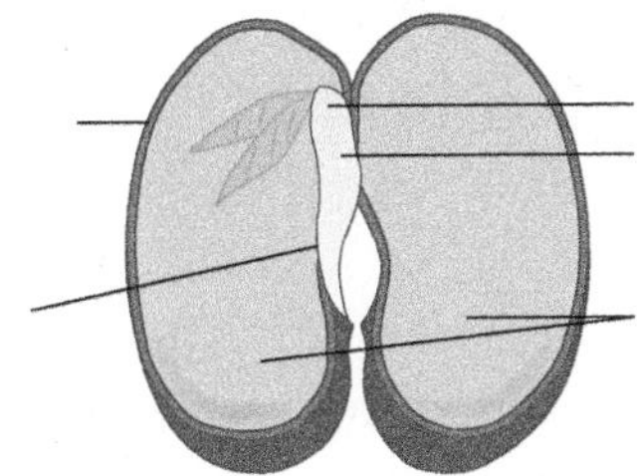

FIGURE 21.6
The structure of a bean seed. Add the labels: *seed coat, cotyledons, radicle, hypocotyl,* and *epicotyl.*

2. Secure a petri dish with germinating *Brassica rapa* seeds. Carefully examine the seeds using a hand lens or the stereoscopic microscope and identify the **seed coat** (which may have been shed), **cotyledons** (how many?), **hypocotyl** (the portion of the plant below the cotyledons), **primary root**, **root hairs**, and **young shoot**, consisting of the epicotyl and young leaves.

Results

1. Label the parts of the bean seed in Figure 21.6.
2. Draw and label several germinating *B. rapa* seeds in the margin of your lab manual.

Discussion

1. What is the function of the seed coat? From what does it develop?

2. What has happened to the endosperm that formed in the embryo sac of the developing ovule observed in Lab Topic 15 Plant Diversity II?

3. How have the cotyledons developed, and from what? (Check your text; see, for example, *Campbell Biology,* Chapter 39.)

4. How does the structure of the seed facilitate the dispersal and survival of the plant?

5. What environmental conditions or cues might be required for seed germination? Consider the variety of seed types (Lab Topic 15 Plant Diversity II) and the range of plant habitats, seasonal growth patterns, and climatic conditions.

Experiment B. Student-Designed Investigation of Seed Germination

Materials

seeds of *Brassica rapa*
various other seeds—beans, corn, radish, okra, lettuce
35-mm film canisters—clear plastic, black, or black with holes punched in the sides
small squares of blotting paper soaking in water
floral foam disks 28 mm diameter, 2–4 mm thick, soaking in water
red, green, and blue light filter sheets
tape
water baths, incubator, refrigerator
grid sheets that fit the lids of petri dishes
wick and grid germination strips
reservoir
forceps
scissors
waterproof pen
agar plates
empty petri dishes
sandpaper
various pH solutions
10% sodium hyperchlorite solution
solutions of gibberellin or other hormones
fertilizer solutions in various concentrations and/or dry fertilizer pellets
hole punch

Introduction

Consider the results of your seed germination experiment and background information provided in this lab and your text as you review the following questions and suggestions for an open-inquiry investigation. If your team chooses to study seed germination for your independent research and report, use the available materials. Design a simple experiment to investigate factors involved in the germination of seeds.

Procedure

1. Collaborating with your research team, read the following potential questions and choose a question to investigate using this list or an original question proposed by your team. You may want to check your text and other sources for supporting information. You should be able to explain the rationale behind your choice of question.
   a. Do seeds need light to germinate? What effect will germination in total darkness have on the process? Is germination better in alternating light and dark, as in nature? Do different species have different light or dark requirements for germination?
   b. Is germination affected by different wavelengths (colors) of light?
   c. Seeds with a hard seed coat can be difficult to germinate. What effect does scarification (scratching the seed coat) have on germination of various seeds? How does scarification occur in nature?
   d. What temperature regimens are optimum for seed germination? Is germination affected by alternating temperatures (as in nature)? What effect does freezing have on seeds from different environments or regions?
   e. Does gibberellin (Exercise 21.3) promote or inhibit seed germination? What about other hormones—for example, cytokinins, abscisic acid, or the recently discovered hormones brassinosteroids or jasmonates?

f. What effect does fertilizer have on seed germination and seedling growth? Is more always better?

g. What effect will salt solutions or acid solutions have on seed germination? Why would questions such as these be of interest?

h. Does seed size have an effect on germination rates?

i. Some crops are limited in their distribution by cold soils that affect seed germination. Will changes in regional climate affect seed germination in okra (a summer plant), broccoli (a fall plant), or varieties of wheat?

2. Design your experiment, proposing hypotheses, making predictions, determining procedures, and recording results as instructed in Exercise 21.4.

EXERCISE 21.2

Plant Growth Regulators: Auxin

Both plants and animals respond to environmental cues. Animals generally respond rapidly via the nervous system or more slowly via hormones secreted from endocrine glands. Plants lack a nervous system, are sedentary, and respond to environmental stimuli via chemical messengers including hormones. Changes in hormones lead to altered patterns of growth and development. **Auxin** is the name given to a complex of substances that promotes stem growth. The natural auxin indoleacetic acid (IAA) is a hormone produced in apical meristems. It migrates down the stem from the zone of production to tissues in the stem, leaves, and roots. If the growing tip is removed, the stem will not elongate, but if the tip is replaced with a paste containing IAA, elongation will continue. At low concentrations, IAA facilitates cell elongation and promotes growth by breaking linkages among cellulose fibers and loosening the cell wall. In this exercise, you will investigate the role of auxin in stem and root curvature in response to light and gravity.

Experiment A. Gravitropic and Phototropic Curvature in *Coleus blumei*

Materials

(on demonstration)
Coleus plant placed on its side
Coleus plant in unilateral light
Coleus plant in upright position

FIGURE 21.7
Phototropism. Plants detect light, which stimulates the redistribution of auxin, resulting in cell elongation and curvature of the stem. Do the cells elongate on the dark or light side of the stem?

Introduction

In this exercise, you will investigate the growth of the stem and root in response to two environmental stimuli, gravity and light. **Gravitropism** is the response of plant organs to gravity. Different plant organs may show positive or negative gravitropism. **Phototropism** is the response of a plant to light (Figure 21.7). A plant organ may grow toward a light stimulus (positive phototropism) or away from a light stimulus (negative phototropism). Three *Coleus blumei* plants are on demonstration in the lab room. One was placed on its side several days ago. Another has been growing in unilateral light for several days. The third, the control, was left undisturbed in the greenhouse until lab time.

Procedure

1. Study gravitropic curvature in *Coleus blumei*.
   a. Carry your lab notebook to the demonstration area and observe the *Coleus* plant placed on its side.
   b. Examine the plant, noting the appearance of different regions of the stems and roots (if visible). What part of each stem has curved? To what degree? How could you measure the amount of curvature in the stem?
   c. Compare this plant with the control plant left in an upright position.
   d. Make a simple sketch of the plant, showing the curvature of stem and roots (if any).
   e. Describe your observations and answer the questions in the Results section.
2. Study phototropic curvature in *Coleus blumei*.
   a. Examine the plant growing in unilateral light, noting the appearance of different regions of the stems.
   b. Compare the plant to the control plant, which has received light from all directions in the greenhouse. What measurements can you make to quantify the differences in the two plants?
   c. Make a simple sketch of the plant, showing the curvature of stem and roots (if any).
   d. Record your observations and answer the questions in the Results section.

Results

1. Describe the appearance of the plant lying on its side. How does this compare with the appearance of the control plant?

2. What is the extent of the response to gravity? How could you measure or quantify the response?

3. Describe the appearance of the plant in unilateral light. Compare this plant with the control plant.

4. What is the extent of the response to the light? How could you measure or quantify the response?

5. What part of the plant specifically is being affected? (If you need to, review the anatomy of the apical meristem in Lab Topic 20 Plant Anatomy.) Can you explain why?

Discussion

1. What type of response would you expect to see if you reoriented the plant lying on its side? Explain.

2. Current research has confirmed that plants detect gravity by the location of **statoliths** (specialized amyloplasts or starch grains). Can you suggest how these might signal a change in orientation of the plant from vertical to horizontal? Consult your text for additional information. (This is not the only mechanism, as plants that lack statoliths still have a weak response to gravity.)

3. Review with your team members the physiological basis for the growth response. How is auxin involved in gravitropism? Where is it produced? How is the action of auxin different in root and stem tissues? Consult your text, if necessary.

4. Is curvature of the stem in response to gravity and light the result of additional cell division or cell elongation? How do you know, or how could you investigate this?

5. What is the role of auxin in phototropism? Is stem elongation stimulated on the shaded side or the brighter side of the stem? Is more auxin produced or redistributed? Use your text and other resources to answer these questions.

Student Media Videos—Ch. 39: Phototropism; Gravitropism

Experiment B. Student-Designed Investigation of Auxins

Materials

auxin solutions in various concentrations
auxin in lanolin paste
lanolin with no auxin
scissors
Coleus plants
glass containers for planting
Brassica rapa seedlings in quads
corn and bean seedlings in pots or flats
lamps
toothpicks
protractor or protractor app downloaded on smartphone
digital camera or smartphone
spray bottles
cotton-tipped applicators
aluminum foil
35-mm black film canisters
floral foam disks, 28 mm diameter, 2–4 mm thick, soaking in water
wick and grid germination strips
forceps
red, green, and blue light filter sheets
Brassica rapa seeds
3-day-old *Brassica rapa* seedlings germinated in petri dishes on wet filter paper

Introduction

Consider the results of your phototropism and gravitropism experiments and the background information provided in this lab and your text as you review the following questions and suggestions for an open-inquiry investigation. Discuss with your team members ways to use *Coleus, Brassica,* or corn or bean seedlings to further investigate these processes. Using the materials available, design a simple experiment to investigate the role of auxin in plant growth or factors involved specifically in phototropism and gravitropism.

Procedure

If spraying solutions with plant hormones or herbicides, isolate plants so that the spray only affects your experimental plants.

1. Collaborating with your research team, read the following potential questions and choose a question to investigate using this list or an original question proposed by your team. You may want to check your text and other sources for supporting information. You should be able to explain the rationale behind your choice of question.
   a. If only some wavelengths stimulate phototropic response, which ones do and which do not?
   b. If the apical meristem is removed, will plants respond to unilateral light?
   c. How does the root respond to gravity (if the cotyledons are kept stationary)?
   d. If the tip of the root is removed, will roots respond to gravity? (Seedlings can be planted close to the wall in glass containers so that root growth can be viewed, or you may use *Brassica rapa* seeds germinating in transparent film canisters.)
   e. Can these tropisms be altered by applying auxin paste to the plant?
   f. What will happen if the tips of the plants (root or stem) are covered with aluminum foil?
   g. How else does auxin affect plants? How does auxin affect apical dominance? Can auxin be used as an herbicide? (In what concentrations? What is the effect on plants?) What concentration of auxin produces the largest roots on stem cuttings? (What horticultural applications would this have?)
   h. Will seedlings growing in the dark respond to auxin applied to the side of the stem? At all locations on the stem?
   i. The herbicide 2,4-dichlorophenoxyacetic acid (2,4-D), used to control weeds, is described as a synthetic auxin. How does 2,4-D affect plant growth? What plants are affected or unaffected by 2,4-D and at what concentrations? Be sure to isolate plants and any equipment when treating with herbicides.
2. Design your experiment, proposing hypotheses, making predictions, and determining procedures as instructed in Exercise 21.4.

EXERCISE 21.3

Plant Growth Regulators: Gibberellins

Gibberellins are another group of important plant growth hormones found in high concentrations in seeds and present in varying amounts in all plant parts. In some plants, gibberellins produce rapid elongation of stems; in others they produce **bolting**, the rapid elongation of the flower stalk. Produced near the stem apex, gibberellins work by increasing both the number of cell divisions and the elongation of cells produced by those divisions. The effects of gibberellins can be induced by artificially applying solutions to plant parts. Not all plants respond to the application of gibberellins, however, and in this exercise, you will investigate the effect of applying gibberellin solutions to normal and to dwarf (mutant) corn seedlings.

Experiment A. Effects of Gibberellins on Normal and Dwarf Corn Seedlings

Materials

2 pots each with 4 tall (normal) corn seedlings
2 pots each with 4 dwarf (mutant) corn seedlings
calculator

Introduction

The seedlings used in this exercise are approximately the same age, but they exist in two phenotypes, tall and dwarf (Figure 21.8). The tall seedlings are wild type, or normal. A genetic mutation produces dwarf plants that lack gibberellins. Each team of students has *four pots, each with four corn seedlings*. Your instructor has previously treated the plants with either water or a gibberellin solution. In this experiment, you will investigate the effects of these treatments on the corn plants.

FIGURE 21.8
Normal corn plants and recessive dwarf mutants.

Procedure

1. Several days ago, your instructor sprayed the plants with either distilled water or a gibberellin solution, as follows:

Control	***Treated***
Pot 1: normal corn, water treatment	*Pot 2*: normal corn, gibberellin treatment
Pot 3: dwarf corn, water treatment	*Pot 4*: dwarf corn, gibberellin treatment

2. Observe the results of the treatments.
3. Measure the height of each of your plants and record these data in Table 21.1.

Results

1. Determine and record the mean height of plants in each category in Table 21.1.
2. Using the mean height for each category of plants, calculate the percentage difference in the mean height of treated normal plants and control normal plants. Then calculate the percentage difference in the mean height of treated dwarf plants and control dwarf plants. Use the given formula for your calculations:

$$\text{Normal \% difference} = \frac{\text{treated} - \text{control}}{\text{control}} \times 100 = \_\_\_\_\_\_\%$$

$$\text{Dwarf \% difference} = \frac{\text{treated} - \text{control}}{\text{control}} \times 100 = \_\_\_\_\_\_\%$$

3. Record your data for the average percentage difference in the mean values for both normal and dwarf plants from Table 21.1 on the class master sheet. Then calculate the average percentage differences for the entire class.

TABLE 21.1 Height of Normal and Dwarf Corn Seedlings with and Without Gibberellin Treatment

Download an Excel version of this table from www.masteringbiology.com in the Study Area under Lab Media.

	Normal Control (Pot 1)	Normal Treated (Pot 2)	Dwarf Control (Pot 3)	Dwarf Treated (Pot 4)
Plant 1 Height				
Plant 2 Height				
Plant 3 Height				
Plant 4 Height				
Mean Height				

	Your Data	*Class Data*
Average % difference: normal	________	________
dwarf	________	________

Discussion

1. How do values for percentage difference compare for dwarf versus normal treated and untreated plants?

2. What is the action of gibberellins? What can you conclude about the mutation that produces dwarfism in corn? Discuss your results with your group, and consult your text or other references.

Experiment B. Student-Designed Investigation of Gibberellins

Materials

normal and dwarf *Brassica rapa* seedlings (*petite* and *rosette* dwarf strains)
normal and dwarf corn seedlings
normal and dwarf pea seedlings (Little Marvel peas, *Pisum sativum*)
Cyocel® solution
solutions of gibberellin
dropper bottles
sprayers
cotton-tipped applicators
supplies to grow *Brassica rapa* seedlings in quads
supplies for seed germination (see Exercise 21.1 Experiment B)

Introduction

Having seen the effect of gibberellins on the growth of normal and dwarf corn seedlings in the preceding experiment, discuss with your team members possible questions for further study of this group of hormones. If you choose to carry out your independent investigation with this system, the questions provided in the following Procedure section will be appropriate for your study. Using the materials available, design a simple experiment to investigate the actions of gibberellins in *plant growth or seed germination.*

Procedure

If spraying solutions with plant hormones, isolate plants so that the spray only affects your experimental plants.

1. Collaborating with your research team, read the following potential questions and choose a question to investigate using this list or an original question proposed by your team. You may want to check your text and other sources for supporting information. You should be able to explain the rationale behind your choice of question.
   a. Plant scientists have discovered a mutant strain of *Brassica rapa* in which plants are dwarf. In these plants, the internodes do not elongate, and plants consist of a cluster of leaves spreading close to the soil. Flowers cluster close to the leaves. How can you determine if these are GA-deficient dwarf plants?
   b. Would other plant hormones produce the same response in dwarf corn seedlings as do gibberellins?
   c. In the demonstration experiment, the gibberellin solution was sprayed on all parts of the plant. If the gibberellin solution were added only to specific regions, such as the roots or apical meristem, would the effect be the same?
   d. Would the results in the corn experiment differ with different concentrations of gibberellin solution?
   e. What effect do gibberellins have on seed germination? Additional supplies for seed germination investigations are provided for Exercise 21.1 Experiment B.

f. What effect do gibberellins have on root growth on cut stems?

g. Is the dwarf condition seen in certain strains of peas (Little Marvel peas) due to the lack of gibberellins?

h. There are two dwarf forms of *Brassica rapa—rosette* and *petite*. Is dwarfism in these mutant strains due to insufficient gibberellin or to some other factor?

i. *Brassica rapa* mutant, *tall* (ein/ein genotype) produces an excess of gibberellins. Can the elongated growth in this mutant be inhibited by the growth regulator Cyocel®? Can growth be further stimulated by applying gibberellins? Research the mode of action for Cyocel®.

2. Design your experiment, proposing hypotheses, making predictions, and determining procedures as instructed in Exercise 21.4.

EXERCISE 21.4

Designing and Performing an Open-Inquiry Investigation

Materials

See each Experiment B materials list in Exercises 21.1, 21.2, and 21.3.

Introduction

Now that you have completed Experiment A in each of the exercises, your research team should select one factor that affects plant growth to investigate. Return to the investigation of your choice and review the suggestions in Experiment B. Use Lab Topic 1 Scientific Investigation as a reference for designing and performing this independent investigation. You will need to think critically and creatively as you ask questions and formulate your hypothesis. As a team, review and modify the procedures, determine any additional required materials, review the techniques and procedures for growing plants and germinating seeds, and assign tasks to all members of your research team. Your experiments will be successful if you plan carefully, think critically, perform techniques accurately and systematically, and record and report data accurately. The following outline will assist you in designing and performing your original investigation.

> You and your lab partner are responsible for the care and maintenance of your plants. Remember to check the water. The success of your investigation depends on the plants' survival.

Procedure

1. **Develop a research question to investigate.** Consider one or two potential questions, then as a team select one question. Suggested questions are included in Experiment B of Exercises 21.1, 21.2, and 21.3 or select an original question of your choice. (Refer to Lab Topic 1, Exercise 1.1, Lab Study A. Asking Questions.)

   Question:

2. **Formulate a testable hypothesis.** (Refer to Lab Topic 1, Exercise 1.1, Lab Study B. Developing Hypotheses.)

   Hypothesis:

3. **Summarize the essential elements of the experiment.** (Use separate paper.)

4. **Predict the results of your experiment based on your hypothesis.** (Refer to Lab Topic 1, Exercise 1.2, Lab Study C. Making Predictions.)

   Prediction: (if/then)

5. **Outline the procedures used in the experiment.** (Refer to Lab Topic 1, Exercise 1.2, Lab Study B. Choosing or Designing the Procedure.)
   a. Review and modify the procedures used in Experiment A of one of the exercises: 21.1 Factors Influencing Seed Germination, 21.2 Plant Growth Regulators: Auxin, or 21.3 Plant Growth Regulators: Gibberellins. Review the General Procedures for Independent Investigations: Germinating Seeds and Growing Plants section in the Introduction to this lab topic. List each step in your procedure in numerical order.
   b. Critique your procedure: check for replicates, level of treatment, controls, duration of experiment, growing conditions, glassware, and age and size of plants. Review measurements and intervals between measurements. Assign team members to check plants periodically and water if needed.
   c. If your experiment requires materials other than those provided, ask your instructor about their availability. If possible, submit requests in advance.
   d. Create a table for data collection. Use examples in this lab topic as a model or design your own. If computers are available, create your table in Excel. Remember to include space for general observations of plant growth.
6. **Perform the experiment,** making observations and collecting data for analysis.
7. **Record results, including observations and data,** in your data table. Make notes about experimental conditions and observations. Do not rely on your memory for information that you will need when reporting your results.
8. **Prepare your discussion.** Discuss your results in light of your hypothesis.
   a. Review your prediction. Did your results correspond to the prediction you made? If not, explain how your results are different from your prediction, and why this might have occurred.
   b. Review your hypothesis. Review your results (tables and graphs). Do your results support or falsify your hypothesis? Explain your answer, using your data for support.
   c. If you had problems with the procedure or questionable results, explain how they might have influenced your conclusion.

d. If you had an opportunity to repeat and expand this experiment to make your results more convincing, what would you do?

e. Summarize the conclusion you have drawn from your results.

9. **Be prepared to report your results to the class.** Prepare to persuade your fellow scientists that your experimental design is sound and that your results support your conclusions.

10. If your instructor requires it, **submit a written laboratory report** in the form of a scientific paper (see Appendix A). Keep in mind that although you have performed the experiments as a team, you must turn in a lab report of *your original writing.* Your tables and figures may be similar to those of your team members, but your paper must be the product of your own literature search and creative thinking.

REVIEWING YOUR KNOWLEDGE

1. Having completed this lab topic, you should be able to define and use the following terms, providing examples when appropriate: *hormones, seed, seedling, seed coat, cotyledon, endosperm, radicle, hypocotyl, epicotyl, germination, dormancy, statoliths, phototropism, gravitropism, apical dominance, auxin, gibberellins, bolting.*
2. Students investigating plant growth and the effects of hormones removed the seeds from developing strawberries and compared the size and time of fruit development. The strawberries failed to enlarge and become red and fleshy in plants where the seeds were removed. Research the roles of auxin and gibberellin and determine which hormone is responsible for fruit development.

3. What adaptive role would positive phototropism play in a natural ecosystem where plants are crowded?

APPLYING YOUR KNOWLEDGE

1. Auxin is directly or indirectly responsible for apical dominance in plants. In this phenomenon, the growth of lateral or axillary buds (described in Lab Topic 20 Plant Anatomy) is inhibited by the auxin that moves down the stem from the apical meristem. It has long been the practice of horticulturists to clip off the apical meristems of certain young houseplants. What impact should this practice have on subsequent plant growth and appearance?

2. Light plays an important role in the growth and development of plants. Plants detect the direction and intensity of light, as well as the length of day and seasonal changes from light stimuli. Phototropin, a kinase protein

embedded in the cell membrane, is a light receptor involved in phototropism by triggering cellular changes that result in auxin redistribution and the curvature of the stem. The *absorption spectrum* for phototropin indicates a peak of light absorption in the blue range of the visible light spectrum. Design an investigation that would provide evidence that phototropin is or is not the light receptor that is mediating phototropism. (*Hint:* refer to your investigation of photosynthetic pigments in Lab Topic 6.)

3. Plant hormones are used in agriculture to enhance crops and regulate the timing of crop development. For each of the following scenarios, suggest the growth hormone (auxin, gibberellin, cytokinin, ethylene, or abscisic acid) involved and the mode of action. Use your textbook or other references to assist with the functions of hormones not investigated in this lab topic.

Thompson seedless grapes are treated with a hormone to increase the size of the grapes and also extend the stem so that the clusters are less compact (more loose and full).

Seed companies treat a field of mature plants of mustard, cabbage, and canola with a growth hormone to stimulate flower growth and seed production in all plants at the same time.

Hormones are used in tissue culture to increase cell division and to promote the differentiation of meristems into mature cells. (More than one hormone is involved.)

Tomatoes are shipped green to the distributor. They are then treated with a hormone that promotes fruit development, changing the tomatoes to the bright red color customers expect.

When seedlings are transplanted, they suffer severe drought stress. Horticulturists spray the plants with a low concentration of a hormone that causes stomata closure and reduces water loss.

4. During the 1960s, a world food shortage was solved by the breeding and cultivation of a race of wheat with higher grain production and short stems that resisted the damage of wind and rain. Norman Borlaug won the Nobel Prize in 1970 for his contributions to this "Green Revolution" that reportedly saved the lives of more than 1 billion people. Based on your knowledge of plant hormones, suggest the hormone that would be deficient in this race of wheat. Explain using the evidence observed in the laboratory.

5. Brassinosteroids are a recently discovered class of plant hormones that are similar in structure to animal steroids. Initially isolated in *Brassica napus* (thus their name), they now have been identified in a wide range of plants. Brassinosteroids are more potent than auxins and gibberellins, requiring much smaller concentrations to effect changes in plant growth. These plant hormones can cause cell elongation, among other effects on plant growth. Scientists have long known that the stems of normal seedlings grown in the dark elongate rapidly and become light colored, a condition called *etiolation*. Recently, scientists have cultured a mutant race of *Arabidopsis*, a model plant for physiology studies, that does not become etiolated in the dark. Utilizing *Arabidopsis* mutant and normal plants, devise an experiment to test the role of brassinosteriods in etiolation. Consider controlled variables, control treatments, replicates, how you will measure plant growth, and the time period for making measurements and observations. State your prediction based on the experiment you design.

STUDENT MEDIA: BioFlix, Activities, Investigations, Videos, and Data Table

www.masteringbiology.com (select Study Area)

Activities—Ch. 39: Flowering Lab

Investigations—Ch. 39: What Plant Hormones Affect Organ Formation?

Videos—Ch. 35: Root Growth in a Radish Seed (time-lapse); Ch. 39: Phototropism; Gravitropism

Data Table—Table 21.1 can be downloaded in Excel format. Look in the Study Area under Lab Media.

REFERENCES

Baskin, C. C. and J. M. Baskin. *Seeds: Ecology, Biogeography, and Evolution of Dormancy and Germination,* 2nd ed. Cambridge, MA: Academic Press, 2014.

Chamovitz, D. *What a Plant Knows: A Field Guide to the Senses*. New York: Scientific American/Farrar, Straus and Giroux, 2012.

Hanson, T. *The Triumph of Seeds: How Grains, Nuts, Kernels, Pulses, and Pips Conquered the Plant Kingdom and Shaped Human History.* New York: Basic Books, 2015.

Evert, R. F. and S. E. Eichhorn. *Raven Biology of Plants*, 8th ed. New York: W.H. Freeman, 2013.

Mauseth, J. D. *Botany. An Introduction to Plant Biology*, 6th ed. Sudbury, MA: Jones and Bartlett Publishers, 2016.

Srivastava, L. M. *Plant Growth and Development: Hormones and Environment*. San Diego, CA: Academic Press, 2002.

Taiz, L., E. Zeiger, I. Moller, and A. Murphy. *Plant Physiology,* 6th ed. Sunderland, MA: Sinauer, 2014.

Urry, L., M. Cain, S. Wasserman, P. Minorsky, and J. Reece. *Campbell Biology,* 11th ed. San Francisco, CA: Pearson, 2017.

WEBSITES

American Society of Plant Biologists website with links for undergraduates and education:
http://www.aspb.org/

Angle Pro app will measure angles and can be downloaded for free:
https://itunes.apple.com/us/app/angle-pro/id750327028?mt=8
https://play.google.com/store/apps/details?id=com.FiveFufFive.AngleProFreeAndroid

Information on Fast Plants and detailed instructions for growing procedures:
http://www.fastplants.org/
http://www.fastplants.org/activities/

Plant Physiology is a journal of the American Society of Plant Biologists; current research in plant physiology:
http://www.plantphysiol.org/

Planting Science; the Botanical Society of America website connecting scientists and students in collaborative research:
http://www.plantingscience.org/index.php

Plants in Motion by Robert Hangarter, University of Indiana, includes time-lapse photos of plant growth and development and instructions for making time-lapse videos:
http://plantsinmotion.bio.indiana.edu/plantmotion/starthere.html

Teaching Tools for Plant Biology, with information on each of the plant hormones and more:
http://www.plantcell.org/site/teachingtools/teaching.xhtml

USDA website with agricultural applications and information on plants, growth, and hormones:
http:www.usda.gov/wps/portal/usdahome

LAB TOPIC 22

Vertebrate Anatomy I: The Skin and Digestive System

Overview of Vertebrate Anatomy Labs (Lab Topics 22, 23, and 24)

In Lab Topic 18 Animal Diversity I and Lab Topic 19 Animal Diversity II, you investigated several major themes in biology as illustrated by biodiversity in the animal kingdom. One of these themes is the relationship between form and function in organ systems. In this and the following two lab topics, you will continue to expand your understanding of this theme as you investigate the relationship between form, or structure, and function in vertebrate organ systems. For these investigations, you will be asked to view prepared slides and isolated adult vertebrate organs, and to dissect a representative vertebrate, the fetal pig. The purpose of these investigations is not to complete a comprehensive study of vertebrate morphology but rather to use several select vertebrate systems to analyze critically the relationship between form and function.

You will explore the listed concepts in the designated exercises.

1. The specialization of cells into tissues with specific functions makes possible the development of functional units, or organs (Exercise 22.1, Histology of the Skin).
2. Multicellular heterotrophic organisms must obtain and process food for body maintenance, growth, and repair (Exercise 22.3, The Digestive System in the Fetal Pig).
3. Because of their size, complexity, and level of activity, vertebrates require a complex system to transport nutrients and oxygen to body tissues and to remove waste from all body tissues (Exercise 23.1, Glands and Respiratory Structures of the Neck and Thoracic Cavity; Exercise 23.2, The Heart and the Pulmonary Blood Circuit; Exercise 24.1, The Excretory System).
4. Reproduction is the ultimate objective of all metabolic processes. Sexual reproduction involves the production of two different gametes, the bringing together of the gametes for fertilization, and limited or extensive care of the new individual (Exercise 24.2, The Reproductive System).
5. Complex animals with many organ systems must coordinate the activities of the diverse parts. Coordination is influenced by the endocrine and nervous systems. Integration via the endocrine system is generally slower and more prolonged than that produced by the nervous system, which may receive stimuli, process information, and elicit a response very quickly (Exercise 24.3, Nervous Tissue, the Reflex Arc, the Sheep Brain, and the Vertebrate Eye).

Laboratory Objectives

After completing this lab topic, you should be able to:

1. Describe the four main categories of tissues and give examples of each.
2. Identify tissues and structures in mammalian skin.
3. Describe the function of skin. Describe how the morphology of skin makes possible its functions.
4. Identify structures in the fetal pig digestive system.
5. Describe the role played by each digestive structure in the digestion and processing of food.
6. Relate tissue types to organ anatomy.
7. Apply knowledge and understanding acquired in this lab to problems in human physiology.
8. Apply knowledge and understanding acquired in this lab to explain organismal adaptive strategies.

Introduction to Tissues

All animals are composed of **tissues**, groups of cells that are similar in structure and that perform a common function. During the embryonic development of most animals, the body is composed of three tissue layers: **ectoderm**, **mesoderm**, and **endoderm**. (Recall from Lab Topic 18 Animal Diversity I, that animals in the phylum Porifera lack true tissue organization and that animals in the phylum Cnidaria have only two tissue layers—ectoderm and endoderm.) It is from these embryonic tissue layers that all other body tissues develop. There are four main categories of tissues: epithelium, connective tissue, muscle, and nervous tissue. Organs are formed from this limited number of tissues, and usually all four tissue types will be found in a single organ.

Tissues are composed of cells and extracellular substances secreted by the cells. These substances are mostly glycoproteins (carbohydrates bonded with proteins), often organized into an extracellular matrix. **Epithelial tissue** has cells in close aggregates with little extracellular substance (see Figure 22.1). The organization and shape of the cells correlates with the tissue function. These cells may be in one continuous layer, or they may be in multiple layers. They generally cover or line an external or internal surface of an organ or cavity. If formed from single layers of cells, the epithelium is called **simple**. If cells are in multiple layers, the epithelium is called **stratified**. If epithelial cells are flat, they are called **squamous**. If they are cube-shaped, they are called **cuboidal**. Tall, prismatic cells are called **columnar**. Thus, epithelium can be stratified squamous (as in skin), simple cuboidal (as in kidney tubules), or other combinations of characteristics. Epithelial layers may be derived from embryonic ectoderm, mesoderm, or endoderm.

In **connective tissue**, cells are widely scattered in an elaborate **extracellular matrix** consisting of a web of fibers and an amorphous foundation that may be solid, gelatinous, or liquid (Figure 22.2). **Loose connective tissue** binds together tissues and organs and helps hold organs in place. Fibers in this tissue are loosely woven in a liquid matrix. **Adipose tissue**, another connective tissue, consists of adipose cells with fibers in a soft, liquid extracellular matrix. Adipose cells store droplets of fat, causing the cells to swell and pushing the nuclei to one side. **Bone** and **cartilage** are specialized connective tissues

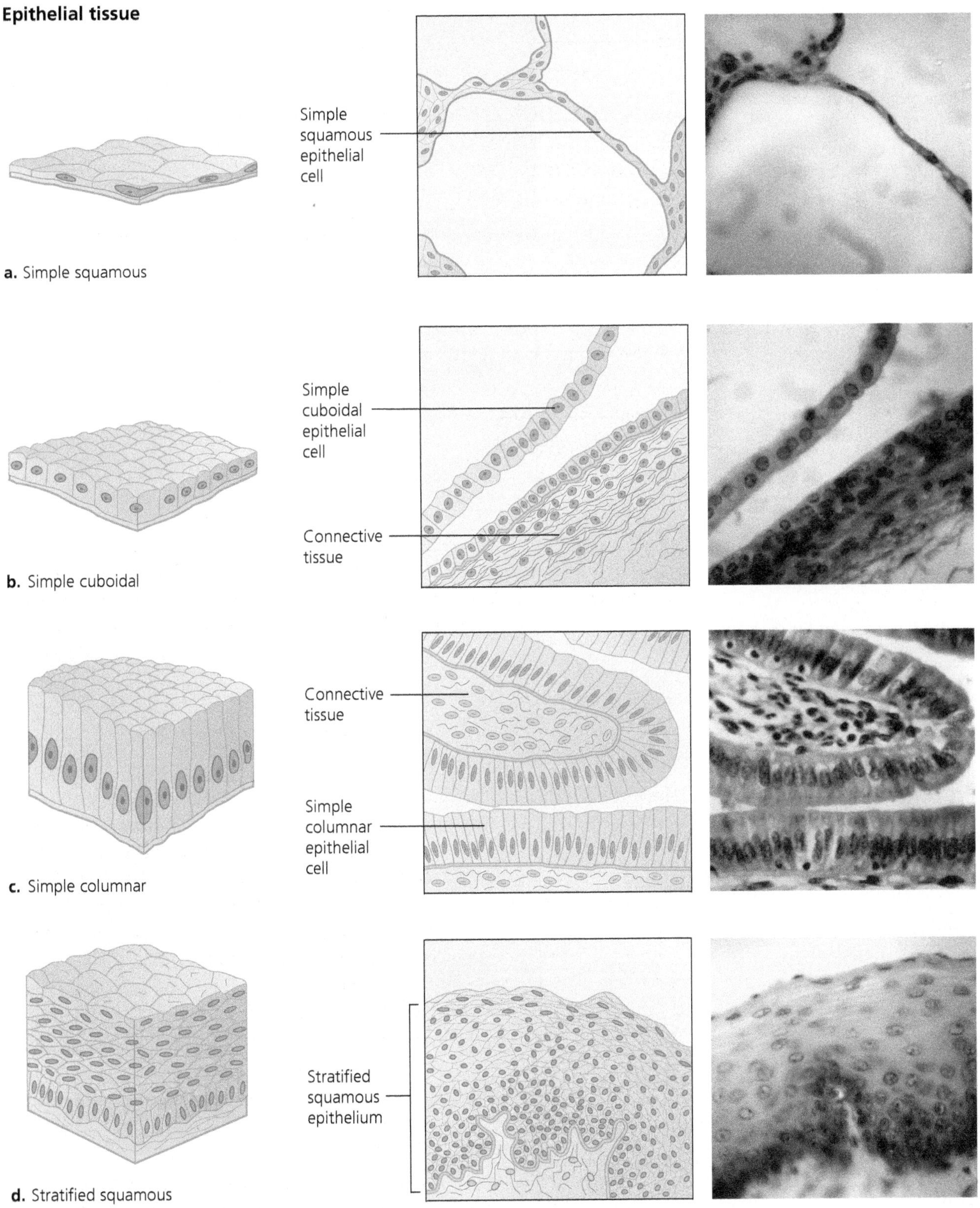

Epithelial tissue

Simple squamous epithelial cell

a. Simple squamous

Simple cuboidal epithelial cell

Connective tissue

b. Simple cuboidal

Connective tissue

Simple columnar epithelial cell

c. Simple columnar

Stratified squamous epithelium

d. Stratified squamous

FIGURE 22.1

Epithelial tissue. Epithelial tissue has closely packed cells with little extracellular matrix. Cells may be (a) squamous (flat), (b) cuboidal (cube-shaped), or (c) columnar (elongated). They may be simple (in single layers) or (d) stratified (in multiple layers).

Connective tissue

a. Loose connective tissue

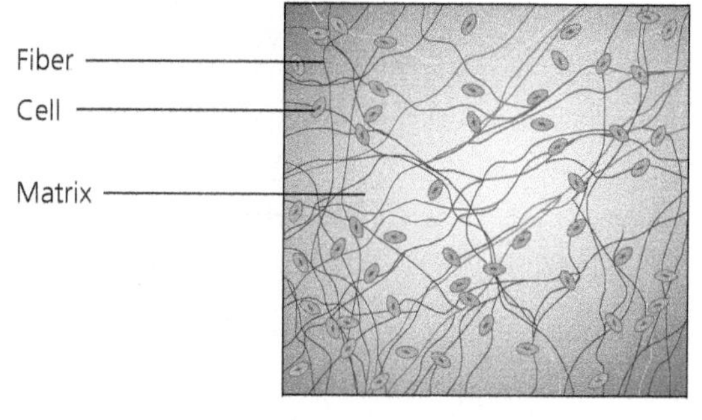

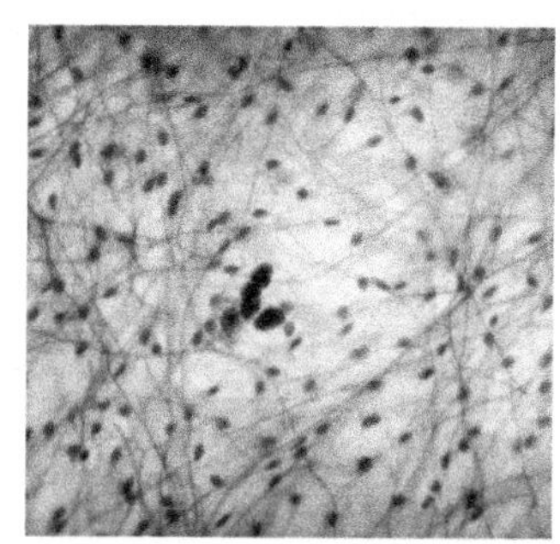

b. Adipose tissue

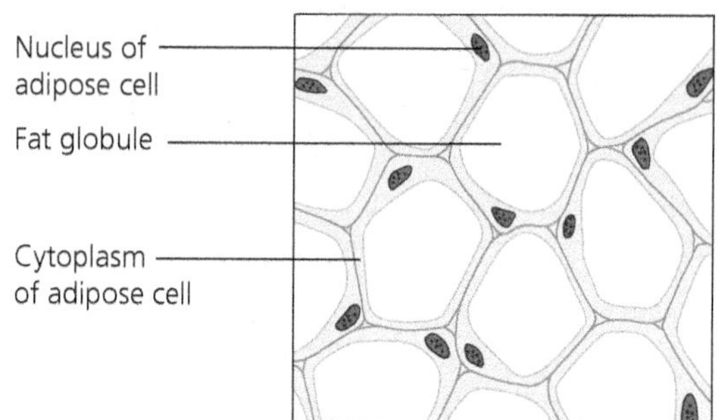

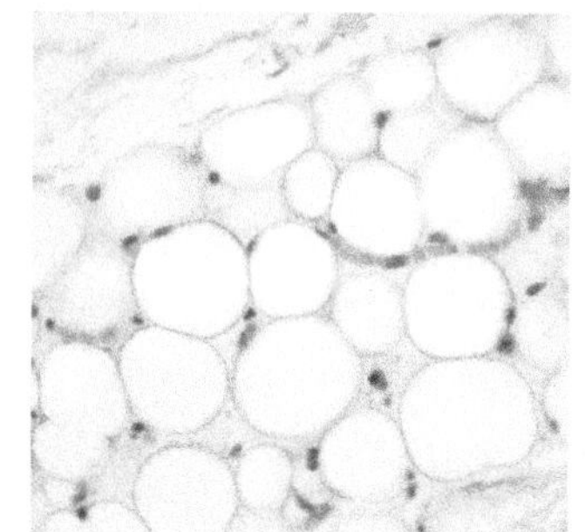

c. Bone

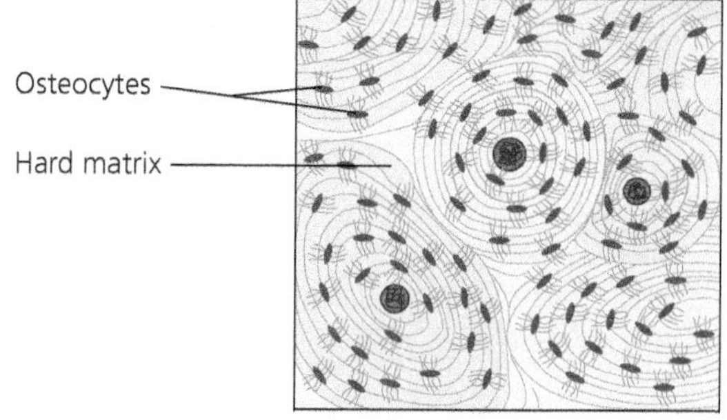

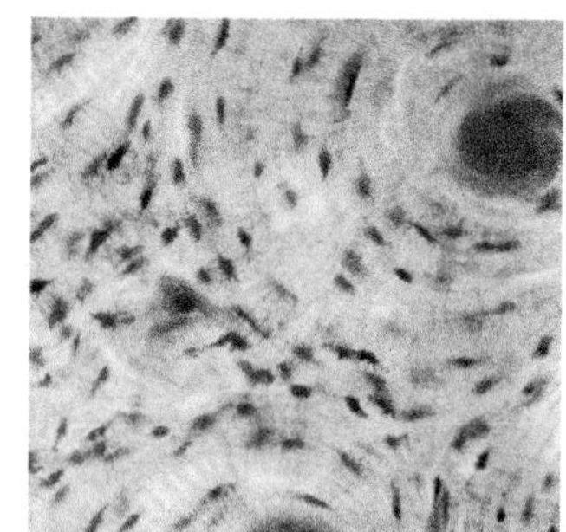

d. Cartilage

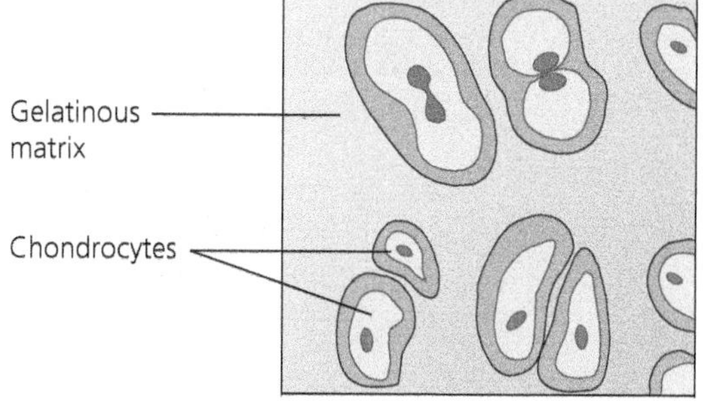

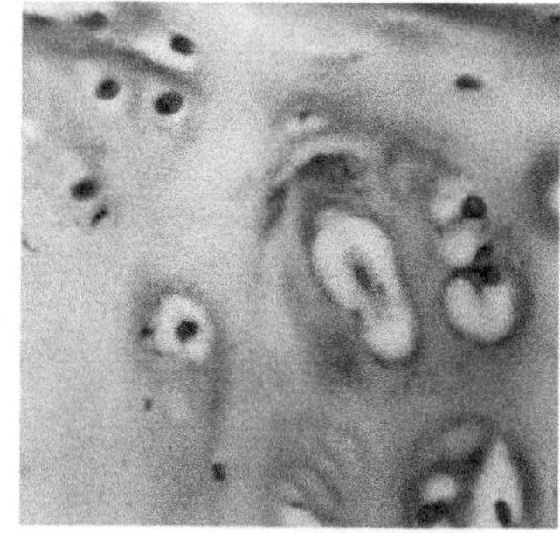

e. Blood

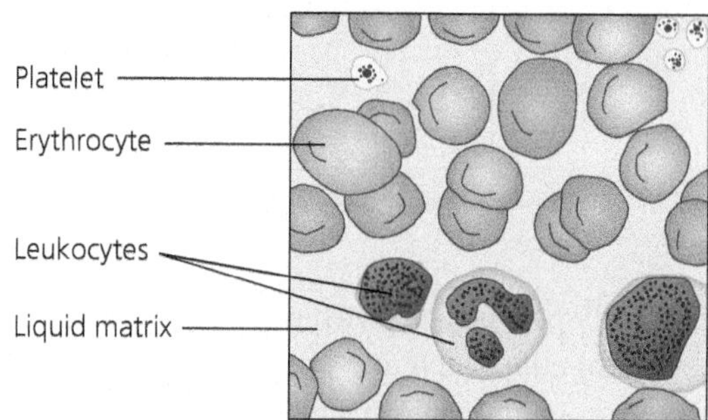

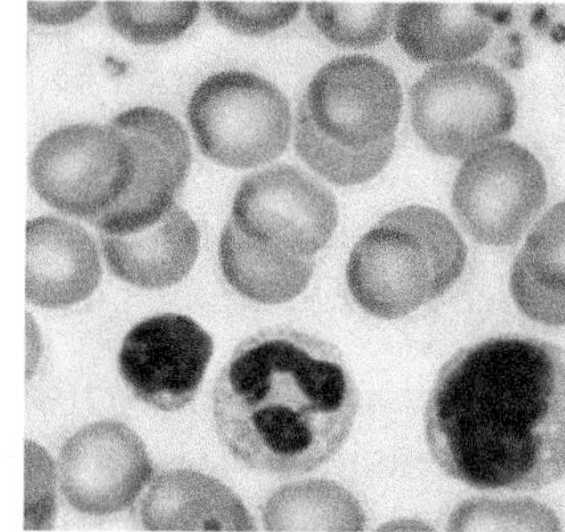

Muscle tissue

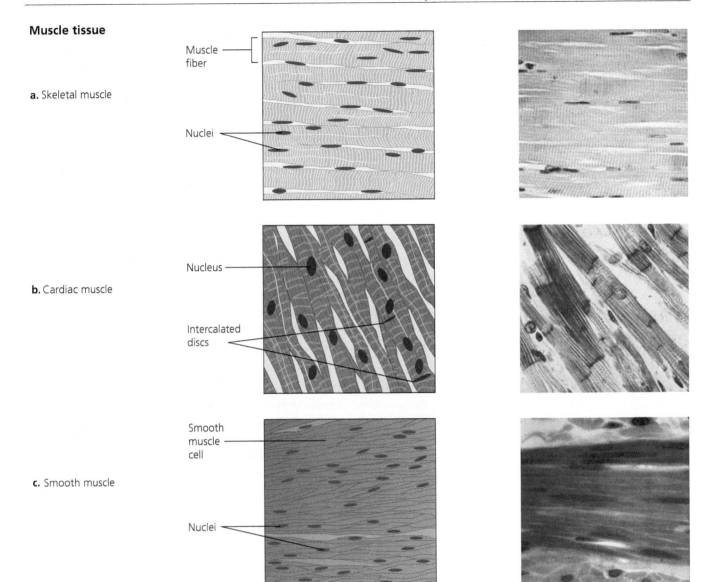

a. Skeletal muscle

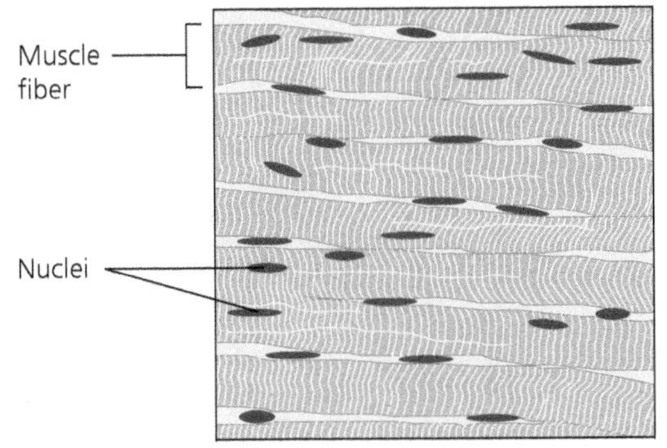

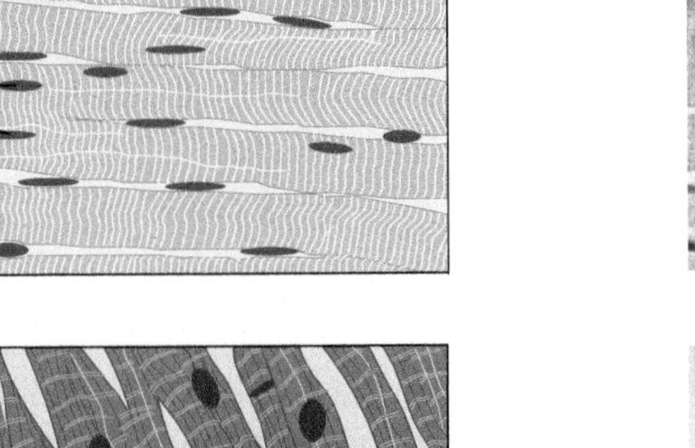

b. Cardiac muscle

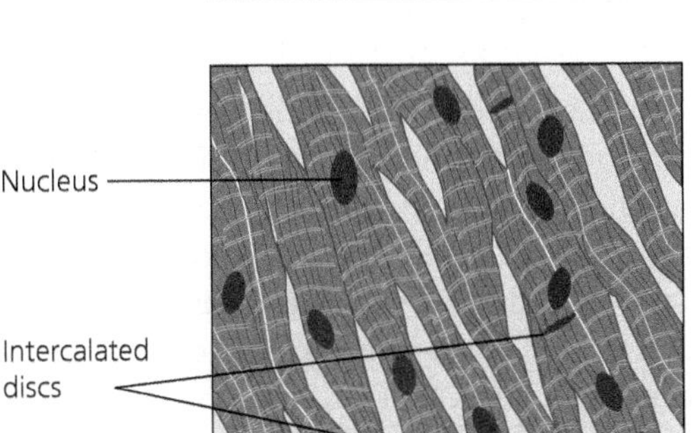

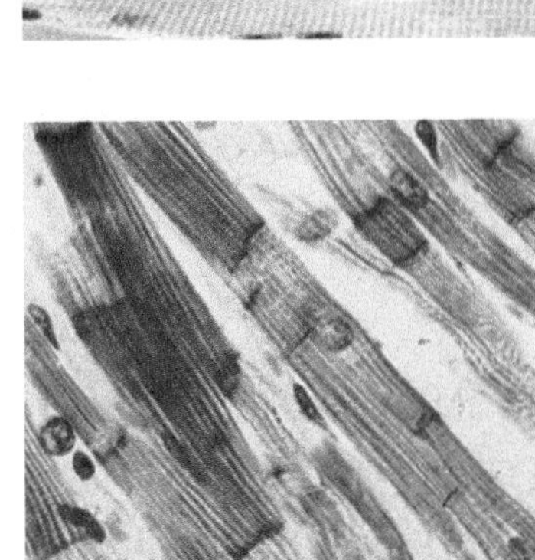

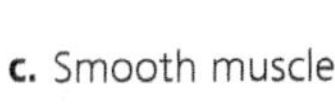

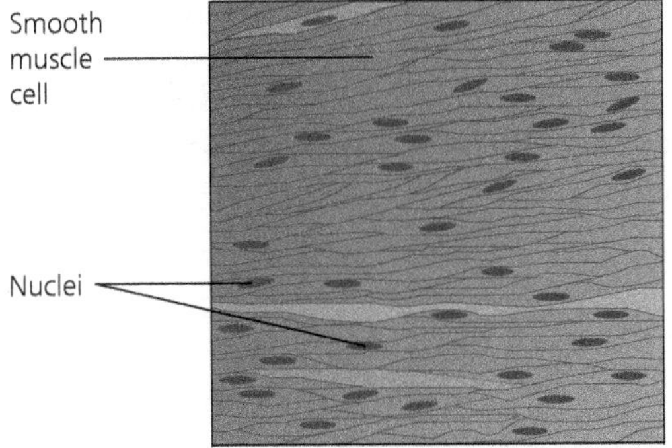

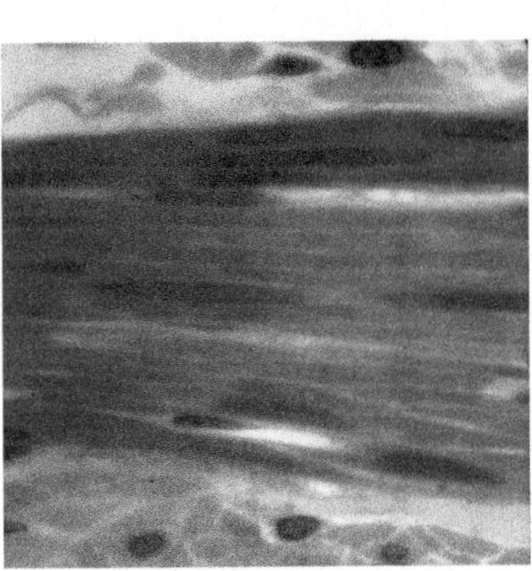

FIGURE 22.2 (at left)
Connective tissue. (a) In loose connective tissue, cells are embedded in a liquid fibrous matrix. (b) Adipose tissue stores fat droplets in adipose cells. (c) In bone, cells are embedded in a solid fibrous matrix. (d) In cartilage, cells are embedded in a gelatinous fibrous matrix. (e) In blood, cells are embedded in a liquid matrix.

FIGURE 22.3 (above)
Muscle tissue. Muscle tissue is either striated or smooth. (a) Skeletal muscle is striated. (b) Cardiac muscle is also striated. (c) Smooth, or visceral, muscle is not striated.

found in the skeleton characterized by cells embedded in, respectively, a hard or a gelatinous extracellular matrix consisting mainly of collagen fibers. In bone, cells called **osteocytes** secrete a rigid mineral component into the matrix. The matrix in cartilage is secreted by cells called **chondrocytes**. **Blood** is a connective tissue consisting of cellular components called **erythrocytes** (red blood cells), **leukocytes** (white blood cells), and **platelets** (cell fragments) in a liquid matrix consisting of water, salts, and dissolved proteins called **plasma**. Other connective tissues fill the spaces between various tissues, binding them together or performing other functions. Connective tissues are derived from the embryonic tissue layer, mesoderm.

Muscle tissue may be **striated**, showing a pattern of alternating light and dark bands, or **smooth**, showing no banding pattern (Figure 22.3). There are

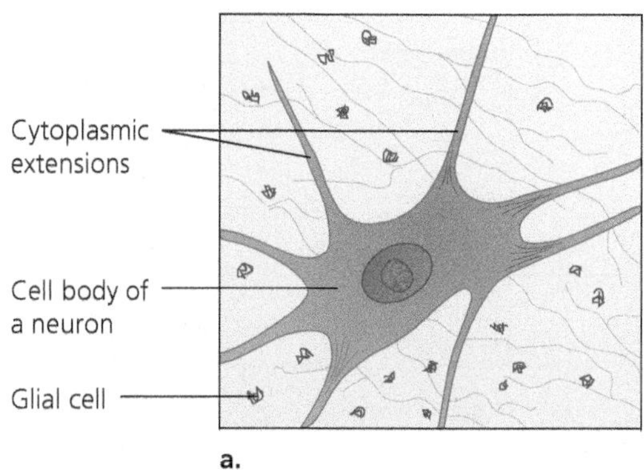

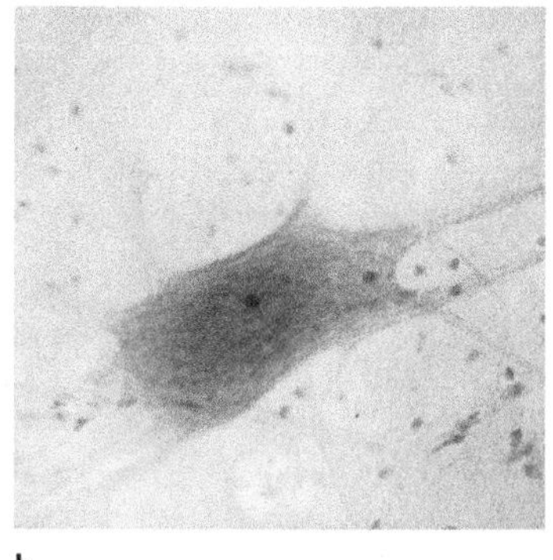

FIGURE 22.4
Nervous tissue. Neurons and glial cells.

two types of striated muscle, skeletal and cardiac. **Skeletal** muscle moves the skeleton and the diaphragm, and is made of muscle fibers formed by the end-to-end fusion of several cells, creating fibers with multiple nuclei. **Cardiac** muscle, found only in the wall of the heart, is also striated, but the cells do not fuse. Cells are attached by **intercalated discs**. **Smooth muscle**, also called **visceral** muscle, is found in the skin and in the walls of organs such as the stomach, intestine, and uterus. Muscle, like connective tissue, is derived from mesoderm.

Nervous tissue is found in the central nervous system (brain and spinal cord) and in the peripheral nervous system consisting of nerves (Figure 22.4). Nervous tissue is found in every organ throughout the body. There are two basic cell types, neurons and glial cells. **Neurons** are capable of responding to physical and chemical stimuli by creating an **impulse**, which is transmitted from one locality to another. **Glial cells** support, insulate, and nourish the neurons. Nervous tissue is derived from ectoderm.

The organs that you will investigate in the following exercises are made up of the four basic tissues. The tissues, each with a specific function, are organized into a functional organ unit.

EXERCISE 22.1

Histology of the Skin

Materials

compound microscope
prepared slide of mammalian skin—pig, monkey, or human

Introduction

Tissues are structurally arranged to function together in **organs**, which are adapted to perform specific functions. Organs are found in all but the simplest animals. The largest organ of the vertebrate body, the skin, illustrates the organization of tissues, each with a specific function, into a functioning organ (Figure 22.5). The skin protects the body from dehydration and bacterial invasion, assists in regulating body temperature, and receives stimuli from the environment. As you work through this exercise, ask how the unique function of each tissue produces the functioning whole—the skin. Review information about each observed tissue type in the Introduction to Tissues (pp. 588–592).

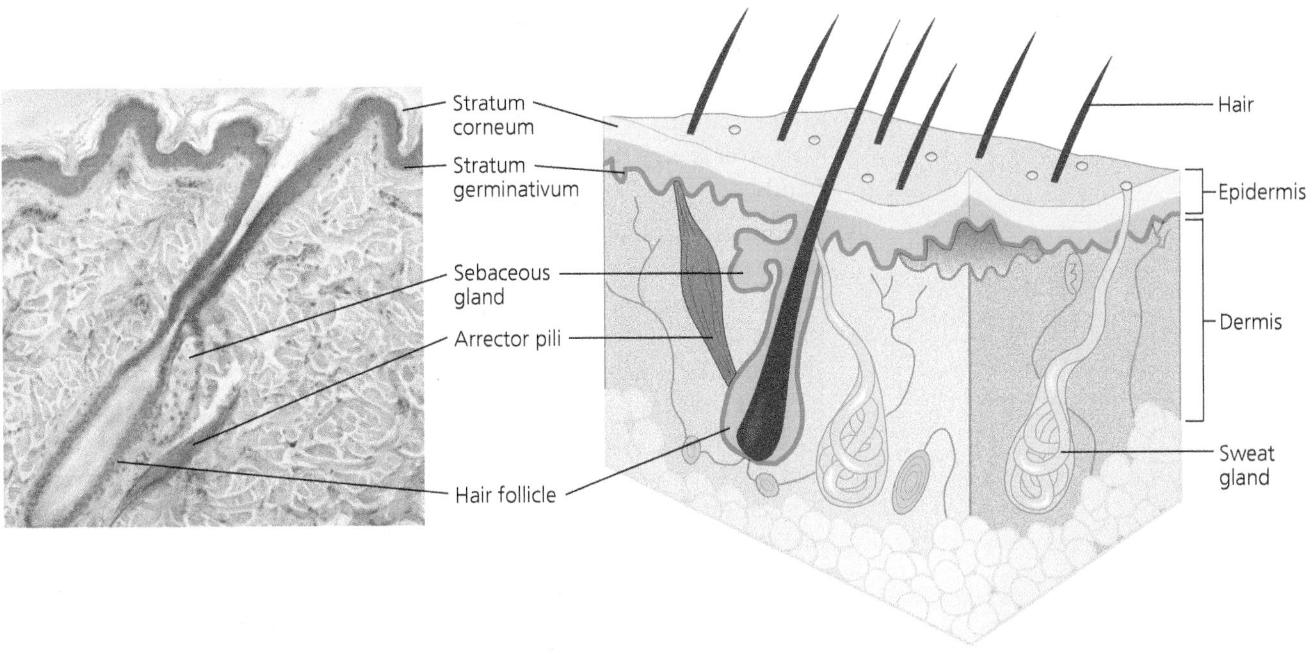

FIGURE 22.5
Mammalian skin structure. (a) Photomicrograph of a cross section of the skin. (b) A diagram detailing the skin structure.

Procedure

1. Obtain a prepared slide of mammalian skin (pig, monkey, or human). View it using the low and intermediate power objectives on the compound microscope.
2. Identify the two main layers of the skin. The thin outer layer, the **epidermis**, consists of **stratified squamous epithelium** (Figure 22.1d); the thicker inner layer, the **dermis**, consists mainly of **dense connective tissue** and scattered **blood vessels**. The dermis merges into layers of adipose tissue, loose connective tissue (Figure 22.2a), and smooth muscle (Figure 22.3c), tissues that are not considered part of the skin but share some of the skin's protective functions.
3. Locate **hair follicles** extending obliquely from the epidermis into the dermis. In the living animal, each follicle contains a hair, but the hair shaft may or may not be visible in every follicle on your slide, depending on the plane of the section through the follicle. The follicle is lined by epithelial cells continuous with the epidermis. Carefully observe several hair follicles. You may be able to find a band of smooth muscle cells attached to the side of a follicle. This muscle, called the **arrector pili**, attaches the hair follicle to the outermost layer of the dermis. When stimulated by cold or fright, it pulls the hair erect, causing "goose bumps." In furry animals this adaptation increases the thickness of the coat to provide additional temperature insulation. Clusters of secretory cells making up **sebaceous glands** (oil glands) are also associated with hair follicles. These are more obvious in monkey and human skin slides than in pig skin slides.

4. Focus your attention on the epidermis and locate the outermost layer, the **stratum corneum**, a layer of dead, keratinized cells, impermeable to water. This layer is continually exfoliated and replaced. The thickness of the stratum corneum varies, depending on the location of the skin. This layer is very thick on the soles of the feet and the palms of human hands.
5. The innermost layer of cells in the epidermis is called the **basal layer** or the **stratum germinativum**. These cells divide mitotically and produce new cells, which, as they mature, are pushed to higher and higher layers of the epidermis, until they fill with keratin and form the stratum corneum. Scattered through the basal layer (although not seen on your slide) are cells called **melanocytes** that produce **melanin**, a pigment that produces brown or black hues in the skin. The melanin is inserted into newly forming epidermal cells as they are pushed outward. Regular exposure to sunlight stimulates melanocytes to produce more melanin, helping protect the body against the potentially harmful effects of sun exposure.
6. In addition to hair follicles, coiled tubular **sweat glands** lined with **cuboidal epithelium** (Figure 22.1b) extend from the epidermis into the dermis. They appear as circular clusters in cross section and may be easily located in pig or human skin but are less numerous or absent in the skin of furry animals, such as monkeys. The tubular secretory portion is convoluted into a ball, which connects with a narrow unbranched tube leading to the skin surface. It is unlikely that you will see an entire intact sweat gland in one section.
7. Look for connective tissue and blood vessels, which often contain red blood cells in the dermis. Look for adipose tissue (Figure 22.2b) below the dermis.

Results

In Table 22.1, list the tissues you have identified in the skin and indicate the specific function of each.

TABLE 22.1 Tissues of the Skin and Their Functions

Tissue	Function

Discussion

1. How does the skin prevent dehydration?

2. How does the skin protect from bacterial invasion?

3. Discuss how each of the following helps regulate body temperature: blood vessels in the dermis, sweat glands, adipose tissue below the dermis, hair, and hair follicles.

EXERCISE 22.2

Introduction to the Fetal Pig

Materials

preserved fetal pig
dissecting pan
disposable gloves
preservative

Introduction

Fetal, or unborn, pigs are obtained from pregnant sows being slaughtered for food. The size of your pig will vary depending on its stage of gestation, a total period of about 112 to 115 days. After being embalmed in a formaldehyde- or phenol-based solution, pigs are stored in a preservative that usually does not contain formaldehyde, although the smell of formaldehyde may remain. Most preserving solutions are relatively harmless; however, they will dry the skin, and occasionally a student may be allergic to the solutions. For these reasons, you should not handle the pigs with your bare hands, and you should perform your dissections in a well-ventilated room.

Wear disposable gloves when handling the pig and other preserved animals.

In this exercise, you will become familiar with the external anatomy of the fetal pig, noting the regions of its body and the surface structures. The skin, just studied in microscopic sections, will be the first organ observed.

Each student, working independently (unless otherwise instructed), will dissect a fetal pig; however, we encourage you to engage in collaborative discussions with your lab partner. Discuss results and conclusions and compare dissections.

Read carefully the rules and techniques for dissection in Appendix E before you begin your study of the fetal pig.

Procedure

1. Obtain a fetal pig and place it in a dissecting pan. Add a small amount of preservative to the pan. Do not allow the pig to dry out at any time. If your pig begins to dry out, use the preservative, not water, to moisten the tissues unless otherwise instructed.

Your pig may have been injected with red and blue latex. The red was injected into arteries through one of the umbilical arteries; the blue was injected into the external jugular vein through an incision in the neck.

2. Using information in Appendix E and collaborating with your lab partner, locate your pig's left and right, dorsal and ventral, and anterior (cranial) and posterior (caudal) regions. Use the terms *proximal* and *distal* to compare positions of several structures.
3. Locate the body regions on your pig. The pig has a **head**, **neck**, **trunk**, and **tail** (Figure 22.6a). The trunk is divided into an anterior **thorax**, encased by ribs, and a posterior **abdomen**. The thoracic and abdominal regions of the body house corresponding cavities that are divisions of the body cavity, or coelom. The **thoracic cavity** is in the thorax, and the **abdominal (peritoneal) cavity** is in the abdomen.
4. Examine the head with its concentration of sensory receptors. Identify the mouth; **external nostrils** on the end of the snout; ears, each with an **auricle**, an external flap supported by cartilage (Figure 22.2d); and the eyes with two eyelids, as in humans, and a third eyelid, the **nictitating membrane**, near the inside corner of each eye.
5. Examine the pig's skin and review its functions. In a fetal pig, an outer embryonic skin layer, the **epitrichium**, lies over the skin. You may find pieces of this layer on your pig. Is hair present on your pig?
6. Locate the cut **umbilical cord** on the ventral surface of the abdomen. Blood vessels pass from the placenta, attached to the wall of the mother's uterus, to the fetal pig through this cord. If your pig's umbilical cord is collapsed, use scissors to make a fresh transverse cut and examine the end more closely. Identify the cut ends of two round, thick-walled **umbilical arteries**; one larger, flattened **umbilical vein**; and one very small, round **allantoic stalk**. You may remember that blood vessels carrying blood

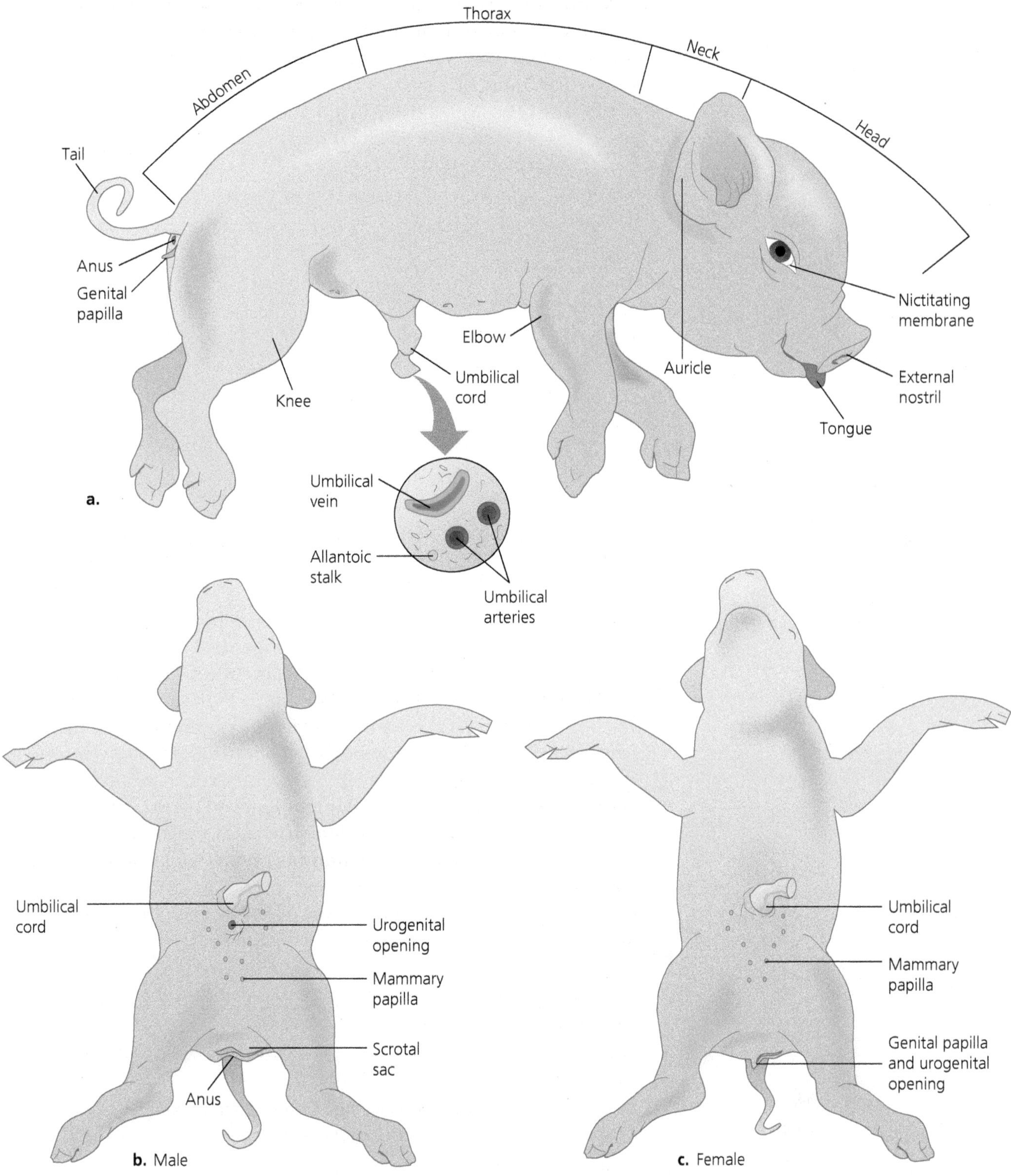

FIGURE 22.6

Fetal pig. (a) Body regions and external structures of the fetal pig with an enlarged cross section of the umbilical cord. (b) Posterior region of male pig. (c) Posterior region of female pig.

away from the heart of the animal are called **arteries**. Blood vessels carrying blood back to the heart are called **veins**. In fetal circulation, the umbilical arteries carry blood from the fetus to the placenta. The umbilical vein carries blood from the placenta to the fetus. The allantois is an extension of the urinary bladder of the fetus into the umbilical cord. Speculate about the nature of blood in the umbilical arteries and the umbilical vein. Explain your conclusions.

a. Which vessel—umbilical artery or umbilical vein—would carry blood *high* in oxygen (oxygen-rich blood)?

b. Which vessel would carry blood *low* in oxygen (oxygen-poor blood)?

c. Which vessel would carry blood high in nutrients?

d. Which vessel would carry blood high in metabolic waste?

7. Look just caudal to the umbilical cord to determine the sex of your pig. If it is a male (Figure 22.6b), you will see the **urogenital opening** in this position. This opening is located below the tail in the female. Locate the **anus** just ventral to the base of the tail in both sexes. In the male, **scrotal sacs** will be present ventral to the anus and caudal to the hind legs. In the female pig (Figure 22.6c), the **urogenital opening** is located ventral to the anus. Folds, or **labia**, surround this opening, and a small protuberance, the **genital papilla**, is visible just ventral to the urogenital opening. The urogenital opening is a common opening from the urinary and reproductive tracts. Notice that **mammary papillae** are present in pigs of both sexes. Locate a pig of the opposite sex for comparison. Having determined the sex of your pig, you are now ready to begin your dissection.

Results

1. List structures observed in the fetal pig that are no longer present in the pig after birth.

2. Modify Figure 22.6 or make a sketch in the margin of your lab manual with any additional details needed for future reference.

Discussion

Review the definition of the term *cephalization*, defined and observed in some of the animals studied in Lab Topic 18 Animal Diversity I, and Lab Topic 19 Animal Diversity II, and describe how the pig demonstrates this phenomenon.

EXERCISE 22.3

The Digestive System in the Fetal Pig

Materials

supplies from Exercise 22.2
dissecting instruments
twine
plastic bag with twist tie and label
stereoscopic microscope or hand lens
compound microscope
prepared slide of jejuno-ileum (small intestine) c.s.

Introduction

Most internal organs, including the entire digestive system, are located in the body cavity, or coelom. A large muscular structure, the **diaphragm**, divides the body cavity into the **thoracic** cavity and the **abdominal (peritoneal)** cavity. The thoracic cavity includes two additional cavities, the **pleural** cavity housing the lungs and the **pericardial** cavity housing the heart. **Coelomic epithelial membranes** line these cavities and cover the surface of all organs. Those epithelial membranes lining the wall of the cavity are called **parietal** (L., pariet, "wall"). Epithelial membranes covering organs are called **visceral** (L., viscera, "bowels"). Each epithelial membrane lining in the coelom is named according to its cavity and location (lining the wall or covering the organ). Thus, the epithelial membrane covering the lungs would be the **visceral pleura**. The epithelial membrane lining the wall of the pleural cavity would be the **parietal pleura**. Likewise, the lining of the abdominal (peritoneal) cavity is **parietal peritoneum**. The epithelium covering organs in the peritoneal cavity is **visceral peritoneum**. What would be the name of the epithelium covering the heart?

Use this convention to complete Table 22.2, naming coelomic epithelial membranes. As you open each body cavity described in these lab topics, return to this table to check your answers.

Recall from studies in Lab Topic 18 Animal Diversity I, and Lab Topic 19 Animal Diversity II that some animals with simple body plans have only one opening to their digestive cavity, and more complex animals have a tubular digestive

TABLE 22.2 Divisions of the Body Cavity and Associated Membranes

Body Cavity	Divisions	Epithelial Membrane Lining the Cavity Wall	Epithelial Membrane Covering the Organs	Organs
Thoracic	Pleural	Parietal pleura	Visceral pleura	Lungs
	Pericardial			
Abdominal	Abdominal (peritoneal)		Visceral peritoneum	Stomach, pancreas, spleen, liver, gallbladder, intestines

system with an anterior opening, the mouth, and a posterior opening, the anus. This pattern in the digestive system allows specialization to take place along the length of the tract, resulting in the development of specific organs that carry out specific functions. As you investigate the digestive system in the fetal pig, ask how each structure or region solves a particular problem in nutrition procurement and processing. Remember that memorizing the names of structures is meaningless unless you understand how the morphology is related to the function of the organ.

Procedure

1. Hold the pig ventral side up in the dissecting pan and use twine to tie the two anterior legs together and the two posterior legs together, leaving enough twine between to slip under the dissecting pan. The pig should be positioned "spread-eagle" in the pan, ventral side up.
2. To expose structures in the mouth cavity, use heavy, sharp-pointed scissors to cut at the corners of the mouth along the line of the tongue. Continue to cut until the lower jaw can be lowered, being careful not to cut into the tissues in the roof of the mouth cavity. Continue to cut, pulling down on the lower jaw until your cuts reach the muscle and tissue at the back of the mouth. Cut through this muscle until lowering the jaw exposes the back of the mouth cavity.
   a. Identify structures in the mouth cavity (Figure 22.7). Locate the **teeth**; the **tongue** covered with **papillae**, which house taste buds; and the

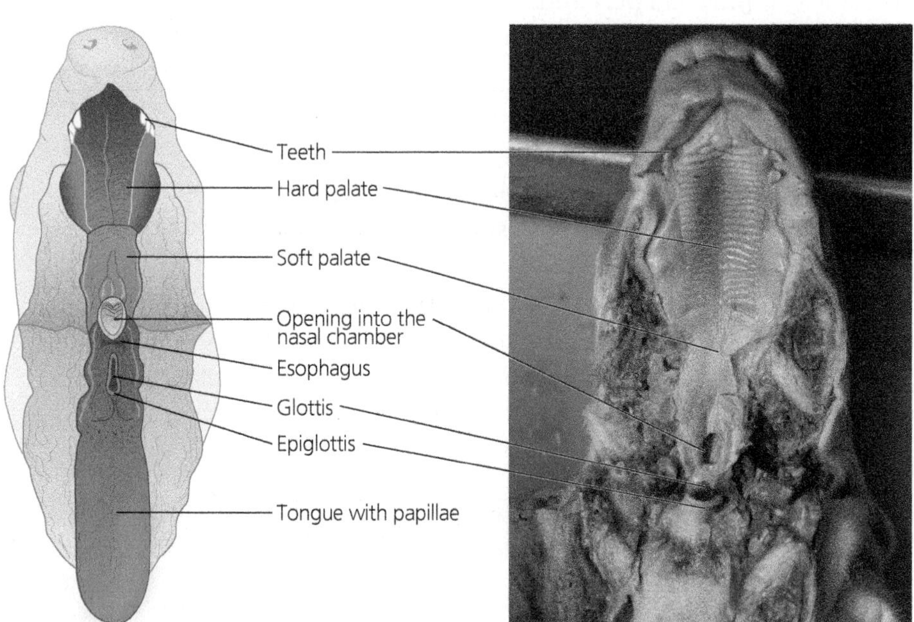

FIGURE 22.7

The mouth cavity. Openings into the pharynx in the rear of the mouth cavity lead to the respiratory system, the digestive system, and the nasal cavity. The tongue occupies the floor of the mouth, and the roof of the mouth consists of the hard and soft palates.

roof of the mouth, composed of the **hard palate** supported by bone (Figure 22.2c) and the **soft palate**. The hard palate is anterior to the soft palate and is marked by transverse ridges.

b. Identify structures and openings at the rear of the mouth cavity: the **glottis**, the space in the beginning of the respiratory passageway; the **epiglottis**, a small flap of tissue supported by cartilage (Figure 22.2d) that covers the glottis when swallowing; the **esophagus**, the beginning of the digestive tube (alimentary canal); and the **opening into the nasal chamber**. All these open into the **pharynx**, the chamber located posterior to the mouth.

c. Insert your blunt probe into the three openings in this region: the opening at the rear of the soft palate that leads into the nasal chamber, the esophagus just posterior and ventral to the opening into the nasal chamber, and the glottis at the beginning of the respiratory passageway leading into the pleural cavity. Notice that the opening into the esophagus lies *dorsal* to the glottis.

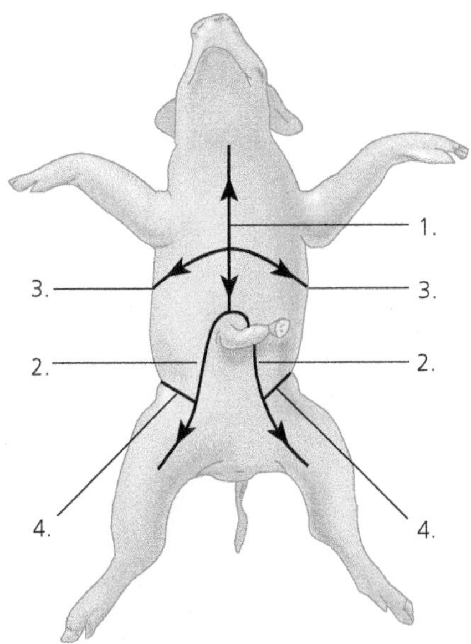

FIGURE 22.8
Incisions to open the abdominal cavity. Incision 1 makes a shallow midventral incision from the base of the throat to the umbilical cord. Incision 2 cuts around the umbilical cord to the medial surface of each leg. Incision 3 cuts through the body wall laterally just posterior to the diaphragm. Incision 4 cuts laterally at the posterior margin of the abdominal cavity.

3. Expose the digestive organs in the abdominal region by opening the posterior portion of the abdominal cavity.

   a. Begin the dissection by *using the scalpel* to make a shallow midventral incision from the base of the throat to the umbilical cord (Figure 22.8, incision 1). Cut lateral incisions (Figure 22.8, incision 2) around each side of the umbilical cord and continue the two incisions, one to the medial surface of each leg.

   b. Now *use the scissors* to cut deep into one of the lateral incisions beside the cord until you penetrate into the abdominal cavity, piercing the parietal peritoneum. At this point, fluid in the cavity should begin to seep out. Use the scissors to cut through the body wall along the two lateral incisions to the legs and around the umbilical cord.

   c. Pull lightly on the umbilical cord. If your dissection is correct, you will see that the umbilical cord and ventral wedge of body wall can be reflected, or pulled back, toward the tail, except for a blood vessel, the **umbilical vein**. This vein passes from the umbilical cord anteriorly toward a large brown organ, the liver. Cut through this vein, leaving a stub at each end. Tie a small piece of twine around each stub so you can find them later. This should free the flap of body wall, which may now be reflected toward the tail, exposing the abdominal organs.

4. Open the anterior portion of the abdominal cavity.

   a. Cut anteriorly through the body wall, deepening the midventral incision until you reach the diaphragm, separating the thoracic and abdominal cavities.

   b. Make four lateral cuts, two adjacent to the rib cage just posterior to the diaphragm (Figure 22.8, incision 3) and two at the posterior margin of the abdominal cavity (Figure 22.8, incision 4). This will produce two flaps of body wall that can be folded back like the lids of a box.

   c. If your specimen contains coagulated blood or free latex from the injection, pull out the latex and rinse the body cavity under running water.

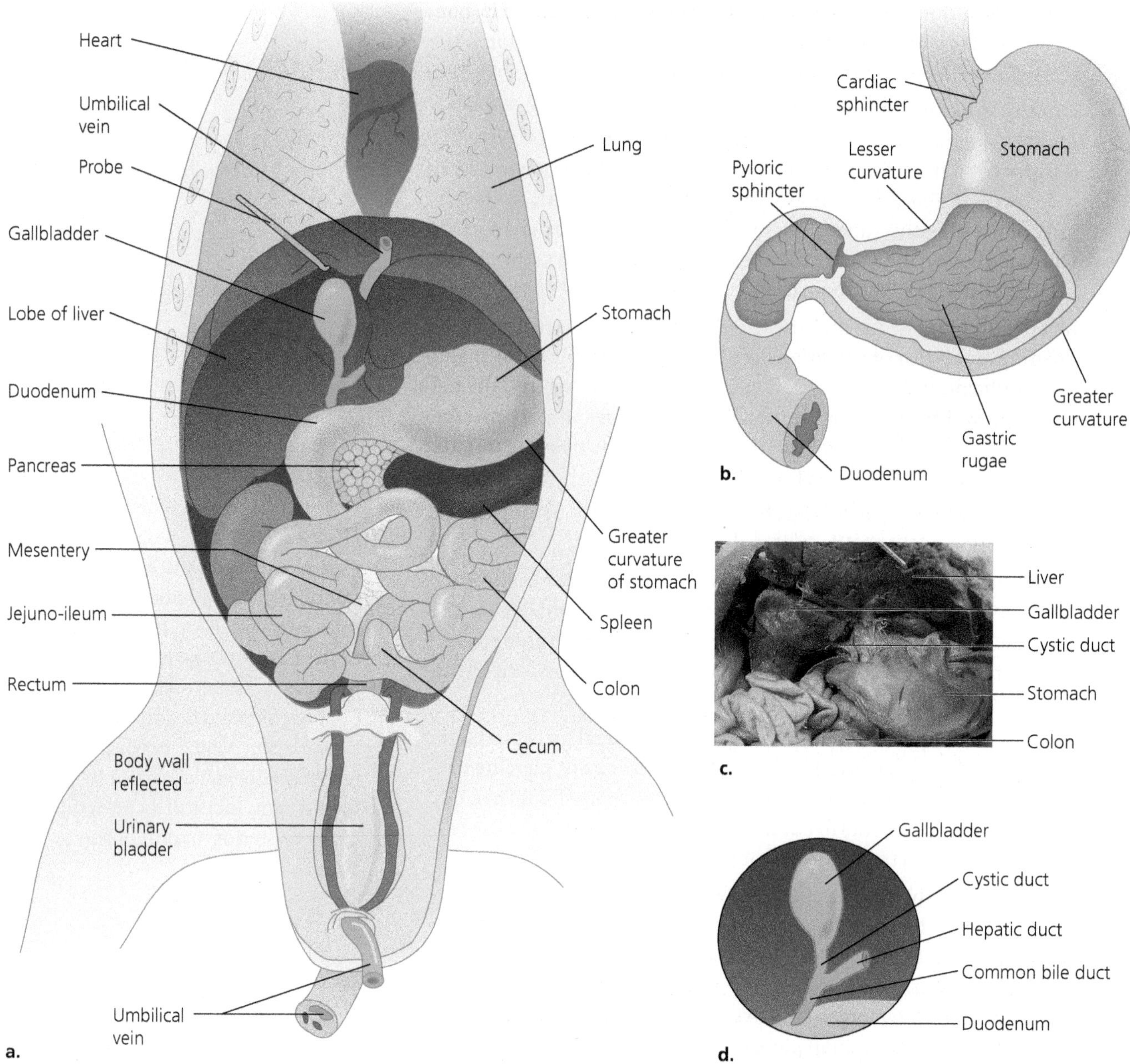

FIGURE 22.9
Digestive organs of the abdominal cavity. (a) The liver is reflected cranially to expose deeper-lying organs. The stomach leads into the duodenum of the small intestine. The jejuno-ileum continues from the duodenum and empties into the colon near the cecum. (b) Cutaway view of stomach showing internal gastric rugae. (c) Photo of gall bladder attached to the under (dorsal) side of the liver that has been reflected cranially. (d) Enlargement of gallbladder and associated ducts.

5. Identify the various structures in the abdominal cavity (Figure 22.9).
   a. Shiny epithelial membranes line the cavity and cover the organs. The **parietal peritoneum** lines the cavity, and the **visceral peritoneum** covers the organs. Speculate about the contents of the space between these two membranes and its function.
   b. The **diaphragm** is a large domed and striated (skeletal, Figure 22.3a) muscle forming the transverse cranial wall of the abdominal cavity, separating this from the thoracic cavity. Only mammals have a diaphragm. The contraction and relaxation of this muscle and muscles between the ribs cause the thoracic cavity to expand and contract, changing the pressure in the cavity and lungs, thus facilitating movement of air into and out of the lungs.

c. The **liver** is the large brown organ that appears to fill the abdominal cavity. Notice that it consists of several lobes. Pull it cranially and locate the **gallbladder**, a small, thin-walled, paddle-shaped sac embedded in the underneath surface near the umbilical vein identified earlier. The liver has many functions, including processing nutrients and detoxifying toxins and drugs. Its main digestive function is the production of **bile**, a substance that emulsifies fats. Bile is stored in the gallbladder until needed.

d. The **stomach** is a large, saclike organ lying dorsal to the liver (Figure 22.9b). Reflect the liver cranially to get a better view. The larger upper-left portion of the stomach tapers down to a narrower portion to the right. The **esophagus** passes through the diaphragm and enters the upper medial border of the stomach. Locate the **spleen**, a dark organ lying along the **greater curvature** of the stomach (the lateral convex border). The spleen filters blood.

e. Cut into the stomach along the greater curvature. Rinse out the stomach contents and identify the numerous longitudinal folds on the inside surface called **gastric rugae**. Speculate about the role played by these folds.

f. Food enters the stomach from the esophagus through the **cardiac valve**, or **cardiac sphincter**. A sphincter valve is a circular band of muscle that encircles an opening. Looking inside the stomach, locate this valve between the esophagus and stomach and speculate about its function.

The stomach mechanically churns food and mixes it with water, mucus, hydrochloric acid, and protein-digesting enzymes.

g. From the stomach, food then passes into the **small intestine**, which joins the stomach at its extreme right narrow portion. Locate another valve, the **pyloric sphincter**, which lies between the stomach and small intestine. This sphincter is closed when food is present in the stomach, preventing food from entering the small intestine prematurely.

h. The **duodenum** is the portion of the small intestine connecting with the stomach (Figure 22.9b). The **pancreas**, an irregular, granular-looking gland, lies in a loop of the duodenum. As food passes through the duodenum, enzymes from the pancreas and the duodenal wall are added to it along with bile, which is produced in the liver and stored in the gallbladder. Using your forceps and a blunt probe to pick away surrounding tissue, locate and separate the **common bile duct**, which enters the duodenum. It is formed by the confluence of the **hepatic duct** passing from the liver and the **cystic duct** leading from the gallbladder (Figure 22.9c and d).

i. The **jejuno-ileum** is the extensive, highly convoluted portion of the small intestine extending from the duodenum to the colon (Figure 22.9a). Whereas the jejunum and the ileum are separate anatomical regions, macroscopically (without a microscope), it is difficult to distinguish between the two in the pig except by position. The duodenum joins the jejunum, which leads to the ileum. The ileum joins the colon.

Most digested food is absorbed into the circulatory system through the walls of the jejuno-ileum. The surface area of the lining of this organ is increased by the presence of microscopic **villi** and **microvilli**, greatly enhancing its absorbing capacity.

Spread apart the folds in the jejuno-ileum and notice the thin membrane called **mesentery**, which supports the folds. Do you see blood vessels in this mesentery? Speculate about the relationship between these vessels and food processing.

j. Cut out a 1-cm-long segment of the jejuno-ileum and cut it open to expose the inner surface. Use a stereoscopic microscope or hand lens to examine this surface. Can you see villi?

k. Follow the ileum to its junction with the large intestine, or **colon**. The diameter of the colon is slightly greater than that of the small intestine, and it is tightly coiled and held together by mesentery. Look for a small outpocketing or fingerlike projection of the colon at its proximal end. This projection is the **cecum**, which is much larger in herbivores than in carnivores. In animals with a large cecum, it probably assists in digestion and absorption. In humans, a **vermiform** (wormlike) **appendix** extends from the cecum.

   One of the important activities in the colon is the reabsorption of water that has been added, along with enzymes and mucus, to the food mass as it passes down the digestive tract. Water conservation is one of the most critical problems in terrestrial animals.

l. The distal portion of the colon is the **rectum**, which passes deep into the caudal portion of the abdominal cavity and to the outside of the body at the anus. Water reabsorption continues in the rectum.

6. After you complete the dissection, use an indelible pen to prepare two labels with your name, lab day, and room. Tie one label to your pig and place the pig in a plastic bag with the label in view. Add preservative from the lab stock and securely close the plastic bag. Tie the second label to the outside of the bag.

7. Using the compound microscope, study a prepared microscope slide of a section through a region of the jejuno-ileum (Figure 22.10a and b). As you study the slide, compare the tissues you see here to the diagrams and photomicrographs of tissues in Figures 22.1, 22.2, and 22.3, near the beginning of this lab topic.

   a. Use the lowest power on the microscope and scan the section, finding the smooth outer surface and then the **lumen**, or central cavity, located within the intestine. Food passing through the intestine passes through the lumen.

   b. The fingerlike **villi** previously observed are now easily discernible, projecting into the lumen of the intestine. Switch to 10× magnification and focus on the simple **columnar epithelial** tissue lining each villus (Figure 22.1c). This tissue functions in the absorption of nutrients into the circulatory system. Capillaries and lacteals (small lymph vessels) are located within each villus. These vessels are not usually visible.

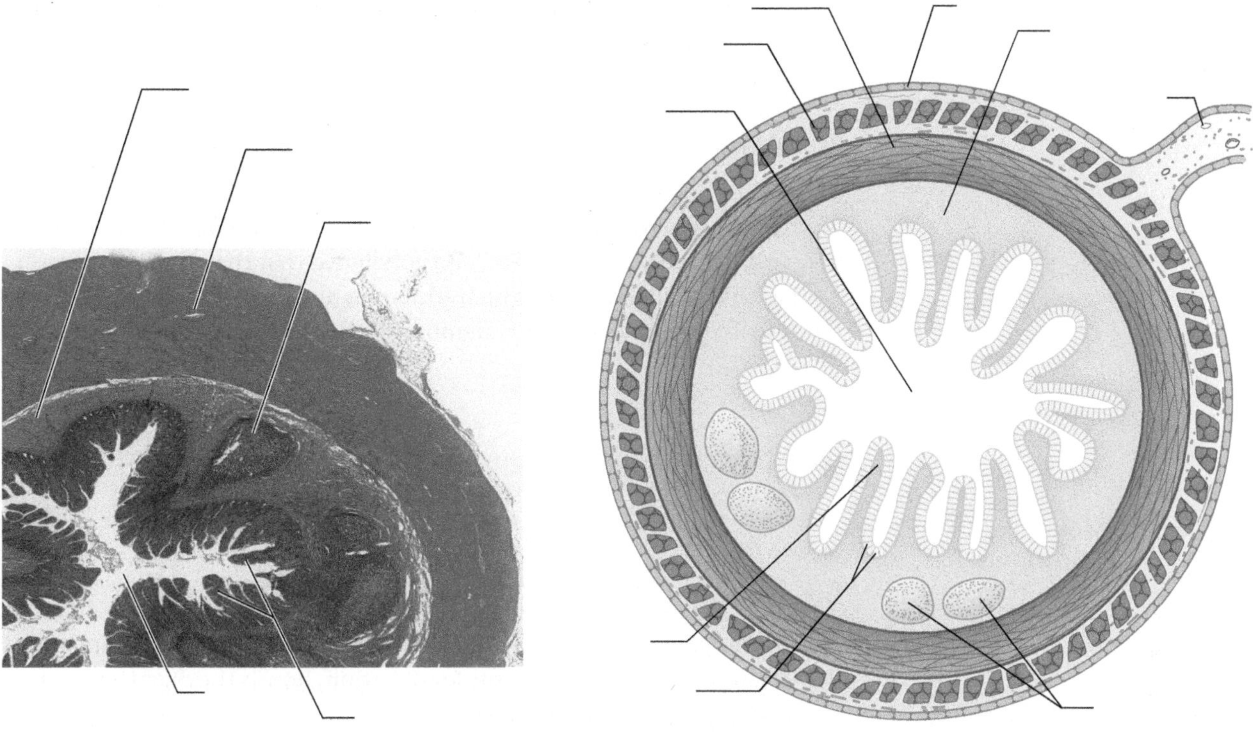

FIGURE 22.10

Cross section through the pig jejuno-ileum. (a) Photomicrograph of a cross section. Locate the lumen, villi, lymph nodules in loose connective tissue, and circular muscle layer. (b) Compare the line diagram with your observation of the prepared slide. Label the cells and structures described in this exercise, step 7. Modify the diagram if needed.

c. Continue your observations, scanning outward toward the surface of the intestine. You will pass through regions with **loose connective tissue** (Figure 22.2a) containing many blood vessels. You may see large masses of cells that are lymphocytes in **lymph nodules**.

d. On the outer surface of the section, locate a thin layer of simple squamous epithelium (Figure 22.1a), called the **visceral peritoneum**, or **serosa**. Two large bands of smooth muscle lie just inside the visceral peritoneum.

e. Locate the outermost muscle layer, composed of smooth muscle fibers extending longitudinally along the intestine. This muscle layer is called the **longitudinal muscle layer**. Because this is a cross section, the longitudinal nature of the fibers will not be apparent. Look just inside the longitudinal muscle layer to see a wide band of muscle, the **circular muscle layer** (appears as in Figure 22.3c). These muscle fibers encircle the intestine.

Imagine each band of muscle contracting as a unit, and speculate about the function of these muscles. Do you know the special name given to the waves of contraction of these muscles?

f. Turn to high power and locate nuclei in both the longitudinal and the circular muscle fibers.

g. Label Figure 22.10.

Results

1. In Lab Topics 18 and 19, you learned that the digestive tract of many invertebrates is described as a "tube within a tube"—a tubular digestive tract within a "tubular" body. You also concluded that the pig digestive tract has this tubular design. Which structures and organs of the pig digestive system develop as tubes or chambers within its tubular digestive tract?

2. Which organs in the digestive system lie outside the "digestive tube" but are important in the digestive process?

Discussion

1. Conservation of water is a critical problem faced by terrestrial organisms. Given the water requirements for digestion, how is the digestive tract anatomy adapted to life on land?

2. Speculate about the outcome if food passes too slowly or too rapidly through the colon.

3. Consult a text and describe the process of food absorption in the small intestine, relating it to the structures observed in this exercise.

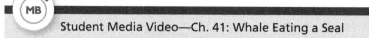

Student Media Video—Ch. 41: Whale Eating a Seal

REVIEWING YOUR KNOWLEDGE

In Table 22.3, beginning with the mouth, list in sequence each region or organ through which food passes in the pig and describe for each the primary digestive functions, the enzymes active in that chamber, and the macromolecules affected, when appropriate. Consult your text if necessary.

TABLE 22.3 Organs of the Pig Digestive Tract and Their Functions

Region/Organ	Function/Macromolecule Digested	Enzymes

APPLYING YOUR KNOWLEDGE

1. The Peachtree Road Race is over, and you have just been awarded the coveted T-shirt. Your body is dripping wet and your skin appears bright red. Explain, from a physiological perspective, what is happening to your body.

2. Tattooing is the risky and unregulated procedure of using a needle to deposit pigment in the skin. Where in the skin would it be necessary to deposit the pigment to make the tattoo permanent?

3. The three most common forms of skin cancer are (1) basal cell carcinoma, (2) squamous cell carcinoma, and (3) malignant melanoma, the most dangerous skin cancer. Using your knowledge of skin histology, predict the skin layers and/or cells that are involved in each of these cancers.

4. Give examples of structures supported by bone and by cartilage observed in this lab topic. What differences in flexibility and function have you noticed in these structures?

5. The illness called celiac disease is an autoimmune disorder that is triggered by eating gluten, a protein found in wheat and other grains. Persons with this disease may appear undernourished, even when fed seemingly adequate diets. Microscopic studies of the digestive tract in affected persons show that villi are flattened and chronically inflamed and damaged.

   a. Using your knowledge of the anatomy and function of different regions of the digestive tract, discuss the impact of gluten on digestion and nutrition in persons with this disease.

   b. Propose a "cure" for the disease.

6. Gallstones are pieces of solid material that form in the gallbladder. Most commonly they consist of crystalized cholesterol, one of the major components of bile. They cause problems when they block the cystic duct.

   a. What effect would you predict that gallstones would have on the digestive process?

   b. Laparoscopic surgery to remove the gallbladder is a common treatment for severe symptoms of gallstones. This process involves inserting a light and surgical tools through several small incisions in the abdominal wall. Review the position of the gallbladder in the abdominal cavity, and imagine all structures, ducts, and membranes involved in the process of gallbladder removal.

7. Traditionally, scientists have described the human appendix as a "relic of the past"—a shrunken portion of the cecum that functioned in human evolutionary history but no longer performs a role in the human body. Recently, scientists have hypothesized that the appendix serves as a reserve for beneficial gut bacteria that are wiped out when prolonged courses of antibiotics are taken. Losing normal bacteria in the gut may open the door for deadly bacteria, for example, *Clostridium difficile* (*C. diff*), to grow quickly and overtake the gut, resulting in diarrhea and abdominal cramping. The *C. diff* infection often recurs after the antibiotic treatment is terminated.

a. Design a *correlation* study to investigate this hypothesis using the medical histories of patients with gut infections.

b. Design an *experiment* as described in Lab Topic 1 Scientific Investigation (with controls, dependent variables, an independent variable, etc.), to investigate this hypothesis.

CASE STUDY

You have just completed a study of the digestive tract, and now you know the names, structures, and basic functions of each part. However, this information alone is just the beginning of understanding the role the digestive system plays in the body. This system functions to provide nourishment for optimum functioning of your tissues and organs. You eat carbohydrates, proteins, and fats, and they are digested in the organs of your digestive system. But are all carbohydrates, proteins, and fats equally useful in providing nourishment? Should your diet include more of one component and less of another?

Over 10 years ago the U.S. Department of Agriculture created a "Food Guide Pyramid" with the goal of giving advice on the best diet for healthy living. In recent years, however, many health professionals have criticized this pyramid, pointing out that it was not based on sound scientific research and that its design was influenced by the food industry. Recently, a "Healthy Eating Plate" has been created with a stronger foundation in scientific evidence linking diet and health.

The "freshman 15" (referring to the number of pounds that many students gain their first year in college) is an indication that many college students are unaware of or choose to ignore the wealth of information now available to help plan a healthy diet. This weight gain may be due to erratic eating patterns, high stress levels, sedentary lifestyles, or any number of factors of college life. The objective of this study is to encourage you to examine your diet and become familiar with the latest guidelines for healthy living.

1. To begin your study, access "The Nutrition Source" at the Harvard School of Public Health website: https://www.hsph.harvard.edu/nutritionsource/healthy-eating-plate/. This website, created by faculty from the Department of Nutrition, presents the latest scientific evidence for the blueprint of a healthy diet. The site opens to a "Healthy Eating Plate." Copy and print the Healthy Eating Plate image for future reference. Note the bulleted list below the image with suggestions for creating healthy, balanced meals. Use this information as you begin to plan and evaluate your specific diet.

2. Compile a personal "meal log" for one or two typical 24-hour periods. Create a table and record each food that you consume, keeping a record of numbers of servings and estimated or actual amounts (volume, e.g., ½ cup; or weight, e.g., one 6-oz. chocolate candy bar, 1 tablespoon [T] oil). Use the Healthy Eating Plate diagram to determine the category for each food that you consumed and record this in your table. Your table might look something like the following:

Time	Item	Number and Amounts of Servings	Category or Recommendation in Healthy Eating Plate
8:00 a.m.	Scrambled egg	1 large	Healthy protein
	Whole-wheat toast	2 slices	Whole grains
10:00 a.m.	Banana	1	Fruits
	Yogurt, plain	½ cup	Limit dairy
12:00 noon	Hamburger	4 oz	Healthy protein
	Lettuce	2 leaves	Vegetable
	Tomato	2 slices	Fruit/vegetable
	Mayonnaise	1 T	Healthy oils
	French fries	½ potato	Don't use

3. Using data from your table, construct a detailed version of your Healthy Eating Plate. Draw the outline of the Healthy Eating Plate image. Annotate each section with the food items from your diet that fit each food category.
4. Evaluate your diet. Compare your diet to the content and proportions of a healthy diet presented in the Healthy Eating Plate. For each food category, indicate if you met the dietary recommendations. Using the website as a reference, describe the changes you should make to improve your diet. Use your data and conclusions to write a paragraph or report. Your instructor may ask you to submit your report as a class assignment.
5. Survey the food available in the food services for your college or university. Based on what you have learned about healthy eating, write a recommended diet for college students from the menu of food available.

STUDENT MEDIA: BioFlix, Activities, Investigations, and Videos

www.masteringbiology.com (select Study Area)

BioFlix—Ch. 41: Membrane Transport

Activities—Ch. 40: Overview of Animal Tissues; Epithelial Tissue; Connective Tissue; Muscle Tissue; Nervous Tissue; Ch. 41: Feeding Mechanisms of Animals; Digestive System Function; Case Studies of Nutritional Disorders

Investigations—Ch. 41: What Role Does Amylase Play in Digestion?

Videos—Ch. 40: Shark Eating Seal; Hydra Eating *Daphnia* (time-lapse)

REFERENCES

Dunn, Rob. "Your Appendix Could Save Your Life," *Scientific American*, March 2012, vol. 306, p. 22.

Fasano, Alessio. "Surprises from Celiac Disease," *Scientific American*, August 2009, vol. 301, pp. 54–61.

Fawcett, D. W. and R. P. Jensh. *Bloom and Fawcett: Concise Histology*, 2nd ed. Cambridge, MA: Oxford University Press, 2002. The successor to Bloom and Fawcett's classic *Textbook of Histology*.

Marieb, E. N. and K. Hoehn. *Human Anatomy and Physiology*, 10th ed. San Francisco, CA: Pearson, 2016.

Smith, D. G. and M. P. Schenk. A *Dissection Guide and Atlas to the Fetal Pig*, 3rd ed. Englewood, CO: Morton Publishing Company, 2011. Includes excellent photographs of fetal pig anatomy.

Urry, L., M. Cain, S. Wasserman, P. Minorsky, and J. Reece. *Campbell Biology*, 11th ed. San Francisco, CA: Pearson, 2017.

Walker, W. F., Jr. *Anatomy and Dissection of the Fetal Pig*, 5th ed. New York: W. H. Freeman, 1998.

Willett, W. C. and P. J. Skerrett. *Eat, Drink, and Be Healthy: The Harvard Medical School Guide to Healthy Eating*. New York: Simon & Schuster, 2005.

WEBSITES

Current information based on scientific evidence for designing a healthy diet:
http://www.hsph.harvard.edu/nutritionsource/healthy-eating-plate/index.html

Photographs of fetal pig anatomy:
home.apu.edu/~jsimons/Bio101/PigDissectionGuide.htm

NOTES

LAB TOPIC 23

Vertebrate Anatomy II: The Circulatory and Respiratory Systems

Laboratory Objectives

After completing this lab topic, you should be able to:

1. Identify and describe the function of the main organs and structures in the circulatory system and trace the flow of blood through the pulmonary and systemic circuits.
2. Identify and describe the function of the main organs and structures in the respiratory system and describe the exchange of oxygen and carbon dioxide in the lungs.
3. Describe how the circulatory and respiratory systems work together to bring about the integrated functioning of the body.
4. Apply knowledge and understanding acquired in this lab to problems in human physiology.
5. Apply knowledge and understanding acquired in this lab to explain organismal adaptive strategies.

Introduction

In Lab Topic 22 Vertebrate Anatomy I, you learned that nutrients are taken into the digestive tract, where they are processed: chewed, mixed with water and churned to a liquid, mixed with digestive enzymes, and finally digested into the component monomers, or building blocks, from which they were synthesized. For an animal to receive the benefits of these nutrients, these products of digestion must pass across intestinal cells and into the circulatory system to be transported to all the cells of the animal's body. Oxygen is necessary for the release of energy from these digested products. Oxygen from the atmosphere passes into the respiratory system of the animal, where it ultimately crosses cells in the lungs (in a terrestrial vertebrate) or gills (in an aquatic vertebrate) and enters the circulatory system for transport to cells of all organs to be utilized in nutrient metabolism. Waste products of cellular metabolism—carbon dioxide and urea—are transported from the tissues that produce them via the blood and are eliminated from the body through the lungs of the respiratory system and the kidney of the excretory system, respectively. Thus, the circulatory, respiratory, and excretory systems function collectively, utilizing environmental materials, eliminating wastes, and maintaining a stable internal environment.

In this and the following lab topic, you will investigate the morphology of the circulatory, respiratory, and excretory systems in the fetal pig. As you dissect, relate the structure and specific function of each system to its role in the integrated body.

EXERCISE 23.1

Glands and Respiratory Structures of the Neck and Thoracic Cavity

Materials

These materials will be used for the entire lab topic.

fetal pig
dissecting pan
dissecting instruments
twine
disposable gloves
plastic bag with twist tie and labels
preservative

Introduction

To study the glands and respiratory structures of the neck, you must first open the thoracic cavity and then remove the skin and muscles in the neck region. This will expose several major glands that lie in the neck region in close proximity to the respiratory structures.

Wear disposable gloves when dissecting preserved animals.

Procedure

1. Begin the dissection by making an incision that extends to the jaw and opens the thoracic cavity, which houses the heart and lungs.
   a. Use scissors to deepen the superficial incision previously made anterior to the abdominal cavity, and continue deepening this incision to the base of the lower jaw.
   b. Cut through the body wall in the region of the thorax, clipping through the ribs slightly to the right or left of the **sternum** (the flat bone lying midventrally to which ribs attach).
   c. Continue the incision past the rib cage to the base of the lower jaw.
2. Using the blunt probe to separate tissues, carefully remove the skin and muscles in the neck region. You will expose the **thymus gland** on each side of the neck (Figure 23.1). This gland is large in the fetal pig and in young mammals, but regresses with age, appearing as a small gland under the sternum in an adult. It plays an important role in the development of the body's immune system.
3. Push the two thymus masses to the side to expose the **larynx** and **trachea** lying deep in the masses. Recall your knowledge about the **glottis**, observed in the dissection of the mouth in Lab Topic 22. The **glottis** leads into the larynx, an expanded structure through which air passes from the mouth to the narrower trachea. The larynx houses the vocal cords.

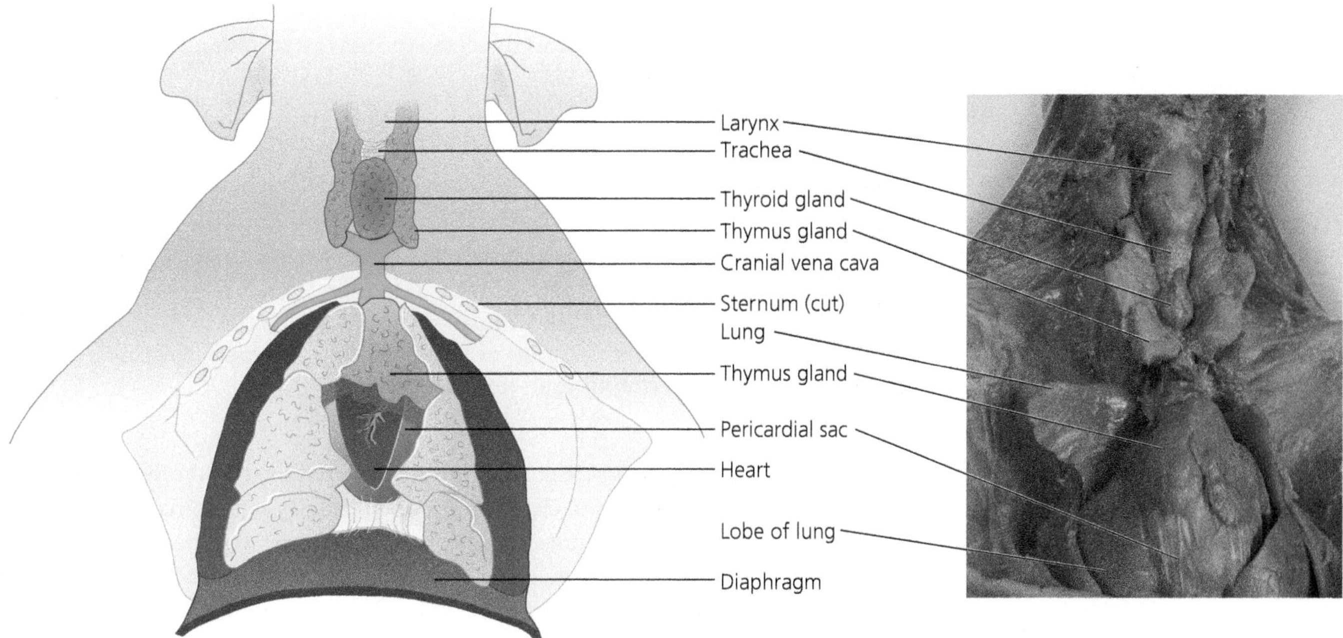

FIGURE 23.1
Ventral view of the anterior region of the pig, showing structures in the neck region and the thoracic cavity. The pericardial sac encloses the heart.

4. A small reddish gland, the **thyroid gland**, covers the trachea. An enlarged thyroid gland is called a goiter, often the result of an iodine-deficient diet. The thyroid gland secretes hormones that influence metabolism. Push this gland aside and observe the rings of cartilage that prevent the collapse of the trachea and allow air to pass to the lungs. Push aside the trachea to observe the dorsally located **esophagus**.
5. Do not continue the dissection of the neck and thoracic regions at this time. To prevent damage to blood vessels, you will complete the dissection of the remainder of the respiratory system (Exercise 23.5) following the dissection of the circulatory system.

EXERCISE 23.2

The Heart and the Pulmonary Blood Circuit

The heart and lungs lie in the **pericardial** and **pleural** (Greek for "rib") cavities, respectively, within the thoracic cavity. In your dissection of the heart and blood vessels, you will distinguish the two circulatory pathways found in mammalian circulation: the **pulmonary circuit**, which carries blood from the heart to the lungs in arteries and back to the heart in veins; and the **systemic circuit**, which carries blood from the heart in arteries to all organs *except the lungs* and back to the heart in veins. This exercise investigates circulation in fetal and adult pig hearts and the pathway of blood to the lungs in the pulmonary circuit.

Materials

isolated adult pig or sheep heart dissected to show chambers and valves, demonstration only
supplies from Exercise 23.1

Procedure

Although, generally, veins contain blue latex and arteries contain red latex, the colors can vary and should not be used as guides to distinguish veins from arteries or vessels carrying oxygen-rich blood from vessels carrying oxygen-poor blood.

1. In the fetal pig, expose the heart lying in the **pericardial cavity** between the two pleural cavities. Gently push open the rib cage, using scissors and a probe to cut through muscle and connective tissue. Another lobe of the thymus gland will be seen lying over the **pericardial sac** housing the heart. The wall of the pericardial sac is a tough membrane composed of two fused coelomic epithelial linings, the **parietal pericardium** and the **parietal pleura**.
2. Cut into and push aside the pericardial sac. Carefully dissect away membranes adhering to the heart until you can identify the four chambers of the heart (Figure 23.2). The walls of heart chambers consist of cardiac muscle (Figure 22.3b, p. 591).

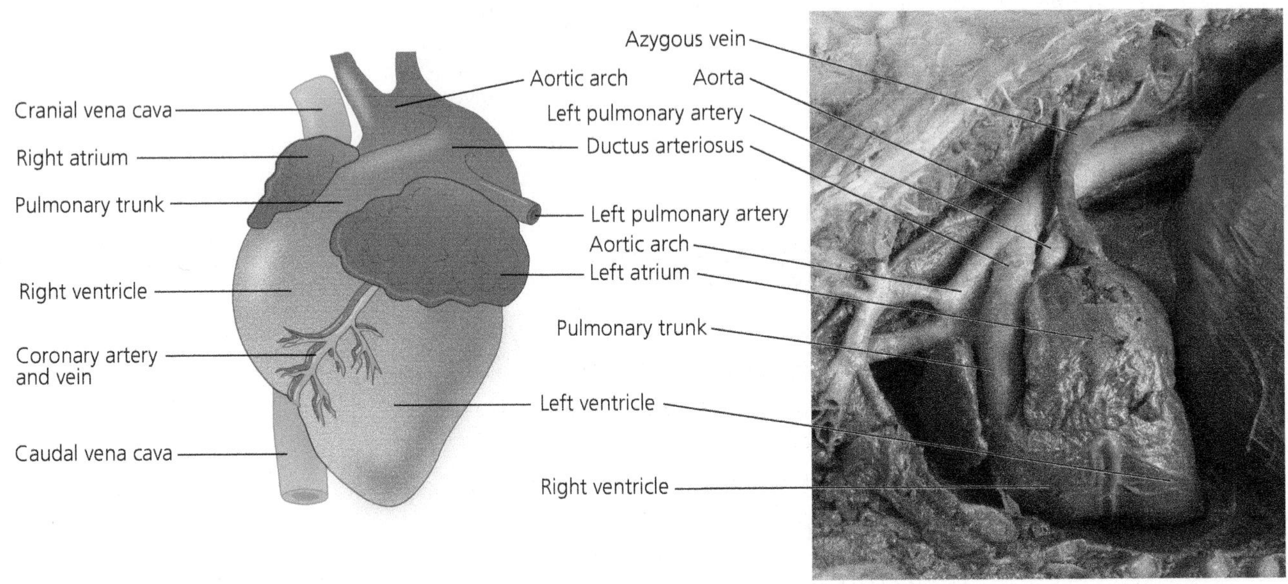

FIGURE 23.2
Fetal pig heart. (a) Ventral view showing the four chambers and the major associated blood vessels. (b) Dissected pig heart with connective tissue around the blood vessels removed, exposing the left pulmonary artery and the ductus arteriosus. The ductus arteriosus carries the greatest volume of fetal blood from the pulmonary trunk into the aorta, bypassing the lungs. Also visible is the azygous vein that carries blood from the ribs back to the heart.

a. The **right atrium** and **left atrium** are small, dark, anteriorly located heart chambers that receive blood from the **venae cavae** and the pulmonary veins, respectively.

b. The **right ventricle** and **left ventricle** are large muscular heart chambers that contract to pump blood. A branch of the **coronary artery** may be seen on the heart surface where the left and right ventricles share a common wall. The coronary artery carries blood to heart tissue.

What is the name of the epithelial lining adhering to the heart surface?

3. Trace the **pulmonary circuit**. As the heart contracts, blood is forced from the right ventricle into the **pulmonary trunk**, a large vessel lying on the ventral surface of the heart. Another large vessel, the **aorta**, lies just dorsal to the pulmonary trunk.

a. Use forceps to pick away tissue around the pulmonary trunk and trace the pulmonary trunk as it curves cranially, giving off three branches: the right and left **pulmonary arteries** and the **ductus arteriosus** (Figure 23.2b).

b. Identify the ductus arteriosus and the **left pulmonary artery** (the right pulmonary artery is not readily visible).

The right and left pulmonary arteries are relatively small at this stage of development. They conduct blood to the right and left lungs, respectively. The ductus arteriosus is the short, large-diameter vessel that connects the pulmonary trunk to the aorta. Because the small right and left pulmonary arteries and compact lung tissue present an extremely resistant blood pathway, the greatest volume of blood in the pulmonary trunk will flow through the ductus arteriosus and directly into the aorta and systemic circulation, bypassing the pulmonary arteries and lungs. At the time of the fetus's birth, when air enters the lungs and the tissues expand, blood will more easily flow into the lungs. The ductus arteriosus closes off and eventually becomes a ligament.

4. Observe the isolated **adult pig or sheep heart** on demonstration and locate the dorsal and ventral surfaces (Figure 23.3).

a. Identify the **right atrium** with associated **cranial** and **caudal venae cavae** and the **left atrium** with associated **pulmonary veins**.

b. Locate the **right** and **left ventricles** and the **atrioventricular valves** between the atria and the ventricles.

c. Locate the **pulmonary trunk**, which carries blood from the right ventricle, and the **aorta**, carrying blood from the left ventricle. The first two small branches of the aorta are **coronary arteries**. Locate these vessels and the **coronary veins** lying on the surface of the heart between the left and right ventricles. Coronary arteries and veins form a short circuit servicing heart tissues.

d. Probe for the **semilunar valves**, both located where major arteries exit the ventricles—one between the right ventricle and the pulmonary trunk, and the other between the left ventricle and the aorta.

Results

Review the heart chambers, blood vessels, and organs in the pathway of the pulmonary circuit in the *adult* heart. To facilitate this review, fill in the blanks

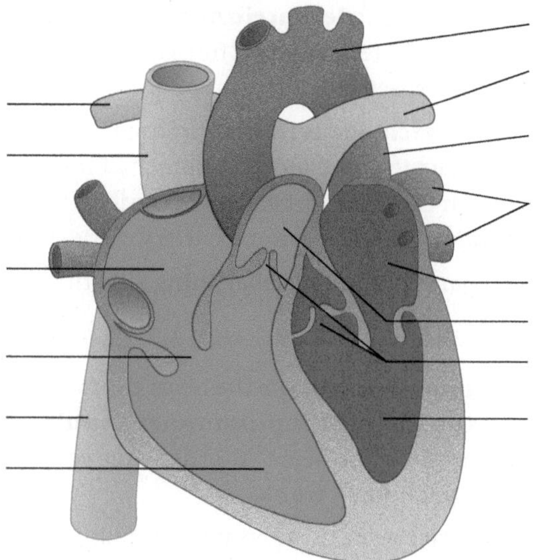

FIGURE 23.3
Internal structures of the adult heart. Label the structures as you trace the pathway of blood flow in the pulmonary circuit.

in the next paragraph. As you fill in the blanks, label the structures in Figure 23.3 and draw arrows showing the pulmonary circuit of blood flow through the adult heart.

Blood entering the heart by way of the cranial and caudal venae cavae passes first into the right atrium. From there it flows through the right ______________ into the right ventricle. When the heart contracts, this blood is forced out of the ventricle through a semilunar valve into the ________ trunk. Branches of this trunk called ______________ carry blood to the lungs. After birth, the blood will become oxygen-rich in the lungs. Blood from the lungs passes back to the heart through ____________, thus completing the pulmonary circuit. It enters the left atrium of the heart and continues on through another atrioventricular valve to the left ventricle. When the heart contracts, blood passes through a second semilunar valve into the aortic arch.

Discussion

1. Define *artery*. Define *vein*.

2. Why would pulmonary arteries be relatively small at the fetal stage of development?

3. Although a pulmonary circuit exists, the heart in amphibians and most reptiles is made up of only three chambers—two atria and one ventricle. The latter receives blood from both atria. Speculate about possible disadvantages to this circulatory pathway.

EXERCISE 23.3

The Heart and the Systemic Circuit in the Thorax

Blood returning from the lungs collects in the left atrium and flows into the left ventricle. When the heart contracts, blood is forced out of the heart into the **aorta**, the origin of which is obscured by the pulmonary trunk. The first branch from the aorta is the small **coronary artery**, previously identified, leading to the heart muscle. The larger volume of blood passes through the aorta to all organs of the body except the lungs. Blood returns to the heart from organs of the body through two large veins, the cranial and caudal venae cavae.

Procedure

1. Identify the venae cavae and their major branches.
   a. Push the heart to the pig's left to see two large veins entering the right atrium; these are the **cranial** and **caudal venae cavae** (Figure 23.4).
   b. Using the blunt probe to separate the vessels from surrounding tissues, follow the cranial vena cava toward the head and identify the two large **brachiocephalic veins**, which unite in the cranial vena cava.
   c. Identify the three major veins that unite to form each brachiocephalic vein: the **external** and **internal jugulars** that carry blood returning from the head, and the **subclavian vein** that drains blood from the front leg and shoulder. Follow the subclavian vein into the front leg. Probe deep into the muscle covering the underside of the scapula (shoulder blade) and you should see the **subscapular vein**, draining blood from the shoulder region. The **axillary vein** carries blood from the front leg, becoming the subclavian vein at the subscapular branch. Occasionally, the subclavian vein is very short, and the subscapular and axillary veins unite close to the brachiocephalic vein. Another vein that is often injected and prominent in the shoulder area is the **cephalic vein**. This vein lies just beneath the skin on the upper front leg. It typically enters the external jugular near its base.
2. Identify branches of the aorta near the heart (Figure 23.5).
   a. Push the pulmonary trunk ventrally and posteriorly to observe the curve of the aorta, the **aortic arch**, lying behind.
   b. Remove obscuring tissue and expose the first two major branches of the arch, which carry blood anteriorly. It may be necessary to remove the veins to do so. The larger of the branches, the **brachiocephalic trunk**, branches off first. The **left subclavian artery** branches off second.
   c. Identify the three major branches from the brachiocephalic trunk: the **right subclavian artery**, which gives off several branches that serve the right shoulder and limb area, and two **common carotid arteries**, which carry blood to the head. The common carotid arteries lie adjacent to the internal jugular veins.

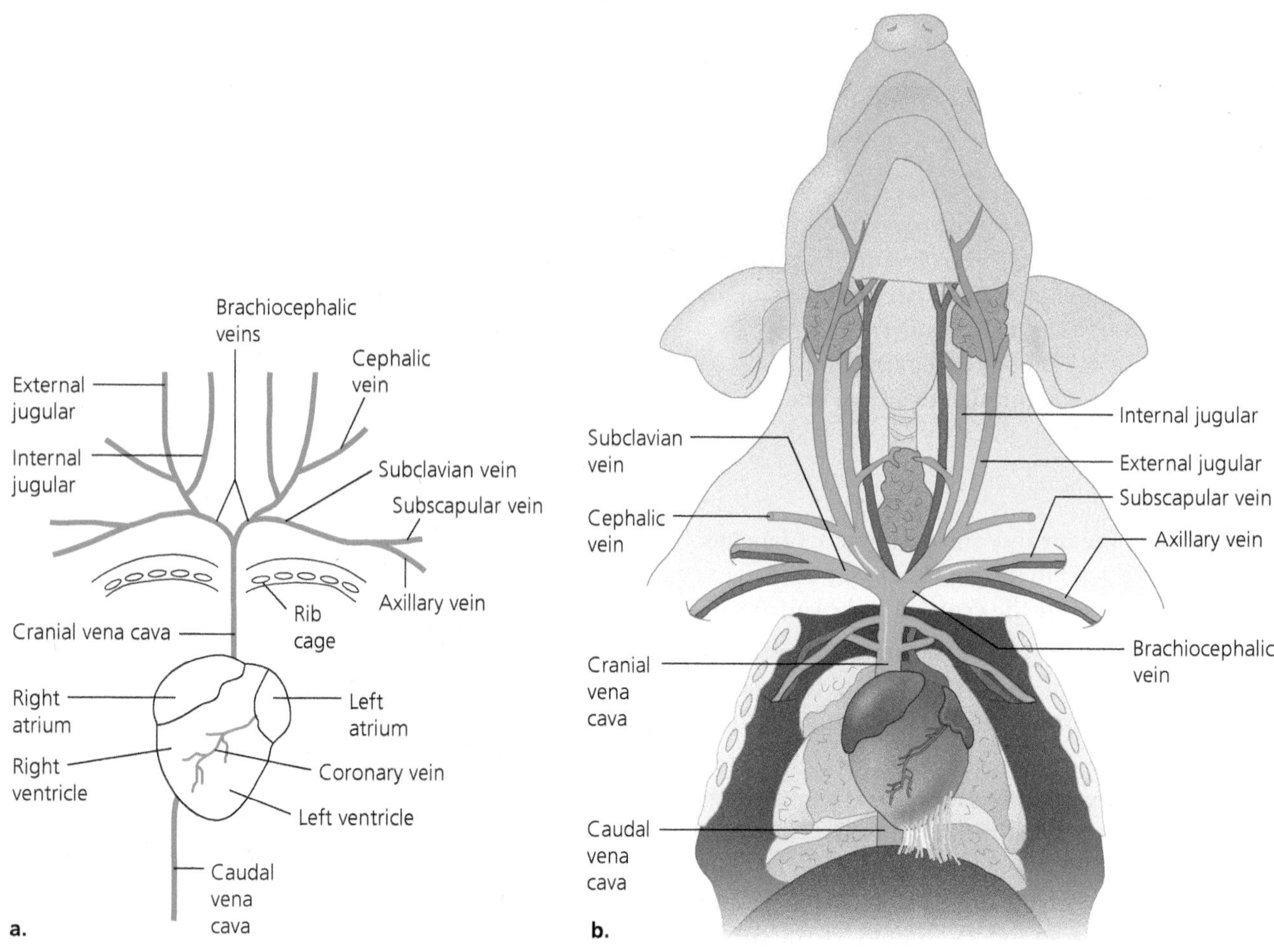

FIGURE 23.4

Veins near the heart. The subclavian vein and the external and internal jugulars carry blood to the brachiocephalic veins, which unite into the cranial vena cava. The caudal vena cava carries blood from the posterior regions of the body.

   d. Trace the branches of the left subclavian artery into the left shoulder and front leg. The branch that passes deep toward the underside of the scapula is the **subscapular artery**. After the subscapular artery branches off, the left subclavian continues into the front leg as the left **axillary artery**. Additional branches of this artery complex may also be visible.

3. Pull the lungs to the pig's right side and trace the dorsal aorta as it extends posteriorly from the aortic arch along the dorsal thoracic wall. Notice again the **ductus arteriosus** connecting from the pulmonary trunk (Figure 23.2b).

4. Note the small branches of the dorsal aorta carrying blood to the ribs. A large conspicuous vein, the **azygos vein**, lies near this region of the aorta. This vein carries blood from the ribs, usually uniting with the cranial vena cava.

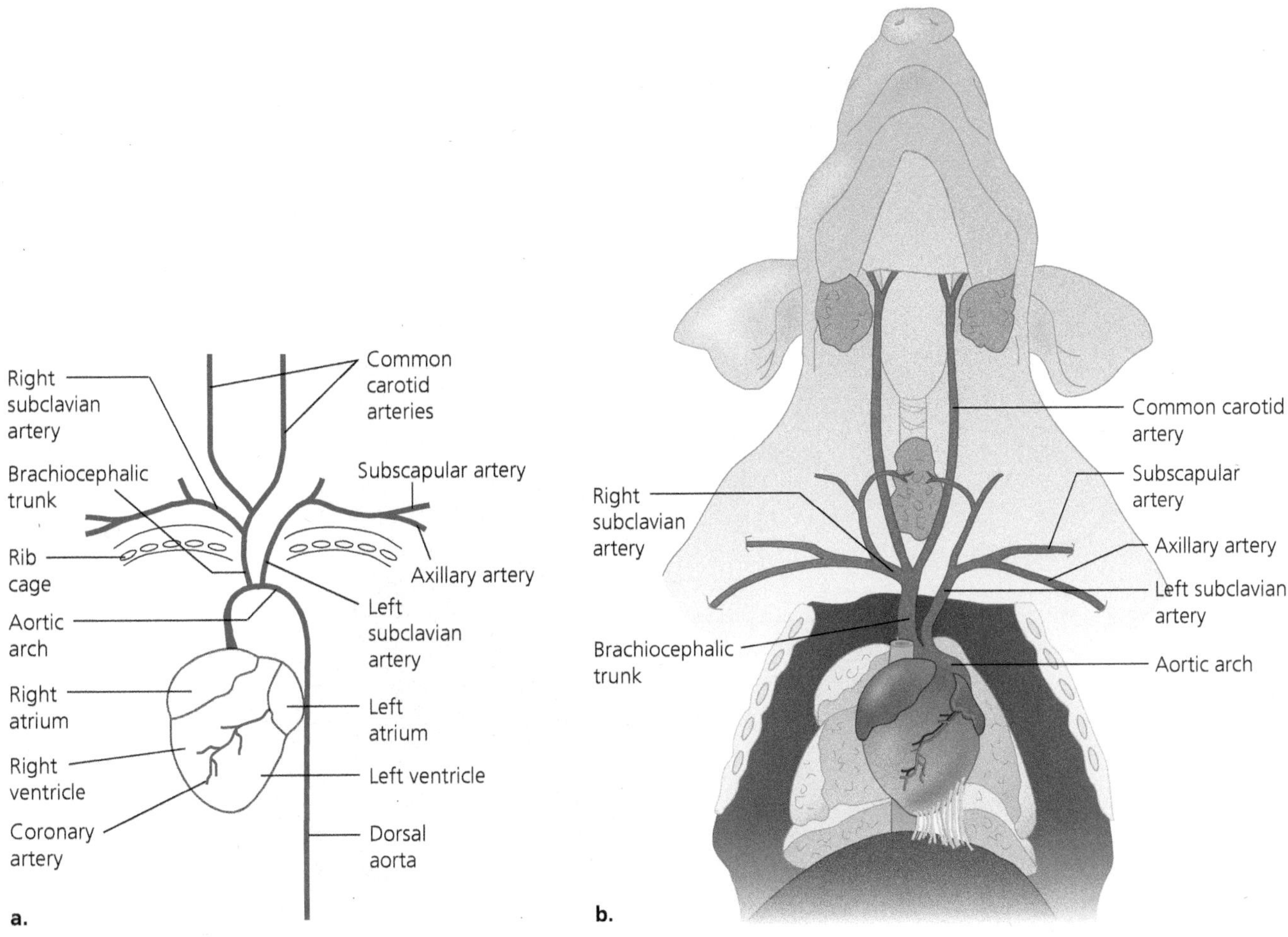

FIGURE 23.5
Branches of the aorta. Branches from the aortic arch carry blood to the head and anterior limbs. The first branch, the brachiocephalic trunk, branches into the right subclavian artery to the right limb and two common carotid arteries to the head. The second branch is the left subclavian to the left limb.

Results

Modify Figures 23.4 and 23.5 or sketch additional details in the margin of your lab manual to indicate particular features of your pig's circulatory system for future reference.

EXERCISE 23.4

The Systemic Circuit in the Abdominal Cavity

The dorsal aorta passes into the abdominal cavity, where it branches into arteries supplying the abdominal organs, the legs, and the tail. In fetal circulation, it also branches into two large umbilical arteries to the placenta. Blood from the legs, tail, and organs collects in veins that ultimately join the caudal vena cava to return to the heart. Blood draining from organs of the digestive system passes through additional vessels in the hepatic portal system before emptying into the caudal vena cava.

Lab Study A. Major Branches of the Dorsal Aorta and the Caudal Vena Cava

In this lab study, you will identify the major blood vessels branching from the dorsal aorta and those emptying into the caudal vena cava.

Procedure

1. Identify branches of the dorsal aorta (Figures 23.6 and 23.8).

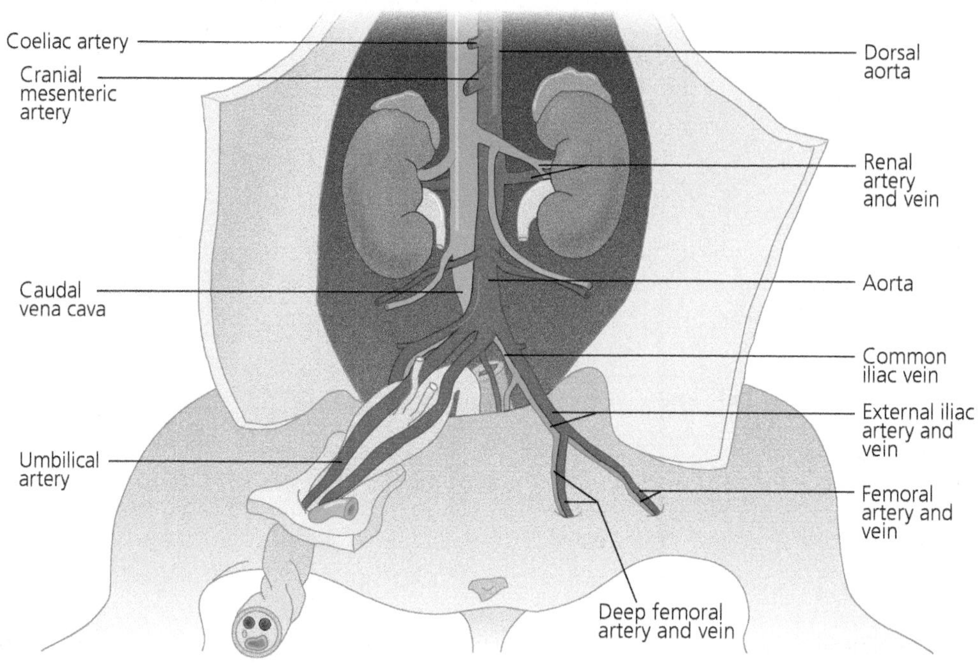

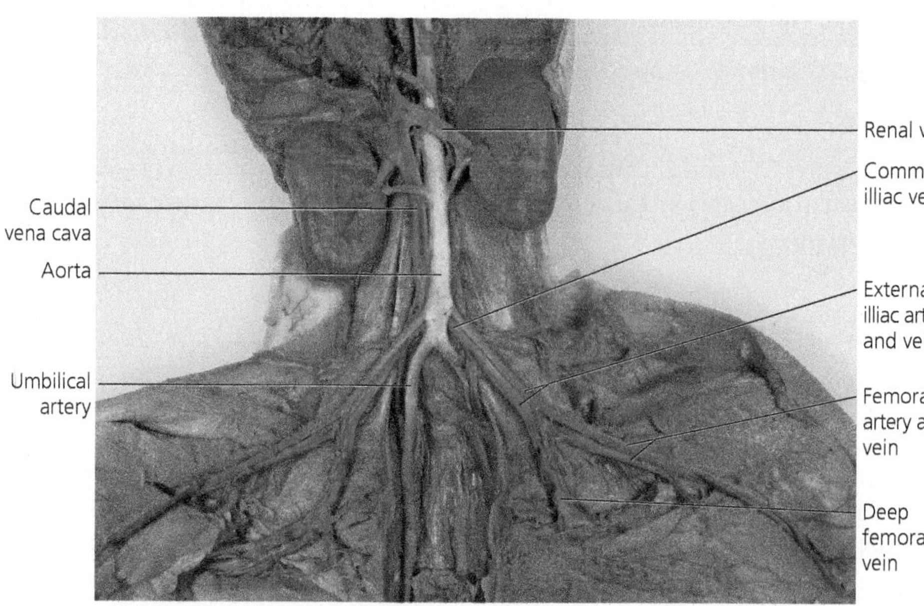

FIGURE 23.6
Branches of the aorta and caudal vena cava in the abdomen. Branches of the aorta supply blood to the stomach (the coeliac artery), the small intestine (the cranial mesenteric artery), the kidney (renal arteries), the hind limbs (iliac arteries), and the placenta (umbilical arteries). Branches of the caudal vena cava drain blood from the kidneys (renal veins) and posterior limbs (common iliac veins).

a. The first large branch of the aorta in the abdominal cavity exits the aorta at approximately the level of the diaphragm. Clip the diaphragm where it joins the body wall, pull all the organs (lungs and digestive organs) to the pig's right, and search for the **coeliac artery**, which carries blood to the stomach and the spleen. You may have to pick away pieces of the diaphragm that are attached to the aorta to see this vessel.

b. Once you have identified the coeliac artery, look for the next branch of the aorta, the **cranial mesenteric artery**, arising slightly caudal to the coeliac artery and carrying blood to the small intestine. The cranial mesenteric artery ultimately branches to the **mesenteric arteries** you observed when you studied the digestive system.

c. Following the dorsal aorta posteriorly, identify the two **renal arteries** leading to the kidneys.

You will observe the posterior branches of the aorta after the dissection of the reproductive system.

d. The dorsal aorta sends branches into the hind legs (the **external iliac arteries**) and to the placenta (the **umbilical arteries**) through the umbilical cord.

e. Separate the muscles of the leg to see that the external iliac artery divides into the **femoral artery** and the **deep femoral artery**. The femoral artery carries blood to the muscles of the lower leg, and the deep femoral artery carries blood to the thigh muscles.

2. Identify branches of the caudal vena cava.

   a. Using Figure 23.6 as a reference, push the digestive organs to the pig's left and trace the caudal vena cava into the abdominal cavity. It lies deep to the membrane lining the wall of the abdominal cavity, the **parietal peritoneum**. Peel off this membrane to see the vena cava, the dorsal aorta, and the kidneys.

   b. Identify **renal veins** carrying blood from the kidneys. **Common iliac veins** (to be identified in Lab Topic 24) carry blood from the hind legs, and **hepatic veins** carry blood from the liver to the caudal vena cava. Hepatic veins are presented in Lab Study B.

Lab Study B. The Hepatic Portal System

In the usual pathway of circulation, blood passes from the heart to arteries, to capillaries in an organ, and to veins leading from the organ back to the heart (Figure 23.7a). In a few rare instances, a second capillary bed is inserted in a second organ in the circulation pathway (Figure 23.7b). When this occurs, the circulatory circuit involved is called a **portal system**. Such a system of portal circulation exists in the digestive system (Figure 23.7c). An understanding of this circulation pathway will increase your understanding of the absorption and processing of nutrients.

FIGURE 23.7
Circulatory pathways.
(a) General circulatory pathway; (b) Circulation in a portal system; and (c) Circulation in the hepatic portal system. Arteries are depicted in red, veins are blue, and portal veins are orange.

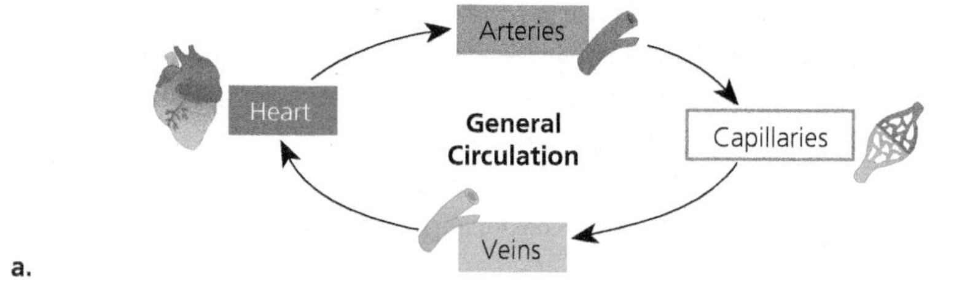

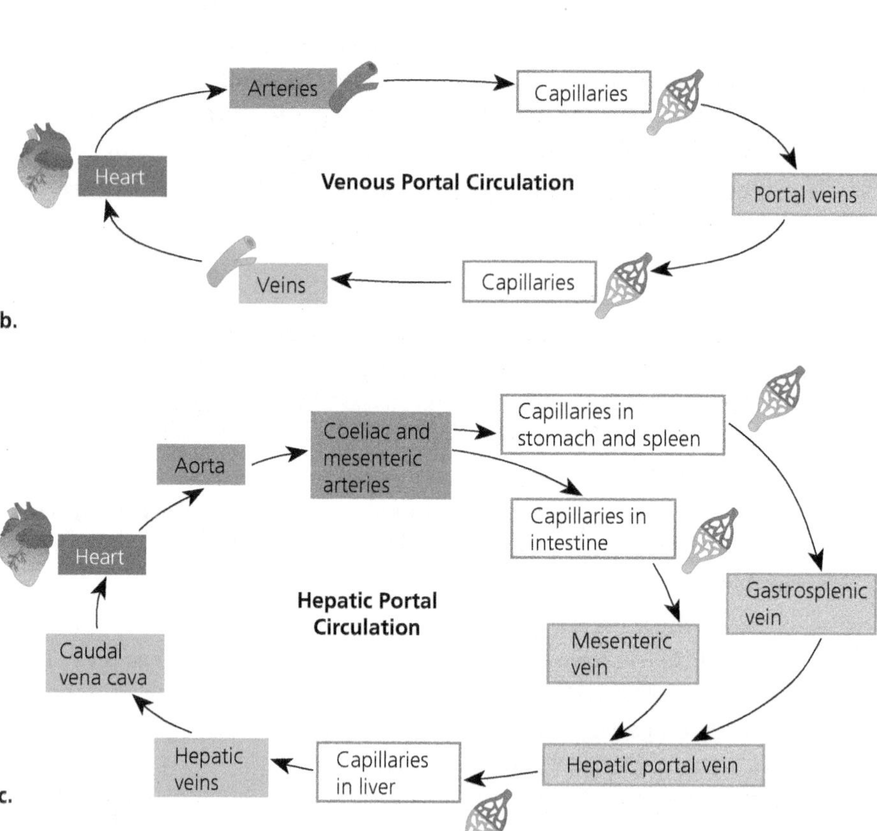

You have previously exposed the coeliac and cranial mesenteric arteries, which send branches to the stomach, spleen, and small intestine. These arteries divide into smaller arteries, to arterioles, and, finally, to capillaries, thin-walled vessels that are the site of exchange between blood and the tissues of the organs. Arteries associated with the small intestine are called **mesenteric arteries**; veins leaving the small intestine are called **mesenteric veins**, and they unite to form one large **mesenteric vein**. Veins from the stomach and spleen unite to form the larger **gastrosplenic vein**. The mesenteric and gastrosplenic veins unite to form the **hepatic portal vein**, which enters the liver (Figure 23.8). In the fetal pig, small branches of the **umbilical vein** join the hepatic portal vein as it enters the liver. However, the greatest volume of blood in the umbilical vein passes directly through the liver into the caudal vena cava.

In the liver, the hepatic portal vein branches into a second capillary bed, where exchange takes place between blood and liver tissue. These capillaries reunite into **hepatic veins**, which join the caudal vena cava. To identify these vessels, begin by dissecting the veins.

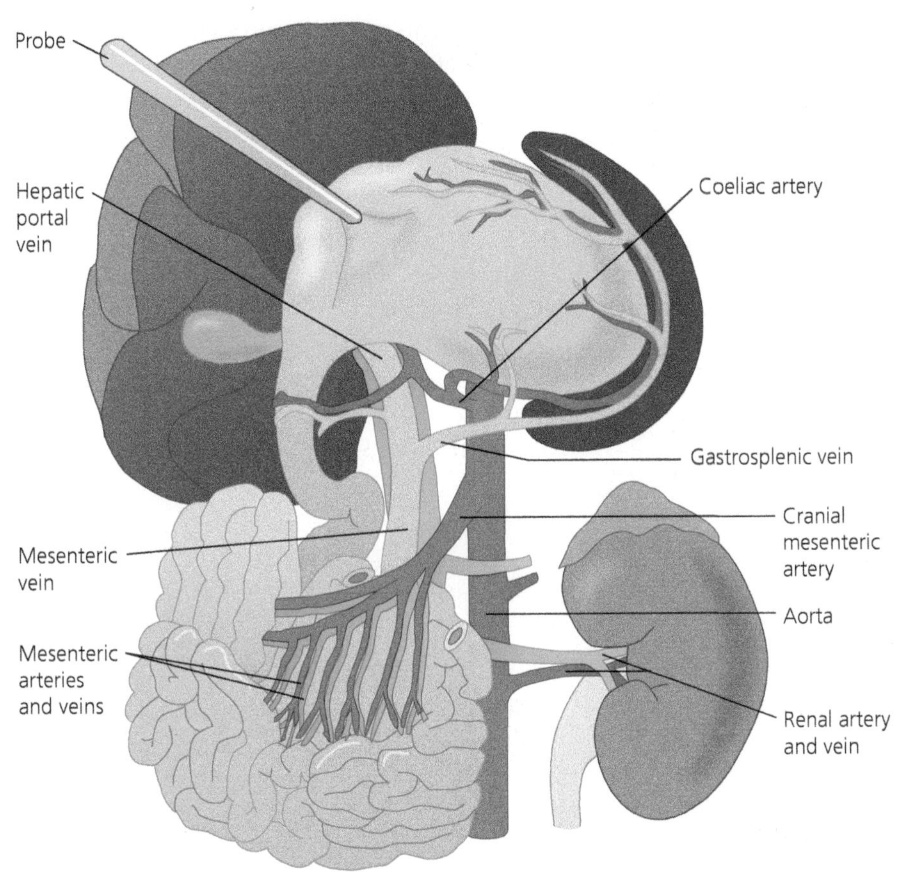

FIGURE 23.8

The hepatic portal system.

Blood from the small intestine passes into the mesenteric vein, which unites with the gastrosplenic vein to form the hepatic portal vein. This vessel leads to the liver, where it breaks into a capillary bed. Blood leaves the liver through the hepatic veins.

Procedure

1. Push the stomach and spleen anteriorly and dissect away the pancreas.
2. Use the blunt probe to expose a vein (it will probably not be injected) leading from the mesenteries of the small intestine. This is the **mesenteric vein**. It is joined by a vein leading from the stomach and spleen, the **gastrosplenic vein**. The two fuse to form the **hepatic portal vein**, which continues to the liver.
3. Review the flow of blood from the mesenteric arteries to the liver.

Results

Review the blood vessels and organs in the pathway of blood through the hepatic portal system of an adult pig with functioning digestive organs. Fill in the blanks in the next paragraph.

Blood that is poor in nutrients is carried from the aorta to the ____________ artery to smaller mesenteric arteries, which divide to a capillary bed in the wall of the ___________, where, in the process of absorption, nutrients enter the blood. This nutrient-rich blood now flows into the ________ vein, which joins with the __________ vein from the spleen and stomach and becomes the __________ vein. This vein now carries blood to the liver, where it breaks into a second capillary bed. Capillaries in the liver converge into the _____ veins, which empty into the caudal vena cava for transport back to the heart.

Discussion

Referring to your text, review the function of the liver in nutrient metabolism and relate this to the function of the hepatic portal system. Include information on digestive products, drugs, and toxins.

EXERCISE 23.5

Blood

Materials

compound microscope
prepared slide, smear of human blood

Introduction

In almost all multicellular animals, the circulatory system has three components: a pump (heart), interconnecting vessels, and a circulatory fluid to carry nutrients, oxygen, and waste throughout the body. You have just investigated two of these components—the heart and the vessels. In this exercise you will observe a slide of **blood**, the circulatory fluid. Recall from Lab Topic 22 that blood is a connective tissue (Figure 22.2e). Blood plasma (the liquid tissue matrix) contains dissolved substances that have many functions, including maintaining the osmotic balance of the body and supplying ions to other tissues. Floating in the blood plasma are cellular components that function not only in transport, but also in fighting infection, developing immunity, and playing a role in blood clotting.

Procedure

1. For now, *set aside your pig* and set up your compound microscope. Obtain a prepared slide of human blood. Observe first on low power, focusing on the cells visible on the slide. Then switch to high power.
2. Scan the slide and identify the most numerous cells, the **erythrocytes** (red blood cells). These cells appear lighter in the center due to their biconcave shape (pinched in at the center). Note that they do not have nuclei. (Erythrocytes in other vertebrates—fish, birds, reptiles, and amphibians—do have nuclei.) As these cells develop from blood-forming cells in bone marrow, they lose their nuclei; consequently, they live only about 120 days and must be replaced throughout life. When mature, these cells carry

hemoglobin, an iron-containing protein that transports oxygen. The biconcave shape increases the surface area of the cell, increasing the efficiency of oxygen diffusion into the cell.

3. Identify **leukocytes** (white blood cells), less numerous and scattered among the erythrocytes. Leukocytes consist of eosinophils, basophils, neutrophils, lymphocytes, and monocytes. All of these cell types have nuclei and the usual organelles. Eosinophils, basophils, and neutrophils can be identified by their lobed nuclei and small stained granules in the cytoplasm. These cells function to fight infection. Lymphocytes are smaller, and their nuclei do not have lobes. These cells function in developing immunity. Monocytes are the largest leukocytes and they are recognized by their kidney-shaped nuclei. These cells also function to fight infection. You may be able to distinguish among some of these cell types on your slide.
4. **Platelets** are the last cellular component that you should identify. Platelets are not complete cells, but are cell fragments that pinch off from specialized bone marrow cells. Using the highest power on your microscope, search among the erythrocytes and leukocytes for small, darkly stained granules that may be clumped together. Platelets function in blood clotting.
5. Label erythrocytes, leukocytes, and platelets in Figure 23.9.

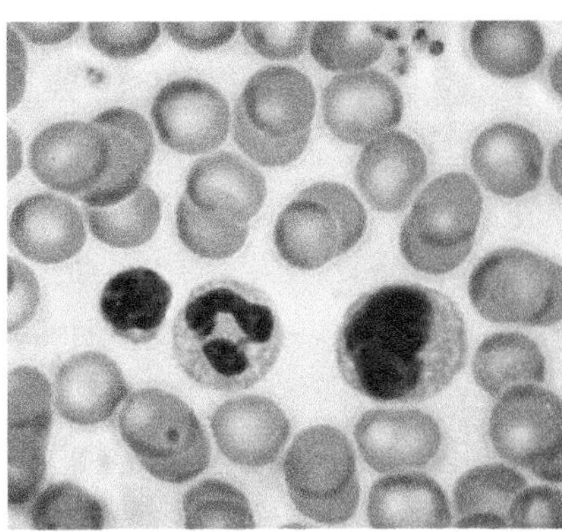

FIGURE 23.9
Mammalian blood. Identify erythrocytes, leukocytes, and platelets.

Return your microscope to storage and continue your dissection of the fetal pig.

EXERCISE 23.6

Fetal Pig Circulation

As you dissected the circulatory system in the fetal pig and observed the adult pig heart, you noted differences between the fetal heart and the adult heart, and you identified blood vessels found in the fetus but not in the adult. In this exercise you will review these vessels and structures, tracing blood flow through the fetal pig.

Procedure

1. Return the umbilical cord to the position it occupied before you began your dissection. Locate again the umbilical vein as it passes from the umbilical cord toward the liver. You cut through this vein when you opened the abdominal cavity. The umbilical vein carries blood from the umbilical cord into the liver. In the liver, small branches of this vein join the hepatic portal vein, passing blood into the liver tissue. However, the majority of the blood passes through a channel in the liver called the **ductus venosus** into the caudal vena cava. Would blood be *high* or *low* in oxygen in the caudal vena cava?

2. Review the anatomy of the fetal pig heart, and retrace the flow of blood through the heart into the dorsal aorta by way of the **ductus arteriosus**. This represents one pathway of blood through the fetal heart.

3. A second pathway of blood through the heart is created by a structure in the fetal heart called the **foramen ovale**. To study this pathway, use your scalpel to open the pig heart by cutting it along a frontal plane, dividing it into dorsal and ventral portions. Begin at the caudal end of the heart and carefully slice along the frontal plane, cutting just through the ventricles, keeping the atria intact. Carefully lift the ventricles and look inside the heart for the wall between the two atria. Using your blunt probe, carefully feel along this wall for an opening between the two atria. This hole is the foramen ovale, which makes possible the second pathway of blood through the heart. How would this hole change the flow of blood through the heart?

   In fact, most blood coming into the heart from the caudal vena cava passes from the right atrium through this hole into the left atrium. After leaving the left atrium, where would blood go next?

4. Follow the dorsal aorta into the abdominal cavity to the umbilical artery branches. These branches pass through the umbilical cord to the placenta. Would blood in these branches be *high* or *low* in oxygen?

Results

Trace fetal blood circulation from the umbilical vein to the umbilical artery by filling in the blanks in the next paragraphs.

Blood from the umbilical vein passes through the liver in a channel called the ____________ and into the ______________, which carries blood into the heart, specifically into the chamber called the __________. In one circuit of blood flow, blood goes from this chamber into the right ventricle and out the ____________. A branch from this vessel, the ______________ (present only in fetal circulation), carries most of this blood into the dorsal aorta. The dorsal aorta passes through the body, giving off branches to all organs of the body. Two large branches located near the tail lead into the umbilical cord and are called the ______________.

An alternate route carries blood from the right atrium through a fetal hole called the ____________ into the heart chamber, the _________. From this chamber, blood next goes into the left ventricle and out the _________. Branches of this vessel lead to the head.

Discussion

1. What is the advantage of the circuit of fetal blood flow through the ductus arteriosus?

2. What is the advantage of fetal blood flow through the foramen ovale?

EXERCISE 23.7

Details of the Respiratory System

You have previously located several of the major structures of the respiratory system (Exercise 23.1). Direct your attention again to the neck region of the pig and complete the study of the respiratory system.

Procedure

1. Identify again the **larynx** and the **trachea** (Figure 23.10a and c).
2. Follow the trachea caudally to the pleural cavities housing the lungs. The trachea branches into **bronchi** (sing., **bronchus**), which lead into the lobes of the **lungs**. It will be necessary to push aside blood vessels to see this. *Take care not to destroy these vessels.*
3. Tease apart lung tissues to observe that the larger bronchi branch into smaller and smaller bronchi. When the tubes are about 1–2 mm in diameter, they are called **bronchioles**. Bronchioles continue to branch and ultimately lead to microscopic **alveoli** (not visible with the unaided eye), thin-walled, blind-ending sacs that are covered with capillaries (Figure 23.10b). It is here that the exchange of oxygen and carbon dioxide takes place—between the blood and the atmosphere.
4. Identify the epithelial lining of the pleural cavity. How would this epithelium be named?

   What is the name of the epithelial lining adhering to the lung surface?

5. After you complete this lab topic, return your pig to its plastic bag. Check that your labels are intact and that your name, lab room, and lab day are legible. Add preserving solution and securely close the bag.

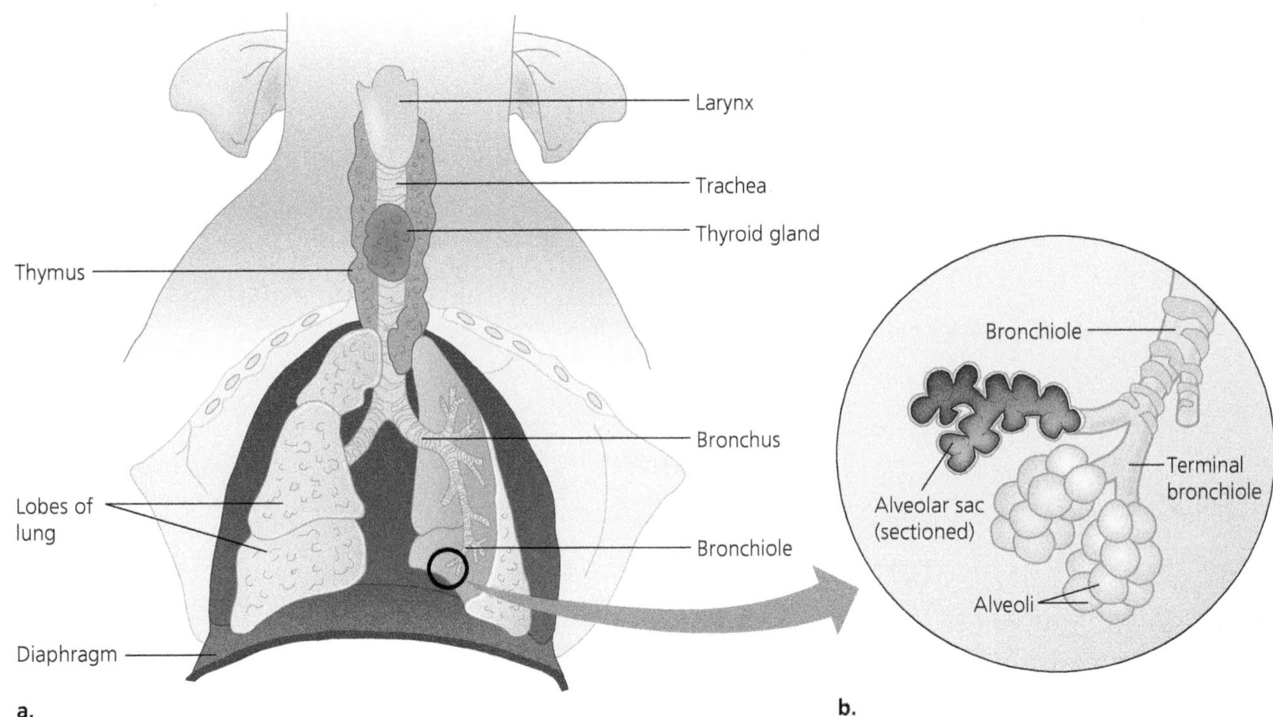

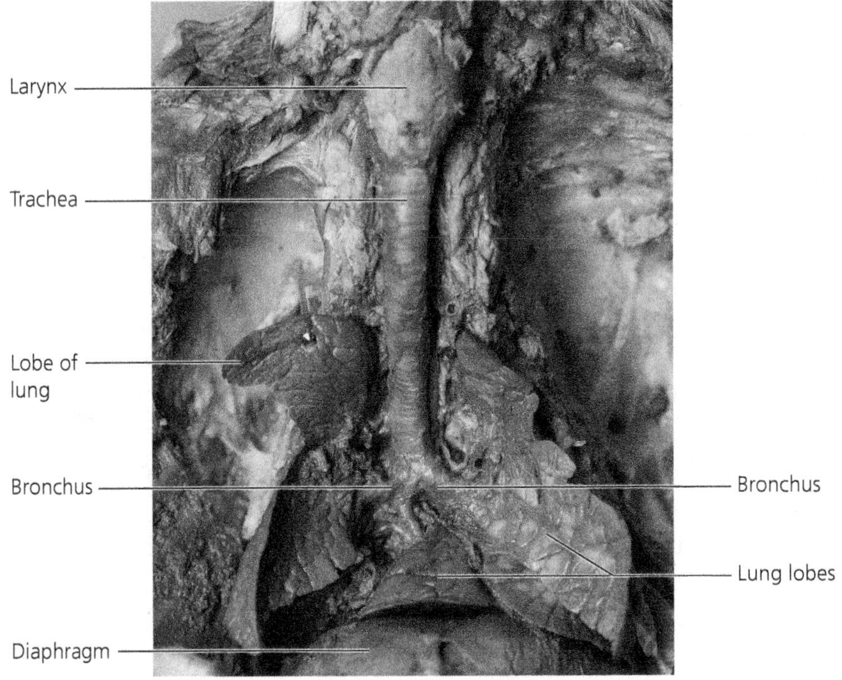

FIGURE 23.10

The respiratory system. (a) In the fetal pig, air passes through a succession of smaller and more numerous tubes: the larynx, trachea, bronchi, bronchioles, and, ultimately, microscopic alveoli. (b) Enlarged view of bronchioles and associated alveoli, the site for oxygen and carbon dioxide exchange. (c) In this photograph, the heart and some lung tissue have been removed to expose the bronchi. Note the diaphragm separating the thoracic and abdominal cavities.

Results

List, in order, the structures, tubes, and cellular barriers through which air passes as it travels from outside the body to the circulatory system of a pig, a terrestrial vertebrate.

Discussion

1. In terrestrial vertebrates, what is the advantage of having the surfaces for oxygen and carbon dioxide exchange embedded deep in lung tissue?

2. The capillaries that lie in close contact with alveoli are branches of what blood vessel?

3. The confluence of these capillaries forms what blood vessel?

4. Compare blood composition in adult circulation with reference to oxygen and carbon dioxide between capillaries approaching alveoli and capillaries leaving alveoli.

REVIEWING YOUR KNOWLEDGE

1. Return to Table 22.2 in the previous lab topic and review the coelomic cavities, the organs contained within them, and the associated coelomic membranes. Which two membranes fuse to form the pericardial sac?

2. Complete Table 23.1 on the next page. List uniquely fetal circulatory structures and vessels, describe the position or location of each, and give the fate of each after birth.
3. Name in order all blood vessels, chambers, and valves through which blood passes in an adult heart from vena cava to aorta.

TABLE 23.1 Names, Positions, and Fate of Uniquely Fetal Circulatory Structures

Fetal Structure or Vessel	Location	Fate After Birth

APPLYING YOUR KNOWLEDGE

1. Using the Web, materials provided in the lab, your text, or library materials, describe the effects and symptoms of the following diseases linked to cigarette smoking.

   Chronic bronchitis

   Emphysema

   Lung cancer

2. What effects does smoking during pregnancy have on the fetus?

3. The trachea is composed of rings of cartilage, whereas the nearby esophagus is composed of muscle and lacks cartilage. How are these structural differences related to the functions of each?

4. In the early 1900s, and in many countries today, iodine deficiency is a serious health problem, as inadequate amounts of iodine in the diet may lead to mental and developmental disabilities. Often, without dietary iodine,

the thyroid gland enlarges, resulting in a conspicuous swelling in the neck called a goiter. How has this problem been solved in the United States? How might recent recommended changes in the American diet lead to an increase in the frequency of goiters?

5. More than any other organ in the body, the liver is most often damaged by excessive intake of substances such as alcohol, anabolic steroids, vitamin A, acetaminophen, and some prescription drugs, a condition called toxic hepatitis. Why would this be?

6. Most mammals and birds are described as *homeothermic endotherms*; that is, they maintain a relatively constant body temperature (homeothermic), and they are warmed mostly by heat generated by metabolism (endothermic). Two features that you have observed in this study of the circulatory system of the pig and present in other mammals are considered to be important adaptations of the mammalian body to support homeothermy and endothermy. Identify these features and explain why they support these conditions.

CASE STUDIES

Extending their knowledge from this lab topic, students may investigate the following medical cases using various texts, the library, or Web resources. They may submit their findings in written or oral reports.

1. During cardiac development in an embryo or fetus, adequate flow of blood through the heart is necessary for normal development. Use your knowledge of blood flow through the fetal heart to predict which chambers and valves of the heart would not develop if the foramen ovale is absent or closes prematurely. Research identified syndromes that result from this condition.
2. Marilyn was admitted to the emergency room with pain in her chest and left shoulder, extending on to the pit of her stomach. After several tests to rule out other conditions, physicians administered nitroglycerin

underneath her tongue and her pain subsided. Did this response give her physician any information to help determine if Marilyn suffered from *angina pectoris* or *myocardial infarction*? How do these two conditions differ? Which of these can be treated with *nitroglycerin*?

3. Bob, a middle-aged male, has been a moderate smoker since his teen years. After many months of procrastination and increasing chest pain, he visited his physician for a physical examination. Tests revealed that he has stage 3B lung cancer. Using Web resources, research the stages of lung cancer. What would be the condition of Bob's lungs in stage 3B? What is the prognosis for his recovery?

STUDENT MEDIA: BioFlix, Activities, Investigations, and Videos

www.masteringbiology.com (select Study Area)

Activities—Ch. 42: Mammalian Cardiovascular System Structure; Path of Blood Flow in Mammals; Mammalian Cardiovascular System Function; The Human Respiratory System; Transport of Respiratory Gases

Investigations—Ch. 42: How Is Cardiovascular Fitness Measured?

REFERENCES

Burggren, W. "And the Beat Goes On—A Brief Guide to the Hearts of Vertebrates." *Natural History,* 2000, vol. 109, pp. 62–65.

Marieb, E. N. and K. Hoehn. *Human Anatomy and Physiology,* 10th ed. San Francisco, CA: Pearson, 2016.

"What You Need to Know About Cancer." *Scientific American,* 1996, Special Issue, vol. 275.

Zimmer, C. "The Hidden Unity of Hearts." *Natural History,* 2000, vol. 109, pp. 56–61.

WEBSITES

A comprehensive report of the consequences of tobacco use in the United States: 50 years of progress: A report of the Surgeon General, 2014. www.hhs.gov

All About Smoking:
http://www.nlm.nih.gov/medlineplus/smoking.html
http://www.cdc.gov/tobacco/quit_smoking/index.htm
http://www.americanheart.org/GettingHealthy

Descriptions of the stages of lung cancer:
http://www.cancerhelp.org.uK/type/lung-cancer/treatment/the-number-stages-of-lung-cancer

For information on smoking, women's health, and pregnancy, search these terms at the following websites:
http://www.lungusa.org
http://www.med.stanford.edu/medicalreview/smrp14-16.pdf
http://www.marchofdimes.org/pregnancy/smoking-during-pregnancy.aspx

For photos of healthy lungs and lungs of smokers, do an online search for: smoking lungs photos.

HHMI Biointeractive lecture on heart function:
http://www.hhmi.org/biointeractive/brave-heart-circle-life

Information on maintaining a healthy heart:
http://www.nhlbi.nih.gov//health/public/heart/other/your_guide/healthyheart.pdf

Photographs of dissected pig, including the heart and blood vessels:
home.apu.edu/~jsimons/Bio101/PigDissectionGuide.htm

Vertebrate Circulatorium | HHMI BioInteractive:
www.hhmi.org/biointeractive/vertebrate-circulatorium

LAB TOPIC 24

Vertebrate Anatomy III: The Excretory, Reproductive, and Nervous Systems

Laboratory Objectives

After completing this lab topic, you should be able to:

1. Identify and describe the function of all parts of the excretory system of the fetal pig, noting differences between the sexes and noting structures shared with the reproductive system.
2. Identify and describe the function of all parts of the reproductive systems of male and female fetal pigs and trace the pathway of sperm and egg from their origins to out of the body.
3. Compare reproductive systems in pigs and humans.
4. Describe the structure of a neuron.
5. Describe the pathway of a simple reflex, relating this to the structure of the spinal cord.
6. Identify the major regions of the mammalian brain, relating each region to a main function.
7. Describe the structure of a representative sensory receptor, the eye.
8. Discuss the role played by the nervous and endocrine systems in integrating all vertebrate systems into a functioning whole organism.

Introduction

In Lab Topic 23 Vertebrate Anatomy II, you saw that, functionally, the excretory system is closely related to the circulatory and respiratory systems. Developmentally, however, the excretory system shares many embryonic and some adult structures with the reproductive system. In the first two exercises of this lab topic, you will investigate form and functional relationships in the excretory and reproductive systems. In the last exercise of this lab topic, you will study the nervous system, which keeps all organ systems functioning appropriately and in harmony.

The action and interaction of organ systems must be precisely timed to meet specific needs within the animal. Two systems in the body, the nervous system and the endocrine system, coordinate the activities of all organ systems. The nervous system consists of a **sensory component**, made up of **sensory receptors** that detect such stimuli as light, sound, touch, and the concentration of oxygen in the blood, and **sensory nerves**, which carry the data to the **central nervous system**. The central nervous system consists of the brain and spinal cord. It integrates information from all stimuli, external and internal, and, when appropriate, sends signals to the motor system. The **motor system** carries impulses along motor nerves to **effectors** such as glands, muscles, and other organs, bringing about the appropriate response. The nervous system provides rapid, precise, and complex control of body activities.

The endocrine system consists of endocrine glands, which respond to stimuli by secreting hormones into the blood to be transported to target tissues in the body. The target tissues then bring about the response. You have already observed several endocrine glands, including the thymus, thyroid, and pancreas. In this lab topic, you will study additional endocrine glands: ovaries and testes. Control mediated by hormones in the endocrine system is slower and less precise than nervous system control. The interaction of the nervous and endocrine systems brings about the coordination of physiological processes and the maintenance of internal **homeostasis**, the steady-state condition in the vertebrate body.

EXERCISE 24.1

The Excretory System

Materials

preserved fetal pig
preserved adult kidney on demonstration
microscope slide of section through the kidney cortex on demonstration
dissecting pan
disposable gloves
dissecting instruments

Introduction

Several important functions are performed by the vertebrate excretory system, including **osmoregulation**, the control of tissue water balance, and **excretion**, the elimination of excess salts and urea, a waste product of the metabolism of amino acids. In terrestrial animals, including most mammals, water conservation is an important function of the excretory system. Studying this system in the pig will reveal the organs and structures involved in producing and eliminating metabolic waste with minimal water loss.

Wear disposable gloves when dissecting preserved animals.

Procedure

1. Locate the blood vessels serving the kidneys, exposed in the dissection of the circulatory system. The arteries branch from the dorsal aorta caudal to the cranial mesenteric artery. Blood enters the kidney through the **renal artery** and exits through the **renal vein**. Identify these vessels and the **kidneys** lying deep to the **parietal peritoneum** lining the abdominal cavity. This membrane was observed and removed in Lab Topic 23 Vertebrate Anatomy II.
2. Dissect the left kidney as follows. Leaving the kidney in the body and attached to all blood vessels and tubes, make a frontal section along the outer periphery, dividing it into dorsal and ventral portions (Figure 24.1a). First, observe the **renal cortex**, **renal medulla**, **renal pyramids**, **renal pelvis**, and **ureter** in your fetal pig. Then observe these parts in the adult kidney on demonstration.

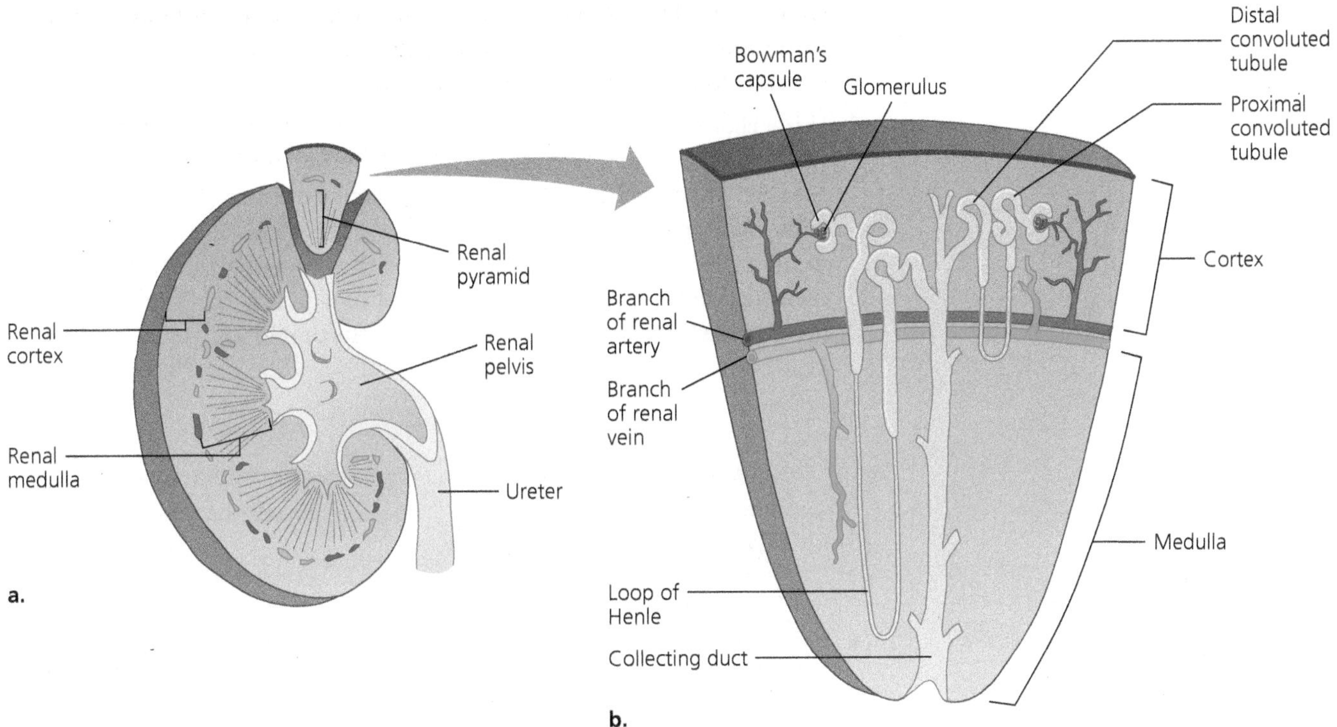

FIGURE 24.1
Structure of the kidney. (a) The kidney consists of three major regions: the cortex, the medulla, and the pelvis. Renal pyramids make up the medulla, and the pelvis is continuous with the ureter. (b) An enlarged wedge of the kidney, including the cortical region over one pyramid. Nephrons—consisting of Bowman's capsule, a proximal convoluted tubule, the loop of Henle, a distal convoluted tubule, and a collecting duct—extend over the cortical and medullary regions. Waste carried in the collecting duct ultimately passes into the pelvis and ureter.

3. Observe on demonstration a microscope slide of tissue taken from the kidney cortex. Microscopic tissues and structures found in the cortex and medulla of the kidney include blood vessels, tubules, and thousands of nephrons (over 1 million in humans). A nephron consists of Bowman's capsule, a proximal convoluted tubule, the loop of Henle, a distal convoluted tubule, and a collecting duct (Figure 24.1b). Cuboidal epithelial cells (see Figure 22.1b, p. 589) line most regions of the nephron. Bowman's capsule, a cup-shaped swelling at the end of the nephron, surrounds a ball of capillaries, the glomerulus (pl. glomeruli). Blood in the glomerulus enters the kidney via the renal artery. Blood is filtered as water and waste pass from the glomerulus into Bowman's capsule. (For details of nephron function, see your text.) Bowman's capsule, the proximal and distal convoluted tubules, and the associated blood vessels lie in the *renal cortex.* Loops of Henle and collecting ducts extend into *renal pyramids,* which make up the *renal medulla.* Both the loop of Henle and the collecting duct play a role in producing concentrated urine, a significant adaptation for terrestrial vertebrates. The hypertonic urine passes into the collecting ducts, which ultimately empty into the renal pelvis, an expanded portion of the ureter, into the kidney. The filtered blood ultimately passes into the renal vein, which exits the kidney and joins the caudal vena cava.

On the slide, identify **cuboidal epithelium**, **Bowman's capsule**, and **glomeruli** inside the capsules. Relate the microscopic anatomy seen on the slide to your observations of the isolated kidney.

4. Using Figure 24.2 as a reference, follow the ureter as it exits the left kidney at its medial border and turns to run caudally beside the dorsal aorta. The ureter then enters the **urinary bladder**. Also locate the ureter draining the right kidney and trace it to the urinary bladder. In the fetal pig, the urinary bladder is an elongate structure lying between the two **umbilical arteries** identified in Lab Topic 23 Vertebrate Anatomy II. It narrows into the small **allantoic stalk** identified in the study of the umbilical cord in Lab Topic 22 Vertebrate Anatomy I.

Do not damage reproductive organs as you expose the structures of the excretory system.

5. Pull on the umbilical cord, extending the urinary bladder, and locate a single tube, the **urethra**, exiting the urinary bladder near the attachments of the ureters. At this stage, you will see only the end of the urethra near the entrance of the ureters. In male pigs (see Figure 24.2a), the urethra leads into the **penis**. This will be visible only after you have dissected the reproductive structures. In female pigs (Figure 24.3a), the urethra joins the **vagina**, forming a chamber, the **urogenital sinus**. You will identify these structures after exposing the reproductive structures.

   In male humans, the urethra is a tube in the penis. In female humans, the urethra does not join the vagina as in the pig, but empties to the outside of the body through a separate opening. The urethra becomes functional after birth when the umbilical cord and allantois wither and fall away. Waste stored in the bladder passes into the urethra, where it is carried to the outside of the body.

Results

Describe the pathway of metabolic waste from the aorta to the outside of the body in the fetal pig.

Discussion

How does the elimination of metabolic waste in the pig change after birth?

EXERCISE 24.2

The Reproductive System

Materials

preserved fetal pig
dissecting instruments
dissecting pan
disposable gloves

Introduction

Reproduction is perhaps the ultimate adaptive activity of all organisms. It is the means of transmitting genetic information from generation to generation. Less complex animals may reproduce sexually or asexually, but in general, vertebrates reproduce sexually. Sexual reproduction, while costly to the organism in terms of energy required for the process, is advantageous because it promotes genetic variation, which is important for species to adapt to changing environments. For evolution to occur, heritable variation must exist in populations. Although mutation is the source of variation, sexual reproduction promotes new and diverse combinations of genetic information. Ultimately, all sexual reproduction involves the production of gametes and the bringing together of gametes to enable fertilization to take place.

Lab Study A. Male Reproductive System

The male reproductive system consists of gonads, ducts, and glands. Testes, the male gonads, produce sperm and secrete testosterone and other male sex hormones. Sperm pass from the testes into the epididymis, where they mature and are stored. When ejaculation takes place, sperm pass from the epididymis through the ductus deferens—also called the *vas deferens* in humans—to the urethra. The urethra leads to the penis, which carries the sperm to the outside of the body. As sperm pass through the male tract, secretions from the seminal vesicles, the prostate gland, and the bulbourethral glands are added, producing semen, a fluid containing sperm, fructose, amino acids, mucus, and other substances that produce a favorable environment for sperm survival and motility.

Procedure

> You will dissect the reproductive system of only one sex. However, you should observe the dissection of a pig of the opposite sex and be able to identify and describe various structures of both sexes.

1. Expose the structures of the male reproductive system (Figure 24.2a). The penis is located in the flap of ventral body wall caudal to the umbilical cord. To prevent damage to this structure, locate it before you make an incision. Hold the flap between your fingers and feel for the cordlike penis just below the skin. Once you locate the penis, using scissors, begin at the **urogenital opening** (identified in Lab Topic 22 Vertebrate Anatomy I) and make a longitudinal incision, extending caudally, just through the skin. Push aside the skin and use the probe to locate and expose the long

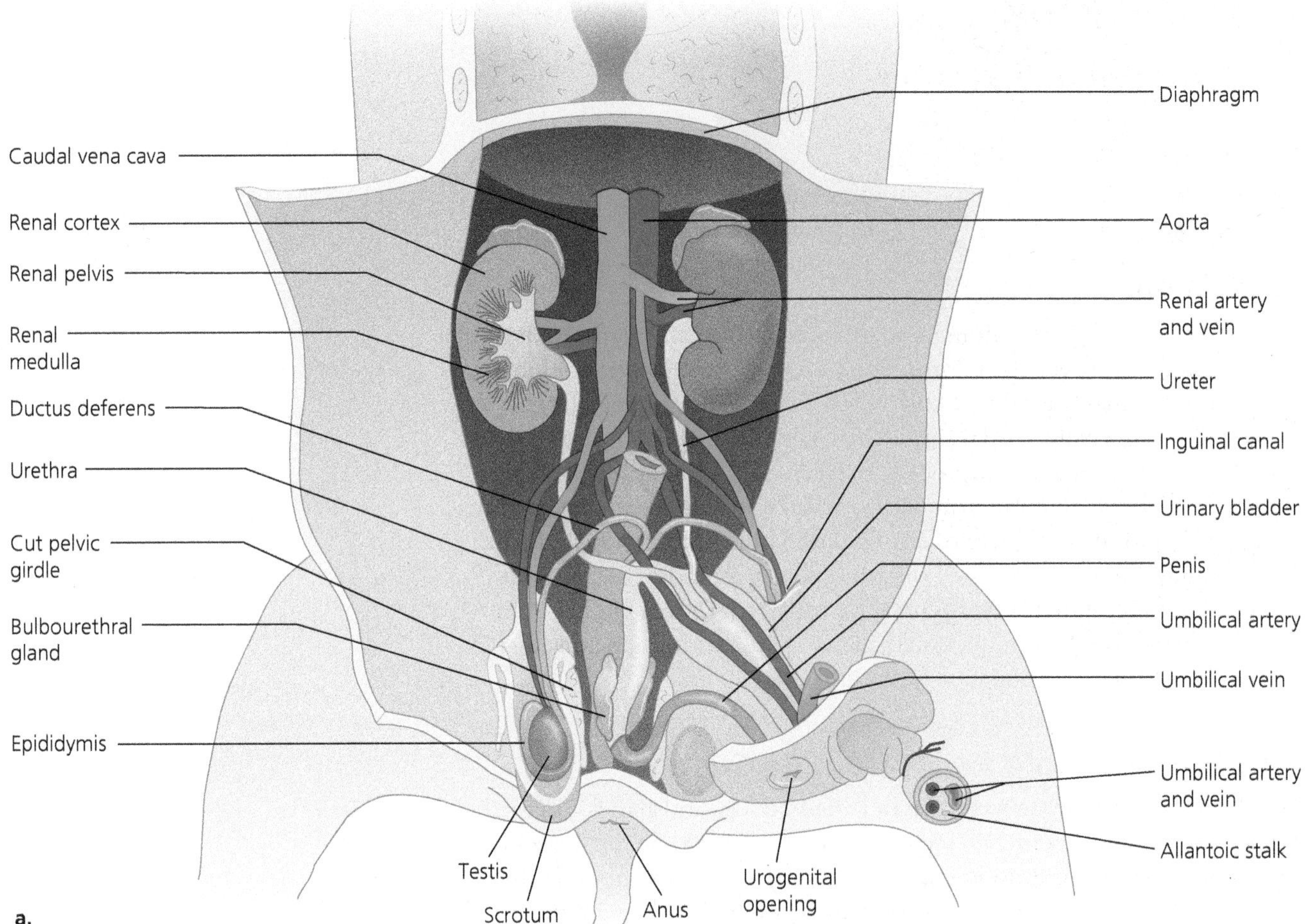

FIGURE 24.2a
Organs of the excretory and reproductive systems in the male fetal pig. The ureters enter the urinary bladder between the umbilical arteries. The urethra exits the urinary bladder and leads to the penis. The penis leads to the urogenital opening to the outside of the body. The testes lie in pouches in the scrotum. Sperm are produced in the testes, stored in the epididymis on the testis surface, and pass to the ductus deferens, which leads to the urethra.

penis from the orifice caudally until it turns dorsally to meet the urethra (see Figure 24.2b and c).

2. Next, begin to expose the testis, epididymis, and ductus deferens. To do this, locate the **ureters** (identified in Exercise 24.1) and observe the right and left **ductus deferentia** (sing., deferens), which loop over the ureters. Follow a ductus deferens outward and caudally to the **inguinal canal** leading into the **scrotum**. Use scissors to cut carefully along the canal to expose the **testis** lying in a membranous sac. Remove this sac and identify the various structures.
   a. Identify the **testis**, a bean-shaped gonad. The testes first develop in the abdominal cavity and descend before birth into the scrotal sacs.
   b. Identify the **epididymis**, a convoluted duct that originates at the cranial end of the testis, extends caudally along one side, then turns and continues cranially as the ductus deferens.

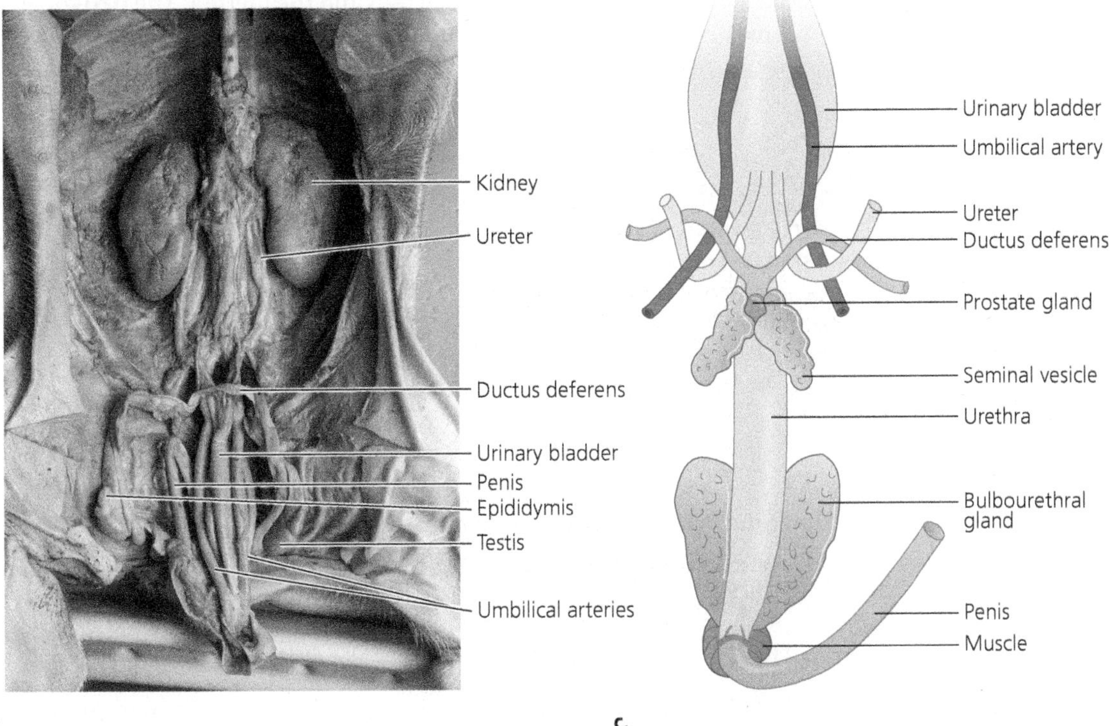

FIGURE 24.2b and c

Male reproductive and excretory organs. (b) Photograph showing ureters extending from the kidneys to the urinary bladder. The urethra, not visible in this photograph, lies dorsal to the urinary bladder. (c) Enlarged **dorsal** view with the penis extended and the urinary bladder elevated, exposing the urethra. (To see this view, pull the umbilical cord with the urinary bladder and umbilical arteries outward, away from the body.) Seminal vesicles lie near the junction of the urethra, urinary bladder, and ductus deferens. Bulbourethral glands lie on either side of the urethra.

c. Identify the **ductus deferens**, a duct that leads away from the epididymis back into the abdominal cavity, where it loops over the ureter and enters the urethra. Also locate the ductus deferens from the other testis.

3. Turn your attention again to the area where the penis turns dorsally to meet the urethra. Push the penis to one side and probe through the muscle between the legs to locate the pubic symphysis, the portion of the pelvic girdle that fuses in a position ventral to several of the reproductive structures and the rectum. *Being careful not to go too deep or to cut the penis,* use heavy scissors to cut the pubic symphysis from posterior to anterior beginning at the bend in the penis. Press the hind limbs apart and trim the ends of the symphysis. Use the probe to remove connective tissue, and expose the **urethra**, which continues anteriorly from the bend of the penis. The urethra continues into the **urinary bladder** lying between the umbilical arteries. Identify the two large **bulbourethral glands** lying on either side of the urethra anterior to its junction with the penis (see Figure 24.2c).

4. Pull on the umbilical cord, reflecting the urethra, and locate a pair of glands, the **seminal vesicles**, that lie on the dorsal surface of the urethra

near the junction of the ductus deferens and the urethra. The **prostate gland** lies between the lobes of the seminal vesicles, but because of its immature stage of development, it is difficult to identify.

At this time, complete the study of the branches of the dorsal aorta (Exercise 23.4, Lab Study A). Identify the **umbilical arteries** and the **external iliac arteries** to the legs and their branches, the **femoral** and **deep femoral arteries.** Also identify the **deep femoral, femoral,** and **common iliac veins,** which drain the legs and empty into the **caudal vena cava.**

5. After you conclude the study of the male pig, find someone with a female pig, and demonstrate the systems to each other.
6. Place your pig in its plastic bag, make sure the labels are legible, add preservative, secure the bag, and store it.

Results

In Table 24.1, list the organs and ducts through which sperm pass from their origin to the outside of the body. Describe what takes place in each organ or duct, and note glandular secretions when appropriate. Refer to your text if needed.

Discussion

1. Vasectomy is the most common form of human male sterilization used for birth control. Describe this process.

TABLE 24.1 Pathway of Sperm

Organ/Duct	Activity and Glandular Secretion

2. What structures identified are common to both the reproductive and excretory systems?

3. The testes develop inside the abdominal cavity and descend through the inguinal canal into the scrotum before birth. Explain the significance of the external scrotum and external testes in mammals. Refer to your text if needed.

Lab Study B. Female Reproductive System

The female reproductive system consists of the ovaries (female gonads), short uterine tubes (also called *fallopian tubes,* or *oviducts* in humans), the uterus, the vagina, and the urogenital sinus. The urogenital sinus is present in the pig but not in the human. In the pig, the uterus consists of a uterine body and two uterine horns in which embryonic pigs develop. In the human female, the uterus does not have uterine horns but consists of a dome-shaped portion, the fundus, which protrudes above the entrance of the fallopian tubes, and an enlarged main portion, the body of the uterus, where embryos develop.

Procedure

1. To study the female reproductive system (Figure 24.3a), use scissors and make a median longitudinal incision, cutting through the skin posterior to the umbilical cord. Push aside skin and muscles and probe in the midline to locate the pubic symphysis, the portion of the pelvic girdle that fuses in a position ventral to many of the female reproductive structures and the rectum. Being careful not to go too deep, use heavy scissors to cut through the muscles and the symphysis. Press apart the hind limbs and trim away the cut ends of the symphysis.
2. Begin observations by locating the **ovaries** in the abdominal cavity just caudal to the kidneys (Figure 24.3a). They are a pair of small, bean-shaped organs, one caudal to each kidney. (When the testes of the male first develop, they are located in approximately the same position in the abdominal cavity as the ovaries; however, the testes later descend, becoming supported in the scrotal sacs.) A small convoluted tube, the **uterine tube**, can be observed at the border of the ovary.
3. The reproductive structures form a long, continuous tract. Follow a uterine tube from one ovary into the associated **horn of the uterus**. Left and right horns join to form the **body of the uterus**. The body of the uterus lies dorsal to the urethra. Push the urethra aside and use the probe to separate the urethra from the uterus. Notice that the urethra and the reproductive structures meet.

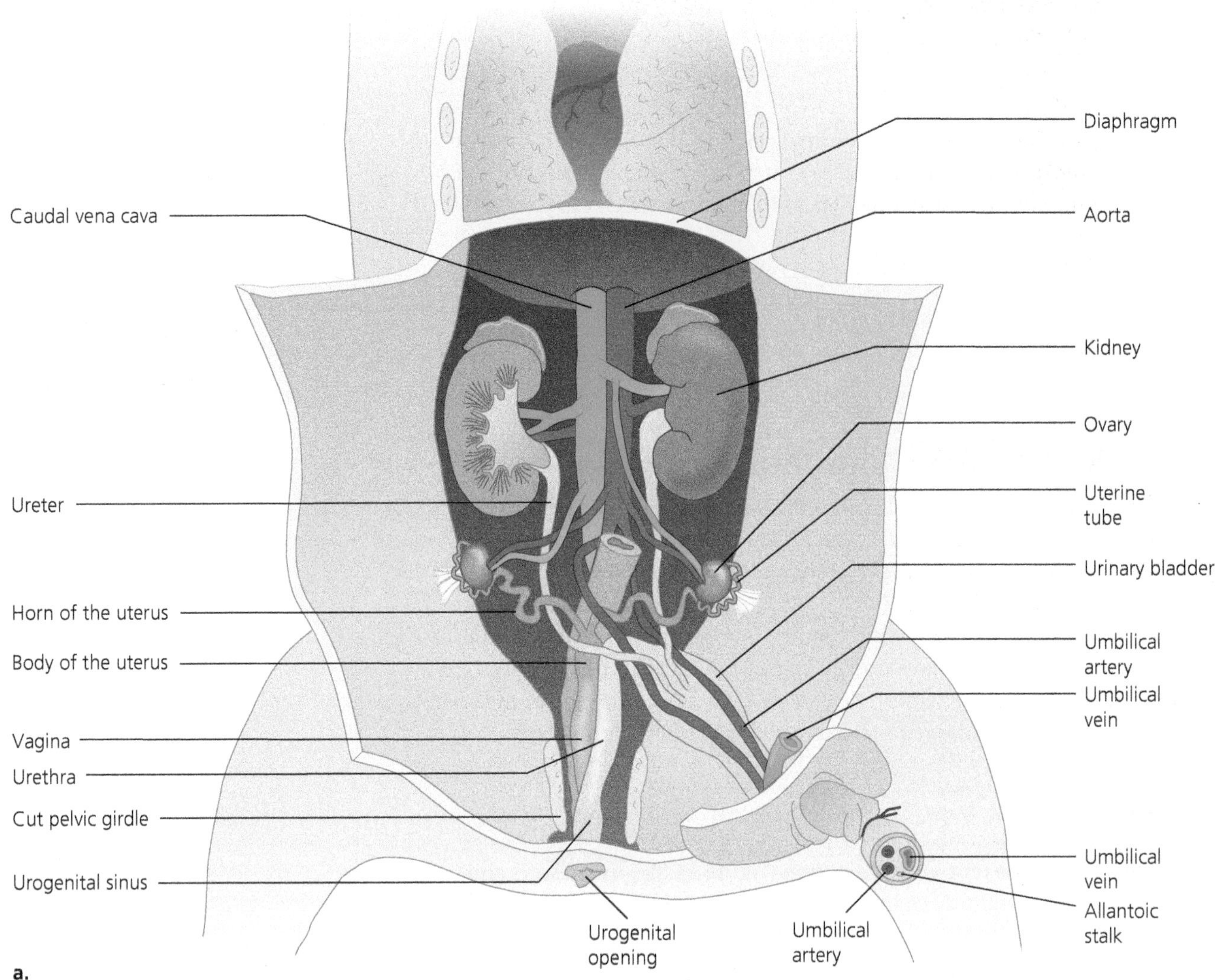

FIGURE 24.3a

Organs of the excretory and reproductive systems in the female fetal pig. The ureters enter the urinary bladder. The urethra exits the urinary bladder and joins the vagina, forming the urogenital sinus.

4. The body of the uterus leads into the **cervix** of the uterus, which leads into the **vagina**. To conclusively identify these regions, you must open the uterus. Without disturbing the junction of the urethra and the reproductive structures, use scissors to make a longitudinal, lateral incision in the reproductive structures and push back the sides, exposing the interior. Your dissection should resemble Figure 24.3c. Now you should be able to identify all parts of the uterus, the vagina, and the opening of the urethra into the reproductive tract. Identify the cervix, easily identified by the presence of internal ridges. The vagina, which joins the cervix, does not have these ridges. The vagina joins the urethra to form a common

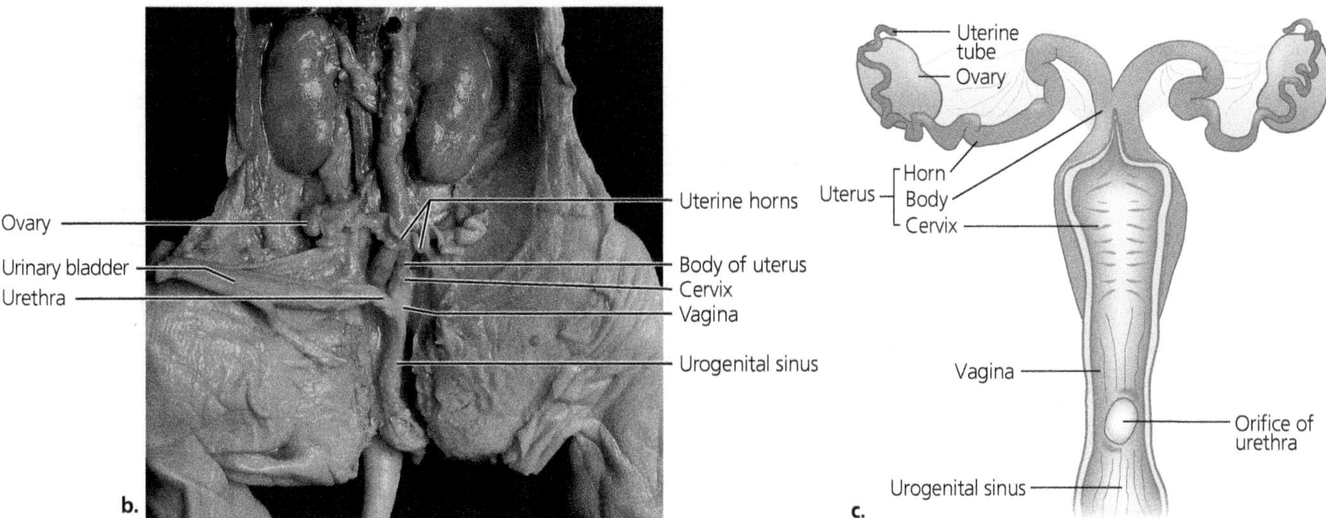

FIGURE 24.3b and c

Female reproductive and excretory organs. (b) Photograph of organs showing the urethra extending from the urinary bladder to the vagina where the two join to form the urogenital sinus. (c) In this enlarged view the cervix and vagina have been opened to show the ridges in the cervix, which are absent in the vagina. The urogenital sinus is the common chamber formed by the confluence of the vagina and the urethra.

chamber, the **urogenital sinus**, leading to the outside of the body. The outer opening is the **urogenital opening**, ventral to the anus (identified in Lab Topic 22 Vertebrate Anatomy I).

At this time, complete the study of the branches of the dorsal aorta (Exercise 23.4, Lab Study A). Identify the umbilical **arteries** and the **external iliac arteries** to the legs and their branches, the **femoral** and **deep femoral** arteries. Also identify the **deep femoral, femoral,** and **common iliac veins,** which drain the legs and empty into the **caudal vena cava.**

5. After you conclude your study of the female pig, find someone with a male pig, and demonstrate the systems to each other.
6. Place your pig in its plastic bag, make sure your labels are legible, add preservative, secure the bag, and store it.

Results

Describe the pathway of an egg from the ovary to the uterus. Then name the structures from the uterus to the outside of the body in a fetal pig, naming regions of organs when appropriate.

Lab Study C. The Pregnant Pig Uterus

On demonstration is an isolated pregnant pig uterus, which should include uterine horns and the body of the uterus. Ovaries and uterine tubes may be attached. Fetal pigs are located in the uterine horns. Each fetal pig is attached to the mother pig by means of the **placenta**, a structure consisting of tissue from the inner lining of the uterus (maternal tissue), and the **chorionic vesicle** (embryonic tissue). These tissues are convoluted, creating interdigitating folds that increase the surfaces where the exchange of nutrients, oxygen, and wastes takes place between mother and fetus. Remember that blood does not flow directly between the mother and fetus, but the close proximity of the tissues allows diffusion across the placental membranes.

Procedure

1. Observe the uterus with one uterine horn partially opened (Figure 24.4). One or more fetal pigs should be visible.
2. If it is not already dissected, using scissors, carefully cut into the **chorionic vesicle**, a saclike structure surrounding each fetal pig. Note that the chorionic vesicle is composed of two fused membranes, the outer **chorion** and the inner **allantois**. Blood vessels are visible lying within the thin allantois. In Lab Topic 22 Vertebrate Anatomy I, you identified the allantoic stalk, a small tube in the umbilical cord extending between the fetal pig's urinary bladder and the allantois. Speculate about the function of the allantois. The blood vessels are branches of which vessels?
3. Observe the **amnion**, a very thin, fluid-filled sac around the fetus. What function do you think this membrane performs?

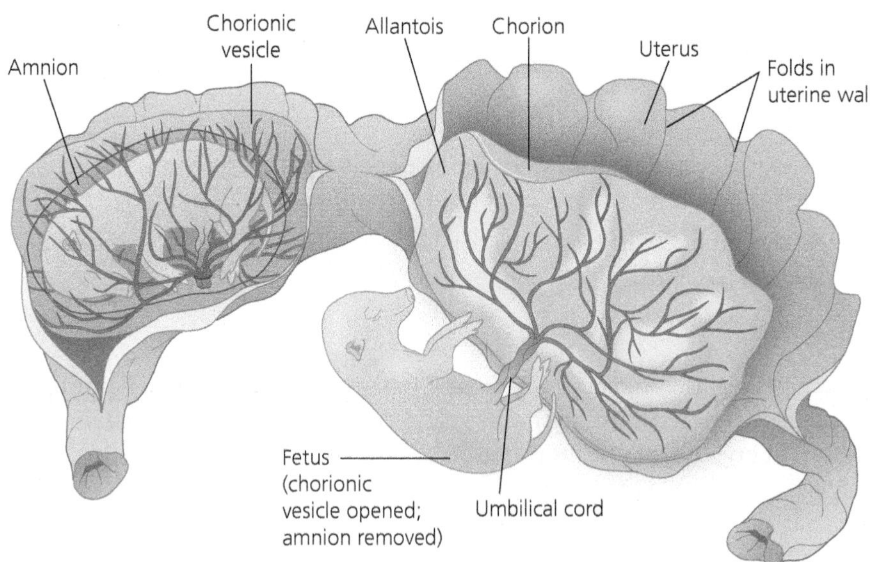

FIGURE 24.4
Section of uterine horn from an adult pig with two fetuses. Two saclike structures, an amnion and a chorionic vesicle, surround the fetus on the left. The chorionic vesicle around the other fetus has been opened and the amnion removed.

4. Open the amnion and see the **umbilical cord** attaching each fetus to the fetal membranes.

Results

Beginning with those membranes closest to the fetal pig, list in order all embryonic and maternal membranes and tissues associated with the fetal pig.

Discussion

1. Using your text if necessary, compare the female reproductive organs in a human and an adult pig with respect to the oviduct and uterus in the human and uterine tube, horns of the uterus, and body of the uterus in the pig. Speculate about the adaptive advantage of the differences.

2. Describe differences in the arrangement of the vagina and urethra in the fetal pig and human.

3. Tubal ligation is a common form of human female sterilization. Describe this process.

EXERCISE 24.3

Nervous Tissue, the Spinal Cord and Reflex Arc, the Sheep Brain, and the Vertebrate Eye

In this exercise, you will study several components of the nervous system: the structure of neurons, the pathway of a reflex arc as it relates to the structure of the spinal cord, the major regions of the sheep brain, and the structure of a sensory receptor, the vertebrate eye.

Lab Study A. Nervous Tissue and the Structure of the Neuron

Materials

compound microscope
prepared slide of nervous tissue

Introduction

To understand the function of nervous tissue, review the structure of the **neuron**, the functional cell of nervous tissue. Neuron structure facilitates nervous impulse transmission. Each neuron has three parts: a **cell body**, which contains cytoplasm and the nucleus; **dendrites**, extensions from the cell body that receive nervous impulses from other neurons and transmit those signals toward the cell body; and an **axon**, an extension that transmits nervous impulses away from the cell body to the next neuron or sometimes to a muscle fiber (Figure 24.5). Neurons are found in the brain and the spinal cord and in nervous tissue throughout the body. You will study the structure of nervous tissue and neurons in the spinal cord.

Procedure

1. Using the intermediate objective on your compound microscope, scan the prepared slide of nervous tissue provided. This is a smear preparation of tissue taken from the spinal cord. You will see hundreds of small, dark dots, which are the nuclei of **glial cells**. Glial cells are nonconducting cells that support and protect neurons.
2. Look for large angular **cell bodies** of **motor neurons** scattered among the fibers and glial cells. Study one of these cell bodies on high power and locate the **nucleus**, usually with a prominent **nucleolus**. Try to identify the two types of **processes** extending from the cell body: the **axon** and **dendrites**. Although it is difficult to be certain, you may be able to differentiate between the single, broader axon and one or more slender dendrites extending from the cell body.

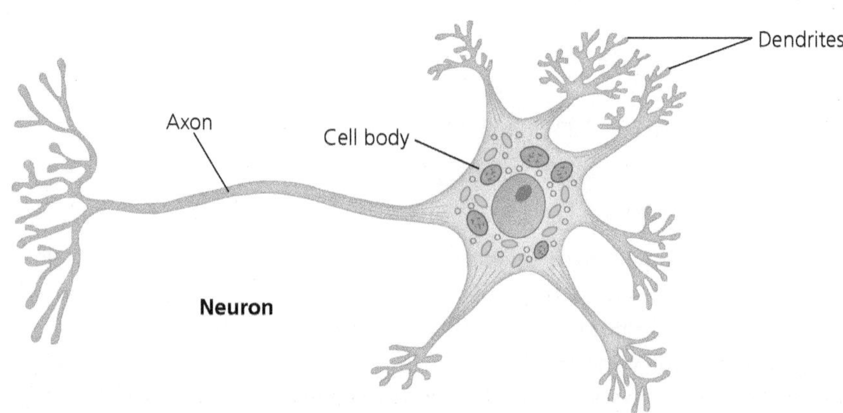

FIGURE 24.5
Structure of a neuron. Dendrites and an axon, cytoplasmic processes, extend from the cell body.

Lab Study B. The Reflex Arc and Structure of the Spinal Cord

Materials

stereoscopic microscope
compound microscope
prepared slide of a cross section (c.s.) of spinal cord

Introduction

This lab study will help you envision the interaction between the two main divisions of the nervous system, the **peripheral nervous system** and the **central nervous system**. Recall from the Introduction to this lab topic that the peripheral nervous system consists of a sensory component with **sensory receptors** and **sensory neurons** (also called afferent neurons), and a motor component with **motor neurons** (called efferent neurons) leading to **effectors**, muscles or glands. Bundles of sensory or motor neurons are called **nerves**. The **spinal cord** and the **brain** make up the central nervous system. In this lab study you will examine the structure of the spinal cord and relate this to a simple **reflex arc**. This will demonstrate the roles of sensory neurons, the spinal cord, and motor neurons in a simple **reflex**—a rapid, involuntary response to a particular stimulus. One example is the knee-jerk reflex.

Procedure

1. Using the stereoscopic microscope, examine a prepared slide of a spinal cord cross section taken at a level that shows **dorsal** and **ventral roots**. The roots are collections of processes of neurons in spinal nerves.
2. Identify the dorsal and ventral surfaces of the spinal cord by locating the **ventral fissure** (Figure 24.6). Recall from Lab Topic 19 Animal Diversity II,

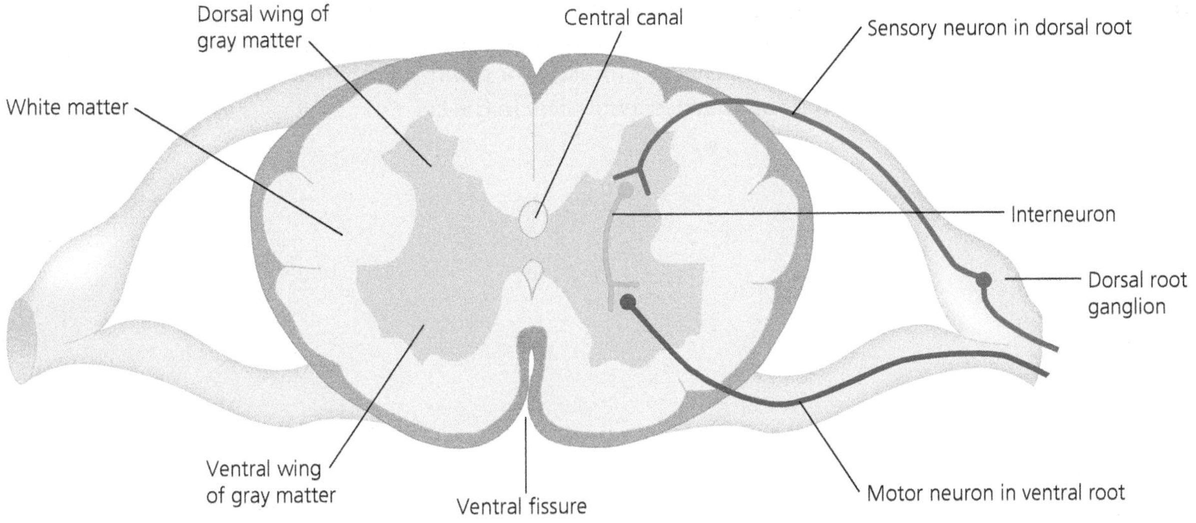

FIGURE 24.6
Cross section of the spinal cord at the level of dorsal and ventral roots. Sensory neurons enter the dorsal wing of the gray matter, and the cell bodies of motor neurons lie in the ventral portion of the gray matter. Interneurons may be present in simple reflex arcs. A simple reflex arc may include two neurons—one motor neuron and one sensory neuron—or three neurons if an interneuron is present.

that vertebrates have a tubular nervous system. This is supported by the presence of the **central canal**, a small channel in the center of the cord.

3. Locate **gray** and **white matter**. In the spinal cord, white matter lies outside the butterfly-shaped gray matter. In sections through the spinal cord at the level where dorsal and ventral roots enter and exit the cord, you will be able to identify the **dorsal root ganglion**, in which cell bodies of sensory neurons lie. Look for the neuronal processes from these cell bodies. These processes continue into the tip of the dorsal "wing" of the gray matter. Sensory neurons receive impulses directly from the environment or from a specific sensory receptor, for example, the eye.
4. Locate cell bodies of **motor neurons** in the ventral "wing" of the gray matter. These are best studied using lower powers on the compound microscope. Many of these cell bodies contain conspicuous nuclei and nucleoli. Whereas the simplest reflex involves only one sensory and one motor neuron, most reflexes involve many **interneurons**, lying between sensory and motor neurons. Motor neurons carry impulses to muscles or glands and bring about a **response**.
5. Careful observations may reveal axons from motor neuron cell bodies coursing through the white matter and into the ventral root of the spinal nerve.

Results

Using information from the study of the spinal cord, list in sequence the structures or neurons involved in the simplest reflex.

Discussion

Most reflexes involve a specific sensory receptor (such as the eye, ear, or pain or touch receptors), several sensory neurons, several interneurons, several motor neurons, and effectors (muscles or glands). Propose a reflex arc that would result if you touched a hot plate in the lab.

Lab Study C. Overview of Sheep Brain Anatomy

Materials

dissecting pan
whole sheep brain with meninges removed
half sheep brain sliced in sagittal section
isolated human brain (optional)
blunt probe
disposable gloves

Introduction

Scientists are using modern tools of molecular biology, biochemistry, immunology, and imaging to reveal the complex functions of the brain. The fields of neurobiology and behavior are developing rapidly and the editors of *Scientific American* have proposed the 21st century as the "Century of the Brain." At all organizational levels of the nervous system, from basic neurons to the brain, there is a clear relationship between structure and function. Understanding the anatomy of the specialized regions of the brain is foundational for investigating the many brain functions, for example, information processing, sensations, emotions, language, and memory, as well as the diagnosis and treatment of disease. Although the anatomy of the brain in humans and other animals has been described in great detail for centuries, there are still surprising and even revolutionary discoveries. In 2015, scientists at the University of Virginia discovered small lymphatic vessels on the brain surface, providing the first evidence that the immune system has direct structural connections to the brain, and generating new questions about diseases such as Alzheimer's and multiple sclerosis. *Science*, one of the leading scientific journals, awarded this discovery one of the top three "Breakthroughs of the Year" (Armitage, 2015).

In this laboratory you will investigate the major specialized regions of the sheep brain, which is similar in structure to the human brain. Working with a partner, you will observe dorsal and ventral surfaces of the whole brain and then a sagittal section through the brain to explore some of the internal structures. You will follow your laboratory investigations by reviewing the functions associated with each of these major regions and structures of the brain.

Procedure

1. Obtain a **whole sheep brain** and determine its dorsal and ventral surfaces and its anterior/posterior axis. Before you proceed, review the definitions of these terms in Appendix E. Place the brain in the dissecting pan ventral side down. As you perform your survey of brain structures, work systematically, *being careful not to cut or dissect the brain tissue.* Use only a blunt probe to locate and demonstrate each structure indicated in bold, and then add labels to Figure 24.7. Inside the skull or cranium (the bony structure covering and protecting the brain), the brain is covered by three layers of tissue. Two of these layers, the dura mater and the arachnoid mater (or simply, the arachnoid), may not be present on the brain you are studying, because they are usually removed with the skull. In Figure 24.7b, the dura mater is still intact, and can be seen as a thin, transparent layer of tissue covering parts of the brain. The dura mater is a tough membrane that is bound to the inside of the skull. The arachnoid mater is attached to the inside of the dura mater. The third membrane is the **pia mater**, a thin vascular membrane directly on the surface of the brain. Cerebrospinal fluid (CSF) flows between the arachnoid mater and the pia mater.

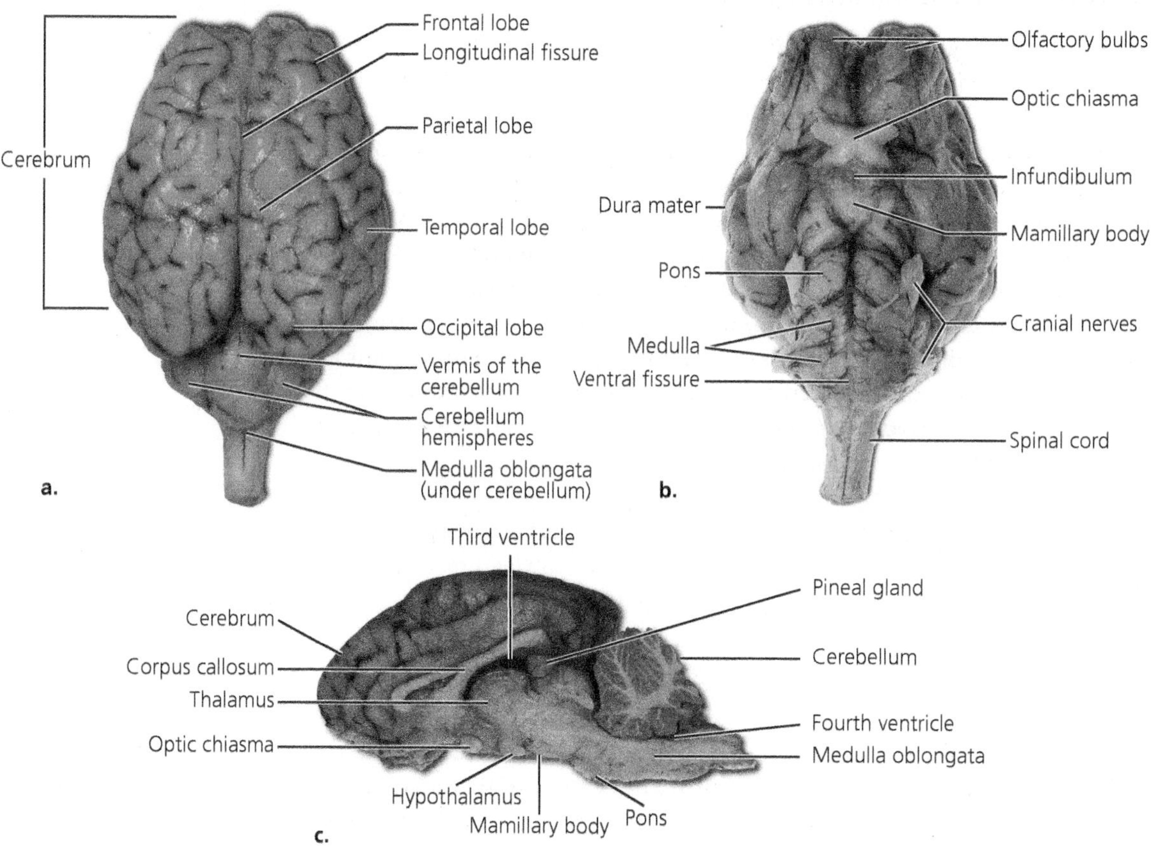

FIGURE 24.7
The sheep brain. (a) Dorsal view. (b) Ventral view. (c) Sagittal section.

2. **Dorsal view of the brain (Figure 24.7a)**

   a. **Locate the cerebrum.** Identify the deep, **longitudinal fissure** (a long narrow groove) separating the largest part of the brain, the **cerebrum**, into two **cerebral hemispheres**. You can see small ridges (called gyri) and grooves (called sulci) on the surface of these hemispheres. The pia mater covers and extends into these ridges and grooves. Each hemisphere is divided into four lobes —the **frontal lobe** is the anterior region of the hemisphere; the **parietal lobe** is posterior to the frontal lobe, with the medial longitudinal fissure separating the two sides; the **temporal lobe** is positioned in the lateral and widest area of the hemisphere; and the **occipital lobe** is in the posterior region of the hemisphere.

   b. **Locate the cerebellum** posterior to the cerebrum. Note the three parts of this structure—the median part, called the **vermis**, (probably because someone thought it looked like a coiled worm) and on either side of the vermis, the two **cerebellum hemispheres**.

   c. **Locate the medulla oblongata** (or simply the medulla) just posterior to the cerebellum. The medulla continues as the spinal cord through a large opening into the skull, the foramen magnum.

3. **Ventral view of the brain (Figure 24.7b)**

(Note that "anterior" "middle" and "posterior" regions described here refer to locations in Figure 24.7b. These do not correspond to the embryonic brain regions—forebrain, midbrain, and hindbrain.)

a. **Observe the anterior region of brain.** Turn the brain over and identify structures on the ventral side, beginning at the anterior end. The anteriormost structures are two large swellings on the ventral brain surface. These are the **olfactory bulbs**. Sensory receptors originate in the epithelium of the nasal cavity and synapse with neurons in the olfactory bulbs. Extending from the posterior end of each bulb are olfactory tracts consisting of nerve fibers (axons) that run beneath the frontal lobes and enter the cerebral hemispheres. *(Note that "nerves" are collections of neuron fibers {axons} in the peripheral nervous system, and "tracts" are bands of axons in the central nervous system.)*

Considering their function, predict the sizes of the olfactory bulbs and tracts in a human, compared with olfactory bulbs in the sheep brain.

b. **Observe the middle region** of the brain. Look just posterior to the olfactory tracts to identify the **optic chiasma**, an X-shaped structure where some of the fibers from the optic nerves originating in the eyes cross over to continue in tracts to opposite sides of the brain. Just posterior to the optic chiasma is the **infundibulum** that attaches the **pituitary gland** (also called the hypophysis) to the **hypothalamus** of the brain, providing an important link between the brain and the endocrine system. The pituitary gland lies in a small cavity in the ventral inside wall of the skull base, just above the posterior region of the nasal cavity. In the isolated sheep brain this gland is usually missing because it is easily torn off as the brain is removed from the skull. It is interesting to consider surgical procedures used to remove a diseased pituitary gland given its location in the brain. What do you think would be the least invasive approach to access the gland?

Just posterior to the infundibulum is a small two-parted swelling, the **mammillary body**, also part of the hypothalamus.

c. **Observe the posterior region** of the brain. Identify the medulla oblongata on the ventral surface, with a median ventral fissure extending its length. Just anterior to the medulla is the **pons**. The **brain stem** refers to the central-posterior portion of the brain, consisting of midbrain structures, the pons, and the medulla, continuing into the spinal cord.

Notice stubs of the **cranial nerves**, 12 pairs in all, designated by Roman numerals. Ten pairs of these nerves emerge from the surfaces of the brain stem. The other two are the olfactory (I) and optic (II) nerves. One cranial nerve, the vagus nerve, (X) originates in the medulla and is the only cranial nerve to extend beyond the head and neck to the thorax and abdomen. Cranial nerves are part of the peripheral nervous system, not the brain proper.

4. **Sagittal view of the brain (Figure 24.7c)**
   a. Obtain a half brain from the demonstration table, and position it with the cut surface up with the anterior end to the left (as in Figure 24.7c) in the dissecting pan. The sagittal section was made separating the brain into right and left halves.
   b. Locate the following structures previously observed in the whole brain: **cerebrum**, **cerebellum**, **medulla**, **pons**, **optic chiasma**, **mammillary body**, and **hypothalamus**. Observe the internal locations and connections of these brain structures and relate these to your observations of their external positions in the whole brain.
   c. The hypothalamus is in the floor of the **third ventricle**. Ventricles are cavities in the brain that originate from the "hollow" characteristic of the vertebrate neural tube. Identify the **thalamus** that appears as an oval structure on the lateral walls of the third ventricle. The thalamus is an important center where sensory neurons from the spinal cord synapse with other neurons on the way to the cerebrum. Look dorsal and posterior to the thalamus, directly posterior to the third ventricle, to locate the **pineal gland**, a small endocrine gland.
   d. Identify the **corpus callosum**, a white band above the third ventricle. This broad tract of nerve fibers located below the longitudinal fissure connects the right and left cerebral hemispheres of the brain. A *corpus callostomy* is a procedure that may be performed in persons having the most extreme forms of epilepsy, not controlled by medications. In this surgery, the corpus callosum is cut to prevent the spread of seizures from one cerebral hemisphere to the other.
   e. There are four **ventricles** in the brain, all connected, and containing cerebral spinal fluid (CSF) produced by a network of blood vessels in each ventricle. CSF is somewhat similar to blood plasma, and plays important roles in providing protection and nourishment to the brain. The first and second ventricles, called lateral ventricles, are cavities within in each cerebral hemisphere. (To keep the cerebral hemispheres intact, we will not observe these ventricles.) Carefully push the cerebellum forward to observe the **fourth ventricle**, located between the cerebellum and the medulla. This ventricle is continuous with the central canal of the spinal cord.
5. As you conclude your observations, review all brain structures and regions with your partner and return the brains to the demonstration table. *If a human brain is available on demonstration, compare the major regions and structures for human and sheep brains.*

Results

Using your text or Web resources, complete Table 24.2, listing a major function for each of the brain structures you have observed. Your instructor may assign this as a take-home assignment.

TABLE 24.2 Major Functions of Brain Structures

Structure or Region	Function
Cerebrum	
Cerebellum	
Medulla oblongata	
Olfactory bulbs and tracts	
Optic chiasma	
Infundibulum	
Hypothalamus	
Mammillary bodies	
Pons	
Ventricles	
Thalamus	
Pineal gland	
Corpus callosum	

Discussion

1. In a previous lab topic you learned four unique chordate characteristics. Which of these characteristics are evidenced by the anatomy of the brain?

2. What is cerebral palsy?

Lab Study D. Dissection of a Sensory Receptor, the Eye

Materials

preserved cow or sheep eye
dissecting instruments
dissecting pan
disposable gloves

Introduction

The goal of this study is not to perform a comprehensive study of eye structure but rather to identify those structures in the eye that enable it to receive stimuli and transmit the signals in sensory nerves to the central nervous system for processing. After processing, the signals are sent through motor nerves to the effector, bringing about the response.

The vertebrate eye is a complex sensory organ containing nervous tissue capable of being stimulated by light to produce nervous impulses. Sensory neurons carry these impulses to the brain, where they are interpreted, resulting in the perception of sight. The light-sensitive, or photoreceptor, cells in the eye are called *rods* and *cones.* They are the sensory part of a multilayered tissue, the retina. Other structures in the eye protect the retina and regulate the amount and quality of light stimulating the photoreceptor cells.

As you dissect an isolated eyeball from a cow or sheep, determine the contribution of each structure to the production of sight.

Procedure

1. Examine the isolated eye and notice that it is covered with fatty tissue and muscle bands except in the region of the **cornea**, a tough, transparent layer that allows light to enter the eye (Figure 24.8a). The cornea serves to protect the eye, and it also plays a role in focusing light entering the eye. It is supplied with many pain receptors, but has no blood supply.

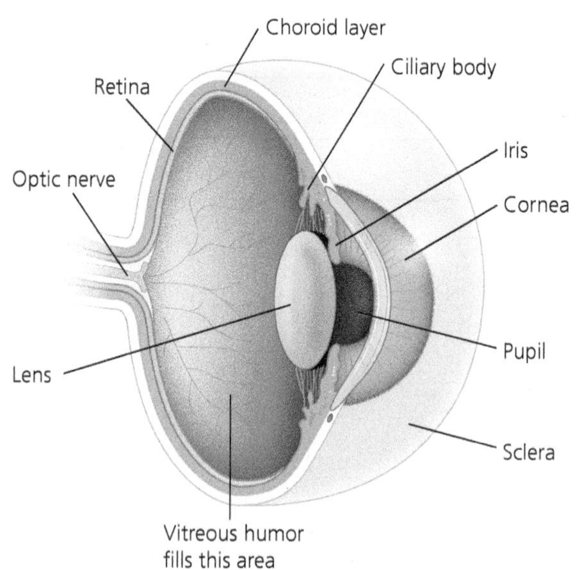

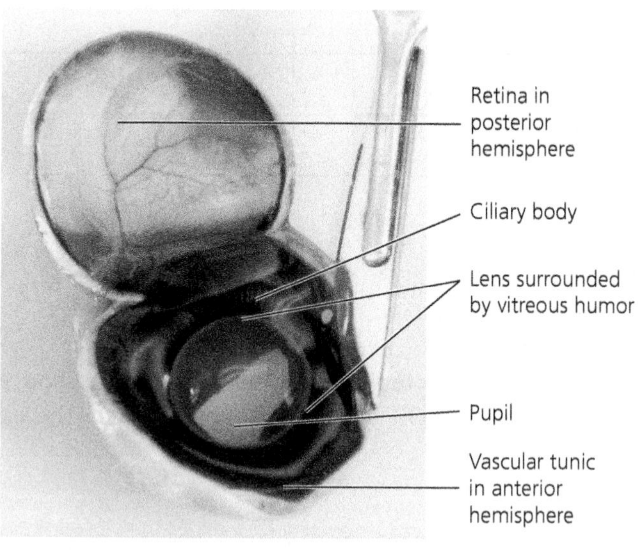

FIGURE 24.8

Anatomy of the eye. (a) Diagram of eye anatomy. (b) Cow eye opened to show structures.

2. Search through the fatty tissue on the eye sphere approximately opposite the cornea and locate the round stub of the **optic nerve** exiting the eyeball (Figure 24.8a). The optic nerve contains axons of sensory cells that transmit sensations from the eye to the brain.
3. Use forceps and scissors to trim away all fat and muscle on the eye surface, taking care to leave the optic nerve undisturbed.
4. Once the fat is removed, you will see that the cornea is the anterior portion of the tough, outer layer of the eyeball, the **fibrous tunic**. The posterior portion of this layer is the white **sclera** (Figure 24.8a). The fibrous tunic protects the internal eye structures.
5. Use scissors to cut the eye in half, making an equatorial incision and separating the anterior hemisphere (with the cornea) from the posterior hemisphere (bearing the optic nerve). Open the eye by placing it, nerve down, in the dissecting tray and lifting off the front hemisphere. Place this hemisphere in the tray with the cornea down. Your dissection should look like Figure 24.8b.
6. Identify the various structures in the anterior hemisphere.
   a. Identify the **lens**, a hard, oval-shaped structure that, together with the cornea, focuses the light on the retina (Figure 24.8a). In life, this is transparent. Surrounding and attached to the lens may be a jellylike clear substance, the vitreous humor, described in step 8.
   b. Identify the **ciliary body**, a dark, ridged, muscular structure surrounding and attached to the lens by thin ligaments. The ciliary body is a component of the second tunic of the eye, the darkly pigmented **vascular tunic**. Contraction of muscle fibers in the ciliary body changes the shape of the lens. What role does this process play in eye function?

7. Carefully remove the lens (and vitreous humor, if attached) and observe that the ciliary body merges anteriorly into another component of the vascular tunic, the **iris**. The iris surrounds an opening, the **pupil**. In the cow or sheep eye, the pupil is more irregular in shape than in the human eye. The pupil allows light to pass through the vascular tunic to the lens. The iris is a sphincter muscle. What is its function?

8. Turn your attention to the posterior hemisphere of the eye. If dissected as described, this hemisphere should hold the **vitreous humor**, which holds the retina in place and is the major internal support of the eye.
9. Using forceps, carefully remove the vitreous humor and identify the pale, delicate **retina**, the sensory layer, the third tunic of the eye. The retina contains microscopic rods and cones. What is the function of the retina?

10. The retina lies on the inside surface of the pigmented **choroid layer**, another component of the vascular tunic. The choroid layer absorbs extraneous light rays passing through the retina. Gently push the retina aside

and notice that it appears to be attached to the choroid layer in only one spot. This point of attachment is actually where processes from sensory neurons exit the retina as fibers of the **optic nerve.** Fibers from the two optic nerves travel to the brain where they enter at the optic chiasma, identified earlier in the study of the sheep brain. Look for a semicircular area of rainbow-colored tissue in the choroid layer. This is the **tapetum lucidum**, a tissue found in the choroid of some animals (but not humans) that enhances vision in limited light.

Results

Examining your dissected eye, list in sequence all tissues and structures in the eye through which light passes to create an image, beginning outside the eye and continuing through to the brain.

Discussion

1. Using your text, library sources, or the Web, describe each listed functional impairment of the eye.
   a. myopia (nearsightedness):

   b. hyperopia (farsightedness):

   c. astigmatism:

   d. detached retina:

   e. cataracts:

   f. presbyopia

   g. Which of the above impairments is (are) most likely to occur as a result of aging?

REVIEWING YOUR KNOWLEDGE

Complete Table 24.3, naming the three tunics of the eye and their subdivisions, if appropriate. Give the function of each subdivision.

TABLE 24.3 Eye Tunics and Their Functions

Tunic	Subdivision	Function

APPLYING YOUR KNOWLEDGE

1. How would hypertrophy (swelling) of the prostate gland (often a symptom of prostate cancer) affect functioning of the excretory system?

2. Normally, a fatty encasement surrounds the kidney, helping to maintain its normal position in the body. In cases of extreme emaciation in humans—for example, as in anorexia—the kidneys may drop to a lower position. Consider the ducts associated with the kidney and propose one side effect to the kidney that could result from severe weight loss.

3. A person who has lost a limb may experience phantom pain, the feeling of pain in the part of the body that is gone. Suggest an explanation for this phenomenon.

4. In the process of transplanting a kidney from a donor (living or cadaver) to a recipient whose kidneys are no longer functioning, the donor kidney is **not** placed in the position of the recipient's diseased kidneys as you might expect, but into the lower right abdominal cavity of the recipient. Refer to Figure 23.6 of the circulatory system. What veins and arteries are available in the recipient abdominal cavity to attach to the donor renal artery and vein?

CASE STUDIES

Extending their knowledge from this lab topic, students may investigate the following medical cases using various texts, the library, or Web resources. They may submit their findings in written or oral reports.

1. Consult a text and write a definition of *homeostasis*. It is said that disease is the failure to maintain homeostatic conditions in the body. Investigate disorders or diseases that may result when homeostasis is disrupted owing to problems in the respiratory, digestive, circulatory, nervous, or excretory systems.
2. Kate has peritonitis, which she was told resulted from a urinary tract infection. What is peritonitis? Why would this result from a urinary tract infection in females but not males? Discuss the occurrence of this disease in relation to the anatomy of the female and male reproductive tracts.
3. Recently scientists have suspected a correlation between Zika virus disease in pregnant women and an increase in the frequency of microcephaly seen in newborn babies, particularly in South America where this disease is somewhat common. Using the Web, research the information that is known about the correlation between microcephaly and Zika virus infection, and its impact on the developing brain.
4. Lasik surgery (laser eye surgery) has become increasingly popular since its approval by the U.S. Food and Drug Administration in the early 1990s. What structures in the eye are modified in Lasik surgery, and how are they modified? What conditions in the eye can be corrected by this surgery? Can presbyopia be corrected by Lasik surgery? What are the potential risks and complications of this surgery?

STUDENT MEDIA: BioFlix, Activities, Investigations, and Videos

www.masteringbiology.com (select Study Area)

BioFlix—Ch. 48: How Synapses Work

Activities—Ch. 44: Structure of the Human Excretory System; Nephron Function; Control of Water Reabsorption; Ch. 46: Reproductive System of the Human Female; Reproductive System of the Human Male; Ch. 48: Neuron Structure; Ch. 49: Neuron Structure; Ch. 50: Structure and Function of the Eye

Investigations—Ch. 46: What Might Obstruct the Male Urethra?

Videos—Ch. 46: Ultrasound of Human Fetus 1; Ultrasound of Human Fetus 2

REFERENCES

Armitage, H. "Lymphatic vessels: The brain's well-hidden secret." *Science*, vol. 350, 2015, p. 1462.

Fawcett, D. W. and R. P. Jensh. *Bloom and Fawcett: Concise Histology*, 2nd ed. Cambridge, MA: Oxford University Press, 2002. The successor to Bloom and Fawcett's classic *Textbook of Histology*.

Marieb, E. N. and K. Hoehn. *Human Anatomy and Physiology*, 10th ed. San Francisco, CA: Pearson, 2016.

Werbin, F. and B. Rosha. "The Movies in Our Eyes." *Scientific American*, April 2007.

Yuste, R. and G. Church. "The New Century of the Brain." *Scientific American*, March 2014, pp. 38–45.

WEBSITES

A Web search for "reflex arc" will yield many informative websites:
http://www.google.com

Cold Spring Harbor Laboratory DNA Learning Center, access the G2C (Genes to Cognition) Online 3D Brain, an interactive brain map that allows users to rotate the brain in three-dimensional space:
https://www.dnalc.org/resources/3dbrain.html

Information on Lasik eye surgery:
http://www.allaboutvision.com/visionsurgery/

Two sites with photographs of fetal pig anatomy. Students may use these photos to review for quizzes:
https://www.biologycorner.com/myimages/fetal-pig-dissection/
http:www.home.apu.edu/~jsimons/Bio101/PigDissectionGuide.htm

Vertebrate eye anatomy:
http://www.webmd.com/eye-health/tc/eye-anatomy-and-function-topic-overview#BM_Topic_Overview

Website for the National Eye Institute with information on eye anatomy, diseases, and disorders:
http://www.nei.nih.gov/health

LAB TOPIC 25

Animal Development

Laboratory Objectives

After completing this lab topic, you should be able to:

1. Describe early development in echinoderms (sea urchin and sea star), an amphibian (frog or salamander), a fish (zebrafish), and a bird (chicken).
2. List the events in early development that are common to all organisms.
3. Compare early development in the organisms studied, speculating about factors causing differences.
4. Relate the events of early development in vertebrates to the formation of a dorsal nerve cord.
5. Discuss the effects of large amounts of yolk on the events of early development.

Introduction

The development of a multicellular organism involves many stages in a long process beginning with the production and fusion of male and female gametes, continuing with the development of a multicellular embryo, the emergence of larval or juvenile stages, growth and maturation to sexual maturity, and the process of aging, and eventually ending with the death of the organism. A range of biological processes functions in development, including **cell division**; **differentiation**, where cells, tissues, and organs become specialized for a particular function; and **morphogenesis**, the development of the animal's shape, or body form, and organization.

Early developmental investigations focused on the description of form, or **morphology**, of animals as they grow. Studies using several model organisms contributed to our understanding of the process of development and the forces involved in morphogenesis. These model organisms included *Drosophila melanogaster,* the fruit flies that you used for genetics studies in Lab Topic 9, the sea urchin and sea star (echinoderms), frogs and salamanders (amphibians), and the chick. More recently, the roundworm, *Caenorhabditis elegans*, the zebrafish, and other organisms have become important subjects for developmental studies. Currently, geneticists and developmental biologists use these same animals to ask questions about the genetic control of development and the processes involved in activating different genes in different cells. Before these questions can be pursued, however, it is important to have a basic understanding of morphology in early development.

In this lab topic, you will use several of the model organisms of classical and current research to investigate major early developmental events common to most animals. These events include **gametogenesis**, the production of gametes; **fertilization**, the union of male and female gametes to form a diploid zygote;

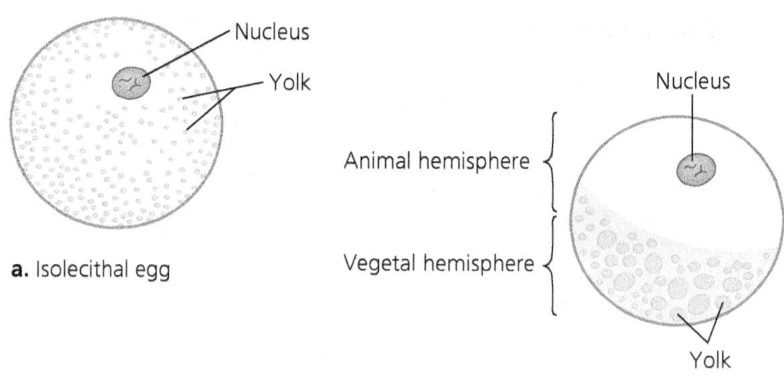

FIGURE 25.1
Egg types based on amount and distribution of yolk. (a) Isolecithal eggs have small amounts of evenly distributed yolk. (b) Telolecithal eggs have large amounts of yolk concentrated at one end.

cleavage, when cells divide by mitosis to produce a multicellular blastula; **gastrulation**, the formation of three primary germ layers—ectoderm, mesoderm, and endoderm; **neurulation**, the formation of the nervous system in chordates; and **organogenesis**, the development of organs from the three primary germ layers. Although you may observe all of these developmental stages, you will study primarily cleavage and blastulation, gastrulation, neurulation, and organogenesis.

Overview of Stages in Early Development

Stage 1: Preparation of the Egg, Fertilization, and Cleavage

Development begins as sperm and egg prepare for fertilization. Sperm develop a flagellum, which propels the cell containing the haploid genetic complement of the paternal parent to the egg, which contains the haploid maternal genetic complement. As an egg matures, sperm-binding receptors develop on its surface. These can only bind to matching receptors in the sperm, ensuring that eggs will be fertilized by sperm of the same species. Within the egg, food reserves called **yolk** accumulate. These reserves are mainly protein and fat and will be utilized by the early embryo.

When egg and sperm unite, their nuclei, each containing a haploid set of chromosomes, combine to form one diploid cell, the **zygote**. The mitotic cell divisions of cleavage rapidly convert the zygote to a multicellular ball, or disc, called the **blastula**. The cells of the blastula are called **blastomeres**. A cavity, the **blastocoel**, forms within the ball of cells. The position of the blastocoel in the developing blastula depends on the amount and distribution of yolk in the egg and developing embryo.

Egg Types

Because early events in development are affected by the amount of yolk present in the egg, the classification of eggs is based on the amount and distribution of yolk. Eggs with small amounts of evenly distributed yolk are called **isolecithal**

FIGURE 25.2 (at right)
Cleavage types based on amount and distribution of yolk. (a) In isolecithal eggs, cleavage is holoblastic, and the blastocoel is centrally located. (b) In moderately telolecithal eggs, cleavage is holoblastic, and the blastocoel develops in the animal hemisphere. (c) In strongly telolecithal eggs, cleavage is meroblastic. Only the active cytoplasm divides, producing a cap of cells, the blastoderm. The blastocoel forms within the blastoderm.

Holoblastic cleavage

a. Isolecithal egg

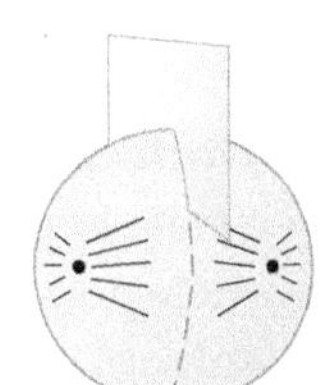
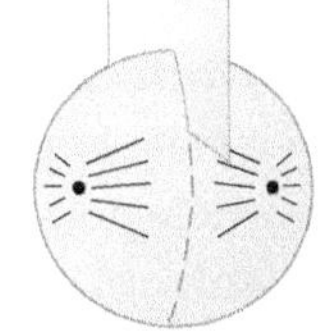

Zygote

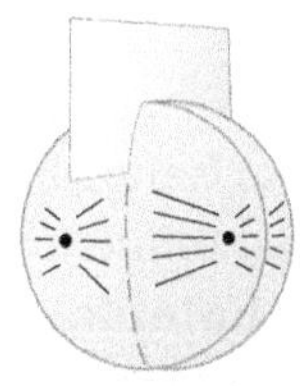

2-cell stage

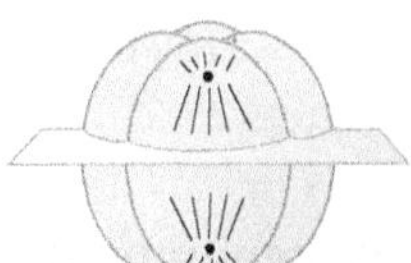

4-cell stage

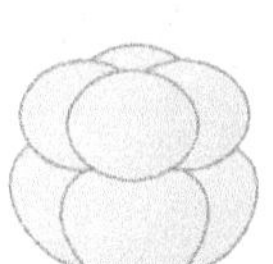

8-cell stage

16-cell stage

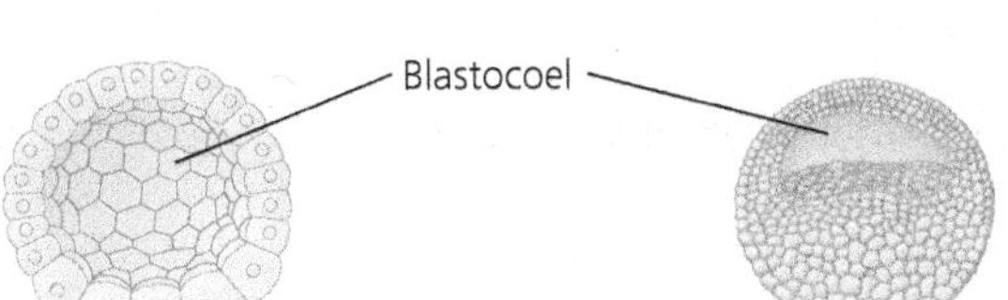

b. Moderately telolecithal egg

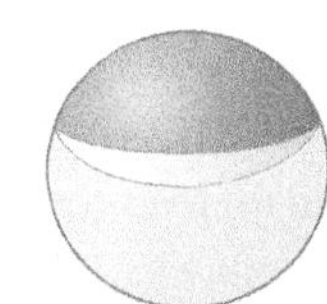
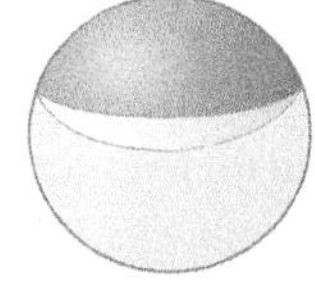

Zygote

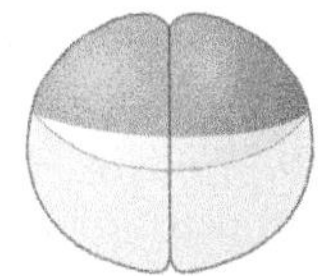

2-cell stage

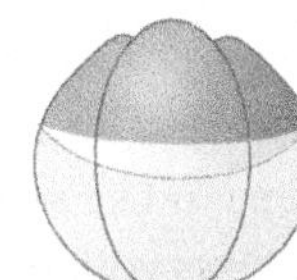

4-cell stage

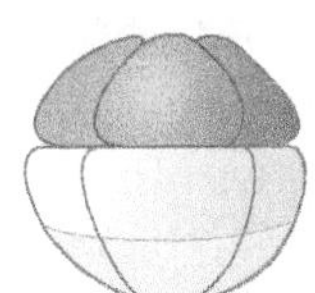

8-cell stage

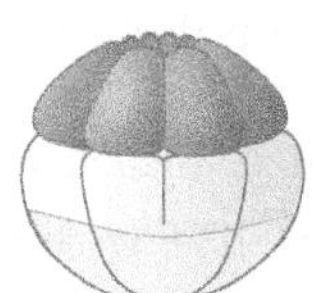

16-cell stage

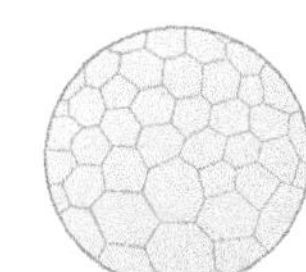

Meroblastic cleavage

c. Strongly telolecithal egg

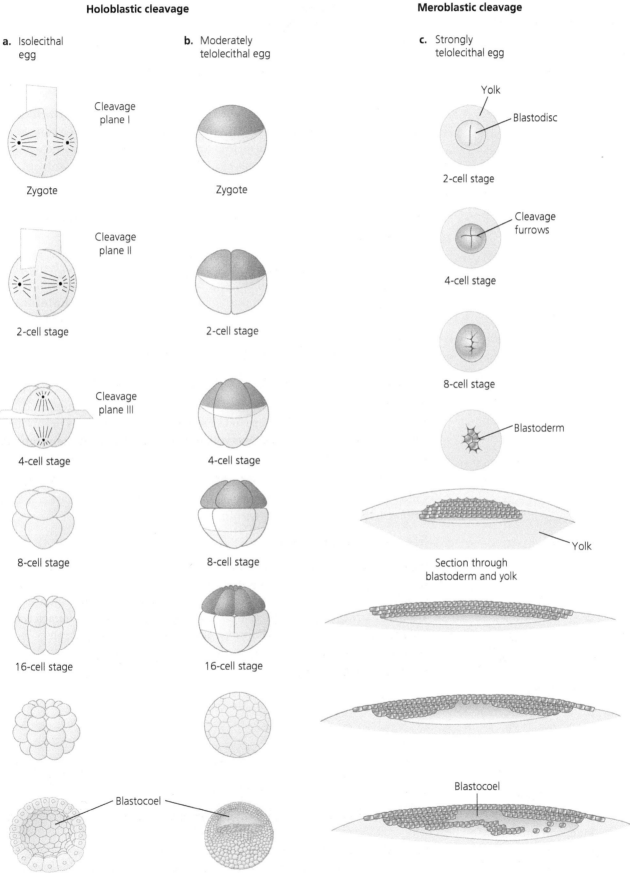

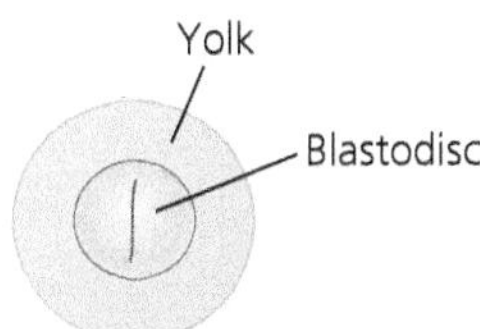

2-cell stage

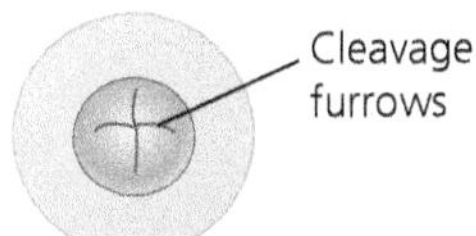

4-cell stage

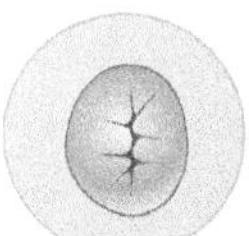

8-cell stage

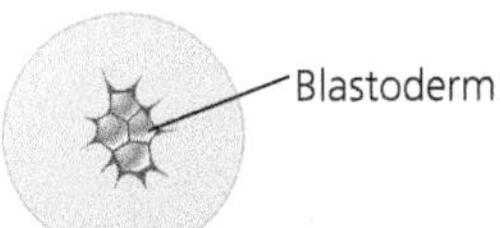

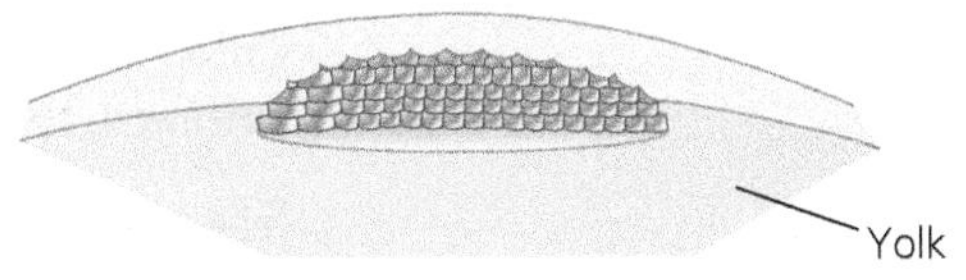

Section through blastoderm and yolk

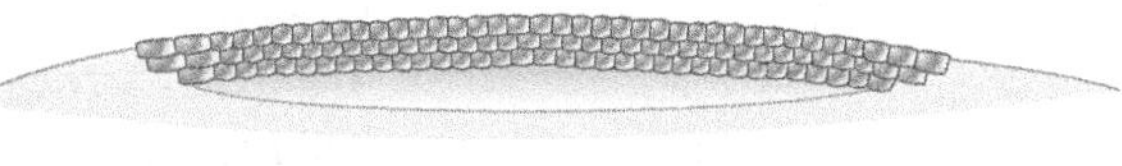

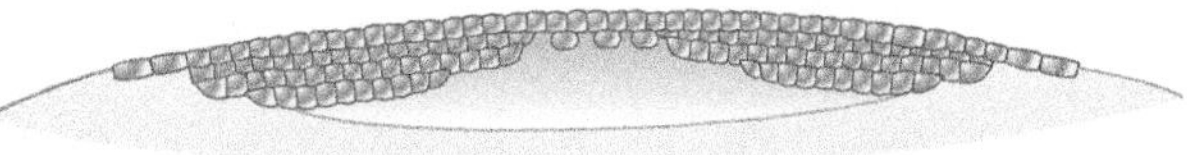

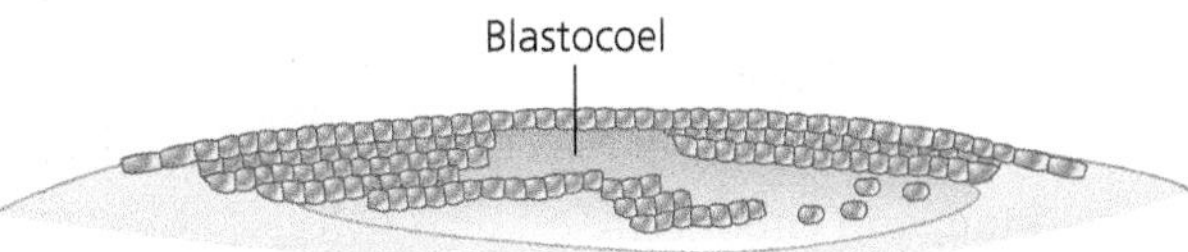

eggs (Figure 25.1a). Eggs containing large amounts of yolk concentrated at one end are called **telolecithal** eggs (Figure 25.1b). Some species are moderately telolecithal, whereas others are strongly telolecithal. Eggs may also be classified as **centrolecithal** (yolk in the center of the egg surrounded by yolk-free cytoplasm, as in insects) or **alecithal** (no significant yolk reserves, as in humans and other mammals). Neither of these conditions will be studied in this lab topic.

In strongly telolecithal eggs, the nucleus is surrounded by **active cytoplasm**, which is relatively devoid of yolk. This nuclear-cytoplasmic region is called the **blastodisc**. The blastodisc is displaced toward the pole of the egg where polar bodies budded from the cell in meiosis. This pole is designated the **animal pole**. The half of the egg associated with the animal pole is the **animal hemisphere**. In these eggs, the yolk is concentrated in the other half of the egg, designated the **vegetal hemisphere**. The pole of this hemisphere is the **vegetal pole**.

Cleavage Types

Although the end result of cleavage, the formation of the blastula, is the same in all organisms, the pattern of cleavage can differ. One factor that influences the pattern of cleavage is the amount of yolk present. In total, or **holoblastic**, cleavage, cell divisions pass through the entire fertilized egg. This type of cleavage takes place in isolecithal eggs, where the impact of yolk is minimal (Figure 25.2a). In these eggs, the early blastomeres are similar in size and the blastocoel forms in the center of the blastula. In moderately telolecithal eggs, the yolk will retard cytoplasmic divisions and affect the size of cells. However, if the entire egg is cleaved, cleavage is considered holoblastic (Figure 25.2b). In this case, the blastocoel develops in the animal hemisphere. Cells in this hemisphere will be smaller and have less yolk than cells in the vegetal hemisphere.

In a strongly telolecithal egg, only the active cytoplasm is divided during cleavage. This process is called **meroblastic** cleavage, and it produces a cap of cells called a **blastoderm** at the animal pole (Figure 25.2c). In meroblastic embryos, the blastocoel forms within the cell layers of the blastoderm.

Stage 2: Gastrulation

Gastrulation transforms the blastula, the hollow ball of cells (in holoblastic cleavage) or cap of cells (in meroblastic cleavage), into a **gastrula** made up of three embryonic, or germ, layers: endoderm, ectoderm, and mesoderm (Figure 25.3). Whereas cleavage is characterized by cell division, gastrulation is characterized by cell movement, including cell *migration* and *invagination* (a layer of cells folds into the blastocoel). Surface cells migrate and invaginate into the interior of the embryo forming a new internal cavity, the **archenteron**.

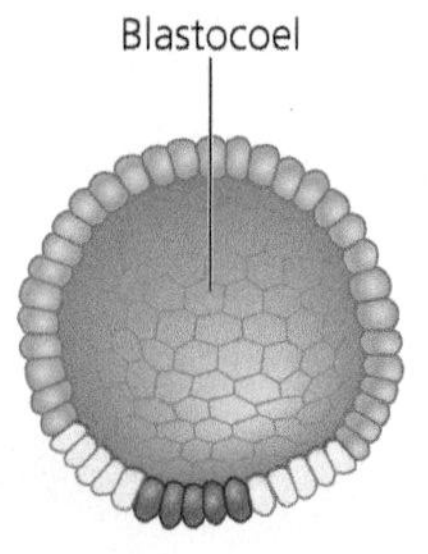

a. Blastula

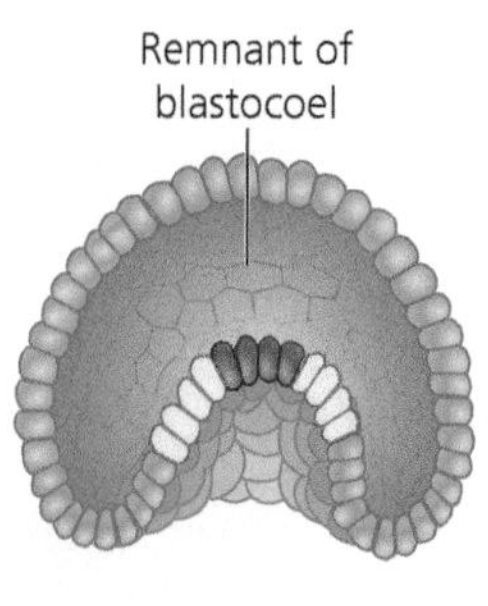

b. Early gastrula

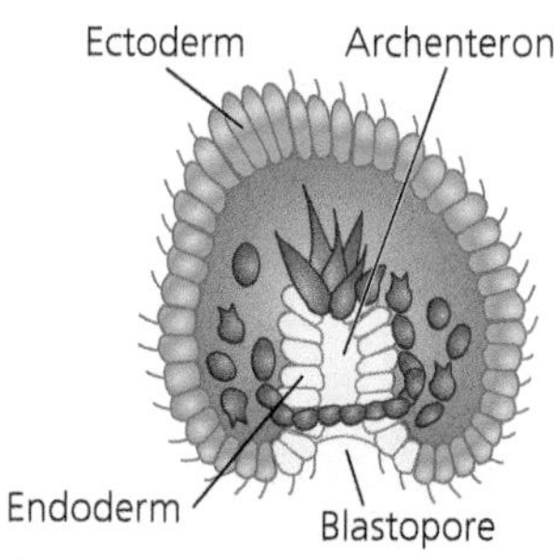

c. Late gastrula

FIGURE 25.3
Gastrulation. The blastula is converted to a three-layer embryo. Ectoderm (blue) and endoderm (yellow) germ layers form first. Mesoderm (red) forms later between the ectoderm and endoderm.

This new cavity is lined by the **endoderm**, the embryonic germ layer that ultimately forms the digestive tract. The archenteron opens to the outside through the **blastopore**, which in deuterostomes becomes the anus. The invertebrate group Echinodermata and all chordates are deuterostomes. In protostome development, the blastopore becomes the mouth. Mollusks, annelids, nematodes, and arthropods are protostomes. The cells that remain on the surface of the embryo become the **ectoderm**. A third layer of cells, the **mesoderm**, develops between ectoderm and endoderm.

Stage 3: Neurulation

Late in gastrulation, neurulation, the formation of a dorsal, hollow neural tube, begins (Figure 25.4). In this *strictly chordate* process, certain ectodermal cells flatten into an elongated **neural plate** extending from the dorsal edge of the blastopore (or from the primitive streak in the chick embryo) to the anterior end of the embryo. The center of the plate sinks, forming a **neural groove**. The edges of the plate become elevated to form **neural folds**, which approach each other, touch, and eventually fuse, forming the hollow **neural tube**. The anterior end of the tube develops into the brain, while the posterior end develops into the nerve (spinal) cord.

Stage 4: Organogenesis

After the germ layers and nervous system have been established, organogenesis, the formation of rudimentary organs and organ systems, takes place. Ectoderm, the source of the neural tube in chordates, also forms skin and associated glands. In chordates, somites and the notochord (an internal supportive rod running along the dorsal side of the body) develop early from mesodermal cells. Later, muscles, the skeleton, gonads, the excretory system, and the circulatory system develop from mesoderm. Nonchordate animals lack somites and the notochord, but their muscles, and organs of the excretory, circulatory, and reproductive systems, develop from mesoderm. The endoderm develops into the lining of the digestive tract and such associated organs as the liver, pancreas, and lungs.

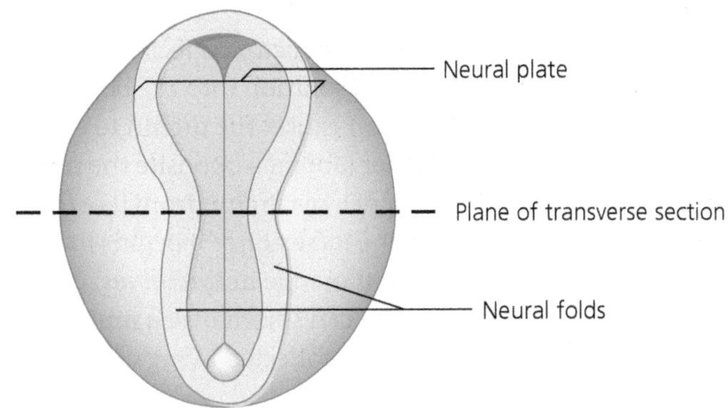

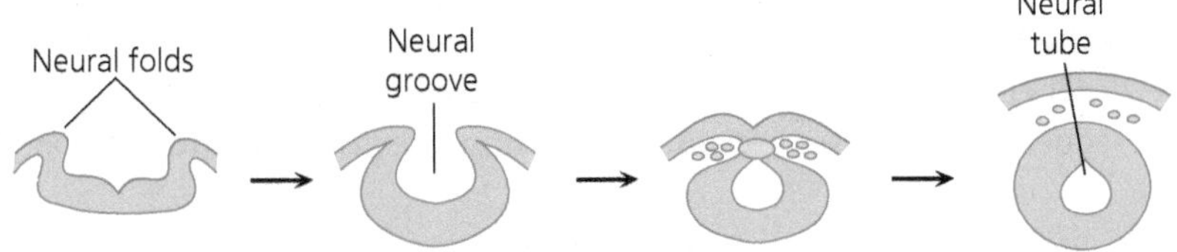

FIGURE 25.4
Neurulation in a developing frog, a chordate. (a) Dorsal view of the entire frog embryo, showing the ectodermal neural plate with edges elevated, forming the neural folds. (b) Seen in transverse section, the neural folds meet and fuse, forming the neural tube.

Today's lab will be a comparative study of early development in five organisms—the sea urchin and the sea star, both echinoderm invertebrates; and three chordates: the salamander, an amphibious vertebrate; the fish, an aquatic vertebrate; and the chick, a terrestrial vertebrate.

EXERCISE 25.1

Development in Echinoderms: Sea Urchin and Sea Star

Sea urchins and sea stars (starfish) are classified in the phylum Echinodermata, the invertebrate group that is phylogenetically closer to chordates than any other. Echinoderms release large numbers of gametes into the sea, and fertilization is external. Early development leads to a larval stage that is free-swimming and free-feeding. In this exercise, you will observe fertilization and early development in living sea urchins, and then you will observe a prepared slide of sea star development.

Lab Study A: Fertilization in Living Sea Urchins

Materials

clean slides and coverslips
sand or glass chips in a small petri dish
transfer pipettes, one labeled "egg," the other labeled "sperm," cut to make a slightly larger bore
small clean test tube labeled "egg" containing a suspension of living eggs from a sea urchin, e.g., *Lytechinus* sp. or *Arbacia* sp.
small clean test tube labeled "sperm" containing a suspension of living sperm from a sea urchin, e.g., *Lytechinus* sp. or *Arbacia* sp.
moisture chamber made from a petri dish containing a piece of moist (not wet) filter paper

Introduction

Before your laboratory began, your instructor collected sperm and eggs from living sea urchins by injecting a KCl solution into the body cavity of the urchin. An injection of this solution causes the urchin to extrude gametes from its genital pore located on its upper (aboral—opposite the mouth) surface. It is not possible to determine the sex of a sea urchin from its external anatomy. However, it is possible to determine its sex once it begins to extrude its gametes. Whereas eggs are colored (e.g., light orange in *Lytechinus*), sperm in all species are whitish. Each female sea urchin can spawn over a million eggs and each male a billion sperm. Your instructor collected living eggs and sperm in sea water.

An unfertilized echinoderm egg is surrounded by a protective *vitelline layer* just outside the plasma membrane and a **jelly coat** that slowly dissolves in sea water. When eggs and sperm come into contact, fertilization takes place and a halo—the **fertilization envelope**—forms around the fertilized egg. (See Figure 25.6.) This envelope helps prevent multiple fertilizations, or *polyspermy*. Two sequential processes prevent polyspermy, the *fast block* and the *slow block*.

When a sperm first fuses with an egg, the permeability of the egg plasma membrane immediately changes, allowing an influx of sodium ions. The sodium ions change the electric potential across the cell membrane and, within a second or so, create a *fast block to polyspermy.* A second block to multiple fertilization takes about 20 to 30 seconds. This *slow block to polyspermy* involves the fusion of egg cytoplasmic vesicles with the egg plasma membrane. These vesicles lie in the cortex, or outer portion of the egg cytoplasm, and are called *cortical granules.* When they fuse with the egg membrane, their contents are expelled to the egg surface. Enzymes from the vesicles break bonds between the vitelline layer and the egg plasma membrane, and water flows between the two layers. The vitelline layer rises up from the egg membrane and becomes the *fertilization envelope.* Some time after the egg and sperm cells fuse, the egg and sperm nuclei, called *pronuclei,* move toward the center of the egg, where they fuse and almost immediately begin to prepare for the first mitotic division of cleavage.

In this lab study you will observe fertilization and early development. In Lab Study B, you will learn more details about the process and early stages of development in the sea star.

Procedure

Place a generous drop of egg suspension on a clean microscope slide. Add a few grains of sand or glass chips and cover with a coverslip. Place the slide on your compound microscope stage and observe the eggs using first 4× and then 10× objectives.

1. Take a drop of sperm suspension and add this to the edge of the coverslip.
2. Working quickly, note the time and immediately begin observing the slide using first the 10× and then the high power objectives. Use phase-contrast microscopy on both powers if your microscope has this capability. Look for sperm swimming toward the eggs.
3. Carefully observe the eggs, watching for the formation of fertilization envelopes (a clear halo surrounding the egg), indicating that fertilization has taken place.
4. After observing fertilization for several minutes, place your slide in a moisture chamber made from a petri dish.
5. At 30-minute intervals until the end of lab, remove your slide from the petri dish, carefully dry off the bottom of the slide, and use the compound microscope to observe developing embryos. After you complete your observations, return the slide to the moisture chamber each time.

Results

1. Describe the events that take place as sperm and eggs are mixed. How long did it take for fertilization envelopes to become visible?
2. Are fertilization envelopes present around all eggs? If not, can you estimate the percent of eggs that have been fertilized?
3. Describe the activity of the sperm. Do active sperm bounce off the egg, or do they become stuck to the jelly coat?

4. Did you observe cleavage? How long after you first observed fertilization? Was the cleavage pattern *holoblastic* or *meroblastic* (Figure 25.2)?

5. At the end of the laboratory period, at what stage of development are most of your embryos?

Discussion

1. What role might the jelly coat surrounding the egg perform in the process of fertilization?

2. What would you predict would happen if sperm were exposed to a solution containing only egg jelly coat? Would these sperm still be capable of fertilization?

Student Media Video—Ch. 47: Sea Urchin Embryonic Development

Lab Study B: Development in the Sea Star

Materials

compound microscope
prepared slide of whole sea star embryos in different stages of development

Introduction

In this lab study, you will observe a prepared slide containing an assortment of whole sea star embryos in various stages of development. You will identify each developmental stage and determine the type of egg and cleavage pattern of the sea star.

Procedure

1. View the prepared slide of sea star embryos using low and intermediate powers on the compound microscope.

Use only low and intermediate powers. Using the high power objective to view this slide will destroy the slide!

2. Find examples of all stages of development. When you find a good example of each of the stages described, make a careful drawing of that stage in the appropriate square in Figure 25.12 at the end of this exercise.

Unfertilized Egg

By the time sea star eggs leave the body of the female, meiosis I and II are completed. The nucleus, called the *germinal vesicle,* is conspicuous because the nuclear envelope is intact (Figure 25.5). A nucleolus is usually distinct. The plasma membrane surrounding the egg cytoplasm closely adheres to a thin external layer known as the *vitelline layer* (not visible at this magnification). Species-specific sperm receptors extend from the egg plasma membrane into the vitelline layer.

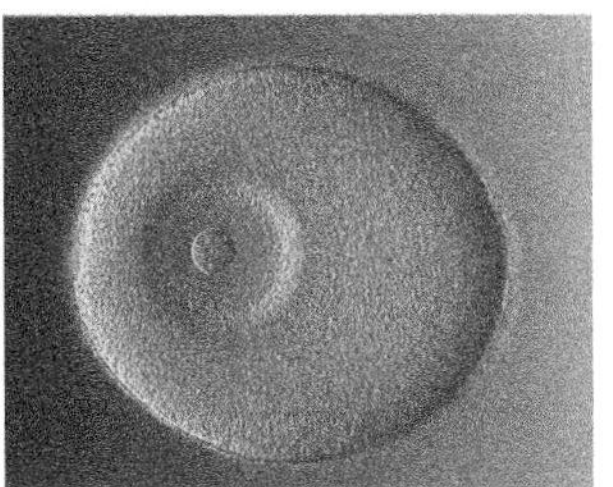

FIGURE 25.5
Unfertilized egg with germinal vesicle.

Fertilized Egg

The fertilized egg, or zygote, has no visible nuclear envelope, giving this cell a uniform appearance (Figure 25.6). Look on the zygote surface for a **fertilization envelope,** most easily seen using phase-contrast microscopy. In Lab Study A you learned that this envelope forms as a result of sperm–egg fusion. The presence of the fertilization envelope and the absence of the visible nuclear envelope will help distinguish fertilized and unfertilized eggs.

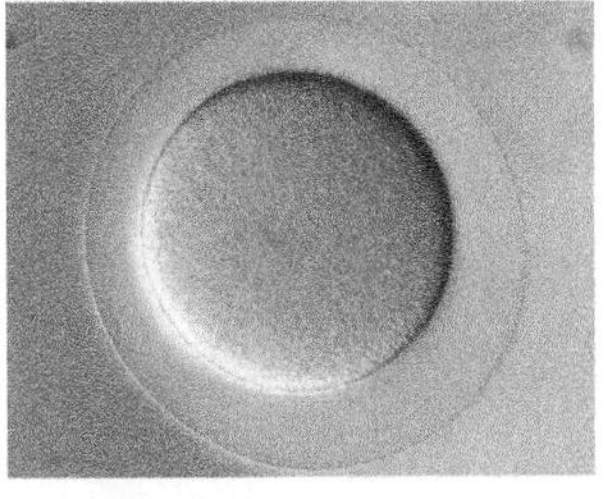

FIGURE 25.6
Fertilized egg.

Early Cleavage

As cleavage begins, the zygote divides by mitosis and cytokinesis (Figure 25.7) and continues to divide as this single cell is converted into a multicellular embryo. The G_1 and G_2 phases of the cell cycle (see Lab Topic 7, Mitosis and Meiosis) are essentially skipped in these mitotic events. Find two-, four-, and eight-cell stages. Is the entire zygote involved in early cleavage?

The fertilization envelope remains intact around the embryo until the gastrula stage. What is happening to the size of the cells as cleavage takes place and cell numbers increase?

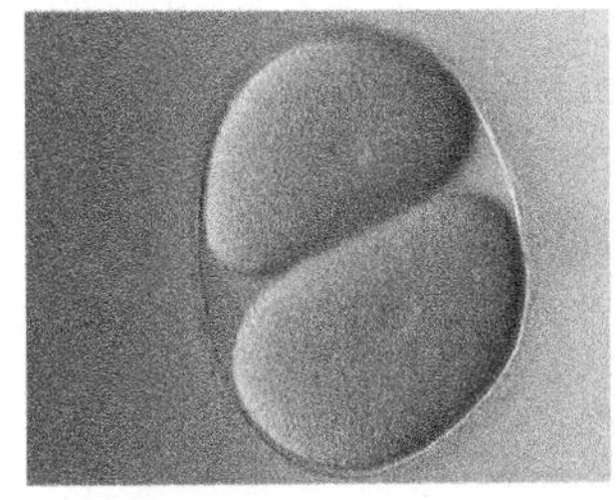

FIGURE 25.7
Early cleavage—two-cell stage.

Late Cleavage

As cleavage continues, a cavity, the **blastocoel**, forms in the center of the cell cluster (Figure 25.8). The end product of cleavage will be a hollow ball of cells, the **blastula**. Locate and study blastulae in early and late stages of cleavage. The late blastula has a thick, dark wall of cells. The lighter blastocoel lies in the center of the blastula. How does the size of individual cells compare with the size of the fertilized egg?

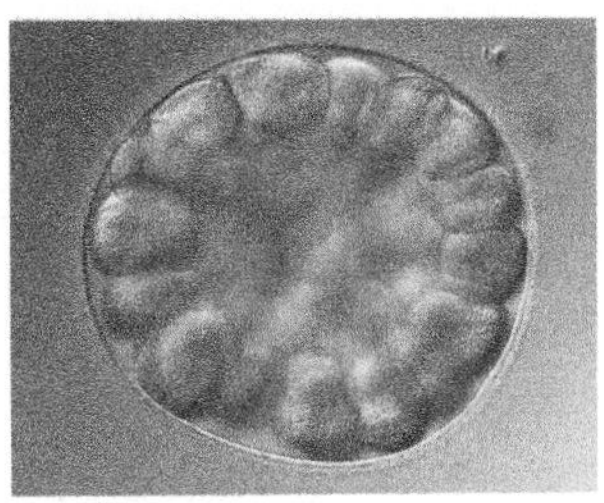

FIGURE 25.8
Later cleavage. Blastula with blastocoel.

How does the overall size of the blastula compare with that of the fertilized egg?

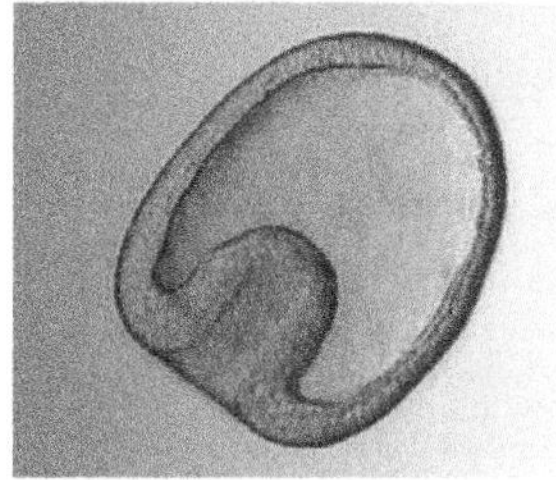

FIGURE 25.9
Early gastrulation.

Early Gastrulation

Gastrulation converts the blastula into the gastrula, an embryo composed of three primary germ layers. The early gastrula can be recognized by a small bubble of cells protruding into the blastocoel (Figure 25.9). These cells push into the blastocoel through a region on the embryo surface called the **blastopore**. As cells continue to *invaginate*, or move inward, a tube called the **archenteron** forms. The archenteron eventually becomes the adult gut. Which embryonic germ layer lines the archenteron?

Middle Gastrulation

The archenteron continues to grow across the blastocoel. It takes on a bulblike appearance as the advancing portion swells.

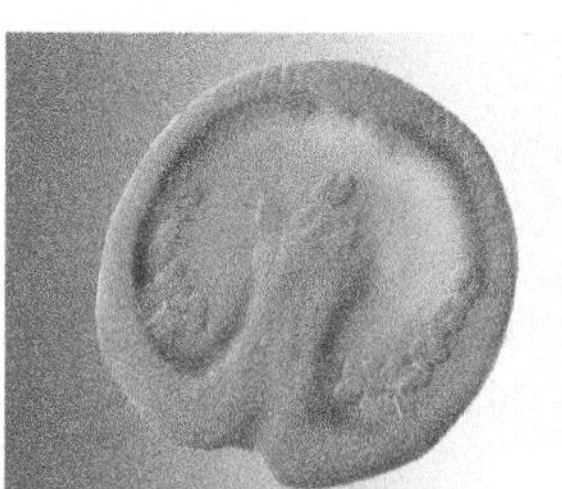

FIGURE 25.10
Late gastrulation.

Late Gastrulation

Cells at the leading edge of the advancing archenteron extend *filopodia* (cellular processes that can extend and contract) that attach to a specific region across the blastocoel (Figure 25.10). As the filopodia make contact, these cells continue to pull the archenteron across the blastocoel. As the tip of the archenteron approaches the opposite wall of the embryo, it bends to one side and fuses with surface cells. The site of fusion will eventually become the mouth of the embryo. What will be formed from the blastopore at the opposite end of the archenteron? (Recall that echinoderms are deuterostomes.)

What is the germ layer of cells on the surface of the embryo called?

The amoeboid cells at the leading edge and surrounding the archenteron are called *mesenchyme cells*. These cells later detach from the archenteron, proliferate, and form a layer of cells within the old blastocoel, now divided by the archenteron. This layer of cells will become the mesodermal germ layer.

Bipinnaria Larval Stage

The archenteron of the gastrula differentiates into a broad **esophagus** leading from the **mouth** to a large oval **stomach** and on to a small, tubular **intestine**. All these structures will be visible in the bilaterally symmetric bipinnaria larva (Figure 25.11). Locate these structures in larvae on your slide. The larva is now self-feeding and begins to grow. It will later be transformed into the radially symmetric adult sea star.

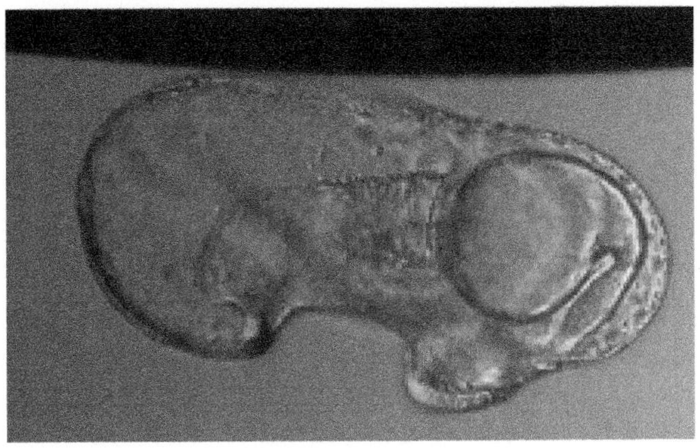

FIGURE 25.11

Bipinnaria larva. Identify the mouth, esophagus, stomach, intestine, and anus.

Results

Draw stages of sea star development in the appropriate boxes in Figure 25.12 on the next page.

Discussion

1. Sea urchins may release their eggs in tidal pool communities that include millions of eggs of different species at a given time. What is the advantage of species-specific sperm receptors in the plasma membrane and vitelline layer of an egg?

2. What type of egg does the sea star have? What evidence have you observed that supports your answer?

3. Describe the pattern of cleavage seen in the sea star and give the name for this type of cleavage.

4. At what stage in the sea star embryonic development did you first observe two layers of cells?

a. Unfertilized egg	**b.** Fertilized egg
c. Early cleavage	**d.** Late cleavage
e. Early gastrulation	**f.** Middle gastrulation
g. Late gastrulation	**h.** Bipinnaria larva

FIGURE 25.12 **Draw early stages of development in the sea star as observed under the microscope.**

EXERCISE 25.2

Development in an Amphibian

Materials

video or film of early development in a salamander or some other amphibian

Introduction

Amphibians are vertebrates that lay jelly-coated eggs in water or in moist areas on land. Common examples include frogs and salamanders. For most species, fertilization is external, with the male depositing sperm over the eggs after the female releases them. Internal fertilization takes place in some amphibians, however, in which cases the young are born in advanced developmental stages. Early development is similar in all species. After fertilization, the zygote begins cleavage followed by gastrulation, neurulation, and organogenesis.

In this exercise, you will study early development in an amphibian by observing a video or film presentation of some species, such as the salamander *Triturus alpestris* or the frog *Xenopus laevis*. The film or video shows dramatic time-lapse photography of cleavage, gastrulation, and neurulation.

Procedure

1. Before viewing the film or video, complete Table 25.1 by defining terms commonly used when describing early embryos. (You may need to refer to your text.)
2. Read the questions in the Results section, view the film or video, and then answer the questions.

TABLE 25.1 Common Terms Used in Embryology

Term	Definition
Animal pole	
Animal hemisphere	
Equator	
Vegetal hemisphere	
Vegetal pole	

Results

1. Would you describe the amphibian egg as isolecithal, moderately telolecithal, strongly telolecithal, or alecithal?

2. Describe the cleavage pattern. Is it holoblastic or meroblastic? Are cleavages synchronous or irregular? Can you detect any particular pattern in the cleavage? Where is the second cleavage plane in relation to the first?

3. Does the size of the embryo change as cleavage progresses?

4. Visually follow surface cells during gastrulation. Note that cells on the surface of the blastula spread out and move through the blastopore to the interior of the embryo, forming the gastrula. Do the surface cells all move at the same rate? Describe gastrulation, comparing the process with that in the sea star. Notice the position of the blastopore and the yolk plug located in the blastopore.

5. During neurulation, do the neural ridges (folds) meet and fuse simultaneously along the entire length of the neural tube or do they close like a zipper?

Discussion

1. Name at least two major differences in early development between the salamander and the sea star and describe factors responsible for these differences.

2. Compare the video of amphibian development with Figure 25.13. Label the following in the appropriate figure: **animal pole**, **vegetal pole**, **blastocoel**, **archenteron**, **yolk plug**, **neural plate**, and **neural folds**.

Student Media Video—Ch. 47: Frog Embryo Development

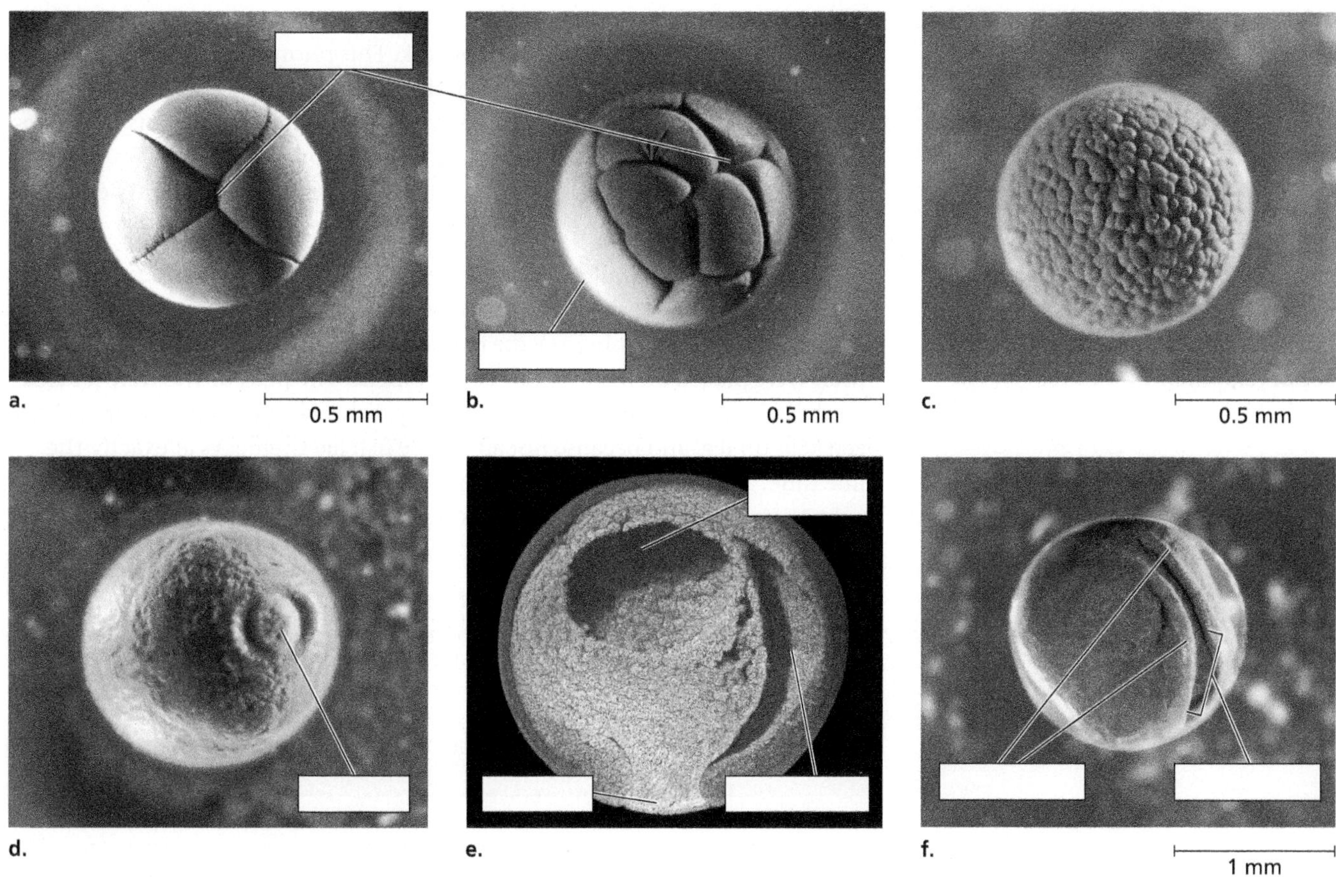

FIGURE 25.13

Early amphibian development. (a) and (b) Early cleavage. (c) Surface view of the hollow blastula in late cleavage. (d) Surface view of a gastrula. (e) Cross section of a gastrula. (f) Surface view of a neurula.

EXERCISE 25.3

Development in the Zebrafish

Materials

small petri dishes
transfer pipettes
depression slides
embryo-rearing solution
clean toothpicks
fish embryos in various developmental stages (some on ice)
stereoscopic microscope
compound microscope

Introduction

In this portion of the laboratory, you will observe living embryos in early developmental stages of a freshwater fish commonly known as the zebrafish, or zebra danio (*Danio rerio*) (Figure 25.14). The natural habitat of these popular aquarium fish is streams in India. Male and female fish are similar in appearance, but the male is generally smaller, with a streamlined body shape. The female is larger and broader than the male, especially when carrying eggs.

FIGURE 25.14
Zebrafish, *Danio rerio.* These fish have become important organisms for developmental studies.

In nature, zebrafish are stimulated to reproduce when days consist of approximately 16 hours of light and 8 hours of dark. This photoperiod corresponds to favorable weather and food supplies for developing embryos. By artificially creating this photoperiod in the laboratory, we can produce conditions that stimulate the zebrafish to spawn. After only 2 or 3 days on a cycle of 16 hours of light and 8 hours of dark, female fish will lay eggs, and male fish will deposit sperm for external fertilization.

The embryos you will observe today were collected from zebrafish on the artificial schedule. Newly spawned embryos were placed in embryo-rearing solution in petri dishes. Some embryos were maintained at room temperature, while others were placed on ice to retard development. The petri dishes were labeled to indicate the approximate stage of development. Because some embryos have been kept on ice, and because not all female fish lay their eggs at exactly the same time, a variety of early developmental stages should be available for your study. Neurulation and organogenesis stages are available from yesterday's spawning.

The approximate schedule of development at 25°C is described in Table 25.2. Use this schedule to predict the approximate stage of development for eggs

TABLE 25.2 Developmental Schedule for Zebrafish at 25°C

Hour	Time	Comments
0	8 A.M.	Lights on. Fish stimulated to spawn. Fish begin to dart back and forth, depositing eggs and sperm (spawning) close to the bottom of the aquarium. Fertilization takes place, forming the zygote. Cleavage begins in 35 minutes.
1	9 A.M.	Cleavage continues. Some embryos are collected and placed on ice to slow development.
2	10 A.M.	Embryos are in midblastula stage (approximately 64 cells).
4	12 noon	Late blastula.
5	1 P.M.	Early gastrula.
6	2 P.M.	Midgastrula.
12	8 P.M.	Gastrulation completed; neurulation taking place.
18	2 A.M.	Neurulation completed; organogenesis beginning.
24	8 A.M.	Second day begins; organogenesis continues. Body axis straightens; circulatory system, pigmentation, and fins develop.
72	8 A.M.	Day 3 begins; embryos begin hatching.

collected at 8 A.M. and maintained at room temperature (approximately 23°C). Remember, however, that different embryos, even those developing from eggs laid at the same time, develop at slightly different rates.

Procedure

1. Obtain a petri dish with an embryo in embryo-rearing solution from the lab supply. The stage of development may be labeled. View it using the stereoscopic microscope. Gently roll the embryo using a toothpick to see the embryo from several angles.
2. Read the description (following) of each stage of development, and determine the stage of the embryo. Remember that these are living embryos, and the stage of development may have changed from that indicated on the petri dish label.
3. Using a pipette, carefully transfer the embryo and rearing solution to a depression slide and view it on the lowest power of the compound microscope.
4. Remembering that these are living embryos, watch carefully to observe cells dividing. You may observe a two-cell embryo changing to a four-cell embryo or a late blastula beginning gastrulation. With careful and patient observations, you may be fortunate enough to see the developmental stages unfolding.
5. Record notes in the margin of your lab manual and modify figures so that you will be able to refer to them later.
6. Using the pipette, return the embryo to the petri dish. Then return the petri dish to the lab supply.
7. Obtain a petri dish with an embryo in a different stage of development and repeat steps 1 to 6. Continue your observations until you have seen all listed stages of development.

Unfertilized Egg

Because most eggs are immediately fertilized after spawning, you may not find any unfertilized eggs. The unfertilized egg is about 0.5 mm in diameter. It is spherical and is filled with evenly distributed yolk granules that will provide nourishment in early development until the fish can feed itself.

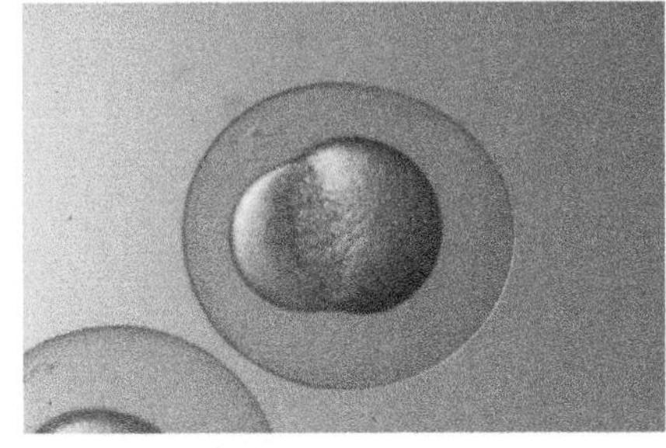

FIGURE 25.15
Fertilized egg, one cell stage.

Fertilized Egg (Zygote)

Most of the eggs that you observe will be fertilized. A thick, clear, protective membrane, the **chorion**, is visible around the egg and will be seen surrounding the embryo as it develops. The yolk granules condense, and the active cytoplasm migrates to the **animal pole**, where it becomes the **blastodisc**, or germinal disc. The blastodisc is visible as a bulge in the otherwise spherical cell. The future embryo will develop from the blastodisc, and the yolk serves as the nutrient supply (Figure 25.15).

Cleavage

Cleavage takes place in the blastodisc. Cells resulting from cleavage are called **blastomeres**. The blastodisc divides into 2 (Figure 25.16), then 4, 8, and 16 cells of equal size.

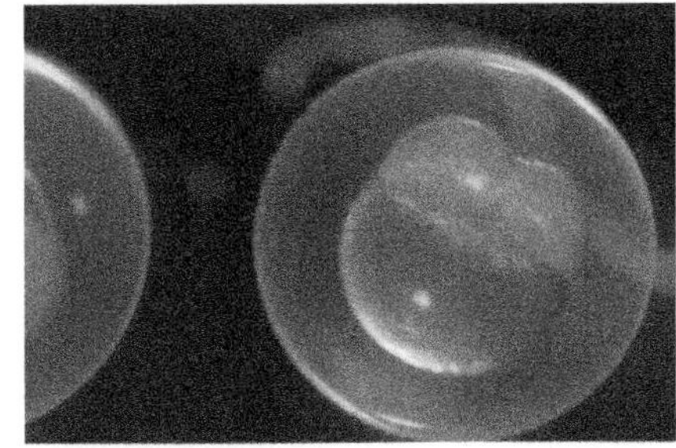

FIGURE 25.16
Two-cell stage.

Blastula

By the 64-cell stage, the blastula appears as a high dome of cells at the animal pole (Figure 25.17). This dome of cells is called the **blastoderm.** As cleavage continues, the cells become more compact, and the interface between the blastoderm and the yolk flattens out (Figure 25.18).

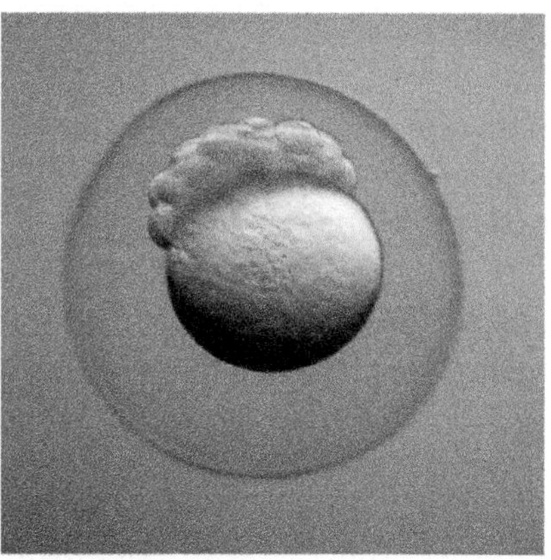

FIGURE 25.17
The 64-cell blastula.

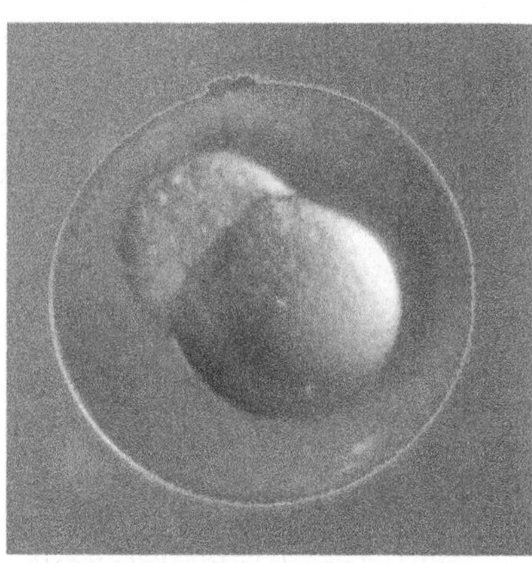

FIGURE 25.18
Later blastula stage.

Gastrula

Gastrulation in the zebrafish is strikingly different from the process in the sea star and the salamander, because cleavage is restricted to the blastodisc. The surface cells of the blastula spread over the entire yolk mass toward the vegetal pole in a process called **epiboly**. The surface cells produce the ectoderm. When the advancing blastoderm cells have covered approximately one-half of the yolk, a type of cell movement called **involution** begins when cells move into the interior of the embryo in a ring at the edge of the advancing blastoderm. These involuting cells will eventually form mesoderm and endoderm. Figures 25.19 and 25.20 illustrate early and late gastrulation.

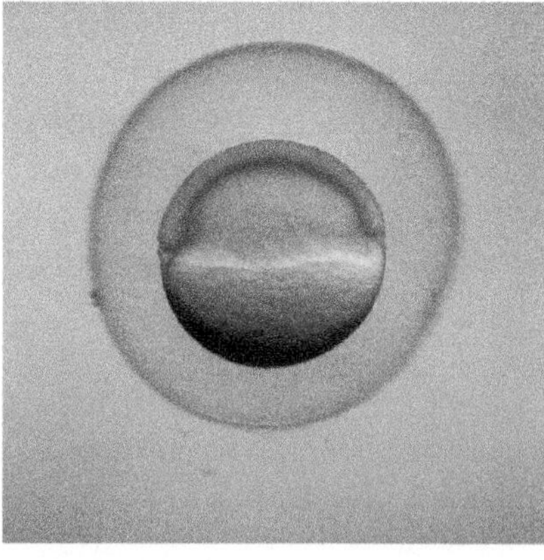

FIGURE 25.19
The early gastrula (epiboly).

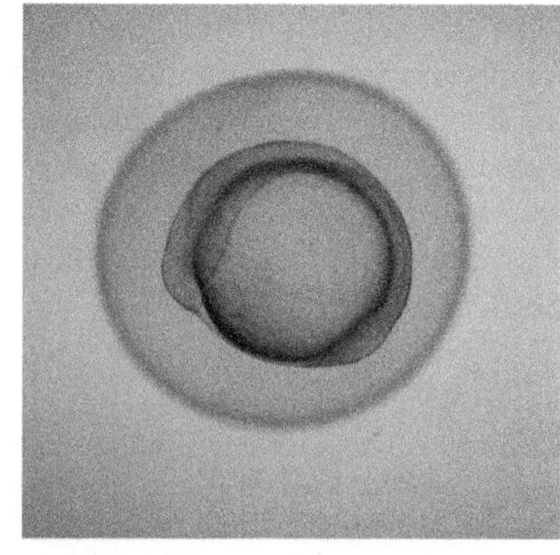

FIGURE 25.20
Late gastrula.

Neurulation

The antero/posterior, dorso/ventral embryonic axes are first obvious during neurulation. As neurulation takes place, ectodermal cells form the neural plate and eventually the neural tube. In the whole embryo, the region of the neural tube appears as a ridge on the dorsal surface. With careful investigation, you can discern that one end of the neural tube is enlarged. This enlarged portion is the **brain** developing at the anterior end of the embryo. The posterior end of the neural tube develops into the **nerve** (spinal) **cord**. As development progresses, a supportive rod, the **notochord**, forms from mesoderm beneath the neural tube. The notochord is later replaced by vertebrae. Blocks of tissue called **somites** form from mesoderm along each side of the nerve cord (Figure 25.21).

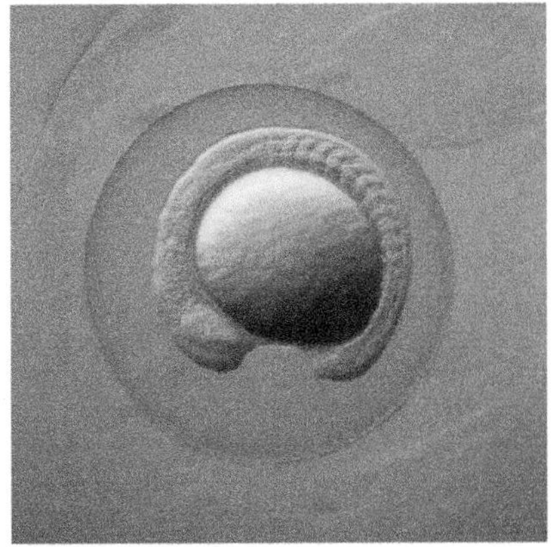

FIGURE 25.21
Neurulation.

Organogenesis

In organogenesis, formation of rudimentary organs and organ systems takes place (Figure 25.22). In the fish, the brain, eyes, somites, spinal cord, and tail bud are visible. The rhythmic beating of the heart and circulating blood may be seen.

Among other things, somites develop into skeletal muscles, and older embryos will be actively twisting and turning in the fertilization membrane. Developing **pigmentation** will be visible in the skin and eye.

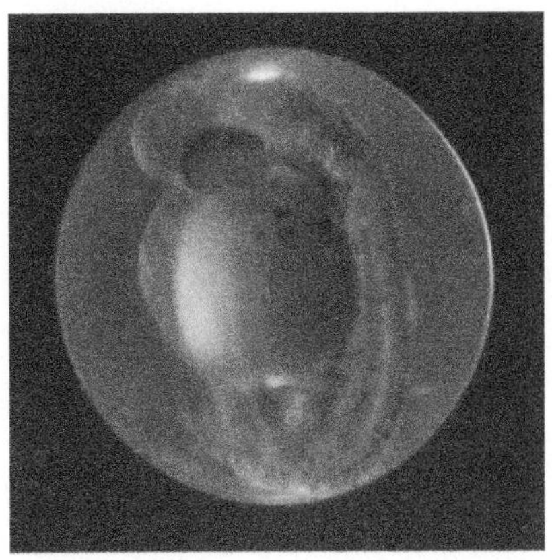

FIGURE 25.22
Organogenesis.

Hatching

As organogenesis continues, embryos grow and develop, and after 3 to 4 days, the embryos hatch, breaking free of the chorion (Figure 25.23).

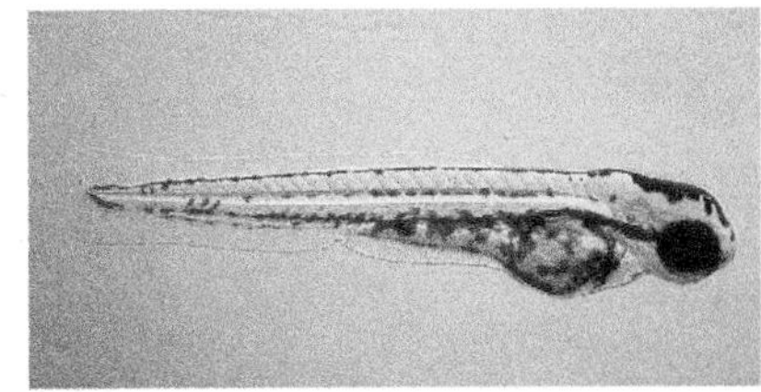

FIGURE 25.23
Newly hatched embryo.

Results

1. As you observe the living embryos, modify the figures to show any differences you observe between the living embryos and those pictured in the figures. Record notes to help distinguish stages. Sketch any additional stages that you observe, for example, 4-cell and 8-cell stages.
2. Label the following on the appropriate figures: *animal pole, vegetal pole, yolk mass, blastomere, somites, developing brain.*

Discussion

1. What type of egg does the zebrafish have (Figure 25.2)?
2. Is cleavage in the zebrafish holoblastic or meroblastic?
3. As the region of active cytoplasm (the blastodisc) undergoes cleavage, it becomes the blastoderm. Compare the size of the blastodisc and the blastoderm, and the size of cells as cleavage takes place.
4. After approximately how many hours and during which developmental stage do somites begin to appear?

EXERCISE 25.4

Development in a Bird: The Chicken

Materials

compound microscope
stereoscopic microscope
unincubated egg (demonstration)
prepared slide of 16-hour chick
prepared slide of 24-hour chick
living egg incubated 48 hours
living egg incubated 96 hours
finger bowl
warm 0.9% NaCl
flat-tipped forceps
sharp-pointed scissors
watch glass
disposable pipette
pipette bulb

Introduction

Immature eggs, or **oocytes**, develop within follicles in the single ovary of the adult female bird. (Two ovaries begin to develop in birds, but the second ovary degenerates.) In the sexually mature bird, hormonal stimulation brings about **ovulation**, the release of oocytes into a single oviduct. An oocyte consists of active cytoplasm, called the **blastodisc**, or germinal disc, floating on a huge amount of food reserve, the yolk, surrounded by a plasma membrane. At the time of ovulation, chromosomes in the large oocyte nucleus have just completed the first maturation division of meiosis (meiosis I). At ovulation in chickens, the oocyte nucleus measures approximately 0.5 mm and the oocyte measures approximately 35 mm in diameter.

Fertilization is internal in birds. If sperm are present in the oviduct at ovulation, they will penetrate each oocyte (one per oocyte), stimulating the completion of meiosis in the oocyte nucleus. The sperm nucleus and the now mature egg nucleus fuse, producing the zygote nucleus, which begins to divide by mitosis followed by cytoplasmic cleavage. As this developing embryo continues its passage down the oviduct, albumin, shell membranes, and, finally, a calcareous shell are deposited on its surface. In chickens, passage down the oviduct takes about 25 hours. This means that a freshly laid chicken egg, if it has been fertilized, has completed about 25 hours of development. The cleaved blastodisc is now called the **blastoderm**, or **blastula**. Development continues in the blastoderm, giving rise to all parts of the embryo, with yolk containing carbohydrates, proteins, lipids, and vitamins serving as the food reserves.

In this exercise, you will observe an unincubated egg and incubated eggs in several stages of development. As you study the embryos, identify developing structures and compare bird development with that of the sea star, salamander or frog, and zebrafish.

Procedure

1. Refer to Figure 25.24 and observe the unincubated chicken egg on demonstration.

   Most eggs sold for human consumption are purchased from commercial egg farms where hens are not allowed contact with roosters. The egg you are studying may or may not have been fertilized.

   a. Observe the broken calcareous **shell** with outer and inner **shell membranes** just inside. The shell and membranes are porous, allowing

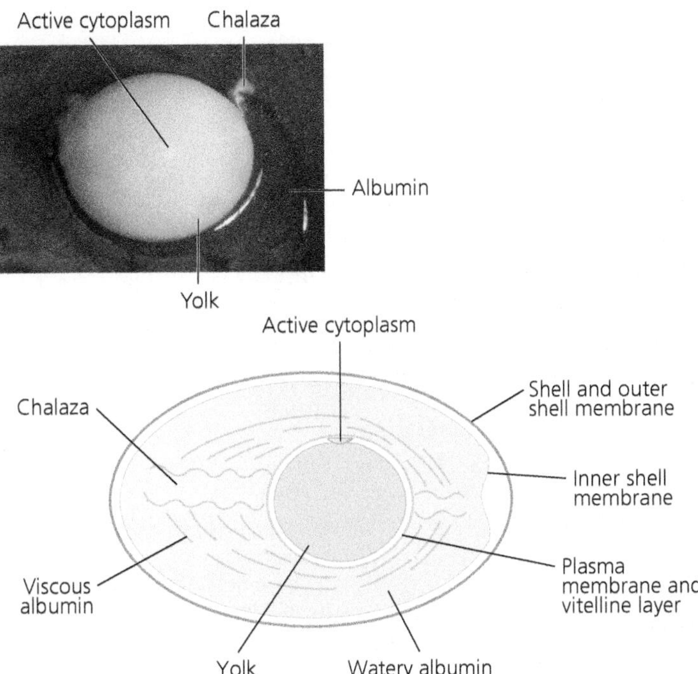

FIGURE 25.24

An unincubated chicken egg. The chalaza suspends the yolk in the albumin. The active cytoplasm (blastodisc) and chalaza associated with the yolk are visible in the inset.

air to pass through to the embryo inside. You have probably noticed an air chamber at one end of a hard-boiled egg, between the two membranes.

b. Observe the watery, proteinaceous **egg albumin** (egg white) and the yellow yolk. The layers of albumin closest to the yolk are more viscous and stringy than the outer albumin. As the yolky egg passes down the oviduct, it rotates, twisting the stringy albumin into two whitish strands on either end of the yolk. Called **chalaza**, these strands suspend the yolk in the albumin.

c. Locate the small whitish disc lying on top of the yolk. This is larger in a fertilized egg because of the development that has taken place. Cleavage takes place in this disc of active cytoplasm, called the *blastodisc*, before cleavage begins and the *blastoderm* after cleavage has begun. If you are studying a fertilized egg, cleavage is completed. Cleavage is restricted to the blastoderm; the yolk does not divide.

   As cleavage takes place, the blastoderm, now a mass of cells, becomes elevated above the yolk. Subsequent horizontal cleavages create three or four cell layers in the blastoderm, and a space, the blastocoel, forms within these layers.

2. Study the prepared slide of the 16-hour egg (the **gastrula**). Refer to Figure 25.25a, a surface view of the embryo.

> Use only low and intermediate powers when viewing this slide. The high power objective will break the slide!

a. Using low and then intermediate powers on the compound microscope, view a prepared slide of the whole chick embryo after about 16 to 19 hours of incubation. At this stage, cells in the blastoderm have

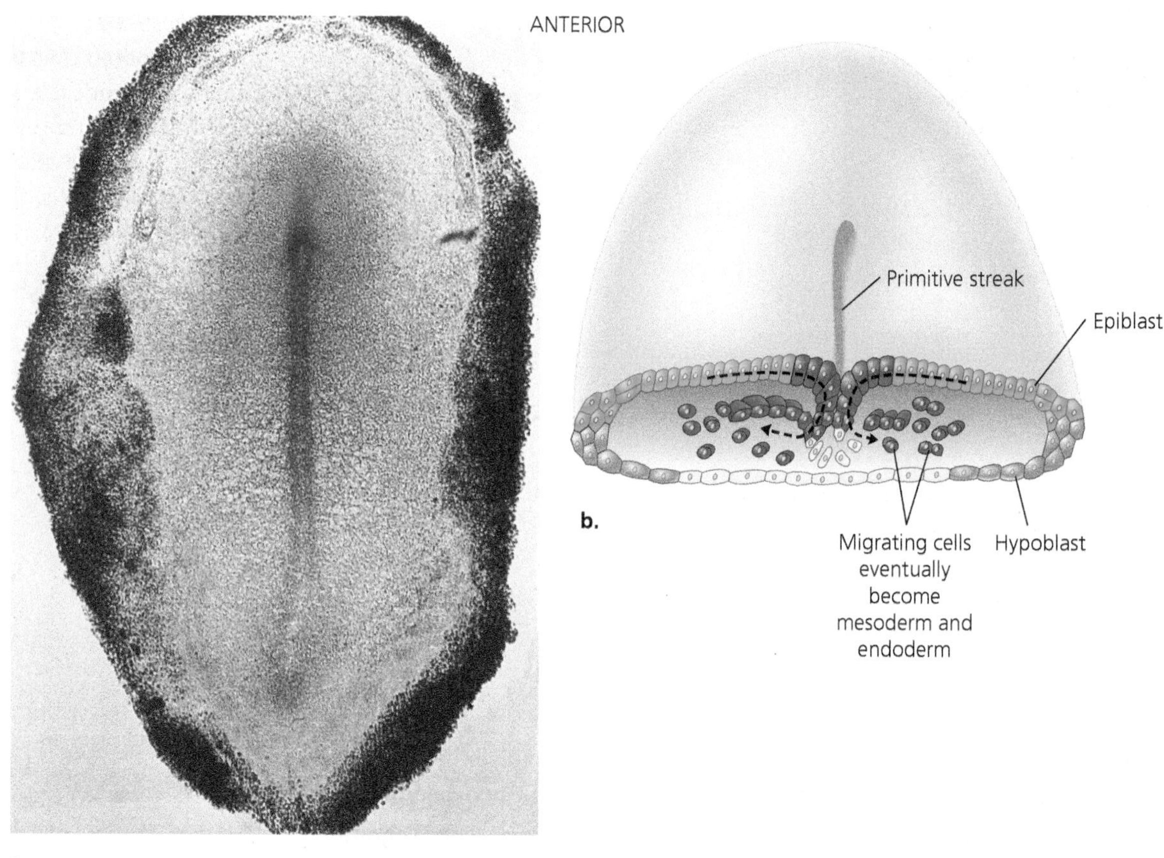

FIGURE 25.25
Chick gastrulation. (a) Surface view of chick blastoderm after 16 to 19 hours of incubation, the gastrula stage. The primitive streak is visible. (b) Cross section of blastoderm after 16 hours of incubation. Cells turn in at the primitive streak, initiating the formation of mesodermal and endodermal tissues.

separated into an upper layer, the *epiblast*, and an inner layer, the *hypoblast*, lying next to the yolk. The cells in the epiblast will eventually form the embryo body. These layers are visible only in sections of the embryo.

b. Locate a dark, longitudinal thickening, the **primitive streak**, equivalent to the blastopore in the amphibian. During gastrulation, cells in the epiblast migrate toward the primitive streak and then turn under, through the primitive streak. By 18 hours, they have spread out and initiated the formation of mesodermal and endodermal tissues, the latter eventually replacing the hypoblast (Figure 25.25b). Epiblast cells that remain on the surface of the developing embryo become the ectoderm.

3. Study the prepared slide of the 24-hour chick (**neurulation**). Refer to Figure 25.26.

Use only low and intermediate powers to view this slide. The high power objective will break the slide!

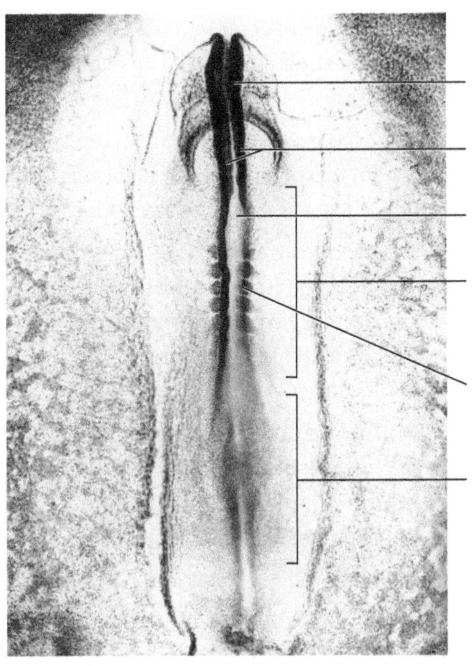

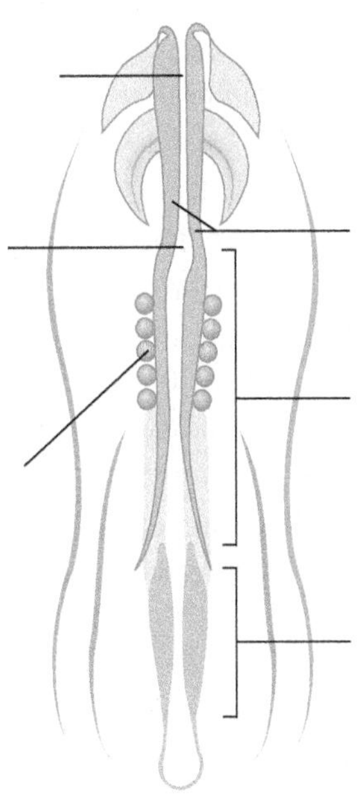

FIGURE 25.26

The chick after 24 hours of incubation (neurulation). Edges of the ectodermal neural plate elevate, forming neural folds. The depressed center is the neural groove. The neural folds eventually fuse, forming the neural tube.

a. Use the low and then intermediate powers on the compound microscope to view the prepared slide of a chick after at least 24 hours of incubation. At this stage, the neural tube is forming anterior to the primitive streak in a process similar to neurulation in fish.

b. Look for the developing longitudinal ectodermal **neural tube** with elevated edges called **neural folds** and a depressed center called the **neural groove**. The margins of the neural folds, which appear as a pair of dark longitudinal bands, become elevated and approach each other until they touch and eventually fuse. This fusion of the folds completes the formation of the neural tube. The anterior end of the neural tube becomes the brain; the posterior end becomes the spinal, or nerve, cord.

c. Observe **somites**, blocks of tissue lying on either side of the **spinal cord region**. Somites develop into body musculature and several other mesodermal organs. To determine the number of hours your chick was incubated, count the number of somites. A chick with five somites was incubated approximately 27 to 30 hours.

d. Label the following on the photo and diagram of Figure 25.26: *neural folds, neural groove, developing brain, spinal cord region* of the *neural tube, somites* and *primitive streak.*

Study the next two stages of development using living chick embryos and the materials listed at the beginning of the exercise. Work in pairs. One student will open the 48-hour chick; the other will open the 96-hour chick. Collaborate as you observe both eggs.

4. Prepare each egg to study the 48- and 96-hour chicks (**organogenesis**).

a. Pour warm (heated to about 38°C) 0.9% NaCl solution into a clean finger bowl.

b. Obtain an egg and carry it to your desk, keeping it oriented with the "top" up as it has been in the incubator. Crack the "down" side of the egg on the edge of the finger bowl. Hold the egg, cracked side down, in the NaCl solution, and carefully open the shell, allowing the egg to slide out of the shell.

c. Using the stereoscopic microscope, observe the embryo on the surface of the yolk. The embryo should either be on top of the yolk or it should float to the top. The embryo is in the center of the vascularized blastoderm.

d. Remove the finger bowl from the microscope and, using broad-tipped forceps and sharp scissors, remove the embryo from the yolk surface by carefully snipping outside the vascularized region of the embryo. Use the forceps to hold the blastoderm as you cut. Do not let go or you may lose the embryo in the mass of yolk.

e. Hold a watch glass under the NaCl solution and carefully pull the blastoderm away from the surface of the yolk into the watch glass. Carefully lift the watch glass and embryo out of the solution and pipette away excess solution until only a small amount remains.

f. Wipe the bottom of the watch glass, place on the stereoscopic microscope, and observe.

5. Study the 48-hour chick (early organogenesis). Refer to Figure 25.27.

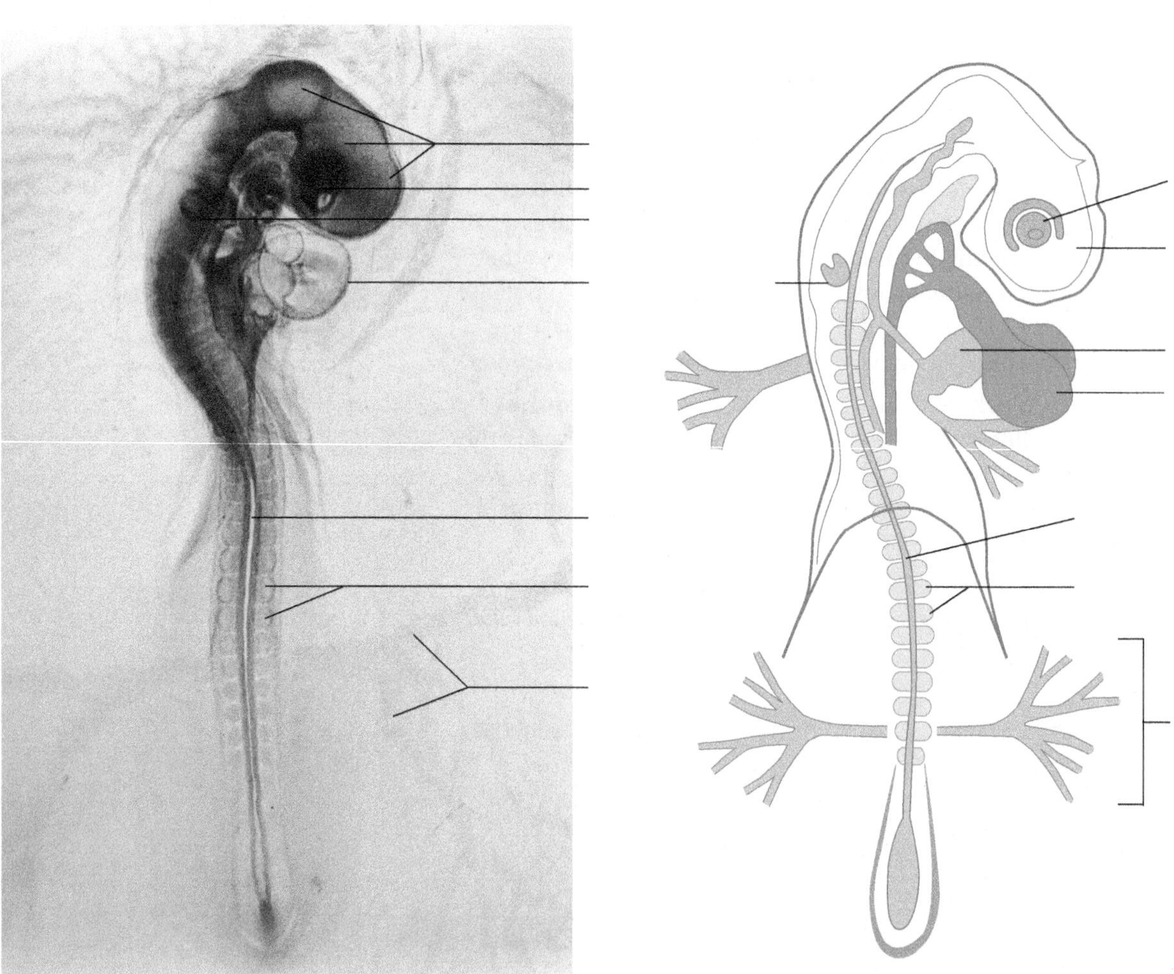

FIGURE 25.27

The chick after 48 hours of incubation. Identify the heart (atrium and ventricle), vitelline blood vessels, the brain, an eye, an ear, the spinal cord, and somites.

a. Identify structures in the circulatory system. In the living embryo, the **heart** is already beating at this stage, pumping blood through the **vitelline blood vessels**, which emerge from the embryo and carry food materials from the yolk mass to the embryo. If the heart is still beating, you should be able to see blood passing from the **atrium** into the **ventricle**. The atrium lies behind the ventricle, which is a larger, U-shaped chamber.

b. Identify structures in the nervous system. The anterior part of the neural tube has formed the **brain**. **Eyes** are already partially formed. You may be able to locate the **developing ear**. Follow the tube posteriorly to the **spinal cord**.

c. Count the number of **somites** lying on either side of the **spinal cord**. A chick with 23 to 24 somites has been incubated about 48 hours.

d. Label Figure 25.27.

e. Swap chick embryos with your lab partner.

6. Study the 96-hour chick (later organogenesis). Refer to Figures 25.28 and 25.29.

a. Notice that the 96-hour chick has a strong **cervical flexure**, bending the body into a C configuration. Several organs are noticeably larger than in the 48-hour chick.

b. Locate the developing **brain**, **eyes**, and **ears**.

c. Identify the conspicuous **heart**.

d. Locate one of the two **anterior limb buds** just behind the heart. Anterior limb buds develop into wings.

e. Locate the **posterior limb buds** near the tail. These limb buds grow into legs.

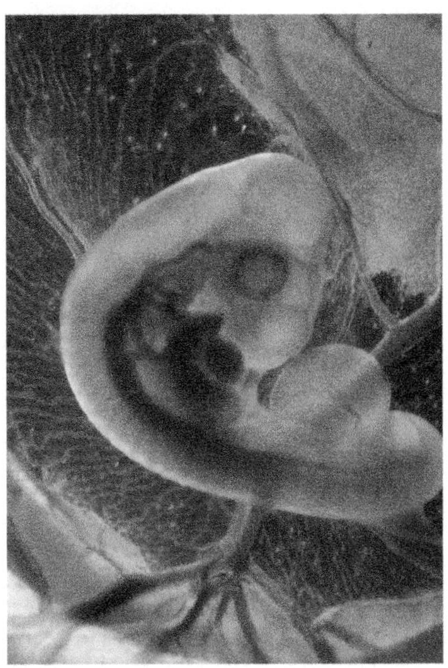

FIGURE 25.28
Chick embryo between 3 and 4 days of development.

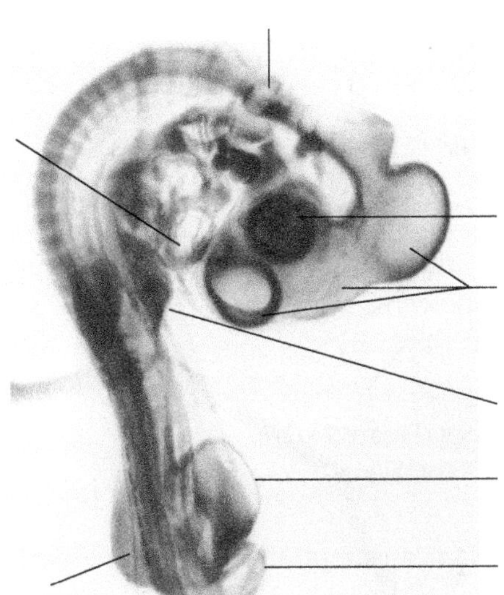

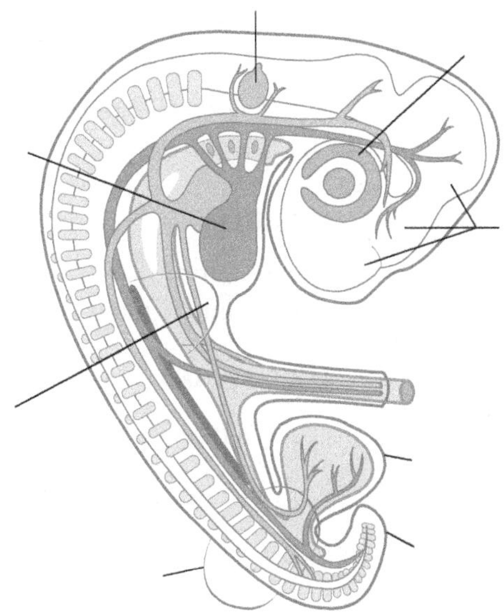

FIGURE 25.29
The chick after 96 hours of incubation. A strong cervical flexure has developed. Limb buds, allantois, ears, and tail are visible.

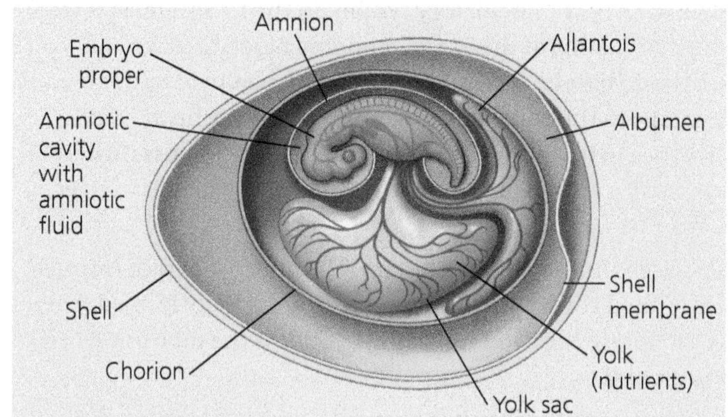

FIGURE 25.30
Extraembryonic membranes in a chick between 96 and 120 hours of incubation. The allantois protrudes from the gut near the posterior limb bud. The amnion surrounds the embryo proper, creating a fluid-filled amniotic cavity; the yolk sac surrounds the yolk; and the chorion grows until it eventually fuses with the shell membrane.

f. Identify one of four **extraembryonic membranes**, the **allantois**, which protrudes outward from the hindgut near the posterior limb bud (see Figures 25.28, 25.29, and 25.30). Be sure you can distinguish between limb bud and allantois. Extraembryonic membranes are derived from embryonic tissue, but are found outside the embryo proper. These membranes are important adaptations for land-dwelling organisms such as reptiles, birds, and mammals. They help solve problems such as desiccation, gas exchange, and waste removal in the embryo. The large saclike allantois will continue to grow until it lies close inside the porous shell. It functions to bring oxygen to the embryo, carry away carbon dioxide, and store liquid wastes.

g. Look for a second extraembryonic membrane, the **amnion**, a thin, transparent membrane that encloses the embryo in a fluid-filled sac.

h. Using the broad-tipped forceps, carefully lift the embryo to observe the yolk stalk and **yolk sac**, a third extraembryonic membrane that surrounds the yolk. The fourth extraembryonic membrane, the **chorion**, is not easily observed.

i. Label Figure 25.29.

j. Swap chick embryos with your lab partner.

Results

Label Figures 25.26, 25.27, and 25.28 and make additional sketches in the margin of your lab manual for future reference.

Discussion

1. What type of egg is the chicken egg (Figure 25.2)?

2. a. Are chicken eggs sold for human consumption fertilized or unfertilized? How can this be controlled?

b. What is added to an egg as it passes from the ovary down the oviduct?

3. A chicken egg undergoes what type of cleavage?

4. Collaborating with your lab partner, describe major differences between the 48-hour and 96-hour chicken embryos.

REVIEWING YOUR KNOWLEDGE

1. Having completed this lab topic, define and describe the following terms, giving examples when appropriate: *cleavage, blastula, gastrula, involution, archenteron, blastocoel, blastopore, isolecithal, telolecithal, alecithal, meroblastic, holoblastic, bipinnaria, epiboly, neurulation, organogenesis, animal pole, vegetal pole, blastodisc, blastoderm, ectoderm, mesoderm, endoderm, neural tube, neural groove,* and *neural fold.*
2. For each of the organs listed below, tell its embryonic germ layer:
   - **a.** sweat glands:
   - **b.** brain:
   - **c.** kidney:
   - **d.** liver:
   - **e.** esophagus:
   - **f.** heart:
3. Define *primitive streak.* At what stage of development in the chick is this structure present? What structure in amphibian development is equivalent to the chick primitive streak?
4. Differentiate between epiboly and involution (see gastrulation in the fish).
5. Review the stages of early development in animals. How are these stages similar in the animals you have studied? How do they differ? Complete Table 25.3.
6. The extraembryonic membranes that you observed in the chick are also found in mammals and reptiles. Hypothesize reasons that these groups have these membranes, but they are not found in embryos of fish and amphibians.

TABLE 25.3 Comparison of Stages of Early Development in the Sea Urchin, Sea Star, Salamander or Frog, Fish, and Chick

Organism	Cleavage	Gastrulation	Neurulation	Organogenesis
Sea urchin and sea star				
Salamander and frog				
Fish				
Chick				

APPLYING YOUR KNOWLEDGE

1. What is yolk? What is its function in development? Do mammals have yolk? Explain.

2. Giving examples from organisms studied, explain how the amount of yolk affects cleavage and gastrulation.

3. How do the differences in development lead to adaptations to the particular lifestyles of these organisms? (Some are aquatic, others terrestrial, and so on.)

4. Sea urchins live in marine communities that include a variety of species living in close proximity. Most of these species, including sea urchins, have external fertilization, a reproductive strategy that potentially has a low probability of sperm and egg contact. List several adaptations of sea urchins that help ensure successful reproduction.

5. Eggs of reptiles, birds, and a few primitive mammals are laid on land, usually away from water and independent sources of food. What features of the land egg (chick) allow this type of early development?

6. The extraembryonic membranes in mammals (including humans) and birds are homologous (share a common ancestry) and develop in a similar way. Use Table 25.4 to compare similarities and differences in the function of these membranes in birds and mammals. Use your text if needed.
7. Using your text or other resources, describe possible differences in the arrangement of the chorion and amnion in the formation of identical twins in humans.

TABLE 25.4 Similarities and Differences in Extraembryonic Membrane Functions in Birds and Mammals

Membrane	Birds	Mammals
Chorion (see Figures 25.30 and 24.4)		
Amnion (see Figures 25.30 and 24.4)		
Yolk sac (see Figure 25.30)		
Allantois (see Figures 25.30, 22.6, Figure 24.2a, and 24.4)		

8. Using information from the Web or other resources, describe the birth defect spina bifida. Which stage of embryonic development studied in this lab topic would be related to this condition?

INVESTIGATIVE EXTENSIONS

Sea urchin, fish, and chicken eggs collected for this lab topic provide an opportunity for student independent projects. Living amphibian eggs also may be available from colleagues or, in season, from streams or ponds in your area. Recent concern about declining populations of sea urchins (for example, *Diadema antillarum*, the black sea urchin once abundant throughout the Caribbean) and the decline of healthy populations of amphibians in North America are just two examples where environmental factors may be having an impact on animal development.

Suggestions follow for possible questions to investigate.

1. What effects, if any, and at what concentration do specific pollutants—pesticides, herbicides (e.g., atrazine), fertilizers—have on the following?
   a. Fertilization rates in sea urchins, amphibians, or fish
   b. Early development in sea urchins, amphibians, or fish (Recently, atrazine, a common herbicide, has been implicated in contributing to the decline and development of abnormalities in amphibian populations.)
2. What effect, if any, does acid rain-induced pH change have on the following?
   a. Fertilization rates in sea urchins, amphibians, or fish
   b. Early development in these organisms
3. Does UV radiation have an effect on the following?
   a. The rate of fertilization in sea urchins
   b. Early development in amphibians or fish

4. In sea urchins, will the addition of a calcium solution bring about the formation of the fertilization envelope in the absence of sperm?
5. Investigate the role of the jelly coat of echinoderm eggs in fertilization. If the jelly coat is removed, does this increase or decrease the rate of fertilization?
6. The chorion is relatively easy to remove from the developing zebrafish embryo. Design an experiment to investigate the effect of water pollution on fertilization and early development in fish embryos with and without a chorion.
7. An age-old process called "egg candling" is used by amateurs working with poultry as well as the industrial poultry business. Using online sources, investigate this process and its applications. Search how to make a homemade "egg candler" and imagine how you might use this process in the design of an investigation for your introductory biology laboratory.

STUDENT MEDIA: BioFlix, Activities, Investigations, and Videos

www.masteringbiology.com (select Study Area)

Activities—Ch. 47: Sea Urchin Development; Frog Development

Investigations—Ch. 47: What Determines Cell Differentiation in the Sea Urchin?

Videos—Ch. 47: Sea Urchin Embryonic Development; Frog Embryo Development; Ch. 20: Cloning

REFERENCES

Beams, H. W. and R. G. Kessel. "Cytokinesis: A Comparative Study of Cytoplasmic Division in Animal Cells." *American Scientist,* 1976, vol. 64, pp. 279–290. Scanning electron micrographs of development in zebrafish.

Gilbert, S. F. *Developmental Biology,* 10th ed. Sunderland, MA: Sinauer Associates, 2013.

Kimmel, C., W. Ballard, S. Kimmel, B. Ullmann, and T. Schilling. "Stages of Embryonic Development of the Zebrafish." *Developmental Dynamics*, 1995, vol. 203, pp. 253–310.

Patten, B. M. *Early Embryology of the Chick,* 5th ed. New York, NY: McGraw-Hill, 1971. The historic textbook describing chick development for the beginning student.

Urry, L., M. Cain, S. Wasserman, P. Minorsky, and J. Reece. *Campbell Biology*, 11th ed. San Francisco, CA: Pearson, 2017.

Westerfield, M. (ed.). *The Zebrafish Book; A Guide for the Laboratory Use of Zebrafish (Brachydanio rerio)*, 5th ed. Eugene: University of Oregon Press, 2007. 4th edition available online. See under Websites.

WEBSITES

A detailed Web page with exercises and procedures for using sea urchins in the teaching laboratory developed by teachers and Stanford University researchers: http://www.stanford.edu/group/Urchin

References for sea urchin fertilization investigations: http://icb.oxfordjournals.org/content/46/3/298.full

Sea urchin laboratory procedures; includes information on gamete collection and fertilization. Search for Urchin: http://www.swarthmore.edu/NatSci/sgilber1/DB_lab/Urchin/urchin_basic.html

The Zebrafish Book online: http://zfin.org/zf_info/zfbook/zfbk.html

Zebrafish early development images. Select zebrafish normal development: http://www.cas.vanderbilt.edu/bioimages/animals/danrer/zfish-devel.htm

ZFIN—The Zebrafish Model Organism Database: http://zfin.org/

NOTES

LAB TOPIC 26

Animal Behavior

Laboratory Objectives

After completing this lab topic, you should be able to:

1. Define *ethology.*
2. Define and give an example of *taxis, kinesis, agonistic behavior,* and *reproductive behavior.*
3. State the possible adaptive significance of each of these behaviors.
4. Propose hypotheses, make predictions, design experiments to test hypotheses, collect and process data, and discuss results.
5. Present the results of your experiments in a scientific paper.

Introduction

Behavior, broadly defined, is the sum of the responses of an organism to stimuli in its environment. In other words, behavior is what organisms do. **Ethology** is the study of animal behavior in the context of the evolution, ecology, social organization, and sensory abilities of an animal (Gould, 1982). Ethologists concentrate on developing accurate descriptions of animal behavior by carefully observing and experimentally analyzing overt behavior patterns and by studying the physiology of behavior.

Explaining a particular behavior in the broad, multivariable context of evolution or ecology can become a complex undertaking. It is often necessary, therefore, to study behavior in animals that have a limited range of behaviors and for which more is known about their evolution, ecology, and sensory abilities. Understanding simple and isolated behaviors is important in unraveling more complex behaviors.

There are two basic categories of behavior—**learned** (depends on experiences) and **innate** (inherited—exhibited by nearly all individuals of a species). Experimental evidence suggests that the basis of both lies in the animal's genes. As with all genetically controlled features of an organism, behavior is subject to evolutionary adaptation. As you study animal behavioral activities in this lab topic, think in terms of both **proximate causes**, the immediate events that led to the behavior, and **ultimate causes**, the adaptive value and evolutionary origin of the behavior. To illustrate, a fiddler crab will respond to human intrusion into its feeding area by running into its burrow. The proximate cause of this behavior might be the vibration caused by footsteps stimulating sensory receptors and triggering nervous impulses. The nervous impulses control muscle contractions in the crab's legs. Ultimate causes are the adaptive value of retreating from predators to avoid being eaten.

Note that you will be asking **causal** questions in your investigations. It is inappropriate to ask **anthropomorphic** questions—that is, questions that ascribe human attributes to the animal. Consider, for example, a behavior that places an animal in its best environment. An anthropomorphic explanation for this behavior would be that the animal makes a conscious choice of its environment. There is no way for us to come to this conclusion scientifically. The causal explanation would be that the animal is equipped with a sensory system that responds to environmental stimuli until the favorable environment is reached.

Ethologists have categorized behavioral patterns based on the particular consequence of that behavior for the organism. **Orientation behaviors** place the animal in its most favorable environment. Two categories of orientation behaviors are **taxis** (plural, **taxes**) and **kinesis**. *A taxis is movement directly toward or away from a stimulus.* When the response is toward a stimulus, it is said to be *positive;* when it is away from the stimulus, it is *negative.* Prefixes such as *photo-, chemo-,* and *thermo-* can be added to the term to describe the nature of the stimulus. For example, an animal that responds to light may demonstrate positive phototaxis and is described as being *positively phototactic.*

A kinesis differs from a taxis in that it is undirected, or random, movement. A stimulus initiates the movement but does not necessarily orient the movement. The intensity of the stimulus determines the rate, or velocity, of movement in response to that stimulus. If a bright light is shined on an animal and the animal responds by moving directly away from it, the behavior is a taxis. But if the bright light initiates random movement or stimulates an increase in the rate of turning with no particular orientation involved, the behavior is a kinesis. The terms *positive* and *negative* and the prefixes mentioned earlier are also appropriately used with *kinesis.* An increase in activity is a positive response; a decrease in activity is a negative response.

Another complex of behaviors observed in some animals is **agonistic behavior**. In this case, the animal is in a conflict situation where there may be a threat or approach, then an attack or withdrawal. Agonistic behaviors in the form of force are called **aggression**; those of retreat or avoidance are called **submission**. Often the agonistic behavior is simply a display that makes the organism look big or threatening. It rarely leads to death and is thought to help maintain territory so that the dominant organism has greater access to resources such as space, food, and mates.

Reproductive behavior can involve a complex sequence of activities, sometimes spectacular, that facilitate *finding, courting,* and *mating* with a member of the same species. It is an adaptive advantage that reproductive behaviors are species-specific. Can you suggest reasons why?

For the first hour of lab, you will perform Experiment A in each of the exercises that follow, briefly investigating the four behaviors just discussed: *taxis in brine shrimp, kinesis in pill bugs, agonistic behavior in Siamese fighting fish,* and *reproductive behavior (courting and mating) in fruit flies.* After completing every Experiment A, your team will choose one of the systems discussed and perform Experiment B in that exercise. Following the outline in Exercise 26.5, you will use the suggested questions in *Experiment B to propose one or more testable hypotheses and design a simple experiment to test your hypotheses*. Then you will spend the remainder of the laboratory period carrying out your experiments.

Near the end of the laboratory period, several of you may be asked to present your team's results to the class for discussion. One part of the scientific process involves persuading your colleagues that your experimental design is sound and that your results support your conclusions (either negating or supporting your hypothesis). Be prepared to describe your results in a brief presentation in which you will use your experimental evidence to persuade the other students in your class.

You may be required to submit a laboratory report describing your experiment and results in the format of a scientific paper (see Appendix A). You should discuss results and come to conclusions with your team members; however, you must turn in an originally written lab report. Your Materials and Methods section and your tables and figures may be similar, but your Introduction, Results, and Discussion sections must be the product of your own library and online research, creative thinking, evaluating, and writing.

Remember, first complete Experiment A in each exercise. Then discuss with your research team a possible question for your original experiment, choosing one of the animals investigated in Experiment A as your experimental organism. Be certain you can pose an interesting question from which you can develop a testable hypothesis. Then turn to Experiment B for your chosen organism and follow Exercise 26.5 to design and execute your experiment.

EXERCISE 26.1

Taxis in Brine Shrimp

Brine shrimp (*Artemia salina*) are small crustaceans that live in salt lakes and swim upside down using 11 pairs of appendages. Their sensory structures include two large compound eyes and two pairs of short antennae (Figure 26.1). They are a favorite fish food and can be purchased in pet stores.

FIGURE 26.1
Brine shrimp (*Artemia salina*) are 8–10 mm in length. A type of fairy shrimp, brine shrimp live in inland salt lakes such as the Great Salt Lake in Utah.

Experiment A. Brine Shrimp Behavior in Environments with Few Stimuli

Materials

brine shrimp
2 large test tubes
black construction paper
1 small finger bowl
salt water
dropper

Introduction

In this experiment, you will place brine shrimp in a test tube of salt water similar to the water of their normal environment. You will not feed them or disturb them in any way. You will observe their behavior in this relatively stimulus-free environment. Notice their positions in the test tube. Are they in groups, or are they solitary? Are they near the top or near the bottom? You should make careful observations of their behavior, asking questions about possible stimuli that might initiate taxes in these animals.

Hypothesis

Hypothesize about the behavior of brine shrimp in an environment with few stimuli.

Prediction

Predict the result of your experiment based on the hypothesis (if/then).

Procedure

1. Place six brine shrimp in a test tube filled two-thirds with salt water. Rest the test tube in the finger bowl in such a way that you can easily see all six shrimp. You may need to use black construction paper as a background.
2. Describe the behavior of the brine shrimp in the Results section; for example, are they randomly distributed throughout the test tube or do they collect in one area?
3. Record your observations in the Results section.

Results

1. By describing the behavior of brine shrimp in an environment with relatively few stimuli, which component of experimental design are you establishing?

2. Describe the behavior of the brine shrimp.

Discussion

On separate paper, list four stimuli that might initiate taxes in brine shrimp and predict the response of the animal to each. What possible adaptive advantage could this behavior provide?

Experiment B. Student-Designed Investigation of Brine Shrimp Behavior

Materials

supplies from Experiment A
piece of black cloth
lamp
ice
magnets
rock salt crystals
salt solutions
dropper bottles—solutions of sugar, egg albumin, acid, and base

Introduction

If your team chooses to perform your original experiment investigating taxes in brine shrimp, return to this experiment after you have completed all the introductory investigations (Experiment A of each exercise). Using the materials available, design a simple experiment to investigate taxes in brine shrimp. If your experiment requires materials other than those provided, ask your laboratory instructor about their availability. If possible, submit requests in advance.

Procedure

1. Collaborating with other members of your research team, read the following potential questions, and choose a question to investigate using this list or an original question proposed by your team. You may want to check your text and other sources for supporting information. You should be able to explain the rationale behind your choice of question.
   a. How do brine shrimp respond to light—are they positively or negatively phototrophic?
   b. Do brine shrimp respond differently to stimuli in cold temperatures and warm temperatures (perhaps reflecting different behaviors in summer and winter)?
   c. Do brine shrimp respond differently to different nutrients, for example, solutions of carbohydrates, proteins, or fats? How does this reflect their diet in nature?
   d. Would a magnetic field change the behavior of brine shrimp?
   e. How would brine shrimp be distributed along a saline gradient?
2. Design your experiment, proposing hypotheses, making predictions, determining procedures, and recording results as instructed in Exercise 26.5.

> Allow a conditioning period of several minutes after the shrimp have been disturbed or stimulated. If you add something to the water in one experiment, begin additional experiments with fresh water and shrimp.

EXERCISE 26.2

Kinesis in Pill Bugs

Kinesis can be studied using a crustacean in the order Isopoda (called *isopods*). These animals are called by many common names, including *pill bugs, sow bugs, wood lice,* and *roly-polies* (Figure 26.2). Although most crustaceans are aquatic, pill bugs are truly terrestrial, and much of their behavior is involved with their need to avoid desiccation. They are easily collected in warm weather under flowerpots, in leaf litter, or in woodpiles. Some species respond to mechanical stimuli by rolling up into a ball.

FIGURE 26.2
Pill bugs measure up to about 15 mm in length. These terrestrial isopods are also called *sow bugs, wood lice,* and *roly-polies.*

Experiment A. Pill Bug Behavior in Moist and Dry Environments

Materials

pill bugs
2 large petri dishes
filter paper
squirt bottle of water

Introduction

In this experiment, you will investigate pill bug behavior in moist and dry environments by observing the degree of their activity, that is, the number of times they circle and turn. As you observe their behavior, ask questions about possible stimuli that might modify this behavior.

Hypothesis

Hypothesize about the degree of activity of pill bugs in moist and dry environments.

Prediction

Predict the results of the experiment based on your hypothesis (if/then).

Procedure

1. Prepare two large petri dishes, one with wet filter paper, the other with dry filter paper.
2. Place five pill bugs in each dish.
3. Place the dishes in a dark spot, such as a drawer, for 5 minutes.
4. After 5 minutes, you will observe the pill bugs in the petri dishes. Before you open the drawer or uncover the petri dishes, assign each of the following procedures to a member of your team.
   a. Count the number of pill bugs moving in each dish.
   b. Choose one moving pill bug in each dish and determine the rate of locomotion by counting revolutions per minute (rpm) around the petri dish.
   c. Determine the rate of turning by counting turns (reversal of direction) per minute for one pill bug in each dish.

Results

Record your results in Table 26.1.

TABLE 26.1 Kinesis in Pill Bugs: Response to Wet and Dry Environments

Environmental Condition	Number Moving	Rate of Locomotion (rpm)	Rate of Turning (turns/min)
Moist			
Dry			

Discussion

1. Kinetic response to varying moisture in the environment is called *hygrokinesis.* What other environmental factors might influence the behavior of pill bugs?

2. On separate paper, list four factors that might initiate kineses in pill bugs and predict their response to each. What possible adaptive advantage could this behavior provide?

3. Some terrestrial arthropods display a behavior where they avoid decaying organisms of their own species, but not other species. Can you suggest an adaptive advantage for this behavior? How would you design an experiment to test this?

Experiment B. Student-Designed Investigation of Pill Bug Behavior

Materials

supplies from Experiment A
white enamel pan
wax pencils
beaker of water
construction paper
manila folder
large pieces of black cloth

Introduction

If your team chooses to perform your original experiment investigating kineses in pill bugs, return to this experiment after you have completed all the introductory investigations (Experiment A of each exercise). Using the materials available, design a simple experiment to investigate kineses in pill bugs. If your experiment requires materials other than those provided, ask your laboratory instructor about their availability. If possible, submit requests in advance.

Procedure

1. Collaborating with your research team, read the following potential questions, and choose a question to investigate using this list or an original question proposed by your team. You may want to check your text and other sources for supporting information. You should be able to explain the rationale behind your choice of question.
   a. Does the amount of moisture have an effect on the behavior of pill bugs in wet environments?
   b. Is the kinetic behavior of pill bugs the same in light and dark?
   c. Do different species of pill bugs respond differently to mechanical stimuli?
   d. Is the behavior of pill bugs different when individual bugs are isolated compared with bugs in groups?
   e. Is the behavior of pill bugs different if there are dead pill bugs in the environment? What if there are dead bugs of different species?
   f. Does temperature affect the behavior of pill bugs?
   g. Do pill bugs respond differently to different nutrients, for example, solutions of carbohydrates, proteins, or fats? How does this reflect their diet in nature?

2. Design and perform your experiment, proposing hypotheses, making predictions, determining procedures, and recording results as instructed in Exercise 26.5.

EXERCISE 26.3

Agonistic Display in Male Siamese Fighting Fish

In this exercise you will investigate agonistic behavior in a species of fish, *Betta splendens*, commonly called Siamese fighting fish. This fish species displays *sexual dimorphism*, a condition where males and females of a species differ in appearance such as color, size, and/or shape. In Siamese fighting fish, the male body is brightly colored with large, brightly colored fins. Usually, the female body is smaller and less colorful than the male, with small, dull-colored fins.

The innate agonistic behavior of the male Siamese fighting fish has been widely studied (Simpson, 1968; Thompson, 1969). The sight of another male *Betta* or even its own reflection in a mirror will stimulate a ritualized series of responses toward the intruder. If two fish are placed in the same aquarium, their agonistic behavior usually continues until one fish is defeated or subordinated.

Experiment A. Display Behavior in Male Siamese Fighting Fish

Materials

one male Siamese fighting fish in a 1- to 2-L flat-sided fishbowl
mirror
timer or clock with second hand

Introduction

The purpose of this experiment is to describe the ritualized agonistic display of a male Siamese fighting fish after being stimulated by its own reflection in a mirror. Before you begin the experiment, become familiar with the fish's anatomy, identifying its dorsal, caudal (tail), anal, pectoral, and ventral (pelvic) fins, and its gill cover (Figure 26.3).

When you begin the experiment, you will be looking for several possible responses: frontal approach (facing intruder), broadside display, undulating

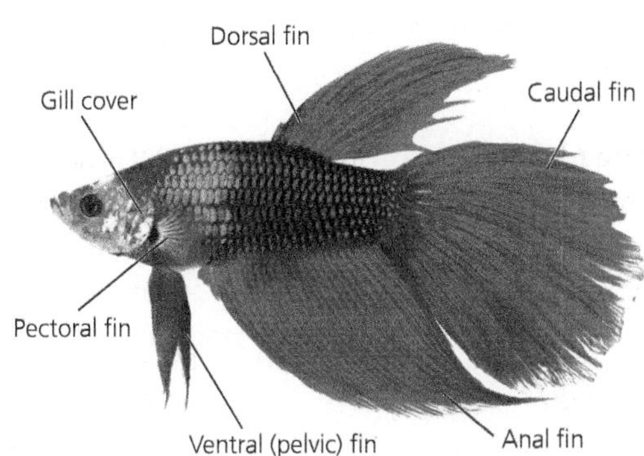

FIGURE 26.3
Male Siamese fighting fish (*Betta splendens*).

movements, increased swimming speed, and enhanced coloration in tail, fin, or body. You should note the duration and intensity of elevation for each fin and changes in the gill cover.

Hypothesis

Hypothesize about the response of the fish to its image in the mirror.

Prediction

Predict the result of the experiment based on your hypothesis (if/then).

Procedure

1. Plan your strategy.
    a. Be ready with your pencil and paper to record your observations. Behaviors can happen very quickly. Your entire team should observe and record them.
    b. Each team member should be responsible for timing the beginning and duration of particular responses (listed in the Introduction). For example, one or two team members will observe the response in one or two specific fins and record their results in Table 26.2 of the Results section. Another team member will observe more general responses, such as "increased swimming speed" or "broadside display for 60 seconds."
2. Place the mirror against the fishbowl.
3. As the fish reacts to its reflection, record your observations in the table or on separate paper.
4. Compare collective results.
5. In the Results section, using the time you present the mirror as time zero, make a *sequential* list of the recognizable responses involved in the display. Be as quantitative as possible; for example, you might record "gill cover extended 90° for 30 seconds," or "broadside display for 60 seconds."
6. Note in the Results section those responses that take place simultaneously.

Results

1. Record your data for each response in Table 26.2. Consider the time when you place the mirror next to the fishbowl as time zero. Then note how many seconds lapse before the specific response and its intensity. Use these symbols as estimates of response intensity: − (none), + (slight response), + + (moderate response), + + + (intense response).

TABLE 26.2 Behaviors, Response Times, and Response Intensity for Siamese Fighting Fish Presented with a Mirror Stimulus

Response	Time Before Response (seconds)	Intensity − (none), + (slight), + + (moderate), + + + (intense)
Fish notices mirror		
Approaches mirror		
Dorsal fin flare		
Caudal fin (tail) flare		
Anal fin flare		
Ventral fin flare		
Pectoral fin flare		
Gill cover flare		

2. Record your sequential list.

3. Record the responses that take place simultaneously.

Discussion

1. Collaborating with your teammates, write a descriptive paragraph, as quantitative and detailed as possible, describing the agonistic display elicited in the Siamese fighting fish in response to its reflection.

2. Biologists have described a behavior called *fixed action pattern*, where a simple stimulus initiates a sequence of unlearned acts that, once initiated, usually lead to the completion of the behavior. Do you conclude that agonistic behavior in Siamese fighting fish is an example of this type of behavior? Why or why not?

3. What is the obvious adaptive advantage of complex agonistic displays that are not followed by damaging fights? Are there advantages that are not so obvious?

4. Name several other animals that demonstrate a strong display that is seldom followed by a damaging fight.

5. Name several animals that do engage in damaging fights.

Experiment B. Student-Designed Investigation of Siamese Fighting Fish Behavior

Materials

supplies from Experiment A
colored pencils and index cards or brightly colored paper
scissors
wooden applicator sticks
transparent tape
fish of different species in fishbowls
female Siamese fighting fish

Introduction

If your team chooses to perform your original experiment investigating agonistic behavior in Siamese fighting fish, return to this experiment after you have completed all the introductory investigations (Experiment A of each exercise). Using the materials available, design a simple experiment to investigate this behavior. If your experiment requires materials other than those provided, ask your laboratory instructor about their availability. If possible, submit requests in advance.

Procedure

1. Collaborating with your research team, read the following potential questions, and choose a question to investigate using this list or an original question proposed by your team. You may want to check your text and other sources for supporting information. You should be able to explain the rationale behind your choice of question.
   a. What is the simplest stimulus that will initiate the response? Is color important? Size? Movement?
   b. Is the behavior "released" by a specific stimulus or by a complex of all the stimuli?
   c. Will another species of fish initiate the response?
   d. Will a female *Betta* fish initiate the response, and, if so, how does the response compare with the response to a fish of a different species?
   e. Is the response all or none—that is, are there partial displays with different stimuli?
   f. Does the fish become "conditioned"—that is, after repeated identical stimuli, does the duration of the display change, or does the display cease?
   g. Could chemical stimulation contribute to the response? (Transfer water from one fishbowl to another.)
   h. Do other species of fish with sexual dimorphism display aggressive behavior similar to that seen in *Betta?* Survey fish available in your local pet store and choose an experimental species where males and females differ in color, size, or shape. For example, you might choose fancy guppies that are highly sexually dimorphic, or some species of cichlids, aquarium fish that may show aggressive behavior. Design an experiment to determine if males of that species behave aggressively toward males or females of their own species or fish of another species.
2. Design your experiment, proposing hypotheses, making predictions, determining procedures, and recording results as instructed in Exercise 26.5.

EXERCISE 26.4

Reproductive Behavior in Fruit Flies

Reproductive behavior in some animals involves a complex sequence of activities that may include courtship and mating. Spieth (1952, described in Marler, 1968) has classified courtship and mating behavior of the fruit fly *Drosophila melanogaster* as being a complex of at least 14 behaviors. These behaviors illustrate four modes of communication in animals that have been described by behavioral biologists. The modes are categorized as visual, chemical, tactile, and auditory. As you observe reproductive behavior in fruit flies, consider which of these four modes is being utilized in each behavior.

Courtship behavior in fruit flies is an example of a *stimulus-response chain*. The response to a stimulus becomes the stimulus for the next behavior. If there is no response, then the behavior changes. Described below are 10 of the most common and easily recognized of these behaviors. Read the list carefully and become familiar with the behaviors you will be required to recognize. Six of the behaviors are seen in males, four in females. The behavior sequence begins as the male orients his body toward the female (Figure 26.4a).

Male Behaviors

1. *Tapping.* The male sees a female and extends a foreleg to strike or tap her abdomen (tactile communication) (Figure 26.4b).
2. *Waving.* The wing is extended and held 90° from the body, then relaxed without vibration (Figure 26.4c).
3. *Wing vibration.* The male extends one or both wings from the resting position and moves them rapidly up and down (producing a courtship song—auditory communication) (Figure 26.4c).
4. *Licking.* The male licks the female's genitalia (on the rear of her abdomen) (Figure 26.4d).
5. *Circling.* The male postures and then circles the female, usually when she is nonreceptive.
6. *Stamping.* The male stamps forelegs as in tapping but does not strike the female.

Female Behaviors

1. *Extruding.* The female releases chemicals that attract the male. A temporary, tubelike structure is extended from the female's genitalia.
2. If receptive, the female allows the male to attempt copulation (Figure 26.4e and f).
3. *Decamping.* A nonreceptive female runs, jumps, or flies away from the courting male.
4. *Depressing.* A nonreceptive female prevents access to her genitalia by depressing her wings and curling the tip of her abdomen down.
5. *Ignoring.* A nonreceptive female ignores the male.

FIGURE 26.4
Mating behavior in the fruit fly, *Drosophila melanogaster*.

Experiment A. Reproductive Behavior in *Drosophila melanogaster*

Materials

stereoscopic microscope
fly vials with 2 or 3 virgin female *D. melanogaster* flies
fly vials with 2 or 3 male *D. melanogaster* flies

Introduction

In this experiment, you will place virgin female *D. melanogaster* flies in the same vial with male flies and observe the behavior of each sex. *Working with one other student,* discuss the behaviors and modes of communication described in the introduction to this exercise and plan the strategy for your experiment. Identify mating behaviors of *D. melanogaster* and record their sequence and duration (when appropriate). As you observe the behavior of the flies, discuss possible original experiments investigating courtship and mating behavior in flies.

Hypothesis

Hypothesize about the presence of flies of the opposite sex in the same vial.

Prediction

Predict the results of the experiment based on your hypothesis (if/then).

Procedure

1. Set up the stereoscopic microscope.
2. Have paper and pencil ready. The behaviors can happen very rapidly. One person should call out observations while the other person records.
3. Obtain one vial containing virgin females and one vial containing males, and gently tap the male flies into the vial containing females.
4. Observe first with the naked eye. Once flies have encountered each other, use the stereoscopic microscope to make observations.
5. Prepare to quantify your observations. To do this, you may consider counting the number of times a behavior takes place or timing the duration of behaviors.

Results

1. Describe, in sequence, the response of the male to the female and the female to the male. Quantify your observations. To do this, you may consider counting the number of times a behavior takes place and timing the duration of behaviors.

2. Describe rejection if this takes place.

3. In the margin of your lab manual, note any behaviors that can be analyzed quantitatively.

Discussion

Speculate about the adaptive advantage of elaborate courtship behaviors in animals.

Experiment B. Student-Designed Investigation of Reproductive Behavior in Fruit Flies

Materials

supplies from Experiment A
fly vials with 2 or 3 virgin females of an alternate fly species (other than *D. melanogaster*)
fly vials with 2 or 3 males of the alternate species

Introduction

If your team chooses to perform your original experiment investigating courtship and mating behavior in fruit flies, continue with this experiment after you have completed all the introductory investigations (Experiment A of each exercise). Using the materials available, design a simple experiment to investigate this behavior. Several questions follow that might provide ideas. If your experiment requires materials other than those provided, ask your laboratory instructor about their availability. If possible, submit requests in advance.

Procedure

1. Collaborating with your research partner, read the following potential questions, and choose a question to investigate using this list or an original question proposed by your team. You may want to check your text and

other sources for supporting information. You should be able to explain the rationale behind your choice of question.

a. Will courtship and mating in another species be identical to that in *D. melanogaster*?

b. Will males placed in the same vial demonstrate courtship behaviors?

c. Will males respond to dead females?

d. What is the response of a male *D. melanogaster* to females of a different species?

e. Do males compete?

2. Design your experiment, proposing hypotheses, making predictions, determining procedures, and recording results as instructed in Exercise 26.5.

EXERCISE 26.5

Designing and Performing Your Open-Inquiry Investigation

Materials

See each Experiment B Materials list in Exercises 26.1, 26.2, 26.3, and 26.4.

Introduction

Now that you have completed all introductory investigations (Experiment A of each exercise), your research team should choose one of the animals and investigate its behavior. Return to the investigation of your choice and review the suggested questions in Experiment B. Use Lab Topic 1: Scientific Investigation as a reference for designing and performing this independent investigation. You will need to think critically and creatively as you ask questions and formulate your hypothesis.

As a team, review and modify the procedures in Experiment A, determine any additional required materials, review the techniques, and assign tasks to all members of your research team. Your experiments will be successful if you plan carefully, think critically, perform lab techniques accurately and systematically, and record and report data accurately. The following outline will assist you in designing and performing your original investigation.

Procedure

1. **Decide on one or more questions to investigate.** Suggested questions are included in Experiment B as a starting point. (Refer to Lab Topic 1, Exercise 1.1, Lab Study A. Asking Questions.)

   Question:

2. **Formulate a testable hypothesis.** (Refer to Exercise 1.1, Lab Study B. Developing Hypotheses.)

   Hypothesis:

3. **Summarize the essential elements of the experiment.** (Use separate paper.)
4. **Predict the results of your experiment based on your hypothesis.** (Refer to Lab Topic 1, Exercise 1.2, Lab Study C. Making Predictions.)
   Prediction: (if/then)
5. **Outline the procedures to be used in the experiment.** (Refer to Lab Topic 1, Exercise 1.2, Lab Study B. Choosing or Designing the Procedure.)
   a. Review and modify the procedures used in Experiment A for your chosen animal. List each step in your procedure in numerical order.
   b. Design at least one table to record the results of your experiment.
   c. Critique your procedure: check for replicates, levels of treatment, controls, time intervals, experimental conditions, and equipment.
   d. If your experiment requires materials other than those provided, ask your laboratory instructor about their availability. If possible, submit requests in advance.
6. **Perform the experiment,** making observations and collecting data for analysis.
7. **Record results, including observations and data,** in a data table.

   Be thorough when collecting data. Make notes about experimental conditions and observations. Do not rely on your memory for information that you will need when reporting your results. Create at least one figure to illustrate your results.
8. **Prepare your discussion.** Discuss your results in light of your hypothesis.
   a. Review your prediction. Did your results correspond to the prediction you made? If not, explain how your results are different from your predictions, and why this might have occurred.
   b. Review your hypothesis. Review your results (notes, tables, and figures).
      Do your results support or falsify your hypothesis? Explain your answer, using your data for support.
   c. If you had problems with the procedure or questionable results, explain how they might have influenced your conclusion.
   d. If you had an opportunity to repeat and expand this experiment to make your results more convincing, what would you do?
   e. Summarize the conclusion you have drawn from your results.
9. **Be prepared to report your results to the class.** Prepare to persuade your fellow scientists that your experimental design is sound and that your results support your conclusions.

10. If required by your instructor, **submit a scientific paper** describing your experiment for evaluation (see Appendix A). Keep in mind that although you have performed the experiments as a team, any report submitted must be *your original writing*. Your tables and figures may be similar to those of your team members, but your Introduction, Results, and Discussion sections must be the product of your own literature search and creative thinking.

REVIEWING YOUR KNOWLEDGE

1. Define, compare, and give examples for each item in the following pairs.
   a. Learned behavior—innate behavior
   b. Proximate cause of behavior—ultimate cause of behavior
   c. Causal explanation for a behavior—anthropomorphic explanation for a behavior
   d. Taxis—kinesis
2. Define the following terms.
   a. Fixed action pattern
   b. Stimulus-response chain
3. Name four modes of communication that have been described for animals, and give an example for each from your observations of reproductive behavior in fruit flies.

APPLYING YOUR KNOWLEDGE

1. Adult male European robins have red feathers on their breasts. A male robin will display aggressive behavior and attack another male robin that invades his territory during mating season. Immature male robins with all brown feathers do not elicit this behavior in the adult robin. How could you explain this behavior? Design an experiment to test your explanation (hypothesis).

2. Bees, ants, and other social insects are able to detect and will remove dead organisms from their colony. Scientists have analyzed decaying insects and found they give off compounds—mostly oleic acid and linoleic acid—as they decay. You know that social insects display some of the most complex behaviors in response to environmental stimuli of any animal groups. In lab today, you learned that it is possible to explain behaviors based on anthropomorphic causes, but that scientists base explanations on proxi-

mate and ultimate causes (see Lab Topic Introduction). Using bees as the experimental animals, in the space below, propose explanations (hypotheses) based on these three perspectives— anthropomorphic, proximate, and ultimate—and then propose a research project for those hypotheses that can be tested scientifically.

a. Anthropomorphic causes:

Possible research project:

b. Proximate causes:

Possible research project:

c. Ultimate causes:

Could an experiment similar to this be used to investigate ancestors of insects and isopods and to answer questions about when these groups diverged?

STUDENT MEDIA: BioFlix, Activities, Investigations, and Videos

www.masteringbiology.com (select Study Area)

Activities: Ch. 51: Honeybee Waggle Dance Video

Investigations: Ch. 51: How Can Pill Bug Responses to Environments Be Tested?; LabBench: Animal Behavior

Videos: Ch. 51: Ducklings; Chimp Cracking Nut; Snake Ritual Wrestling; Albatross Courtship Ritual; Blue-footed Boobies Courtship Ritual; Chimp Agonistic Behavior; Wolves Agonistic Behavior; Giraffe Courtship Ritual

REFERENCES

Alcock, J. *Animal Behavior: An Evolutionary Approach,* 9th ed. Sunderland, MA: Sinauer Associates, 2009.

Gould, J. L. *Ethology: The Mechanisms and Evolution of Behavior.* New York, NY: W. W. Norton & Company, 1982.

Greenspan, R. J. "Understanding the Genetic Construction of Behavior." *Scientific American,* 1995, vol. 272, no. 4, p. 72.

Johnson, R. N. *Aggression in Man and Animals.* Philadelphia, PA: Saunders College Publishing, 1972.

Marler, P. "Mating Behavior of *Drosophila.*" In *Animal Behavior in Laboratory and Field,* editor A. W. Stokes. San Francisco, CA: Freeman, 1968.

Reebs, S. "Death Whiff." *Natural History,* October 2009. (pill bugs' response to their dead)

Raham, G. "Pill Bug Biology." *The American Biology Teacher,* 1986, vol. 48, no. 1.

Simpson, M. J. A. "The Threat Display of the Siamese Fighting Fish, ." *Animal Behavior Monograph,* 1968, vol. 1, p. 1.

Thompson, T. "Aggressive Behavior of Siamese Fighting Fish," in *Aggressive Behavior,* editors S. Garattini and E. B. Sigg. Proceedings of the International Symposium on the Biology of Aggressive Behavior. New York, NY: Wiley, 1969.

Urry, L., M. Cain, S. Wasserman, P. Minorsky, and J. Reece. *Campbell Biology,* 11th ed. San Francisco, CA: Pearson, 2017.

Waterman, M., and E. Stanley. *Biological Inquiry: A Workbook of Investigative Cases.* San Francisco, CA: Pearson, 2014. In this supplement to *Campbell Biology,* 11th ed., see "Back to the Bay," a case study applying principles of animal behavior to an environmental problem.

WEBSITES

Animal Behavior Society Website. The education section describes experiments submitted by college professors suitable for independent projects. Select Education and Behavior:
http://animalbehaviorsociety.org

Behavior experiments using cockroaches, fruitflies, isopods, and crickets:
www.ableweb.org/volumes/vol-3/7-larsen.pdf

LAB TOPIC 27

Ecology I: Terrestrial Ecology

Laboratory Objectives

After completing this lab topic, you should be able to:

1. Describe the trophic levels of an ecosystem and provide examples from field experience.
2. Describe the environmental factors that are important components of the ecosystem.
3. Calculate density, frequency, dominance, and species diversity.
4. Describe the relative importance of particular species in the ecosystem as determined by their density, size, or role in the ecosystem.
5. Work as a collaborative team, setting up sample plots, collecting data, and making field observations.
6. Construct a model of the ecosystem trophic relationships.

Introduction

Scientists at Hubbard Brook Experimental Forest in New Hampshire have been studying the ecological structure and function of the forest system since 1955, and their Forest Ecosystem Study has focused on the many aspects of forest **ecology** (the study of the relationship of organisms and their biological and physical environments). They have one of the longest continuous and complete data sets of forest science in the world. Hubbard Brook continues systems-level research that investigates interactions among components of the forest to understand this complex ecological system. For example, scientists are studying the effects of air pollution and climate change on soil properties and plant diversity, and the effect of seed and nut production on the population dynamics of rodents. This kind of long-term experimental research at the ecosystem level is essential to understanding the effects of a warming climate, as well as predicting changes in ecosystems in the future.

The National Science Foundation (NSF) has recently funded NEON (National Ecological Observation Network) to establish 81 monitoring sites in diverse ecological regions across the continent. Scientists will use standardized experimental designs and procedures to collect data for air quality, water quality, soil temperature, precipitation, plant presence and cover, as well as invertebrates and small mammals. In this laboratory, you will be collecting similar types of data to characterize a local ecosystem. You can also contribute to the NEON database by participating in Project BudBurst, collecting records of first flowering dates across the continent (http://www.neonscience.org/learn-experience/project-budburst). NEON will provide long-term data that can be utilized by scientists and managers across disciplines to pursue questions at the systems level.

Ecological investigations may address questions at several hierarchical levels: **individuals**, **populations** (organisms of the same species that share a common gene pool and occur in the same area), **communities** (populations of different species that inhabit the same area), and **ecosystems** (the community of plants and animals plus the physical environment). A physiological ecologist studies the effects of the environment on individual organisms. A population ecologist might be interested in questions about the reproductive biology of a population of endangered plant species. The community ecologist might investigate the sequence of species composition changes in a forest following a disturbance. And the ecosystem ecologist might question how biological diversity or the interactions among members of different trophic levels (feeding levels) are influenced by environmental conditions.

An ecosystem can be divided into **biotic** components (living organisms) and **abiotic** components (physical features). In a forest ecosystem, the abiotic components include the climatic factors, soil type, water availability, and landscape features. The biotic components of the forest include trees, shrubs, wildflowers, squirrels, foxes, caterpillars, eagles, spiders, millipedes, fungi, and bacteria on the forest floor and in the soil (Figure 27.1). The biotic components can be further characterized based on **trophic structure**, the ecological role of organisms in the food chain. Plants and some protists are categorized as **primary producers** (autotrophic organisms) capable of transforming light energy into chemical energy stored in carbohydrates through the process of photosynthesis. *The amount of energy available for all other trophic levels is dependent on the photosynthetic ability of the primary producers.* **Consumers** are animals and heterotrophic protists, which literally consume the primary producers or each other or both. They may be divided into **primary consumers** (**herbivores**, which consume plants) and **secondary** and **tertiary consumers** (**carnivores**, which eat other consumers). Rarely do ecosystems support additional levels of consumers. In general, ecosystems have a **trophic efficiency**, the energy available at one trophic level that is transferred to the next trophic level, of only 10–20%. The number of organisms that can be supported at subsequent higher trophic levels is limited by the available energy captured and stored by primary producers. **Detritivores** obtain their nutrients and energy from dead organisms and waste materials, and **decomposers** (fungi and bacteria) absorb nutrients from nonliving organic material.

With the assistance of your lab partner, for each of the following trophic levels provide an example of an organism(s) from the forest system just described. Remember that the trophic level reflects who is eating whom (Figure 27.1).

Primary producers:

Primary consumers:

Secondary consumers:

Tertiary consumers:

Detritivores:

Decomposers:

Note that some organisms fit into more than one trophic level—because, like many humans, some are omnivores, eating whatever is available.

FIGURE 27.1
Forest ecosystem. A simplified forest system showing trophic (feeding) relationships among primary producers, consumers, detritivores, and decomposers.

Designing and Organizing Your Investigation

In this lab topic, you will investigate the structure and function of a local ecosystem. The exercise is designed for a forest ecosystem but can be adapted easily for use in grasslands or even a weedy urban lot. (An outline for adapting the lab for use in a weedy field on or near your campus is included at the end of the lab topic.) The study site has been selected in advance by your instructor, who may have prepared an introductory description of the site.

This lab topic provides diverse resources and methodologies to pursue questions about the trophic structure of a forest ecosystem, or to perform a comparative study of two sites, for example, with different records of disturbance (cutting, herbicides, or invasive species). As another option, you might choose to study biological diversity for only one component of the ecosystem, for example, the litter layer. *As a class, discuss the possible questions that you will pursue. Record a brief description of your study site, and then state the question.*

Study Site

Question

Organization of Teams

Based on your question, select the components of the ecosystem (see Table 27.1) that you will be investigating. Next, determine the number of replicate plots; this will be the number of teams. Each team, composed of six to eight students, will sample one plot, thus providing three or four replicate samples, depending on the number of teams. Within teams, each student will have specific assigned responsibilities, as suggested in Table 27.1 and fully described in the following exercises. If you choose to focus on only one or two components of the ecosystem, then you may be able to sample more replicates or compare two sites.

Following the field sampling, you will share results with other teams, make calculations, pool and analyze data, and develop a model of the ecosystem. Students should read assignments for all field studies.

TABLE 27.1 Suggested Organization of Student Teams

Exercise	Sampling	No. of Students
Exercise 27.1	**Biotic Components**	
A	Trees	2
B	Shrubs, saplings, and vines	2
C	Seedlings and herbaceous vegetation	2*
D	Macroinvertebrates	2*
E	Microinvertebrates	2**
F	Microorganisms	2*
G	Other forest animals	All
Exercise 27.2	**Abiotic Components**	**2****

*Two students can complete Field Studies C, D, and E.

**Two students can complete both Exercise 27.2 and Field Study F in Exercise 27.1.

EXERCISE 27.1

Biotic Components

As you begin your field observations of the forest, note the vertical layers of the forest from the forest floor up to the tallest and largest trees, which form a canopy. The forest vegetation can be subdivided arbitrarily into categories according to the structural pattern of the forest (Figures 27.2 and 27.7): forest **trees** (woody plants with a diameter at breast height (DBH) of >10 cm); **shrubs**, **saplings**, and **vines** (DBH 2.5–10 cm); and **seedlings** and **herbaceous plants** (DBH <2.5 cm). *(Note that DBH is measured at a height of 1.5 m from the base of the tree.)* **Microorganisms** and **small animals** can be sampled from the forest floor, **litter** (fallen leaves, for example), and **soil**, with **large animals** observed directly or indirectly by animal signs (nest sites, feces, tracks).

In your student teams, you will sample part of the study site using circular plots to estimate the abundance of plants and animals in each of the categories just described. The size of the circular plots varies according to the size and abundance of the organisms (Table 27.2). Thus, trees are sampled using the largest plot, whereas the herbs of the forest floor are sampled in small plots. Imagine the difficulty of sampling trees in plots of only 1 m^2 or counting all the non-woody plants in a plot with a diameter of over 10 m^2!

FIGURE 27.2
Vertical stratification of the forest. The forest can be divided into layers: the trees (DBH >10 cm); shrubs, saplings, and vines (DBH 2.5–10 cm); and seedlings and herbaceous plants (DBH <2.5 cm). See also Figure 27.7.

TABLE 27.2 Sampling Design for Determining the Biotic Components of a Forest Ecosystem

Organisms	Plot Size
Trees (DBH >10 cm)	100 m^2
Shrubs, saplings, and vines (DBH 2.5–10 cm)	50 m^2
Seedlings and herbs (DBH <2.5 cm)	0.50 m^2
Litter macroinvertebrates	0.50 m^2
Microinvertebrates from litter and soil samples	—
Microorganisms from litter and soil samples	—
Large animals observed by all	—

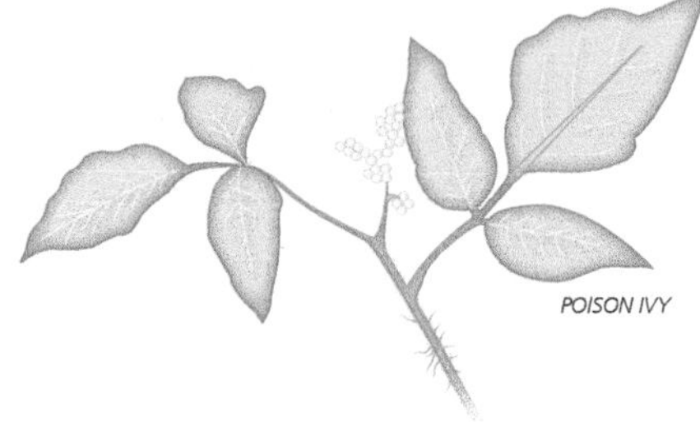

Avoid contact with poison ivy. Your instructor will identify the plant for you. After completing the fieldwork, thoroughly wash all areas of exposed skin with soap and cold water. Avoid contact with your clothing if you think you have come into contact with poison ivy. Notify your instructor if you are allergic to bee stings or other plants and animals that you might encounter outdoors.

Field Study A. Trees

Materials

3 lines, 5.64 m long, with a clip on one end and a stake on the other end of each
center post
mallet
DBH measuring tape
permanent marker
re-sealable plastic bags

Introduction

Trees form the uppermost layer, or **canopy**, of the forest. They influence the physical environment, such as the quality and quantity of light reaching all other vegetation, water availability, and temperature. A forest description is usually based on the largest and most abundant species of trees, for example, oak-hickory forests of the Southeast.

Procedure

1. Determine the sample plot location based on recommendations from your instructor. The plot center point may have been flagged in advance.
2. Locate the center of the sample plot and use the mallet to hammer the center post into the ground at that location.
3. Clip one end of a line to the center post. Extend the line, keeping it straight and taut, and hammer the stake into position (Figure 27.3, line 1).
4. Attach and secure lines 2 and 3 at an angle, forming two wedge-shaped sections, A and B (see Figure 27.3). Each wedge should be equal to approximately one-fifth of the plot.

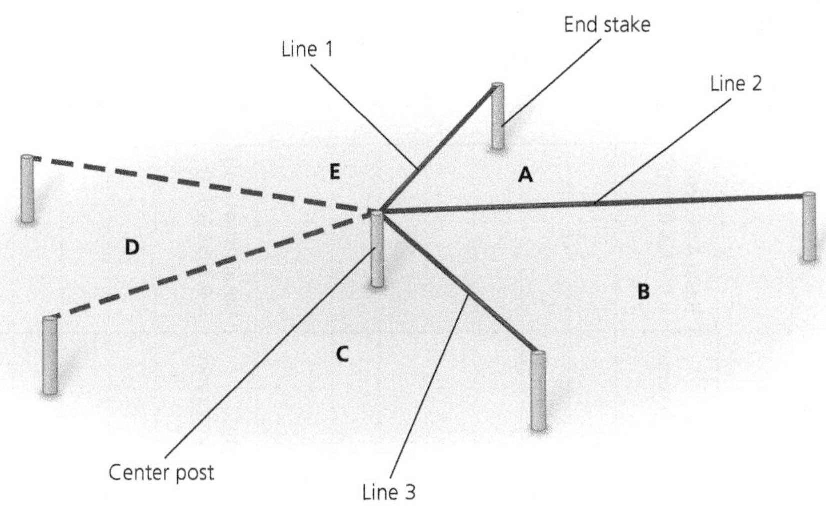

FIGURE 27.3
Establishing and sampling a circular plot. Locate the center post and attach three lines (1, 2, 3) to form two wedge-shaped sections, A and B. After sampling A and B, move line 2 to form section C. After sampling C, move line 3 to form sections D and E.

5. Identify and measure DBH (>*10 cm*) for all trees in sections A and B (see Figure 27.3). The number of DBH measurements will also provide a record of abundance. If a tree is on the outer perimeter of the plot, it should be counted in the plot if at least 50% of the tree is within the plot.

> Diameter at breast height (DBH) is measured 1.5 m above the base of the tree. Measure or estimate this height on your body to ensure that you determine DBH in a consistent fashion. If DBH measuring tapes are not available, measure trees with a circumference of >32 cm.

6. If you cannot identify a tree, assign a number to that type of tree. Collect a leaf sample and place in a numbered re-sealable plastic bag for later identification. If leaves are not available, write a brief description and sketch or photograph.
7. *Before moving any lines*, check with students of the shrub, sapling, and vine group to be sure they have completed their sampling. Their plot is smaller and lies within the boundary of your tree plot.
8. Do not move line 1. Once all groups have completed sampling sections A and B, create section C. Detach line 2 and move it into position to form the boundary between sections C and D (the first dotted line in Figure 27.3).
9. Identify and measure DBH for all trees in section C (see Figure 27.3).
10. Detach line 3 and move it into position to form the boundary between sections D and E (the second dotted line in Figure 27.3).
11. Identify and measure DBH for all trees in sections D and E (see Figure 27.3).

Results

1. Record your data in Table 27.3 on the next page.
2. In Exercise 27.3, Data Analysis, you will calculate:

   DBH (if measurements were taken of circumference rather than DBH)

   Density and relative density

   Frequency and relative frequency

   Basal area, dominance, and relative dominance

   Importance value

TABLE 27.3 Sampling Results for **Trees** (DBH >10 cm)

Download an Excel version of this table from www.masteringbiology.com. Look in the Study Area under Lab Media.

Locality: ______________ **Plot ID #:** ______________ **Plot size:** ______________ **Date:** ______________

Students: ______________________________ **Instructor:** ______________________________

SPECIES:			SPECIES:			SPECIES:		
Circumference (cm)	DBH (cm) $= \frac{\text{Circumference}}{\pi}$	Basal Area (cm^2) $= 0.7854\,(\text{DBH})^2$	Circumference (cm)	DBH (cm) $= \frac{\text{Circumference}}{\pi}$	Basal Area (cm^2) $= 0.7854\,(\text{DBH})^2$	Circumference (cm)	DBH (cm) $= \frac{\text{Circumference}}{\pi}$	Basal Area (cm^2) $= 0.7854\,(\text{DBH})^2$

SPECIES:			SPECIES:			SPECIES:		
Circumference (cm)	DBH (cm) $= \frac{\text{Circumference}}{\pi}$	Basal Area (cm^2) $= 0.7854\,(\text{DBH})^2$	Circumference (cm)	DBH (cm) $= \frac{\text{Circumference}}{\pi}$	Basal Area (cm^2) $= 0.7854\,(\text{DBH})^2$	Circumference (cm)	DBH (cm) $= \frac{\text{Circumference}}{\pi}$	Basal Area (cm^2) $= 0.7854\,(\text{DBH})^2$

SPECIES:			SPECIES:			SPECIES:		
Circumference (cm)	DBH (cm) $= \frac{\text{Circumference}}{\pi}$	Basal Area (cm^2) $= 0.7854\,(\text{DBH})^2$	Circumference (cm)	DBH (cm) $= \frac{\text{Circumference}}{\pi}$	Basal Area (cm^2) $= 0.7854\,(\text{DBH})^2$	Circumference (cm)	DBH (cm) $= \frac{\text{Circumference}}{\pi}$	Basal Area (cm^2) $= 0.7854\,(\text{DBH})^2$

Field Study B. Shrubs, Saplings, and Vines

Materials

circular plot established for trees in Field Study A; the radius lines should be flagged conspicuously at 4 m for the SSV plot
re-sealable plastic bags
calipers (or DBH tape)
permanent marker

Introduction

Lower vertical layers of the forest are inhabited by young trees of the types seen in the canopy, trees and shrubs unique to this layer, and some vines. The shrub, sapling, and vine (SSV) plot is located within the tree plot (Figure 27.4). Therefore, each line should be boldly marked to indicate the radius for 50 m^2.

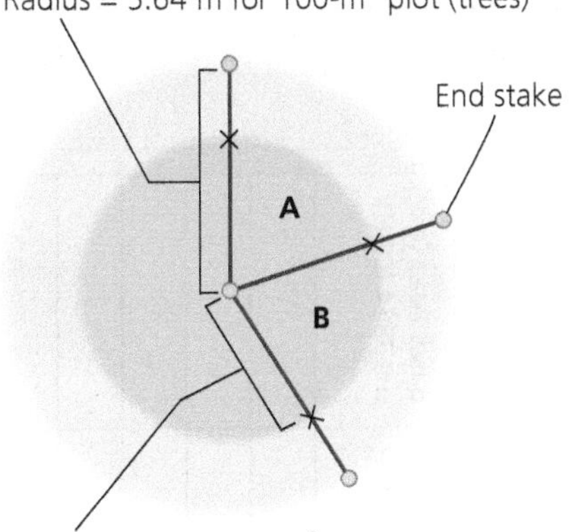

FIGURE 27.4
Placement of circular plot for shrubs, saplings, and vines.
A circular plot with a smaller diameter is marked within the boundary of the large tree plot. Radius lines are clearly marked to indicate the outer perimeter for the plot (Xs).

Coordinate your sampling with the tree group, Field Study A.

Procedure

1. Locate the circular plot established by your team members who are sampling trees. Restrict your sampling to the smaller plot located within the tree plot. The radius of the smaller plot should be marked clearly with flagging or tape along each line. Sample vines and other woody plants with a *DBH of 2.5–10 cm* within the SSV plot.
2. Begin sampling in section A (see Figure 27.3). Identify and measure each individual in the SSV category for all sections of the plot. If available, use calipers, rather than a meter tape, to determine DBH.

Diameter at breast height (DBH) is measured 1.5 m above the base of the woody plant. Measure or estimate this height on your body to ensure that you determine DBH in a consistent fashion. If DBH measuring tapes are not available, measure plants with a circumference of 8–32 cm.

3. If you cannot identify a woody plant, assign a number to that type of plant. Collect a leaf sample and place in a numbered re-sealable plastic bag for later identification. If leaves are not available, write a brief description and sketch or photograph.

Results

1. Record your data in Table 27.4 on the next page.
2. In Exercise 27.3, Data Analysis, you will calculate:

   DBH (if measurements were taken of circumference rather than DBH)
   Density and relative density
   Frequency and relative frequency
   Basal area, dominance, and relative dominance
   Importance value

TABLE 27.4 Sampling Results for **Shrubs, Saplings,** and **Vines** (DBH 2.5–10 cm)

Download an Excel version of this table from www.masteringbiology.com. Look in the Study Area under Lab Media.

Locality: ________________ **Plot ID #:** ________________ **Plot size:** ________________ **Date:** ________________

Students: ________________________________ **Instructor:** ________________________________

SPECIES:

Circumference (cm)	DBH (cm) $= \frac{\text{Circumference}}{\pi}$	Basal Area (cm^2) $= 0.7854\ (\text{DBH})^2$

SPECIES:

Circumference (cm)	DBH (cm) $= \frac{\text{Circumference}}{\pi}$	Basal Area (cm^2) $= 0.7854\ (\text{DBH})^2$

SPECIES:

Circumference (cm)	DBH (cm) $= \frac{\text{Circumference}}{\pi}$	Basal Area (cm^2) $= 0.7854\ (\text{DBH})^2$

SPECIES:

Circumference (cm)	DBH (cm) $= \frac{\text{Circumference}}{\pi}$	Basal Area (cm^2) $= 0.7854\ (\text{DBH})^2$

SPECIES:

Circumference (cm)	DBH (cm) $= \frac{\text{Circumference}}{\pi}$	Basal Area (cm^2) $= 0.7854\ (\text{DBH})^2$

SPECIES:

Circumference (cm)	DBH (cm) $= \frac{\text{Circumference}}{\pi}$	Basal Area (cm^2) $= 0.7854\ (\text{DBH})^2$

SPECIES:

Circumference (cm)	DBH (cm) $= \frac{\text{Circumference}}{\pi}$	Basal Area (cm^2) $= 0.7854\ (\text{DBH})^2$

SPECIES:

Circumference (cm)	DBH (cm) $= \frac{\text{Circumference}}{\pi}$	Basal Area (cm^2) $= 0.7854\ (\text{DBH})^2$

SPECIES:

Circumference (cm)	DBH (cm) $= \frac{\text{Circumference}}{\pi}$	Basal Area (cm^2) $= 0.7854\ (\text{DBH})^2$

Field Study C. Seedlings and Herbaceous Vegetation

Materials

circular 0.50-m^2 plot
compass
re-sealable plastic bags
permanent marker

Introduction

Tree seedlings and herbaceous plants appear near the forest floor. These plants may be difficult to count, and, in some cases, it may be almost impossible to determine what is actually one individual. Because of this, the abundance of these plants is estimated based on the percent cover (percent of the sample area covered by the plant).

The seedling and herb plot is not located within the larger plots but should be placed near the tree plot at a predetermined position.

Procedure

1. Place a 0.50-m^2 plot just outside the tree plot. Determine the exact location for the plot before beginning. For example, it might always be 1 m away from the tree plot at a compass heading of due north.
2. Identify and estimate the abundance of each species by estimating the percent of the plot area covered by the species. Sketching the plot may help determine cover and ensure consistency in your estimates.
3. If you cannot identify a plant, assign a number to that type of plant. Collect a leaf or flower sample if possible and place in a numbered re-sealable plastic bag for later identification. If leaves or flowers are not available, write a brief description and sketch or photograph.

Results

1. Record results in Table 27.5 on the next page.
2. In Exercise 27.3, Data Analysis, you will calculate average percent cover, frequency, and relative frequency.

TABLE 27.5 Sampling Results for **Seedlings** and **Herbaceous Vegetation** (DBH <2.5 cm)			
Download an Excel version of this table from www.masteringbiology.com. Look in the Study Area under Lab Media.			
Locality: _______ **Plot ID #:** _______		**Plot size:** _______ **Date:** _______	
Students: _______		**Instructor:** _______	
Species	**Percent Cover**	**Species**	**Percent Cover**

Field Study D. Macroinvertebrates

Materials

circular 0.50-m^2 plot
thin plastic sheet
forceps
dissecting probes
vials with 70% alcohol
labeling tape
permanent marker
compass
re-sealable plastic bags

Introduction

Large invertebrates—for example, grasshoppers, ants, and spiders—can be surveyed by sifting through the dead leaves or litter on the forest floor. These animals may fit into any of the consumer trophic levels. For example, grasshoppers are primary consumers, spiders are secondary consumers, and ants are omnivores.

TABLE 27.6 Sampling Results for **Consumers, Detritivores,** and **Decomposers**

Download an Excel version of this table from www.masteringbiology.com. Look in the Study Area under Lab Media.

Locality: ______ **Plot ID #:** ______ **Date:** ______

Students: ______ **Instructor:** ______

Macroinvertebrates		Microinvertebrates
Species/Group	**No. of Individuals**	**Groups Observed**
Microorganisms:		
Vertebrates:		

Procedure

1. Carefully place a 0.50-m^2 circular plot on the ground in an undisturbed spot outside and near the large vegetation plot. This should be a preselected spot; for example, it might always be 1 m away from the large plot at a compass heading of due south. Avoid disturbing the animals present. Quickly cover the plot with plastic to discourage "escapees."
2. Carefully remove all animals from the vegetation and litter at ground level.
3. Sort animals into similar groups in the field. Place the organisms into vials of 70% alcohol according to their group. If you cannot identify an animal, assign a number to that type of animal. Preserve the animal in alcohol and label the sample using tape and a permanent marker.

Results

1. Record results in Table 27.6.
2. In Exercise 27.3, Data Analysis, you will calculate density for the groups.

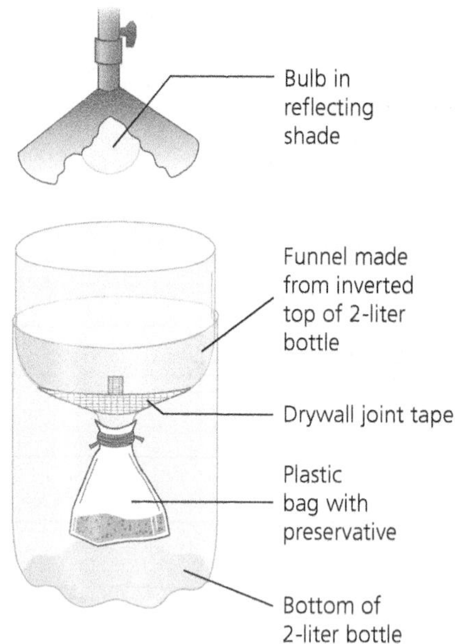

FIGURE 27.5

Berlese-Tullgren funnel used for the extraction of microinvertebrates from litter and soil. The samples are placed in a funnel with the narrow opening covered with two crossed pieces of drywall tape, adhesive side down, or fine mesh window screening. The light and heat drive the animals down into the small sampling bag containing preservative. (Setup modified from Kingsolver [2006].)

Field Study E. Microinvertebrates

Materials

circular 0.50-m^2 plot
index cards
self-sealing plastic bags
Berlese-Tullgren funnels
light sources
soil sample collected in Exercise 27.2, Field Study A
cheesecloth

Introduction

Various invertebrates, including mites and springtails, inhabit the litter layer and the upper layers of the soil. Many of these animals are not easily seen with the naked eye and would be impossible to collect by inspecting samples. These consumers and detritivores can be sampled by collecting soil and litter and placing these in Berlese-Tullgren funnels (Figure 27.5). The animals move away from the light and heat source into a small sampling vial. The animals can then be identified in the laboratory using a dissecting microscope. You may be asked to collect samples for extraction later in the lab, or your instructor may have already collected the samples and started the extractions.

Procedure

1. Place the 0.50-m^2 plot in an undisturbed location. Carefully remove half the litter (discard sticks) and place it into the bag provided. Include a card with the plot number in your sample bag. Close the bag securely.
2. Upon return to the lab, extract the microinvertebrates using a Berlese-Tullgren funnel. A soil sample will also be collected by the soil group (see Exercise 27.2, Field Study A), and microinvertebrates will be extracted using the same procedure.
   a. Place the soil or litter sample on the mesh tape in a large funnel under a light bulb. *The soil samples can be wrapped in a layer of cheesecloth to prevent soil particles from sifting into the collection bags.*
   b. The light and heat force the animals down the funnel into the collecting bag containing 70% alcohol. (See Figure 27.5 and the Preparation Guide for extraction funnel setup.)
3. After 24 to 48 hours, the organisms can be sorted and identified using the dissecting scope. Refer to the illustrated key to common microinvertebrates in Figure 27.6.

Results

1. Record in Table 27.6 the microinvertebrates present.
2. Sketch the most common organisms in the margin of your lab manual if they are not represented in the illustrated key.

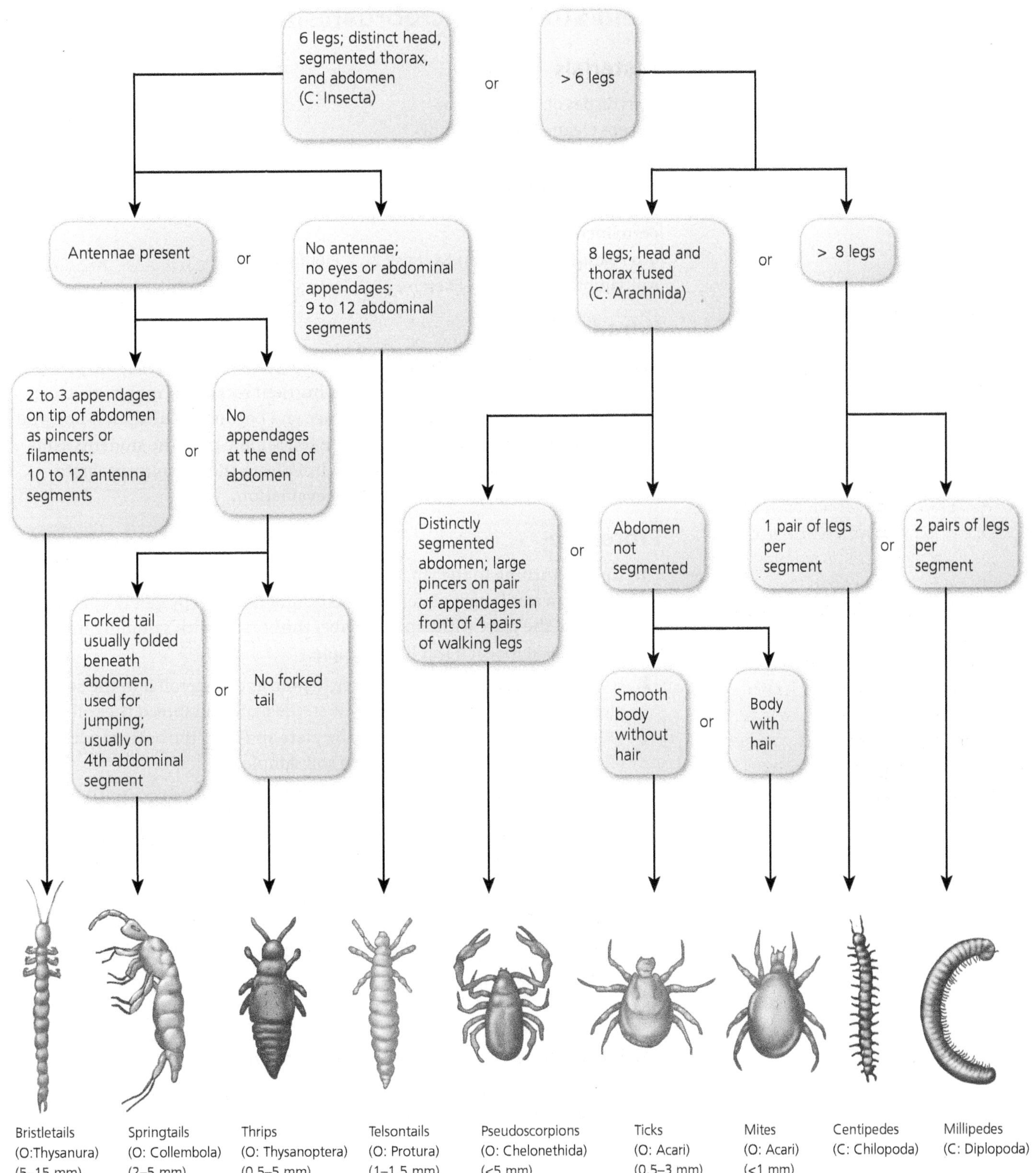

FIGURE 27.6
Picture key to microinvertebrates commonly found in litter and soil samples.
C: class; O: order.

Field Study F. Microorganisms

Materials

petri dishes of nutrient agar
forceps cleaned with alcohol
sterile swabs
test tube with sterile water
tape
permanent marker
litter sample collected in Exercise 27.1, Field Study E
soil sample collected in Exercise 27.2, Field Study A

Introduction

Microorganisms in the litter and soil are important decomposers, the primary functions of which are soil building and nutrient recycling. Your instructor may have collected the samples in advance and begun the cultures. If not, you will need to obtain a small portion of the soil sample from the students sampling the abiotic components and some litter from the microinvertebrate group. You will prepare cultures for later evaluation.

Procedure

1. Upon returning to the lab, use clean forceps to place a small sample of leaf litter on a sterile agar plate. Avoid exposing the plate by barely opening the top. Seal the plate with tape and label the bottom with sampling information. Repeat for each leaf litter sample.
2. For each soil sample, moisten a sterile swab and then roll it in the soil sample. Carefully open the top of a new sterile plate and spread the soil sample onto the surface of the agar. Seal the plate and label the bottom with sampling information. Repeat for each soil sample.
3. Incubate the plates at room temperature for a couple of days until bacteria colonies and fungal growth are abundant, and then refrigerate to prevent further growth.
4. Observe the diversity of colonies present after 2 or 3 days.

Results

Record the presence of different bacterial colony types and fungi in Table 27.6. Refer to Lab Topic 12 Bacteriology for help with determining the microbial diversity using colony characteristics.

Field Study G. Other Forest Animals

Materials

binoculars
field guides

Introduction

You will not trap forest vertebrates in this field study, but all students will need to make observations of these larger but often secretive animals. Observe the activities of these animals as well as their calls, tracks, nest sites, and feces. Think about their role in the ecosystem.

Procedure

1. After completing your field sampling, survey the surrounding forest for birds, mammals, reptiles, and amphibians.
2. Record observations of animals sighted, plus any tracks, nests, burrows, carcasses, feces, owl pellets, songs, or calls. Field guides are provided to assist with identification. See also suggestions for apps and field guides to assist in identification in References and Websites at the end of this lab topic.

Results

Record your observations in Table 27.6.

EXERCISE 27.2

Abiotic Components

Abiotic components include the physical features of the environment that influence the biotic components and in turn may be influenced by the organisms inhabiting the area. You will record the climatic conditions both for the local environment and for the smaller-scale microenvironment within the ecosystem. Two students in each team will measure and record all abiotic features.

Field Study A. Soil

Materials

soil thermometer
soil auger
self-sealing plastic bags
permanent marker
trowel
diagram of soil profile
soil map of the area
soil test kit

Introduction

The soil is a biotic component of the ecosystem with living organisms and dead organic matter. The soil is also abiotic, characterized by minerals, atmosphere, water, and other inorganic materials. In this exercise, you will determine the soil type and collect samples to be used in extracting microinvertebrates (Exercise 27.1, Field Study E) and microorganisms (Exercise 27.1, Field Study F).

Procedure

1. Read about soil temperature (Field Study A) and air temperature (Field Study B); then place thermometers in appropriate locations to equilibrate while you begin your other work.
2. Remove the vegetation from a small undisturbed area near the tree plot. Place the soil thermometer in the soil and allow it to equilibrate for 10 to 15 minutes before recording the temperature.
3. Carefully remove the leaf litter. Using the soil auger, remove two soil cores approximately 30–45 cm long, if possible.
4. Sketch or photograph the profile of the soil layers. (See an example of a soil profile in your materials.) Note changes in color and texture. How much sand or clay or dark organic matter appears to be present?
5. Place each soil sample in a separate plastic (self-sealing) bag. Follow the directions provided with the soil test kit. Determine the soil pH and any other characteristics suggested by your instructor for one sample. Instructions are included with the soil test kit. Save the other sample to share with your team.

6. Remove a third sample of soil to be used for determining the presence and types of microinvertebrates and microorganisms. Collect this sample by using a trowel to dig some of the upper loose dirt. Seal this sample in a third plastic bag. Label it "organisms" with the plot number. The organisms will be extracted as described in Exercise 27.1, Field Studies E and F.
7. Study the soil association map of this general area and determine the type of soil typical of this region.

Results

1. Record the soil temperature and pH in Table 27.7.
2. Sketch or photograph the soil profile and indicate the layers and soil colors.
3. Record in Table 27.7 the soil type and other pertinent information from soil maps.

Field Study B. Climatology

Materials

NOAA climatological data sheets
sling psychrometer (with directions)
min/max thermometer (with directions)
gun-style infrared thermometers
wind anemometer (with directions)
light meter
thermometers

Introduction

The climatic conditions of the environment can be viewed on the local or regional level or on a smaller scale (microclimate) within the forest ecosystem. Not only does the microclimate affect the organisms that are present, but the process is mutual: Organisms will influence the microclimatic conditions as well. For example, the air temperature above bare ground and open grassland may differ significantly owing to differences in water loss and the shading effect of the vegetation.

Procedure

1. *Local climate.* For general climate patterns in your area, consult the local climatological data sheets provided or find local weather information by entering your zip code on the Weather Service website at http://w2.weather.gov/climate/local_data.php?wfo=ffc. In Table 27.7, record the average annual rainfall for this month, temperature averages, and the relative humidity at 1300 hours (1 P.M.).
2. *Microclimate.* Variation in microclimate due to such factors as elevation, slope, and shade can result in temperatures, humidities, and light intensities quite different from those of surrounding areas only a few meters away. For example, the microclimate conditions near ground level will be different from those a few meters above the surface. Measure the following microclimate conditions and record your results in Table 27.7.
   a. *Air temperature.* Record the temperature at the soil surface and 2 m or so above the ground for a sunny spot and a shady spot. Allow 5 to 10 minutes for the thermometer to equilibrate at each position. Record the minimum and maximum temperature for the previous 24 hours if a min/max thermometer was left in the forest overnight.

TABLE 27.7 Characteristics of the Physical Environment

Locality: ____________________ **Plot ID #:** ____________ **Date:** ____________

Students: ________________________ **Instructor:** ______________________

Soil

Temperature (°C):

pH:

Other:

Description:

Local Climate

Annual rainfall:

Annual average temperature:

Relative humidity:

Microclimate

Temperature

Soil surface: sun: ______ shade: ______

2 m above surface: sun: ______ shade: ______

Min ____________

Max ____________

Light

Forest ____________

Full sun ____________

% full sun ____________

Relative humidity ____________

Wind speed ____________

Topography

Disturbance

Comments

b. *Light.* Using a light meter, measure the light intensity in the forest and in full sunlight. Calculate the percentage of full sun that reaches the forest floor.

c. *Humidity.* Humidity refers to the amount of water vapor in the air. It has important biological effects on respiration, transpiration, and evaporation. **Relative humidity** (the most common measurement) is the actual amount of water vapor in the air divided by the total possible water vapor (or saturation vapor) in the air at its temperature. Refer to the directions provided and measure relative humidity with a **sling psychrometer**. *Do not hit anyone with the psychrometer!*

d. *Wind.* Using the **wind anemometer**, measure the wind speed in the sampling area. What factors can you suggest that affect wind speed?

Results

Record in Table 27.7 the results of your measurements for air temperature, light, humidity, and wind.

Field Study C. Topographic Features

Materials

field notebook
pencil
USGS topographic map for study site

Introduction

Before completing your study of the abiotic components, you should stand back and observe the general features of the landscape that are difficult to measure but that may influence other physical factors and the organisms as well.

Procedure

Observe the topography of the sampling site. Walk to the highest and/or lowest point in your sampling area. Is the area sloping, eroded, on a hillside, or cut by a stream or ditch? Record in Table 27.7 indications of past or present disturbance—for example, scarring from past fires, cut stumps, old fences, or terracing.

Results

1. Record the results of your observations in Table 27.7.
2. In the margin of your lab manual, sketch the topographic features that appear to influence your sample site.

EXERCISE 27.3

Data Analysis

Materials

summary of all student data
calculator

Introduction

Students will need to pool data with team members and other teams. You may determine the *abundance* (density or cover), *distribution* (frequency), overall *size* (dominance), and *importance* (the sum of relative density, frequency, and dominance) for each species in the sampling categories (trees, SSV, and so on).

Density, the number of individuals per unit area, provides a summary of abundance by species. (**Percent cover** is the measure of abundance used for seedlings and herbaceous plants.) However, the density of a species does not necessarily reflect the distribution of the species on the landscape. For example, you might sample 20 white oaks in one plot or 5 white oaks in each of four plots. The number of white oaks will be the same, but the distribution of the white oaks is very different. **Frequency** provides information on the distribution of a species and is calculated as the percent of plots sampled that have at least one individual of the species present. **Dominance** is a measure of the influence of a species based on the size of individuals. (Dominance is determined from the area, called basal area, calculated from DBH measurements.) Adding the relative values for each of these measures provides an estimate of overall **importance**, which includes abundance, distribution, and size. The equations for calculating these parameters are provided in Table 27.8 on the next page.

Procedure

Trees, Shrubs, Saplings, and Vines

1. If circumference was measured for trees and shrubs, saplings, and vines, then calculate DBH for individuals in each sample plot and record the results in Tables 27.3 and 27.4, respectively.
2. Calculate basal area for each tree and shrub, sapling, and vine. Record your results in Tables 27.3 and 27.4, respectively.
3. Pool the data for all sample plots and record for each species the total number of individuals, number of plots in which a species was present, and total basal area. Summarize tree data in Table 27.9, and SSV data in Table 27.10.
4. Calculate density. Total the densities for all species and calculate the relative density of each species.
5. Calculate frequency. Total the frequencies for all species and calculate the relative frequency of each species.
6. Calculate dominance. Total the dominance for all species and calculate the relative dominance of each species.
7. Calculate importance value by totaling the relative density, relative frequency, and relative dominance for each species.

TABLE 27.8 Calculating Density, Frequency, Dominance, and Importance Values for the Biotic Components of the Ecosystem

DBH $= \dfrac{\text{circumference}}{\pi}$	**Basal area** $= 0.7854\ (\text{DBH})^2$
Dominance $= \dfrac{\text{total basal area}}{\text{total area sampled}}$	
Relative dominance $= \dfrac{\text{dominance for a species}}{\text{total dominance for all species}} \times 100$	
Density $= \dfrac{\text{no. of individuals}}{\text{total area sampled}}$	
Relative density $= \dfrac{\text{density for a species}}{\text{total density for all species}} \times 100$	
Frequency $= \dfrac{\text{number of plots in which species recorded}}{\text{total number of plots sampled}}$	
Relative frequency $= \dfrac{\text{frequency for a species}}{\text{total frequency for all species}} \times 100$	
Average percent cover $= \dfrac{\text{total percent cover}}{\text{total number of plots sampled}}$	
Importance value = relative density + relative dominance + relative frequency	

Seedlings and Herbs

1. Pool the data for all sample plots and record in Table 27.11 the percent cover for each species and the number of plots in which a species was present.
2. Calculate average percent cover.
3. Calculate frequency. Sum the frequencies for all species and calculate the relative frequency of each species.

Macroinvertebrates

1. Pool the data for all sample plots and record in Table 27.12 the total number of individuals for each species and the number of plots in which a species was present.
2. Calculate density. Sum the densities for all species and calculate the relative density of each species.
3. Calculate frequency. Sum the frequencies for all species and calculate the relative frequency of each species.

Abiotic Components

Pool the data for all sample plots and record it in Table 27.13.

Results

1. Record all summary results for *trees* in Table 27.9 and those for *shrubs, saplings,* and *vines* in Table 27.10.
2. List in the spaces provided the three most important tree species and the three most important shrub, sapling, and vine species.

   Trees:

   Shrubs, saplings, and vines:

3. Record the summary results of *seedlings* and *herbs* in Table 27.11.
4. List in the space provided the three most common seedling and herb species based on their abundance.

   Seedlings and herbs:

5. Record summary results for *macroinvertebrates* in Table 27.12.
6. List in the space provided the three most common macroinvertebrate species based on abundance. Indicate, if possible, the appropriate trophic level: *primary* (1°) or *secondary* (2°) *consumers,* or *detritivores* (D) by placing the appropriate letter or number by each. Remember that some invertebrates may occupy more than one trophic level. To make these determinations, observe mouthparts or other body structures. Consult reference books or handouts provided in the laboratory or from the library. See "Who Eats What" (Hogan, 1994).

   Macroinvertebrates:

TABLE 27.9 Summary of Results for **Trees**

Download an Excel version of this table from www.masteringbiology.com. Look in the Study Area under Lab Media.

Locality: ____________ **Size class:** Trees **Date:** ____________

No. of plots sampled: ____________ **Total area sampled:** ____________

Species	Total No. of Individuals	Density	Relative Density	No. of Plots Present	Frequency	Relative Frequency	Total Basal Area	Dominance	Relative Dominance	Importance Value
Totals			100			100			100	

TABLE 27.10 Summary of Results for **Shrubs, Saplings,** and **Vines**

Download an Excel version of this table from www.masteringbiology.com. Look in the Study Area under Lab Media.

Locality: ____________ **Size class:** SSV **Date:** ____________

No. of plots sampled: ____________ **Total area sampled:** ____________

Species	Total No. of Individuals	Density	Relative Density	No. of Plots Present	Frequency	Relative Frequency	Total Basal Area	Dominance	Relative Dominance	Importance Value
Totals			100			100			100	

TABLE 27.11 Summary of Results for **Seedlings** and **Herbaceous Vegetation**

Download an Excel version of this table from www.masteringbiology.com. Look in the Study Area under Lab Media.

Locality: ____________ **Size class:** S/Herb **Date:** ____________

No. of plots sampled: ____________ **Total area sampled:** ____________

Species	Average Percent Cover	No. of Plots Present	Frequency	Relative Frequency
Totals				100

TABLE 27.12 Summary of Results for **Macroinvertebrates**

Download an Excel version of this table from www.masteringbiology.com. Look in the Study Area under Lab Media.

Locality: ____________ **Size class:** Macroinvertebrates **Date:** ____________

No. of plots sampled: ____________ **Total area sampled:** ____________

Species	Total No. of Individuals	Density	Relative Density	No. of Plots Present	Frequency	Relative Frequency
Totals			100			100

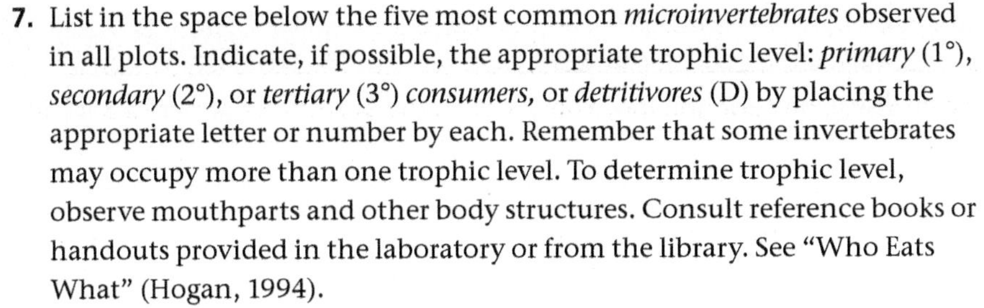

7. List in the space below the five most common *microinvertebrates* observed in all plots. Indicate, if possible, the appropriate trophic level: *primary* (1°), *secondary* (2°), or *tertiary* (3°) *consumers,* or *detritivores* (D) by placing the appropriate letter or number by each. Remember that some invertebrates may occupy more than one trophic level. To determine trophic level, observe mouthparts and other body structures. Consult reference books or handouts provided in the laboratory or from the library. See "Who Eats What" (Hogan, 1994).

Microinvertebrates:

8. What types of microorganisms grew in your agar plates? Did you observe fungal hyphae (filaments)? Did you observe bacterial colonies? Describe them briefly. (Refer to Lab Topic 12 Bacteriology.)

9. Record the summary results for all *abiotic* components of the physical environment in Table 27.13.

10. List the abiotic features that appear to be important influences in the ecosystem.

11. Describe any summary information not included in the tables.

TABLE 27.13 Summary of Results for **Physical Environment**

Download an Excel version of this table from www.masteringbiology.com. Look in the Study Area under Lab Media.

	Plot				
	1	2	3	4	5
Soil					
Description					
pH					
Temperature					
Climate					
Annual rainfall					
Annual avg. temp.					
Microclimate					
Temperature					
°C soil surface					
°C 2 m above					
°C min					
°C max					
Light intensity					
Forest					
Full sun					
% full sun					
Relative humidity					
Wind speed					
Topography					
Comments					

FIGURE 27.7
Profile diagram of a typical forest system. Using your results, label the important species in each of the vertical layers of the forest, from trees to herbs on the forest floor.

Discussion

1. Prepare ecosystem profiles. Using the results from your study of the forest, label the profile diagram provided in Figure 27.7. Label the trees, shrubs, saplings, vines, and herbs to illustrate the species composition of each vertical layer of the forest ecosystem. If you are not investigating a forest system, construct a profile diagram that will illustrate the specific composition and patterns of your study site. Refer to Figure 27.7.
2. Incorporating all components of the ecosystem, complete the compartmental model of the ecosystem in Figure 27.8, indicating the trophic levels and interactions observed in the forest. Provide examples for each trophic level, from producers to top carnivores to decomposers.
3. Write a one-page discussion of your results. Characterize the forest by vegetation type. Describe trophic levels and provide examples, using your model as an illustration. Propose features of the physical environment that appear to influence or be influenced by the biotic community.

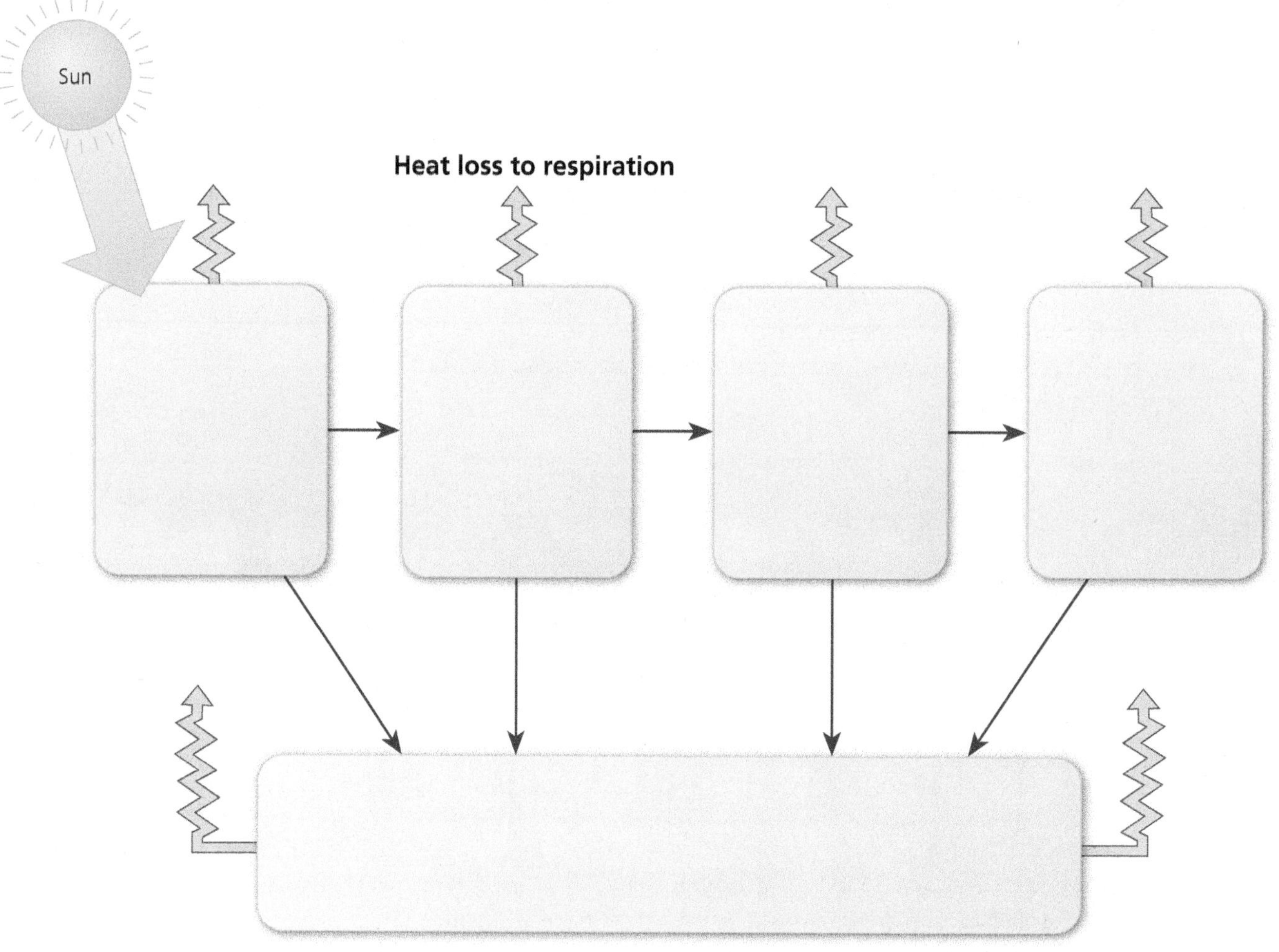

FIGURE 27.8

Compartmental model for trophic relationships in an ecosystem. Heat losses to respiration are represented by yellow arrows. *Label the trophic levels and add arrows to represent energy flow from trophic level to trophic level. For each trophic level, list in each box examples from organisms observed in this laboratory.*

REVIEWING YOUR KNOWLEDGE

1. What are the factors that determine the importance of species in the tree category?

2. Are the same species important in the tree and the shrub, sapling, and vine categories?

3. What are the advantages and disadvantages of using cover as a measure of abundance for herbaceous plants?

4. Which group of consumers, primary or secondary, had the highest density?

5. What are the top carnivores in this ecosystem?
   Explain the relatively low density of carnivores (for example, hawks, owls, or snakes) in this ecosystem.

6. Do your results from this ecosystem analysis adequately represent the forest ecosystem you studied? Explain.

APPLYING YOUR KNOWLEDGE

1. To investigate the structure and function of ecosystems, ecologists may construct a microcosm using organisms and materials from the ecosystem. Properly constructed, these model systems should be self-sustaining.
   a. Once the system is established, what will happen if you remove half the primary producers from the microcosm. Would you predict that your model would continue to be self-sustaining? Explain.

   b. If you remove the decomposers and detritivores, would the microcosm be self-sustaining? Explain.

2. Using your knowledge of ecosystem structure and function, compare the trophic structure of a desert to that of a temperate hardwood forest. Include the relative number of organisms and energy availability for the different trophic levels.

3. Sudden oak death, the rapid decline of red oaks and tan oaks, was first described for central California forests in 1994. The disease is caused by the pathogen *Phytopthora ramorum,* which infects and kills tan oak seedlings and trees. However, mature red oaks are more susceptible, developing large lesions on their stems, with clusters of trees rapidly dying. The death of saplings could have very different effects on the forest community compared with the death of mature trees. How might forest structure and regeneration be affected by the loss of mature oaks that dominate the canopy? How might the forest ecosystem be affected by the loss of young saplings?

4. Wolves did not inhabit Yellowstone National Park from the 1930s until 1995, when 14 gray wolves were introduced by the National Park Service and U.S. Fish and Wildlife Service. In 1995–1996 an additional 17 wolves were released into the park. In 2004, there were 16 packs with about 10 wolves per pack. Some of the changes in the ecosystem were expected, for example, the decline of elk, a major food source for the wolves. However, changes in other animals and vegetation over 10 years provide a study of trophic structure and ecological connections. Between 1930 and 1990, there was no growth of young aspen, cottonwood, and willows along creeks, but now young aspen saplings are thriving. No beavers were seen in these areas, but some have now returned. Draw a simple model of interactions among these components in the Yellowstone ecosystem, including wolves, elk, aspen, willows, and beavers.
   Explain how the introduction of a top carnivore could affect vegetation in an ecosystem.

   After 20 years, the results of the wolf introductions are complicated. Wolf and elk populations have stabilized, and aspens are thriving, however, the willows have not recovered, and beavers are still scarce. Willows require wet flooded soils, and beavers require larger willows to dam streams forming ponds. What factors do you think could be influencing the return of beavers and willows?

5. Design a sampling regimen to answer the following question: Are there differences in the abundance and species composition of microinvertebrates found in the tree litter from forests of different ages? Include the types of plots, organisms, and physical features to be measured and the selection of sample sites.

OPTIONAL EXERCISE

Investigating Weedy Fields

If a forest system is not accessible for study, small and more numerous plots can be used to study weedy fields, which are ubiquitous in both urban and rural environments. The scale for environmental factors can also be reduced and will require sampling at intervals that correspond to the smaller plot sizes. If the weedy field has woody vegetation, then refer to the plot sizes recommended earlier for the forest ecosystem. If only a few woody plants are present, consider recording the presence of these in the general description of the study site, but do not sample them. If the field has only seedlings or nonwoody (herbaceous) vegetation (or both), then use 0.50-m^2 plots to sample the vegetation and invertebrates.

It may be necessary to modify your field studies to correspond to the ecosystem being studied. Additional resources are provided at the end of this lab topic.

Students should work in teams of four: two for seedlings, herbaceous plants, and macroinvertebrates, and two for microinvertebrates and abiotic factors (optional). A class of 24 could sample 6 to 12 plots.

Suggested Outline for Weedy Fields

Exercise 27.1, Biotic Components
- Field Study C. Seedlings and Herbaceous Plants
  - D. Macroinvertebrates
  - E. Microinvertebrates
  - F. Microorganisms
  - G. Other Forest Animals

Exercise 27.2, Abiotic Components (optional)
- Field Study A. Soil
  - B. Climatology

Exercise 27.3, Data Analysis

STUDENT MEDIA: BioFlix, Activities, Investigations, Videos, and Data Tables

www.masteringbiology.com (select Study Area)

Activities—Ch. 52: Terrestrial Biomes; Ch. 54: Interspecific Interactions; Food Webs; Primary Succession; Ch. 55: Energy Flow and Nutrient Cycling

Investigations—Ch. 30: How Are Trees Identified by Their Leaves? Ch. 52: How Do Abiotic Factors Affect Distribution of Organisms? Ch. 54: How Are Impacts on Community Diversity Measured?

Data Tables—Tables 27.3, 27.4, 27.5, 27.6, 27.9, 27.10, 27.11, and 27.12 can be downloaded in Excel format. Go to www.masteringbiology.com. Look in the Study Area under Lab Media.

REFERENCES

Boyce, R. "Life Under Your Feet: Measuring Soil Invertebrate Diversity." *Teaching Issues and Experiments in Ecology,* vol. 3, Ecological Society of America, 2005. Downloadable soil invertebrate key accessed at http://www.esa.org/tiee/vol/v3/experiments/soil/abstract.html

Cain, M., W. E. Bowman, and S. D. Hacker. *Ecology,* 3rd ed. Sunderland, MA: Sinauer Associates, 2013.

Cox, G. W. *Laboratory Manual of General Ecology,* 8th ed. Columbus, OH: McGraw Hill, 2001.

Groffman, P. M. et al. "Long-Term Integrated Studies Show Complex and Surprising Effects of Climate Change in the Northern Hardwood Forest." *Bioscience,* 2012, vol. 62, pp. 1056–1066.

Gurevitch, J., S. M. Seheiner, and G. A. Fox. *Ecology of Plants,* 2nd ed. Sunderland, MA: Sinauer Associates, Inc., 2006.

Hogan, K. "Who Eats What." *Eco-Inquiry,* Appendix A. Dubuque, IA: Kendall-Hunt Publishing, 1994, pp. 355–382. Available to download at http://www.caryinstitute.org/sites/default/files/public/downloads/lesson-plans/Appendix_A-_Who_Eats_What_Guide.pdf

Holmes, R. T. and G. E. Likens. *Hubbard Brook: The Story of a Forest Ecosystem.* New Haven: Yale University Press, 2016.

Kingsolver, R. W. *Ecology on Campus.* San Francisco, CA: Pearson, 2006.

National Audubon Society Regional Guides. This series provides an overview of ecosystems and natural areas by region: Pacific Northwest, Southeast, Southwest, New England, Rocky Mountain, California, Florida, etc.

Rizzo, D. M. and M. Garbelotto. "Sudden Oak Death: Endangering California and Oregon Forest Ecosystems." *Frontiers in Ecology and the Environment,* 2003, vol. 1(5), pp. 197–204.

Robbins, J. "Lessons from the Wolf." *Scientific American,* 2004, vol. 290(6), pp. 76–81.

Smith, T. M. and R. L. Smith. *Elements of Ecology,* 9th ed. San Francisco, CA: Pearson, 2014.

Vodopich, D. *Ecology Laboratory Manual.* Dubuque, IA: McGraw Hill, 2009.

Waterman, M. and E. Stanley. *Biological Inquiry: A Workbook of Investigative Cases.* San Francisco, CA: Pearson, 2017.

WEBSITES

Audubon Guides — Field Guides to Birds, Mammals, Wildflowers, and Trees can be downloaded as apps: http://www.audubonguides.com/field-guides/birds-mammals-wildflowers-trees-north-america.html

Discover Life Nature ID Guides: http://www.discoverlife.org/mp/20q

Encyclopedia of Life Field Guides; find or create your own using the tools and apps: http://fieldguides.eol.org/

EPA Ecosystems web page: http://www.epa.gov/research/ecoscience/

Finding and ordering USGS topographic maps: http://topomaps.usgs.gov/ordering_maps.html

Hubbard Brook Ecosystem Study; review the research questions that are pursued in ecosystem studies of terrestrial and aquatic ecosystems: http://www.hubbardbrook.org/

Leafsnap, an online nature guide to trees that can be downloaded as an app: http://leafsnap.com/

NEON (National Ecological Observation Network), overview and videos on design and data collection http://www.neonscience.org/about

Project Bud Burst: http://www.neonscience.org/learn-experience/project-budburst

Soil Biology Primer: http://soils.usda.gov/sqi/concepts/soil_biology/biology.html

Tobin, K. "Did Wolves Help Restore the Trees to Yellowstone?" *PBS News Hour*, September 4, 2015: http://www.pbs.org/newshour/rundown/wolves-greenthumbs-yellowstone/

Tree of Life Project; includes current information on relationships of all major groups of living organisms: http://tolweb.org/tree/phylogeny.html

The world's biomes (and interesting information on all organisms): http://www.ucmp.berkeley.edu/exhibits/biomes/index.php

vTree app for tree identification and fact sheets from Virginia Tech: http://dendro.cnre.vt.edu/

NOTES

NOTES

LAB TOPIC 28

Ecology II: Computer Simulations of a Pond Ecosystem

Laboratory Objectives

After completing this lab topic, you should be able to:

1. Develop a computer model to investigate a pond ecosystem.
2. Calibrate the model using information from field investigations.
3. Describe the importance of models for investigating large-scale complex systems.
4. Answer questions and test hypotheses using a computer model.
5. Evaluate the effects of disturbance on the model ecosystem.
6. Apply the results of computer simulations to predict the results in real ecosystems.

Introduction

Changing climate patterns are affecting the ecological relationships and functional dynamics of ecosystems. Evidence from field research and ecological modeling reveal several areas where notable changes are occurring: the timing of life cycles, shifts in species ranges, disruption of food webs, rise in pathogens and disease, and an increase in extinction rates (Environmental Protection Agency, 2016). In 1980, the National Science Foundation established the Long Term Ecological Research Network (LTER), which currently supports long-term ecosystem studies at 26 sites across the continental United States, Alaska, Antarctica, and the Caribbean and Pacific islands—including lakes, coral reefs, prairies, forests, tundra, deserts, and urban areas. This research is providing the scientific evidence needed to understand climate change and to model the effects of loss of species, changes in predator/prey relationships, and the impact of invasive species at the ecosystem level. To understand these large complex systems and to make predictions concerning their responses to disturbance (both natural and human), ecologists often depend on computer models that correspond to the particular ecosystem. Information obtained in field research is used to develop the structure of the model, the appropriate interactions among ecosystem components, and the actual values used in the model.

You have already worked with models in other laboratory topics, including the bead models of cellular reproduction and population genetics, the diagrammatic models in terrestrial ecology, and computer simulations in population genetics. Models generally provide a simplified view of a complex phenomenon. In this lab topic, you will actually construct a computer model of an aquatic ecosystem, the pond. As you develop your model, you can simulate a variety of conditions, including the effects of natural disturbance and human disturbance (fishing).

If you have completed Lab Topic 27 Ecology I: Terrestrial Ecology, you should be able to construct a compartmental model of the trophic structure of a forest ecosystem and develop a general model based on the trophic levels present in most ecosystems. Before continuing this lab, use your knowledge of ecology to complete Table 28.1, providing definitions and examples. Refer to Lab Topic 27 and your textbook if necessary.

TABLE 28.1 Definitions and Examples of Ecological Terms

Terms	Definitions	Examples
Primary producer		
Consumer		
Trophic level		
Ecosystem		
Biotic component		
Abiotic component		
Food chain		

The Pond Ecosystem

The primary producers in a pond are generally the algae floating in the surface layers (phytoplankton) and some plants growing at the pond edge, or margin. These autotrophic organisms convert the energy of sunlight into chemical energy stored in organic compounds. The producers, in turn, are consumed by primary consumers (herbivores such as aquatic invertebrates and zooplankton), which are ingested by secondary consumers (carnivores such as sunfish) and even tertiary consumers (carnivores such as bass) (Figure 28.1). The movement of energy through these systems is called **energy flow,** because energy *does not cycle*, rather, at each trophic level the amount of energy is reduced. All organisms expend energy in the activities of life (work, growth, movement, reproduction, and more), and ultimately this energy is dissipated as heat. The **trophic efficiency**, that is, the transfer of energy from one trophic level to another, is only on average about 10%. Biological systems require a constant input of energy, which in most systems comes from the sun and sustains ecosystem structure and function.

The rate at which light energy is converted to chemical energy is called **primary productivity**. Productivity is measured as the amount of **biomass**, or organic matter, added to the system per unit area per unit time (for example, $kg/m^2/yr$). In tropical rain forests, for example, the primary productivity is 2.2 $kg/m^2/yr$; in temperate grassland, 0.6 $kg/m^2/yr$; and in lakes and streams, 0.25 $kg/m^2/yr$ (*Campbell Biology,* 11th ed.). **Secondary productivity** is the rate at which consumers and detritivores accumulate new biomass from organic matter that was consumed. In the computer model of the pond, the changes in each trophic level will be measured as *biomass*.

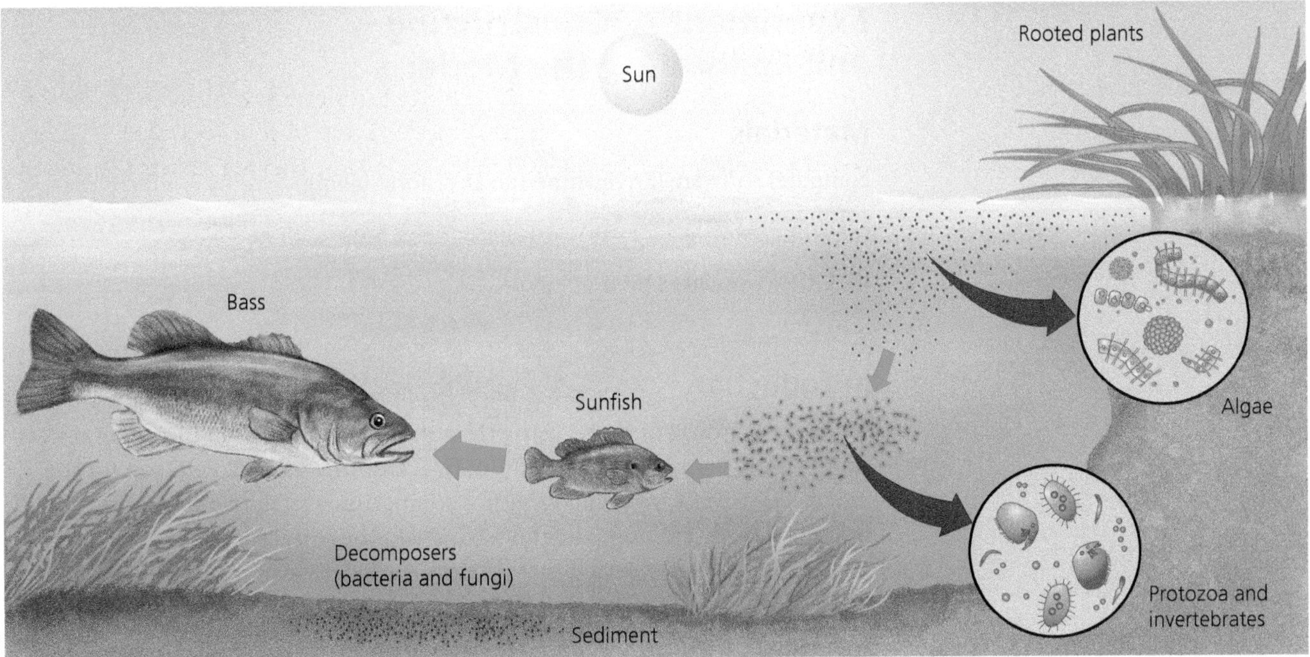

FIGURE 28.1
The food web of a pond ecosystem. The primary producers (algae) are consumed by primary consumers (protozoa and invertebrates), which in turn are consumed by secondary consumers (sunfish and bass). In this pond, bass also may be tertiary consumers.

Refer to the food web in Figure 28.1, and in the margin of your lab manual, *sketch a compartmental model of a pond ecosystem.* (See also the example of a compartmental model in Figure 27.8.) For each trophic level (producers, consumers, etc.) draw a box (compartment) and then use arrows to indicate the flow of energy and biomass.

EXERCISE 28.1

Computer Model of the Pond Ecosystem

The computer model of an ecosystem is based on observations of the natural ecosystem in the field. Your sketch of a pond ecosystem compartmental model, including biotic components and interactions, represents the first step in constructing a model. You must also determine the biomass associated with each compartment based on data collected from field investigations. These data are used to **calibrate** the model—that is, to set the starting values for each component.

In this exercise, you will construct and calibrate a computer model for a pond ecosystem. You will run an initial simulation to determine the **steady-state**, or stable, values for your pond, then compare your results with those expected from a field study. In the process, you will practice changing components of the model, graphing, and printing your results. Once you have determined that your model adequately represents a natural ecosystem, you can begin to ask questions, formulate hypotheses and predictions, and test these using the model of the pond ecosystem. This process will continue as you adjust the model, analyze your results, and simulate conditions.

Experiment A. Constructing and Calibrating the Model

Materials

computer software (Environmental Decision Making)
computer
printer
USB drive (provided by students)

Introduction

Working in groups of three, using the computer software Environmental Decision Making, written by E. C. Odum, H. T. Odum, and N. S. Peterson for the BioQUEST project, you will develop a compartmental model for a pond ecosystem. You will calibrate the model and run the simulation to determine the steady-state values, the values at which the model stabilizes.

The 1-hectare (= 2.47 acres) pond that you are modeling is inhabited by a variety of **pond life**, including such small organisms as algae, micro- and macroinvertebrates, plants, animals, and microbes that inhabit the pond bottom. Two species of fish are present: **sunfish**, which feed primarily off the pond life, and **bass**, which eat the sunfish. Each component of the ecosystem is represented by an icon (Figure 28.2).

Procedure

Environmental Decision Making is available in both Mac and Windows™ versions. Depending on the version you are using, the screens may be slightly different.

1. **Open the program** by double-clicking on the icon Environmental Decision Making (EDM), then double-click on *Pond Worksheet.* (Double-click on *Pondwork* in the Windows version.) Click on the *Extend* screen. You should see an empty window with the plotter icon in the corner with the label "Quantity." Note: You can also open the model file from the Extend LT application by clicking on *Open* in the File menu.
2. **Construct the model** to incorporate pond life, sunfish, and bass.
   a. Choose *BioQUEST Library* from the Library menu, and then choose *BIOQULIB.LIX* from the Library menu. Select *Sunlight* from the submenu.
   b. When you select an icon and hold down the mouse button, the pointer changes to a hand that allows you to move the icon. Move the sunlight icon to the left of the window.
   c. Choose *Pond Life* from the BioQUEST Library submenu and position it to the right of the sunlight icon.

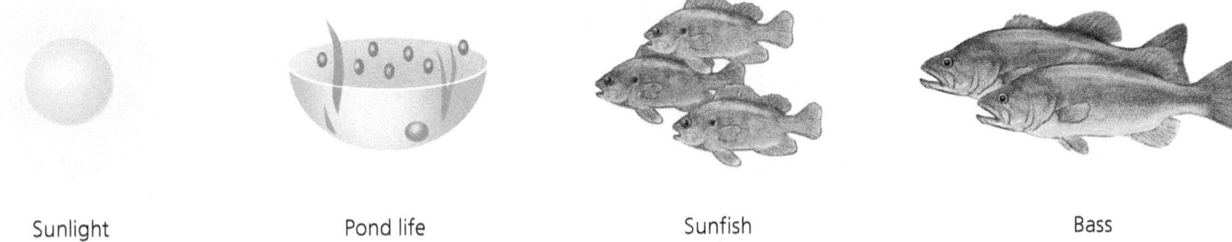

FIGURE 28.2
Icons for the components of the model ecosystem.

d. To connect the sunlight and pond life components, draw a line from one icon to another, connecting the small boxes on each icon. To do this, move the pointer over the small dark box by the sunlight icon. The pointer changes to a pen. Drag the pen from the dark box of the sunlight icon to the open box of pond life (Figure 28.3). Release the button. The solid line should connect the dark box of the sunlight icon (indicating flows out of this component) with the open box of the pond life icon (indicating flows into the second component). For example, sunlight provides the energy input to the ecosystem through pond life. Therefore, the dark box of sunlight is connected to the open box of pond life.

e. Connect the pond life to the plotter (Figure 28.4) by drawing a line connecting the upper dark box of the pond life icon to the first open box on the plotter. The plotter can connect up to four components.

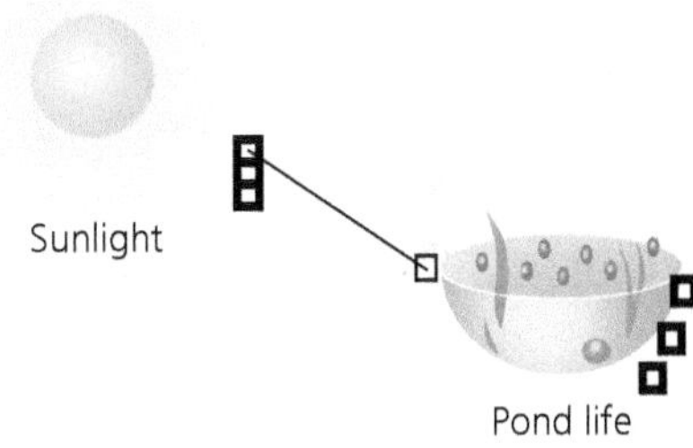

FIGURE 28.3
Drawing lines with the mouse to connect icons. Position the mouse over the dark box, then click and hold. A pen will appear, and you can draw a connecting line.

To erase an icon, move the pointer to the icon, and a hand should appear. Select the icon and press *Delete*.

To erase a line, move the pointer to the line and click it to select it. Press *Delete*.

f. Choose *Sunfish* from the BioQUEST Library and position it to the right of the pond life icon. Connect the open box of the sunfish icon to a dark box of the pond life icon. Connect the upper dark box of the sunfish to the second box of the plotter.

g. Choose *Bass* from the BioQUEST Library and position the icon to the right of the sunfish icon. Connect one of the open boxes of the bass icon to one of the dark boxes of the sunfish icon. Connect the upper

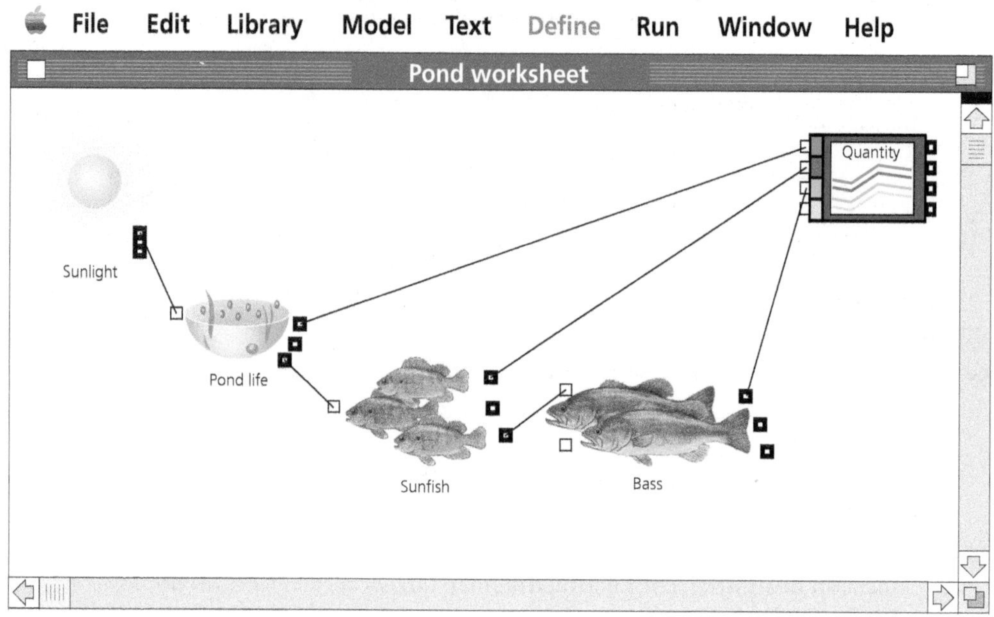

FIGURE 28.4
An example of the model ecosystem.

dark box of the bass icon to the third box of the plotter. Beginning on the left side of the screen, you should have sunlight connected to pond life; pond life should in turn be connected to sunfish, which consume the pond life. Finally, the bass should be connected to the sunfish. All the organisms should be connected to the plotter (Figure 28.4). What trophic level is represented by the bass?

3. **Calibrate the model** by determining the starting values for the components of the ecosystem: sunlight, pond life, sunfish, and bass.
   a. Refer to the map of solar radiation for the United States (Figure 28.5). Determine the solar energy for your location in kilocalories per meter squared. If your location is not included on the map, ask your instructor for appropriate values to use.
   b. Double-click on the sunlight icon. A *dialog box* should appear on the screen. Calibrate the model by entering the appropriate energy values in the box and click *OK*.
   c. Double-click on the pond life icon. Enter *1,000 kg/ha (kilograms per hectare)* for your starting value. Click *OK*.
   d. Double-click on the sunfish icon and enter the total biomass of sunfish in kilograms in the dialog box. Begin with 500 sunfish as a starting number for a 1-ha pond. *Assume 10 sunfish per kilogram, or 50 kg/ha.* Click *OK*.
   e. Double-click on the bass icon and enter the total biomass of bass in kilograms in the dialog box. *The starting value for bass might be 10 large bass, 1 kg each, or 10 kg/ha.* Click *OK*.
   f. Record the starting values for all components in Table 28.2.

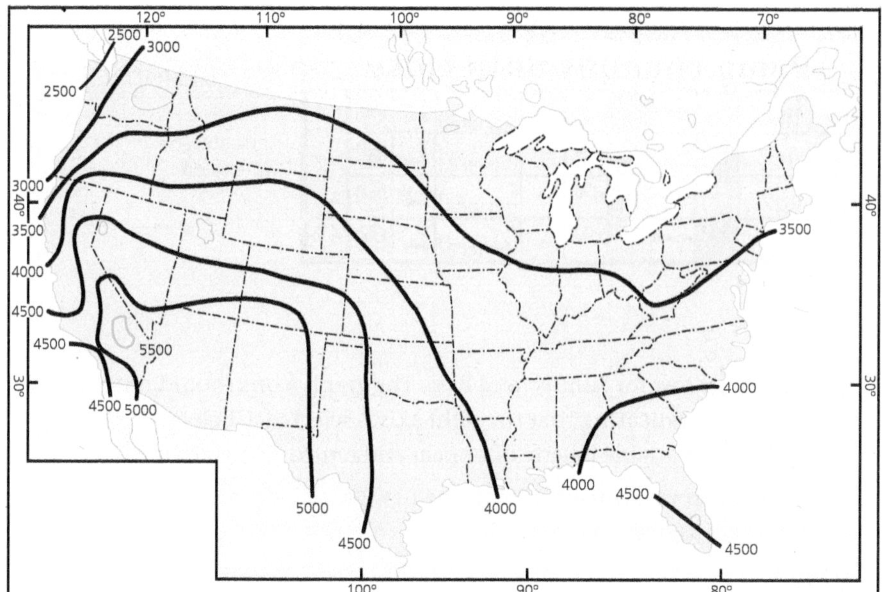

FIGURE 28.5
Solar radiation for regions of the United States. Select the kilocalories of solar radiation for your location.

4. To **calibrate the plotter**, choose *Simulation Setup* from the Run menu. In the time dialog box that appears, change the time to end the simulation to 2 years (730 days). Enter *730* in the first box and click *OK*.
5. **Run the simulation:** Choose *Run Simulation* from the Run menu. The simulation should run and the simulation graph, similar to the screen shown in Figure 28.8, should be plotted.
   a. *Check the lines on the graph.* There should be three, representing pond life, sunfish, and bass. If not, the icons on the model are not correctly attached to the plotter or each other. Return to the model by closing the Pond System window. Correct the model and run the simulation again.
   b. *Check the axes of the graph.* You can modify the graph and table, if needed, to accommodate all the components of the ecosystem and to scale axes appropriately. To change the values on the axes, click on the lower or upper value for the axis. Type in the new number and press *Enter.* To modify the labels, click in the space above the axis and type in the appropriate label. Press *Enter.*
   c. *Label the axes for your graph.* The controls for changing the features of the graph appear in a bar at the top of the screen. You may want to experiment with these in a later simulation. (Refer to Figure 28.6.) Click on the icon in the first box of the display bar. A palette showing the variables in the graph will appear. *Pond life* should appear in the first box, *Sunfish kg/ha* in the second, and *Bass kg/ha* in the third (Figure 28.7). Click and type in the box to change the variables. To read the small values for the sunfish and bass, you will use a different scale on the right axis (Y2). If the right axis is not selected, click on the graph icon (eighth column) on

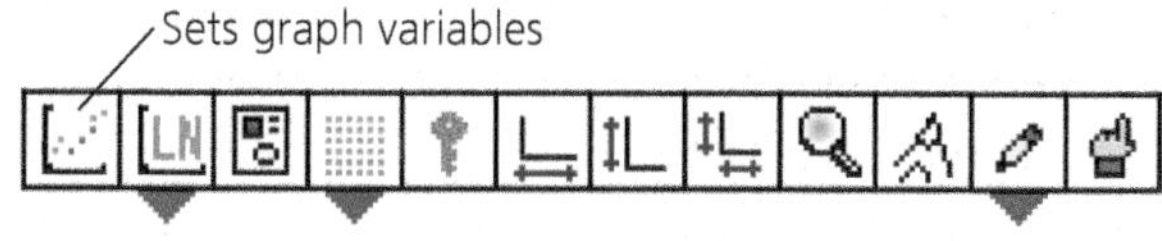

FIGURE 28.6
Display bar for graph. Select the graph icon to adjust the axes and labels for the simulation graph.

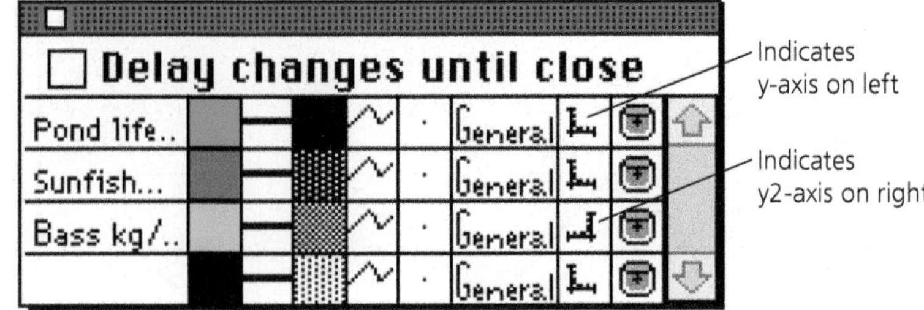

FIGURE 28.7
Example of palette for modifying graph variables. Modify the axes and labels by entering information in this palette.

the same row for Sunfish and Bass. The graph icons should now be reversed, indicating that the right axis is selected. Click on *Close box* to make the variable palette disappear and return to the graph.

d. The graph will run again with the new graph features. Make additional changes as needed.

Results

1. On the simulation graph, note the point at which biomass levels off for the ecosystem components. These various points indicate the steady-state values for the energy levels used in your model, showing that the system has stabilized. *To read values from the graph, move the cursor to the point of interest (for example, the steady-state point) and read the corresponding biomass in the table at the bottom of the graph (Figure 28.8).* The top row of the table should show the time and biomass at the position on the graph indicated by the cursor line. By moving along the graph, you can follow the values on the top row, indicating increases, decreases, and relative stability. Other values and times are listed in the table.

2. Record the steady-state values and the time it took to reach steady state in Table 28.2.

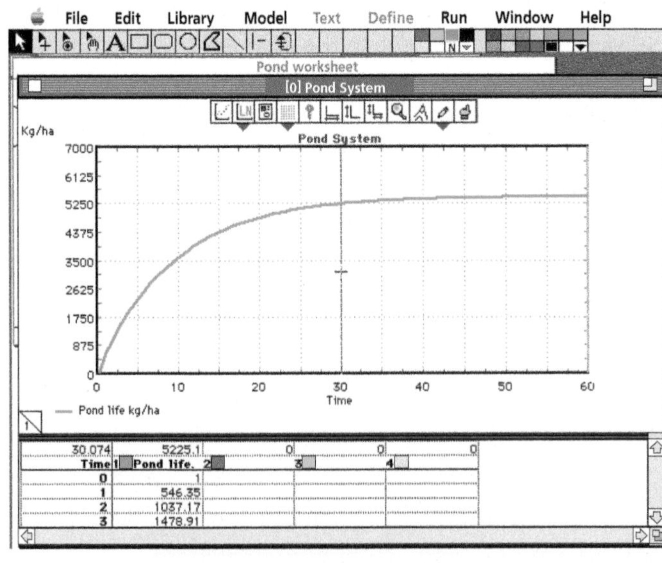

FIGURE 28.8
Example of reading values from a simulation graph. To determine the values at any point on the graph, move the cursor to the position. Read the values on the top row of the table below the graph. The cursor line is at 5,225 kg/ha of pond life on day 30.

TABLE 28.2 Starting and Steady-State Values for Computer Simulation, Experiment A

Component	Starting Value	Steady-State Value
Bass		
Time to steady state:		

3. Choose *Print Plotter* from the File menu. If you want to print only the graph, then click on the *Top Plot Only* button. If *Plot Data Tables* is selected (a ✓ appears in the box), then click on this box to remove the table option. Click *Print.*
4. Close the simulation graph window.

Discussion

1. Describe the changes in the biomass of each component of the ecosystem over time.

   Pond life:

Sunfish:

Bass:

2. Did the highest and lowest values for the three components occur at the same time? Explain why or why not.

3. What do you predict would happen to the steady-state values if you increased or decreased the amount of sunlight available?

Experiment B. Steady-State Model

Materials

computer software (Environmental Decision Making)
computer
printer
USB drive (provided by students)

Introduction

Your initial calibration was based on data from field observations and investigations. Having run the model, you should have estimates for the *steady-state* values, the biomass that can be maintained by this system given the energy available. This is referred to as the **carrying capacity** for each component of the ecosystem. In this simulation, you will calibrate and run the model using the steady-state values from Experiment A.

Hypothesis

State a hypothesis for changes in the components of the ecosystem at steady state.

Prediction

Based on your hypothesis, predict the appearance of the simulation.

Procedure

1. Calibrate the model. (Refer to Experiment A, Procedure step 3.) Double-click on each of the icons and set the starting value for each component at the steady-state value based on results in Table 28.2.
2. Record the new starting values in Table 28.3.
3. Run the simulation.

TABLE 28.3 Starting and Steady-State Values for Computer Simulation, Experiment B

Component	Starting Value	New Steady-State Value
Sunlight		
Pond life		
Sunfish		
Bass		
Time to steady state:		

Results

1. Compare the graph of this simulation with the results from the first simulation (Experiment A). If the starting values for this experiment are not similar to the ending steady-state values from the first simulation, check the starting value for each component. Rerun the simulation if necessary.

2. Record your results in Table 28.3.
3. Print the graph. Choose *Print* from the File menu.
4. Label each line on your graph.

Discussion

1. Did your results match your predictions? If not, explain.

2. Were the values you set for this simulation actually at steady state? If not, consider using new steady-state values from this simulation and repeat the simulation.
3. The steady-state value for each component represents the biomass that can be sustained by this ecosystem (carrying capacity). At carrying capacity, how many bass can be sustained by this pond system? (*Hint:* Refer to Experiment A; the values used in the model are biomass, and you must convert to numbers of bass.) Why is knowing the carrying capacity of the pond important to ecologists and fisheries biologists?

4. Does the biomass increase or decrease as you move from producers to the secondary and tertiary consumers? Would you ever expect the pond biomass to be the same in all the components? Explain.

EXERCISE 28.2

Investigating the Effects of Disturbance on the Pond Ecosystem

Materials

computer software (Environmental Decision Making)
computer
printer
USB drive (provided by students)

Introduction

In this exercise, you will discuss questions that interest your group and design experiments to determine the effects of disturbance on the components of the pond ecosystem. You will run a series of three simulations.

In groups of three, discuss with your teammates possible options for *simulating disturbance of the pond ecosystem.* Using Environmental Decision Making, you can change a number of factors. Consider the following options as you discuss your ideas:

1. What happens to biomass and steady-state values for a pond in another region that receives more or less sunlight (see Figure 28.5)?
2. What would happen to each trophic level if reproduction or mortality was affected so that more (or fewer) sunfish (or bass) inhabited the pond?
3. What changes would occur to the biomass of the sunfish and bass if another fish (e.g., gar) were introduced to the pond? (Gar eat sunfish, but not bass.)
4. What changes would you expect in each trophic level if sunfish (or bass) were eliminated due to a massive fish kill (contamination by toxins, disease, or low levels of oxygen in a summer pond)?
5. Design a question based on local conditions that you can model. Locate your watershed at the EPA site, "How's My Waterway," or use the downloadable app to determine local conditions of lakes and ponds in your area and across the nation: https://watersgeo.epa.gov/mywaterway/mywaterway.html.

The questions you pose should have a biological basis; that is, they should be meaningful questions based on your understanding of ecosystems, particularly pond ecosystems. Just because you can change the model does not mean the question is appropriate. If necessary, consult your textbook or additional readings supplied by your instructor.

List the questions of interest to your group.

Questions:

You will run two computer simulations of your own design. In the *first simulation* you may pursue one or more of the questions generated in your discussion and listed previously. With a computer model, you can make any number of changes easily, but, depending on what you do, you may have difficulty in stating predictions and interpreting your results. *It is best to begin with one change at a time, increasing the level of complexity in subsequent simulations.*

In the *second simulation* (Experiment B), you will investigate the effects of a human disturbance, *fishing,* on the components of the pond ecosystem. In the *third simulation,* Experiment C, you are presented with a case study about the introduction of an exotic fish species. Your team will develop a model to investigate the effects of this invasive species on the pond trophic structure.

Your group should be prepared to describe one of your questions, explain the model, and present your results at the end of the lab period.

Experiment A. Open-Inquiry Computer Simulation of Disturbance

Based on your discussion of factors that might disturb the steady state of the model pond ecosystem, select one question to pursue using the computer simulation. Formulate a hypothesis and make a prediction for your computer simulation.

Question

Hypothesis

Formulate a hypothesis about the effects of disturbance on the pond ecosystem.

Prediction

State a prediction based on your hypothesis. (This is an if/then statement that predicts the results of changes in the steady-state model in Exercise 28.1.)

Test your prediction using the computer simulation model.

Procedure

1. In Table 28.4 on the next page, briefly state the disturbance problem you have chosen to simulate. Record the selected starting values for your model and indicate, by circling, the values that differ from those of the steady-state model.
2. Note any other changes in the model—for example, changes in interactions, solar radiation, or addition or removal of components.

3. Calibrate the model based on your changes. Modify the model according to your problem. *You should be able to make changes in the model based on your experience in Exercise 28.1, Experiment A, Procedure step 2. For additional help, refer to the EDM manual provided with the software.*
4. Run the simulation until steady-state values are reached. At steady state, the biomass for each component should change only slightly or not at all.
5. Modify the graph features if necessary.

TABLE 28.4 Starting and Steady-State Values for Exercise 28.2, Disturbance Model

Problem:		
Component	**Starting Value**	**Steady-State Value**
Sunlight		
Pond life		
Sunfish		
Bass		
Time to steady state:		

Results

1. Describe the problem and then record the new steady-state values and the time to steady state in Table 28.4.
2. Print the simulation graph.
3. If instructed to do so, save your model on a USB. Refer to the following instructions for assistance.

To Save a Model

a. Insert your USB. Click on *File* and pull down to *Save Model As.* Enter the name of the file, which might be, for example, your name and the experiment number (e.g., Stef1.A). Select the USB. Click on *Save.* The file is saved on your USB.

b. If the file has been saved previously, click on *File* and *Save.* The current version of the file will be saved with the existing name. Save your files to a USB for future use. Save each model for Experiments A to C with a different name.

Discussion

1. Describe the interactions among the trophic levels in the pond ecosystem.

2. Do your results match your predictions? Discuss any differences between results and predictions.

3. How did the disturbance simulated in this experiment affect the pond ecosystem? How might these results be applied to a natural ecosystem?

Experiment B. Open-Inquiry Simulation of Fishing Activities

Although we often think of our human activities as being separate from natural ecosystems, we can have a dramatic effect on these systems. Ponds support not only carnivores such as bass, but also humans, who fish selectively. In this simulation, you will add fishing as a component to the model. Your objective might be to allow fishing at levels that will maintain a stable pond ecosystem. You might choose to challenge the system by increasing fishing on one or more species. You do not have to begin with the original steady-state model but can choose to add fishing to one of your previous models. Consider what type of fish will be caught—sunfish, bass, or gar. Is it possible to catch more than one type of fish? Does fishing compound disturbance factors that you may have introduced in previous simulations? State the question of interest below.

Question

Hypothesis

State a hypothesis about the effect of fishing on the pond ecosystem.

Prediction

State a prediction based on your hypothesis.

Test your prediction using the computer simulation model.

Procedure

1. In Table 28.5, describe the problem you have chosen to simulate. Record the selected starting values for your model.
2. Record in the space provided the proposed changes to the model. What fish will be caught? Are any other interactions being changed?
3. Modify the model. Select *Fishing* from the Library. Connect the open box for fishing to the dark box for the fish of choice. Connect the dark box for catch to the fourth plotter line.
4. Calibrate the model based on your changes. The level for fishing is set at 1 hour of fishing per day. Begin with this level. You may adjust the model according to your hypothesis.
5. Run the simulation until steady-state values are reached. At steady state, the biomass for each component should change only slightly or not at all.
6. You will need to modify the graph features. *Fishing* should now be displayed on the right (Y2) axis. Select the first box on the display bar over the graph. In the fourth row, type in *Fishing kg/ha/d* and select the right axis by clicking in the eighth column. Return to the graph and adjust the right axis values as needed. The right axis label is now *Catch/day* (for fishing) and *kg/ha* (for fish). Refer to the Procedure section for the first simulation for additional assistance. (See also Figures 28.7 and 28.8.)

Results

1. Record the new steady-state values and the time to steady state in Table 28.5.
2. Print the simulation graph.

TABLE 28.5 Starting and Steady-State Values for Exercise 28.2, Fishing Model

Problem:		
Component	**Starting Value**	**Steady-State Value**
Sunlight		
Pond life		
Sunfish		
Bass		
Fishing		
Time to steady state:		

3. Save your model to a USB drive. Refer to the instructions in Experiment A for assistance.

Discussion

1. Describe the interactions among the trophic levels in the pond ecosystem.

2. Do your results match your predictions? Discuss any differences between results and predictions.

3. Discuss the effects of fishing on the pond ecosystem. How might these results be applied to a natural ecosystem?

Experiment C. Computer Simulation—Invasive Species Case Study

Problem

Jake, a fish farmer, counted the number of dead striped bass in his net. Twenty more today! He skillfully cut one open, exposing intestines and the mottled mass of muscles riddled with worms. "I'm not sure I can even harvest enough stripers to cover my expenses, much less make a profit. At this rate, it looks like the worm has turned. They're eatin' my fish instead of the other way 'round." Bud, the fishery's technician, suggested, "Can't you just kill those worms? You know, worm the fish." Jake countered, "I can't kill the worms because they reproduce in the snails. I can't kill the snails, because every chemical I could use would also kill the fish and contaminate my fish farm." "How about a biological control? You know—we use black carp to keep the plants down in the pond?" suggested Bud. Jake thought about this idea. He knew that black carp would eat the snails. No snails, no worms. But he remembered that black carp is considered an invasive species when it escapes from ponds into streams and lakes. Maybe he should find out more before investing in snail-crunching black carp. Jake is alarmed after finding a video on the Web about Asian carp (http://www.invasivespeciesinfo.gov/aquatics/asiancarp.shtml) and reading a news item on black carp.

News Focus from Science:

"Will Black Carp Be the Next Zebra Mussel?" by Dan Ferber, vol. 292, pp. 203–204, April 13, 2001.

There's good reason to worry about black carp, says ichthyologist Jim Williams of the U.S. Geological Survey (USGS) Caribbean Research Center in Gainesville, Florida. In a detailed 1996 risk assessment of the fish, Williams and USGS colleague Leo Nico concluded that black carp would survive and reproduce in U.S. rivers, consuming native mollusks and competing with native mollusk-eating fish such as redear sunfish and freshwater drum.

Additional Information

If Jake adds black carp to his pond and they escape to a natural pond, there could be problems. The structure and function of the pond ecosystem would change. Black carp will eat the snails (part of the *pond life* in the computer model). The bass prefer to eat the sunfish, because carp are too bony. Therefore, black carp would be competing with the sunfish for food, but with little or no predation by bass. (Assume no bass predation.) Any model is based on the connections established by the scientist. Review the trophic interactions among the elements of the pond ecosystem now that you have added the black carp.

Questions

What might be the consequences if black carp were to escape to a nearby natural pond? What do you think would happen to the pond life, sunfish, and bass in a natural pond?

Hypothesis

Formulate a hypothesis about the effects of introducing black carp to the pond ecosystem.

Test your hypothesis using the computer simulation model.

Prediction

State a prediction based on your hypothesis. (This is an if/then statement that predicts the results of changes in the model.)

Test your prediction using the computer simulation model.

Procedure

1. In Table 28.6, describe the components of the model you will simulate. Record the selected starting values for your model. The starting value for black carp should be small relative to sunfish. For example, if sunfish is 30 kg/ha, then black carp might be 10 kg/ha.
2. Note below any other changes to the model, such as changes in interactions or additions or removal of components.
3. Calibrate the model based on your changes. Modify the model according to your problem. Select a new fish icon from the Library menu to represent the black carp. (Black carp does not appear as an icon, so you should select the gar icon or add another sunfish icon.) Connect the dark box of pond life to the open box of the new fish icon. Connect the new icon to the plotter. Double-click on the new icon to set the starting value for black carp (Table 28.6). Remember, the starting value for carp should be less than that for sunfish. (You may want to check the starting values for all components at this time.) Should the carp be connected to bass? Why or why not? This is a question about who eats whom and will have large effects on the results of your model.
4. Run the simulation until steady-state values are reached.
5. You will need to modify the graph features. Select the first box on the display bar over the graph. In the fourth row, type in *Black carp kg/ha* and select the right axis by clicking in the eighth column. Return to the graph and adjust the right axis as needed. Refer to the Procedure section of the first simulation for additional assistance. (See also Figures 28.7 and 28.8.)

Results

1. Describe the problem you are simulating, then record the new steady-state values and the time to steady state in Table 28.6.
2. Print the simulation graph.
3. Save your model on your USB drive. Refer to the instructions in Experiment A for assistance.

TABLE 28.6 Starting and Steady-State Values for Exercise 28.2, Invasive Species Model

Problem:		
Component	**Starting Value**	**Steady-State Value**
Sunlight		
Pond life		
Sunfish		
Bass		
Black carp		
Time to steady state:		

Discussion

1. Describe the interactions among the trophic levels in the pond ecosystem.
2. Do your results match your predictions? Discuss any differences between results and predictions.
3. Do you think that Jake should add black carp to his pond? Why or why not? Remember that Jake wants to increase his yield of striped bass (after all, he is a fish farmer). Should the federal government ban the sale of black carp, given your prediction of what will happen if black carp invade natural aquatic ecosystems?

REVIEWING YOUR KNOWLEDGE

1. Define the following terms and provide examples if appropriate: *producers, consumers, decomposers, biomass, primary productivity, secondary productivity, energy flow, trophic efficiency, food web, carrying capacity, steady state, calibration.*
2. Define *model* and provide examples from several areas in biology. Consider areas of biology other than those studied in the laboratory.

3. Critique the computer model used in this lab topic. In what ways is it an appropriate model of the pond ecosystem? In what ways does it fail to model the pond adequately?

4. Even when you adjust the starting values for components in the model, the steady-state values remain the same. However, if you change the energy from the sun, new steady-state values will emerge. Explain.

5. Which disturbance factors had the greatest effect on the pond ecosystem? How are you measuring the effect?

APPLYING YOUR KNOWLEDGE

1. In 1997, nonnative and invasive Asian swamp eels were collected in Florida for the first time at two sites near Tampa and Miami. By 2008, the eel had also been found in a lake in New Jersey. These fish are extremely adaptable to a wide range of freshwater habitats, from wetlands to streams and ponds. They are predators that feed on worms, insects, crayfish, frogs, and other fishes, including bluegill and bass. Swamp eels have the ability to gulp air, which allows them to survive in only a few inches of water and to move over land to a nearby body of water. Scientists are tracking their movements and increasing numbers in the Southeast. In one pond, several species of fish have been completely eliminated.

   Based on your understanding of the pond ecosystem, predict the effect of introducing swamp eels on the following components of the pond.

   Bluegill:

   Bass:

   Pond life:

2. Using your knowledge of ecosystem structure and function, propose a plan of action for eliminating the swamp eels (described in question 1) from the pond before they eliminate the other organisms. You cannot use toxins, as the local anglers fish in this pond.

3. Global warming (increase in global mean temperatures by 0.4°C over the last 150 years) has been greater in the Arctic, with twice the rate and a decline in minimum sea ice cover by 45,000 km^2/yr (about twice the size of New Jersey) over the past two to three decades. Changes in the ecological systems in the Arctic are rapid and complex, with serious effects on plants, animals, and their relationships. During the Fourth International Polar Year, ecological studies revealed the complex dynamics of climate change (Post et al., 2009).

   Over the last decade, Arctic plants have tracked warm temperatures in early spring by growing and flowering earlier. However, caribou continue to calve at the same time of year, which means that calves are born after the peak of plant resources. This has contributed to reduced reproduction and survival of caribou calves. In another example, the loss of snow cover in the Arctic has resulted in the crash of small rodent (lemmings and voles) population cycles, and these populations remain low. Small rodents are the food source for predators such as snowy owls and Arctic foxes.

   For these two examples, state the *trophic level* for each organism: flowering plants, caribou, foxes, owls, lemmings, and voles. In the margin of your lab manual, sketch a food web that reflects the interactions among the organisms in this ecosystem.

   Describe the changes in caribou populations and flowering plant populations given this mismatch.

   Sketch the changes in plant, lemming, and snowy owl populations that you might expect as the lemmings crash and then remain low (no rebound).

4. A local television news bulletin urges you not to eat fish caught in nearby Lake Ketchum because water levels of the pollutant PCB have reached 0.0001 part per million (ppm). With such a small reading, why should you be concerned?

5. Eliminating top predators from an ecosystem can have effects that "cascade" down the trophic levels, with serious results for the survival and interactions of many species. Over the last 35 years, large predatory sharks, including sandbar, blacktip, bull, dusky, tiger, and scalloped hammerhead sharks, have been effectively eliminated from the world's oceans in the hunt for fins and meat. Over the same time period, northwestern Atlantic marine systems suffered severe losses of the economically important bay scallops, as well as clams and oysters. The populations of rays and smaller sharks (blacknose, bonnethead, and finetooth sharks; cownose, lesser devil, spiny butterfly, smooth butterfly, and spotted eagle rays; plus several skates) have increased over the same time period, with cownose ray populations increasing by 20-fold (Myers et al., 2007).

   Given the changes in population sizes for these three groups of consumers (great sharks, smaller rays and sharks, and bivalves), sketch a model of trophic relationships that would best explain the data. What do you think ecologists mean by the "cascade effect"?

INVESTIGATIVE EXTENSIONS

Environmental Decision Making has two other ecosystem models, *Grasslands* as well as *Forestry and Logging*, and an option for creating your own model using general symbols. These programs can be used to pursue additional topics, including ecosystem dynamics and management.

STUDENT MEDIA: BioFlix, Activities, Investigations, and Videos

www.masteringbiology.com (select Study Area)

Activities—Ch. 54: Food Webs; Ch. 55: Pyramids of Production; Energy Flow and Chemical Cycling

Investigations—Ch. 54: How Are Impacts on Community Diversity Measured? Ch. 55: How Do Temperature and Light Affect Primary Production?

REFERENCES

Environmental Protection Agency (EPA), 2016, "Climate Change: Impacts and Adaptations." https://www3.epa.gov/climatechange/impacts/

Estes, J. A. et al. "Trophic Downgrading of Planet Earth." *Science*, 2011, vol. 333, pp. 301–306.

Groffman, P. M. et al. "Long-Term Integrated Studies Show Complex and Surprising Effects of Climate Change in the Northern Hardwood Forest." *Bioscience*, 2012, vol. 62, pp. 1056–1066.

Mack, R. N., D. Simberloff, W. M. Lonsdale, H. Evans, M. Clout, and F. Bazzaz. "Biotic Invasions: Causes, Epidemiology, Global Consequences and Control." *Issues in Ecology,* 2000, no. 5, Spring.

Myers, R. A., J. K. Baum, T. D. Shepard, S. D. Powers, and C. H. Peterson. "Cascading Effects of the Loss of Apex Predatory Sharks from a Coastal Ocean." *Science*, 2007, vol. 315, pp. 1846–1850.

Odum, E. C., H. T. Odum, and N. S. Peterson. *Environmental Decision Making, The BioQUEST Library Volume VI.* San Diego, CA: Academic Press, 2002. Lab written to correspond to software by permission of publisher.

Odum, H. T. and E. C. Odum. *Computer Minimodels and Simulation Exercises,* Gainesville, FL: Center for Wetlands, Phelps Laboratory, University of Florida, 1989. Examples using BASIC with program listings for Apple II, PC, and Macintosh.

Post, E. et al. "Ecological Dynamics Across the Arctic Associated with Recent Climate Change." *Science,* 2009, vol. 325, pp. 1355–1358.

Simberloff, D. *Invasive Species: What Everyone Needs to Know.* Oxford: Oxford University Press, 2013.

Smith, T. M. and R. L. Smith. *Elements of Ecology*, 9th ed. San Francisco, CA: Pearson, 2014.

SOS for America's Streams (video). Gaithersburg, MD: Izaak Walton League of America, 1990. See the Preparation Guide for ordering information. The Preparation Guide can be downloaded at masteringbiology.com in Instructor Resources, Instructor Guide for Supplements.

Urry, L., M. Cain, S. Wasserman, P. Minorsky, and J. Reece. *Campbell Biology*, 11th ed. San Francisco, CA: Pearson, 2017.

Waterman, M. and E. Stanley. *Biological Inquiry: A Workbook of Investigative Cases*. San Francisco, CA: Pearson, 2017.

WEBSITES

The BioQUEST Curriculum Consortium: http://bioquest.org

BioQUEST Library Online. Environmental Decision Making is now available for free download: http://bioquest.org/BQLibrary/library_result.php

EPA Climate Change: Impacts and Adaptations includes an arctic ecosystem model: https://www.epa.gov/climate-impacts/climate-impacts-ecosystems

EPA Ecosystems Web page with terrestrial and aquatic ecosystems and information on watersheds across the United States:
http://www.epa.gov/research/ecoscience/
https://www.epa.gov/hwp

EPA, How's My Waterway. Website and downloadable app for local conditions of lakes, streams, rivers and other waterways in your area and across the nation: https://watersgeo.epa.gov/mywaterway/mywaterway.html

NREL Solar Radiation Data for the U.S., provides monthly and annual solar radiation data: http://rredc.nrel.gov/solar/old_data/nsrdb/1961-1990/redbook/atlas/Table.html

Simulation video of global climate change from 1870–2100 using CCSM: http://www.youtube.com/watch?v=d8sHvhLvfBo

University Corporation for Atmospheric Research (UCAR) and National Corporation for Atmospheric Research (NCAR)—Understanding Climate Change: Multimedia Gallery; includes several simulations (global surface warming, Arctic sea ice changes) based on the NCAR-based Community Climate System Model (CCSM): http://www2.ucar.edu/news/understanding-climate-change-multimedia-gallery#future

USDA Invasive Species Information Center: http://www.invasivespeciesinfo.gov/

USDA National Invasive Species Information Center; information on Asian carp, including informative and entertaining video: http://www.invasivespeciesinfo.gov/aquatics/asiancarp.shtml#.UOJhekJNGfQ

U.S. Long Term Ecological Research Network with links to 26 sites: http://www.lternet.edu/sites/

NOTES

Appendix A

Scientific Writing and Communication

For the scientific enterprise to be successful, scientists must clearly communicate their work. Scientists share their ideas and results with other scientists, encouraging critical review and alternative interpretations from colleagues and the entire scientific community. Communication, both oral and written, occurs at every step along the research path. While working on projects, scientists present their preliminary results for comments from their collaborators at laboratory group meetings and in written research reports. At a later stage, scientists report the results of their research activities as a poster or oral presentation at a scientific meeting. Then the final report is prepared in a rather standard scientific paper format and submitted for publication in an appropriate scientific journal. At each stage in this process, scientists encourage and require critical review of their work and ideas by their peers. The final publication in a peer-reviewed journal generally promotes additional research and establishes this contribution to current knowledge.

One of the objectives of every lab topic in this manual is to develop your writing skills. You will generate and write hypotheses, observations, answers to questions, and more, as one way of learning biology. Also, you will practice writing in a scientific paper format and style to communicate the results of your investigations. The scientific process is reflected in the design of a scientific paper and the format you will use for your laboratory papers.

A scientific paper usually includes the following parts: a **Title** (statement of the question or problem), an **Abstract** (short summary and preview of the paper), an **Introduction** (background and significance of the research), a **Materials and Methods** section (report of exactly what you did), a **Results** section (presentation of data), a **Discussion** section (interpretation and discussion of results), and **References Cited** (books and articles used). **Acknowledgments** (recognition of assistance) may also be included. When scientists prepare a scientific paper for publication, they select a journal and then check its website for "Instructions for Authors." Every scientific journal has specific required elements that must be followed for publication. These instructions describe the components of the paper and the required formatting from the font to the figures and tables, and even the reference citations. You may consider this appendix as the "Instructions to Authors."

The Writing Program in this laboratory manual encourages you to develop scientific writing skills through a series of writing assignments. *We propose that you practice writing throughout the biology laboratory program by submitting individual sections of a scientific paper before writing a complete scientific paper.* Your instructor will determine which sections you will write for a given lab topic and will evaluate each of these sections, pointing out areas of weakness and suggesting improvements. By the time you have completed these assignments, you will have submitted the equivalent of one scientific paper. Having practiced writing each section of a scientific paper in the first half of the laboratory program, *you will then write one or two complete laboratory papers in scientific paper format during the second half of the laboratory program,* reporting the results of experiments, preferably those that you and your research team have designed and performed.

Successful Scientific Writing

Scientific writing is concise and informative with the objective to report findings to other scientists. Scientists limit the use of adjectives and adverbs as they present their research results and make a logical argument for their conclusions based on evidence. The following notes for success apply to writing throughout all sections of a scientific paper. This section will be helpful to review before you begin writing. After completing your paper, review these points again as a check on your work.

- Your writing should be **clear and concise** to communicate effectively with your scientific audience. Write objectively in direct and informative sentences that explain what you mean with a minimum of words. **Avoid adjectives and adverbs** that can contribute to flowery language and have limited use in describing your work.
- **Write in short and logical, but not choppy, sentences.** Avoid run-on sentences and use grammatically correct English. Avoid long introductions. (See Chapter 5, "Revision," in Knisely [2013] and Pechenik [2016] for basic rules of writing and practice in editing for common errors.)
- **Write for your audience:** other student-scientists and your professor. Thinking about your audience is key to knowing how general or specific you need to be in writing the Introduction and Discussion. You can assume general scientific knowledge and background, so including definitions of terms (e.g., osmosis) is not appropriate. Considering your audience, scientists rather than your lab group, will help you understand the difference between a scientific paper and a lab report. Scientists read scientific papers to be current in the field, find background information on a topic, and develop or improve their own research ideas. Scientific writing should be informative, rather than entertaining.
- **Support your writing with evidence.** Your explanation and interpretation of your work in the Discussion section will be based on the results of your research presented in the Results section. In the Introduction and Discussion sections you will provide support for your ideas by describing and referencing work of other scientists.
- **Locate sources related to your work early** in the research project. Read the sources carefully to understand the work. *To avoid issues with plagiarism, do not copy source materials verbatim or cut and paste sections of another paper. As you read the reference material, take time to make notes and summarize the work in your own words.* Make a complete record of the reference citation information to use in the References Cited section of your paper.
- **Avoid using quotations in scientific writing.** Unlike writing in other disciplines, scientific writing seldom includes direct quotes. You should summarize and explain the work of others in your own words. Then provide a reference citation to the work. *Do not use footnotes.*
- **Do not plagiarize.** When you restate the words and ideas of others, you must credit that work by including the source in a reference citation in the text. Simply paraphrasing by changing a word or two in the sentence is still plagiarism. See Pechenik (2016) and Knisely (2013) for specific examples and suggestions for avoiding plagiarism.
- **Use the past tense** in the Abstract, Materials and Methods, and Results sections. Also use the past tense in the Introduction and Discussion sections when referring to *your* work. **Use the present tense** when relating the background information as you refer to other investigators' published

work. Previously published research is considered established in the present body of knowledge.

- **Use the active voice** when possible. Doing so makes the paper easier to read and more understandable. *However, in the Materials and Methods section you may use the passive voice* so that the focus of your writing is the methodology, rather than the investigator.
- When referring to the **scientific name** of an organism, **the genus and species should be in italics or underlined.** The first letter of the genus is capitalized, but the species is written in lowercase letters, for example, *Drosophila melanogaster.* After the first use of a scientific name you may abbreviate the genus name, for example, *D. melanogaster*. However, when beginning a sentence with a scientific name always write out the genus.
- **Pay particular attention to the rules for numbers.** Use metric units for all measurements. Use numerals when reporting measurements, counts, percentages, decimals, and magnifications. When beginning a sentence, write the number as a word. Numbers of ten or less that are not measurements are written out. Numbers greater than ten are given as numerals. *Decimal numbers less than one should have a zero in the one position* (e.g., 0.153; not .153).
- **Include a heading for each section of the scientific paper** (except the Title Page), placing the heading of the section against the left margin on a separate line. *Each section does not begin a new page but continues in order.*
- **Revise, revise, and then revise!** For suggestions and examples of how to revise your work, see Chapter 5, "Revision," in Knisely (2013) and Chapter 6, "Revising," in Pechenik (2016).
- **A few points to remember.** Note the word *data* is plural, for example, "The data were analyzed graphically." Remember that the results cannot "prove" the hypothesis, but rather they may "support" or "falsify" the hypothesis.
- **Carefully proofread** your work even if your word processor has checked for grammatical and spelling errors. These programs cannot distinguish between "your" and "you're," for example.
- Always **save a copy** of your work on your computer, USB drive, or the Cloud, and **print a copy** of your paper before turning in the original.
- **Begin writing early, leaving time for** searching the literature, thinking, analyzing, revising, and final proofreading. Synthesizing and writing take time and thought.

Locating Appropriate References

References are important to provide a context and evidence from the work of other scientists to support your explanations and interpretations. Scientific papers in general rely on **primary references,** *reports of original research that present the work of scientists in such a way that it can be repeated. Primary references are journal articles that have been peer-reviewed by other scientists and the journal editor.* In addition to articles in journals (for example, *American Journal of Botany, Cell, Ecology,* and *Science*), primary references include conference papers, dissertations, and technical reports. Many scientific journals are available in a full-text version online; these are still primary references. (Websites are not primary references, because they are not required to participate in the peer-review process.)

As you begin your study and literature review, it may be helpful to start with **secondary references.** Textbooks, review articles, and articles from popular science magazines are secondary references, *which generally provide a summary* *and interpretation of research* (for example, *Annual Review of Genetics, Science News,* and *Scientific American*). Secondary references can be particularly helpful in developing an understanding of terminology, methods, and the current state of knowledge in a field. For example, if your project is on plant hormones, you might consult a plant physiology textbook or search in the *Annual Review of Plant Physiology*. These secondary references do not provide the primary evidence needed for your scientific paper. However, they usually include a reference section with carefully selected and useful primary research articles that you can read and incorporate into your Introduction or Discussion sections. (Also see the References section for each lab topic in this manual.)

Given the enormous number of scientific papers and journals available, searching for primary references on your research topic may seem to be a daunting task. The key to success is to use effective **search strategies** with **online databases, indexes, and search engines** that specifically target scientific literature (for example, *PubMed, Google Scholar, Biosis Previews, Science Direct, JSTOR, Science Citation Index,* and *Web of Science*). You can compare these online resources and read about search strategies in Pechenik (2016) and Knisely (2013). Most databases and search engines provide a tutorial for searching under "Help" or "Advanced Searches." Consult with a reference librarian for help with developing search terms and strategies and using the online resources at your institution.

As you read scientific papers, stop after each section of the paper and summarize the essential points in your own words. Do not record quotes, as scientific writing does not use direct quotations. When writing your paper, you will be confident that your notes are in your own words, and you are not plagiarizing. Record the complete citation information for any references at the time you read the article. Any information from these sources must be acknowledged with a citation in the text and inclusion in the References Cited section. *Refer to the citation format in the References Cited section of this appendix.* Your instructor will indicate the number of primary references required for your paper. Pechenik (2016) and Knisely (2013) provide useful suggestions for how to read and evaluate scientific papers.

Plagiarism

Plagiarism is the use of someone else's words or ideas without acknowledging the source or author. Plagiarism is a serious offense and one that can and must be avoided. All sources of information must be acknowledged even if you write about the work in your own words. Although you will work collaboratively on your laboratory investigations, *students must write and submit their papers independently*. Because performing the experiment will be a collaborative effort, you and your teammates will share the results of your investigation. A scientific paper must be the product of your own personal literature research and creative thinking. If you are not certain about the level of independence and what constitutes plagiarism in this laboratory program, ask your instructor to clarify the policy. *In the most extreme case of plagiarism, a student presents another student's report as his or her own. However, representing another person's ideas as your own without giving that person credit is also plagiarism and is a serious offense.* Many instructors are using Internet and software resources to detect plagiarism. The penalties for plagiarism are severe, ranging from a failing grade on the work to expulsion. For best practices to avoid plagiarism see Purdue's OWL website (2016), and to identify plagiarism see Frick's (2016) website.

Writing a Scientific Paper

The sections of a scientific paper in order of appearance are: Title Page, Abstract, Introduction, Materials and Methods, Results, Discussion, and References Cited. However, most scientists do not follow that sequence in the actual writing of the paper, but rather begin with the methodology. See Table A.1 for a suggested plan to organize and manage writing a scientific paper.

TABLE A.1 Plan for Writing a Scientific Paper

Order of Writing	Section	Notes
Begin 1st	Materials and Methods	The first draft of this section can be written before all the results are completed. Remember to review and carefully edit after completing all work. (See Materials and Methods section, p. 797.)
2nd	Results	First, construct the **tables and figures.** Review criteria for tables and figures in Lab Topic 1 Scientific Investigation. **Compose the text** for the Results section stating the important findings and trends depicted in the tables and figures. (See Results section, p. 797.)
3rd	Literature Search—Continued	In developing your investigation, you reviewed some references. However, you will need to locate and review additional primary and secondary references for use in the Discussion and Introduction sections for background information and interpretation of results. (See Locating Appropriate References section, p. 793, and References Cited section, p. 799.)
4th	Discussion and References	Write the **Discussion** section and begin the **References Cited** section. (See Discussion section, p. 798, and References Cited section, p. 799.) Most scientists prefer to write the Discussion before the Introduction. Both sections require references for background information and a clear understanding of the results of the work. If you write the Discussion first, remember to return to this section to carefully check and revise.
5th	Introduction and References	Develop the **Introduction** section and complete the **References Cited** section. (See Introduction section, p. 796, and References Cited section, p. 799.)
6th	Abstract and Title	**Write the Title** and **Abstract**. (See pp. 795-796.) The Abstract is a short summary of all components of the paper, and therefore must be written last. The title is a succinct statement of the objective of the research. Refer to the instructions for creating and formatting the Title Page and the essential elements of the Abstract.
7th	Revise and Proofread	Leave the paper for a short time and then reread carefully. **Revise** to be clear, concise, and well written. Review your use of evidence and evaluate your sources. Revise accordingly. **Proofread carefully** for grammatical, punctuation, and word processing errors. Review a checklist, if available, before preparing the final version of the paper. Check for all formatting rules and suggestions.

Title Page and Title

The **Title Page** is the first page of the paper and includes the **title** of the paper, **your name**, the **course title**, your **lab section**, your **instructor's name**, and the **due date** for the paper. *The title should be as short as possible and long as necessary to describe the objective and significance of your research topic.* For example, if you are asking a question about the inheritance patterns of the gene for aldehyde oxidase production in *Drosophila melanogaster,* a possible title might be "Inheritance of the Gene for Aldehyde Oxidase in *Drosophila melanogaster.*" Something like "Inheritance in Fruit Flies" is too general, and "A Study of the Inheritance of the Enzyme Aldehyde Oxidase in the Fruit Fly *Drosophila melanogaster*" is too wordy. The words "A Study of the" are superfluous, and "Enzyme" and "Fruit Fly" are redundant. The suffix *ase* indicates that aldehyde oxidase is an enzyme, and most scientists know that *Drosophila melanogaster* is the scientific name of a common fruit fly species. However, it is appropriate to include in the title both common and scientific names of lesser-known species.

The Title Page has a specific format. Place the title about *7 cm from the top* of the Title Page. Place "by" and your name in the *center of the page,* and place the course name, lab section, instructor's name, and due date, each on a separate centered line, at the bottom of the page. *Leave about 5 cm below* this information.

Abstract

The **Abstract** is placed at the beginning of the second page of the paper, after the Title Page. *The Abstract concisely summarizes the question being investigated in the paper, the methods used in the experiment, the results, and the conclusions drawn.* The reader should be able to determine the major topics in the paper so that they can decide if the paper is relevant to their research and if they want to read further. The Abstract should be no more than 250 words, and fewer if possible. Compose the Abstract after the paper is completed.

Introduction

The **Introduction** *has two functions: (1) to provide the context for your investigation and (2) to state the question asked and the hypothesis tested in the study.* Begin the Introduction by providing background information that will enable the reader to understand the objective of the study and the significance of the problem, relating the problem to the larger issues in the field. Include only information that directly prepares the reader to understand the question investigated. Most ideas in the Introduction will come from synthesizing information from a variety of sources, such as scientific journals, review articles, or books dealing with the topic you are investigating. Briefly describe and connect the study to your investigation as you develop the background and context. All sources of information must be referenced and included in the References Cited (or References) section of the paper, but the Introduction must be in your own words. Refer to the references to provide evidence and support your ideas. Unless otherwise instructed, place the author of the reference cited and the year of publication in parentheses at the end of the sentence or paragraph relating the idea; for example, (Finnerty, 1992). Additional information on citing references is provided in the References Cited section, p. 799. *Do not use citation forms utilized in other disciplines. Do not use footnotes and avoid the use of direct quotes.*

The Introduction begins broadly, providing background and context from other studies, and then becomes more focused on your specific research question. As you describe your investigation, state the question and hypothesis that you finally investigated. Briefly describe the experiment performed and the outcome predicted for the experiment. Although these items are usually presented after the background information near the end of the Introduction, you should have each clearly in mind before you begin writing the Introduction. It is a good idea to write down each item (question, hypothesis, prediction) before you begin to write your Introduction, so that you can clearly connect the background information with your investigation.

Materials and Methods

The **Materials and Methods** *section describes your experiment in such a way that it can be repeated. This section should be a narrative description that integrates the materials with the description of the procedures used in the investigation.* The Materials and Methods section is often the best place to begin writing your paper. The writing is straightforward and concise, and you will be reminded of the details of the work. The following notes apply to writing the Materials and Methods section.

- Write the Materials and Methods section concisely in paragraph form in the past tense.
- *Do not list the materials and do not list the steps of the procedure.*
- Be sure to include levels of treatment, numbers of replications, and controls.
- If you are working with living organisms, include the scientific name and the sex of the organism.
- Describe the analysis of data, including any statistical analyses or computer software used.
- *What to include and what not to include:* The most difficult part of writing the Materials and Methods is determining which details are essential for another investigator to repeat the experiment. However, do *not* include details about standard laboratory practices and equipment. For example, if in your experiment you incubated potato pieces in different concentrations of sucrose solution, it would not be necessary to explain that the pieces were incubated in plastic cups labeled with a permanent marker or to provide the numbers of the cups. In this case, the molarity of the sucrose solutions, the size of the potato pieces and how they were obtained, and the amount of incubation solution are the important items to include.
- Do not include failed attempts or information on troubleshooting your procedures. Do not try to justify your procedures in this section.
- Results should not be included in the Materials and Methods.

Results

The **Results** section is the central and most important section of a scientific paper. It consists of at least four components: *(1) one or two sentences reminding the reader about the nature of the research, (2) one or more paragraphs that describe the results, (3) figures (graphs, diagrams, pictures), and (4) tables.*

- First, prepare the tables and figures; only then can you write the narrative portion of the Results section. *See Lab Topic 1 Scientific Investigation for instructions on creating and formatting figures and tables and their presentation.* Do not repeat the same information in both a table and figure. The data included in tables and graphs should be summarized and emphasized in a narrative paragraph written in past tense.

- Draw the reader's attention to the results that are important.
- Describe trends in your data and highlight specific results that you will later discuss.
- Report your data as accurately as possible without explanation or discussion. Do not report what you expected to happen in the experiment nor whether your data supported your hypothesis.
- Do not critique the results.
- *Any data you plan to include in the Discussion section must be presented in the Results. If you collected data or made measurements described in the Materials and Methods section, those results must be included in the Results section.*
- Write the Results section before attempting the Discussion section. This will ensure that the results of your investigation are clearly organized, logically presented, and thoroughly understood before they are discussed.
- Remember to number figures and tables consecutively throughout the paper. Refer to figures and tables within the paragraph as you describe your results, using the word "Figure" or "Table," followed by its number—for example, (Figure 1 and Table 1). If possible, place each figure or table at the end of the paragraph in which it is cited.
- If you have performed a statistical analysis of your data, such as chi-square, include the results in this section.

Discussion

In the **Discussion** *section, you will analyze and interpret the results of your experiment.* The Discussion section begins with an explanation of the evidence from your research, then develops more broadly to present a logical argument using the work of other scientists as evidence to support your interpretations and conclusions. Simply restating the results is not interpretation. The Discussion must provide a context for understanding the significance of the results. Your results will either support or confirm your hypothesis or will negate, refute, or contradict your hypothesis; but the word *prove* is not appropriate in scientific writing. If your results do not support your hypothesis, you must still state why you think this occurred. Review the related work of other scientists and provide a summary of their research. Use this evidence to support your interpretation of the results of your study. State your conclusions in this section.

Complete the Results sections before you begin writing the Discussion. The figures and tables in the Results section will be particularly important as you begin to think about your discussion. The tables allow you to present your results clearly to the reader, and graphs allow you to visualize the effects that the independent variable has had on the dependent variables in your experiment. Studying these data will be one of the first steps in interpreting your results. As you study your data in the Results section, write down relationships and integrate these relationships into a rough draft of your discussion.

The following steps may be helpful as you *begin to outline your discussion* and *before you write the narrative:*

- Restate your question, hypothesis, and prediction.
- Summarize the trends and specific data, including results of statistical tests.
- State whether your results did or did not confirm your prediction and support or falsify your hypothesis.

- Write down what you know about the biology involved in your experiment. How do your results fit in with what you know? What explanations for your results can you suggest based on current knowledge? Do you need to do an additional literature search?
- Locate and describe the essentials of the approach, results, and conclusions from relevant published research by other scientists. How do your results support or conflict with previous work? Summarize the findings and include references to this work.
- Clearly state your **conclusions.**
- If you were writing a laboratory report describing your work in progress to your laboratory team, you might include all the weaknesses and problems that you speculate could affect future experiments. However, remember that other scientists are your audience, and this is a scientific paper describing your research. *Therefore, include one or two sentences only if you identified problems that affected the results.* Remember that the focus of the Discussion is to convey the significance of the results.
- You are now ready to write the narrative for the Discussion. Integrate all of the above information into clear and concise paragraphs. Discuss the results; do not simply restate the data. Refer to and describe work by other scientists to support your conclusions.

References Cited (or References)

A **References Cited** section lists only those references cited in the paper. A References section (bibliography), on the other hand, is a more inclusive list of all references used in producing the paper, including books and papers used to obtain background knowledge that may not be cited in the paper. Most references will be cited in the Introduction and Discussion sections of your paper. References may also be included in the Materials and Methods when acknowledging the source for a procedure or method of analysis. For your paper you should have a References Cited section that includes only those references cited in the paper. For additional assistance, see “Locating Appropriate References” on p. 793.

Constructing the References Cited Section

The **format for the References Cited** section differs slightly from one scientific journal to the next. How does an author know which format to use? Every scientific journal provides “Instructions to Authors” that describe specific requirements for this important section and all other aspects of the paper. You may use the format used in this lab manual and provided in the following examples, select the format in a scientific journal provided by your instructor, or use another accepted format for listing your references. Your instructor may provide additional instructions. Be sure to read the references that you cite in your paper.

If a journal article is published in a print journal, use the standard reference citation below, even if you located the article on the Internet. However, some scientific journals are published exclusively online. These articles may have a URL that is included in the citation. Recently, some sources are including a **DOI** (Digital Object Identifier) for these articles, rather than the URL, which can change for online materials. If available, this number should be included in place of or in addition to the URL. Additional sources for how to cite Internet materials are located at the end of this appendix.

Journal article, one author:
Gould, S. J. "Is a New and General Theory of Evolution Emerging?" *Paleobiology*, 1980, vol. 6, pp. 119–130.

Journal article, two or more authors:
Jinek, M., K. Chylinski, I. Fonfar, M. Hauer, J. A. Doudna, and E. Charpentier. "A Programmable Dual-RNA-guided DNA Endonuclease in Adaptive Bacterial Immunity." *Science*, 2014, vol. 337(6096), pp. 816–21.

Journal article, published in an online journal:
Atwood, T. C., E. Peacock, M. McKinney, K. Lillie, R. Wilson, D. Douglas, S. Miller, and P. Terletzky. "Rapid Environmental Change Drives Increased Land Use by an Arctic Marine Predator." *PLoS One*, 2016, vol. 11(6) e0155932. doi:10.1371/journal.pone.0155932.

Journal article, published in an online journal (no DOI):
Browning T. "Embedded Visuals: Student Design in Web Spaces." *Kairos: A Journal for Teachers of Writing in Webbed Environments,* 1997, 3(1). http://english.ttu.edu/kairos/2.1/features/browning/bridge.html. Accessed May 15, 2010.

Book:
Darwin, C. R. *On the Origin of Species.* London: John Murray, 1859.

Chapter or article in an edited book:
Funk, D. J. "Investigating Ecological Speciation," in *Speciation and Patterns of Diversity,* eds. R. K. Butlin, J. R. Bridle, and D. Schluter. Cambridge, UK: Cambridge University Press, 2009, pp. 195–218.

Government publication:
Office of Technology Assessment. *Harmful Non-indigenous Species in the United States.* Publication no. OTA-F-565. Washington, D.C.: U.S. Government Printing Office, 1993.

Citing References in Text

Scientific writing uses a different format than you may have used in other disciplines for acknowledging sources of information in the body of the paper. After referring to and describing the work of another scientist in your own words, place the author's name and the date in parentheses at the end of the passage. This date and year citation format is not the only one used in scientific writing, but it is the one used in this laboratory manual and many scientific journals. *Do not use citation formats from other disciplines.*

Example for one author:
The innate agonistic behavior of the male Siamese fighting fish has been widely studied (Simpson, 1968).

For two authors use both names:
Telomere terminal transferase was identified as the enzyme that adds nucleotide repeats at the telomeres (ends of chromosomes) during eukaryotic replication (Grieder and Blackburn, 1985).

For three or more authors use first author plus "et al.":
Taxol, a bioactive compound extracted from the Pacific yew, inhibits cell replication at the metaphase/anaphase stage. The blocking of mitosis occurs in tissue culture cells as a result of stabilizing the microtubules so that the chromosomes cannot move to the poles of the cell (Jordan et al., 1993).

Information Sources and the Internet

The Internet can provide access to online search engines and databases, including *Biological Abstracts, Current Contents, Medline,* and *Annual Reviews,* among many others. These search tools provide access to a wide range of published papers, some of which may be available online as full-text journals. For suggestions and examples of how to locate sources using the Internet, see Pechenik (2016) and Knisely (2013). Scientific papers published in professional journals have gone through an extensive review process by other scientists in the same field. Most scientific articles have been revised based on comments by the reviewers and the editors. Sources of information that lack this critical review process do not have the same validity and authority.

The Internet is an exciting, immediate, and easily accessible source of information. However, unlike traditional bibliographic resources in the sciences, the Internet includes websites with material that has not been critically reviewed. *Your instructor may prefer that you use the Internet only for locating peer-reviewed primary references or as a starting point to promote your interest and ideas. Consult your instructor concerning use of Internet information.*

If you do use the Internet to locate information, you should be prepared to evaluate these sites critically. Remember always to record the online address (URL) for any site you use as a reference. For more information and a checklist on critically evaluating sources on the Web, see the University of Maine Library website listed at the end of this appendix.

The following is a model format, along with examples, for citing Internet sources in the References Cited section of your paper. Other formats may be suggested by your instructor or librarian.

Model:
Author's last name and initials. Date of Web publication. Document title. URL or DOI. Date of access.

Examples:

Professional site:
[CBE] Council of Biology Editors. 2010. CBE home page. http://www.councilscienceeditors.org. Accessed May 15, 2010.

Government publication:
Food and Drug Administration, 1996, Sep. "Outsmarting Poison Ivy and Its Cousins." *FDA Consumer Magazine.* http://www.fda.gov/fdac/features/796_ivy.html. Accessed May 15, 2010.

Note: If using a Web search engine such as Google Scholar, *do not include the entire search path.* Cite only the actual URL or DOI in the reference citation.

Oral and Poster Presentations

Oral Presentations

Scientists regularly present their work at scientific meetings in oral paper presentations. They can share preliminary findings and receive critical questions and comments from others in their field that are essential to scientific research. In this laboratory program you may be asked to give an oral report on your experiments in the laboratory or in a research symposium. For your 10- to 15-minute presentation, you need to capture the interest of your audience and make a convincing presentation of your results and conclusions. **Following are suggestions for preparing oral presentations:**

- Include the components of a scientific paper, with a brief Introduction, essential Methods, Results, Discussion, and Conclusions.
- In the Introduction, provide background information, state your hypothesis, give a brief description of your experiment, and state your predicted results.
- Briefly describe the experimental design and only essential procedures.
- Emphasize your results with bold and clearly labeled figures and tables that have visual impact. Be sure to take time to explain your tables and figures to the audience.
- Discuss and interpret your results, providing supporting evidence from primary references.
- State your conclusions.
- Include references to sources of information and for all images.
- Be prepared to answer questions.
- Use simple, bold visual aids that are easily visible from the back of the room. For information on preparing PowerPoint slides, see Knisely (2013).

Following are suggestions for delivering effective presentations:

- Slides should have a small number of bullets (six or so), not extensive text.
- Use a simple and consistent template or theme for slides.
- Minimize distracting transitions and sounds.
- Do not read the slides. Be prepared to describe the points presented in each slide.
- Practice so that you are well prepared and can talk to the audience (not the screen). Speak slowly and clearly, projecting your voice so all members of the audience can hear you.
- Keep your objective in mind—to clearly communicate your ideas.

Instructions for preparing successful oral presentations, tips for effective communication, and common errors may be found in Knisely (2013) and Pechenik (2016). A copy of an evaluation form for oral presentations can be downloaded at http://sites.sinauer.com/knisely4e/.

Poster Presentations

Another form of scientific communication that has become popular in recent years is the **scientific poster**, a document usually created on a single, large sheet. If a large printer is not available, a poster may be created on a multiple-paneled mat board. Most scientific societies are now organizing poster sessions at their annual meetings, often featuring undergraduate and graduate student research. Your instructor may ask you to model this form of communication by preparing a poster about your student-designed investigations or extended research projects as part of a poster symposium.

In many situations, posters may be a more effective method of presentation than writing a scientific paper or giving an oral report. A poster may be prominently displayed for a wider audience for an extended period. The presenters are then available at designated times to explain the research and engage in meaningful discussion.

Following are suggestions for preparing and presenting a poster:

- Present the essentials of your research with *minimal text and maximum visual impact.* Studies show that you have less than a minute to grab the attention of interested persons.
- Include the sections of a scientific paper—Title banner, Introduction, Materials and Methods, Results, Discussion, and brief References. For an alternative way to organize a poster see Chapter 11 "Preparing Talks and Poster Presentations" (Pechenik, 2016).
- Use a large typeface for headings and the content in abbreviated format under each heading. Use fonts that are easily readable from 3 to 6 feet. Title fonts should be 72 point, section headings 28 point, and text should be no smaller than 24 point (Knisely, 2013).
- Include simple, bold, and clearly labeled figures and tables.
- Before creating your poster, sketch the layout on paper, blocking out sections in two or three columns. Consider the location of text, images, tables, and figures. You can print the elements of the layout on sheets of paper and then move these around to create the layout that has the greatest impact.
- Consider the use of fonts, color, and images to highlight your poster and direct the viewer's eye.
- Proofread for typos and errors before printing or constructing the poster.
- Be prepared to answer questions and discuss your work.

You can design your poster in an electronic form using PowerPoint. This allows you to edit the text, images, and overall design. You can import sections from electronic files and insert digital images. You can effectively collaborate with your research team as you modify electronic versions of your poster. If your institution or community has a printing service, the electronic file can be sent to it for printing on large poster-sized paper. For detailed instructions, see Chapter 7 in Knisely (2013).

Your instructor may give specific guidelines for your poster, or you may follow guidelines from other sources. There are several excellent websites with instructions for preparing posters. One of the most complete and well-designed sites is http://colinpurrington.com/tips/poster-design. This website includes a poster template, suggested layouts, and examples of good and mediocre posters.

REFERENCES

The following sources are recommended to give additional help and examples in scientific writing.

Gopen, G. D. and J. A. Swan. "The Science of Scientific Writing." *American Scientist,* 1990, vol. 78, pp. 550–558.

Knisely, K. *A Student Handbook for Writing in Biology,* 4th ed. Sunderland, MA: Sinauer Associates, 2013.

Pechenik, J. A. *A Short Guide to Writing about Biology,* 9th ed., San Francisco, CA: Pearson, 2016.

Style Manual Committee, Council of Biology Editors. *Scientific Style and Format: The CBE Manual for Authors, Editors and Publishers,* 7th ed. Cambridge, MA: Cambridge Univ. Press, 2006.

WEBSITES

Best practices to avoid plagiarism from the Purdue Online Writing Lab. "OWL:Avoiding Plagiarism," 2016: http://owl.english.purdue.edu/owl/resource/589/03/

"Reader Expectations, Interview with Steve Vogel (Duke University)." A great discussion of how to read and write scientific papers: https://www.youtube.com/watch?v=eYBMp5sYchA

Checklists and Evaluation forms for papers, posters and oral presentations: http://sites.sinauer.com/knisely4e/

Frick, T. 2016. "Understanding Plagiarism." Indiana School of Education. Available at: https://www.indiana.edu/~tedfrick/plagiarism/

How to evaluate Web sources: http://library.berkeley.edu/TeachingLib/Guides/-Internet/Evaluate.html

"Evaluating Web Resources," a checklist and information on critical evaluation of web material provided by the University of Maine Library: http://usm.maine.edu/library/checklist-evaluating-web-resources

Purrington, C. B. 2016. Advice on designing scientific posters: http://colinpurrington.com/tips/poster-design

"Searching the World Wide Web: Overview," 2016. Includes best practices, hints, and how to evaluate information: https://owl.english.purdue.edu/owl/resource/558/1/

"Student's Guide to Writing in the Life Sciences," provides excellent information for scientific writing: http://writingproject.fas.harvard.edu/files/hwp/files/life_sciences.pdf

Appendix **B**

The Metric System

Metric Prefixes:	10^9 = giga (G)	10^{-2} = centi (c)	10^{-9} = nano (n)
	10^6 = mega (M)	10^{-3} = milli (m)	10^{-12} = pico (p)
	10^3 = kilo (k)	10^{-6} = micro (μ)	10^{-15} = femto (f)

Measurement	Unit and Abbreviation	Metric Equivalent	Metric-to-English Conversion Factor	English-to-Metric Conversion Factor
ength	1 kilometer (km)	= 1,000 (10^3) meters	1 km = 0.62 mile	1 mile = 1.61 km
	1 meter (m)	= 100 (10^2) centimeters = 1,000 millimeters	1 m = 1.09 yards 1 m = 3.28 feet 1 m = 39.37 inches	1 yard = 0.914 m 1 foot = 0.305 m
	1 centimeter (cm)	= 0.01 (10^{-2}) meter	1 cm = 0.394 inch	1 foot = 30.5 cm 1 inch = 2.54 cm
	1 millimeter (mm)	= 0.001 (10^{-3}) meter	1 mm = 0.039 inch	
	1 micrometer (μm) (formerly micron, μ)	= 10^{-6} meter (10^{-3} mm)		
	1 nanometer (nm) (formerly millimicron, mμ)	= 10^{-9} meter (10^{-3} μm)		
	1 angstrom (Å)	= 10^{-10} meter (10^{-4} μm)		
rea	1 hectare (ha)	= 10,000 square meters	1 ha = 2.47 acres	1 acre = 0.405 ha
	1 square meter (m^2)	= 10,000 square centimeters	1 m^2 = 1.196 square yards 1 m^2 = 10.764 square feet	1 square yard = 0.8361 m^2 1 square foot = 0.0929 m^2
	1 square centimeter (cm^2)	= 100 square millimeters	1 cm^2 = 0.155 square inch	1 square inch = 6.4516 cm^2
Mass	1 metric ton (t)	= 1,000 kilograms	1 t = 1.103 tons	1 ton = 0.907 t
	1 kilogram (kg)	= 1,000 grams	1 kg = 2.205 pounds	1 pound = 0.4536 kg
	1 gram (g)	= 1,000 milligrams	1 g = 0.0353 ounce 1 g = 15.432 grains	1 ounce = 28.35 g
	1 milligram (mg)	= 10^{-3} gram	1 mg = approx. 0.015 grain	
	1 microgram (μg)	= 10^{-6} gram		
olume (solids)	1 cubic meter (m^3)	= 1,000,000 cubic centimeters	1 m^3 = 1.308 cubic yards 1 m^3 = 35.315 cubic feet	1 cubic yard = 0.7646 m^3 1 cubic foot = 0.0283 m^3
	1 cubic centimeter (cm^3 or cc)	= 10^{-6} cubic meter	1 cm^3 = 0.061 cubic inch	1 cubic inch = 16.387 cm^3
	1 cubic millimeter (mm^3)	= 10^{-9} cubic meter = 10^{-3} cubic centimeter		
olume (liquids and gases)	1 kiloliter (kL or kl)	= 1,000 liters	1 kL = 264.17 gallons	
	1 liter (L or l)	= 1,000 milliliters	1 L = 0.264 gallon 1 L = 1.057 quarts	1 gallon = 3.785 L 1 quart = 0.946 L
	1 milliliter (mL or ml)	= 10^{-3} liter = 1 cubic centimeter	1 mL = 0.034 fluid ounce 1 mL = approx. ¼ teaspoon 1 mL = approx. 15–16 drops (gtt.)	1 quart = 946 mL 1 pint = 473 mL 1 fluid ounce = 29.57 mL 1 teaspoon = approx. 5 mL
	1 microliter (μL or μl)	= 10^{-6} liter (10^{-3} milliliter)		
emperature			°F = $\frac{9}{5}$°C + 32	°C = $\frac{5}{9}$ (°F − 32)

Appendix C

Instrumentation and Techniques

Using an Electronic Balance

1. **Weighing Materials or an Object**
   a. Place the balance on a flat, stable surface. Press the power switch. The weight display screen will light up. If zeroes do not appear in the screen, press the "Tare" or "Zero" button.
   b. Place an empty container to hold the material being weighed on the balance pan. The container may be a plastic tray or other container, for example, a glass beaker. In some cases, you may use weighing paper squares or pieces of aluminum foil to hold the material.
   c. Press the "Tare" or "Zero" button to automatically deduct the weight of the container or paper tray from future calculations. Zeroes will again appear on the screen.
   d. Carefully add the material to be weighed to the container. The weight of the material will appear on the screen. Depending on the balance model used, it may weigh to either two or three decimal places (0.00 or 0.000). Fluctuations in the third decimal place (the thousandths, or 0.000) are typical, as air movement in the room may cause fluctuation.
2. **Weighing a Specific Amount of a Substance, Such As a Dry Chemical**
   a. Place the empty container to hold the substance on the balance pan.
   b. Tare the balance (see step c above).
   c. For dry substances, use a spatula or scoopula to carefully add the substance to the container until the desired amount is added.
   d. To weigh a liquid, use a pipette to carefully add the liquid to the container.
   e. If too much material is added, use the spatula or pipette to carefully remove the excess and dispose of it as directed by your instructor. *Do not* return the excess to the stock bottle.

Using a Vernier Caliper

A **caliper** is a tool that is used by scientists, engineers, and precision woodworkers to measure the distance between two opposite sides of an object. Two versions of calipers that may be used in introductory biology laboratories are the digital caliper (described in Lab Topic 3 Diffusion and Osmosis) and the vernier caliper. When a digital caliper is used, the measurement is read from a digital display, and the instrument accurately measures to 0.01 millimeter. A **vernier caliper** consists of a ruler with a stationary arm and a movable arm with an attached vernier scale. A typical vernier caliper can accurately measure to 0.1 millimeter. The instructions that follow are for operating a vernier caliper.

1. Identify the following parts of the caliper: **stationary arm**, **movable arm**, **ruler**, **vernier scale** (see Figure C.1a).

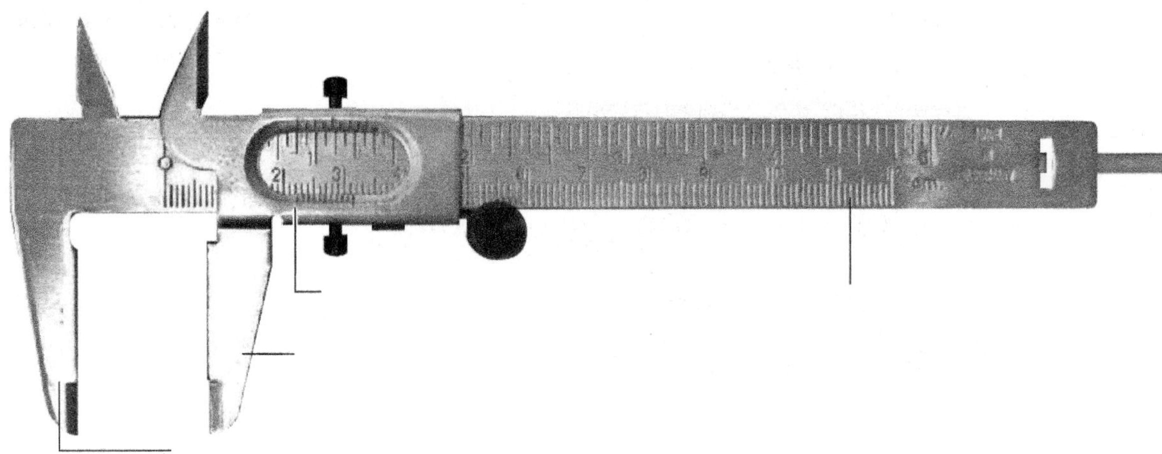

FIGURE C.1a
Vernier caliper. Identify the stationary arm, movable arm, ruler, and vernier scale.

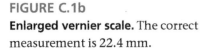

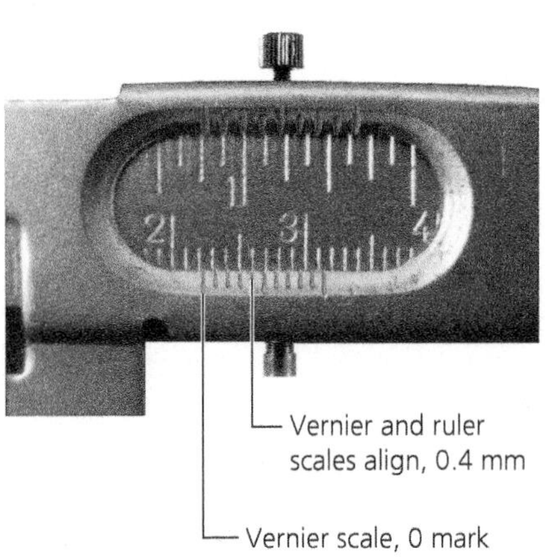

FIGURE C.1b
Enlarged vernier scale. The correct measurement is 22.4 mm.

2. Place the object to be measured between the two arms, adjusting the movable arm until both arms just touch the object.
3. Note the 0 mark on the vernier scale (Figure C.1b). The graduated line on the lower edge of the *ruler* just to the *left* of the 0 mark on the vernier scale is the distance between the caliper arms measured in whole millimeters. In Figure C.1b, that number is 22.
4. Look at the graduated lines between 0 and 10 on the vernier scale. Note the line on the vernier scale that *exactly* matches with a line on the ruler. That line on the vernier scale is the measurement in tenths of a millimeter. In Figure C.1b, that number is 4. Add this to the whole-millimeter reading. What is the final measurement?

Using a Spectrophotometer

A spectrophotometer is an instrument that measures the amount or proportion of light of different wavelengths that is absorbed or, conversely, transmitted by a solution. A light beam inside the instrument is separated into its component wavelengths and then a specific wavelength can be selected to investigate. In this laboratory manual, this instrument is used to measure the amount of light of different wavelengths that is absorbed or transmitted by various pigments.

1. **Thermo Scientific Spectronic 20D+ (Figure C.2).**
   a. Turn on the machine (**power switch** C) at least 15 minutes before beginning.
   b. Using the **wavelength control knob** (A), select the wavelength to be measured.
   c. Press the **mode selection** button to select "Transmittance."
   d. Zero the instrument by adjusting the control knob (the same as power switch C) so that the meter reads 0% transmittance. *There should be no cuvette in the instrument, and the* ***sample holder cover*** *must be closed.*

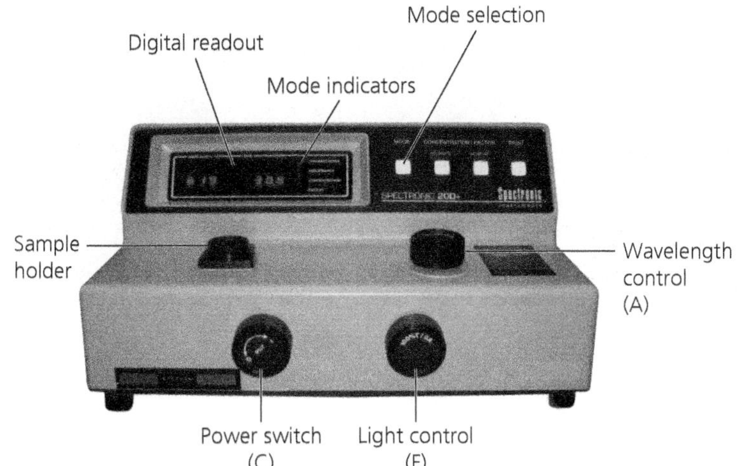

FIGURE C.2
The Thermo Scientific Spectronic 20D+

e. You will use at least two cuvettes, here called A and B. Cuvette A is filled with the solution to be tested and the other is a **blank (B)**, that is, a solution that is identical to the test solution, but that lacks the experimental component.

f. Begin taking a measurement with the blank (B). Cover the cuvette tightly with Parafilm and invert it to mix the reactants. Wipe the cuvette with a Kimwipe to remove prints or smudges, remove the Parafilm, and insert the cuvette into the **sample holder**. Be sure you align the etched mark on the cuvette with the line on the sample holder. Close the cover. Adjust the **light control** (F) until the meter reads 100% transmittance. The blank corrects for differences in transmittance due to substances other than the test material in the solution.

g. Remove cuvette B. You are now ready to test the experimental cuvette (A).

h. Cover the opening to cuvette A with Parafilm and invert it to mix. Wipe with a Kimwipe. Remove the Parafilm and insert the cuvette into the sample holder with lines aligned and record the percent transmittance.

i. To take the next reading, change the wavelength, if specified in the experiment, and repeat the procedure. *Before each new reading*, insert the blank (cuvette B) into the sample holder and adjust to 100% transmittance. Then invert the test solution to mix it again, wipe and insert the cuvette, and record the percent transmittance.

2. **Thermo Scientific Spectronic 200 (Figure C.3)**
The Spectronic 200 is the latest digital model of spectrophotometer offered by Thermo Scientific, replacing all Spectronic 20 models used for teaching laboratories.?

a. Before you begin, locate the following:
   - the **power switch** on the back of the machine
   - the color graphics **display** where you will see instructions
   - two **chambers** to hold samples
   - **sample compartment** covered with a blue hinged lid

   There are two receptacles in the sample compartment to accommodate either 10-mm square cuvettes or test tube cuvettes up to 25 mm in diameter (Figure C.3a). The power switch should be *off* before plugging in the power cable.

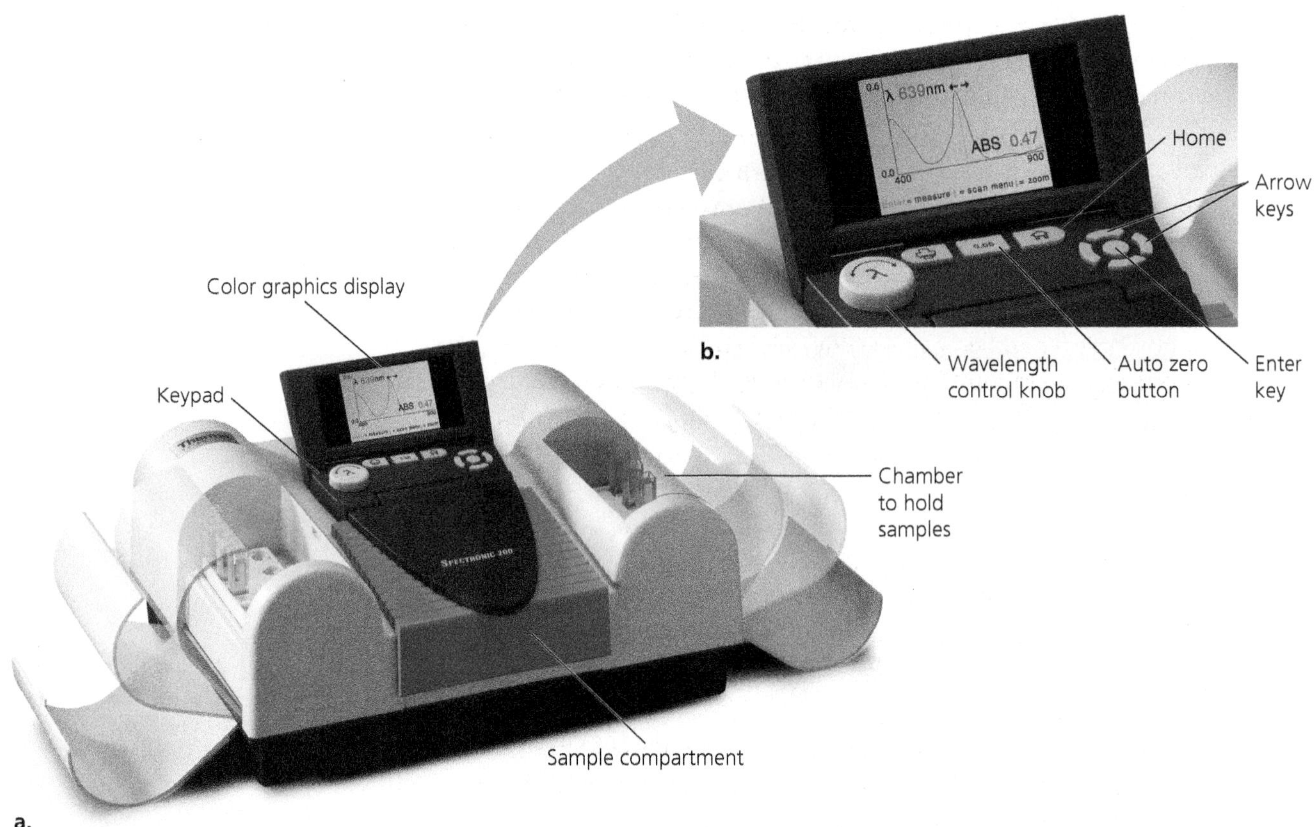

FIGURE C.3

The Spectronic 200. (a) **The digital spectrophotometer.** Locate the *color graphics display* above the *keypad*, the *sample compartment*, and the two *chambers to hold samples*. (b) **Keypad and color graphics display**. Locate the *enter key, arrow keys, home key, auto zero button*, and the *wavelength control knob* (λ).

b. Switch on the power. You should see the display screen where you will be prompted to *remove any cuvettes and close the sample compartment.*

c. Press the **enter key** to continue (Figure C.3b). The machine will perform a self-diagnosis and initialization. After the initialization is completed (this may take several minutes), the main menu is displayed and the system is ready for operation. (Note: Your instructor may have already performed this step.) The Home menu will appear.

d. Press the enter key. You will see *Application: Live Display* on the screen. This application is used to measure and display absorption or transmission values in real time, and will be the application most frequently used for activities in this laboratory manual.

e. Note the **arrow keys** around the enter key. Use the down arrow key to select Measurement Mode, and then the right/left arrow keys to toggle between **%T (transmittance)** and **Abs (absorbance)**. Once you have selected either %T or Abs, use the down arrow to select *Measurement* (wavelength).

f. Use the left and right arrow keys or the **wavelength control knob** (λ) to select the desired wavelength. The knob has a coarse/fine feature. Turn it normally to alter the wavelength in units of 10 nm. Press it down and turn to alter the wavelength in 1 nm. Use the **down arrow**

key to select *GO*, then press the enter key to start the *Live Display* mode. A triangle marker and white line on the spectrum at the bottom of the screen show the color of the light for the selected wavelength.

g. Prepare two cuvettes, A and B. *(Do not write on the cuvettes!)* Cuvette A will be the solution to be tested. Cuvette B is the **blank**, a solution identical to the test solution, but lacking the experimental component. Use a Kimwipe to wipe the cuvette containing the blank (B) solution to remove residue or fingerprints. Locate the etched line on the cuvette and the line on the sample holder, and place the cuvette into the sample holder chamber with the lines aligned. Press the 0.00 button (**auto zero**) and wait until 0.00 Abs (or 100% T) appears on the screen.

h. Once the screen shows 0.00 Abs (or 100% T), remove the blank (B) and use a Kimwipe to wipe cuvette A containing the solution to be tested. Insert the cuvette into the sample holder and align the lines. The reading for transmittance (or absorbance, whichever you selected) for the test solution will be shown on the screen. The digital display may fluctuate. To freeze the display, press the **enter key**.

i. You can measure up to four wavelengths in a single measurement.

   Procedure:
   - Press the **home** button on the instrument keypad.
   - Press the enter key and the right arrow key. *Multi* λ will appear.
   - Press the down arrow key to select Measurement Mode, and then the left or right arrow keys to select %T or Abs.
   - Use the down arrow to select *GO* and press the enter key. Four possible wavelengths will appear.
   - Use the wavelength control knob to choose the wavelength for λ1. Press the down arrow key. Use the knob again to choose the wavelength for λ2, and so on.
   - Use the down arrow key and select *GO*. The absorption or transmittance at each wavelength will be shown.

Unlike the Spectronic 20 D+, this instrument performs a 0%T measurement at startup and retains this information in internal memory. Similarly, when you place a blank cuvette in the sample stage and press the auto zero button on the instrument keypad, the instrument measures the spectrum and records the 100%T value at all wavelengths. This means that it is never necessary for the user to manually set 0%T, and after measuring 100%T at one wavelength, it is not necessary to measure it again if the user changes wavelengths.

Useful online information: http://www.nanodrop.com/library/BR51930_E%200610M_Spec200Educ_L.PDF

Pipetting

A *pipette* is a laboratory tool used to measure and transport a volume of liquid. *Macropipetting* is a process used to transport volumes greater than 1 mL. *Micropipetting* is used to transport volumes of liquids in a range from less than 1 microliter (μL) to 1000 microliters (1mL).

1. **Macropipetting**

   Several types of pipettes are used to measure and/or transport volumes greater than 1mL. **Pasteur pipettes** are usually glass tubes tapered to a narrow point. They are also known as droppers or eye droppers. They are usually used with a small latex or rubber bulb that is attached to the larger opening and is compressed to draw the liquid into the pipette. **Transfer pipettes** are plastic Pasteur pipettes with one end of the pipette closed and expanded into a bulb. Pasteur pipettes and transfer pipettes are not used to transfer *exact* volumes of liquids. These pipettes must be used in a vertical or near-vertical position with the bulb on top and the tip of the pipette pointed down. The liquid should not flow into the bulb.

Graduated pipettes are marked with a series of lines to indicate different volumes and are used when it is important to transfer a given volume of liquid. Note that the volume and units are printed on the top (larger round opening) of the pipette. *Choose a pipette that closely matches the volume you are measuring.* **Pipette fillers** (pipettors) should be used with graduated pipettes. To use a pipette filler:

a. Select an appropriate pipette filler that corresponds to the size of the pipette and place securely on the top of the pipette.
b. Insert the tip of the pipette into the liquid to be measured and, using the thumb wheel, slowly draw the liquid up into the pipette. When using a graduated pipette, as the liquid is drawn into the pipette it will form a curved surface against the glass, called the *meniscus*. Draw the liquid into the pipette up to the *bottom* of the meniscus aligned with the desired level.
c. Be careful to keep the pipettor upright to prevent the liquid from contaminating the pipettor.
d. Move the tip of the pipette to the new container and deliver the correct volume. Press the release lever on the side to *completely* empty the pipette.

2. **Micropipetting**

A **micropipettor** is used to measure small volumes of liquid in a range from less than 1 microliter (1µL) to 1000 µL (1 mL). Micropipettors come in different sizes and each size can be adjusted to deliver an exact volume of liquid within the range of that micropipettor. Typically, three different micropipettors are used to measure three ranges of volumes: a **P20** = 0.5–20 µL, a **P200** = 20–200 µL, and **P1000** = 200–1000 µL. *Never exceed the upper or lower limits of the micropipettor.*

Before using the micropipettor, identify the following items in bold, and locate parts of your pipettor. Refer to Figure C.4, and modify the figure to correspond to your pipettor. Note that the volume adjustment knob and the plunger are the same knob on the model in Figure C.4. This may not be true for the model you are using.

a. Determine the amount of liquid that you need to transfer and find the pipettor that delivers in the range of that amount of liquid. Locate the **volume adjustment knob** that rotates a **display** in a **digital window** (Figure C.4a and C.4b). This display indicates the volume to be delivered. Rotate the knob until the correct volume appears in the display window. Turning the knob clockwise increases volume, and counterclockwise decreases volume. For example, if you would like to transfer 01.5 µL of a solution, use a P20 micropipettor. In the display window of a P20, whole microliters are in black and fractions of a microliter are in red. Turn the knob until the numbers in the window read **01** in black and **5** in red (**01.5**). To measure 15 μL, turn the knob until the numbers read **15** in black and **0** in red. To measure 25 μL, choose the P200 micropipettor and rotate the knob to read **025**.

b. Micropipettors are used in conjunction with **disposable plastic tips** that attach to the narrow tip of the pipettor, color coded for the specific micropipettor. Locate the tip that corresponds to the pipettor you are using. *Never use a pipettor without first attaching the tip.* Attach the tip and check that it is securely in place. Rotate the knob to the designated amount you will transfer. You are now ready to transfer the liquid.

c. *Before placing the tip into the liquid,* depress the pipette **plunger** with your thumb to the *first stop* to eject any air. Place the tip into the solution and slowly release the plunger. This results in the desired amount of solution being taken into the tip. Remove the tip from the liquid.

d. Locate the **microtube** where you will deliver the liquid. Insert the pipette tip into the microtube so that the tip is close to the bottom of the tube. Touch the tip to the side of the tube and slowly press the plunger down to the first stop, and then *continue to press all the way down to the second stop* to release all the liquid from the tip.

e. *Do not release the plunger yet!* Remove the pipettor and tip from the microtube. When the tip is completely out of the tube, slowly release the plunger to the starting position. Never release the plunger inside the tube or you could withdraw an unknown amount of solution back into the tip.

f. To transfer the liquid in the microtube to a well in a gel, brace your elbows on the lab bench and hold the micropipettor with two hands—one hand to deliver the sample and the other to stabilize the end. Be sure that the sample is all the way down in the tip of the pipette and that there is no air between the sample and the tip. Carefully place the tip just inside the well, but not piercing the side or bottom of the well. Slowly release the liquid into the well. Do not release the plunger until the pipette tip is out of the well.

g. Discard the tip, using the **tip ejector button** on the pipettor. Use a new tip each time you use the pipettor, even if you are pipetting the same liquids.

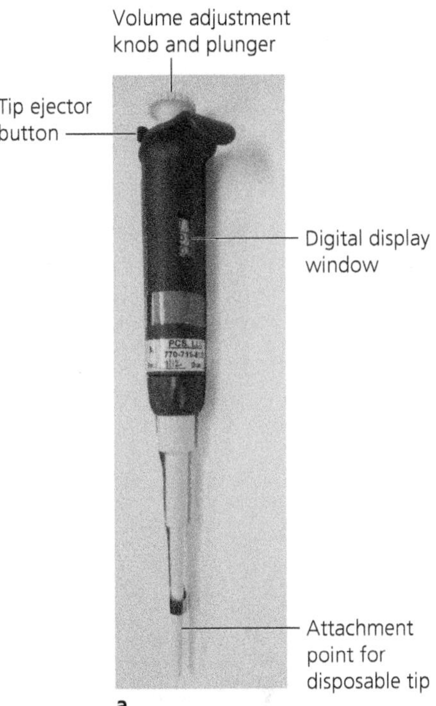

FIGURE C.4

Micropipettor. (a) Identify the *volume adjustment knob, plunger, digital display window, tip ejector button,* and the *attachment point* for a disposable tip. (b) The digital volume indicator in this display window indicates 10 μL to be transferred.

Aseptic Technique

When working with bacteria and some fungi, it is very important to practice certain aseptic techniques to ensure that the cultures being studied are not contaminated by organisms from the environment and that organisms are not released into the environment. Always use the following procedure when working with bacteria. Use aseptic techniques when studying some fungi, if directed by your instructor.

1. Wear a lab coat, a lab apron, or a clean old shirt over your clothes to lessen chances of contamination accidents.
2. Wipe the lab bench with disinfectant before and after the lab activities.

3. Wash your hands before and after performing an experiment. If directed by your instructor, use disposable gloves.
4. Using an alcohol lamp or Bunsen burner, flame all *nonflammable* instruments used to manipulate bacteria or fungi before and after use.
5. Place any swabs or other *flammable* materials in a specially designated container immediately after use. *Never place one of these used items on the lab bench*!
6. Open the lids of cultures only partially when making transfers, to limit exposure to air.

Appendix D

Chi-Square Test

Chi-square is a statistical test commonly used to compare observed data with data we would expect to obtain according to a specific scientific hypothesis. For example, if, according to Mendel's laws, you expect 10 of 20 offspring from a cross to be male and the actual observed number is 8 males out of 20 offspring, then you might want to know about the "goodness of fit" between the observed and the expected. Were the deviations (differences between observed and expected) the result of chance, or were they due to other factors? How much deviation can occur before the investigator must conclude that something other than chance is at work, causing the observed to differ from the expected? The chi-square test can help in making that decision. The chi-square test is always testing what scientists call the **null hypothesis**, which states that there is no significant difference between the expected and the observed result.

The formula for calculating chi-square (χ^2) is:

$$\chi^2 = \sum (o - e)^2/e$$

That is, chi-square is the sum of the squared difference between observed (*o*) and expected (*e*) data (or the deviation, *d*), divided by the expected data in all possible categories.

For example, suppose that a cross between two pea plants yields a population of 880 plants, 639 with green seeds and 241 with yellow seeds. You are asked to propose the genotypes of the parents. Your scientific hypothesis is that the allele for green is dominant to the allele for yellow and that the parent plants were both heterozygous for this trait. If your scientific hypothesis is true, then the predicted ratio of offspring from this cross would be 3:1 (based on Mendel's laws), as predicted from the results of the Punnett square (Figure D.1). The related null hypothesis is that there is no significant difference between your observed pea offspring and offspring produced according to Mendel's laws. To determine if this null hypothesis is rejected or not rejected, a χ^2 value is computed. To calculate χ^2, first determine the number expected in each category. If the ratio is 3:1 and the total number of observed individuals is 880, then the expected numerical values should be 660 green and 220 yellow:

$$({}^3/_4 \times 880 = 660; {}^1/_4 \times 880 = 220)$$

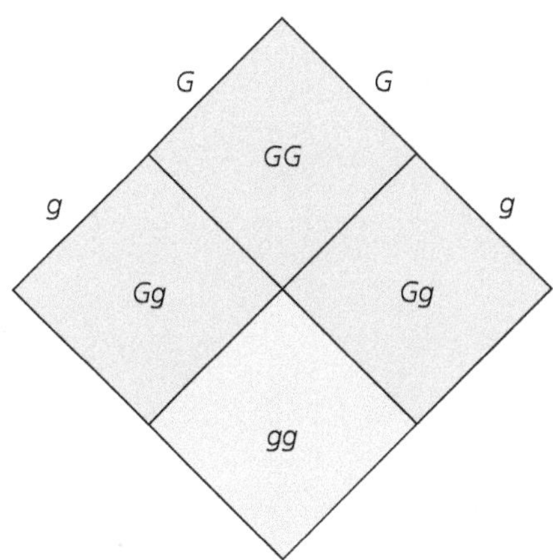

FIGURE D.1

Punnett square. Predicted offspring from cross between green- and yellow-seeded plants. Green (*G*) is dominant (¾ green; ¼ yellow).

Chi-square analysis requires that you use numerical values, not percentages or ratios.

Then calculate χ^2 using the formula, as shown in Table D.1. Note that we get a value of 2.673 for χ^2. But what does this number mean? Here's how to interpret the χ^2 value:

1. Determine **degrees of freedom** (df). Degrees of freedom can be calculated as the number of categories in the problem minus 1. For example, if we had four categories, then the degrees of freedom would be 3. In our example, there are two categories (green and yellow); therefore, there is 1 degree of freedom.
2. Determine a relative standard to serve as the basis for rejecting the hypothesis. Scientists allow some level of error in their decision making for testing their hypotheses. The relative standard commonly used in biological research is $p < 0.05$ The ***p* value** is the *probability* of rejecting the null hypothesis (that there is no difference between observed and expected), when the null hypothesis is true. In other words, the p value gives an approximate value for the error of falsely stating that there is a significant difference between your observed numbers and the expected numbers, when there is *not* a significant difference. *When we pick $p < 0.05$, we state that there is less than a 5% chance of error of stating that there is a difference when, in fact, there is no significant difference.* Although scientists and statisticians sometimes select a lower significance value, in this manual we will assume a value of 0.05.
3. Refer to a chi-square distribution table (Table D.2). Using the appropriate degrees of freedom, locate the value corresponding to the p value of 0.05, the error probability that you selected. If your χ^2 value is *greater* than the value corresponding to the p value of 0.05, then you reject the null hypothesis. You conclude that the observed numbers are significantly different from the expected. In this example, your calculated value, $\chi^2 = 2.673$, is not larger than the table χ^2 value of 3.84 (df = 1, $p = 0.05$). *Therefore, your observed distribution of plants with green and yellow seeds is not significantly different from the distribution that would be expected under Mendel's laws. Any minor differences between your offspring distribution and the expected Mendelian distribution can be attributed to chance or sampling error.*

Step-by-Step Procedure for Testing Your Hypothesis and Calculating Chi-Square

1. State the hypothesis being tested and the predicted results.
2. Gather the data by conducting the relevant experiment (or, if working genetics problems, use the data provided in the problem).
3. Determine the expected numbers for each observational class. Remember to use numbers, not percentages.

Chi-square should not be calculated if the expected value in any category is less than 5.

4. Calculate χ^2 using the formula. Complete all calculations to three significant digits.

TABLE D.1 Calculating Chi-Square for the Pea Experiment

	Green	Yellow
Observed (*o*)	639	241
Expected (*e*)	660	220
Deviation (*o* – *e*)	–21	21
Deviation² (d^2)	441	441
d^2/e	0.668	2.005
$\chi^2 = \sum d^2/e = 2.673$		

TABLE D.2 Chi-Square Distribution

Degrees of Freedom (df)	Probability (*p*)										
	0.95	**0.90**	**0.80**	**0.70**	**0.50**	**0.30**	**0.20**	**0.10**	**0.05**	**0.01**	**0.001**
1	0.004	0.02	0.06	0.15	0.46	1.07	1.64	2.71	3.84	6.64	10.83
2	0.10	0.21	0.45	0.71	1.39	2.41	3.22	4.60	5.99	9.21	13.82
3	0.35	0.58	1.01	1.42	2.37	3.66	4.64	6.25	7.82	11.34	16.27
4	0.71	1.06	1.65	2.20	3.36	4.88	5.99	7.78	9.49	13.28	18.47
5	1.14	1.61	2.34	3.00	4.35	6.06	7.29	9.24	11.07	15.09	20.52
6	1.63	2.20	3.07	3.83	5.35	7.23	8.56	10.64	12.59	16.81	22.46
7	2.17	2.83	3.82	4.67	6.35	8.38	9.80	12.02	14.07	18.48	24.32
8	2.73	3.49	4.59	5.53	7.34	9.52	11.03	13.36	15.51	20.09	26.12
9	3.32	4.17	5.38	6.39	8.34	10.66	12.24	14.68	16.92	21.67	27.88
10	3.94	4.86	6.18	7.27	9.34	11.78	13.44	15.99	18.31	23.21	29.59
	Nonsignificant								**Significant**		

Source: R. A. Fisher and F. Yates, *Statistical Tables for Biological, Agricultural, and Medical Research*, 6th ed., Table IV, Longman Group, UK Ltd., 1974.

5. Use the chi-square distribution table to determine the significance of the value.
   a. Determine the degrees of freedom, one less than the number of categories. Locate that value in the appropriate column.
   b. Locate the χ^2 value for your significance level ($p = 0.05$ or less).
   c. Compare this χ^2 value (from the table) with your calculated χ^2.
6. State your conclusion in terms of your hypothesis.
   a. If your calculated χ^2 value is greater than the χ^2 value for your particular degrees of freedom and p value (0.05), then *reject the null hypothesis* of no difference between expected and observed results. *You can conclude that there is a significant difference between your observed distribution and the theoretical expected distribution* (for example, under Mendel's laws).
   b. If your calculated χ^2 value is less than the χ^2 value for your particular degrees of freedom and p value (0.05), then *fail to reject the null hypothesis* of no difference between observed and expected results. *You can conclude that there does not seem to be a significant difference between your observed distribution and the theoretical expected distribution* (for example, under Mendel's laws). *You can conclude that any differences between your observed results and the expected results can be attributed to chance or sampling error.* (Note: It is incorrect to say that you "accept" the null hypothesis. Statisticians either "reject" or "fail to reject" the null hypothesis.) The chi-square test will be used to test for the goodness of fit between observed and expected data from several laboratory investigations in this lab manual.

REFERENCES

Moore, D., W. Notz, and M. Fligner. *Basic Practice of Statistics*, 7th ed. New York: W. H. Freeman, 2015.

Motulsky, M. *Intuitive Biostatistics,* 3rd ed. New York: Oxford University Press, 2013.

Triola, M. *Elementary Statistics Update,* 11th ed. San Francisco, CA: Pearson, 2012.

Whitlock, M. and D. Schluter. *The Analysis of Biological Data*, 2nd ed. New York: W. H. Freeman, 2014.

WEBSITES

HHMI Biointeractive, "Statistics and Math;" resources that cover fundamentals of statistics, including chi-square: http://www.hhmi.org/biointeractive/statistics-and-math

Electronic Statistics Textbook (Tulsa, OK: Stat Soft); a relatively easy-to-understand free online statistics textbook:
http://www.statsoft.com/textbook/

Appendix E

Terminology and Techniques for Dissection

Orientation Terminology

The terms defined here are used with bilaterally symmetrical animals, both invertebrates and vertebrates (Figures E.1 and E.2). Use these terms as you describe animals studied in this laboratory manual. Note that texts may use different terminology for those animals called bipeds (for example, humans) and those called quadrupeds (for example, the fetal pig). Terminology used exclusively with bipeds is not included in this appendix.

Right/left: always refer to the animal's right or left, not yours.

Anterior, cranial: toward the head.

Posterior, caudal: toward the tail.

Dorsal: backside; from the Latin *dorsum*, meaning back.

Ventral: bellyside; from the Latin *venter*, meaning belly.

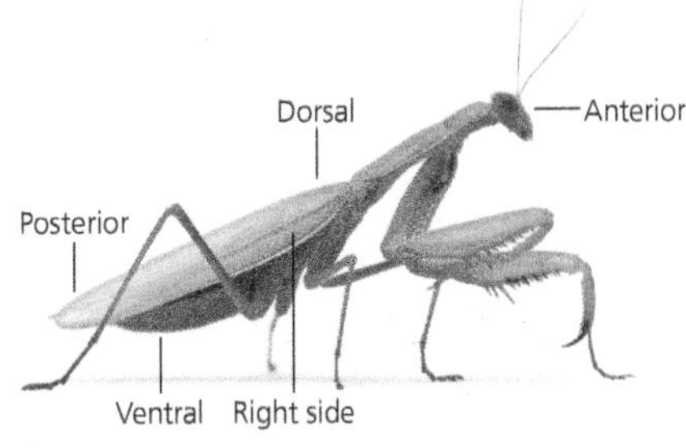

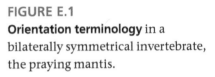

FIGURE E.1
Orientation terminology in a bilaterally symmetrical invertebrate, the praying mantis.

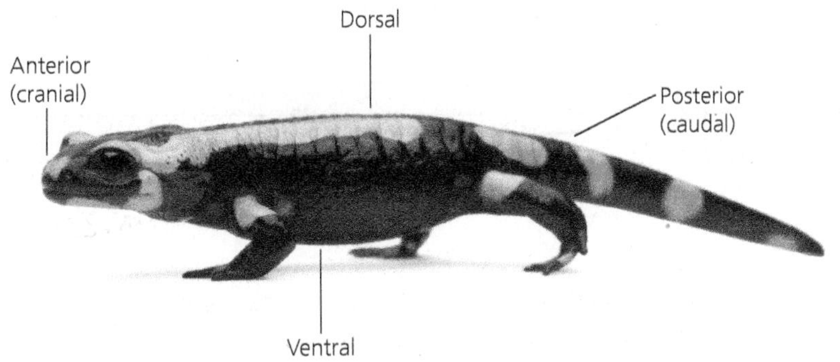

FIGURE E.2
Orientation terminology in a quadruped vertebrate, a fire salamander.

Terms Relating to Position in the Body

Proximal: near the trunk, attached portion, or point of reference; for example: "The elbow is *proximal* to the wrist."

Distal: farther from the trunk, attached portion, or point of reference; for example: "The toes are *distal* to the ankle."

Superficial: lying on top or near the body surface; for example: "The dermis of the skin is *superficial* to a layer of fatty tissue called the hypodermis."

Deep: lying under or below; for example: "Tough connective tissue wrapping skeletal muscles lies *deep* to the hypodermis."

Planes and Sections

A **section** is a cut through a structure. A **plane** is an imaginary line through which a section can be cut. Anatomists generally refer to three planes or sections (Figure E.3).

- **Sagittal section:** This divides the body into left and right portions or halves. This is a longitudinal or lengthwise section from anterior to posterior.
- **Frontal section:** A longitudinal or lengthwise section from anterior to posterior, this divides the body into dorsal and ventral portions or halves.
- **Transverse section:** Also called a **cross section,** this divides the body into anterior and posterior portions or cuts a structure across its smallest diameter.

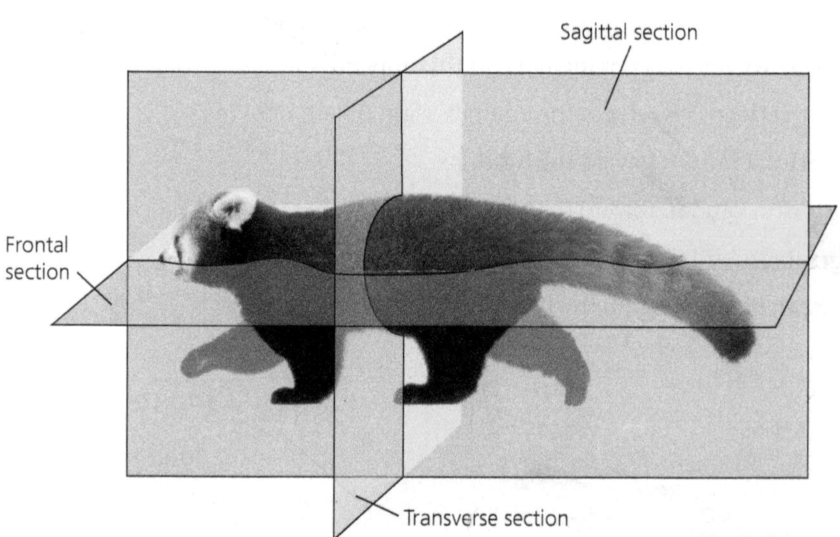

FIGURE E.3
Sections of a bilaterally symmetrical animal. A young red panda.

Dissection Techniques

When studying the anatomy of an organism, the term **dissection** is perhaps a misnomer. *Dissection* literally means to cut apart piece by piece. In lab, however, it is usually more appropriate to *expose* structures rather than dissect them. Initial incisions do require that you cut into the body, but after body cavities are opened, you will usually only separate and expose body parts, using dissection rarely. Accordingly, you will use the scalpel when you make initial incisions into the body wall of large animals, but seldom when studying small animals or the organs of large animals.

Scissors are used to deepen initial cuts made by the scalpel in large animals and to cut into the bodies of smaller animals. When using scissors, direct the tips upward to prevent gouging deeper organs. Once the animal's body is open, use forceps and the blunt probe to carefully separate organs and to pick away connective tissue obstructing and binding organs and ducts. Needle probes are only minimally useful. Never cut away an organ or cut through a blood vessel, nerve, or duct unless given specific instructions to do so.

Producing a good dissection takes time and cannot be rushed. As you study the anatomy of animals, your goal should be to expose all parts so that they can be easily studied and demonstrated to your lab partner or instructor.